Advances in Intelligent Systems and Computing

Volume 393

Series editor

Janusz Kacprzyk, Polish Academy of Sciences, Warsaw, Poland
e-mail: kacprzyk@ibspan.waw.pl

About this Series

The series "Advances in Intelligent Systems and Computing" contains publications on theory, applications, and design methods of Intelligent Systems and Intelligent Computing. Virtually all disciplines such as engineering, natural sciences, computer and information science, ICT, economics, business, e-commerce, environment, healthcare, life science are covered. The list of topics spans all the areas of modern intelligent systems and computing.

The publications within "Advances in Intelligent Systems and Computing" are primarily textbooks and proceedings of important conferences, symposia and congresses. They cover significant recent developments in the field, both of a foundational and applicable character. An important characteristic feature of the series is the short publication time and world-wide distribution. This permits a rapid and broad dissemination of research results.

More information about this series at http://www.springer.com/series/11156

Ryszard Jabłoński · Tomas Brezina
Editors

Advanced Mechatronics Solutions

Springer

Editors
Ryszard Jabłoński
Warsaw University of Technology
Warsaw
Poland

Tomas Brezina
Brno University of Technology
Brno
Czech Republic

ISSN 2194-5357 ISSN 2194-5365 (electronic)
Advances in Intelligent Systems and Computing
ISBN 978-3-319-23921-7 ISBN 978-3-319-23923-1 (eBook)
DOI 10.1007/978-3-319-23923-1

Library of Congress Control Number: 2015950432

Springer Cham Heidelberg New York Dordrecht London

Printed on acid-free paper

Springer International Publishing AG Switzerland is part of Springer Science+Business Media (www.springer.com)

Preface

Mechatronics is in fact the combination of enabling technologies brought together to reduce complexity through the adaptation of interdisciplinary techniques in production.

Focusing on the most rapidly changing areas of mechatronics, this book discusses signals and system control, mechatronic products, metrology and nanometrology, automatic control & robotics, biomedical engineering, photonics, design manufacturing and testing of MEMS. It is reflected in the list of contributors, including an international group of 302 leading researchers representing 12 countries.

The book is intended for use in academic, government and industry R&D departments, as an indispensable reference tool for the years to come. Also, we hope that the volume can serve the world community as the definitive reference source in Mechatronics.

Through this publication we hope to establish and provide an international platform for information exchange in the mechatronics fields mentioned above.

The book comprises carefully selected 93 contributions presented at the 11th International Conference Mechatronics 2015, organized by Faculty of Mechatronics, Warsaw University of Technology, on September 21–23, in Warsaw, Poland. It is the fifth volume in series, following the editions in 2007, 2009, 2011 and 2013.

We would like to thank all authors for their contribution to this book.

Ryszard Jabłoński

Contents

Part III Mechatronics Product: Design, Manufacturing and Testing

Part VI Photonics, Vision Systems and Image Processing

Part VII Robotics

Part I
Automatic Control

The weak isolability of the structure of binary residuals of multiple faults

Michał Bartyś

Institute of Automatic Control and Robotics, Warsaw University of Technology,
św. A. Boboli 8, 02-525 Warsaw, Poland
{bartys@mchtr.pw.edu.pl}

Abstract. The chosen keypoints of the theoretical framework of the multiple fault isolability assessment are briefly outlined in this paper. The set of multiple fault isolability metrics have been proposed. Introduced distinctiveness matrix as well as corresponding multiple fault metric, allow for characterization of multiple fault isolability of technical systems. This is particular important in respect to functional safety systems were multiple faults must be taken into account. Finally, a new definition of a weakly isolating structure of residual sets in respect to the multiple faults is proposed. This definition can be considered as an extension of Gertler's theoretical framework of a structural approach to fault isolation.

Keywords: weakly isolating structure, multiple fault isolation, binary diagnostic matrix, fault isolabilty metric, fault distinguishability, fault distinctiveness matrix

1 Introduction

In this paper, we will give a new perspective on the problem of dimensioning of multiple fault isolability with the use of binary valued structures of residual sets. The necessity of definition of multiple fault isolability metrics arises, among others, as a result of challenging demands for optimization of diagnostic systems e.g. for solving constrained sensor placement problems [21, 22] or support for solving the problem of diagnostic coverage in the functional safety systems. The multiple fault isolation is a commonly recognized problem having serious theoretical meaning and substantial practical impact [1–3, 6, 7, 10, 11, 15, 18, 20, 23].

In structured residuals based approaches [8, 10], faults are recognized indirectly by analysis of binary evaluated results of tests or observations that are sensitive to these faults.

The fault isolability from the perspective of structural analysis for early determination of fault detectability was discussed in [7]. It has been shown how different levels of knowledge about faults result in isolability properties. Basseville in [5] fixed the isolation as a simultaneous multiple-hypothesis testing problem and formulated three isolation criteria based on Kullback distance.

The main contribution of this paper is the proposition of a new isolability metric and generalizing the definition of a weakly isolating structure introduced by Gertler [8].

R. Jabłoński and T. Brezina (eds.), *Advanced Mechatronics Solutions*,
Advances in Intelligent Systems and Computing 393,
DOI 10.1007/978-3-319-23923-1_1

2 Preliminaries

The set of following assumptions have been adopted for further discussion:

1° the bi-valued diagnostic matrix of the diagnosed system is known;

2° the diagnostic matrix contains exclusively non all zeroes columns;

3° the signature of multiple faults of single faults is the logical alternative of single fault signatures.

Consider a diagnostic system in which all residuals are binary valued. Let the finite set of single faults in this system be F_1:

$$F_1 = \{f_i : i = 1..n\} \tag{1}$$

and finite set of diagnostic signals be S:

$$S = \{s_j : j = 1..m\} \tag{2}$$

Diagnostic signal values are derived from residuals by the application of appropriate evaluation processing approaches. Let relation R_{FS} be the Cartesian product of sets F_1 and S.

$$R_{FS} \subset F_1 \times S. \tag{3}$$

According to the geometrical interpretation of the Cartesian product, the relation is the set of $n \cdot m$ points of the plane defined in F_1 and S coordinates. It is possible to spread out a three-dimensional mesh over the $F_1 \times S$ plane by attributing diagnostic signal values $v_{j,i}$ of all diagnostic signals s_j for all f_i faults. The three-dimensional mesh is easily transformable into the form of a two-dimensional $m \cdot n$ matrix V_1 of the diagnostic signal values $v_{j,i}$. Hence, $V_1 = [v_{j,i}]_{m \times n}$. In the simplest case, the diagnostic signals are bi-valued, i.e. the relation faults-symptoms has the form of a binary structure of residual sets [8] or Binary Diagnostic Matrix [10, 14]. In this paper, we will limit our considerations exclusively to the case of binary valued structures of residual sets. Each fault $f_i \in F_1$; $\forall i \in \{1..n\}$ is associated with one and only one i-th column $V_{1,i}$ of the binary diagnostic matrix V_1:

$$V_{1,i} = [v_{1,i}, v_{2,i}, ...v_{m,i}]^\top. \tag{4}$$

Specific vector $V_{1,i}$ consisting of diagnostic signals associated with a particular single fault is referred to as the single *i-th* fault signature. For simplicity of notion, the diagnostic matrix V_1 will be further represented comprehensively as a block matrix of single fault signatures $V_{1,i}$

$$V_1 = [V_{1,i}]_{[n \times 1]} \tag{5}$$

In order to achieve fault distinguishability in the structure, it is beneficial if fault signatures will be unique. In practice, this requirement does not necessarily hold. The problem of single fault distinguishability in the binary structure of residual sets was considered in many publications [8, 10, 22]. Here, we will undertake an attempt towards assessment of multiple fault distinguishability.

Let us denote the fault multiplicity by k. Let us now generate $q = \binom{n}{k}$ subsets V_k^q, each consisting of k non repetitive single fault signatures $V_{1,i}$:

$$V_k^q = \{V_{1,r}^1, V_{1,s}^2, \dots, V_{1,z}^k\}; \quad r \neq s \neq \dots \neq z; \quad r, s, \dots z \in \{1..n\}. \tag{6}$$

According to assumption 3° of Sect. 2, all signatures of binary evaluated multiple faults are assumed as logical alternatives of single fault signatures. Hence, any subset V_k^q of power k represent the alternative signature of any k-multiple fault. Therefore, any q-th signature of any k-multiple fault equals:

$$V_{k,q} = \bigvee_{p=1}^{k} V_k^q = V_{1,r}^1 \vee V_{1,s}^2 \vee \dots \vee V_{1,z}^k; \qquad k \in \{1, \dots, n\}; \; q \in \{1, \dots, \binom{n}{k}\} \tag{7}$$

Definition 1. *The diagnostic matrix of k-multiple faults V_k is a matrix of ordered k-multiple fault signatures:*

$$V_k = [V_{k,1}, V_{k,2}, \dots, V_{k,q}] \tag{8}$$

Now we will define the set of all signatures of all single and multiple faults by generalization of formula (5).

Definition 2. *The multiple fault binary diagnostic matrix is a block matrix consisting of ordered diagnostic matrices of all multiple faults:*

$$\mathbf{V} = [[V_1], [V_2], \dots, [V_n]]_{[1 \times n]} = [V_i]_{[1 \times n]} \tag{9}$$

The disitinguishability of multiple faults in a given isolation structure should be considered as a feature which characterizes at least the uniqueness (isolability or distinction) of any signature of any fault independently of its multiplicity.

Definition 3. *The multiple faults are isolable in a binary diagnostic matrix if their signatures are unique within isolation structure.*

Therefore, the problem is how to determine and measure a distinction between any fault signatures in a given multiple fault binary diagnostic matrix? This problem was considered for single fault binary isolation structures in [8, 22]. Firstly, we obtain the partitioned vector of multiple fault signatures from (9) by concatenating all V_k.

$$\mathbf{v} = [V_{1,1}, V_{1,2}, \dots V_{1,q_1}, V_{2,1}, V_{2,2}, \dots V_{2,q_2}, \dots, V_{n,1}, V_{n,2}, \dots V_{n,q_n}]_{[1 \times N]} \tag{10}$$

where: $q_k = \binom{n}{k}$ is the number of k-multiple faults
Next, we will introduce a matrix of multiple fault distinctiveness referred also to as matrix of discernibility. Principally, it expresses in binary terms, whether any pair of multiple fault signatures is disjunctive or not. To do this, let us define the disjunction of binary matrices.

Definition 4. *Disjunction of binary matrices* $\mathbf{A}_{[m \times n]}$ *and* $\mathbf{B}_{[n \times p]}$ *be denoted as* $\mathbf{A} \oplus \mathbf{B}$ *and defined as the matrix* $\mathbf{C}_{[m \times p]}$*:*

$$\mathbf{C} = \mathbf{A} \oplus \mathbf{B} = \begin{bmatrix} C_{11} & C_{12} & \cdots & C_{1p} \\ C_{21} & C_{22} & \cdots & C_{2p} \\ \vdots & \vdots & \ddots & \vdots \\ C_{m1} & C_{m2} & \cdots & C_{mp} \end{bmatrix} \tag{11}$$

where: $C_{ij} = \bigvee_{k=1}^{n} A_{ik} \oplus B_{kj}$ and $\oplus$ is the Boolean disjunction operator.

3 The multiple fault distinctiveness matrix

In order to define the distinctiveness of multiple fault signatures, we keep in mind that any multiple fault is associated with its own specific signature. Thus, the distinctiveness of signatures of multiple faults might be represented by mutual disjunction of signatures of all multiple faults. Let us define now the multiple fault distinctiveness matrix.

Definition 5. *The multiple fault distinctiveness matrix is the tensor disjunction of partitioned vectors of multiple fault signatures.*

$$\mathbf{D} = \mathbf{v}^{\top} \oplus \mathbf{v} \tag{12}$$

Remark 1. The multiple fault distinctiveness matrix is a square matrix, since the number o rows of $\mathbf{v}^{\top}$ and number of columns of $\mathbf{v}$ vectors is equal.

Corollary 1. *Each* D_{ij} *entry of square multiple fault distinctiveness matrix is the only disjunction of signatures of i-th and j-th multiple fault.*

Proof. According to (11), each D_{ij} entry of the square multiple fault distinctiveness matrix $\mathbf{D}$ is an alternative of disjunctions of binary signatures of i-th and j-th multiple faults over m i.e. ($D_{ij} = \bigvee_{k=1}^{n} V_{ik} \oplus V_{kj}$). Since the number m of columns of transposed partitioned vector of multiple fault signatures $\mathbf{v}^{\top}$ equals 1 then:

$$D_{ij} = V_i \oplus V_j. \tag{13}$$

It is worth to mention that property of symmetry allows to speed up calculations of the multiple fault distinctiveness matrix and should be considered as beneficial in the practice.

4 The binary isolability metric

Corollary 2. *Any pair of multiple faults* $\langle f_i, f_j \rangle$ *is mutually weakly isolating if corresponding entry* D_{ij} *of multiple fault distinctiveness matrix is different from zero for any* $(i \neq j)$.

Proof. In accordance with definition of weak isolability [8], the structure is weakly isolating if the fault signatures are different and nonzero. This apply independently whether faults are single or multiple.

Remark 2. Any pair of multiple faults $\langle f_i, f_j \rangle$ is indistinguishable also if multiple fault distinctiveness matrix entry is all-zeroes vector ($D_{ij} = 0$) or multiple fault distinguishability matrix entry ($\mathcal{D}_{ij} = 0$) for any ($i \neq j$).

The distinctiveness matrix $\mathbf{D}$ might be presented in the form of a block matrix of n fault distinctiveness matrices d_k, each referring to k-th fault multiplicity:

$$\mathbf{D} = [[d_1], [d_2], ..., [d_n]]_{[1 \times n]} \,. \tag{14}$$

Here, the d_i is a matrix of the size $[N : q_i]$. The distinctiveness matrix $\mathbf{D}_k$ of the all k-multiple faults is a block matrix consisting of k ordered matrices d_i:

$$\mathbf{D}_k = [[d_1], [d_2], ..., [d_k]] \,. \tag{15}$$

Now we introduce a measure of a binary multiple fault isolability. Firstly we define the Boolean conjunction of all entries of upper triangular part of distinctiveness matrix $\mathbf{D}$ located above main diagonal in its any i-th column.

$$\mathfrak{i}_i = \bigwedge_{j=1}^{i-1} d_{ij} \,; \forall i \in \{2, N\} \,. \tag{16}$$

Remark 3. Since distinctiveness of any pair of identical faults is always zero, then the upper bound in the Boolean conjunction in formula (16) is $(i - 1)$.

Let the measure of isolability of k-multiple faults be denoted as $\mathfrak{I}_k$.

Definition 6. *The binary isolability metric $\mathfrak{I}_k$ of k-multiple faults be a Boolean conjunction of all entries of multiple distinctiveness matrix $\mathbf{D}_k$ which are located above main diagonal:*

$$\mathfrak{I}_k = \bigwedge_{i=i_k}^{\sum_{k=1}^{k} q_k} \mathfrak{i}_i \tag{17}$$

where: $i_k = 2 + \sum_{k=1}^{k-1} q_k$.

Remark 4. The binary isolability metric of k-multiple faults allows to judge either the k-multiple faults are mutually isolated or not. The k-multiple faults are mutually indistinguishable if ($\mathfrak{I}_k = 0$). Obviously, if this condition holds then multiple faults are not weakly isolating.

Definition 7. *Given the binary isolability vector $\mathfrak{I}^k[1 : k]$ of all single, double, ... , and k-multiple faults:*

$$\mathfrak{I}^k = [\mathfrak{I}_1, \mathfrak{I}_2, .. , \mathfrak{I}_k] \tag{18}$$

Definition 8. *The binary isolability metric $\mathfrak{I}$ of multiple binary fault isolation structure be a Boolean conjunction of all entries of isolability vector $\mathfrak{I}^k$:*

$$\mathfrak{I} = \bigwedge_{i=1}^{n} \mathfrak{I}_i \tag{19}$$

The binary isolability metric $\mathfrak{i}$ indicates isolability of all single, double, ... and n-multiple faults within the isolating structure. Now, we will generalize the definition of a weakly isolating structure introduced by Gertler [8].

Definition 9. *The binary isolating structure is weakly isolating if it does not contain any all-zeros column and if its binary isolability metric $\mathfrak{I} \neq 0$.*

It should be mentioned, that weak isolability might be considered in context of fault multiplicity. In this case:

Definition 10. *The binary isolating structure is weakly isolating k-multiple faults if it does not contain any all-zeros column and if its binary isolability metric $\mathfrak{I}_k \neq 0$.*

5 Illustrative example

Let us consider an example of a column-canonical unidirectional and bidirectional strong isolating binary valued structure of residual sets counting $m = 4$ rows and $n = 3$ columns presented in Tab. 1a. According to the assumption 3° in section 2, a signature of multiple fault is the logical alternative of single fault signatures. The signatures of double and triple faults are depicted respectively in Tab. 1b and Tab. 1c. The partitioned block matrix of multiple fault signatures

Table 1. An example of the binary valued structures of residual sets representing: a) matrix $\mathbf{V}_1$ of single fault signatures; b) matrix $\mathbf{V}_2$ of double fault signatures ; c) matrix $\mathbf{V}_3$ of triple fault signatures.

(a)

S/F	f_1	f_2	f_3
s_1	1	1	0
s_2	1	0	0
s_3	0	0	1
s_4	0	1	1

(b)

S/F	f_1f_2	f_1f_3	f_2f_3
s_1	1	1	1
s_2	1	1	0
s_3	0	1	1
s_4	1	1	1

(c)

S/F	$f_1f_2f_3$
s_1	1
s_2	1
s_3	1
s_4	1

$\mathbf{V} = [\mathbf{V}_1, \mathbf{V}_2, \mathbf{V}_3]$ is depicted in Tab. 2. In this table, vertical lines are used to separate k-multiple fault signature matrices $\mathbf{V}_i$. Each matrix counts respectively $q = [q_1, q_2, q_3] = [3, 3, 1]$ columns and $m = 4$ rows. The distinctiveness matrix $\mathbf{D}$ is depicted in Tab. 3. Fault distinctiveness matrix $\mathbf{D}$ show either any pair of multiple faults is distinguishable or not. For example double fault $\{f_1 f_3\}$ is indistinguishable with triple fault $\{f_1 f_2 f_3\}$. All other faults are distinguishable.

The results obtained from Tab. 3 could be interpreted as follows:

Table 2. The partitioned block matrix of multiple fault signatures $\mathbf{V}$ of the structure of residual sets depicted in Tab. 1a.

S/F	f_1	f_2	f_3	f_1f_2	f_1f_3	f_2f_3	$f_1f_2f_3$
s_1	1	1	0	1	1	1	1
s_2	1	0	0	1	1	0	1
s_3	0	0	1	0	1	1	1
s_4	0	1	1	1	1	1	1

Table 3. The multiple fault distinctiveness matrix $\mathbf{D}$ for isolating structure of residual sets depicted in Tab. 1a.

	f_1	f_2	f_3	f_1f_2	f_1f_3	f_2f_3	$f_1f_2f_3$
f_1	0	1	1	1	1	1	1
f_2	1	0	1	1	1	1	1
f_3	1	1	0	1	1	1	1
f_1f_2	1	1	1	0	1	1	1
f_1f_3	1	1	1	1	0	1	0
f_2f_3	1	1	1	1	1	0	1
$f_1f_2f_3$	1	1	1	1	0	1	0
$\mathfrak{i}_i$		1	1	1	1	1	0
$\mathfrak{I}_k$		1			1		0

1. All single faults are isolable because $\mathfrak{I}_1 = 1$;
2. All single and double faults are isolable because $\mathfrak{I}_2 = 1$;
3. Triple faults are not isolable because $\mathfrak{I}_3 = 0$;
4. Triple fault $\langle f_1f_2f_3\rangle$ and double fault $\langle f_1f_3\rangle$ are indistinguishable because $D_{4,7} = 0$;
5. Not all multiple are isolable because $\mathfrak{I} = 0$
6. Matrix shown in Tab. 1a is weakly isolating for single and double faults because $\mathfrak{I}_1 = 1$ and $\mathfrak{I}_2 = 1$ and is not weakly isolating for triple fault.

6 Final remarks

The main contribution of this paper is a proposition of a new formal definition of a weakly isolating structure of the binary residual sets. This definition can be considered as an extension of Gertler's [8] theoretical framework of a structural approach to fault isolation.

The proposed definition is based on binary isolability metric introduced in this paper. This metric allows for quantitative analysis of indistinguishability of single and multiple faults within the binary valued structure of residual sets. Searching for a weakly isolating structure plays an important role in the fault indistinguishability design and analysis of diagnostic systems. It is extremely useful for automatized searching for optimal fault isolating structures within the given constrains. It allows for a time inexpensive judgement either multiple faults of a given multiplicity k are distinguishable or not within a given binary fault isolation structure.

The functional safety standard IEC 61508 distinguishes the following types of failures: dangerous detectable failures λ_{DD}, dangerous undetectable failures λ_{DU}, safe detectable failures λ_{SD} and safe undetectable λ_{SU} failures. The total intensity of failures λ is defined as the sum of intensities of all types of failures.

$$\lambda = \lambda_{DD} + \lambda_{DU} + \lambda_{SD} + \lambda_{SU} \tag{20}$$

Diagnostic system must be designed in such a way that it should detect and isolate at least dangerous failures. A measure of the quality of detection of dangerous failures is the factor referred to as diagnostic coverage (DC). It is interpreted as the relative reduction of the probability of dangerous failures resulting from application of automated diagnostic tests:

$$DC = \frac{\lambda_{DD}}{\lambda_{DD} + \lambda_{DU}} \tag{21}$$

Clearly, undetectable faults are unisolable and indistinguishable. Therefore, the fault isolation influences λ_{DD} and λ_{SD} intensities only. Hence, each measure allowing for achieving better fault distinguishability enhance DC factor. In this context, the fault isolability metrics introduced in this paper can support searching for optimal diagnostic coverage in diagnostic systems at design stage. So far, the introduced metrics have not been already used in a diagnostic coverage in the real functional safety system design.

References

1. Bartyś, M.: Chosen Issues of Fault Isolation - Theory, Practice, Applications. PWN, Warszawa (2014)
2. Bartyś, M.: Generalised reasoning about faults based on diagnostic matrix, International Journal of Applied Mathematics and Computer Science, 23(2), 407–417 (2013)
3. Bartyś, M.: Multiple Fault Isolation Algorithm Based on Binary Diagnostic Matrix. In: Korbicz, J., Kowal, M., (eds.) Intelligent Systems in Technical and Medical Diagnosis: Advances in Intelligent Systems and Computing, Springer-Verlag Heilderberg, New York, Dordrecht, London, vol. 230, pp. 441–452 (2013)
4. Basseville, M.: On fault detectability and isolability. European Journal of Control, 7(6), 625–637 (2001)
5. Basseville, M.: Information Criteria for Residual Generation and Fault Detection and Isolation. Automatica, 33(5), 783–803 (2001)
6. Blanke, M., Schröder, M.: Diagnosis and fault tolerant control. Vol. 115, Springer (2003)
7. Blanke, M., Staroswiecki, M.: Structural design of systems with safe behaviour under single and multiple faults. In: Proc. of IFAC Symposium Safeprocess 2006, Beijing, PR China, pp. 511–515 (2006)
8. Gertler, J.: Fault Detection and Diagnosis in Engineering Systems. New York, Marcel Dekker Inc., New York, Basel, Hong Kong (1998)
9. Isermann, R.: Fault Diagnosis Systems. An Introduction from Fault Detection to Fault Tolerance, Springer-Verlag New York (2006)

10. Korbicz, J., Kościelny, J.M., Kowalczuk, Z., Cholewa, W., (eds.): Fault Diagnosis. Models, Artificial Intelligence, Applications, Springer-Verlag Berlin, Heildelberg, New York (2004)
11. Kościelny, J.M., Syfert, M. Application properties of methods for fault detection and isolation in the diagnosis of complex large-scale processes, Bulletin of the Polish Academy of Sciences Technical Sciences, 62(3), 571–582 (2014)
12. Kościelny, J.M., Bartyś, M., Rzepiejewski, P., Sá da Costa, J.M.G. Actuator fault distinguishability study of the DAMADICS benchmark problem, Control Engineering Practice, Pergamon–Elsevier, 14(6), 645–652 (2006)
13. Kościelny, J.M., Bartyś, M., Syfert, M.: The practical problems of the fault isolation in large scale industrial systems, semi–plenary paper, In Proc. of 6th Symposium IFAC Safeprocess 2006, Beijing, Peoples Republic of China, 13–24 (2006)
14. Kościelny, J.M.: Fault Isolation in Industrial Processes by Dynamic Table of States Method, Automatica, 31(5), 47–753 (1995)
15. Krysander, M., Frisk, E.: Sensor placement for fault diagnosis, IEEE Transactions On Systems, Man and Cybernetics, IEEE, 38(6), 1398–1410 (2008)
16. Ligęza, A., Kościelny, J.M.: A new approach to multiple fault diagnosis. Combination of diagnostic matrices, graphs, algebraic and rule–based models. The case of two–layer models. International Journal of Applied Mathematics and Computer Science, 18(4), 465–476 (2008).
17. Nyberg, M.: A Fault Isolation Algorithm for the Case of Multiple Faults and Multiple Fault Types. In: Proc. of 6th IFAC Symposium Safeprocess 2006, Beijing, PR China, pp. 679–689 (2006)
18. Nyberg, M., Krysander, M.: Combining AI, FDI, and statistical hypothesis–testing in a framework for diagnosis. In: Proc. of 5th of IFAC Symposium Safeprocess, Washington, D.C., USA, pp. 891–896 (2003)
19. Patton, R., Frank, P., Clark, R., (eds.): Issues of fault diagnosis for dynamic systems, Springer-Verlag Berlin, Heidelberg, New York, (2000)
20. Reiter, R.A.: Theory of Diagnosis from First Principles. Artificial Intelligence, (32), 57–95 (1987)
21. Rosich, A., Sarrate, R., Puig, V., Escobet, T.: Efficient optimal sensor placement for model-based FDI using an incremental algorithm. In: Proc. of 46th IEEE Conference on Decision and Control, 2590–2595 (2007)
22. Rostek, K. Measure of fault isolability of diagnostic system. In: 25th International Workshop on Principles of Diagnosis (DX), September 8-11, Gratz, Austria (2014)
23. Travé–Massuyés, L.: Bridges between diagnosis theories from control and AI perspectives. In: Korbicz, J., Kowal, M., (eds.) Intelligent Systems in Technical and Medical Diagnostics, Springer-Verlag Heidelberg, New York, Dordrecht, London, (230), 3–28,(2013)

Operational limits in vibration diagnostics

Tomas Budik[1], Robert Jankovych[2], Milos Hammer[3],

[1] Faculty of Mechanical Engineering, Brno University of Technology, Brno, Czech Republic
y125952@fme.vutbr.cz
[2] Faculty of Mechanical Engineering, Brno University of Technology, Brno, Czech Republic
jankovych@fme.vutbr.cz
[3] Faculty of Mechanical Engineering, Brno University of Technology, Brno, Czech Republic
hammer@fme.vutbr.cz

Abstract. The paper discusses some practical problems occurring at ISO 10816-1 application in technical practice. Technical diagnosticians find out frequently that the zone boundaries proposed by the standard are not sensitive adequately to the changes in vibrations during operation resulting from a change of technical condition of the machine. The principle of statistical determination of individual operational limits from the measured overall level of RMS vibration velocity (overall level) with utilization of statistical regulation of individual values is also described in the paper. The alarm limit for the overall level (boundary between the B and C zones) in the range 10 Hz to 1,000 Hz is specified in the paper on the basis of measuring of vibrations on the electric motor of Bosch Rexroth hydraulic unit.

Keywords: Operational limits; vibration diagnostics; zones boundary.

1 Introduction

It is assumed in technical practice that measuring of vibrations on the non-revolving components is very often suitable for the purpose of assessment of technical condition for a number of machineries. We most frequently measure and assess overall level the significant changes of which indicate progressive changes of technical condition of rotational parts resulting e.g. from imbalance, misalignment, mechanical backlash of bearings in mounting, structure resonances, insufficiently rigid foundations, bent shaft, excessive wear of bearings, etc. relatively reliably.
Four typical zones (A, B, C and D) for assessment of the level of vibrations are defined in the standard [1] and in other related references (such as [4], [5], [6], [8], [10], etc.) for the purpose of evaluation of intensity of vibrations of a given machine and provision of the guidance for potential maintenance measures. We mostly find out at practical application of the ISO 10816-1 standard to the conditions of operation of a specific machine that the boundaries of the mentioned zone proposed by this standard are not adequately sensitive to changes of vibrations resulting from a change of technical condition of the machine [3].

R. Jabłoński and T. Brezina (eds.), *Advanced Mechatronics Solutions*,
Advances in Intelligent Systems and Computing 393,
DOI 10.1007/978-3-319-23923-1_2

2 Assessment of vibrations of the hydraulic unit electric motor

Verification of the above mentioned assumptions was carried out on HA 0070-120/008-050/050 hydraulic unit (hereinafter the hydraulic unit, Fig. 1) manufactured by Bosch Rexroth. The hydraulic unit is specified as a source of pressure oil for a didactic stand for lessons of hydraulics in our Institute for Production Machines, Systems and Robotics. The hydraulic unit (Fig. 1.) represents a set of hydraulic elements arranged functionally on an aluminium tank positioned in a valve table made of aluminium profiled components.

Fig. 1. Hydraulic unit with Micrologue CMXA 48

The hydraulic circuit of the unit consists of two pressure circuits. A combination of two PV7 variable vane pumps produced by Bosch Rexroth (Fig. 1, item 1) with geometrical capacity of 12 and 7.5 cm^3/rev. controlled to a constant pressure and driven with Siemens 2.2 kW electric motor (Fig. 1, item 2) with synchronous speed of 1,000 rpm represents a source of pressure energy.

The Micrologue CMXA 48 apparatus (Fig. 1, item 3) developed by SKF was used for evaluation of technical condition of the hydraulic unit. This portable vibration data collectors and analyzer has a good simultaneous triaxial or four-channel vibration measurement capabilities over a range of 0.16 Hz to 80 kHz.

Electric motor vibrations were measured using three CMSS 2111 accelerometers mounted to the flange of the electric motor with permanent magnets approximately in the horizontal direction (hereinafter H), in the vertical direction (hereinafter V) and in the axial direction (hereinafter A), see Fig. 1, item 4. The CMSS 2111 is a small footprint accelerometer that includes an integrated 2 m cable along with a magnetic mount (CMSS 908-LD).

We understand the value of the measured variable proportional to the total vibration energy (strength or intensity of vibrations) related to a considered time record as overall level of mechanical vibration o. The overall level can be calculated from the time record according to the following formula:

$$o = \frac{\sqrt{\sum_{i=1}^{m}(y_i)^2}}{\sqrt{m}}, \tag{1}$$

where y_i is an amplitude of the determining quantity of mechanical vibration in the time t_i and m is a number of discrete points in the corresponding time record (in which the y value is ascertained).

By ascertaining of the value of overall level o and its comparison with the normal level, we will obtain some of possible information on the condition of the machine or any of its parts. When comparing the measured values, it should be verified whether the measurements of both values (the original one which represents information on the normal status and the new one through which current technical condition of the machine is assessed) have been made in the same frequency range (e.g. 10-1,000 Hz) and expressed in the same way (e.g. the peak value). If the value of total vibrations is higher than the normal level, we can assume that there is a problem which caused these higher values.

Such method of assessment of technical condition is recommended also by ISO 10816-1 standard. In this standard, individual types of machines are classified either in the group of "small machines" such as electric motors up to power of 15 kW or in the group of "bigger machines".

In our case, an asynchronous motor (2.2 kW) drives a pair of pumps. The first pump with a geometric capacity of 12 cm3/rev. was loaded using a throttle valve to a constant value of flow rate of 7 l/min at pressure of 5 MPa during vibration measuring. The other pump was hydraulically alleviated. It results from these loading parameters that the electric motor is loaded with approximately 0.6 kW, i.e. at about 30% of its nominal power. And that is why the machine is classified in the group of "small machines".

Values of measuring in accordance with ISO 10816-1 standard were set on Microlog CMXA 48 the lower frequency limit was set to 10 Hz while the upper one to 1,000 Hz. Number 10 was selected for averaging process, overlapping was 0% and measurements were repeated 30 times in order to obtain a sufficiently representative statistical sample of vibration values.

As the power of an electric motor was lower than 15 kW, it results from the standard that overall level for the zone boundary of the A/B zones can be specified to the value of 0.71 mm/s. Similarly, the value 1.8 mm/s is recommended for the alarm limit (boundary of B/C zones) while the value of 4.5 mm/s for the trip limit.

Fig. 2 shows the measured values of vibrations in all three directions (H, V, A) with the illustrated zone boundaries in accordance with the standard [1]. It is possible to say on the basis of Fig. 2 that the alarm and operation shutdown limits are very probably too far from the measured values. The alarm value specified using this method would not probably allow timely warning of the operator in case of a significant rise of vibrations.

It means that it is necessary to define new substantiated operational limits, e.g. using suitable statistical tools, in practice when establishing a system of predictive maintenance.

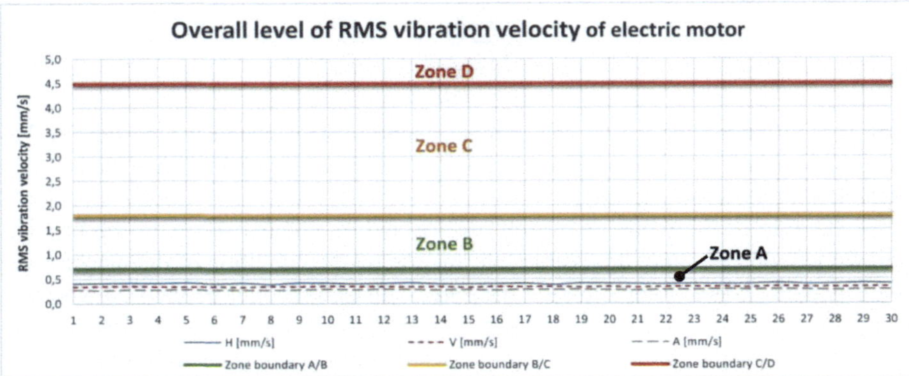

Fig. 2. Measured overall levels and zones for their assessment

3 Statistical determination of operational limits

We will identify overall level calculated from i-th time record with o_i symbol while the mean value of all measured overall levels for the past monitored period will be identified with $\bar{o}$ symbol. Moving range of overall levels between two successive measurements will be identified with R_{O_i} symbol:

$$R_{o_i} = |o_i - o_{i-1}|, \quad (2)$$

The alarm limit for overall level which should not be exceeded is given by the following formula:

$$UCL_{o_i} = \bar{o} + 2.66\bar{R}_o, \quad (3)$$

where R_{O_i} represents the mean value of the moving ranges of overall levels from all previous measurements.

The alarm limit for moving range of overall levels can be calculated according to the formula:

$$UCL_{R_{o_i}} = 3.267\bar{R}_o. \quad (4)$$

The formulas (3) and (4) apply only when the assumed risk of unnecessary error signal (risk α, type I error) equals to 0.27 % (for more details please see [2]).

Both limit values are in the distance of 3 sample standard deviations from the mean values $\bar{o}$ of overall levels and mean values $\bar{R}_O$ of their moving ranges.

Using the formulas mentioned above, the alarm limits for the hydraulic unit electric motor were determined on the basis of completed measurements of vibrations; these alarm limits are listed in Table 1. The mentioned electric motor has been used in non-demanding operation for about 10 years. It is in a good technical condition so that measuring of overall levels can be considered to be the assessment of the status of vibrations which correspond to the zone B in accordance with the standard [1]. On the basis of the above mentioned facts, we can consider the limit value calculated according to the formula (3) to be the boundary between the B and C zones (alarm limit).

We can conclude on the basis of assessment of the measurements that the maximum vibration values were measured in the horizontal direction as expected and for that reason, we will use them for illustration of the proposed procedure of determination of the alarm limit.

Fig. 3 illustrates the measured overall levels in the horizontal direction from which the alarm limit for this direction was calculated (0.438 mm/s).

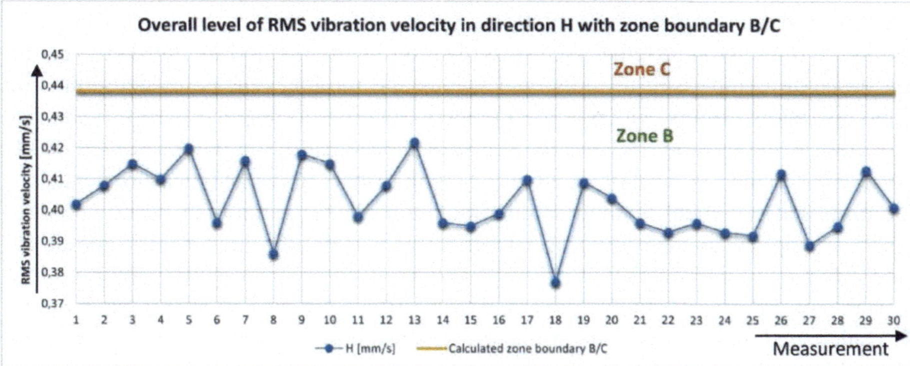

Fig. 3. Overall level of RMS vibration velocity in direction H

Fig. 4 illustrates the corresponding moving ranges of overall levels and the calculated alarm limit of moving range for the horizontal direction (0.044 mm/s).

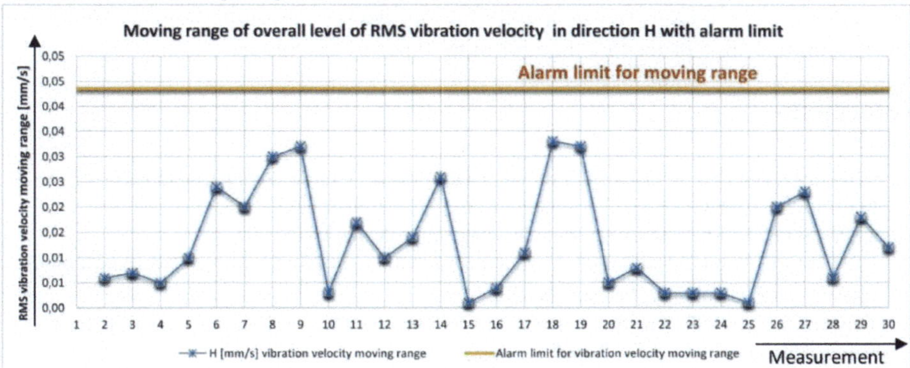

Fig. 4. Moving range of overall level of RMS vibration velocity in direction H

Table 1 contains the calculated alarm limits for overall level all three measured directions.

Tab. 1. Calculated alarm limits for overall level and moving range

Alarm limits:	H [mm/s]	V [mm/s]	A [mm/s]
For overall levels:	0.438	0.361	0.299
For a moving range of overall levels:	0.044	0.029	0.027

4 Conclusion

It happens very frequently in practice that the general standards for assessment of vibrations of machines and equipment during operation provide only approximate setting of an alarm limit and an operation shutdown limit (trip limit) which must be made more accurate progressively on the basis of experience and evaluation of the real course of measurements of vibrations.

The paper presents an original idea of utilization of the method of statistical regulation based on comparison of individual values and moving ranges for determination of the operational limits. The proposed procedure allows to respond to significant rise of overall levels in time. Beyond the scope of the standard, the proposed method offers also assessment of change of vibrations with respect to the last measurement.

The proposed statistical method of determination of the operational limits of vibrations was applied to measuring of vibrations of a hydraulic unit electric motor for three directions (horizontal-H, vertical-V and axial-A). The measured data represent the status of vibrations which correspond to the B zone according to the standard [1]. The calculated limit values are proposed as the boundary between B and C zone (alarm limit). When at least one of the calculated limits is exceeded (overall level or moving range), it is necessary to arrange for a suitable maintenance intervention immediately.

Identical method can be used for determination of the substantiated value of the boundaries between the A and B or C and D zones according to the standard [1].

The authors of the paper offer cooperation to all persons from the industry interested in verification of this procedure also for other types of machines.

References

1. ČSN ISO 10816-1 + Amd. 1 Vibrace - Hodnocení vibrací strojů na základě měření na nerotujících částech - Část 1: Všeobecné směrnice. Praha, ČNI 1998.
2. ČSN ISO 8258 Shewhartovy regulační diagramy. Praha, ČNI 1994.
3. ČSN ISO 13373-1 Monitorování stavu a diagnostika strojů - Monitorování stavu vibrací – Část 1: Obecné postupy. Praha, ČNI 2003.
4. Jankových R., "Vibrační diagnostika odměrových mechanismů tanků", Habilitační práce, Vojenská akademie, Czech Republic, Brno, 2004.
5. Hammer M., "Metody umělé inteligence v diagnostice elektrických strojů", BEN – technická literature, Czech Republic, Praha, 2009.
6. Berry J. E., "Proven Method for Specifying Spectral Band Alarm Levels and Frequencies Using Today's Predictive Mainetenance Software Systems", USA, Charlotte, 1997.
7. Broch T.J., "Mechanical Vibration and Shock Measurements", K. Larsen & Son, Denmark, Naerum, 1984.
8. Hines J.W., and all., "Study Shows Shaft Misalignment Reduces Bearing Life" In: Maintenance Technology (4/1999), pages 11-17.
9. Tůma J., "Zpracování signálů získaných z mechanických systémů užitím FFT", Sdělovací technika, Czech Republic, Praha, 1997.
10. White G.D., "Introduction to Machine Vibration", DLI Engineering Corp., USA, Bainbridge Island, 1997.

Securing communication layer of uBUS protocol

Juraj Ďuďák[1], Gabriel Gašpar[2], Štefan Šedivý[1]

[1] Research Centre University of Žilina, Žilina, Slovakia
{juraj.dudak, stefan.sedivy, ludovit.dvorsky}@rc.uniza.sk
[2] Faculty of materials and technology in Trnava, Slovak university of technology in Bratislava
ggaspar@gmail.com

Abstract. The essential component of all control and sensory systems is a communication layer. The MODBUS protocol is used mainly in automation and sensory industries and defines a simple format of communication with the master/slave principle. uBUS protocol derived from MODBUS brings advanced features for today's industrial requirements. The implementation of secure transmission among nodes on the bus is a part of the protocol and can be prohibited or enforced. As an encryption method was chosen RSA asymmetric encryption and the encryption strength is given by the size of its keys (32 bit - 1024 bit).

Keywords: MODBUS protocol extension, communication protocol, secure communication.

1 Introduction

The essential component of all the control and sensory systems is a communication layer. There are several communication standards for specific industries such as CAN for automotive, FIELDBUS for automation (factory automation environment), M-bus for sensory systems. The MODBUS protocol is used mainly in automation and sensory industries and defines a simple format of communication with the master/slave principle. Since it was developed in 1979, the possibilities of its use are limited to equipment which was used at the time of its creation. For new types of terminals such as smart sensors and intelligent control modules was defined uBUS protocol derived from the MODBUS[6] protocol, which retains the principle of communication and the specification of the physical and data link layer. Certain modifications were made in the application layer, where were added function codes for new types of terminal nodes. uBUS has been utilized in a variety of applications from simple data collection from connected sensors, through smart sensors, to use in control applications.

Communication protocols generally do not solve security issues against attacks and planted false information. Such an attack could in critical applications cause significant material and environmental damage.

R. Jabłoński and T. Brezina (eds.), *Advanced Mechatronics Solutions*,
Advances in Intelligent Systems and Computing 393,
DOI 10.1007/978-3-319-23923-1_3

The issue of of securing communication is in MODBUS protocol elaborated for MODBUS over TCP/IP. The security of MODBUS protocol is discussed in the sources [1] and [2]. There is not defined any mechanism for ensuring communication using MODBUS over serial line. The issue of the lack of secure communication on the level of serial communication is to some extent due to the nature of the end devices on MODBUS bus.

The article describes the uBUS protocol, which is derived from the MODBUS protocol. The reasons for the need of developing a modification of the MODBUS protocol were thoroughly defined. One of the main reasons for this modification is to add secured communication at RS-485 serial bus.

2 Communication protocol uBUS

2.1 MODBUS protocol

MODBUS is a protocol of the application-layer of ISO/OSI reference model, which provides client/server communication between devices connected on different types of buses or networks.

Devices on the MODBUS bus system communicate using master/slave method (client-server), which may be initiated only by one communication device and it is always the master (client). Other devices (slave, respectively server) respond by sending the required data to the master device or only by an action specified in the received message. Slave devices are peripheral devices (I/O converter, valve or other measuring device) which are able to process the information and send the output to the MODBUS master device via the bus.

MODBUS application protocol defines the protocol data unit (PDU - Protocol Data Unit), which is independent of the lower layers of the communication. For the implementation of the MODBUS protocol for a specific bus or network, are to PDUs added additional data fields. Node that initiates MODBUS serial line transaction creates enhanced MODBUS frame ADU (Application Data Unit), where are added additional fields (Fig.1).

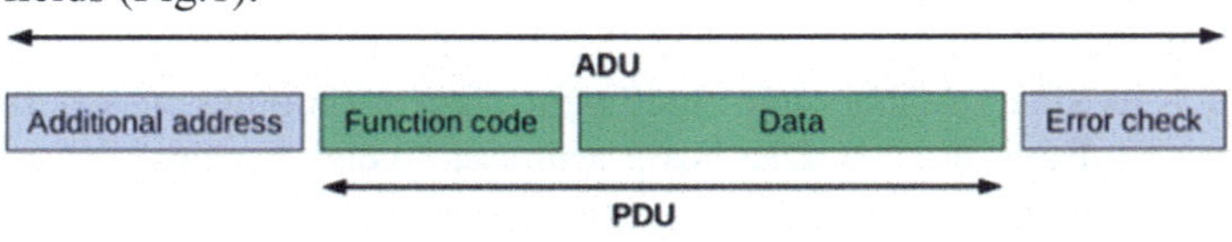

Fig. 1. MODBUS data frame

2.2 uBUS protocol

Currently, the original specification of the MODBUS application layer does not fit the new control and sensory systems containing smart sensors [3]. uBUS protocol uses the identical specification of communication layer on the 1. and 2. level of ISO/OSI reference model. The differences are only in the application layer. In the data frame format, the difference is only in the address part. Table 1 shows a comparison of the changes.

Table 1. Comparison of the address space of the MODBUS and uBUS protocols

Protocol	Address length	Properties
MODBUS	1 Byte	The number of addressable nodes is 255, 247 real addresses (according to the original protocol specification)
uBUS	2 Byte	The number of addressable nodes is 65535, as specified for MultiSlave it is 4095 terminal nodes.

3 uBUS application protocol specification

In the following text will be listed the basic functions, respectively commands in the uBUS application protocol. Data frame is shown in Fig. 1. The commands are divided as follows: diagnostic, configuration, data access and special.

3.1 MultiSlave extension

In the uBUS application protocol has the Address field 2 bytes. This change is related to the following facts:

- *Advanced features of uBUS protocol.* Part of the uBUS specification is MultiSlave mode, where one physical slave device can act as multiple logical terminal nodes. These logic devices are implemented in software.
- *Extension of the address space.* Using MultiSlave terminal nodes radically reduces address space. At the capacity of 16 logical devices for a MultiSlave device, the address space is narrowed to 16 devices (4-bit for the device address and to 4-bit for the address of the logical device within MultiSlave)

The motivation for the proposal of MultiSlave extension was to use throughput capacity of devices on which is implemented a slave node. In many cases the sensor is implemented in special hardware which performs the actual measurement (eg. using the A/D converters) and the subsequent preprocessing of the measured value. Since this hardware is a part of uBUS bus, there has to be an implemented protocol UBUS in the device. In order to use computing capacity of such hardware, there can be implemented multiple slave nodes or special equipment slave (MultiSlave) on a given hardware, which in one device defines the so-called virtual slave nodes. The principle of addressing is as following:

- MultiSlave node address: 12bit,
- address of virtual slave within MultiSlave: 4 bit.

In overall, the address of the terminal slave is 16bit (2 Byte), of which the last 4bits define the address of the virtual slave node within MultiSlave.

3.2 Securing communication of uBUS protocol

MODBUS protocol does not define any security encryption of the communication, it means that the communication is always transmitted in a clear, readable form. An attacker can quite easily enter the communication in a server role and gain the full control over all slave devices. He can set new values, new parameters, even reboot. This vulnerability is known as “Man-in-the-middle”.

The goal of uBUS protocol is to eliminate this apparent disadvantage of MODBUS protocol. The implementation of secure transmission among nodes on the bus can be prohibited or forced. The terminal MultiSlave device decides whether to use encryption or not at the communication initialization. This is based on its configuration. For the used communication encryption were defined following conditions:

1. Encryption must be strong enough so that the immediate decryption is not possible (without the knowledge of the decryption key).
2. Encryption must be extremely simple to be implementable on the slave devices, with a low computing power.
3. Data frame format shall not be altered.

To achieve these conditions we propose the following solutions:

A) RSA asymmetric encryption will be used. This type of encryption ensures ease of implementation at the MultiSlave devices, while providing strong encryption on the principle of a pair of keys - encryption and decryption. Encryption strength is given by the size of the key (32 bit - 1024 bit). The key length is not limited to the actual hardware of the terminal MultiSlave device, where the encryption is implemented.

B) The proposed uBUS extension must be backwards compatible with MODBUS, respectively uBUS. It must therefore be guaranteed that it will work even when communication does not have encryption. It must be clearly identifiable when it comes to unencrypted and when to the encrypted type of communication.

For added security, there is an implemented "floating code" mechanism, whereby the encryption key is only valid for a certain period (of the order of minutes). After this time will be generated new encryption keys. For the distinction of encrypted and unencrypted communication is used byte order in CRC16 checksum. The length of CRC part is 2B, so it can be written as:

$$CRC = CRC_{hi}\,CRC_{lo} \tag{1}$$

where CRC_{hi} is the first byte of the checksum and CRC_{lo} is the second byte of the checksum. In the case of encrypted communication, the right cyclic shift is applied to CRC. According to the original specification of the MODBUS application protocol, after receipt of a message by a slave node the first operation to make is the CRC checksum. If an error is detected when calculating the CRC checksum, the message is marked as damaged and is not further processed. While using the uBUS protocol, the message can be encrypted. This means that the last 2 bytes (CRC sum) is shifted right.

3.3 Implementation of 32-bit encryption on the MultiSlave device

MultiSlave device is implemented on a single physical hardware and can contain up to 16 virtual slave end devices. For the MultiSlave implementation was selected 32-bit ARM microcontrollers, specifically STM32F030. The microcontroller enables performance scaling. For testing purposes, it was set to max clock frequency, i.e. f=48MHz. Communication interface USART has been configured to 19200 Bd. The time required to encrypt responses for different lengths of messages is shown in Table 2.

Table 2. Typical encryption delays measured with STM32F0

Number of bytes for encryption	Encryption time
2	0.79 ms
4	1.5 ms
8	3.9 ms
16	7.9 ms

At the speed of USART serial interface used for communication with a host computer, the time required to encrypt the message is approximately the same as the time of transmission.

4 Implementation of uBUS protocol in real systems

In the following text are examples of deployment of uBUS protocol in real systems.

4.1 Solar water heating system

uBUS protocol uses a variety of physical media for communication. In solar systems which may be relatively large (from the collector to a point of consumption can be tens of meters), cabling costs play a not insignificant role and therefore the trend is to make use of existing physical media, e.g. existing powerlines, HF communications in the public zone etc. The uBUS communication protocol has been proposed so that the threat of compromising transmitted information was minimized or eliminated. For small applications such as solar system with low power, the risk lays particularly in hijacking telemetry data. Actuators (pumps and valves) are usually placed close to the control unit, in larger systems should be considered the secured control of components such as solenoid valves, pumps, motors and the like. In our solar system are installed 3 solar collectors in series with sensory and control elements [5]. All the devices have uBUS protocol implemented on-board.

4.2 System for acquisition of environmental data via wireless sensors

For the needs of monitoring the temperature of biological materials stored outdoors arose the problem of monitoring the course of temperature in such a mass of material. In the case of woodchips (wood pellets) the improper storage can cause slow autoignition. To monitor the temperature of such a biomass, has been designed sensory system consisting of several measuring probes and a number of receivers. The number of measuring probes is in the tens and the number of receivers in the order of units. The final number of the receivers depends on several factors such as the actual size of the monitoring site, its topography and the presence of interfering elements such as metal construction and wireless transmission. To transfer data from the receivers of the radio signal to a host computer is used uBUS protocol.

5 Conclusions

This paper presents securing of a uBUS protocol communication layer against planted false information in a wide range of applications. uBUS protocol extends the address space, allows to utilize the computing capacity of the microcontroller used in the sensor using MultiSlave extension where one physical device slave can contain multiple logical devices. MODBUS protocol does not contain any encryption mechanism, but using the uBUS protocol allows adding the encryption layer. uBUS protocol implementation optionally provide secured transmission between nodes on the bus. For securing communication we use asymmetric cipher on the RSA principle.

Acknowledgement
The research is supported by the European Regional Development Fund and the Slovak state budget by the projects "Research Centre of the University of Žilina" - ITMS 26220220183.

References

1. Dao-gang Peng ; Coll.: Design and Realization of Modbus Protocol Based on Embedded Linux System In: Embedded Software and Systems Symposia, 2008. ICESS Symposia '08. International Conference. ISBN: 978-0-7695-3288-2 , pp. 275-280
2. Hayes, G.: Securing modbus transactions using hash-based message authentication codes and stream transmission control protocol. In: Communications and Information Technology (ICCIT), 2013 Third International Conference. ISBN: 978-1-4673-5306-9, pp. 179-184
3. Dudak, J., Gaspar, G., Kebisek, M.: Application of modified MODBUS protocol for the needs of modern measurement systems. In Advances in Mechatronics 2011: Proceedings of the 6th International Conference on Advances in Mechatronics 2011 (AIM′11), Brno, Czech Republic, pp. 9–14 University of Defence, Brno (2011)
4. Giovanni A. Cagalaban, Yohwan So, Seokso Kim: SCADA Network Insecurity: Securing Critical Infrastructures through SCADA Security Exploitation. In: Journal of Security Engineering. Vol. 6, No. 6, 2009. ISSN: 1738-7531
5. Gaspar, G., Pavlikova, S., Masarova, R.: Distributed control system of solar domestic hot water heating using open-source. In Mechatronics 2013: 10th International conference. Brno, Czech Republic, pp. 749–756. Springer International Publishing (2014)
6. MODBUS APPLICATION PROTOCOL SPECIFICATION V1.1b. Modbus. [Online] 28.4.2015. http://www.modbus.org/docs/Modbus_Application_Protocol_V1_1b.pdf.

Application of Neural Networks in Fault Classification of a kind of Clutch Mechanism Retainer

Ebrahim Ebrahimi[1], Saman Bahrami[1], Nasrolah Astan[1], Maziar Mahdipour Jalilian[1]

[1] Department of Mechanical Engineering, Kermanshah Branch, Islamic Azad University, Kermanshah, Iran
ebrahimi@iauksh.ac.ir
e.ebrahimi57@gmail.com

Abstract. The purpose of this study is to introduce an intelligent method for fault classification of farm machinery with vibration analysis using the ANFIS. Test conditions include safe, roller fault, seal fault and axis friction on the experimental setup of retainer and clutch mechanism of Massey Ferguson. The time-domain vibration signals with normal and defective modes processed for feature extraction. Three features of the time domain (T_{11}), frequency domain (A_8) and phase angle (A_9) data as premium features were selected. Data are classified into eight experimental models and also loaded in the ANFS network. In all models, appropriate membership functions were selected. For ANFIS training 1000 epoch was considered. The statistical indicators and the result of ANFIS prediction of evaluation models were presented. Total classification accuracy was 100% in both models. The results showed that ANFIS can be used as a powerful tool for intelligent fault classification of tractor mechanisms.

Key words: vibration signals, ANFIS, premium features, fault classification.

1 Introduction

Agricultural machines should be available in a timely condition. Early diagnosis can stop car suddenly. Condition monitoring is used to find faults at early level [1]. Maintenance procedures are often provided by the equipment manufacturers to inhibit major fault occurrences, leading to possible shutdowns [2]. Maintenance of rotary machineries are the essential components of abnormal event management, that it attracts more attention. The components such as shafts, bearings, and gears are important machine elements for any rotary machine [3]. In recent years the complexity of mechanical systems has been increased and the maintenance of such systems has become an important and inevitable task in industry. Visual inspection and physical assessment alone no longer provide adequate early warning to any emerging problem in a complex system which strict down the time, permanency of cost and increase of damage. Vibration signal analysis has emerged as an extremely useful and essential for early warning technique to predict onset of defect in its nascent form thus giving adequate indication and time to plan preventive permanency. The relations among vibration amplitude and frequency or time, and etc. obtained from the recorded vibra-

R. Jabłoński and T. Brezina (eds.), *Advanced Mechatronics Solutions*,
Advances in Intelligent Systems and Computing 393,
DOI 10.1007/978-3-319-23923-1_4

tion signal of machine system are known as machine signature [4]. The suitable methods for processing measured data contain the frequency domain technique, time domain technique, and time– frequency domain technique [5]. This feature of time-frequency analysis meets the requirements for analysis vibration signals that are non-stationary [6]. Numerous vibration feature extraction techniques have been developed to date. Time-frequency analysis has been applied to machinery fault diagnosis due to its advantages in the representation of signals in both the time and frequency; a feature extraction and evaluation method was proposed for fault diagnosis of rotating machinery. Based on the central limit theory an extraction procedure is presented to compute statistical features with the help of existing signal processing tools. Such statistical features are close to normal distributions.

The raw features which are directly computed with suitable feature extraction methods that compared with the statistical one by using them as the inputs for ANN and SVM based on fault classifiers [7]. Fast Fourier transform (FFT) analyzers became effective for general applications and its cost. The raw signatures were acquired through a vibration sensor needed further processing and classification of the data for any meaningful surveillance of the condition of the system being monitored [5]. They largely increase the reliability of fault detection and diagnosis systems. ANFIS is a hybrid model which combines the ANNs adaptive capability and the fuzzy logic qualitative approach [8]. ANFIS maps inputs through input membership functions and associated parameters, and then through output membership functions to outputs. The initial membership functions and rules for the fuzzy inference system can be designed by employing human expertise about the target system to be modeled [9]. The ANFIS learns features in the data set and adjusts the system parameters according to a given error criterion [10]. ANFIS adopts the combination of the neural and FL approaches to exploit the advantages of both, for example the simple learning procedures and computational power of ANN and reasoning of fuzzy systems. The result offers an appealingly powerful framework for tackling practical classification problems. Networks (ANNs) and fuzzy logic have been successfully applied to automated detection and diagnosis of machine conditions. They largely increase the reliability of fault detection and diagnosis systems [11]. The research has shown that fault detection in machines is classified for using vibration signals and ANFIS [12-15]. The purpose of this study is to introduce an intelligent method for fault classification of farm machinery with vibration analysis using the ANFIS.

2 Material and method

A method is introduced for intelligent fault classification of farm machinery with vibration analysis and using ANFIS. Experiments has performed on the set up of the holder clutch mechanism of Massey Ferguson tractor in a single laboratory in Iran as shown in Figure 1.

Fig. 1. Laboratory set up

After assembling Laser, Digital Tachometer (D- T2234- B) and industrial Dimmer were used to supply different input shaft speeds. Electric motor Stream type (YCA90S-2, 0.75 kW), was coupled to the main shaft and applied the load mechanism.

Data acquisition was done in 4 modes includes safe, roller fault, seal fault and axis friction. An accelerometer (VMI Ltd, VMI-102 Sweden)was installed in two positions: vertical and horizontal and connected to the analog to digital converter (APC-40). Data was acquired in different speeds 1000, 1500 and 2000 rpm, 24 states was selected. For each case, 130 samples were obtained. That is for the safe mode there were 780 cases and 2340 cases for defective modes and totally 3120 samples are observed. a SONY VAIO laptop and ARMA software were applied. Data for each sample was recorded. 10240-voltage, data at 4 seconds interval and was saved in an Excel file.

Which; A is safe mode, B is failure mode, C is location of sensor and D is speed (rpm). Specifications of each failure were includes the inner seal is turn and the coil is damage in seal failure, shaft diameter is reduced up to 10 percent in shaft friction and three roller balls and the inner ring are destroyed in roller failure.

3 Signal Processing and Feature Extraction

For signal processing and feature extraction we need a code in a software e.g. MATLAB & Simulink Release in 2010. It is the most important part of the monitoring process. Some of outstanding features and a lot of troubleshooting are closely related to each other and others are prominent. This Excel program contains vibration data and using the filter applied, the data filtered the noises that eliminated, finally the time and frequency statistical parameters calculated. Thus, some conversion features of Fourier series (FFT) with 10240 points were calculated for each signal. Also, the power spectrum density (PSD) and phase angle (FFT) of vibration signals calculated for using MATLAB as illustrated.

4 ANFIS Modeling

Because with ANFIS network output cannot be a linguistic variable, data classification and ANFIS modeling were used for each of the modes that defined codes .

For ANFIS modeling characteristics of 2000 rpm were considered. Table 3 shows classification of modes and ANFIS input to the models.

Because Gaussian and bell-shaped membership functions are smoothly and concisely, they are the most popular membership functions that associated with fuzzy sets. So, in models number 1, 2, 3 and 4, a Gaussian membership function and in models 5, 6, 7 and 8 a bell-shaped membership function has been selected for inputs and linear membership function has selected for outputs.

5 Discussion and Conclusions

After data acquisition and signal processing, according to the structure of fuzzy rules shown in Figure 2,show number of three vectors feature were used as input to the ANFIS network.

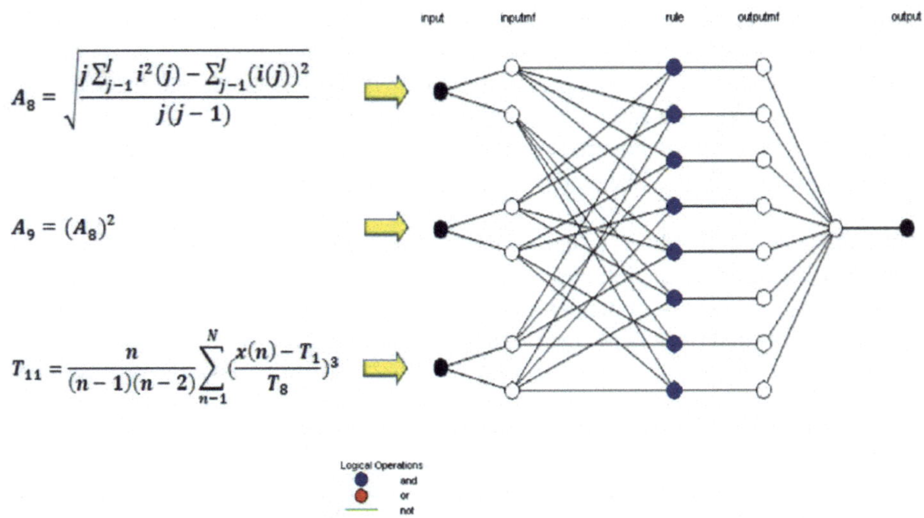

Fig. 2. Fuzzy rules structure

For training of the system, number of 1000 epochs were considered, and ahybrid learning approach was selected. ANFIS results for each model include statistical index (R) and thepredicted values by the ANFIS. To assess the accuracy of each model that related to the correlation coefficient, ‘R’ was used as shown in Table 5. A correlation coefficient expresses the degree of correlation among the results predicted by the model and the data. The equation is as follows.

Where: y_{act}actual observed values,$\bar{y}_{act}$ average actual observed values, y_{act}estimatedvalues ofoutput, $\bar{y}_{ast}$Average estimate of output in model.

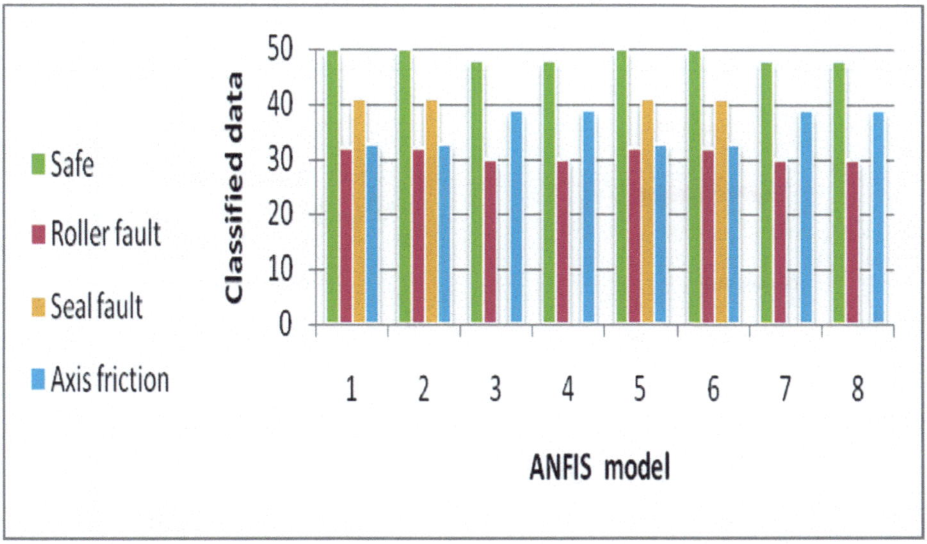

Fig. 3. Number of data classified in each model

ANFIS results were observed values by the amounts which recognized with the graphs shown in Figure 4. It was seen that in models 1, 2, 5 and 6, distribution of detected responses was high. But in models 3, 4, 7, and 8, which seal failure data was not classified, and predict values was more correctly and less fragmentation was observed. Thus, it’s based on the results the fuzzy neural inference system and could not detect seal failure that was detected very well before. While in all models failure in roller bearing detected with 100% accuracy. Based on ANFIS results of the all models, the correlation coefficient between of actual values and predicted values closer to 1. R values have been shown in Table 5 and the results can be seen in the graphs in Figure 8 as well as models that have a higher R, could detect failures better than other models. Then correct and incorrect responses were detected by ANFIS counted. Confusion matrix was formed and the performance of each model was assessed. The total classification accuracy of the models is shown in figure 4. Models that have higher total classification accuracy can be used as desirable models for fault diagnosis.

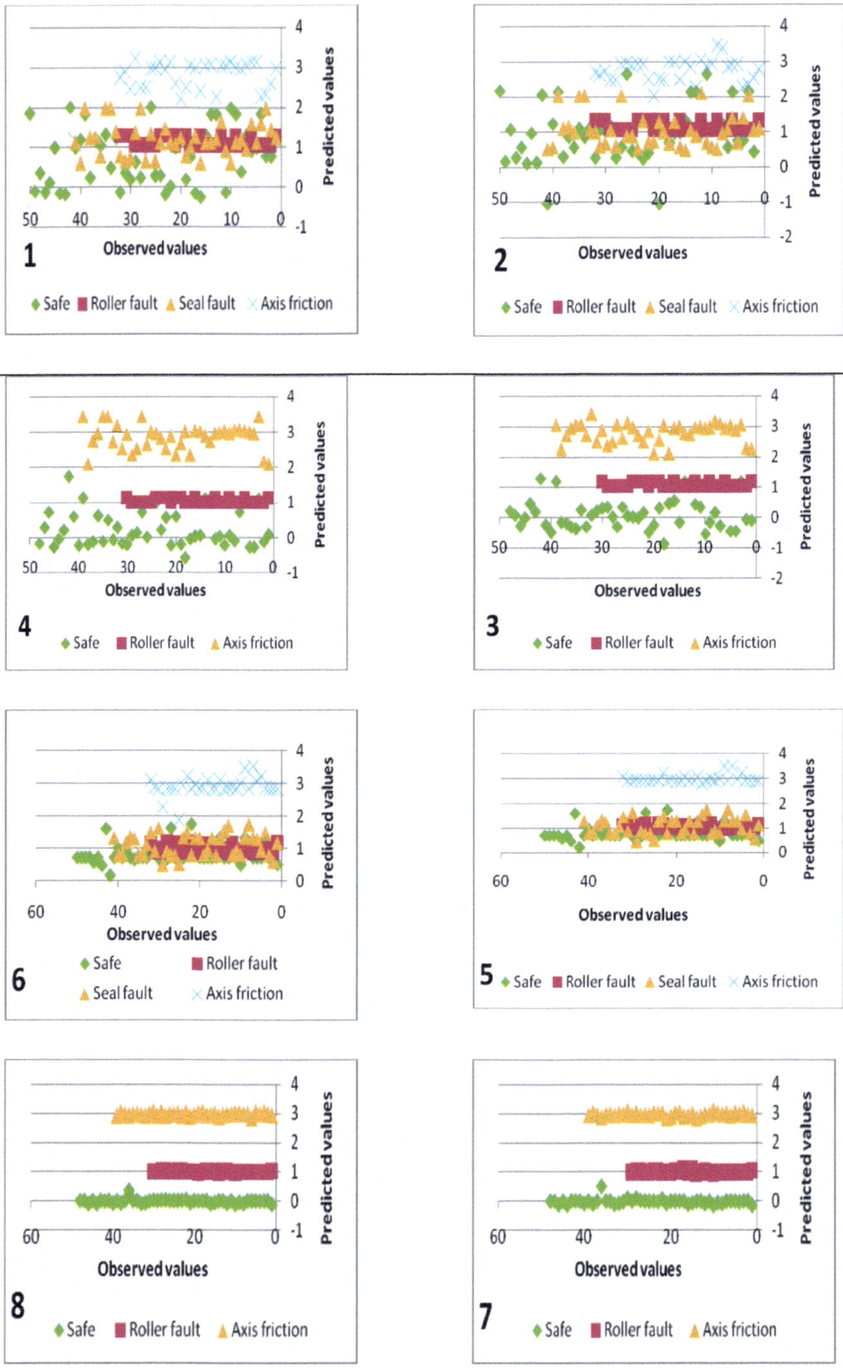

Fig. 4. Fault classification result with ANFIS

The results showed that the use of premium features A9 The and A_8 as vectors in ANFIS is an effective mechanism for detecting defects of holder Clutch in Massey Ferguson Tractor. Finally, based on the results, the total accuracy classification of the models number 7 and 8, were 100% as shown in Figure 5. Other models showed high diagnostic accuracy. The failure of sample holder clutch mechanism was identified with high accuracy. The neural fuzzy inference system can be used as a suitable and effective tool for the fault detection and classification of defects in farm machinery.

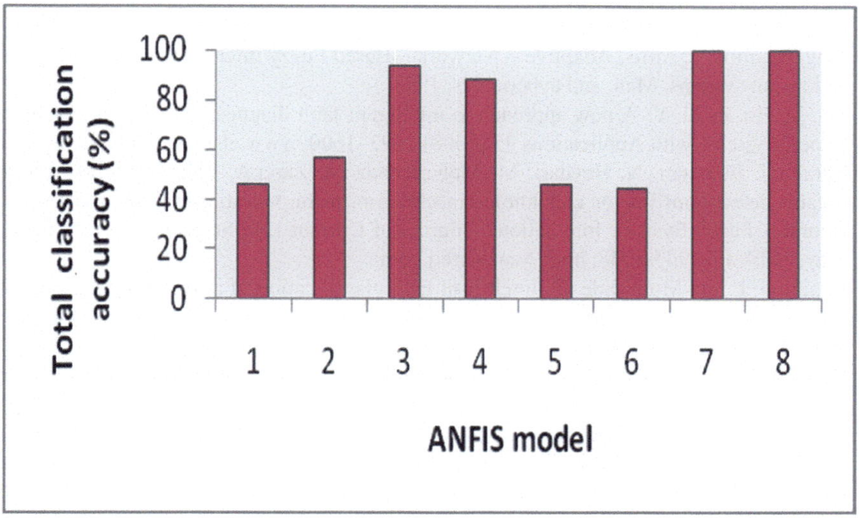

Fig. 5. Total classification accuracy of the models

References

1. Sakthivel N.R., Binoy B. Nair, M. Elangovan , Sugumaran V., Saravanmurugan S., Comparison of dimensionality reduction techniques for the fault diagnosis of mono block centrifugal pump using vibration signals, Engineering Science and Technology, an International Journal 17 (2014) 30- 38, http://ees.elsevier.com/jestch/default.asp.
2. Salahshoor. K , Kordestani. M , Khoshro.M, Fault detection and diagnosis of an industrial steam turbine using fusion of SVM (support vector machine) and ANFIS (adaptive neuro-fuzzy inference system) classifiers, Energy 35 (2010) 5472-5482,www.elsevier.com/locate/energy.
3. Indira V., Vasanthakumari R., Sugumaran V., Minimum sample size determination of vibration signals in machine learning approach to fault diagnosis using power analysis, Expert Systems with Applications 37 (2010) 8650–8658, www.elsevier.com/locate/eswa.
4. Chakravorty. P.K,signal processing in vibration analysis-advanced techniques, Def Sci J, Vol41, No 3, July 1991, pp 241-249.
5. Saravanan N., Cholairajan S., Ramachandran K.I., Vibration-based fault diagnosis of spur bevel gear box using fuzzy technique, Expert Systems with Applications 36 (2009) 3119–3135,www.sciencedirect.com.

6. Yang, Q. 2004. Model-based and data driven fault diagnosis methods with applications to process monitoring. PhD Thesis. Case Western Reserve University.
7. Li. W, Zhu. Z, Jiang. F, Zhou, Guoan Chen. G, Fault diagnosis of rotating machinery with a novel statistical feature extraction and evaluation method, Mechanical Systems and Signal Processing 50-51 (2015) 414–426, www.elsevier.com/locate/ymssp.
8. Lei. Y, Hea. Z, Zia.Y, Hu. O, Fault diagnosis of rotating machinery based on multiple ANFIS combination with Gas, Mechanical Systems and Signal Processing 21 (2007) 2280–2294,www.sciencedirect.com.
9. Lei. Y, Lin. J , He. Zand Kong. D, A method based on multi-Sensor data fusion for fault detection of planetary gearboxes, Sensors 2012, 12, 2005-2017; doi:10.3390/s120202005, www.mdpi.com/journal/sensors.
10. Jang. R, shing. J, Anfis, Adaptive – Network – Based Fuzzy inference system, IEEE transactions on systems, Man, and cybernetics, 1993.
11. Lei. Y, He. Z, Zi. Y, A new approach to intelligent fault diagnosis of rotating machinery, Expert Systems with Applications 35 (2008) 1593–1600, www.elsevier.com/locate/eswa.
12. Ebrahimi. E, Astan. N, Heydari. M, Mohammadi. M, Zarei.A, Effectsof sensorsposition onfault detectionof tractor clutchholdermechanism, using vibrationanalysis and Inference Neural – Fuzzy System, International Journal of Current Life Sciences - Vol.4, Issue, 5, May, 2014. Pp 1943-1948. http://www.bretj.com.
13. Ebrahimi. E and Mollazade. K, Intelligent fault classification of a tractor starter motor using vibration monitoring and adaptive neuro-fuzzy inference system, Insight Vol 52 No 10 October 2010. Pp 561-567.
14. Hussain. S, and. Gabbar. H, vibration analysis and time series prediction for wind turbine gearbox prognostics, International Journal of Prognostics and Health Management, ISSN 2153-2648, 2013 014. Pp 1-11.
15. Li. H, Zhang. X, and Xu. F, Experimental investigation on centrifugal compressor blade crack classification using the squared envelope spectrum, Sensors 2013, 13, 12548-12563; doi:10.3390/s130912548. Pp 12934-12963. www.mdpi.com/journal/sensors.

Model-based Predictive Control of IC Engine Involving an Algebraic Link between Plant Inputs and Outputs

Martin Florian[1], Vit Dolecek[1], Zbynek Sika[1], Pavel Steinabauer[1]

[1] CTU in Prague, Faculty of Mechanical Engineering
{martin.florian,pavel.steinbauer,zbynek.sika}@fs.cvut.cz

Abstract. The paper deals with the predictive control of a supercharged IC engine, based on a real-time predictive model of the engine. This allows for using signals in the feedback loop are not actually measured, just estimated by the predictive model. New approach to the control law computation is proposed, involving a direct algebraic link between the inputs and outputs of the controlled plant. The control is tested using both the MIL and HIL approach.

Keywords: IC engine control, model-based predictive control, MPC, predictive models, dynamic system identification, LOLIMOT

1 Introduction

At present, there is a general trend to increase the energetic efficiency of mechanical (mechatronic) devices and appliances, and to minimize their negative impact on the living environment at the same time. The very tendency may be perhaps best observed in the development of transportation means, especially their driving units. This is related to the ever-increasing number of vehicles in operation, globally rising oil price (in the long term), and also general desire to limit the production of waste energy and pollutants.

In case of internal combustion (IC) engines, the stated reasons stand behind the development of advanced control units able to optimize fuel consumption and emission production. Such a development is mostly carried out by engine manufacturers themselves, and is a part of their know-how, the details of which are rarely published.

The goal of the present work is to propose a universal control framework that could be used for various control tasks related to IC engines. Such a demand calls for the application of modern control methods, which would treat the considered dynamic system as one complex multi-input-multi-output (MIMO) unit.

Model-based predictive control (MPC) [2.,3.,4.] is a natural choice here, since discrete predictive models are very well suited to the "clockwork" nature of IC engines. A reliable predictive model of the controlled plant is crucial for the proper control function, so one of the important tasks is finding a suitable method of creating such a model, i.e. of the system identification.

One of the most promising and universal methods of dynamic systems identification is LOLIMOT [1.]. It can produce robust predictive models that have the advantage of being piecewise linear, so that linear predictive control may be based on them.

R. Jabłoński and T. Brezina (eds.), *Advanced Mechatronics Solutions*,
Advances in Intelligent Systems and Computing 393,
DOI 10.1007/978-3-319-23923-1_5

2 Predictive model of an IC engine

Predictive models have an ability to predict the behaviour of a dynamic system in the (near) future, therefore they are used for the calculation of the predictive control law. A predictive model is usually a discrete state model of the system in fact.

2.1 LOLIMOT

One of the most promising methods of predictive model generation is an algorithms called LOLIMOT [1.]. Its features are robustness (if well applied), real-time capability and versatility – it does not depend on the fact whether a simulation model is available of a given dynamic system. What is however necessary to have for the proper identification of the system, are comprehensive training data sets, i.e. records of the dynamic responses of the system (or its model) on a number of input excitations.
LOLIMOT produces a combination of local input-output models that approximate the nonlinear behaviour of an identified system by linear functions valid in particular subregions of the whole domain of definition. The mathematical models of a system identified with the LOLIMOT are called 'LOLI-models' (LLMs).
The basic principle of LOLIMOT is the approximation of the generally non-linear multivariable input-output function of a system by the scalar product of the vector of linear input-output functions and the vector of validity functions. Each linear function approximates the system output in sub-region determined by a relevant validity function. The output of the model can be then written as:

$$
\begin{aligned}
\tilde{y} = \sum_{i=1}^{M} \tilde{y}_i \Phi_i(\mathbf{u}_L) = \sum_{i=1}^{M} \Big(& w_{i,0} + w_{i,1}u_1(k) + w_{i,2}u_1(k-1) + \ldots + w_{i,n_1+1}u_1(k-n_1) + \\
& + w_{i,n_1+2}u_2(k) + w_{i,n_1+3}u_2(k-1) + \ldots + w_{i,n_1+n_2+2}u_2(k-n_2) + \\
& \ldots + w_{i,n_1+n_2+\ldots+n_{p-1}+p}u_p(k) + w_{i,n_1+n_2+\ldots+n_{p-1}+p+1}u_p(k-1) + \ldots + w_{i,n_1+n_2+\ldots+n_p+p}u_p(k-n_p) + \\
& + w_{i,n_1+n_2+\ldots+n_p+p+1}y(k-1) + w_{i,n_1+n_2+\ldots+n_p+p+2}y(k-2) + \ldots + w_{i,n_1+n_2+\ldots+n_p+n_y+p}y(k-n_y) \Big) \Phi_i(\mathbf{u}_L)
\end{aligned} \tag{1}
$$

where M is the number of LLMs, $\tilde{y}_i$ is the output of the i-th LLM, $\mathbf{u}_L = [u_1(k), u_1(k-1), \ldots, u_1(k-n_1), u_2(k), \ldots\ u_p(k-n_p), y(k-1), \ldots, y(k-n_y)]^{\mathrm{T}}$ is the vector of inputs, $\Phi_i(\mathbf{u}_L)$ is a validity function for the i-th LLM (designed as a normalized orthogonal Gaussian function), $\mathbf{w} = \left\lfloor w_{i,0}, w_{i,1}, \ldots, w_{i,n_1+n_2+\ldots+n_p+n_y+p} \right\rfloor^{\mathrm{T}}$ is the vector of the parameters of the i-th LLM.
The process of computing LLMs parameters, i.e. the identification of a given dynamic system, is called 'training of LLMs', and the computation is based on training signals.

2.2 Generating the predictive model of an IC engine

LOLIMOT was utilized to produce piecewise-linear predictive model of the engine. LLMs were trained using two independent sources of dynamic response data sets – a simulation model (for a model-in-the-loop simulation) and a real test-bed engine (for a hardware-in-the-loop arrangement).

2.2.1 Tested Engine

A six-cylinder diesel engine IVECO was used for the purpose of this work, equipped with a common rail fuel injection system. The original turbocharger was replaced by another one with a variable-geometry turbine (VGT), which is adjustable online. The original ECU of the engine was replaced by the control unit rCube2 (provided by an industrial partner), which is fully open and modifiable.

2.2.2 Simulation model

For the development and testing of the predictive model and controller, a fast running model (FRM) was generated by the simplification of a detailed 1-D model created in GT-Power, where especially manifold pipes were substituted by 0-D volumes. FRM had to be further simplified to gain real-time capability (enlarging sample time, etc.), while maintaining an acceptable level of accuracy.

2.2.3 Predictive model based on LOLIMOT

Although calculated LLMs generally constitute a nonlinear predictive system, their big advantage is that they can be directly transformed into a discrete state-space description with *locally* constant state matrices **A**, **B**, **C**, **D**:

$$\begin{aligned} \Delta \boldsymbol{x}_{k+1} &= \mathbf{A}\Delta \boldsymbol{x}_k + \mathbf{B}\Delta \boldsymbol{u}_k \\ \Delta \boldsymbol{y}_k &= \mathbf{C}\Delta \boldsymbol{x}_k + \mathbf{D}\Delta \boldsymbol{u}_k \end{aligned}, \tag{2}$$

where $\Delta \boldsymbol{x}$ are system state variables, $\Delta \boldsymbol{u}$ are system inputs (control variables) and $\Delta \boldsymbol{y}$ are system outputs (i.e. variables to be controlled).

Three quantities were designated the states – engine speed, turbocharger speed, and boost pressure. LLMs were subsequently calculated of the state variables. Each LLM had five input signals (two actual control inputs and the past values of three state variables themselves). Another LLM was computed for the engine torque, which was (together with the boost pressure) chosen as the controlled quantity.

3 Model-based predictive control of an IC engine

A model-based control scheme utilizes a concurrently running numerical model as a basis for the application of the control law. One of the benefits of this approach is the possibility to replace measurements by computations, which considerably reduces demands on the instrumentation of the whole control system. This may have a big impact on both the price and the reliability of the system. Moreover, the simulation model may provide for some data that are not measurable using standard means.

Rewriting the state model (2) for N subsequent steps, while getting rid of the incremental form of the states, inputs and outputs (i.e. writing $\boldsymbol{x}$, $\boldsymbol{u}$, and $\boldsymbol{y}$ instead of $\Delta \boldsymbol{x}$, $\Delta \boldsymbol{u}$, and $\Delta \boldsymbol{y}$), one gets the formula sequence (3). The inclusion of the direct algebraic link between the plant input and output variables (via the matrix **D**) within the control framework is in fact rather novel (not seen in the available sources). It naturally resulted from the practical application demands.

$$
\begin{aligned}
&(\boldsymbol{y}_k = \mathbf{C}\boldsymbol{x}_k + \mathbf{D}\boldsymbol{u}_k)\\
&\boldsymbol{x}_{k+1} = \mathbf{A}\boldsymbol{x}_k + \mathbf{B}\boldsymbol{u}_k\\
&\boldsymbol{y}_{k+1} = \mathbf{C}\boldsymbol{x}_{k+1} + \mathbf{D}\boldsymbol{u}_{k+1} = \mathbf{C}\mathbf{A}\boldsymbol{x}_k + \mathbf{C}\mathbf{B}\boldsymbol{u}_k + \mathbf{D}\boldsymbol{u}_{k+1}\\
&\boldsymbol{x}_{k+2} = \mathbf{A}\boldsymbol{x}_{k+1} + \mathbf{B}\boldsymbol{u}_{k+1}\\
&\boldsymbol{y}_{k+2} = \mathbf{C}\boldsymbol{x}_{k+2} + \mathbf{D}\boldsymbol{u}_{k+2} = \ldots = \mathbf{C}\mathbf{A}^2\boldsymbol{x}_k + \mathbf{C}\mathbf{A}\mathbf{B}\boldsymbol{u}_k + \mathbf{C}\mathbf{B}\boldsymbol{u}_{k+1} + \mathbf{D}\boldsymbol{u}_{k+2}\\
&\ldots\\
&\boldsymbol{y}_{k+N} = \mathbf{C}\boldsymbol{x}_{k+2} + \mathbf{D}\boldsymbol{u}_{k+2} = \ldots = \mathbf{C}\mathbf{A}^N\boldsymbol{x}_k + \mathbf{C}\mathbf{A}^{N-1}\mathbf{B}\boldsymbol{u}_k + \mathbf{C}\mathbf{A}^{N-2}\mathbf{B}\boldsymbol{u}_{k+1} + \ldots + \mathbf{C}\mathbf{B}\boldsymbol{u}_{k+N-1} + \mathbf{D}\boldsymbol{u}_{k+N}
\end{aligned}
\tag{3}
$$

Using the sequence (3), the output variables may be written in a special complex matrix form (suitable for the control law derivation) as follows:

$$\hat{\boldsymbol{y}} = \boldsymbol{f} + \mathbf{G}\boldsymbol{u}, \tag{4}$$

where

$$
\boldsymbol{f} = \begin{bmatrix} \mathbf{C}\mathbf{A}^1 \\ \mathbf{C}\mathbf{A}^2 \\ \ldots \\ \ldots \\ \mathbf{C}\mathbf{A}^N \end{bmatrix} \boldsymbol{x}_k, \quad
\mathbf{G} = \begin{bmatrix}
\mathbf{CB} & \mathbf{D} & \mathbf{0} & \ldots & \ldots & \mathbf{0} \\
\mathbf{CAB} & \mathbf{CB} & \mathbf{D} & \ldots & \ldots & \mathbf{0} \\
\ldots & \ldots & \ldots & \ldots & \ldots & \mathbf{0} \\
\mathbf{C}\mathbf{A}^{N-2}\mathbf{B} & \mathbf{C}\mathbf{A}^{N-3}\mathbf{B} & \ldots & \ldots & \mathbf{D} & \mathbf{0} \\
\mathbf{C}\mathbf{A}^{N-1}\mathbf{B} & \mathbf{C}\mathbf{A}^{N-2}\mathbf{B} & \mathbf{C}\mathbf{A}^{N-3}\mathbf{B} & \ldots & \mathbf{CB} & \mathbf{D}
\end{bmatrix}
\begin{bmatrix} \boldsymbol{u}_k \\ \boldsymbol{u}_{k+1} \\ \boldsymbol{u}_{k+2} \\ \ldots \\ \boldsymbol{u}_{k+N} \end{bmatrix}
\tag{5}
$$

The control is derived from the optimization of a quadratic performance index J_k [2.]:

$$
\begin{aligned}
J_k &= \varepsilon\left\{(\hat{y} - w)^T Q(\hat{y} - w) + u^T pu\right\} = \\
&= \varepsilon\left\{(Gu + f - w)^T Q(Gu + f - w) + u^T pu\right\}
\end{aligned}
\tag{6}
$$

The performance index is optimized in the step k using the prediction $\hat{\boldsymbol{y}} = [\boldsymbol{y}_{k+1} \quad \boldsymbol{y}_{k+2} \quad \ldots \quad \boldsymbol{y}_{k+N}]^{\mathrm{T}}$ for the sequence of output vectors. ε is a mean value operator, N is the prediction horizon, $\boldsymbol{y}$ is the output vector, $\boldsymbol{w}$ is the desired output vector, $\boldsymbol{Q}$ is a penalization matrix for the outputs, $\boldsymbol{p}$ is the penalization of the inputs, and $\boldsymbol{u} = [\boldsymbol{u}_k \quad \boldsymbol{u}_{k+1} \quad \ldots \quad \boldsymbol{u}_{k+N}]^{\mathrm{T}}$ is the sequence of output vectors.

From the requirement of the minimization of the performance index

$$J_{k,\mathrm{opt}} \overset{!}{=} \min_{\boldsymbol{u}}(J_k), \tag{6}$$

the control law may be derived:

$$\boldsymbol{u} = \left(\mathbf{G}^{\mathrm{T}}\mathbf{Q}\mathbf{G} + \mathbf{p}\right)^{-1}\mathbf{G}^{\mathrm{T}}\mathbf{Q}(\boldsymbol{w} - \boldsymbol{f}). \tag{7}$$

Only the first element of the vector $\boldsymbol{u}$ is used for the nearest control action.

Speaking in terms of the actual work, the following quantities were designated the system variables (relevant for the control):

- 2 system inputs $\boldsymbol{u}$: 1. fuel mass per cycle, 2. rack position of the VGT;
- 3 state variables $\boldsymbol{x}$: 1. engine speed, 2. turbocharger speed, 3. pressure;
- 2 system outputs $\boldsymbol{y}$: 1. engine torque, 2. intake manifold (boost) pressure.

The engine is supposed to work in a mode of prescribed engine speed, which is being set externally, while controlled variables are the two above mentioned system outputs. While the choice of the desired engine torque is arbitrary (within certain limits), a function (in form of a 2-D lookup table) is imposed on the online calculation of the boost pressure setpoint, which takes engine speed and torque setpoint into account, and its goal is to provide optimum combustion conditions.

4 Simulation results

Model-in-the-Loop simulation scheme was firstly built. It was in fact a co-simulation process, where the fast running model of the engine was run in GT-POWER, while the controller model was implemented in Simulink, including an integrated predictive model of the engine. Sample simulation results are shown in the following pictures. A part of an exhaust emission regulation test (WHTC) was run to test the control system capability. The test is defined by the given time series of the engine speed and torque. The engine speed sequence was set in the simulation model, while the torque was used (together with a boost pressure) as a setpoint for the controller – see Fig.1.

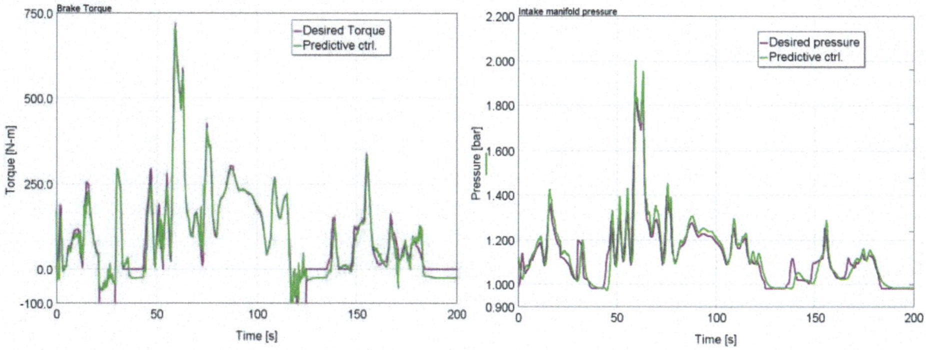

Fig.1. Simulation results: engine torque (left) & intake manifold pressure (right)

The simulation demonstrated a reasonably good agreement between the desired and realized values of the outputs. The deviations from setpoints had several reasons, among them the abrupt changes in engine speed (which in fact act as disturbances from the control viewpoint) and the limited accuracy of the predictive model, which could be bettered using more complete training data during the LLMs training phase.

5 Experimental results

The controller was further tested on a real IC engine in a Hardware-in-the-Loop fashion. The control algorithm was compiled and implemented in the available CPU. The engine was loaded by a dynamometer brake, working in the mode of prescribed engine speed: random speed changes were applied this time, again acting as disturbances, which had to be tackled by the controller. A sample result of the test-bed measurements is shown in the Fig.2 – it shows the comparisons of the realized engine torque and boost pressure vs. their setpoints. There were more sources of errors and disturbances this time, so the agreement between the desired and realized values is a bit worse, but the controller still worked acceptably. There will be an effort to increase the accuracy and stability of the control in the future (again more thorough LLMs training phase, for them to become more robust and precise)

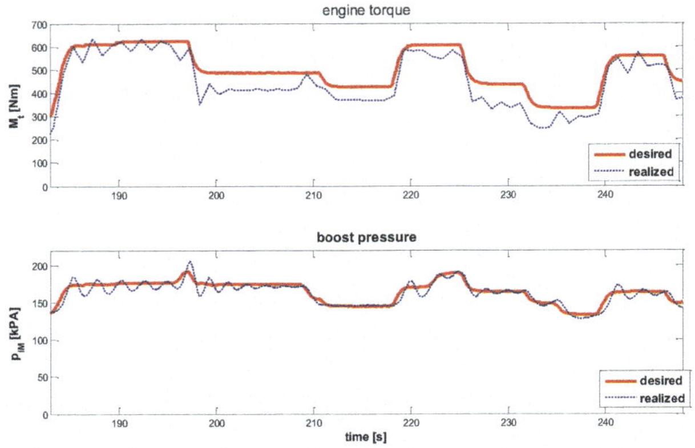

Fig.2. Test-bed results: engine torque (left) & intake manifold pressure (right)

6 Conclusions

Model-based predictive control of combustion engines has been investigated. The utilized predictive models are locally linearized and used as a basis for the linear predictive control law calculation. Own predictive control approach was developed, newly involving a direct algebraic link between the inputs and outputs of the controlled system. The control scheme was tested both in the MIL and HIL arrangement, with the results having turned out to be promising. It is a step forward to the practical application of a versatile and robust multi-input-multi-output control.

Acknowledgment

This research has been realized using the support of EU Regional Development Fund in OP R&D for Innovations (OP VaVpI) and The Ministry of Education, Youth and Sports, Czech Republic, project # CZ.1.05/2.1.00/03.0125 Acquisition of Technology for Vehicle Center of Sustainable Mobility. This research has been further realized using the support of The Ministry of Education, Youth and Sports program NPU I (LO), project # LO1311 Development of Vehicle Centre of Sustainable Mobility. This support is gratefully acknowledged.

References

1. Nelles, O.: Nonlinear system identification with local linear fuzzy-neuro models. Automatisierungs technik, Shaker Verlag, Aachen, 1999.
2. Camacho, E.F., Bordons, C.: Model Predictive Control. Springer Verlag, Berlin, 1999.
3. Valasek, M. et al.: Model Based Predictive Control of Combustion Engine with Constraints, Review of Automotive Engineering of Japan SAE2005, Vol.26, No.3, pp.349-356.
4. Sika, Z., Valasek, M., Florian, M., Macek, J., Polasek, M.: Multilevel Predictive Models of IC Engine for Model Predictive Control Implementation. SAE Technical Paper 2008-01-0209, doi: 10.4271/2008-01-0209.

Modeling of Kinematics, Dynamics and Design of Electronics Control Unit for an Experimental Robot with Hybrid Locomotion

Grepl, R.[1], Čoupek, P[1]., Radoš, J[1]., Konvičný, J[1]., Krejčiřík, M.[1]

[1] Mechlab, Faculty of Mechanical Engineering, Brno University of Technology
{grepl}@fme.vutbr.cz

Abstract.
This paper presents particular design and construction of robot with hybrid locomotion. First, the state-of-the-art situation of wheeled/walking robots designs is described. Hybrid locomotion is chosen as to gain benefits of both robot designs. The resulting mechatronic solution is presented, including the associated modelling of kinematics and dynamics, including used algorithms. Mechanical design is then introduced, consisting of model produced by 3D laser sintering technology, fitted with necessary control and power electronics.

Keywords: mobile robot · hybrid locomotion · walking robot · statics · kinematics · SimMechanics

1 Introduction

In the development of mobile robots, wheeled robots still dominate, having a range of benefits (much simpler design, lower power consumption during movement) when compared to walking robots. But walking robots have significantly higher potential of maneuverability and throughput through difficult terrain.

An interesting area of research and development of robots are robots that combine the characteristics of wheeled and walking robots.

In this paper we will briefly describe an experimental study of such a robot.

1.1 Robots with hybrid locomotion

Primary problem in this area lays in twofold demand that is placed on the robot design. On one hand, in good terrain, robots are capable of relatively high speed movement, using mainly wheels, which from the mechatronic point of view does not present that many obstacles for the design and control of the robot. But such relatively easy solution fails when it comes to various less favorable terrains. Number and variety of obstacles in such conditions pose a serious challenge in terms of actual movement and also in terms of behavior of the robot. In such conditions, designs of walking robots are (often inspired by insects, mammals and by human) the answer to this problem. But

R. Jabłoński and T. Brezina (eds.), *Advanced Mechatronics Solutions*,
Advances in Intelligent Systems and Computing 393,
DOI 10.1007/978-3-319-23923-1_6

even if successful, these solutions provided by walking robot designs suffer from lack of speed in favorable terrain, as the technical solution of the actual "walk" often does not allow high speed movement due various reasons. Moreover, these designs have significantly higher demands on construction and energy consumption as well. Therefore, motivation and faith in the prospects of walking robots is based on prerequisite of continuing progress in actuators and energy sources.

There were some attempt in this field to incorporate the speed, high maneuverability and climbing ability. One of such designs is Roller Skating Robot. This robot [1] moves in irregular terrain like four-legged robot. For this movement, uses lateral side of its wheels. For moving on plane solid surface, it uses its passive wheels. Another example is AZIMUT robot [2]. It can move by walking and driving equally. For walking uses end tip of its leg. For driving robot has two options: wheels and cogged belt. The Halluc II is very sophisticated robot [3] from Japan. Operator can choose from three types of movement: driving using active wheels, insect walking and animal walking. Next construction of robot [4] is inspired by another insect, by the move of cockroach. Cockroach slow movement is based on usual pattern. On the other side, when running, cockroach moves its legs round its hip joint (almost plane rotation) and uses them like wheels. That allows very fast movement independent on terrain. Yet another design is robot called PAW [5]. Robot is equipped with four, individually actuated wheels fixed on rotational limbs.

As demonstrated above, there are many approaches, how to solve the movement problem in the robot design.

1.2 Goals of the project

The aim of the project was to design, construct and revive experimental design of a robot that can combine driving and walking and thus achieve high maneuverability. The desired travel speed is 8 km/h, climbing ability is 30 deg. and ability to overcome stair with a height of 0,1 meters.

When designing and controlling the robot, it was necessary to use advanced tools and algorithms for modeling of kinematics and dynamics. Also, an important part of this experimental study was the styling. 3D printing technology was used for the production of the robot with significant related advantages [12].

2 Design of the experimental robot

2.1 Study of design variants and modelling of kinematics and dynamics

We have designed several variations of the robot configuration topologies with different numbers of joints (actuators). The aim was to minimize the number of joints while maintaining the minimum required mobility [8]. During the development and selection of variants we took into account the impact of design [12]. The final selected arrangement of the robot consists of four motors for travel and 6 actuators (including 2 in the central joint) (Fig. 1). The robot thus has substantially fewer joints than e.g. dog AIBO robot, which has 18 controlled joints.

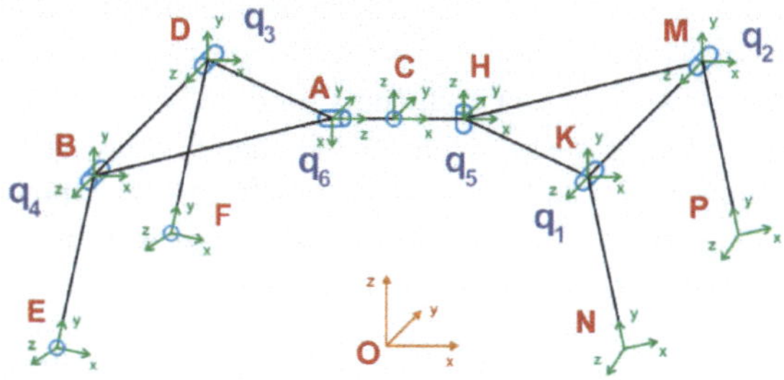

Fig. 1. Design concept (left); Joints and coordinate systems(right)

Models and simulations carried out during the design of the robot can be divided into four categories:

- **Kinematic model to determine the mobility of structure variants** – SimMechanics tool was used [13]. When determining the properties of the movement space, three feet of the robot were bound by spherical joint with the base [14], the fourth moved freely and was excited by random vector forces. Thus, we got the idea of the resulting leg workspace and hence of the robot mobility and were able to optimize eg. size of the particular parts of the robot.
- **Static analysis of propulsion strain** – gradually, individual movement phases of the robot were simulated (Fig. 2 top) and appropriate moments in the joints determined (Fig. 2 bottom). For the contact of the legs with the terrain, spherical joint was used again.
- **Dynamic simulation of robot motion** – this type of simulation was the most challenging, for modeling of the contact between the foot of the robot and the ground was used unilateral link created through virtual spring with shock-absorber [9].
- **The kinematic model for the control system** – this model must be implementable in embedded hardware. The model was designed based on the description of the homogeneous transformation matrices and inverse problem was solved by the Levenberg-Marquardt method.

The algorithm of inverse kinematics is based on the differential relationship:

$$\Delta \mathrm{x} = \mathrm{J}\Delta \mathrm{q}, \qquad \mathrm{q} = [q_1,..q_6, C_x, C_y, C_z, C_R, C_P, C_Y] \qquad (1)$$
$$\mathrm{x} = [x_{A,} x_{B,} x_{C,} x_{D,} x_{E,} x_{F,} x_{G,} x_{H,} x_{K,} x_{M,} x_{N,} x_P]$$

where $q_1,..q_6$ are the actuator joint coordinates, C_x, C_y, C_z are translations and

C_R, C_P, C_Y are the rotations of point C and x_i are Cartesian coordinates of points according to Fig. 1 right. The inversion of the Jacobian is then replaced by damped-least squares:

$$\Delta \mathrm{q} = \mathrm{J}^T (\mathrm{J}\mathrm{J}^\mathrm{T} + \lambda^2 \mathrm{I})^{-1} \Delta \mathrm{x} \qquad (1)$$

2.2 Mechanical design and styling

The key parts of the robot prototype have been made with SLS (Selective Laser Sintering) technology. Mechanical properties of the resulting product are approximately 90% of the characteristics of the material (polyamide). The used technology allowed the organic shaping and easy customizing of the shape based on the used actuators, sensors and electronics (Fig. 3 left). For drive, small DC gear motors were used, for actuators in the joints of the robot, modified Hitec servos were used. Installation into the structure of the robot is shown in Fig. 3 on the right.

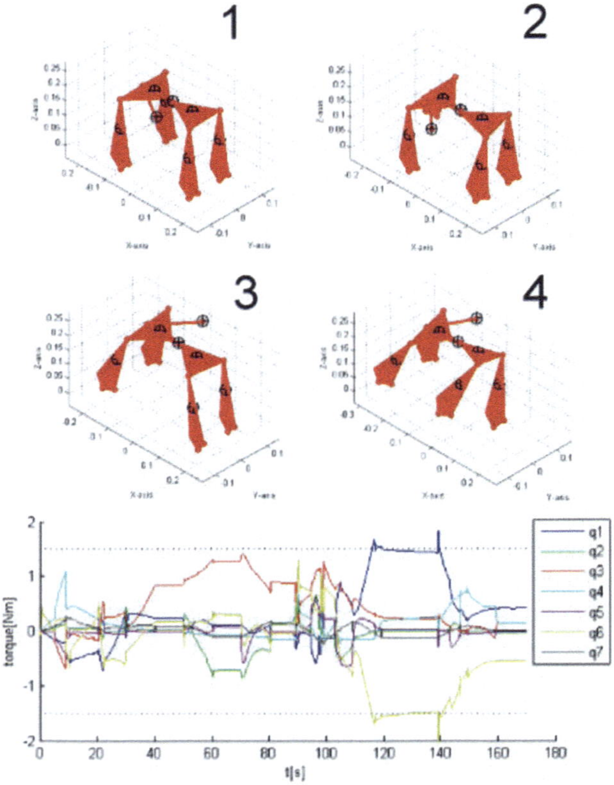

Fig. 2. Visualization of movement sequence for static analysis (top); Simulated load in robot joints resulting from (quasi) static analysis (bottom)

2.3 Sensors and distributed control

For walk control and hybrid robot movement is necessary to use a relatively large number of sensors. To ensure the basic movement, incremental encoders are used for each loop motor, current sensors for servodrives in joints, strain gauges for measurement of the robot leg load and MEMS accelerometers [10,11] and gyros for stabilization.

The temperature of the individual drives and battery status are also measured, and infrared collision sensors are used. The control structure is distributed (5x ARM microcontroller), communication takes place via the CAN bus, HMI (actuators) and UART (parent control unit Eyebot with camera) (Fig. 4).

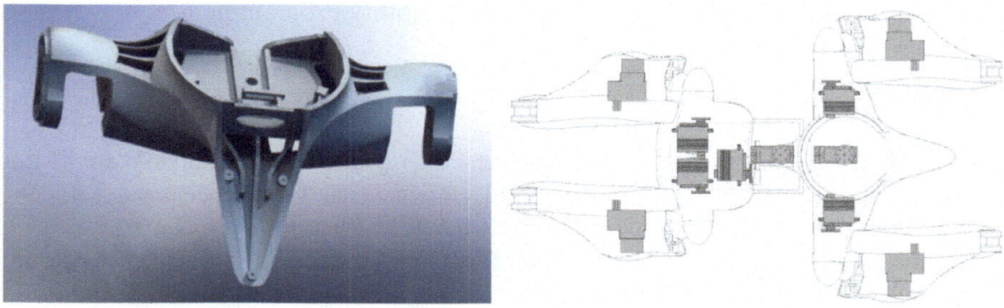

Fig. 3. Render of the part of robot body manufactured using SLS (left); Placement of motors and servodrives in the robot construction (right)

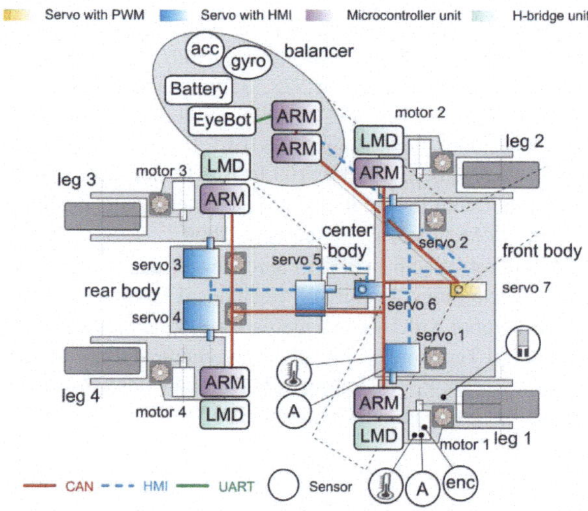

Fig. 4. Structure of sensory and control systems

3 Conclusion

In the paper, the project of experimental robot with a hybrid way of movement that combines driving and walking was briefly presented. For the development of the robot, modern design techniques of mechatronic systems were used by means of Model Based Design methodology. The control unit has been programmed using the automatically generated code from Simulink. A substantial part of the structure of the robot was made using 3D printing [13]. The design of the robot is listed in national Patent and Utility Model database under nr. 35057 – 2010.

Acknowledgement

Published results acquired with the support of project FSI-S-14-2204, by NETME Centre, Reg. No. CZ.1.05/2.1.00/01.0002 and, in the follow-up sustainability stage, through NETME CENTRE PLUS (LO1202) by financial means from the Ministry of Education, Youth and Sports under the „National Sustainability Programme I.

References

1. Endo, G.; Hirose, S., "Study on Roller-Walker - Energy efficiency of Roller-Walk -," Robotics and Automation (ICRA), 2011 IEEE International Conference on , vol., no., pp.5050,5055, 9-13 May 2011, doi: 10.1109/ICRA.2011.5980144
2. Michaud, F.; Letourneau, D.; Arsenault, M.; Bergeron, Y.; Cadrin, R.; Gagnon, F.; Legault, M.; Millette, M.; Pare, J.-F.; Tremblay, M.-C.; Lapage, P.; Morin, Y.; Bisson, J.; Caron, S., "AZIMUT, a leg-track-wheel robot," Intelligent Robots and Systems, 2003. (IROS 2003). Proceedings. 2003 IEEE/RSJ International Conference on , vol.3, no., pp.2553,2558 vol.3, 27-31 Oct. 2003, doi: 10.1109/IROS.2003.1249254
3. Halluc II: Futuro robotic Technology Center, Japan, http://furo.org/en/works/halluc2.html
4. Adachi, H.; Koyachi, N., "Development of a leg-wheel hybrid mobile robot and its step-passing algorithm," *Intelligent Robots and Systems, 2001. Proceedings. 2001 IEEE/RSJ International Conference on* , vol.2, no., pp.728,733 vol.2, 2001
5. Smith, J.A.; Sharf, I.; Trentini, M., "PAW: a hybrid wheeled-leg robot," Robotics and Automation, 2006. ICRA 2006. Proceedings 2006 IEEE International Conference on , vol., no., pp.4043,4048, 15-19 May 2006, doi: 10.1109/ROBOT.2006.1642323
6. Radoš, J.: Mobile robot with hybrid undercarriage: engineering design, Brno University of Technology, Faculty of Mechanical Engineering, 2009
7. Konvičný, J.: Mobile robot with hybrid undercarriage: design and testing of servodrives. Brno University of Technology, Faculty of Mechanical Engineering, 2009.
8. Krejčiřík, M.: Mobile robot with hybrid undercarriage: kinematics and programming of control. Brno University of Technology, Faculty of Mechanical Engineering, 2009
9. Grepl, R., Vlach, R., Krejčí, P.: Modelling of Unilateral Constraints for Virtual Prototyping in SimMechanics IEEE International Conference on Mechatronics 2007, Kumamoto, Japan, 2007
10. Łuczak, S.: Specific Measurements of Tilt with MEMS Accelerometers. In Jabłoński R., Brezina T. (eds.) Mechatronics. Recent Technological and Scientific Advances, pp. 705-711. Springer, Berlin Heidelberg (2011)
11. Łuczak, S.: Guidelines for Tilt Measurements Realized by MEMS Accelerometers. Int. J. Precis. Eng. Manuf. 15, 489-496 (2014)
12. Coupek, P.; Skaroupka, D.; Palkova, J.; et al., Development of Experimental Mobile Robot with Hybrid Undercarriage, 9th International Conference on Mechatronics, Warsaw, 2011, RECENT TECHNOLOGICAL AND SCIENTIFIC ADVANCES, p.405-411
13. Grepl, R., Simulation of unilateral constraint in MBS software SimMechanics, International Conference on Mechatronics Location: Warsaw Univ Technol, Warsaw, 2007, RECENT ADVANCES IN MECHATRONICS, Pages: 318-322
14. Bagiński K., Jasińska-Choromańska D., Wierciak J.: Modelling and simulation of a system for verticalization and aiding the motion of individuals suffering from paresis of the lower limbs. Bulletin of the Polish Academy of Sciences, Technical Sciences, vol. 61, nr 4, 2013, pp. 919–928

Sensorless Speed Control of BLDC Motor using EKF with Computed Inputs and Disturbance

Hrbáč, Z.[1], Sova, V.[1], Grepl, R.[1]

[1] Mechlab, Faculty of Mechanical Engineering, Brno University of Technology
{sova,grepl}@fme.vutbr.cz

Abstract. This paper presents the application of Extended Kalman Filter to the speed control of a BLDC motor. The inputs to EKF are computed based on the measured data as well as the disturbance (an external mechanical load). The non-linear model of the system is used for the simulation and later implemented on the dSPACE HW to obtain experimental results. The resulting solution sufficiently control the speed from 50 rpm to the nominal speed with the presence of disturbance and with the low torque ripple.

Keywords: BLDC motor · sensorless control · Extended Kalman Filter · experimental results · disturbance rejection.

1 Introduction

BLDC motors have a great variety of possible applications. Due to their high power-to-weight ratio, high reliability and easy controllability, they can be found in from small devices, such as disc drives, to quite large and heavy-duty applications, in example driving industrial fans, powering cars or taxying airplanes.

Many applications require a sensorless control of the motor. Mostly used Hall-effect position sensors would withstand in an environment with high temperatures or humidity. Or if there is just a demand for low cost of an application, sensorless algorithm can be implemented in a controller, needed for BLDC control application anyway.

1.1 Sensorless control of BLDC

There are quite a few developed and widely used sensorless control algorithms. The most well-known are perhaps the Back-EMF zero crossing detection and the Back-EMF integration [1]. Both methods detect a point on a non-powered phase of the motor, where Back-EMF crosses zero. Crossing appears exactly in one half of the 60° long non-powered phase period.

In the crossing detection technique, once this crossing is sensed, a delay of 30° electrical is realized and all phases are commutated. Time-wise, the delay changes with motor speed. Exact delay is constantly maintained by the control algorithm (variable counter, etc.) [5].

R. Jabłoński and T. Brezina (eds.), *Advanced Mechatronics Solutions*,
Advances in Intelligent Systems and Computing 393,
DOI 10.1007/978-3-319-23923-1_7

In the integration technique, the delay of 30° is done by an integrator. The integrated area from detected zero crossing to the actual commutation stays approximately the same at all motor speeds. When integrator approaches pre-defined threshold value, all phases are commutated and integrator is held in reset state until next crossing occurs [1].

Both algorithms improve BLDC motor behavior, because they more precisely detect the actual commutation, compared to conventional Hall-effect position sensor information.

To further improve BLDC motor sensorless control, an algorithm using Extended Kalman Filer, which had shown lower output torque ripple and greater RPM controllability spectrum compared to above mentioned techniques, is proposed in this paper.

Since sensorless control algorithm's main purpose is to omit presence of a position sensor mounted on a rotor, the Extended Kalman Filter (EKF) is used to estimate rotor's electrical angle by continuously measuring phase voltages and line currents.

2 Implementation of sensorless control using EKF

2.1 Model of BLDC motor

For the EKF algorithm a known model of a BLDC motor has been used [6]. Fig. 1 shows winding diagram of a BLDC motor, with following convention: u is phase voltage, i line current, R winding resistance, L winding inductance, M mutual inductance and e back-EMF.

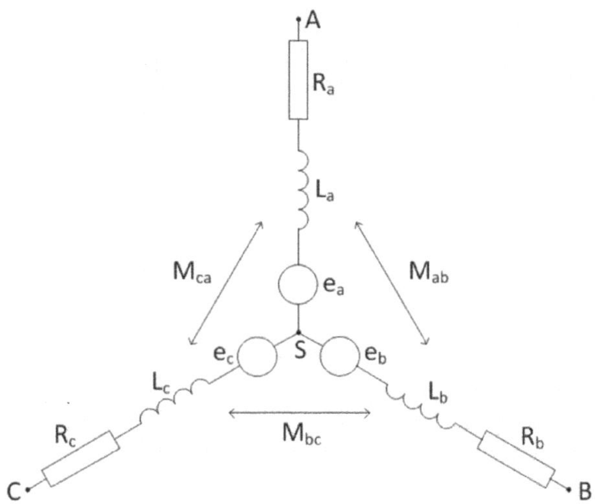

Fig. 1. Wye wound BLDC motor winding diagram

Symmetrical motor is supposed, thus all three winding resistances, winding inductances and mutual inductances equal the same value (R, L and M respectively). Also, the total inductance substitution $L_s = L - M$ had been used during development.

2.2 Nonlinear state space model

From the above described motor model, nonlinear state space representation can be derived (relation between input vector $\underline{u}$ and state vector $\underline{x}$: $\dot{\underline{x}} = f(\underline{x}, \underline{u})$):

$$\dot{\underline{x}} = \begin{bmatrix} -\frac{R}{L_s}x_1 - \frac{\lambda}{L_s}f_a(x_5)x_4 + \frac{1}{L_s}u_1 \\ -\frac{R}{L_s}x_2 - \frac{\lambda}{L_s}f_b(x_5)x_4 + \frac{1}{L_s}u_2 \\ -\frac{R}{L_s}x_3 - \frac{\lambda}{L_s}f_c(x_5)x_4 + \frac{1}{L_s}u_3 \\ \frac{\lambda}{J}f_a(x_5)x_1 + \frac{\lambda}{J}f_b(x_5)x_2 + \frac{\lambda}{J}f_c(x_5)x_3 - \frac{C}{J}x_4 - \frac{1}{J}u_4 \\ px_4 \end{bmatrix} \tag{1}$$

Where $\underline{x} = [i_a, i_b, i_c, \omega_m, \varphi_{el}]'$,$\underline{u} = [u_{as}, u_{bs}, u_{cs}, \tau_z]'$, ω_m is rotor speed, φ_{el} rotor electrical angle and τ_z output torque. Back-EMF prototype function is represented by $f(\varphi_{el})$. This function is dependent of rotor's electrical angle and has maximal amplitude of 1.

Relation between output vector $\underline{y}$ and state vector $\underline{x}$ ($\underline{y} = \boldsymbol{C}\underline{x} + \boldsymbol{D}\underline{u}$):

$$\begin{bmatrix} i_a \\ i_b \\ i_c \end{bmatrix} = \begin{bmatrix} 1 & 0 & 0 & 0 & 0 \\ 0 & 1 & 0 & 0 & 0 \\ 0 & 0 & 1 & 0 & 0 \end{bmatrix} \begin{bmatrix} i_a \\ i_b \\ i_b \\ \omega_m \\ \varphi_{el} \end{bmatrix} + \begin{bmatrix} 0 & 0 & 0 & 0 \\ 0 & 0 & 0 & 0 \\ 0 & 0 & 0 & 0 \end{bmatrix} \begin{bmatrix} u_{as} \\ u_{bs} \\ u_{cs} \\ \tau_z \end{bmatrix} \tag{2}$$

2.3 Rotor angle EKF estimation

Fig. 2 shows design concept of the whole algorithm:

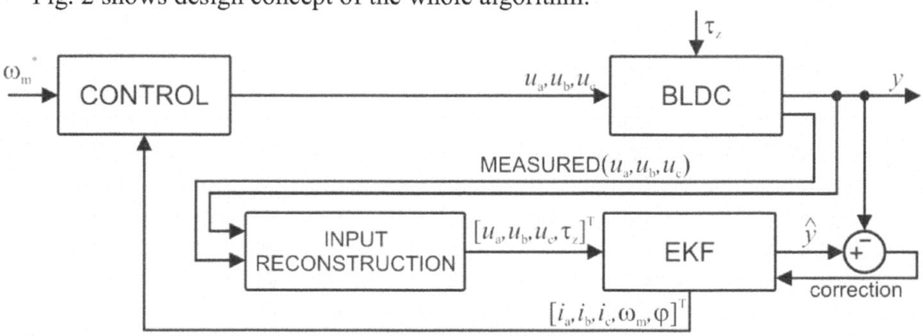

Fig. 2. Design concept

Phase voltages, which, together with output torque, create input vector $\underline{u}$ for both the EKF and the BLDC, were measured on the BLDC's motor input terminal. Output torque was determined from total inputting current to the power electronics, multiplied by torque constant. Output of both the EKF and BLDC were line currents vectors (estimated $\hat{\underline{y}}$ and measured $\underline{y}$). EKF correction was done based on comparison of these two vectors. Line currents were physically measured in the power electronics. Most importantly, the estimated rotor electrical angle $\widehat{\varphi_{el}}$ was read directly from EKF's state vector. Estimated Hall-effect position sensor signal, needed for the actual BLDC motor control, was derived from this estimated electrical angle information.

One of the effective ways to linearize a nonlinear BLDC motor model is to use a Taylor series. Taylor series has to be then recalculated at every time sample [3]. Jacobian matrix implemented in the EKF:

$$\underline{J} = \begin{bmatrix} -\frac{R}{L_s} & 0 & 0 & -\frac{\lambda}{L_s} f_a(x_5) & -\frac{\lambda}{L_s} x_4 \frac{df_a(x_5)}{dx_5} \\ 0 & -\frac{R}{L_s} & 0 & -\frac{\lambda}{L_s} f_b(x_5) & -\frac{\lambda}{L_s} x_4 \frac{df_b(x_5)}{dx_5} \\ 0 & 0 & -\frac{R}{L_s} & -\frac{\lambda}{L_s} f_c(x_5) & -\frac{\lambda}{L_s} x_4 \frac{df_c(x_5)}{dx_5} \\ \frac{1}{J} f_a(x_5) & \frac{1}{J} f_b(x_5) & \frac{1}{J} f_c(x_5) & -\frac{B}{J} & \frac{\partial f_4}{\partial x_5} \\ 0 & 0 & 0 & p & 0 \end{bmatrix} \quad (3)$$

Where λ is voltage constant depending on motor's design, J rotor inertia and B viscous friction coefficient. Partial derivative of f_4 with respect to x_5:

$$\frac{\partial f_4}{\partial x_5} = \frac{\lambda}{J}\left(\frac{df_a(x_5)}{dx_5} x_1 + \frac{df_b(x_5)}{dx_5} x_2 + \frac{df_c(x_5)}{dx_5} x_3\right) \quad (4)$$

Derivatives of $f(x_5)$ with respect to x_5:

φ_{el}	$\partial f_a/dx_5$	$\partial f_b/dx_5$	$\partial f_c/dx_5$
0π÷1/3π	6/π	0	0
1/3π÷2/3π	0	0	-6/π
2/3π÷π	0	6/π	0
π÷4/3π	-6/π	0	0
4/3π÷5/3π	0	0	6/π
5/3π÷2π	0	-6/π	0

2.4 Simulation and experimental results

Proposed method of sensorless control was first tried in Simulink simulations. Known model of a BLDC motor has been used [6], which has served as a tool, where the algorithm can be verified and sorted out. Simulated BLDC motor's outputs were rotor angular speed and electrical angle, which were compared to estimated values outputting the EKF. Simulations were done with white noise variance of $0.1V^2$ (noise variance based on real-measured variance on the dSPACE modular HW input). Following figure shows comparison of simulated end estimated rotor speed:

Experiments were done with dSPACE modular hardware platform (DS 2202 HIL I/O Board, DS1006 Processor board), Microchip power electronics (MC1L) and Fulling BLDC motor (FL86BLS125). Control algorithm was developed in Matlab/Simulink, compiled and loaded to dSPACE [7]. While running, the algorithm was commanded via dSPACE ControlDesk and Real-Time Interface. Sensorless control block diagram and experimental setup is outlined in Fig. 4:

EKF setup remained the same as it was during simulation runs. Covariance matrices had to be adjusted to better comply with the real-world environment, especially with the noise disturbance. The only exact mechanical output form the BLDC was the measured Hall-effect position sensor output, which was then compared with estimated one from EKF. Fig. 5 shows comparison of one of the three measured and estimated position sensor outputs.

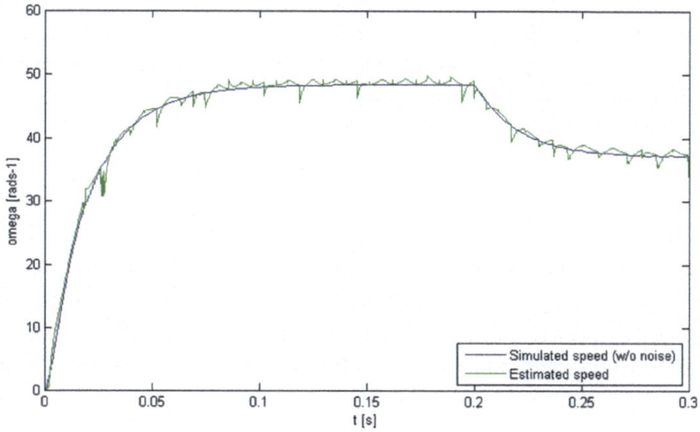

Fig. 3. Estimated and simulated speed comparison

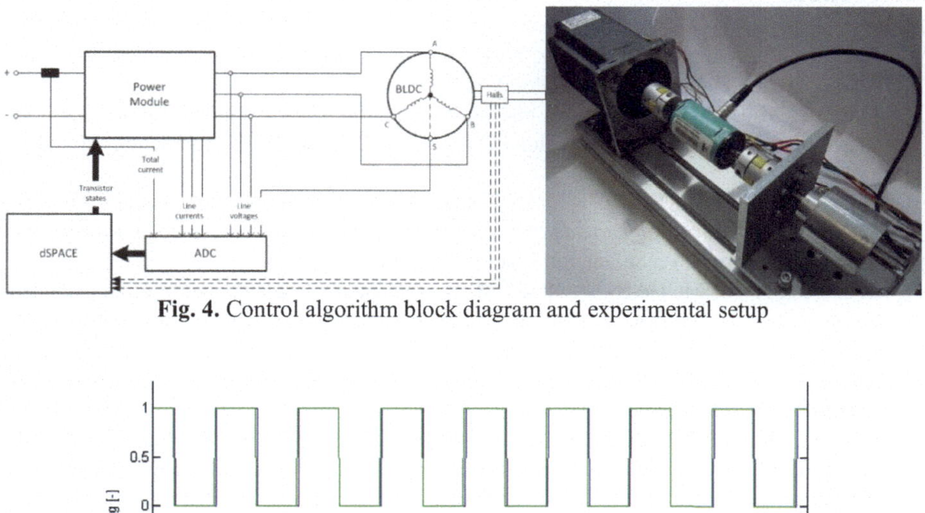

Fig. 4. Control algorithm block diagram and experimental setup

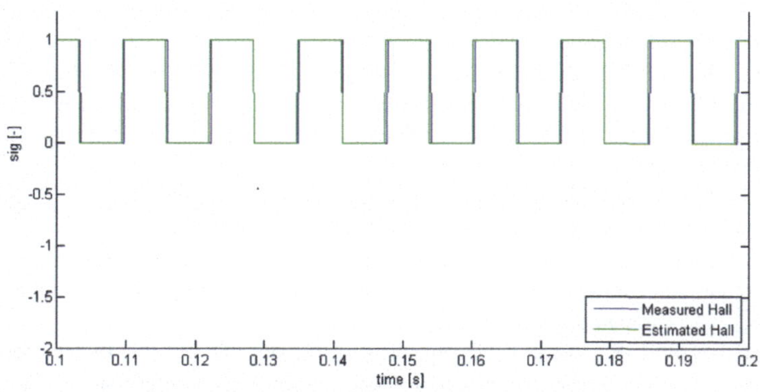

Fig. 5. Estimated and measured position sensor comparison

3 Conclusion

This paper presented simulation and experimental results of proposed sensorless control algorithm of BLDC motor using Extended Kalman Filter. Method showed sufficient results from 50 RPM up to nominal speed, both with and without torque load (up to nominal torque of the motor). Compared to other two sensorless methods mentioned in paper's introduction proposed algorithm presented lower output torque ripple at all speeds and greater controllability range.

Acknowledgement

This work was supported by the European Commission within the FP7 project Efficient Systems and Propulsion for Small Aircraft "ESPOSA", contract No. ACP1-GA-2011-284859-ESPOSA, by NETME Centre, Reg. No. CZ.1.05/2.1.00/01.0002 and, in the follow-up sustainability stage, through NETME CENTRE PLUS (LO1202) by financial means from the Ministry of Education, Youth and Sports under the „National Sustainability Programme I.

References

1. Gamazo, J.C., Vázquez, E., Gómez, J. Position and Speed Control of Brushless DC. Valladolid, Spain : University of Valladolid, 2010
2. TERZIC, B.; JADRIC, M.: Design and implementation of the extended Kalman filter for the speed and rotor position estimation of brushless DC motor. IEEE Transactions on Industrial Electronics. 2001.
3. LENINE, D.; RAMI REDDY, B.; VIJAY KUMAR, S.: *Estimation of speed and rotor position of BLDC motor using Extended Kalman filter.* International Conference on Information and Communication Technology in Electrical Sciences. 2007.
4. DHAOUADI, R.; MOHAN, N.; NORUM, L.: Design and implementation of an extended Kalman filter for the state estimation of a permanent magnet synchronous motor. IEEE Transactions on Power Electronics. 1991.
5. Iizuka, K., Uzuhashi, H., Kano, M., Endo, T., Mohri, K., *Microcomputer Control for Sensorless Brushless Motor*, 1985.
6. Sova, V.; Grepl, R.: Hardware in the Loop Simulation Model of BLDC Motor Taking Advantage of FPGA and CPU Simultaneous Implementation, 10th International Conference on Mechatronics, Brno, 2013, MECHATRONICS 2013: RECENT TECHNOLOGICAL AND SCIENTIFIC ADVANCES, p. 135-142
7. Grepl, R., Real-time control prototyping in MATLAB/simulink: Review of tools for research and education in mechatronics, IEEE ICM 2011, pp. 881-886.
8. Sova V., Chalupa J., Grepl R.: Fault Tolerant BLDC Motor Control for Hall Sensors Failure, 21st International Conferenceon Automation & Computing, University of Strathclyde, Glasgow,UK, 2015 (prepared for publication)
9. Chalupa, J., Grepl, R., Sova, V., Design of Configurable DC Motor Power-Hardware-In-the-Loop Emulator for Electronic-Control-Unit Testing, 21st International Conferenceon Automation & Computing, University of Strathclyde, Glasgow,UK, 2015 (prepared for publication)

Step response identification of inertial model for oscillating system

J. Kurek [1]

[1] Institute of Automatic Control and Robotics, Warsaw University of Technology, Warsaw, Poland
jkurek@mchtr.pw.edu.pl

Abstract. There is presented a method for calculation of inertial model for oscillating system based on system step response. The method is simple and can be easily used in automatic control practice. Identified model can be applied in control algorithms more advanced than PID controller, for instance in predictive control.

Keywords: oscillating system, identification, step response.

1 Introduction

Simple identification methods of dynamical systems based on system step response is frequently used in industrial automatic control practice, e.g. for PID controller settings [2]. There are proposed models for (i) inertial system with dead-time and (ii) integral (astatic) system with dead-time. These simple models can be used also in more advanced control algorithms, e.g. for design of predictive controller.
In this note we consider inertial oscillating system

$$G(s) = \frac{k}{T^2 s^2 + 2T\xi s + 1} G_s(s) \tag{1}$$

where $\xi \in (0,1)$ is damping ratio and $G_s(s)$ is asymptotically stable system without dead-time, for instance

$$G_s(s) = \frac{1}{(T_s s + 1)^n}$$

We show how to calculate model for system (1) in the form

$$G_m(s) = \frac{k_m}{T_m^2 s^2 + 2T_m \xi_m s + 1} \frac{1}{T_{1m} s + 1} \tag{2}$$

based on system step response in very simple way.

R. Jabłoński and T. Brezina (eds.), *Advanced Mechatronics Solutions*,
Advances in Intelligent Systems and Computing 393,
DOI 10.1007/978-3-319-23923-1_8

2 Oscillating model identification

Step response of system (2) can be estimated as response of inertial system for sinusoidal input generated by oscillating system, fig. 1. On the output we have sinusoidal signal with the same frequency and smaller magnitude and phase shift.

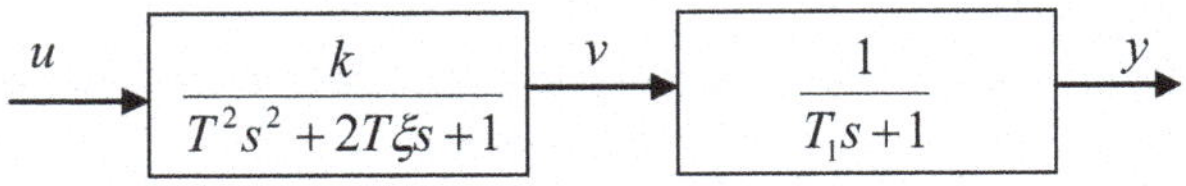

Fig. 1. Serial connection of oscillating and inertial systems

Based on system step response, fig. 2, one can calculate frequency of sinusoidal signal

$$\omega_{osc} = \frac{2\pi}{T_{osc}}$$

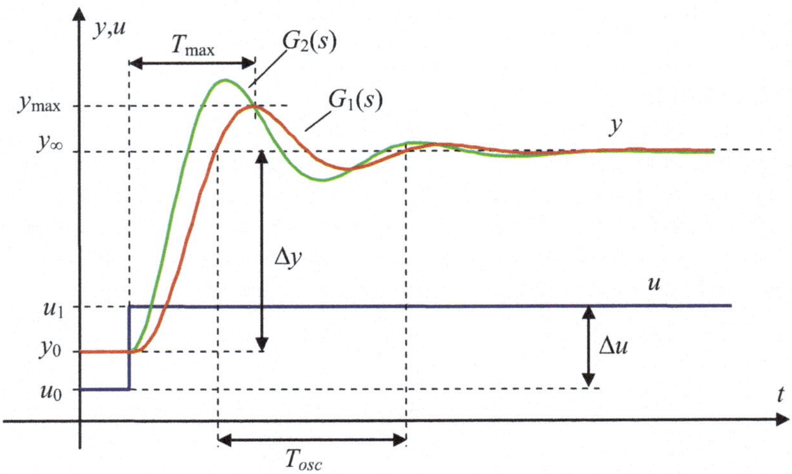

Fig. 2. Step response of inertial oscillating system $G_1(s)$ and oscillating system $G_2(s)$.

It is known, [3], that step response of oscillating system has maximum after time $T_{max} = \frac{T_{osc}}{2}$ and in the inertial system we have phase shift $\varphi(\omega) = -\arctan(T_{1m}\omega)$. Thus, for inertial system, fig. 1, one has for sinusoidal input generated by oscillating system $v(t) \approx k\sin(\omega_{osc}t)$

$$v_{max} \approx k\sin(\omega_{osc}t_{max}) = k\sin\left(\omega_{osc}\frac{T_{osc}}{2}\right)$$

and

$$y_{\max} \approx m(\omega_{osc}) k \sin[\omega_{osc} T_{\max} + \varphi(\omega_{osc})]$$

From the above one finds

$$\omega_{osc} \frac{T_{osc}}{2} \approx \omega_{osc} T_{\max} + \varphi(\omega_{osc})$$

and one has

$$\left(T_{\max} - \frac{T_{osc}}{2}\right)\omega_{osc} \approx -\varphi(\omega_{osc}) = \arctan(T_{1m}\omega_{osc})$$

Thus, time constant of inertial system can be approximated as follows

$$T_{1m} \approx \frac{1}{\omega_{osc}} \tan\left[\left(T_{\max} - \frac{T_{osc}}{2}\right)\omega_{osc}\right] \tag{3}$$

Unfortunately, tangent is nonlinear function which for small inexactness of the argument, e.g. $T_{\max}$, gives large errors. However, for small argument one can calculate the following approximation: $\tan(\alpha) \approx \alpha$, e.g. $\tan\left(\frac{\pi}{4}\right) = 1 \approx \frac{\pi}{4} = 0.7854$, obtaining

$$\left(T_{\max} - \frac{T_{osc}}{2}\right)\omega_{osc} \approx -\varphi(\omega_{osc}) \approx T_{1m}\omega_{osc}$$

and one has

$$T_{1m} \approx T_{\max} - \frac{T_{osc}}{2}$$

Next one can calculate oscillating gain of inertial system

$$m(\omega_{osc}) \approx \frac{1}{\sqrt{1 + T_{1m}^2 \omega_{osc}^2}}$$

Then, we correct value of $y_{\max}$ and oscillation magnitude of the oscillating system as follows

$$\bar{y}_{\max} = \frac{y_{\max}}{h_i(T_{\max})} = \frac{y_{\max}}{1 - e^{-\frac{T_{\max}}{T_{1m}}}}$$

where h_i denotes the step response of the inertial system.
Then, one can calculate coefficients of oscillating system knowing that inertial system dumps oscillating response of the whole system

$$\xi_m = \sqrt{\frac{\ln^2\left(\dfrac{\bar{y}_{\max} - y_\infty}{m(\omega_{osc})y_\infty}\right)}{\pi^2 + \ln^2\left(\dfrac{\bar{y}_{\max} - y_\infty}{m(\omega_{osc})y_\infty}\right)}} = \sqrt{\frac{\ln^2\left[\left(\dfrac{\bar{y}_{\max}}{y_\infty} - 1\right)\dfrac{1}{m(\omega_{osc})}\right]}{\pi^2 + \ln^2\left[\left(\dfrac{\bar{y}_{\max}}{y_\infty} - 1\right)\dfrac{1}{m(\omega_{osc})}\right]}}$$

and

$$T_m = \frac{\sqrt{1-\xi_m^2}}{2\pi} T_{osc}$$

It is easy to see that for dead time system $G(s) = G_0(s)e^{-T_0 s}$ one has to use $T_{\max 0} = T_{\max} - T_0$ instead of $T_{\max}$. Referring to (3) it should be pointed up that dead-time T_0 usually is estimated not exactly.

Then, based on step response of the system, fig. 2, one can calculate gain of system model (2) finding

$$k_m = \frac{\Delta y}{\Delta u}$$

Example 1.

Consider oscillating system described by the following transfer function

$$G(s) = \frac{2.5}{5^2 s^2 + 2\cdot 5\cdot 0.3s + 1}\frac{1}{5s+1}\frac{1}{3s+1}e^{-3s} = \frac{2.5}{25s^2 + 3s + 1}\frac{1}{5s+1}\frac{1}{3s+1}e^{-3s}$$
$$= \frac{k}{T^2 s^2 + 2T\xi s + 1}\frac{1}{T_1 s + 1}\frac{1}{T_2 s + 1}e^{-T_0 s}$$

For the system we look for the model (2).

Step response of the system is presented in fig. 3. One read from the figure

$$\Delta u = 1,\quad \Delta y = 2.5,\quad T_0 = 12,\quad T_{0m} = 17,\quad T_{osc} = 33,\quad y_{\max} = 2.95,\quad T_{\max} = 34.5$$

and finds

$$k_m = \frac{2.5}{1} = 2.5$$

Next, calculating simple oscillating model with dead-time one finds [3]

$$G_{mo}(s) = \frac{k_m}{T_{mo}^2 s^2 + 2T_{mo}\xi_{mo}s + 1}e^{-T_{0m}s} = \frac{2.5}{4.6101^2 s^2 + 2\cdot 4.6101\cdot 0.4791s + 1}e^{-17s}$$

It is easy to see, that there is quite big difference between damping ratio of model and system (160%), this will give us a difference in response of the model and system.

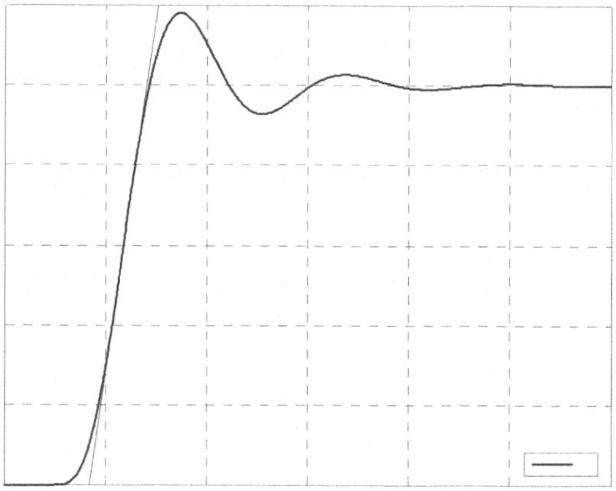

Fig. 3. Step response of the oscillating system *G*(*s*).

Next calculating model of oscillating system with inertia (2) with dead-time one finds

$$\omega_{osc} = \frac{2\pi}{33} = 0.1904$$

and

$$T_{1m} = 34.5 - 12 - \frac{33}{2} = 6$$

and

$$m(\omega_{osc}) = \frac{1}{\sqrt{1 + 6^2 \cdot 0.1904^2}} = 0.6587$$

Then after correction of maximal output of the system

$$\bar{y}_{\max} = \frac{2.95}{1 - e^{-\frac{34.5-12}{6}}} = 3.0210$$

one has

$$\xi_m = \sqrt{\frac{\ln^2\left[\left(\frac{3.0210}{2.5} - 1\right)\frac{1}{0.6587}\right]}{\pi^2 + \ln^2\left[\left(\frac{3.0210}{2.5} - 1\right)\frac{1}{0.6587}\right]}} = 0.3439$$

and

$$T_m = \frac{\sqrt{1-0.3439^2}}{2\pi}\cdot 33 = 4.9317$$

Thus, we have obtained the following model (2) for the system

$$G_{mio}(s) = \frac{k_m}{T_m^2 s^2 + 2T_m \xi_m s + 1}\frac{1}{T_{1m}s+1}e^{-T_0 s} = \frac{2.5}{(6s+1)(4.9317^2 s^2 + 2\cdot 4.9317\cdot 0.3439s+1)}e^{-12s}$$

In fig. 4 there are presented step and impulse responses of the system, simple oscillating model and oscillating model with inertia. It is easy to see that the responses are similar, however, there are differences. It is easy to see that the oscillating model with inertia is better one.

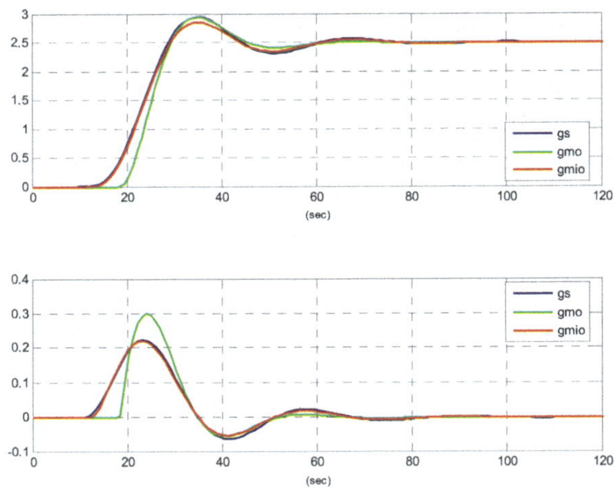

Fig. 4. Step and impulse responses of the oscillating system (gs), simple oscillating model (gmo) and oscillating model with inertia (gmio).

3 Concluding remarks

The presented model for oscillating with inertia is quite good approximation for real oscillating system. The model can be easily calculated based on system step response. The model should improve control in the case of oscillating systems particularly in more advanced control algorithms than PID, for instance in the predictive control.

References

1. Findeisen Wł. (ed.), Handbook for Control Engineer, WNT, Warsaw, 1973 (in Polish).
2. Iserman R., Digital Control, Springer Verlag, Bonn, 1990.
3. Kurek J., Fundamentals of Automatic Control, Warsaw University of Technology, Warsaw, e-book (in Polish).

An Improved Extraction Process of Moving Objects' Silhouettes in Video Sequences

Tomasz Posłuszny and Barbara Putz

Warsaw University of Technology, Faculty of Mechatronics, Warsaw, Poland
{tefposluszny@gmail.com, bputz@mchtr.pw.edu.pl}

Abstract. In this paper we propose a new method for extracting silhouettes of moving objects in images acquired with a static camera. First the background subtraction algorithm with an adaptive Gaussian mixture model is used to obtain moving regions. The output binary mask is then refined using a region-filtering algorithm based on an adaptive fast-scanning segmentation algorithm. Next, the resulting mask is morphologically processed in order to prepare the input for the GrabCut algorithm. Finally, the GrabCut algorithm leverages spatial and color relationships between pixels in order to improve the background subtraction result. We show through experiments that for certain types of video sequences our approach can perform better than state-of-the-art methods as regards the mask accuracy.

Keywords: motion detection, silhouette extraction, background subtraction, Gaussian mixture model, GrabCut

1 Introduction

Moving-object detection is an important part of many computer vision systems. Along with object classification, it can serve as a tool for collecting data about the objects that appear in the camera's viewport. The data can be aggregated in real time and used in automatic control and decision-making systems, intelligent transportation systems, data mining, and monitoring systems. Detecting moving objects is often referred to as background subtraction or foreground detection and can also be used in video editing software.

Background subtraction is always executed based on a background model. Recently in the literature have appeared many new background models including advanced statistical models, fuzzy models, discriminative and mixed subspace learning models, robust subspace models, subspace tracking, low rank minimization, sparse models, and domain transform models [1]. In this work we focus on improving the results of a background subtraction algorithm based on a Gaussian mixture model [2], a widely used approach for background modeling [3], through additional processing steps. The process of improving and refining the moving-object mask (foreground mask) is often called silhouette extraction. Different techniques of

R. Jabłoński and T. Brezina (eds.), *Advanced Mechatronics Solutions*,
Advances in Intelligent Systems and Computing 393,
DOI 10.1007/978-3-319-23923-1_9

obtaining better foreground masks can be found in the literature, e.g., morphological processing [4] or defining and applying additional mask models [5].

In the next section we describe a process that creates a background model using a mixture of Gaussians, performs background subtraction, refines the moving objects mask, and improves it with the GrabCut algorithm [6]. Automatic preparation of a trimap as the input for GrabCut algorithm has been described in many works. In [7] the input for the GrabCut algorithm was prepared using depth camera images. In [8] a trimap was prepared using person and face detectors. Using a background subtraction algorithm with GrabCut to get accurate moving objects' masks, as proposed in this paper, seems to be a novel approach.

In section 3, the experimental results of key processing steps are presented and, in particular, benchmarking results for two up-to-date video datasets. In section 4, the conclusions are drawn and scene characteristics, for which the process yields best results, are described.

2 Extraction Process

The process of silhouette extraction was designed for static camera images like the images from monitoring systems. It improves the segmentation result of a motion detection algorithm using GrabCut. The extraction process consists of several algorithmic modules—processing steps (Fig. 1). In the first step, moving regions are determined using background subtraction with a Gaussian mixture model (GMM). In the second step, small connected regions are removed from the binary image of moving regions. The third and last step extracts moving objects' masks with enhanced accuracy using the GrabCut algorithm, preprocessed by preparation of the trimap needed as input.

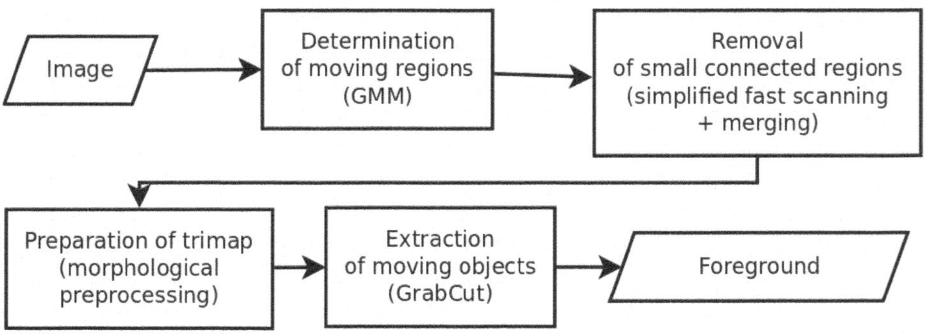

Fig. 1. The block diagram of the silhouette extraction process

1.1 Determination of Moving Regions

During the initialization, a background model is created for the static camera image without moving objects (for n frames, process tested for $n = 50$). Then it is used to get the mask of objects appearing in the viewport. GMM (Gaussian mixture model) is used for background modeling, as described in [2].

A background subtraction algorithm using GMM is encapsulated in OpenCV library in the BackgroundSubtractorMOG2 class. The running time of this class implementation was optimized for the case when learning rate equals zero by avoiding recalculation of Gaussians for every pixel. That change, being the result of the work on the proposed extraction process, has been merged with the main branch of the OpenCV library [9].

Learning rate (LR) is a parameter that indicates how quickly the GMM model is updated, e.g., when lighting conditions change or a new object ceases to move and becomes background. If the learning rate is zero, the background model is not updated. Higher values of learning rate result in a process that adapts to environmental changes but, as a downside, slow-moving objects may be partly incorporated into the background model and their mask may not be determined correctly.

1.2 Removal of Small Connected Regions

Moving objects of interest usually have a specific size range. The process leverages this by filtering out small regions. This removes noise and leaves only large regions that include objects of interest.

Implementation of the algorithm uses the concept of fast scanning [10] to label connected regions. Labeled regions are then evaluated. If a region is too small, it is removed, i.e., it adopts the color of its surroundings. The algorithm of region merging is applied until all small regions are removed.

For optimal time complexity, a label-merge array is maintained. The indexes of the array are labels (referred to as index-labels). The values of the array are labels of regions with which index-labels were merged or −1 if the index-label was not merged. It is easily concluded that at the beginning all values in the label-merge array are equal to −1.

The proposed merging routine is as follows:

1. label regions using a fast scanning algorithm [10] with the following simplification: since we have a binary image, we do not decide on creating new regions by comparing mean pixel values because every region has connected pixels with the same value—black or white. The pixel is added to a region if it has the same value as all pixels in the region and is adjacent to it.
2. create neighborhood graph of regions
3. visit all white regions and merge neighboring black regions if they are too small
4. visit all black regions and merge neighboring white regions if they are too small

5. repeat steps (1)–(4) until all regions have a larger area than the demanded threshold (S_R parameter). The choice of removal threshold is arbitrary and dependent on the characteristics of the moving objects that need to be detected.

1.3 Extraction of Moving Objects

The last processing step is the GrabCut segmentation algorithm [6]. The algorithm uses GMM to model foreground and background. Background and foreground models consist of K components and every pixel has one component from the background and one component from the foreground assigned to itself in vector $\mathbf{k} = \{k_1, ..., k_N\}$ with $k_i \in 1, \ldots, K$.

A good segmentation $\alpha = (\alpha_1, \ldots, \alpha_i, \ldots, \alpha_N)$ for each of N pixels corresponds to an energy function minimum ($\alpha_i = 1$ means that pixel i belongs to foreground; $\alpha_i = 0$, to background). The energy function, also known as Gibbs energy, is minimized in accordance with image data matrix **z**. Energy (1) consists of element U that encourages segmentation values best fitting to the GMM and element V that encourages coherence in regions of similar color:

$$E(\alpha, \mathbf{k}, \theta, \mathbf{z}) = U(\alpha, \mathbf{k}, \theta, \mathbf{z}) + V(\alpha, \mathbf{z}) \tag{1}$$

The Gaussian mixture model θ is represented by (2):

$$\theta = \{\pi\,(\alpha, k), \mu\,(\alpha, k), \Sigma(\alpha, k),\ \ \alpha = 0, 1,\ \ k = 1 \ldots \mathrm{K}\}\ , \tag{2}$$

where π, μ, and Σ are means, weights, and covariances (respectively) of the $2K$ Gaussian components for the background and foreground distributions [6].

In the designed process the GrabCut implementation from OpenCV library is used. GrabCut input is often referred to as a trimap and consists of foreground, background, and mixed pixel regions. For each pixel in the mixed region, GrabCut determines whether it belongs to the foreground or background. In the proposed process of extraction, the trimap for the GrabCut algorithm is prepared in three preprocessing steps of morphological operations. They are performed in regard to the binary image being the result of the removal of small connected regions:

1. the binary image is eroded and the result is labeled as belonging to the foreground,
2. the binary image is dilated and the pixels beyond the result mask are labeled as belonging to the background,
3. the difference between dilated and eroded binary images consists of an uncertain region and its pixels may be labeled as possibly belonging to the background or possibly belonging to the foreground.

The kernel used in steps (1) and (2) has variable size (k_{GC} parameter). The best value of this parameter depends on scene and moving objects' characteristics.

3 Experimental Results

The results of successive steps of the proposed process are presented in Fig. 2. A human silhouette is a moving object that appeared in the camera's viewport after preparing the background model. The colors of the silhouette are similar to the colors of the background (image 1), what results is many imperfections in the resulting foreground mask (image 2).

After removing small connected regions, only two regions remain—one is background and the other is foreground (image 3). The foreground region is jagged and needs refining. The subsequent preprocessing step creates a stripe at the border between foreground and background (image 4). Next, the pixels' affiliations in the stripe are defined using the GrabCut algorithm and the final mask is applied to the input (image 5). Undoubtedly, the use of the GrabCut algorithm enhances background subtraction results, which is clearly shown in Fig. 3.

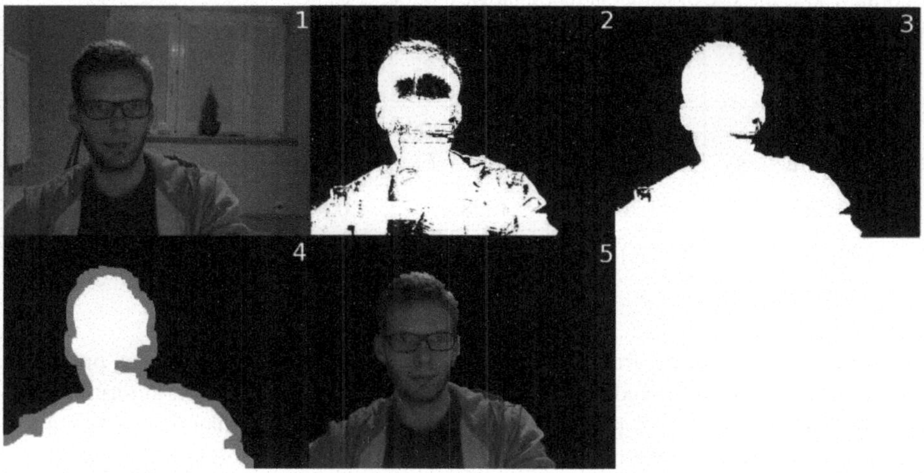

Fig. 2. Input image (1), partial processing results (2–4), and the output image (5)

Fig. 3. Process result without (left) and with (right) GrabCut algorithm

The processing was benchmarked using videos from the [11] dataset described in [3]. The benchmarking results are shown in Tables 1 and 2. The values of the process parameters that gave the best result are given in Table 1. The results were evaluated using the number of true positive (TP), false positive (FP), false negative (FN) pixels, and the following statistical measures:

(1) Recall (RC), $RC = \frac{TP}{TP+FN}$

(2) Precision (PR), $PR = \frac{TP}{TP+FP}$.

The recall measure is the percentage of ground truth foreground pixels that were also classified as foreground by the analysis algorithm. The precision measure is the percentage of pixels classified as foreground by the algorithm that also belong to the ground truth foreground pixels.

Table 1. Experimental results from [11] dataset; S_R, minimum region size in region filtering algorithm; k_{GC}, the size of kernel in morphological preprocessing

Problem Type	Bootstrap	Camouflage*	Foreground Aperture*	WavingTrees
Process parameters	$S_R = 3$ $k_{GC} = 0$	$S_R = 2000$ $k_{GC} = 7$	$S_R = 30$ $k_{GC} = 5$	$S_R = 2000$ $k_{GC} = 7$
Recall	0.70	0.98	0.56	1.00
Precision	0.61	0.97	0.77	0.88

* with shadow detection

In Table 2, the resulting frames were compared with ground truth. The proposed process resulted in improved masks from the background-subtraction algorithm that already had good quality. It provides substantially better results than the background-subtraction algorithm in the case of scenes with dynamic background and camouflage effects. Both camouflage and foreground problems result in false negative blobs, which can be removed to some extent if they are not too large. The false negative blob on the Foreground Aperture sequence (Table 2) result creates two disjointed foreground blobs that have not been connected by the process.

Table 2. Resulting masks from [11] dataset

Sequence	Bootstrap	Camouflage	Foreground Aperture	Waving Trees
Test image				
Ground truth				
MOG2+ morphological processing+ GrabCut (this paper)				

The second dataset [12] is described in [13]. Average precision and recall across all frames in two videos were calculated as well as the following parameters:

(1) False Positive Rate (FPR), $FPR = \frac{FP}{FP+TN}$

(2) False Negative Rate (FNR), $FNR = \frac{FN}{TN+FP}$

(3) Percentage of Wrong Classifications (PWC), $\mathrm{PWC} = \frac{100(FN+FP)}{(TP+FN+FP+TN)}$

(4) Shadow Errors Rate (SER), $SER = \frac{ShadowErrors}{FP}$ (shadow errors: false positive detections due to the classification of shadowed background as foreground).

The first video comes from highway monitoring and the second comes from overpass monitoring (Table 3). The highway sequence belongs to a "baseline" category, which has average recall = 0.94 and precision = 0.93 for the top four background subtraction methods presented in [13]. The process performs worse because of filtering regions representing cars that are far away or are disjoint and offer no shadow detection.

The overpass sequence belongs to the "dynamicBackground" category, which has average recall = 0.85 and precision = 0.86 for the top four methods as above. The proposed process performs better for this particular sequence as regards precision, because moving regions belonging to the dynamic background are filtered.

In Table 4, the examples of resulting masks from above sequences were compared with ground truth.

Table 3. Experimental results from changedetection.net dataset [12], no shadow detection

Video sequence	Process parameters	FPR	FNR	PWC	SER	Recall	Precision
Highway sequence	$S_R = 100$ $k_{GC} = 3$ LR= 0.009	0,0049	0,0128	1,67	0,75	0,796	0,91
Overpass sequence	$S_R = 80$ $k_{GC} = 9$ LR= 0.002	0,0005	0,0021	0,26	0,13	0,843	0,96

Table 4. Examples of resulting masks from changedetection.net dataset [12]. Gray pixels on ground truth images are shadow or do not belong to the tested region of interest (ROI).

Sequence	Highway	Overpass
Test image		
Ground truth	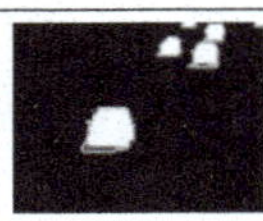	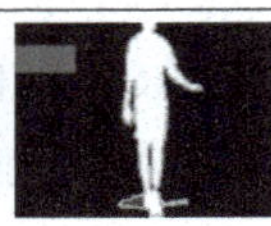
MOG2 + morphological processing + GrabCut (this paper)	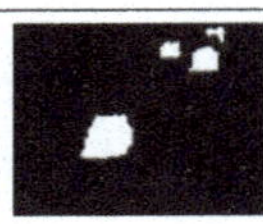	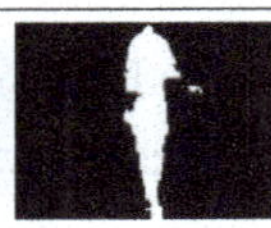

4 Conclusions

We have shown that the use of the GrabCut algorithm enhances background subtraction results. Both background subtraction and GrabCut methods use GMM to model the background. Merging these algorithms so that they share a common GMM for background can result in better performance of the presented process.

The extraction process can be used in tasks that require high-quality moving objects' silhouettes for video sequences acquired with a static camera. It can be useful in moving-object classification tasks, where the moving objects' minimum size can be estimated. The moving-object silhouettes can also improve recognition-task effectiveness.

The proposed process improves the foreground mask quality when the resulting mask from the background-subtraction algorithm is good enough, without many false positive or false negative values. Moreover, this process is most suitable when there is small amount of large moving objects that cover a substantial part of the viewport.

References

1. Bouwmans, T.: Traditional and recent approaches in background modeling for foreground detection: An overview. Computer Science Review 11, 31-66 (2014)
2. Zivkovic, Z., van der Heijden, F.: Efficient adaptive density estimation per image pixel for the task of background subtraction. Pattern recognition letters 27(7), 773-780 (2006)
3. Bouwmans, T., El Baf, F., Vachon, B.: Background Modeling using Mixture of Gaussians for Foreground Detection - A Survey. Recent Patents on Computer Science, Bentham Science Publishers, 1(3), 219-237 (2008)
4. Kim, H., et al.: Robust Silhouette Extraction Technique using Background Subtraction. 10th Meeting on Image Recognition and Understand (MIRU), Japan, 1-6 (2007)
5. Konushin, V., Konushin, A:. Improvement of background subtraction by mask constraints. Proc. GraphiCon., 96-99 (2010)
6. Rother, C., Kolmogorov, V., Blake, A.: Grabcut - Interactive Foreground Extraction using Iterated Graph Cuts. ACM Transactions on Graphics (TOG), 23(3), 309-314 (2004)
7. Gulshan, V., Lempitsky, V., Zisserman, A.: Humanising GrabCut: Learning to segment humans using the Kinect. Workshop on Consumer Depth Cameras in Computer Vision, (ICCV), 1-7 (2011)
8. Hernández-Vela, A., Reyes, M., Ponce, V., Escalera, S.: GrabCut-based Human Segmentation in Video Sequences. Sensors 12 (11), 15376-15393 (2012)
9. Tomasz Posłuszny among OpenCV patches' contributors (2014), `http://opencv.org/opencv-3-0-beta.html`
10. Ding, J.-J., Kuo C.J., Hong, W.C.: An efficient image segmentation technique by fast scanning and adaptive merging. CVGIP, 1-9 (2009)
11. Wallflower Test Images, `http://research.microsoft.com/users/jckrumm/WallFlower/TestImages.htm`
12. CDnet: a video database for testing change detection algorithms, `http://www.changedetection.net`
13. Wang, Y., et al.: CDnet 2014: An Expanded Change Detection Benchmark Dataset. Computer Vision and Pattern Recognition Workshops (CVPRW), 393-400 (2014)

Optimal Sensor Placement for Fault Information System

Kornel Rostek*

Institute of Automatic Control and Robotics,
Faculty of Mechatronics, Warsaw University of Technology, Poland
k.rostek@mchtr.pw.edu.pl

Abstract. In this paper a method for solving the optimal sensor placement problem for diagnostic purposes is presented. The proposed approach maximizes isolability under budgetary constraint. The proposed strategy is based on a Fault Information System. The considered isolability measure distinguishes weak and strong isolability. The method described in this paper leads to Mixed Integer Linear Programming problem, which is then solved with branch-and-bound algorithm. The sensor placement method is tested on Three Tank System.

Keywords: Fault Detection, Fault Isolation, Fault Information System, Optimal Sensor Placement, Mixed Integer Programming

1 Introduction

The performance of a fault diagnosis system for a given industrial process is strongly dependent on available measurements. A designer of Fault Detection and Isolation (FDI) systems often wants new sensors to be installed. However, it is vital to achieve the best possible FDI system performance with minimal additional costs. The problem of optimal sensor selection can be understood as a combinatorial problem of selecting the optimal set of new measurements.

In recent years, numerous papers were published, devoted to different problems of the optimal sensor placement. The model-based FDI considers faults as deviations from normal values of process parameters or as unknown process inputs. Faults are detected when modeled and measured signals behave differently. This differences are called a residuals. They are often found using Analytical Redundancy Relations (ARRs). In [10] a method for finding the optimal sensor set based on ARRs is proposed. First, all ARRs are found under the assumption that all sensor candidates are installed. Then, the sensor set is selected that minimizes the cost while satisfying detectability and isolability requirements. However, this solution is computationally expensive. A modified, incremental approach, using Minimal Structurally Overdetermined (MSO) sets, was proposed in [4]. In [8] the Binary Integer Programming is used to find the optimal sensor set using the set

* This work was partially supported by the Warsaw University of Technology, Faculty of Mechatronics Deans Grant 504/01536.

R. Jabłoński and T. Brezina (eds.), *Advanced Mechatronics Solutions*,
Advances in Intelligent Systems and Computing 393,
DOI 10.1007/978-3-319-23923-1_10

of all possible MSO sets. FDI requirements were ensured with non-linear constraints. The resulting problem is computationally difficult to solve. This method was further improved in [2] and [3]. There, FDI requirements were specified as linear constraints. The cost function was also linear, so the problem was Binary Integer Linear Programming (BILP). It can be efficiently solved with branch-and-bound algorithm with standard Linear Programming (LP) solver. Those methods were thoroughly compared in [6]. Budgetary constraints were analysed in [7]. Branch-and-bound algorithm was used to obtain the optimal solution.

This paper presents a new method of solving the optimal sensor placement problem for a FDI system based on Fault Information System (FIS). The main contribution of this paper is solving a fault isolability maximization problem, using the isolability measure proposed in [5] as cost function. This approach allows to formulate a Mixed Integer Linear Programming (MILP) problem with a budgetary constraint using a FIS. Then, the optimization problem can be efficiently solved. To illustrate the proposed method, the model of Three Tank System is used.

The paper is organized as follows. In Section 2, the theoretical background is given. Fault Information System is defined. Then the measure of fault isolability is introduced. Section 3 describes a Mixed Integer Linear Programming problem formulation. Section 4 describes the application of the proposed method to a Three Tank System. In Section 5 some remarks and conclusions are given.

2 Theoretical background

2.1 Fault Information System

Fault Information System FIS was defined in [1]. It is a polyvalent, discrete form of notation of a symptoms - faults relation. Each diagnostic signal s_j has the set of possible values V_j. For each fault f_k a subset $V_{j,k}$ of V_j is assigned. When a fault occurs, one of those values appear. The vector $[V_{1,k}, \dots, V_{m,k}]^T$, where m is a number of signals, is called a fault f_k signature. Lets call the actual combination of single values of every signal an alternative signature. Each fault can have many different alternative signatures. An example of FIS is shown in Table 1.

Table 1. FIS example.

	f_1	f_2	V_j
s_1	{-1, 1}	1	{-1, 0 1}
s_2	0	1	{0, 1}

Assuming that sensitivities and dynamics of diagnostic signals are different, after a fault occurrence, non-zero values may not appear simultaneously. Therefore, Definition 1 can be given. In Table 1 fault f_2 has only one alternative signature and it excludes fault f_1. Fault f_1 has two alternative signatures and only one of them ($[-1 \quad 0]^T$) excludes f_2.

Definition 1. *The alternative signature of a fault f_i is excluding a fault f_k if signatures are different and any alternative signature of f_k cannot be obtained from the alternative signature of f_i by changing zeros into non-zero values.*

2.2 The measure of fault isolability for Fault Information System

In [5] a new measure of fault isolability for polyvalent diagnostic systems was proposed. It is calculated in two steps.

1. Calculate for every ordered pair of faults:

$$D(f_i, f_j) = \frac{card(S_{i,j})}{card(S_i)} \tag{1}$$

   where S_i is the set of all possible alternative signatures of f_i, and $S_{i,j}$ is the set of alternative signatures of f_i excluding f_j. If faults are unconditionally isolable then $D(f_i, f_j) = 1$, if they are unconditionally unisolable then $D(f_i, f_j) = 0$, otherwise $D(f_i, f_j) \in (0, 1)$.
2. Calculate the measure as:

$$\frac{1}{(n-1)n} \sum_{i=1}^{n} \sum_{\substack{j=1 \\ j \neq i}}^{n} D(f_i, f_j). \tag{2}$$

3 Optimisation Problem Formulation

Let S be the set of all possible new diagnostic signals. Denote by $\bar{s_k} \in \bar{S}$ variable indicating if signal $s_k \in S$ is available in a FDI system ($\bar{s_k} = 1$ if true, 0 otherwise). Using (2) it is possible to construct an optimization problem for finding the set of sensors offering the best isolability.

$$\begin{aligned} \underset{x}{\text{maximize}} \quad & \frac{1}{(n-1)n} \sum_{i=1}^{n} \sum_{\substack{j=1 \\ j \neq i}}^{n} D(f_i, f_j). \\ \text{s.t.} \quad & c^T x \leq b \\ & x_i \in \{0, 1\} \end{aligned} \tag{3}$$

Where: x is a vector of decision variables, $x_i = 1$ when i^{th} sensor is chosen, c is cost vector and b is available budget. The value of $D(f_i, f_j)$ can be calculated in the following way:

$$D(f_i, f_j) = \sup_{\bar{s}}(\bar{S}_{i,j}). \tag{4}$$

$\bar{S}_{i,j} \subset \bar{S}$ is a set of indicating variables for diagnostic signals sensitive to a fault f_i and not sensitive to a fault f_j. Similarly, indicating variable $\bar{s_k}$ can be calculated as:

$$\bar{s_k} = \underset{x}{inf}(X_{\bar{s_k}}), \tag{5}$$

where $X_{\bar{s_k}}$ is a set of new measurements necessary for a signal s_k.

The optimization problem (3), substituted with (4) and (5) is non-linear and difficult to solve. There are techniques that will allow us to transform it into a linear problem, which can be easily solved.

Lemma 1.
Problem $\underset{x}{maximize}\ min\{x_1, \ldots, x_k\}$ *has the same optimal solution as linear, constrained problem:*

$$\begin{aligned} &\underset{x}{maximize} && x_{k+1}, \\ &s.t. && x_{k+1} \leq x_1, \\ & && \vdots \\ & && x_{k+1} \leq x_k. \end{aligned} \tag{6}$$

Proof. The solution of (6) is the biggest lower bound (infimum) of the set $\{x_1, \ldots, x_k\}$. For finite sets it is always equal to a minimum, which is also the solution of the original problem. □

Lemma 2.
BILP problem $\underset{x}{maximize}\ max\{x_1, \ldots, x_k\}\ s.t. x_i \in \{0, 1\}$ *has the same optimal solution as:*

$$\begin{aligned} &\underset{x}{maximize} && min\{x_1 + x_2 + \cdots + x_k, 1\} \\ &s.t. && x_i \in \{0, 1\}. \end{aligned} \tag{7}$$

Proof. When x_i is binary i.e. $x_i \in \{0, 1\}$, $\max\{x_1, \ldots, x_k\} = 0$ iff $x_1 = x_2 = \cdots = x_k = 0$. In such a case $x_1 + x_2 + \cdots + x_k = 0$. □

Using Lemma 1 and 2 it is possible to construct a higher dimensional, linear equivalent of (3) by substituting $D(f_i, f_j)$ and $\bar{s_k}$ with constrained new control variables. In this way, MILP problem, equivalent to (3), can be constructed.

3.1 Solving problem with Branch-and-Bound algorithm

In order to solve the original problem this algorithm solves a series of relaxed LP problems. Binary constraints $x \in \{0, 1\}$ are replaced with $0 \leq x \leq 1$. If a solution to the relaxed problem is not feasible in the original problem (x_i is not integer), two new LP problems (nodes) are created. One with a constraint $x_i = 0$ and the other with $x_i = 1$. Both problems are then iteratively solved. This operation is called branching. If there are multiple infeasible variables, only one, the most infeasible, is used for branching. A variable feasibility is calculated as $abs(0.5 - x)$.

The solver retains current best integer solution. A solution to relaxed problem is always equal or better than solution to the original problem. Therefore, if optimal solution to a relaxed problem is worse than current best, such a node can be discarded. A node is also discarded if after branching LP problem is no longer feasible. Those operations are called bounding.

4 Three Tank System example

The proposed method was tested on the example of a FIS for Three Tank System (Fig. 1). In total 16 faults were considered (Table 2). Eight proposed new sensor locations with cost estimation are presented in (Table 3).

Table 2. Considered faults.

Fault	Description
f_1 - f_5	measurement chain faults
f_6	control signal fault
f_7, f_8	valve seat and actuator faults
f_9	pump fault
f_{10}	not enough water
f_{11} - f_{13}	leaks from tanks
f_{14}, f_{15}	clogging between tanks
f_{16}	clogged output flow

Table 3. Sensor locations.

Symbol	Measurement	Cost
F_1	Input flow	5
P_v	Valve position	2
L_1	Level in T1	1
L_2	Level in T2	1
L_3	Level in T3	1
CV_v	Control signal	0
p_{zp}	Pressure before pump	1
n	Pump rotational speed	2

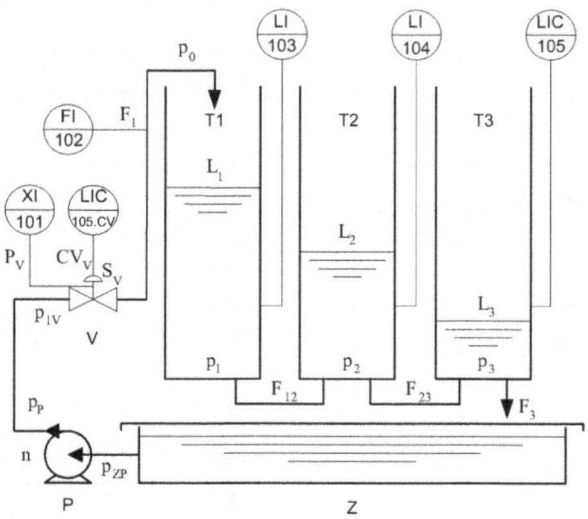

Fig. 1. Three tank system.

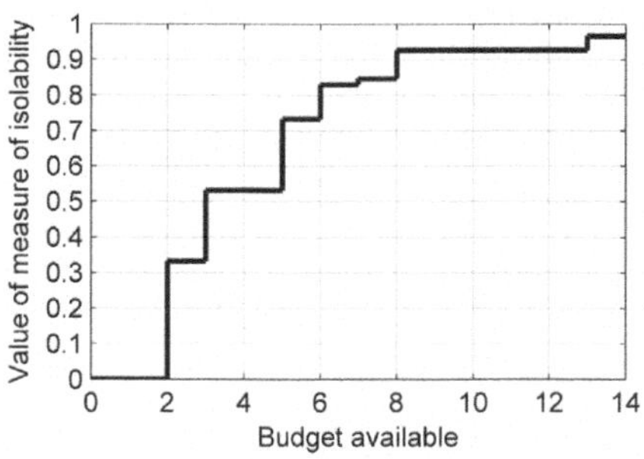

Fig. 2. Fault isolability trade-off.

Using method presented in [9] 25 possible diagnostic signals were generated, using sensors from Table 3. Trivalent diagnostic signals were analysed using information about directions of changes caused by faults. The method proposed in this paper was tested for all reasonable budgets. The values of isolability measure for diagnostic systems with different maximum total costs are presented in Fig. 2.

5 Conclusion

In this paper the sensor placement problem was addressed. A key contribution of this work is introduction of a new measure of fault isolability for Fault Information System as an objective function to an optimisation problem. A strategy of introducing new variables which allows to obtain MILP problem was presented. This allows to use efficient linear optimisation tools to find the optimal sensors set. Results obtained with the method described in this paper depend on a set of possible diagnostic signals considered during the optimization procedure. If this set is incomplete, then results may not be optimal.

The method used in this paper considers all diagnostic signals uniformly. Additional research on including uncertainty of fault detection by given signals should be considered.

References

1. Korbicz, J., Kościelny, J., Kowalczuk, Z., Cholewa, W.: Fault diagnosis: models, artificial intelligence, applications. Springer Science & Business Media (2004)
2. Nejjari, F., Sarrate, R., Rosich, A.: Optimal sensor placement for fuel cell system diagnosis using bilp formulation. In: Control & Automation (MED), 2010 18th Mediterranean Conference on. pp. 1296–1301. IEEE (2010)
3. Rosich, A., Sarrate, R., Nejjari, F.: Optimal sensor placement for fdi using binary integer linear programming. In: 20th International Workshop on Principles of Diagnosis (2009)
4. Rosich, A., Sarrate, R., Puig, V., Escobet, T.: Efficient optimal sensor placement for model-based fdi using an incremental algorithm. In: Decision and Control, 2007 46th IEEE Conference on. pp. 2590–2595. IEEE (2007)
5. Rostek, K.: Measure of fault isolability of diagnostic system. In: 25th International Workshop on Principles of Diagnosis (2014)
6. Sarrate, R., Nejjari, F., Rosich, A.: Model-based optimal sensor placement approaches to fuel cell stack system fault diagnosis. Fault Detection, Supervision and Safety of Technical Processes, Volume# 8— Part# 1 (2012)
7. Sarrate, R., Nejjari, F., Rosich, A.: Sensor placement for fault diagnosis performance maximization under budgetary constraints. In: 2nd International Conference on Systems and Control (2012)
8. Sarrate, R., Puig, V., Escobet, T., Rosich, A.: Optimal sensor placement for model-based fault detection and isolation. In: Decision and Control, 2007 46th IEEE Conference on. pp. 2584–2589. IEEE (2007)
9. Sztyber, A.: Model based diagnosis using causal graph. Pomiary, Automatyka, Robotyka 17, 83–88 (2013)
10. Travé-Massuyes, L., Escobet, T., Olive, X.: Diagnosability analysis based on component-supported analytical redundancy relations. Systems, Man and Cybernetics, Part A: Systems and Humans, IEEE Transactions on 36(6), 1146–1160 (2006)

A comparison of industrial installations modelling methods for the purpose of HAZOP support

Bartłomiej Fajdek[1], Michał Syfert[1], Jan Maciej Kościelny[1]

[1] Warsaw University of Technology, Faculty of Mechatronics, Institute of Automatic Control and Robotics
{b.fajdek,m.syfert,jmk}@mchtr.pw.edu.pl

Abstract. The paper presents a comparison of modelling methods that can be used as a tools supporting the conduction of hazard and operability study (HAZOP). The paper presents the following industrial modelling techniques: quantitative modelling using PExSim package, qualitative modelling using the GP process graph and qualitative modelling using DFM graph methodology. As an example of industrial installation a fragment of gasoline reforming section for chlorides removal was presented. A brief description of each method and results obtained from the simulation of the sample installation are included.

Keywords: qualitative modelling, quantitative modelling, process graph GP, DFM graph, HAZOP.

1 Introduction

One of the major hazards that currently exists in developed countries is a hazards of serious industrial accidents. In Poland, the risk of serious industrial accidents is very high. The hazards posed by enterprises that have products of petroleum distillation and flammable substances are especially dangerous. An important element in industrial failure prevention is a problem of early detection and elimination of potential sources of hazard.

There are known many methods to identify sources of hazards that uses formal techniques of different degrees of advancement. The methods can be divided into the following classes [1]:

- comparative methods – based on knowledge of safety analyses of similar installations (index method, checklists),
- inspection methods – that are based on the systematic study of all known potential sources of hazards, e.g. HAZOP, PHA, FMEA,
- analytical methods – allow to detect sources of hazards, to evaluate of accident scenarios and estimate numerically the risk size (FTA, ETA, CCA).

R. Jabłoński and T. Brezina (eds.), *Advanced Mechatronics Solutions*,
Advances in Intelligent Systems and Computing 393,
DOI 10.1007/978-3-319-23923-1_11

2 Description of industrial installation example

In order to perform a comparative analysis of various modelling techniques that could be potentially used as a tool for HAZOP analysis support, a fragment of gasoline reforming section for chlorides removal was presented. Reforming is a process of preparation of high-octane gasoline or aromatic hydrocarbons from light petroleum distillates, mainly from low-octane gasoline

Figure 1 shows a simplified diagram of the analysed industrial installation for reforming process.

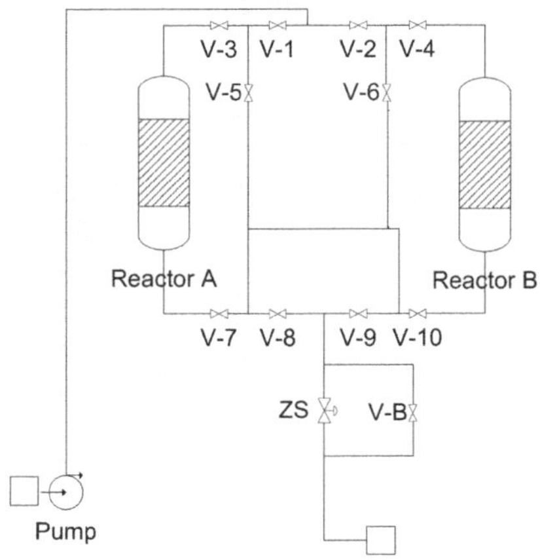

Fig. 1. A simplified installation diagram

3 Modelling of example installation using PExSim package

As one of the possible techniques for HAZOP analysis support is to use quantitative modelling. It involves the implementation of an accurate physical model of the analysed process. In this case the DiaSter system was used for modelling and simulation of exemplary process. It is possible to freely design processing circuits of individual system (process) variables in the DiaSter system. This task is performed by the PExSim module in a manner very similar to the Matlab Simulink package. The user can use a number of blocks performing basic mathematical and logic operations, input / output operations, signals flow control, integration and differentiation operators, filtration and many others. Moreover, thanks to an open architecture, it is possible to easily extend the system functionality. List of standard libraries is freely expandable through the plug-ins mechanism and can be easily adapted to user needs. There is also the possibility to create user libraries, as plug-ins, implementing specific (designed and programmed by the user) functionality. User blocks, from the functionality point of view, are the same as the "original" PExSim ones. Each of the function blocks has a number of configuration parameters determining how the specific functions are realized by given block.

The model of the analyzed system was implemented in the PExSim environment using blocks defined in the platform. Next, the simulation of flow reduction at the output of the installation was performed. It was realized by incomplete opening of the control valve. In figure 2a, 2b, 2c time series that illustrates presented situation are presented.

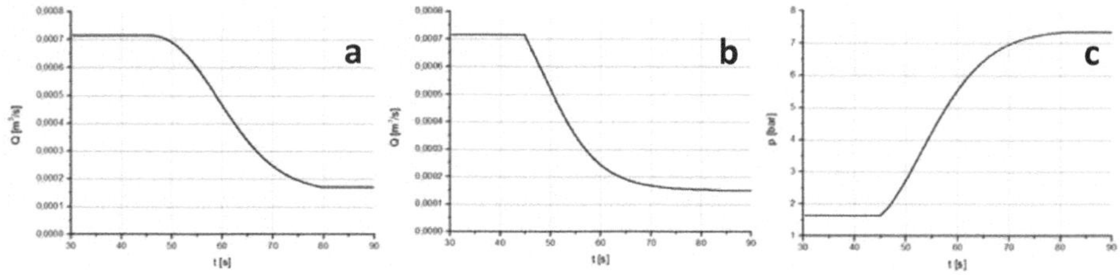

Fig. 2. The waveforms: a - flow after the ZS valve, b – outflow from the pump, c – pressure after the ZS valve

Presented time series clearly shows that, as a result of failure, the system production capacity was reduced by decreasing output flow from the system and the outflow from the pump . The presented approach allows a "very accurate" mapping of the system behaviour. It is possible to use complex or simplified blocks. Unfortunately, the analytical models, based for example on physico-chemical equations, require high expert knowledge during its creation. Furthermore, modelling of complex industrial installation is time consuming. This approach allows to perform only inductive analysis (from causes to effects), it is not possible to perform the deductive analysis. It is also impossible to check the completeness of HAZOP analysis. It seems that quantitative modelling can be used to analyse only the behaviour of individual system components, e.g. furnaces, reactors, etc.

4 Modelling of example installation using process graph

The process graph (GP) is a qualitative model of the process. It represents the cause–effect relationships between state variables, controls and measurements, taking into account the impact of failures on these variables. The relationships between inputs and corresponding outputs are presented in a graphical manner. The analysis of these relationships can be useful in many practical problems. Just a basic knowledge of the physical phenomena occurring in the process is required to create a GP model. Additional source of knowledge can be the operators and process engineers experience. The GP graphs can be used as a tool to support HAZOP analysis. Using the GP graph of the industrial process it is possible to determine which variables affect a given variable. Thus, it is possible to analyse potentially emergency situation in the industrial installation. First, the GP graph of the example industrial installation was implemented.

As the modelled variable the output flow after the control valve ZS (Q_ZS) was selected. Then, all the process variables that affect the modelled variable were selected and presented in the form of TM tree. The TM is tree with highlighted root directed from leaves to the root. The root is a variable that represents a modelled variable, while remaining nodes represent variables that have influence on modelled variable. The tree stores only information about case – effect relationships between input variables of individual components, and modelled variable. Figure 3 shows a fragment of TM tree for Q_ZS modelled variable.

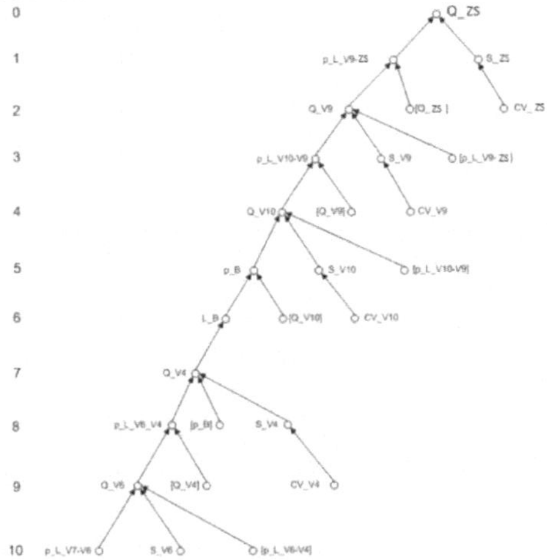

Fig. 3. The TM tree for Q_ZS modelled variable

The TM tree for a particular process variable allows to analyse the relationships between modelled variable and other process variables. The use of designed methodology as a tool for HAZOP analysis support will allow ensuring the completeness of its implementation. This approach allows to perform both inductive and deductive analysis. The disadvantage of that approach is not taking into account the timing relationships between process variables.

5 Modelling of example installation using DFM graph

The Dynamic Flowgraph Methodology (DFM) [6][7] is a two-character directed graph that allows implementation of a qualitative model of the process. It reflects the relationship between process variables similarly to the GP graph. It also takes into account the timing relationships between process variables. The quality of the implemented model strongly affects the quality of performed analyses. One must specify the physical variables and variables associated with the control system that are necessary to reconstruct the behavior of the system in order to build a model of analyzed system. Variables are represented in the model in the form of variable blocks (variable nodes). Then, the nodes are connected together via a transform blocks to create a network of cause–effect relationships. In the next phase of creation of the qualitative model it is necessary to specify correct and uncorrect operating states of individual components. They are represented in the form of state blocks which are connected to the transformation blocks. After connecting all required blocks it is necessary to enter discrete values of variables blocks, state blocks and set decisions tables in transformation blocks. It is possible to carry out two basic types of analysis: inductive and deductive based on this kind of qualitative model.

Deductive analysis due to the method of processing (from effects to causes) is the main tool to support HAZOP analysis. As a result of analysis one obtain a list of main implicants, which are the sources of the main event (top event). The first step in the analysis is to identify the main event. Then, using the model, the successive states of the various system components are determined, which lead as a result to the top event (failure of the system). This type of analysis is, in many aspects, very similar to the method based on fault tree. Fault tree is a graphical representation of possible combinations of failures that lead to creation of the major event. In the case of classical fault tree Boolean algebra can be used in order to determine the tree minimum cross section (MCS). The minimum cross section of the tree is the minimum set of elementary events whose simultaneous occurrence leads to a top event. In the case of analysis using the fault tree, the elementary events are binary encoded. For this reason multivalent fault tree is applied. It is also possible to obtain major implicants using decision tables included in each components of the system. The method is based on the creation of a single critical decision table. Then the table is subjected to merge and absorption operations in order to obtain complete set of major implicants.

Inductive analysis is the type of reasoning known as inference "from particular to general". It can be used primarily to verify the correctness of the model of analyzed system. The first step in the analysis is to determine the states of input variables blocks and state blocks. Then, as a result of simulation, other values of variables are determined automatically.

The following variation was simulated "less at the outlet of the V7 valve", i.e. flow reduction in given installation point. Results obtained from the deductive analysis are shown in Figure 4.

```
For the top event:

At time 0  ,  Q_V7 = Low    (Low flow)

There are 4 prime implicants

Prime Implicant #1
      At time 0  ,       n = Low     (Low speed)        AND
      At time 0  ,    Pump = Normal  (Normal)           AND
      At time 0  ,     CV1 = Opened  (Opened)           AND
      At time 0  ,     CV3 = Opened  (Opened)           AND
      At time 0  ,     CV7 = Opened  (Opened)

Prime Implicant #2
      At time 0  ,    Pump = Normal  (Normal)           AND
      At time 0  ,      L1 = Leakage (Leakage)          AND
      At time 0  ,     CV1 = Opened  (Opened)           AND
      At time 0  ,     CV3 = Opened  (Opened)           AND
      At time 0  ,     CV7 = Opened  (Opened)

Prime Implicant #3
      At time 0  ,    Pump = Normal  (Normal)           AND
      At time 0  ,     CV1 = Opened  (Opened)           AND
      At time 0  , L_V1-V3 = Leakage (Leakage)          AND
      At time 0  ,     CV3 = Opened  (Opened)           AND
      At time 0  ,     CV7 = Opened  (Opened)

Prime Implicant #4
      At time 0  ,    Pump = Normal  (Normal)           AND
      At time 0  ,     CV1 = Opened  (Opened)           AND
      At time 0  ,     CV3 = Opened  (Opened)           AND
      At time 0  ,       A = Leakage (Leakage)          AND
      At time 0  ,     CV7 = Opened  (Opened)
```

Fig. 4. The results of the deductive analysis of reduced flow through the V7 valve

As a result of deductive analysis one obtain a set of equilent implicants. Each of the prime implicants is caused by several simultaneous reasons, e.g. implicant 1 – in order to Q_V7 flow was low at t=0 the following reasons must occur simultaneous: n = Low, Pump=Normal, CV1 = Opened, CV7 = Opened.

The presented approach is functionally similar to the method based on GP graph. It allows to perform both inductive and deductive analysis. It is also possible to take into account timing relationships between process variables. The main disadvantage of the presented method is a large number of prime implicants in the case of more complex industrial installations. It seems reasonable to develop a mechanism for reduction of implicants obtained from deductive analysis.

6 Summary

The paper presents three different techniques that can be used as support tools to conduct HAZOP analysis. Each of presented approach has advantages and disadvantages. There is no universal method. It seems that quantitative modelling in the slightest degree is the right approach, mainly due to the large time-consuming. It requires a lot of time and expert knowledge to implement particular components of an analyzed industrial installation. The accuracy of obtained model is too accurate for conducting HAZOP analysis. More appropriate tool to support HAZOP analysis is qualitative modelling using GP process graph and DFM graph. Both approaches are very similar. The main advantage of DFM methodology is inclusion of timing between process variables. However the main disadvantage is a large number of applicants for the deductive analysis of complex industrial installations.

References

1. Głodek, W.: The risk of industrial accidents, BMP Chemia Przemysłowa 4, Warszawa (2002) (in Polish)
2. Vincoli, J.W.: Basic Guide to System Safety, Wiley-Interscience (2006)
3. Ericson, C.A.: Hazard Analysis Techniques for System Safety, Wiley-Interscience (2007)
4. Ostasz, A., Kościelny, J. M.: Application of GP process graph in diagnostics of industrial processes, Pomiary Automatyka Kontrola, 51 (2005) (in Polish)
5. Tabor, Ł.: Construction of causal graphs of processes using archival signal time series, Pomiary Automatyka Robotyka 14, pp. 65-68 (2010) (in Polish)
6. Motamed, M. et al.: Development of tools for safety analysis of control software in advanced reactors, U.S. Nuclear Regulatory Commission Report, NUREG/CR-6465 (1996)
7. Yau M. et al.: The Dynamic Flowgraph Methodolo-gy: A methodology for assessing embedded system software safety, In: Proceedings of the 2nd International Conference on Probabilistic Safety Assessment and Management, San Diego (1994)

The concept of public cargo transporting pipeline system

Mateusz Turkowski[1], Łukasz Michalak[2]

[1] Warsaw University of Technology, professor at Faculty of Mechatronics,
m.turkowski@mchtr.pw.edu.pl
[2] Warsaw University of Technology, Ph.D. student at Faculty of Mechatronics
lmichalak82@gmail.com

Abstract. One of the options of replacing traditional transport by the sustainable one is the concept of transporting the consumer goods in pipelines with the use of capsules. The paper covers the wide range of technical issues that have to be resolved before the construction of such systems, especially design of the terminals, air locks, propelling and guiding the capsules. The methodology for these purposes has been proposed, including theoretical part - modelling the capsules dynamics moving in the pipe and experimental part.

Keywords: sustainable transport · underground freight transport · capsule pipelines

1 Introduction

The necessity of the sustainable transport development has been discussed in numerous publications and is now beyond dispute [4]. The present trends in transport are not sustainable and fundamental changes in the technology are needed.
One of the options of replacing traditional transport by the sustainable one is the concept of transporting the consumer goods (food, drinks, electronics, clothes, cleaners etc.) in pipelines with the use of special capsules.
Pipeline transport is secure and independent of weather conditions. The pros and contras of pipeline transport systems for consumer goods are discussed in [2, 3] and presented in the form of SWOT analysis in [3]. The authors conclude that there are opportunities for the use of freight pipelines, but research is yet required to fully understand the supply chain, logistics and other related activities that technology may influence.
The encouraging analyses are presented in [5]. The author claims that 92% of the energy used to transport consumer goods is spent on moving the vehicles and only 8% to move the goods. Replacing heavy vehicles with lightweight capsules running in pipelines will save billions of litres of fuel and limit the CO_2 added to the atmosphere.
The systems for various purposes, such as transporting municipal solid waste, mail and parcels, delivering goods on pallets, dispatching containers from seaports to an station are described in [6, 7, 8]. According to the presented concept called pneumatic

R. Jabłoński and T. Brezina (eds.), *Advanced Mechatronics Solutions*,
Advances in Intelligent Systems and Computing 393,
DOI 10.1007/978-3-319-23923-1_12

capsules pipelines (PCP) the stationary linear induction motors (LIM) are used as a prime mover. Similar concept is adopted in Foodtubes project [11].
The concepts of transport of goods in motorized capsules moving in the tunnel is presented by Stein [9] and wide research has already been done in Germany. The system called CargoCap resembles a small subway system.
In some concepts the use of magnetic levitation (maglev) is planned. It is possible to incorporate both maglev and linear electric motor functions. Such motor would extend over the whole pipeline length. These are Pipe§net system [1] and ET3 [10] systems, they envisage the use of vacuum pumps to remove air from the tubes which practically eliminates aerodynamic drag forces. The authors claim that the capsules could achieve velocities as high as 1 500 km/h (Pipe§net) and even more (ET3).

2 General assumptions for the system

At present, prevailing and most advanced ideas is the system of pneumatic wheeled capsules pipelines, i.e. Foodtubes [11] or systems presented in [6, 7, 8]. There is, however a variety of detailed solutions. The first questions is the shape of the tunnel – circular or other, square or rectangular.
The system should be cost competitive with trucks and trains. It means, it should adopt the proven, optimized and economical existing technologies, such as pipeline construction technologies, enabling the construction of even 2 m diameter pipelines e at relatively low price. In this respect circular shape is optimal.
In some concepts each capsule is provided with the motor, i.e. the system presented in [9]. The capsules move like the subway or a train in a tunnel. It means that the capsule should transport the motor. In case of battery powered motors both motor and battery should be transported, limiting the cargo load. In case when the motor is powered from the conductor rail, the construction of pipeline is much more complicated.
Another concept, preferred by the authors, consist of the pipeline with stationary linear induction motors situated every, say, several hundred meters (Fig. 1).

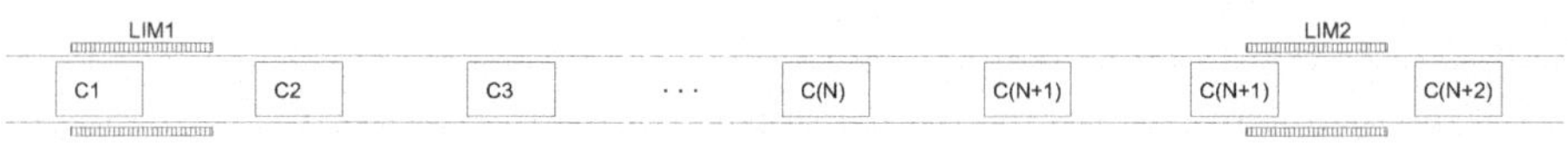

Fig. 1. General idea of PCP

Between the motors the capsules (C2 … CN, CN+2) move thanks to pneumatic cushions between them. The capsule that at the moment is in the area of LIM (C1, CN+1) pushes the capsules situated between LIMs.

3 The concept and optimal geometry of capsules

The dimensions of the capsules should enable connection of PCP to the other logistic chains. It seems that compatibility with euro palettes will be of great value from this point of view. All the existing infrastructure for materials handling, i.e. forklift trucks, high storage warehouses etc. can then be used without modifications.

Fig. 2 shows the load space on euro palettes in pipelines of various diameters. Standard euro palette is 80 cm wide, 120 cm long and 14.4 cm high.

Fig. 2 shows maximal height of the load for various capsule diameters *D*. The most convenient values of *h/D* ratio lie in the diameter range 1–1.8 m, but for diameters smaller than 1.2 m the load space can be too small for some wares. The pipes of outer diameter 1500 - 1600 mm would be most appropriate.

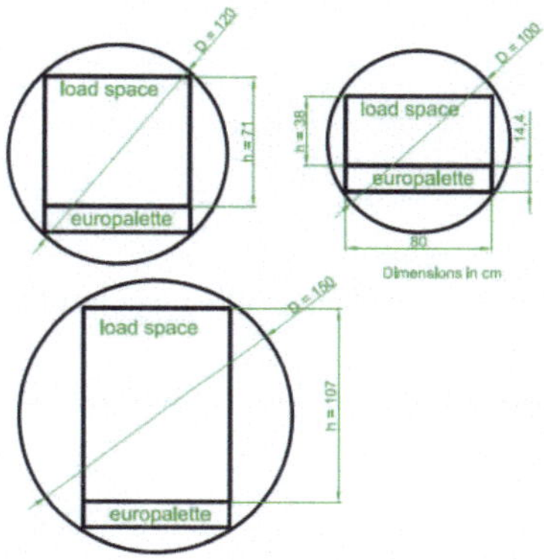

Fig. 2. Dimensions of euro palette and capsule load space for various capsule diameters

The length of the capsule about 2.5 – 3.6 m should be enough for 2 – 3 capsules.

To produce maximum drag and minimize the leakage of air the capsule should be provided with the sealing flanges. There is a question – one or more sealing, in the rear or in the front. The seals should be smaller in diameter than the inner diameter of the pipe to avoid contact between seal and pipe.

4 Modelling the system dynamics

As shown in Fig. 1 the system consists of the pipeline with stationary linear induction motors at distance of several hundred meters.

Between the motors the capsules move thanks to the pneumatic cushions between them and inertia forces.

The capsules that at the moment are inside the LIM act as a pneumatic piston and push the capsules situated between LIMs. For the clarity reasons the system presented in Fig. 3 is simplified to the case of 2 capsules. At a time t_0 the velocity of capsules is zero. Let us also suppose that the force is exerted by tubular LIM only on the ring at the front of the capsule and that the LIM force is constant.

When $x_1 <$ LSI, the first capsule is propelled through LIM with the force F_{LIM} and the second one with the pneumatic cushion exerting the force due to pressure difference $\Delta p = p_{1\text{-}2} - p_{atm}$.

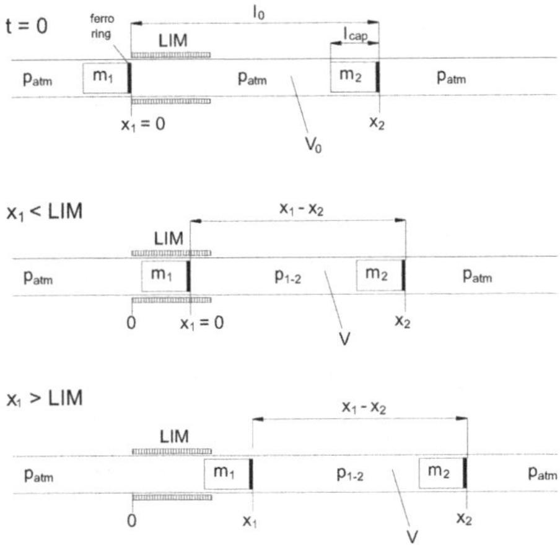

Fig. 3. The sketch for evaluation of equations of capsules motion

Both capsules are influenced by forces due to inertia, and viscous friction. The equations of motion will take form:

for capsule 1 for capsule 2

$$m_1 \frac{d^2x_1}{dt^2} + L\frac{dx_1}{dt} + m_1 gT \operatorname{sgn}\left(\frac{dx_1}{dt}\right) = F_{LIM} - A\Delta p \quad m_2 \frac{d^2x_2}{dt^2} + L\frac{dx_2}{dt} + m_1 gT \operatorname{sgn}\left(\frac{dx_2}{dt}\right) = A\Delta p \quad (1)$$

When $x_1 > LSI$, both capsules are influenced by pneumatic cushion, (first capsule in positive direction, second – in negative). For this case equations are as follows:

for capsule 1 for capsule 2

$$m_1 \frac{d^2x_1}{dt^2} + L\frac{dx_1}{dt} + m_1 gT \operatorname{sgn}\left(\frac{dx_1}{dt}\right) = -A\Delta p \quad m_2 \frac{d^2x_2}{dt^2} + L\frac{dx_2}{dt} + m_1 gT \operatorname{sgn}\left(\frac{dx_2}{dt}\right) = A\Delta p \quad (2)$$

To close the system of equations, one still needs the equation describing how the pressure between capsules depends on capsules coordinates, in simplified form the isothermal process can be assumed. The dry friction coefficient *T* depends on guiding wheels construction and the viscous friction coefficient *L* can be calculated from the equations describing the aerodynamics of the system, i.e. with the use of Computational Fluid Dynamics.

This simplified model should be extended for greater number of capsules. The dynamical model will enable the prediction of system behavior at various mass of capsules, distance between capsules, LIM characteristics, pipe inclinations etc.

5 Terminals

Construction of terminals depends strongly on the logistics of the system. In the simplest case the system will be built and operated through one operator for his own purposes, i.e. courier company or supermarket chain to connect the logistic center with individual stores. The relatively simple terminals should enable alternately reception of capsules or expedition of empty capsules with an empty packaging back to the logistic center. In such cases single bi-directional pipelines will be sufficient to build the system. The means of controlling the speed of the capsules and braking them when they have moved outside the pipeline and entered the terminal should be worked out.

In case of a public available system, the terminals should allow to connect every point where a freight shipment or receiving is needed. The pipelines net will be much more complicated and additional equipment is necessary, such as automatic switches sending capsules into branches, and means for controlling the track of each capsule.

The grand challenge will be the design of terminals enabling capsules insertion to the pipeline between the capsules moving already in the pipe. The right moment, when there is enough space between capsules, must be chosen and the velocity of the capsule to insert and velocity of the capsules already moving must be synchronized.

Another necessary elements are airlocks. The system should control the airlocks in a way that the volume of pneumatic cushions between capsules already moving does not change largely. The airlock design will be the compromise between high speed of opening/closing, low cost of manufacturing, low maintenance cost and low energy use during operation.

6 Propelling of capsules

The capsules will be propelled with the use of tubular LIMs. But perhaps in terminals mechanical, hydraulic or pneumatic propulsion will be more convenient, so both systems would function in parallel.

The motors situated along the pipe will accelerate all the already moving capsules (probably several dozen) but momentum increase for each capsule will be small, only to compensate friction and aerodynamic forces.

Another question is bringing up to speed the capsules in the expedition terminal in the case, when the capsules should be inserted between other capsules already moving. Only one capsule at a time will be accelerated, but in this case from zero velocity to the velocity of already moving capsules. Perhaps pneumatic, hydraulic or mechanical propulsion will be more convenient for this case.

7 Guiding the capsules in the pipe

The capsule should be guided in the pipeline smoothly and without rotation. The simplest, but not necessarily best from the economical point of view, way to achieve this is the installation of rails along the pipeline. The advantage of using two rails is a convenient automatic control of the path of the capsules and possibility of the use of standard rail switches for controlling the path of the capsules when they have moved outside the conduit and entered the terminal.

The possible concepts of rails are shown in Fig. 4. Typical railway wheels and rails should be supported by rolls for stabilizing the trajectory and transferring centrifugal forces in curves. In the simplified version only one rail can be used.

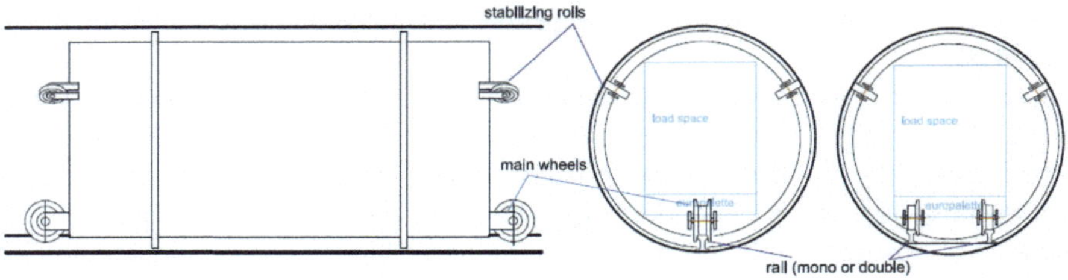

Fig. 4. Possible concepts of capsule guiding with the use of rails

From the point of view of pipeline builders, the rails make the construction much more difficult. An ideal solution would be to eliminate any rails. In the case of no-rails concept, the steering rolls can be used, controlled by the sensors of angular position of the capsule (gravitational or gyroscopic).

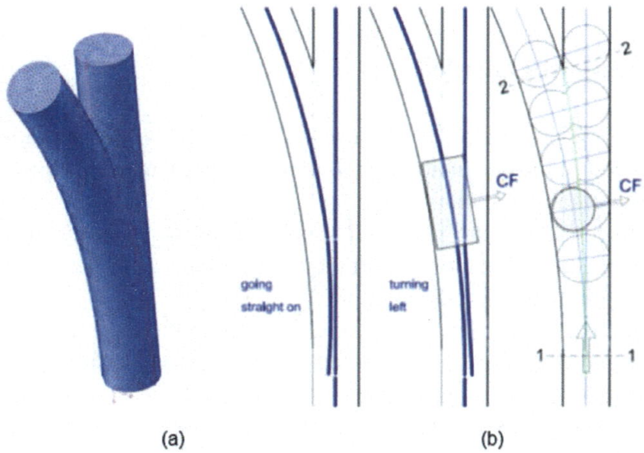

Fig. 5. Turnout shape (a) and centrifugal force in the turnout (b), capsule turning left, mono rail version

If the system is intended to be public, it should have a high number of terminals located as close to customers as possible. A lot of switches (turnouts) to the local terminals should be built then. The switch will have the form shown in Fig. 5. It will be difficult to overcome the problem of centrifugal forces in the switches without the use of 2 rails. Only by the use of two rails and railway switches technologies the problems of centrifugal force and switching the way of the capsule can be eliminated. Between the sections 1 – 1 and 2 – 2 of the turnout, in the case of no rails solution the pipe system has complicated shape composed of two circles. The stabilizing rolls lose contact with the pipe wall and capsule trajectory control is no longer possible. The turnout construction must be, therefore, much more complicated.

8 Other elements necessary to safe operation of the system

The capsule should be equipped with distance sensors and an automatic, autonomous system controlling the distance between the capsules. Distance control can be achieved through by-pass or with the use of compressed air vessel, filled at the terminal, during loading/unloading the capsule (Fig. 6).

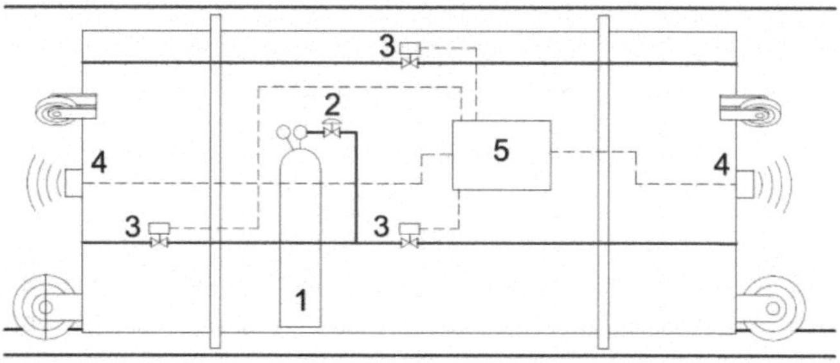

Fig. 6. The concept of distance control with the use of compressed air and by-pass valve. 1 – compressed air bottle, 2 – pressure regulator, 3 – solenoid valves, 4 – sonar or radar, 5 – controller

The system should detect failures such as increase of motion resistance or blockage of the capsule i.e. through pressure measurements. In case of increased friction forces or capsule blockage the pressure drop across the capsule will increase.
The efficient system of the control of capsules distance is crucial if the high linefill (space in the pipe occupied by capsules) has to be achieved. Low linefill limits the throughput of the system. The better and more precise control the lower distance between capsules can be maintained, so the higher linefill. For the and maintenance purposes special service capsules should be designed.

9 Conclusions

Besides rather simple issues, such as optimizing the capsule geometry, determining minimal bends radius, design of seals, mathematical models (capsules dynamics and CFD simulation of capsule aerodynamics), automatic control or maintaining safe distance between capsules, more difficult problems have to be resolved. These are problems of propelling and guiding the capsules and design of terminals and air locks. Propelling the capsules with the use of LIMs, should be examined in details. The question is, if LIM is adequate both for propelling capsules already moving in the pipe and for rapid acceleration of capsule in the expedition terminal.

Most economical would be guiding the capsules in the pipe without any rails. But problems of avoiding rotation and stabilizing the trajectory of the capsules, especially in the switches, can also be a difficult task.
The design of the air locks, switches and terminals can be difficult when the no-rails concept will be adopted, especially for side terminals enabling insertion of subsequent capsules between others, already moving.
The research should base on experiments supported by the theoretical analysis. The experiments should be conducted on the model at a scale about 1:10, so the pipe diameter in the model would be about 150 mm.
The construction of the functioning model of the system at reduced scale can reduce costs and accelerate the research, enabling the testing of many various versions of crucial system elements. Such functioning system will enable to encourage potential investors to build and test the experimental loop in full, 1:1 scale.

References

1. Cotana F., Rossi F., Marri A., Amantini M.: Pipe§net: Innovation in the transport through high rate small volume payloads. 5th International Symposium on Underground Freight Transportation by Capsule Pipelines and Other Tube/ Tunnel Systems ISUFT 2008, Arlington, March 20-22, 2008
2. Egbunike O., Potter, A., Found, P. and Packianather, M.: Are capsule pipelines a pipe dream? a critical review (2008), Proceedings of the 13th Logistics Research Network Conference, Liverpool,10th-12th September, s.563-568.
3. Egbunike O. N., Potter A.: Are freight pipelines a pipe dream? A critical review of the UK and European perspective. Journal of Transport Geography 19 (2011) 499–508
4. Greene D.L, Wegener M.: Sustainable transport. Journal of Transport Geography, Vol. 5, 1997. s. 177-190.
5. Hodson N.: Foodtubes – Energy saving pipeline capsule goods transport. 5th International Symposium on Underground Freight Transportation by Capsule Pipelines and Other Tube/ Tunnel Systems ISUFT 2008, Arlington, Texas, March 20-22, 2008.
6. Liu H., Lenau C.W.: An Advanced Electromagnetic Capsule Pipeline System for Conveying Minerals and Mine Wastes, 4th International Symposium on Underground Freight Transport (ISUFT), Shanghai, China, October 20-21, 2005 12 s.
7. Liu H. Transporting freight containers by pneumatic capsule pipelines: security and other issues, Global Pipeline, November, 2006
8. Liu H.: Feasibility of Using Pneumatic Capsule Pipelines in New York City for Underground Freight Transport. ASCE Pipeline Conference, San Diego, California, Aug.1-4, 2004
9. Stein D. Cargocap pipes through urban congestion. Mobility, issue 19/2011, p. 137 – 139
10. www.et3.com, 2013
11. www.noelhodson.com/index_files/foodtubes-project-team.htm, 2013

Assessment of the measurement method precision in interlaboratory test by using the robust "Algorithm S"

Eugenij Volodarsky[1], Zygmunt Warsza[2], Larysa Kosheva[3], and Adam Idzkowski[4]

[1] Science National Technical University of Ukraine "KPI", Department of Automation of Experimental Studies, Kiev, Ukraine
vet-1@ukr.net

[2] Industrial Research Institute of Automation and Measurement (PIAP), Warsaw, Poland
zlw@op.pl

[3] Department of Biocybernetics and Aerospace Medicine, Kiev, Ukraine
l.kosh@ukr.net

[4] Bialystok University of Technology, Faculty of Electrical Engineering, Bialystok, Poland
a.idzkowski@pb.edu.pl

Abstract. The application of robust statistical methods to assess the precision (uncertainty) of the results of interlaboratory testing of measurement method is presented. These results may include outliers. An usual rejection of such data reduces the reliability of evaluation, especially for small samples. And the robust methods take into consideration all the sample data (also with outliers). The use of two robust methods are provided for international rules of comparative studies in accredited laboratories. However, there is a lack of instructions on how to proceed them in practice. Evaluation of the precision of results, using the same method for homogeneous objects in nine laboratories, was presented. By traditional calculations, the estimate of the standard deviation was 1.5 times higher without rejection of outlier in comparison to one with rejection of outlier. Then, after using the robust method "Algorithm S", the obtained value was near the lower of two mentioned above, and with greater reliability.

Keywords: robust statistics · outliers · measurement uncertainty · interlaboratory comparisons.

1 Introduction

During the process of certification and verification of test methods and their procedures, the dependence of the accuracy on the changing conditions of measurement and on the specific organization of the experiment in each laboratory must be taken into account. A solution is to carry out the same measurements of homogenous objects in several accredited laboratories and to calculate the mean precision of all results. These interlaboratory comparisons are an experimental implementation of a physical model by specific test procedures in certain conditions. This model is created on the basis of the measurement results obtained in the laboratories of a similar essen-

R. Jabłoński and T. Brezina (eds.), *Advanced Mechatronics Solutions*,
Advances in Intelligent Systems and Computing 393,
DOI 10.1007/978-3-319-23923-1_13

tial level of competence, which are specialized in a particular type of testing. The measurement result is represented by the following statistical model [1]

$$y = m_{\bar{\bar{y}}} + B + e \,. \tag{1}$$

where: $m_{\bar{\bar{y}}} = \mu + \delta$ - mean value of the measurement results from all laboratories; δ - component of the correctness of the result, i.e. moving average value (bias) due to the imperfections of the test procedure; B - validation results component (under reproducibility conditions); e - random measurement error component (under repeatability conditions).

The relationship between parameters of the model (1) is shown in Fig. 1. This model formalizes the test procedure. Reproducibility variance σ_R^2 represents the results of the interlaboratory test, conducted with the controlled measuring method, according to certified procedures. It is the sum of variances which defines repeatability, i.e.

$$\sigma_R^2 = \sigma_L^2 + \sigma_r^2 \,. \tag{2}$$

Component σ_L^2 is a between-laboratory variance which describes a dispersion of the measurement results for the homogeneous objects in individual laboratories when the same measuring procedure is used. These spreads result from the permissionable differences in the organization of laboratory process. Repeatability variance σ_r^2 is a component which expresses repeatability of the scattering of results. It also describes the average impact of changes in size of the quantities which randomly interact within the limits allowed by the applicable standards.

Usually, the statistical data analysis is based on the assumption that the scattering of data is normally distributed. It is also the basis for making a decision in statistical inference. Significant percentage of measurement results in practice can include data outliers. In particular this concerns the datasets with a small number of samples.

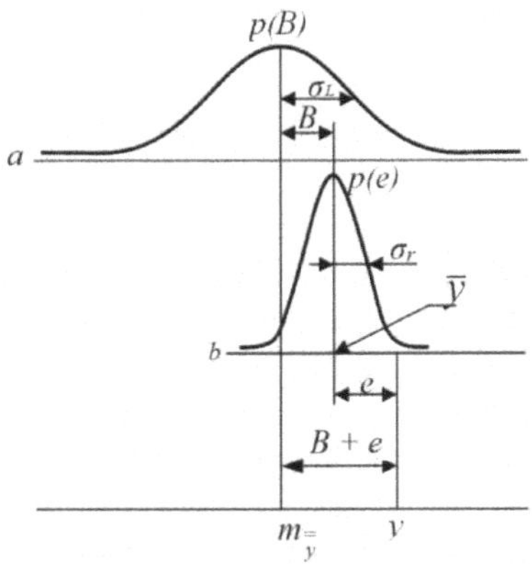

Fig. 1. Basic statistical model of the measurement result [1]

The reason of outlier values in datasets are: failure of measuring instruments, non-compliance with the principles of an experiment, errors in the estimation of results, the impact of external factors. Rejecting these data in the calculations can significantly affect accuracy and reliability of the statistical evaluation of research precision.

The classic parametric evaluation of the experimental results based on normal distribution as well as the theory of statistical inference are firmly settled in practice. Cancellation of this approach would have been inadequate. Thus, a need of adaptation of the "old" model to the new challenges emerged. It can be realized by developing such methods of estimation which, under certain conditions, include "data outliers" or allow sufficiently to assess the parameters of results on the basis of acquired data. Several methods, named as robust, were developed by Tukey, Huber and others [2], [4], [7]. Some of them are applied in the accredited laboratory practice and inter-laboratory comparisons [1-part 5], [5-7].

2 Rules of the scattering data assessment by using of robust methods

In practice, the most commonly used statistical analysis procedures assume a normal distribution of data. However, they are quite sensitive to minor variations in the parameters of this distribution. It particularly applies for the estimation of variance. In fact, we often have to deal with distributions that differ from the ideal normal distribution. However, recently in the processing of the experimental data, the robust methods are increasingly used [3-5]. As the term robust is meant an insensitivity of determined parameters to the different deviations in data samples and heterogeneity of a scatter of elements as well. They emerge from unknown reasons.

The basic model used in the robust method is not based on single normal distribution, it is mixed. Different samples of the central part of the actual scatter of data can be modeled by using of the same normal distribution. And the lower areas of a real distribution curve (tail areas) are less stable and more stretched than ones for the central area of a normal distribution. Measuring observations in the tail areas are less common. Some of them, especially for small samples, can be detected as outliers and pseudo-outliers of the central area of distribution. This approach preserves the traditionally accepted hypothetical assumption of homogeneity of the general population, as a basis for statistical evaluations. Some deviations differing from the normal distribution for the central area are permitted in the tail areas of a real distribution. However, for the tail areas some limitations are assumed. They are modeled by a normal distribution with other parameters, or by other statistics.

Approach proposed by Tukey is often used [2]. He assumed that there are a large number n of measurement data, as accidentally mixed "good" and "bad" observation x_i from a population with a mean value μ, respectively, with probability (1-ε), where ε is a low number. Both types of observations x_i have different normal distributions, i.e. the first - $N(\mu, \sigma^2)$ and the second - $N(\mu, 9\sigma^2)$, but with the same mean value μ – Fig. 2. The standard deviation of the "bad" is 3 times higher than "good". Assuming that all values x_i are independent, the following joint distribution can be expressed as

$$F(x) = (1-\varepsilon)\cdot\Phi\left(\frac{x-\mu}{\sigma}\right) + \varepsilon\cdot\Phi\left(\frac{x-\mu}{3\sigma}\right), \quad (3)$$

where

$$\Phi(x) = \frac{1}{\sqrt{2\pi}}\int_{-\infty}^{x} e^{-\frac{y^2}{2}}\,dy .$$

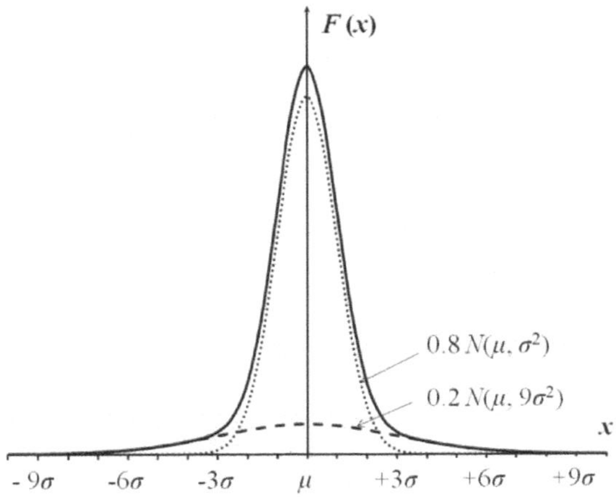

Fig. 2. Joint distribution $F(x) = (1- \varepsilon)\ N(\mu, \sigma^2) + \varepsilon\ N(\mu, 9\sigma^2)$ for $\varepsilon = 0.2$.

Among the robust methods and algorithms the approach of Huber is widely spread [4] and it is currently regarded as classical. He introduced k value which depended on the degree of "contamination" of the general population. It defines the boundaries of the central area of the measurement data histogram, i.e. difference between the upper and lower quartiles modeled by the normal distribution – Fig. 3 [6], [10].

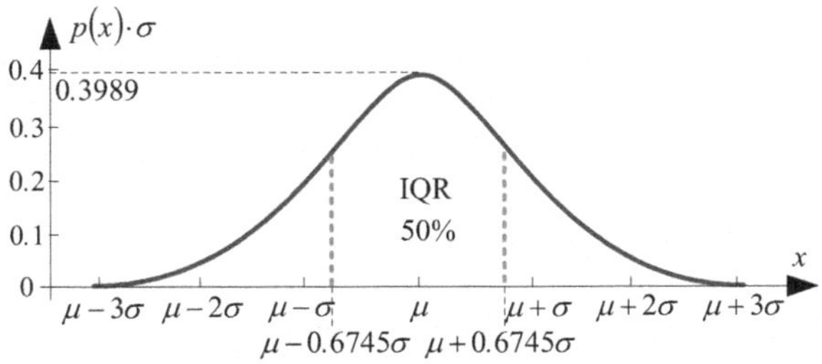

Fig. 3. The inter-quartile range (IQR) of a probability density function (pdf) of normal distribution $N(\mu, \sigma^2)$ used for the central part of sample data with outliers.

Observations are less common in the lateral areas and in one of the criteria they can be considered as outliers. In the method IRLS (iteratively reweighted least squares) extreme observations are subject to winsorizing, i.e. pulling them on the borders of the central area. It follows a change in the mean value and standard deviation of the new set of observations, and constriction of the central area. Therefore customizing the extreme data should be repeated. This process is iterated until changes become negligible.

The application of this robust method (to assess: the result obtained with a measurement method, proficiency testing for laboratory using small samples of data and the occurrence of outliers) was presented in [10]. The difference between the average values designated in the interlaboratory study is utilized to assess the reproducibility of the result. The basis of applied robust algorithms in these works is high stability of inter-quartile range (Fig. 3) with the "pollution" reaching up to 50%.

3 Robust analysis "Algorithm S"

The aim of the interlaboratory comparison study is to estimate and to standardize the variance describing the repeatability of particular method on its results obtained in several accredited laboratories. Therefore, it is necessary to determine a joint probability distribution of such variances obtained by individual laboratories participating in this joint experiment.

For the tests conducted with this method it is allowed to take into consideration the impact of possible combinations of changes in conditions (within acceptable limits). In many cases in practice, it is needed to make separate estimates for different limitations (e.g. the cost of experiment, duration of the experiment, destructive testing) only on the basis of a small number of measurement observations. Normally, their values are asymmetrically distributed and may differ significantly from a Gaussian distribution. According to the Cochran's C test, some of these observations would be regarded as the outliers. Therefore, they should be removed from the statistical processing. Such an approach would be acceptable when the average value was sought.

However the goal of an interlaboratory experiment discussed here is to assess the acceptable scattering of results from laboratories on the basis of obtained experimental data. The assessment is used to standardize the repeatability of the testing procedures performed with the use of method controlled in this experiment. The use of robust methods, as based on all the available experimental data, gives a more reliable estimate of the actual statistical dispersion of results. To obtain a stable estimate of repeatability variance (i.e. precision), the most suitable method is based on robust "Algorithm S" [1-part 5], [7].

The implementation condition of this algorithm is that the bias estimate of robust standard deviation of results from laboratories should be equal to zero. For real experimental data at each j-th step of iteration, this assessment is closer to the standard deviation σ of the normal distribution. Adjustment factor ξ is introduced to estimate a variance shift. The condition should be provided

$$E\left\{ \left(\xi s^{*}\right)^{2} \right\} = \sigma^{2}. \tag{4}$$

Robust standard deviation s^{*} should be stable with some probability, i.e. it should be within specified limits. Therefore, the maximum deviation $\eta\sigma$ of the preferred distribution is limited

$$P\left\{ s^{*} > \eta\sigma \right\} = \alpha, \tag{5}$$

where: σ - standard deviation of a normally distributed population, which corresponds to the experimental assuming their "pure" normal distribution; η - limit factor dependent on the number of data in the sample; $P = (1\text{-}\alpha)$ – probability of fulfilling a condition of the limiting of acceptable standard deviation s^{*} for the expected normal distribution.

The values of adjustment factor ξ and limit factor η are usually determined for $\alpha = 0.1$. It is made by intersecting of cumulative curves of one-modal distributions near the point where the probability equals 0.9. This approach should be examined analytically and its effectiveness should be assessed. Factor η corresponds to the upper value $(1\text{-}\alpha)$ 100% of distribution describing the scattering of robust standard deviation s^{*}. Standard deviation of this distribution may be used to assess the scattering. For a number of elements n in the sample it is dependent on the number of degrees of freedom $\nu = n - 1$. It is included by multiplying both sides of equation (4) by ν

$$E\left\{ \nu \left(\frac{s^{*}}{\sigma} \right)^{2} \right\} = \frac{\nu}{\xi^{2}} \quad \text{or} \quad E\left\{ \chi^{2}_{\nu\ robust} \right\} = \frac{\nu}{\xi^{2}}. \tag{6}$$

According to (5) the probability P of the upper limit of χ^{2} variable is equal to

$$P\left\{ \chi^{2}_{\nu\ robust} > \nu \cdot \eta^{2} \right\} = \alpha. \tag{7}$$

A tail area of χ^{2} distribution containing $\alpha \cdot 100\%$ of the value of a random variable can be approximated by a uniform distribution with density $\nu \cdot \eta^{2}$

$$\int_{\nu \cdot \eta^{2}}^{+\infty} x \cdot p_{\chi_{\nu}}(x) dx = \alpha \cdot \nu \cdot \eta^{2}. \tag{8}$$

From Pearson distribution tables [8], [9] a value $\chi^{2}_{\nu, P=1-\alpha}$ can be found and then limit factor η for which the condition (4) occurs

$$\eta^2 = \frac{\chi^2_{\nu,\,P=0,1}}{\nu}. \tag{9}$$

Starting from the relation $P\left(\chi^2_{\omega} \le \nu \cdot \eta^2\right) = 1 - \alpha$, for the main part of the distribution, z value corresponding to the value of probability P can be found from the tables and

$$\xi = \frac{1}{\sqrt{z + 0.1\eta^2}}. \tag{10}$$

It is an adjustment factor ξ for the selected limit factor η, which assures that robust estimate will not be shifted. s_j^* is a robust standard deviation calculated for the j-th step of iteration. In the iterative calculation the value s_j^* is updated as follows

$$\psi_j = \eta\, s_j^*. \tag{11}$$

In the ordered series of variances of results from laboratories participating in the experiment, a median is selected as an initial assessment of the standard deviation of the predicted normal population

$$s_0^{*2} = Me\{\, s_i^{*2} \,\}, \tag{12}$$

where $i = 1 .. n$ - number in an ordered series of laboratories.
Then the laboratory standard deviations are changed according to formula

$$s_{ij}^* = \begin{cases} \psi_j & \text{when } s_i > \psi_j \\ s_i & \text{in other cases} \end{cases} \qquad j = 0, 1, \ldots . \tag{13}$$

On the basis of the value ψ_j which is found in the current step, the values of deviations s_{ij}^* in the dataset are modified and the new values are calculated from

$$s_{j+1}^* = \xi \sqrt{\sum_{i=1}^{n} \frac{(s_{ij}^*)^2}{n}}, \tag{14}$$

where - s_{ij}^* robust standard deviation in the j-th step of iteration, for the i-th laboratory participating in the joint experiment (n - the total number of laboratories). Robust estimate s_{j+1}^* is used to establish a new limit ψ_{j+1}. Iterative procedure is continued until all standard deviations of the laboratories involved in this experiment converge within the ranges of current limit.

4 Example of using "Algorithm S"

Nine laboratories with extensive experience in this type of research were selected for the experiment. In each of them two homogeneous physical objects were examined. Absolute differences in the results in the i-th laboratory were

$$w_i = |x_{i1} - x_{i2}|,\ i = \overline{1, n}$$

where: x_{i1}, x_{i2} - the results of two experiments in i-th laboratory. Standard deviation (range) values w_i for all ($n = 9$) laboratories were as follows:

$w_1 = 0.28$; $w_2 = 0.49$; $w_3 = 0.40$; $w_4 = 0.00$; $w_5 = 0.35$; $w_6 = 1.98$; $w_7 = 0.80$; $w_8 = 0.32$; $w_9 = 0.95$.

The variance (squared standard deviation) of the difference of two results from the i-th laboratory was $s_i^2 = \frac{1}{2}|x_{i1} - x_{i2}|^2$. The assessment of the repeatability is examined for $\sum_{i=1}^{n} w_i^2$. The mean squared range equalled to $w_0 = \sqrt{\frac{1}{9}\sum_{i=1}^{9} w_i^2} = 0.827$.

Analyzing the absolute values of the differences w_i it can be noticed that the value $w_6 = 1.98$ is significantly different from the other. The hypothesis of a statistical outlier in a laboratory no 6 (value $w_6 = 1.98$) were tested using the Cochran's C test [8], [9]: $G_p = \frac{1.98^2}{6.1663} = 0.636$.

From table of this distribution [1-part 2, table 4], [8], [9] the critical values are $G_{kr}(5\%) = 0.638$ and $G_{kr}(10\%) = 0.754$. Thus G_p for w_6 is below the lower limit of this range and w_6 should be treated as a quasi-outlier. According to the rules of the traditional approach, the value $w_6 = 1.98$ should be omitted in further data processing. Then for $n = 8$, the "more precise" standard deviation $w_0' = 0.530$ is obtained. It is much smaller ($w_0' = 0.64 w_0$) than the value w_0 obtained when all data ($n = 9$) is considered in calculations. On the basis of these two estimates it is clear that the exclusion of only one difference from source data lying slightly outside a line that separate outliers, has a significant impact on the outcome of the analysis. It has influence on the standard deviation assessment of scatter of measurement procedure.

Presently, one of the robust methods will be considered. As it has been already mentioned, in these methods all the experimental data is used, also outliers. However, the data is modified. A robust method "Algorithm S" [1, part -5], [7] allows to evaluate precision of the control interlaboratory measurement procedures on the basis of the results obtained in all ($n = 9$) laboratories. Basic relationships for its implementation were above mentioned. The number of degrees of freedom is $\nu = 1$. The values of the adjusting and limit factors, according to (10) and (9), equal ξ=1.097 and η=1.645, respectively. Below, the iterative procedure is presented.

In the first step of iteration $\psi_1 = w_5^* \eta = 0.40 \cdot 1.645 = 0.658 \approx 0.66$ is determined. This is a limit value for this step. From the raw data $w_{70}^*, w_{80}^*, w_{90}^*$ should be modified because they are greater than ψ_1. New set of differences w_{i1}^* is obtained. In the first step of iteration $w_1^* = \xi \sqrt{\frac{1}{9} \sum_{i=1}^{9} \left(w_{i1}^*\right)^2} = 1.097 \cdot 0.47 = 0.52$ is calculated on the basis of the obtained values. This gives a "new" limiting value $\psi_2 = 1.645 \cdot 0.52 \approx 0.86$ and so on. In the fourth step, we already obtain robust value $w_4^* = 0.68$ which differs from $w_3^* = 0.66$ by $\frac{0.02}{0.66} \cdot 100 \approx 3\%$. As the final result $w^* = 0.68$ can be assumed. This value is between the two estimates calculated conventionally, i.e. the mean value w_0 =0.827 of the precision results for all 9 laboratories and $w_0^{'} = 0.530$ for only 8 results - after rejection of a value recognized as quasi-outlier. Finally, the common robust standard deviation should be taken $s_r = \frac{1}{\sqrt{2}} w^* = \frac{1}{\sqrt{2}} 0.68 = 0.48$.

This robust estimation of precision of the measurement method tested in this inter-laboratory experiment is more reliable than the traditional one which is based on the results from 8 laboratories, i.e. after rejection of the outlier. It was obtained from the results in all 9 laboratories.

5 Summary

The method of determining precision of a measurement method is briefly presented. If the full model is not known then tests are conducted on homogeneous objects by the same procedure in several laboratories with similar competencies. It can be assumed that the scattering is modeled by random variable with normal distribution. On the basis of the results of this research a statistical model is created and its accuracy is determined. In practice the outliers in results may occur. Rejection of them from further processing, when there is a small number of experimentally acquired data, diminishes the credibility of the assessment. Thus a robust statistical method should be applied.

For illustration, a numerical example was presented. The standard deviation as a result of research in the one of nine laboratories was an outlier. An evaluation test of precision in a conventional way with outlier rejection and robust method called "Algorithm S" [1], [7] was executed. The method uses all the experimental data. A joint assessment of the standard deviation for all results was achieved. It was slightly larger than the traditional assessment (with rejection of the outlier), however more statistically reliable.

Evaluation of reproducibility of a particular method carried out by a specific procedure (assessment of precision) is derived from results of research of interlaboratory tests. If in this study heterogeneous experimental data (with outliers) is obtained then their evaluation should be estimated using the robust "Algorithm S". It is more reliable than traditional methods.

References

1. International Organization for Standardization, Accuracy (trueness and precision) of measurement methods and results ISO 5725-2, ISO 5725-5:2002
2. Tukey, J.W.: Exploratory Data Analysis. Addison-Wesley (1978)
3. Willinik, R.: What is robustness in data analysis. Metrologia 45, 442-447 (2008)
4. Huber, P.J., Ronchetti, E.M.: Robust Statistics 2nd edition. Wiley (2011)
5. Wilrich, P.T.: Robust estimates of the theoretical standard deviation to be used in interlaboratory precision experiments. Accreditation and Quality Assurance 12 (5), 231-240 (2007)
6. Volodarsky, E.T., Warsza, Z.L., Application of two robust methods on the example of inter-laboratory comparison. In: Pavese, F., Bremser, W., Chunovkina, A.G., Fischer, N., Forbes, A.B. (eds.) Advanced Mathematical and Computational Tools in Metrology and Testing X. Series on Advances in Mathematics for Applied Sciences volume 86, World Scientific Publishing Company, 385-391 (2015)
7. International Organization for Standardization, Statistical methods for use in proficiency testing by inter-laboratory comparisons (IDT) attachment C2, ISO 13528:2005
8. Zieliński, R.: Tablice statystyczne (Statistical Tables). PWN, Warszawa (1972)
9. Farrant, T.J.: Practical statistics for the analytical scientist: A bench guide. Royal Society of Chemistry (1997)
10. Volodarsky, E., Warsza, Z., Kosheva, L, Idzkowski, A.: Evaluating the precision of interlaboratory measurements using robust S-algorithm. Problems and Progress in Metrology - PPM 2015, Proceedings of Commission Metrology of Polish Academy of Sciences in Poland (division Katowice), Series: Conferences no 20, 53-59 (2015)

Innovative car back door actuator

Michael Valášek[1], Marián Musil[2], Jan Zavřel[1], Jan Vích[1], Pavel Steinbauer[1] and Lukáš Hrnko[2]

[1] CTU in Prague, Faculty of Mechanical Engineering, Prague, Czech Republic
{Michael.Valasek,Jan.Zavrel,Jan.Vich,Pavel.Steinbauer}@fs.cvut.cz
[2] BRANO a.s., Hradec nad Moravicí, Czech Republic
{Marian.Musil,Lukas.Hrnko}@brano.eu

Abstract. Regular back door actuators in vehicles are currently quite fragile devices. Cruel handling by drivers causes frequent damage of both electronic and mechanical parts. The movement of the actuator is also associated with the possibility of injury. The paper presents use of mechatronic and inventive design methods for development of novel robust solution, which however still fits into current vehicle design and ensures safe operation.

Keywords: car; door; actuator; safety; mechatronic design; simulation

1 Introduction

The purpose of development of actuator (strut) was to find a design solution, while maintaining the most similar installation dimensions as a competitive actuator, which will have a built-in safety mechanism which ensures immediate interruption of the tailgate the car in a collision with surrounding objects. In the same time the product has to meet strict parameters required both by general (ISO) standards and also by automotive producer standards on the other hand.

The design problem is truly mechatronic as it requires holistic approach comprising of inventive thinking design methods in early stage of the project, multidomain modelling, simulations and optimization, software design, rapid prototyping and testing in later phases [1],[2].

Selected solutions were checked again by patent search for novelty and further processed. It started with detailed multi-domain modelling and parameter optimization. Detailed 3D models and overall design was created. It enabled direct manufacturing of novel parts so that functional prototypes could be rapidly assembled.

In parallel, the test bed simulating vehicle back door was built. Also control software analyses were carried out and software prototype using RAD chain Matlab/Simulink/Arduino was developed. It was also necessary to design a test plan and that apply to strut including its control software.

The mechanical strut was tested and software reliability also. The mechanical strut was tested especially in terms of durability and safety.

R. Jabłoński and T. Brezina (eds.), *Advanced Mechatronics Solutions*,
Advances in Intelligent Systems and Computing 393,
DOI 10.1007/978-3-319-23923-1_14

2 Inventive methods

The first step was problem specification and functional decomposition. Specification was based on related standards (ISO) and reference automotive producer standard, which is even more strict in terms of noise, safety etc.

Competitive benchmarking was used to determine key features and parameters of competitor's products. Especially patent research was useful to determine existing competitor's solutions and functions.

Functional decomposition approach [3] was finally used to determine ten principal product functions. They were further used for concept generation in the first, divergente phase of design process [1]. The many concepts for each function realization were elicitated using patent and literature search, brainstorming and mainly Zwicky morfological table [4] (Fig. 1).

Function	Realization	A	B	C	D	E	F	G	H	I	J
Lifting mechanism drive - force source		Electric motor + moving screw	Electric motor + push chain	Pneumatics	Electric motor hydraulic pump hydraulic cylinder	[illegible]				New types of magnets	[illegible]
Energy acumulation, force source support							Gas link (energy acumulation into gas compression)	Mechanical spring	Controlled damper		
Manual override of the lifting mechanism drive		motor clutch	locking daisy wheels on the screw	Two segment nut	Locking wheel with belt	Friction nut	safety valves	Hydraulic or pneumatic bypass			
Opening mechanism		telescopic brace	articulated mechanism	[illegible]	tooth mechanismus comb	cam mechanism	[illegible]	Complaint mechanism	Variator - friction mechanism		[illegible]
Collision detection (during lifting)	Mechanical	Mechanical bump stop									
	Electronical - sensoric	Photo cell - space surveylance	camera	Electric current in the motor detection	ultrasound sensor	x	x	x	x	x	x
Upper position lock - gravity compensation		Spring mechanism	Integrated in the lifting mechanism by flip	Mechanical lock - latch	Friction locking	Magnet	Friction clutch				
Closing mechanism drive - force source		Gravity	Reversion of F1								[illegible]
Manual override of the closing mechanism drive		same as F2									
Closing mechanism		same as F3									
Colision detection (during closing)											
Overloading handling (safety features against mechanism destruction due to overload) during closing		switching mechanism controlled by force applied on the lid									

Fig. 1. Zwicky morphological table.

Convergent design phase started with elaboration of several promising concepts (from many aroused) into 3D sketches and simple kinematic and energy-flow models, The concept evaluation was based on several selected criteria, like energy consumption, space requirements, estimated durability, noise generation etc.

3 Design process

The main design was based on the strut dimensions that are associated with significant and specific car model. Based on the installation dimensions force analysis was conducted, from which is based in the design of the solution.

3.1 Requirements for the design

According to the actual installation dimensions to design a car has been calculated transmission of forces in the support during the opening and closing of the door (**Fig. 2**). Because the relationship between the force in the strut and the length of the strut is strongly nonlinear, it is necessary to propose a strut in such a way as to force acting on the strut in accordance with the best use of the required energy. This applies when opening and closing the door. It should also be based on power option, which is limited. The requirements arising from the geometry you need to add features that will increase the usefulness of the strut. Requirements to increase the functionality are two. The first is the overload possibility. Competitive models do not allow overload of the strut. It leads to failure or destruction. The proposed strut should have a safety feature which enables such overload. The second requirement is to ensure safety in terms of the possibility of injury during the movement of the strut. The problem arises when door closes. This movement can lead to injury

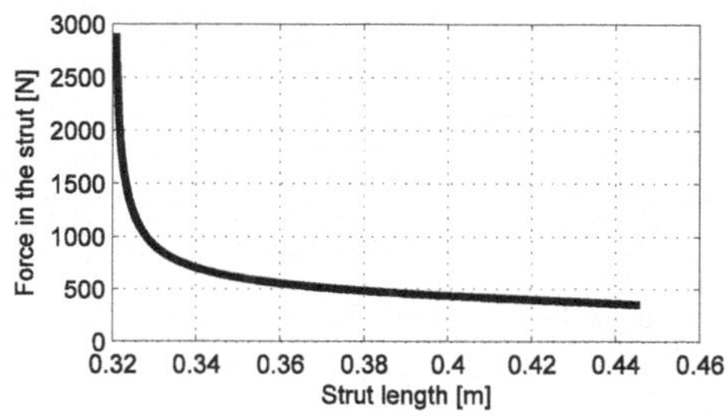

Fig. 2. The force in the strut, depending on its length when opening or closing the door.

3.2 Concept evaluation

Before proposing the final designs was to be ruled out solutions that are patent protected.

There were designed many structural variants and types of possible solutions, especially of safety mechanism. Based on the evaluation of the feasibility of the proposed solutions decommissioning drive strut at a given installation dimensions, it was decided to implement the following variant (Fig. 3). The decisive criteria for the choice of this variant were lower requirements for installation dimensions than other proposed solutions, design and manufacturing simplicity of the mechanism and the expected simpler installation compared to other proposed solutions.

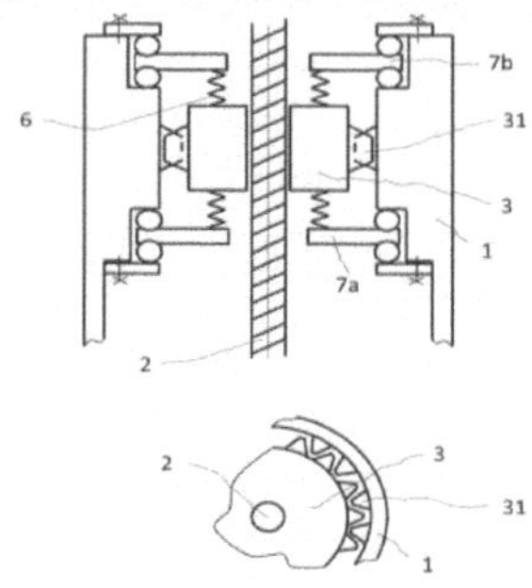

Fig. 3. Structural design of security feature of the strut.

3.3 Modelling and optimization

Before the main design and manufacture was necessary to verify the functionality of the proposed solution [6]. For the proposed system were built simulation models, which were developed for the purpose of obtaining information about the energy performance and functionality mainly of the security element. Models were developed in MATLAB-Simulink (Fig. 5). The selected result of the simulation is in Fig. 4.

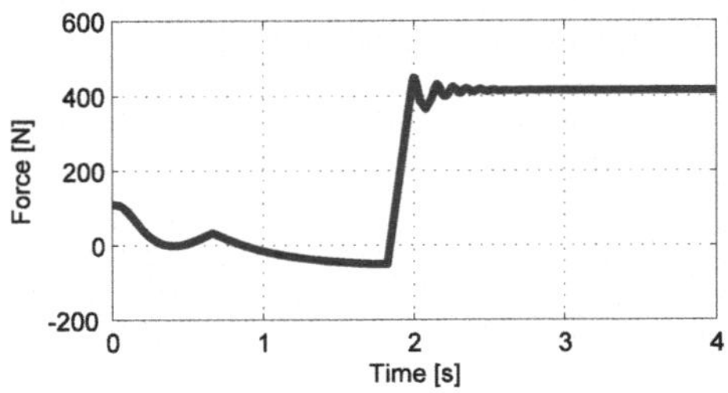

Fig. 4. Force curve when the door opens.

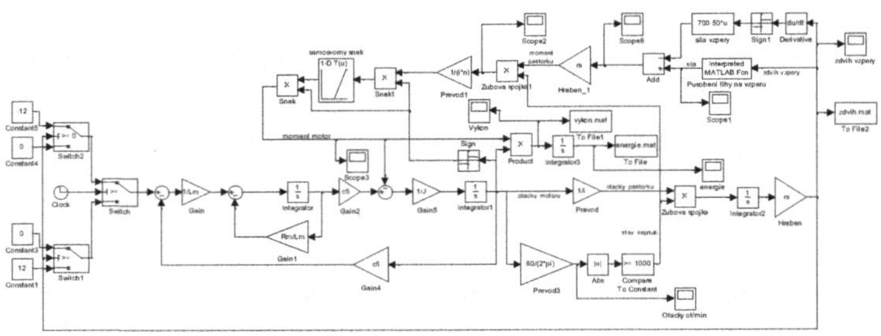

Fig. 5. The simulation model of the strut with the inclusion of a security element.

3.4 Control system and software design

Software of control system has been developed on the basis of the functional logic of the system of opening and closing the back door. According to the sequence of states was designed a flowchart. This scheme was implemented on the control system. The resulting algorithm was tested for possible mistakes of functionality. They were generated random sequences of commands and will evaluate the correct results.

Although the strut seems to be simple system, it interacts with many sensors actuators (especially lock actuators and sensors) and ECUs. So careful analyses of control system dynamic behavior was necessary using state transition diagram.

Verifying of proposed concept consists of two parts. The first is a strut, which is composed of mechanical parts and electric motor. The second part is the electronic control system.

Each part requires a specific approach to design of experiments, including functional verification. The strut will be tested in functional principle of movement, safety element and durability. Control system will be tested to fulfill the required control of the strut and if it ensures trouble-free operation, including security safety functions.

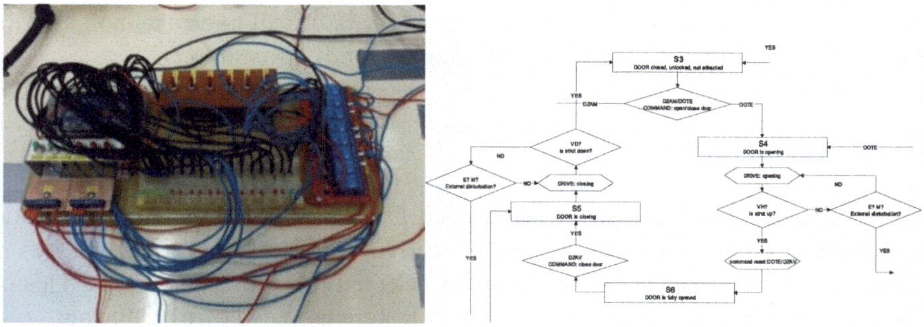

Fig. 6. Control system with ARDUINO hardware and a part of the implemented flowchart.

For controlling, testing and debugging was chosen ARDUINO platform (Fig. 6). This platform is fully sufficient for the proposed concept. The control part is composed of the concept of the physical hardware and the software part. In the automotive industry, all functions of the hardware development will be finally transferred into industrial control system for the specific car.

Part of the control system, which is common to both developmental hardware and automobile control unit is an operational flowchart. Thus, control logic of the strut and its security functions. It is also necessary to focus on the logical order and ensure the proper functioning of the software.

4 Final Concept – Prototype

The proposed elimination mechanism is shown in **Fig. 7**.

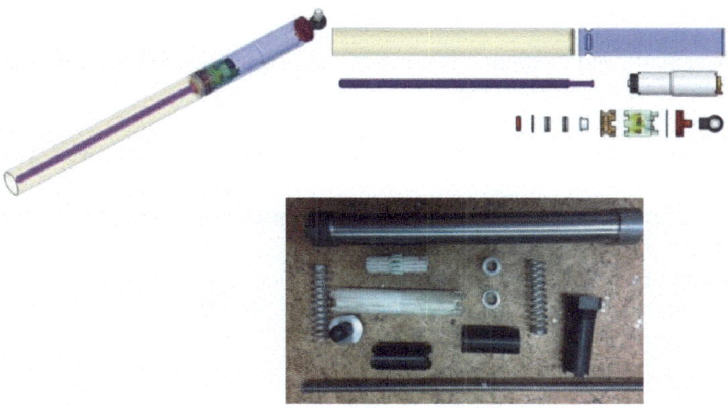

Fig. 7. Modelled strut decomposed into parts and made prototype.

It consists (Fig. 3) of an outer casing (1) with internal involute splines (31), a bearing (7a & 7b), the compression spring (6), axially floating movement nut with an internal thread (3) and a motion screw with trapezoidal thread (2). Function described mechanism consists in the fact that the axially floating movement of the matrix is maintained by compression springs in the initial position. Due to additional forces in the bolt axis indicating a collision with an obstacle, the door will move the motion matrix from a starting position and thus stops the movement of the strut.

5 Conclusion

The novel back door actuator was developed [5], patented and functional prototype manufactured. Using mechatronic and inventive thinking design methods, several solutions could be found. One of them was selected for the real solutions. This one was implemented in physical strut with safety feature provided by movable nut.
The prototype of the strut was manufactured and successfully verified about functionality of the safety mechanism. So industrial partner of the project (Brano, a.s.) has now novel product with significant comparative advantage.
The project also demonstrates successful story of cooperative design between industrial and academic partners.

Aknowledgement: This research has been realized using the support of EU Regional Development Fund in OP R&D for Innovations (OP VaVpI) and The Ministry of Education, Youth and Sports, Czech Republic, project # CZ.1.05/2.1.00/03.0125 Acquisition of Technology for Vehicle Center of Sustainable Mobility.
This research has been realized using the support of The Ministry of Education, Youth and Sports program NPU I (LO), project # LO1311 Development of Vehicle Centre of Sustainable Mobility. This support is gratefully acknowledged.

References

1. Liu, Y-C., A. Chakrabarti, and T. Bligh. "Towards an 'ideal'approach for concept generation." Design Studies 24.4 (2003): 341-355.
2. Valášek, M.: Mechatronika, ČVUT v Praze, 1995.
3. Pahl, G. and Beitz W., Engineering Design : a systematic approach, Berlin ; New York : Springer-Verlag, 1988.
4. Fritz Zwicky, Entdecken, Erfinden, Forschen im Morphologischen Weltbild, Verlag Baeschlin, Glarus, 1966, 2. (reprint) Auflage, 1989.
5. PATENT CZ 304676 B6
6. Ullman, D.G.: The mechanical design process, McGraw Hill, New York, 1992.

Part II
Biomedical Engineering

Magnetic Resonance quantification of myocardial perfusion reserve using Fermi function model: comparison to visual qualification

Tomasz Kubik[1], Konrad Werys[2,3], Krzysztof Mikołajczyk[1], Mateusz Śpiewak[3], Joanna Petryka-Mazurkiewicz[3], and Jolanta Miśko[3]

[1] Institute of Metrology and Biomedical Engineering, Warsaw University of Technology, Poland,
[2] Institute of Radioelectronics, Warsaw University of Technology, Poland,
[3] Institute of Cardiology, Warsaw, Poland

Abstract. Quantitative perfusion assessment using magnetic resonance imaging (MRI) is one of the methods used in Coronary Artery Disease (CAD) diagnosis. In this study we investigated the possibility of performing perfusion quantification using Fermi function modeling on the data acquired in the way typical for visual perfusion assessment with high dose of contrast agent. 25 patients with CAD symptoms have been examined by MRI in stress and rest condition. Image processing and numerical model have been implemented to quantify myocardial perfusion reserve (MPR). The results have been compared with the visual perfusion reserve qualification by an experienced physician. Accuracy of quantitative method in classification perfusion in myocardial segments was below 80%, meaning this method in this setup is not yet ready for clinical applications.
Keywords — Cardiac Magnetic Resonance. Myocardial Perfusion Reserve. Fermi deconvolution. Coronary Artery Disease.

1 Introduction

The aim of this study is to investigate the possibility of the use of the first pass perfusion Dynamic Contrast Enhanced Magnetic Resonance Imaging (DCE-MRI) data, acquired with saturation recovery sequence and with high contrast agent dose (0,1 mmol/kg) to quantify myocardial perfusion reserve by Fermi function deconvolution method. This type of data is typically used for visual perfusion qualification.

In this article we used first-pass DCE-MRI data acquired for visual evaluation. In typical first pass perfusion MRI two image series are acquired, one during vasodilation (stress) and one during normal heart work (rest). In stress, after effect of vasodilation is observed (higher heart rate and higher pressure), bolus of gadolinium contrast agent is injected (usually 0.1 mmol/kg). Series of images is acquired and contrast agent transit through heart chambers and myocardium can be observed. Usually 3 short axis slices are imaged, namely basal,

R. Jabłoński and T. Brezina (eds.), *Advanced Mechatronics Solutions*,
Advances in Intelligent Systems and Computing 393,
DOI 10.1007/978-3-319-23923-1_15

medial and apical positions. After imaging, stress and rest time series are visually evaluated.

One way of DCE-MRI perfusion evaluation is visual interpretation. It is one of the most popular methods in clinical practice nowadays and its reliability was proven in the literature [8]. Usually physician is looking at all imaged planes obtained both under stress and rest condition, observing contrast distribution, looking for areas with perfusion defect (lower signal).

An emerging evaluation method is a full qualitative analysis. Initial reports show better sensitivity and specificity then the one described above [7]. Time intensity curves are acquired both in blood and in myocardium regions. Deconvolution of blood and tissue curves estimates impulse response curve. Most of the quantitative methods described in literature involve special procedure of contrast agent injection, different from the one used in qualitative evaluation. Some groups proposed low contrast dose, for example: Christian et al [3] used 0.025mmol/kg. Other groups suggested dual-bolus technique, using first small contrast dose to obtain time intensity curve in blood, than big dose to obtain intensity curve in the tissue. Christian et al [4] injected 0.01mmol/kg as first bolus and 0.1 mmol/kg as second bolus. Only Costa et al [5] presented good results on data obtained with typical for visual evaluation, single bolus high dose (0.1 mmol/kg) contrast agent injection. Our goal was to confirm his findings.

2 Materials and methods

This study was performed on cardiac MRI data from 25 patients with CAD symptoms. Patients age ranged from 40 to 82 (average age: 59.3). Among patients there were 21 males and 4 females. Images were acquired during gadolinium first pass perfusion MRI performed in our institute. Contrast agent (Gadovist, Schering AG, Germany) dose was 0.1 mmol/kg both in stress and rest and was followed by saline flush.

Ground truth used in this paper was acquired as follows. Physician with 10 years of experience in cardiac MRI was given full image data from the study, including LGE-MRI images, and full medical record of the patient. Heart was divided into 16 segments according to American Heart Association (AHA) standards [2]. The physician marked each segment either as healthy, or as having defective perfusion.

In other articles reference data was taken from FFR procedure [7], coronarography or PET perfusion imaging [10]. Though those methods give quantitative values, they have a disadvantage. FFR procedure and coronarography gives information about the stenosis, not about the perfusion.

3 Theory and Calculation

All calculations, image processing and modeling described in this section have been performed on a software we developed in Java programming language.

Quantitative assessment involves determination of a parameter called MPR (Myocardial Perfusion Reserve). MPR is a ratio of maximum blood flow to the baseline [6], its quantification can be performed using Fermi function model - described in details in [9].

According to [9], some assumptions have to be made to perform modeling correctly:

1. Pixel intensity is proportional to the contrast agent concentration in tissue.
2. Relationship signal to contrast agent concentration is the same within a ventricle as well as in a muscle.
3. Signal measured in the left ventricle is considered as an input to the analyzed structure.
4. The course of the impulse response can be approximated using the Fermi function [1].

The Fermi function is defined as it follows:

$$R_F(t) = F\left[1 - \int_0^t h(s)ds\right] = F\left[\frac{1}{e^{(\tau - \tau_0 - \tau_d{}^k} + 1}\right]\theta(t - \tau_d). \tag{1}$$

F is an initial response at time $t = 0$ and is proportional to the cardiac blood flow, so the MPR can be calculated as the quotient of the $R_F(t = 0)$ values from the stress and rest tests.

Prior to modeling, some image analysis is needed to obtain signal curves. The process can be divided into three stages: automatic motion correction, manual segmentation and intensity curves modeling.

Because MR first pass perfusion imaging is a dynamic study, there is a need to consider heart motion (e.g. due to patient breathing). Acquisition is ECG gated, so usually there are no heart shape changes between frames. Without applying motion correction it is impossible to precisely define location of different heart muscle sectors on the subsequent frames.

To quantify perfusion reserve the following structures need to be precisely indicated: heart muscle and left ventricle. Segmentation was performed manually, individually for each image. The example of MRI short axis segmentation according to AHA 16 sectors model can be seen in Figure 1. Manual segmentation is a time consuming process, it takes approximately 1h to analyze data from one patient.

Before curves modeling, elimination of the DC signal component has been performed. Having measured the signal before contrast agent's injection, the signal's zero level can be designated. Because the signal contains noise, the average value of the pre-contrast signal has been calculated and subtracted from the signal curves. The modeling process sets the function (1) parameters, so that the convolution of the left ventricle signal with the myocardium signal would fit the Fermi function at most. To obtain this, a Levenberg-Marquard optimization algorithm has been used.

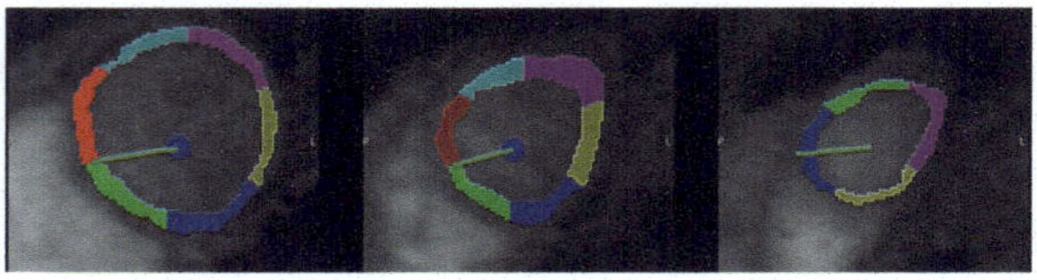

Fig. 1. Example of AHA model short axis image division. On each layer the green line links the middle of a left ventricle with the middle of interventricular septum.

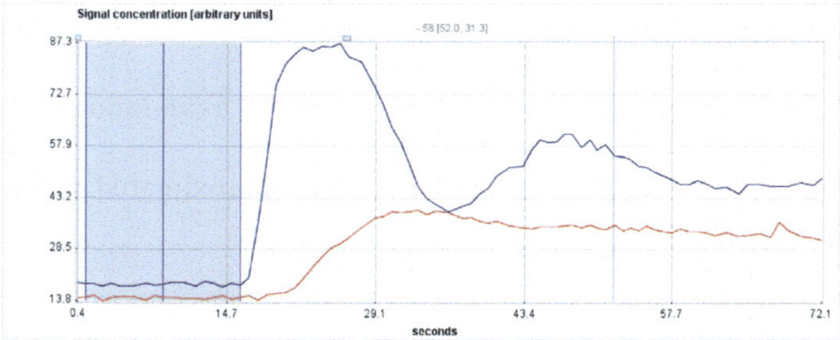

Fig. 2. The blue line represents the signal taken from the left ventricle- AIF. Flat top of the AIF curve can be observed. The green line represents the signal from one of the myocardium regions. The light blue area shows a pre-contrast time

4 Results

All results were calculated in all 16 left ventricle segments according to AHA standards [2], further in this text we refer to this as per segment results. Each segment was assigned to one of the main coronary artery territory (left anterior descending - LAD, right coronary artery - RCA and left circumflex coronary artery - LCX). If a perfusion loss was detected in one of the segments of coronary artery territory, whole territory was selected as having perfusion loss both in visual and in qualitative evaluation. Further in this text we refer to this as per coronary artery territory results. Receiver Operator Characteristic curve (ROC), illustrating performance of binary classifier system as MPR threshold is varied,

Table 1. Results with MRP threshold chosen from ROC curve (ground truth vs Fermi quantification with MPR=1.3). TP - true positives, FP - false positives, FN - false negatives, TN - true negatives, Sens. - sensitivity, Spec. - specificity, Acc. - accuracy

	TP	FP	FN	TN	Sens.	Spec.	Acc.
AHA segment analysis	35	36	46	283	43,2%	88,7%	79,5%
Coronary artery supply territory analysis	12	6	12	45	50%	88,2%	76%

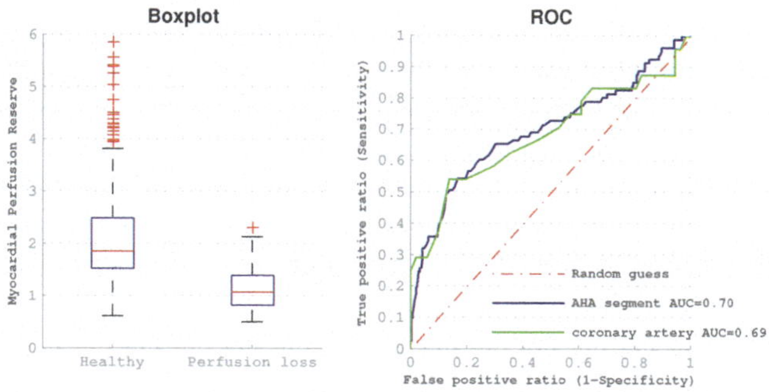

Fig. 3. Left: Box plot with classification of healthy and perfusion loss segments. Right: Receiver operating characteristic curve showing impact of chosen threshold on sensitivity and specificity

is presented in the Figure 3. The highest conformance to the reference results is obtained when MPR value equals 1.3 (chosen by calculating maximum square root of sum of squares of TPR - true positive ratio and TNR - true negative ratio), the results are presented in the Table 1.

Segment data was divided into healthy and perfusion-loss categories (based on ground truth) and presented versus MPR (Figure 3). On each box the central mark is the median, the edges of the box are the 25th and 75th percentiles, the whiskers extend to the most extreme data points not considered outliers and outliers are plotted individually.

5 Discussion

By applying Fermi deconvolution method on DCE-MRI with 0.1mmol/kg of contrast agent to quantify perfusion, we obtained worse results than those presented in literature [5], [7]. With MPR=1.3 cutoff, sensitivity is low and there are as many true positive regions as false positive ones. Manipulation of MPR threshold gives even worse results, as can be observed on ROC curve (Figure 3).

We observed that AIF never exceeds certain value, which results in flat top of the curve. In flow qualification integral of AIF curve is calculated, so AIF underestimation is one of the reasons of MPR error. There are few possible explanations of this phenomenon. High dose of contrast agent significantly shortens longitudinal relaxation time (T_1) and causes full recovery of blood signal, observed as flat top of AIF curve. Another source of error could be the assumption of linear dependence between signal intensity and contrast agent concentration. This assumption could be false with applied contrast agent doses. These problems were addressed in literature and special acquisition techniques with dual bolus injection [4] and blind AIF estimation [11] were proposed. Another important problem during signal analysis is patients breathing motion, which is caused by a long data acquisition time (at least 1 minute in the used MRI sequence). Even though motion correction was implemented and segmentation was performed with care, this may introduce source of error to calculations.

We performed quantitative assessment of cardiac perfusion performed on typical DCE-MRI perfusion data with 0.1mmol/kg of contrast agent using Fermi model deconvolution method. Results from this method in this setup poorly agree with experienced physician assessment (accuracy less than 80%). We conclude that this method in described setup is not yet ready for clinical application.

References

1. Axel, L.: Tissue mean transit time from dynamic computed tomography by a simple deconvolution technique. Invest Radiol 18(1), 94–99 (Feb 1983)
2. Cerqueira, M.D.: Standardized myocardial segmentation and nomenclature for tomographic imaging of the heart: A statement for healthcare professionals from the cardiac imaging committee of the council on clinical cardiology of the american heart association. Circulation 105(4), 539–542 (Jan 2002)
3. Christian, T.F., Bell, S.P., Whitesell, L., Jerosch-Herold, M.: Accuracy of cardiac magnetic resonance of absolute myocardial blood flow with a high-field system. JACC: Cardiovascular Imaging 2(9), 1103–1110 (Sep 2009)
4. Christian, T.F., Rettmann, D.W., Aletras, A.H., Liao, S.L., Taylor, J.L., Balaban, R.S., Arai, A.E.: Absolute myocardial perfusion in canines measured by using dual-bolus first-pass MR imaging1. Radiology 232(3), 677–684 (Sep 2004)
5. Costa, M.A., Shoemaker, S., Futamatsu, H., Klassen, C., Angiolillo, D.J., Nguyen, M., Siuciak, A., Gilmore, P., Zenni, M.M., Guzman, L., Bass, T.A., Wilke, N.: Quantitative magnetic resonance perfusion imaging detects anatomic and physiologic coronary artery disease as measured by coronary angiography and fractional flow reserve. Journal of the American College of Cardiology 50(6), 514–522 (Aug 2007)
6. Cullen, J., Horsfield, M., Reek, C., Cherryman, G., Barnett, D., Samani, N.: A myocardial perfusion reserve index in humans using first-pass contrast-enhanced magnetic resonance imaging
7. Desai, R.R., Jha, S.: Diagnostic performance of cardiac stress perfusion MRI in the detection of coronary artery disease using fractional flow reserve as the reference standard: A meta-analysis. American Journal of Roentgenology 201(2), W245–W252 (Aug 2013)
8. Giang, T., Nanz, D., Coulden, R., Friedrich, M., Graves, M., Alsaadi, N., Luscher, T., Vonschulthess, G., Schwitter, J.: Detection of coronary artery disease by magnetic resonance myocardial perfusion imaging with various contrast medium doses: first european multi-centre experience. European Heart Journal 25(18), 1657–1665 (Sep 2004)
9. Jerosch-Herold, M., Wilke, N., Stillman, A.E., Wilson, R.F.: Magnetic resonance quantification of the myocardial perfusion reserve with a fermi function model for constrained deconvolution. Medical physics 25(1), 7384 (1998)
10. Lee, D.C., Klocke, F.J.: Magnetic resonance approaches and recent advances in myocardial perfusion imaging. Curr Cardiol Rep 8(1), 59–64 (Feb 2006)
11. Schabel, M.C., DiBella, E.V.R., Jensen, R.L., Salzman, K.L.: A model-constrained monte carlo method for blind arterial input function estimation in dynamic contrast-enhanced MRI: II. *In vivo* results. Physics in Medicine and Biology 55(16), 4807–4823 (Aug 2010)

Impact of breathing mechanics, body posture and physique on heart rate variability

Marcel Młyńczak[1], Wiktor Niewiadomski[2], and Gerard Cybulski[1,2]

[1]Warsaw University of Technology, Faculty of Mechatronics, Institute of Metrology and Biomedical Engineering, Boboli 8, 02-525 Warsaw, Poland
mlynczak@mchtr.pw.edu.pl
[2]Polish Academy of Sciences, Mossakowski Medical Research Centre, Department of Applied Physiology, Pawinskiego 5, 02-106 Warsaw, Poland

Abstract. The aim of this work was to assess the impact of breathing mechanics, body posture and subject's physique on heart rate variability (*HRV*). 13 healthy students performed series of breathing with different rates and depths, while supine, sitting and standing. Impedance pneumography (*IP*) and electrocardiography signals were registered. Repeated ANOVA was used to estimate the factor effect on 4 parameters proposed. For standard deviation one (**SD**), ratio of **SD** and mean heart rate (**SDM**), and for maximum value of cross-correlation (**COR**, which describing the correspondence between *IP* and *HRV* signals in term of shapes), depth and rate of breathing and body posture had statistically significant impact. Slow, deep breathing caused increase of *HRV*.

Keywords: heart rate variability, impedance pneumography, breathing

1 Introduction

It is known that variability of heart rate is observed in stable physiological conditions and is considered as a predictor of cardiovascular morbidity and mortality [1]. Heart rate variability (*HRV*) is one of the key parameters in the cardiac and autonomic nervous systems activity analyses [2]. The natural aging is associated with a loss of complex variability in RR intervals, however if the cardiac cycle equalizes over time, there might be a disorder in regulation between the action of the sympathetic and parasympathetic systems [3]. In this way *HRV* shows off the degree of coupling between these systems.

HRV could be influenced by a number of diseases, e.g. cancer, chronic fatigue syndrome, sleep apnea or septic shock. Circadian effect seems to be other natural factor. *HRV* indexes are larger in patients with higher functional capacity. No correlation was noted between *HRV* and *BMI* [4].

It seems that current body posture should be also considered. The force of gravity and the mutual arrangement of organs would change the heart activity in natural way. The nonlinear study of *HRV* time series showed highly significant differences between supine and standing postures [5].

R. Jabłoński and T. Brezina (eds.), *Advanced Mechatronics Solutions*,
Advances in Intelligent Systems and Computing 393,
DOI 10.1007/978-3-319-23923-1_16

In our previous research, we have verified that impedance pneumography method allows the registration of the rate of breathing, as well as its depth. Joint together they could reflect the dynamics of breathing, interesting factor varying *HRV*, which was overlooked in the analyses or was regarded as the effect, but without its diversity [6]. For example, deep breathing is taken into account in the maneuvers that make up the "Ewing battery", but it is standardized [7]. Several studies have shown, that slow, controlled breathing resulted in a decrease of systolic, diastolic and mean arterial blood pressure, and also in the increase of *HRV* [8].

The aim of this study was to check the impact of the current body posture, rate and depth of breathing, sex and subject's physique (understood as *BMI* index) on heart rate variability. The work is a pilot one, which was to examine the possibility of conducting association studies of heart rate variability with impedance pneumography as a non-invasive method of quantitative determination of the parameters characterizing the activity of the respiratory system.

2 Method

13 healthy students (9 males, aged 20-24, *BMI*: 21.5 ± 1.6, and 4 females, aged 21-26, *BMI*: 22.5 ± 2.3, without any reported respiratory and cardiac diseases) were participated in this study. We informed all subjects about the aims and they gave the written informed consent to take part in it.

The *ECG* signal was measured by the cardiomonitor FX2000P, manufactured by Emtel, Poland, in single-lead configuration. Electrodes were mounted on the chest and in the upper abdomen in the standard Einthoven's positions. Impedance signal was registered using our own pneumograph prototype - Pneumonitor [9]. *IP* measurements were performed using the tetrapolar method. Electrode placement configuration was chosen as proposed by *Seppa et al.* [10]. The receiving electrodes were positioned on the midaxillary line at about 5th rib level. The application electrodes were mounted on the proximal side of the arm on the level of receiving ones. In both cases, standard spot *ECG* electrodes were used. The outputs were connected to the WinAcq ADC converter, which sampled the signals at 200Hz and stored to data files.

Every subject was asked to perform 8-10 normal and deep breaths with 6, 10 and 15 breaths/minute rates in three body postures - supine, sitting and standing. Between every series we proposed short breaks in order to establish heart rate. In that way we carried out 18 series of measurements for each subject. As the *IP* signal relates to the volume changes in terms of shape [11], it was used without any pre-calculations, only drift was removed.

We also subtracted the baseline drift from *ECG* signal (low-pass filtered with $f_{pass} = 0.5\ Hz$ and $f_{stop} = 5\ Hz$) and we automatically detected R waves and their locations in time. Heart rate was calculated as the interpolated first differentiate (using second order finite difference) of R wave locations.

We proposed the parameters estimated from (*HRV*) to show the diverse impact of the body posture, breathing rate and depth on the *HRV*:

– standard deviation of the *HRV* signal (abbreviation - **SD**)
– the ratio of **SD** and the mean *HR* value (**SDM**)
– the maximum value of cross-correlation for shifts in range $\pm 5s$ (**COR**)
– the shift value for which the cross-correlation maximum was gained (**DEL**), what was indirectly understood as a value describing whether *HRV* signal was leading or lagging the breathing one.

In order to estimate the impact of body posture, breathing rate and breathing depth, we firstly calculated basic summary of parameters for specific series and presented them in the box-plot figures. Secondly we performed repeated ANOVA analysis, with body posture, breathing rate and breathing depth as within-subject effects. Sex and *BMI* was considered as between-subject effects. For those factors, for which ANOVA states statistical significance, we performed post-hoc Tukey HSD tests.

We also performed correlation test for all pairs of parameters in order to estimate the correlation/dependance between parameters and statistical significance. All signal processing was carried out with MATLAB and statistical analysis using R.

3 Results

Sample *IP* and *ECG* signals, with corresponding *HRV* curve was presented in the Fig. 1. This example was taken from the measurements of third subject, in supine body posture, during deep breathing at 10 breaths per minute rate.

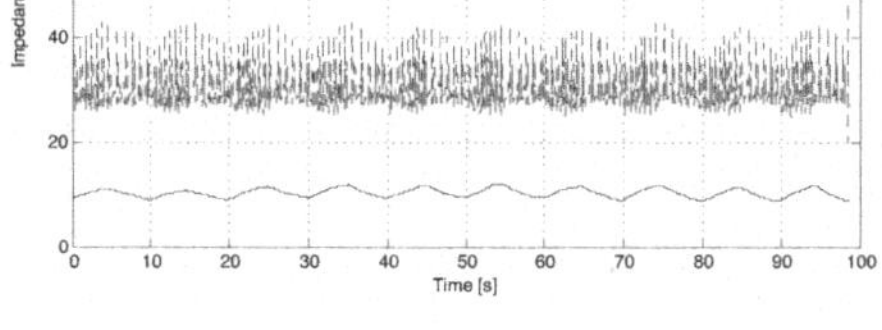

Fig. 1. The example of the impedance pneumography, electrocardiography and corresponding heart rate variability signals for third subject, supine body posture, deep breathing at 10 breaths per minute rate.

We also performed exploratory data analysis on the four described parameters, before ANOVA. The box-plot figures, presenting the impact of sex, body posture, breathing rate and depth for all parameters (**SD**, **SDM**, **COR** and **DEL**, respectively) was showed in the Fig. 3.

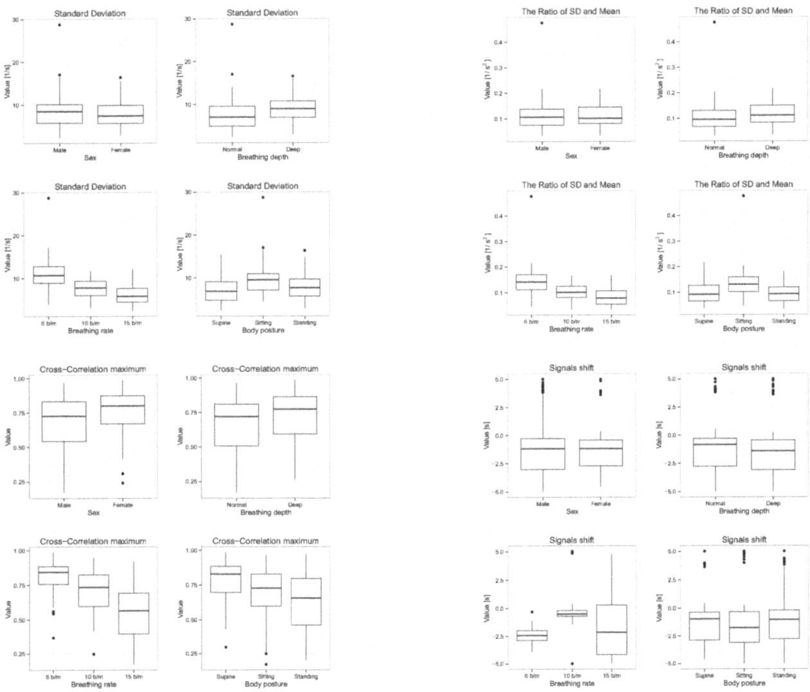

Fig. 2. The comparison of the impact of the sex, body posture, breathing rate and depth for all considered parameter: Standard Deviation in the upper left, Ratio of Standard Deviation and Mean in the upper right, Maximum Cross-Correlation Value in the lower left and Signal Shift for Maximum Correlation in the lower right.

The ANOVA analysis results (*p-values*) was collected in the Table 1.

Post-hoc tests performed for those situations, for which *p-value* was under the significance level, showed that all pairs were statistically significant, except the supine-standing for **SDM** parameter and sitting-standing for **COR** one.

As an additive information, only first and second parameter are correlated each other ($r = 0.948$, $p\text{-}value < 10^{-5}$) and the average tidal volume during normal breathing was 1036 *ml*, and during deep breathing was 2374 *ml*.

4 Discussion

In our study we only analyzed time-related parameters and did not consider frequency-related ones, due to the observation, that during controlled breathing its component of the *HRV* spectrum was covering the other parts and strongly influencing the analysis.

In the natural way, slow, deep breathing influenced on the larger **SD** values. It seems that during such breathing, respiration affects the control and regulation of heart activity in the greatest way. Relatively interesting and intriguing remark is

Table 1. The *p-values* of the ANOVA analysis for all considered parameters and factors effecting the values of parameters

	SD	**SDM**	**COR**	**DEL**
Sex	0.60	0.94	0.06	0.85
BMI	0.36	0.63	0.36	0.92
Breathing Rate	<10-5 ***	<10-5 ***	<10-5 ***	<10-5 ***
Breathing Depth	0.001 **	0.026 *	<10-3 ***	0.76
Body Posture	<10-4 ***	<10-4 ***	<10-5 ***	0.86

that sitting posture caused the largest **SD** parameters, besides every considered body postures were "static".

As the correlation between **SD** and **SDM** was strong, it seems that division by mean value does not change the relation with **SD**. It drew the finding, that standard deviation of *HRV* is not related with HR mean value. Instead of using MANOVA, we decided to use simple ANOVA two times to treat these two parameters separately and show off their impact and make the inference as simple in terms of physiological conclusions as possible.

The average values of maximum of cross-correlation coefficient, approximately 0.75, suggest breathing affecting on *HRV* in terms of signals shape. The results for **COR** parameter seems to match the preliminary expectations, that the slower inspiration and expiration allows the reaction of the sympathetic and parasympathetic systems, respectively. Depth of breathing improves the effect by changing the geometry of the thorax more.

The $\pm 5s$ distraction to **DEL** calculations was added in order to exclude these shifts, which did not correspond to the possible physiological effect. Due to the laboratory conditions signals was recurrent, quasi-sinusoidal. However, still the locations of the cross-correlation maximum differed. The average value indicated that the *HRV* slightly overtakes the breathing signal, because the largest value of heart rate occurred in the beginning of inspiration and the rate was decreasing at the end of inspiration. This suggests, that breathing might be considered as the cause for the *HRV* signal, however the sequence of the signals in terms of shapes was reverse. In our opinion, the influence of other parameter is compensated, because the maximums of cross-correlation were sometimes obtained for positive shifts and sometimes, for negative ones. It seems necessary to estimate causality of the breathing signal on heart activity using other nonlinear techniques.

5 Conclusion

It was spotted, that both breathing mechanics and body posture had statistically significant impact on the heart rate variability. Either subject's sex and *BMI* index did not. The greatest changes of *HRV* existed for deep and slow breathing.

Due to the fact that both respiration and body posture seem to be an important factors, they should be taken into account in the physiological studies considering heart activity.

It appears that the impedance pneumography method is the interesting non-invasive and portable-ready solution for performing respiration signal registration for that purpose.

Breathing seems to be the cause for heart rate changes, however simple, linear method do not assess it in the proper way. Some more research on that topic is needed.

References

1. Moura-Tonello S.C.G., Takahashi A.C.M., Francisco C.O., Lopes S.L.B., Del Vale A.M., Borghi-Silva A., Leal A.M.O., Montano N., Porta A., Catai A.M.: Influence of type 2 diabetes on symbolic analysis and complexity of heart rate variability in men. Diabetology and Metabolic Syndrome. 6:13, 1–11 (2014)
2. Kox M., Pompe J.C., van der Hoeven J.G., Cornelia W., Hoedemaekers C.W., Pickkers P.: Influence of different breathing patterns on heart rate variability indices and reproducibility during experimental endotoxaemia in human subjects. Clinical Science 121, 215–222 (2011)
3. Aysin B., Aysin E.: Effect of Respiration in Heart Rate Variability (HRV) Analysis. In: Proceedings of the 28th IEEE EMBS Annual International Conference, pp. 1776–1779. New York City, USA (2006)
4. Antelmi I., Silva De Paula R., Shinzato A.R., Peres C.A., Mansur A.J., Grupi C.J.: Influence of Age, Gender, Body Mass Index, and Functional Capacity on Heart Rate Variability in a Cohort of Subjects Without Heart Disease. Am J Cardiol 93, 381-–385 (2004)
5. Radhakrishnaa R.K.A., Narayana Dutta D., Vikram Kumar Yeraganib: Nonlinear measures of heart rate time series: influence of posture and controlled breathing. Autonomic Neuroscience: Basic and Clinical 83, 148–158 (2000)
6. Brown T.E., Beightol L.A., Koh J., Eckberg D.L.: Important influence of respiration on human RR interval power spectra is largely ignored. J Appl Physiol 75, 2310–2317 (1993)
7. Ewing D.J., Campbell I.W., Clarke B.F.: Assessment of cardiovascular effects in diabetic autonomic neuropathy and prognostic Implications. Ann Intern Med 92, 308–311 (1980)
8. Howorka K., Pumprla J., Tamm J., Schabmann A., Klomfar S., Kostineak E., Howorka N., Sovova E.: Effects of guided breathing on blood pressure and heart rate variability in hypertensive diabetic patients. Autonomic Neuroscience: Basic and Clinical 179, 131—137 (2013)
9. Mlynczak M., Niewiadomski W., Żyliński M., Cybulski G.: Ambulatory impedance pneumography device for quantitative monitoring of volumetric parameters in respiratory and cardiac applications. In: Computing in Cardiology Conference 41, pp. 965–968 (2014)
10. Seppa V.P., Hyttinen J., Uitto M., Chrapek W., Viik J.: Novel electrode configuration for highly linear impedance pneumography. Biomed Tech 58/1, 35–38 (2013)
11. Mlynczak M., Niewiadomski W., Żyliński M., Cybulski G.: Verification of the Respiratory Parameters Derived from Impedance Pneumography during Normal and Deep Breathing in Three Body Postures. In: MBEC IFMBE Proceedings 45, pp. 881–884, Dubrovnik (2015)

Dependence of sleep apnea detection efficiency on the length of ECG recording

Agata Pietrzak[1], and Gerard Cybulski[1]

[1] Institute of Precision and Biomedical Engineering, Department of Mechatronics, Warsaw University of Technology, Warsaw, Poland
agata.pietrzak89@gmail.com, G.Cybulski@mchtr.pw.edu.pl

Abstract. Our computer program allows the calculations of commonly accepted six heart rate variability (HRV) parameters in time domain. Those parameters, obtained from long-time one-channel ECG signal recordings, were used for detection of sleep apnea. The classification model was based on the Support Vector Machines (SVM) method using the discriminative Radial Basis Function (RBF) kernel. The aim of study was to check how the length of analyzed single channel ECG overnight recording influences on accuracy of sleep apnea detection.

Keywords: Sleep apnea detection · Support Vector Machines · ECG · respiratory disorders

1 Introduction

Abnormal pauses in breathing or episodes of abnormally low breathing during sleep is called sleep apnea. Pauses can last from a several seconds to a few minutes. They may occur 5 to 30 or even more times per hour [1]. Usually sleep apnea is associated with the following symptoms: morning headaches, difficulty concentrating, memory or learning problems, unstable emotional states (irritation, depression, or mood swings), or urination at night. Untreated sleep apnea can lead to the degeneration of some parts of the central nervous system and/or increase the risk of hypertension, stroke, obesity, heart attack, and diabetes. It may also increase the risk of arrhythmias and heart failure and lead to increasing the probability of having work-related accidents or driving accidents (falling asleep at the wheel). Sleep apnea symptoms may be present for years without identification causing the daytime sleepiness and fatigue associated with significant levels of sleep disturbance [1, 2].

It was estimated that in middle-age as many as 24 percent of men and 9 percent of women and were affected, undiagnosed, and untreated [1, 2, 3]. Thus, the early detection of sleep apnea occurrence is an important clinical challenge. Although polysomnography is the gold standard in detection of all types of sleep apnea, the search continues for a simple method for detecting these episodes based on a limited source of information (e.g. single ECG recording).

R. Jabłoński and T. Brezina (eds.), *Advanced Mechatronics Solutions*,
Advances in Intelligent Systems and Computing 393,
DOI 10.1007/978-3-319-23923-1_17

The sleep apnea identification methods based on single channel ECG analysis could be divided into two groups. In first one it is performed the extraction of the respiratory signal from ECG trace [4, 5] and the time localization of the event is performed. In the second one the identification of episode's occurrence is based on analysis of indices characterizing long-time ECG recording without episode's time localization [6, 7].

In this study we applied our computer program [8] enabling detection of sleep apnea occurrence using Support Vector Machines (SVM) classification performed on one lead ECG signal. The aim of this paper was to check how the length of analyzed single channel ECG overnight recording influences on accuracy of sleep apnea detection. We intended to establish the shortest set of data which might be analyzed to obtain the acceptable accuracy of the classification.

2 Material and methods

The computer program for detection of sleep apnea using a one-channel ECG signal with Graphical User Interface was prepared in the MatLab environment (v. 2013a) [8]. Program was used to analyze ECG sample recordings taken from Physionet database [9, 10].

2.1 Computer program features

The program consists of four modules enabling:

- data acquisition and preparation
- RR-interval calculation basing on QRS complex detection according to Pan-Tompkins algorithm [9],
- extraction of basic heart rate variability (HRV) indices [10],
- signal classification using the Support Vector Machine (SVM) method [11, 12].

2.2 Data description

Data were imported from the Physionet webpage's "Apnea-ECG Database (apnea-ecg)" [13] in European Data Format (EDF) [14]. The database contains 70 recordings of one-channel ECG lasting 7-10 hours. Data were recorded with a sampling frequency of 100Hz and a resolution of 16 bits, where 1 bit denotes 5μV. The analyzed signals were arranged in records of 2000, 4000, 6000, 8000, and 10,000 seconds. The signals were classified by providers according to the Apnea–Hypopnea Index (AHI), which represents the number of apnea and hypopnea events per hour of sleep. There are four categories of signals according to AHI: normal (23 records), when AHI is between 0 and 4, mild sleep apnea (3 records, AHI=5…14), moderate sleep apnea (13 records, AHI=15…30), and severe sleep apnea (31 records, AHI>30).

2.3 Heart rate variability indices

Based on the RR-interval lengths series, the commonly accepted heart rate variability parameters in time domain [10] are calculated: the mean value of the RR interval (MeanNN), the standard deviation of the successive differences between adjacent NN's (SDNN), the root mean square of successive differences (RMSSD), the number of pairs of successive NN's that differ by more than 50 ms (NN50), the proportion of NN50 divided by total number of NN's (pNN50), and the standard deviation of successive differences (SDSD). Their definitions are presented below according to the recommendations [13]:

$$MeanNN = \overline{RR} = \frac{1}{N}\sum_{i=1}^{N} RR_i \quad (1)$$

$$SDNN = \sqrt{\frac{1}{N-1}\sum_{i=1}^{N}(\overline{RR} - RR_i)^2} \quad (2)$$

$$RMSSD = \sqrt{\frac{1}{N-1}\sum_{i=1}^{N-1}(RR_{i+1} - RR_i)^2} \quad (3)$$

$$NN50 = \sum_{i=1}^{N-1} f_i \quad (4)$$

$$pNN50 = \frac{NN50}{N-1} \times 100\% \quad (5)$$

$$SDSD = \sqrt{\frac{1}{N-1}\sum_{i=1}^{N-1}(RR_{i+1} - RR_i)^2} \quad (6)$$

where

$$f_i = \begin{cases} 1\ gdy\ (RR_{i+1} - RR_i) > 50ms \\ 0\ gdy\ (RR_{i+1} - RR_i) \leq 50ms \end{cases} \quad (7)$$

N = number of records and RR_i = value of ith RR interval length in [ms].

2.4 Classification with Support Vector Machine method

It was used the Support Vector Machine (SVM) method, described by Vapnik [11], to identify sleep apnea occurrence basing on HRV descriptive, time domain indices. In the program it was adopted procedure developed by Kris De Brabanter et al. [12]. The procedure is based on Least Squares SVM using the discriminative Gaussian RBF as a kernel function in the learning process for separation of "occurrence" from "normal activity" for two sets of data: learning and testing. The Gaussian RBF kernel describes the boundary between classes separated nonlinearly. It was determined using the following formula:

$$\mathrm{K(xi, xj)} = e^{-\frac{||\mathrm{xi}-\mathrm{xj}||^2}{2\sigma^2}} \quad (8)$$

where $||\mathrm{xi} - \mathrm{xj}||^2$ represents the squared Euclidean distance between the two feature vectors and σ is a parameter [12].

The learning process was performed on 35 of 70 signals from database [13] for the records lasting 2000, 4000, 6000, 8000, and 10000 seconds.

The efficiency of the obstructive sleep apnea detection software was evaluated using the Receiver Operating Characteristic (ROC) method by comparing the automatically derived results with true results. The program draws the ROC curve, calculates the area under the curve (AUC) and the cost-effective cut-off point (CUT). The AUC is equal to the probability that a classifier will rank a randomly chosen positive instance higher than a randomly chosen negative one (assuming “positive” ranks higher than “negative”). The CUT is the point on the ROC curve the shortest distance from the upper left corner. Assuming the cut-off point’s optimal level, the confusion matrix containing the true positive (TP), true negative (TN), false positive (FP), and false negative (FN) was determined. The following formula of the classification procedure was evaluated using accuracy (ACC) defined as:

$$ACC = (TP + TN)/(TP + TN + FP + FN)\,. \tag{9}$$

3 Results

Table 1 presents the Area Under the Curve (AUC) and Accuracy for the recordings of different length. It was found that the highest accuracy (at the level of 91.4%) was obtained for the signal of the length of 10000 seconds. The AUC for that analysis was 0.93. The relatively high Accuracy (85.7%) was also observed for the analysis performed on 8000 seconds time recording, but the AUC, parameter expressing the level of reliability of that analysis was relatively low (0.74).

Table 1. The Area Under the Curve (AUC) and Accuracy (ACC) for the recordings of different length.

Time length of the signal [s]	Area Under the Curve (AUC)	Accuracy (ACC) [%]
2000	0.97	74.3
4000	0.89	86.7
6000	0.75	82.9
8000	0.74	85.7
10000	0.93	91.4

4 Discussion and conclusions

We applied computer software [8] enabling calculation of heart rate variability time domain indices. The system also allows to use HRV indices for sleep apnea classification based on a Support Vector Machine. The quality of the classification was evaluated using AUC and ACC parameters calculated when ROC analysis was performed and confusion matrix was constructed.

We wanted to check how the length of analyzed single channel ECG overnight recording influences on accuracy of sleep apnea detection. We intended to establish the shortest set of data which might be analyzed to obtain the acceptable accuracy of the classification.

The preliminary results suggests that our software [8] classifies signals as ‘with apnea’ and ‘without apnea’ with accuracy higher than 90% only for the signal length of 10000 seconds. For that period we achieved accuracy at the level similar to the high-

est accuracies found in literature [4, 15, 16]. For the shorter periods of recordings the accuracies were on level below 90%.
The work has some limitations. Among them, two seems to be important. First of all, the analysis was performed on relatively small amount of data. Secondly, there were only 3 recordings with mild sleep apnea. Despite those limitations, it seems that the program can be used as a base for further development of sleep apnea detection using ECG signals on longer sets of data.

References

1. Lee, W., Nagubadi, S., Kryger, M.H., Mokhlesi, B.: Epidemiology of obstructive sleep apnea: a population-based perspective. Expert Rev. Respir. Med. 2 (3), pp. 349–364 (2008)
2. Young T., Palta, M., Dempsey, J., Skatrud, J., Weber, S., Badr, S.: The occurrence of sleep-disordered breathing among middle-aged adults. The New England Journal of Med. 328 (17), pp. 1230–5 (1993)
3. Young, T., Peppard, P.E., Gottlieb, D.J.: Epidemiology of obstructive sleep apnea: a population health perspective. American Journal of Respiratory and Critical Care Med. 165 (9), pp. 1217–1239, (2002)
4. Chazal, P., Penzel, T., Heneghan, C.: Automated Detection of Obstructive Sleep Apnoeaa at Different Time Scales Using the Electrocardiogram. Physiol. Meas. 25(4), pp. 967-983 (2004)
5. Yilmaz, B., Asyali, M., Arikan, E., Yektin, E., Ozgen, F.: Sleep Stage and Obstructive Apneaic Epoch Classification Using Single- Lead ECG. Biomedical Engineering Online, 9, pp. 39 (2010)
6. Khandoker, A. H., Palaniswami, M., Karmakar, C.: Support vector machines for automated recognition of obstructive sleep Apnea syndrome from ECG recordings. IEEE Trans. Inf. Technol. Biomed., vol. 13 (1), pp. 37–48 (2009)
7. Almazaideh, L., Elleithy, K., Faezipour, M.: Obstructive sleep apnea detection using SVM-based classification of ECG signal features. Proc. of the Annual International Conference of the IEEE EMBS, pp.4938-4941 (2012)
8. Pietrzak, A., Cybulski, G.: New indices for sleep apnea detection in long-time ECG recordings (accepted for publication in conference proceedings), Computing in Cardiology (2015)
9. Pan, J., Tompkins, W.J.: A real-time QRS detection algorithm". IEEE Transactions on Biomedical Engineering, BME-32 (3), pp. 230-236 (1985)
10. Heart Rate Variability: Standards of Measurement, Physiological Interpretation, and Clinical Use, Task Force of the European Society of Cardiology, the North American Society of Pacing Electrophysiology, Circulation, 93, pp. 1043-1065 (1996)
11. Cortes, C., Vapnik, V.: Support-Vector Networks. Machine Learning, 20 (3), pp. 273-297 (1995)
12. De Brabanter, K., De Brabanter, J., Suykens, J.A.K., De Moor, B.: Approximate confidence and prediction intervals for least squares support vector regression, IEEE Transactions on Neural Networks. vol. 22 (1), pp. 110-120 (2011)
13. PhysioNet, http://www.physionet.org/physiobank/database/apnea-ecg/
14. European Data Format, http://www.edfplus.info
15. Manrique, Q., Hernandez, A., Gonzalez, T., Pallester, F., Dominquez C.:Detection of obstructive sleep apnea in ECG recordings using time-frequency distributions and dy-

namic Features. In: Proceedings of the IEEE International Conference on Engineering in Medicine and Biology Society, pp. 5559-5562 (2009)
16. Mendez, M., Ruini, D., Villantieri, O., Matteucci, M., Penzel, T., Bianchi, A.: Detection of sleep apnea from surface ECG based on features extracted by an autoregressive model.: In Proceedings of the IEEE International Conference on Engineering in Medicine and Biology Society (EMBS 2007), pp. 6105-6108 (2007)

Sensitivity Analysis of the Material Parameters of the Ceramics on the Inner Radius of the Hip Joint Endoprosthesis Head

Vladimir Fuis[1] and Premysl Janicek[1]

[1] Institute of Thermomechanics, ASCR, v.v.i. – branch Brno, Czech Republic
{fuis, janicek}@fme.vutbr.cz

Abstract. The paper globally deals with calculation of the parameters of ceramic material from a set of destruction tests of ceramic heads of total hip joint endoprosthesis. It is not possible to use the standard way to calculate the material parameters, because the specimens cut-outed from the heads are smaller than the norm required (the material parameters derived from them would exhibit higher strength values than those which the given ceramic material really has). On that score, a special testing jig was made, in which 40 heads were destructed. From the measured values of circumferential strains of the head's external spherical surface under destruction, the state of stress in the head under destruction was established using the final elements method (FEM). From the values obtained, the sought for parameters of the ceramic material were calculated using Weibull's weakest-link theory. The analysis of the sensitivity of the obtained material parameters on the value of the inner radius in the head's hole is realized in detail.

Keywords: Hip joint endoprosthesis, ceramic head, material parameters, Weibull's weakest-link theory, sensitivity analysis.

1 Introduction

A problem that is being solved is the destruction of the ceramic heads of total hip joint endoprostheses in vivo, that had occurred in a series of Czech hospitals [2], [4], [10], [12] (Fig. 1). The ceramic heads are made of Al_2O_3 and put on conical stem made of austenitic steel [16-19]. The implant's failure of the „ceramic head destruction" type" has always traumatic consequences for the patient, since a part of or even the whole endoprosthesis has to be re-operated, after which must again follow reconvalescence and rehabilitation. Hence, it is desired to reduce the number of implant reoperations to the minimum. The reliability of the ceramic component is based on Weibull weakest-link theory [1], [3] and the failure probability depends on three Weibull's parameters [6], [8], [11]. These parameters are obtained from the statistical analysis of the set of the destructed specimens which are subjected to 3- or 4- point bending [13], [15]. In case of ceramic heads of endoprosthesis, the snag is that the head's dimensions are too small and do not allow to cut out from them a specimen

R. Jabłoński and T. Brezina (eds.), *Advanced Mechatronics Solutions*,
Advances in Intelligent Systems and Computing 393,
DOI 10.1007/978-3-319-23923-1_18

that would comply with the standard for 3- or 4-point bending. One of the possible solutions to this problem is not to cut out the specimens from the heads, but to calculate the material parameters directly from the destruction of whole ceramic heads [7].

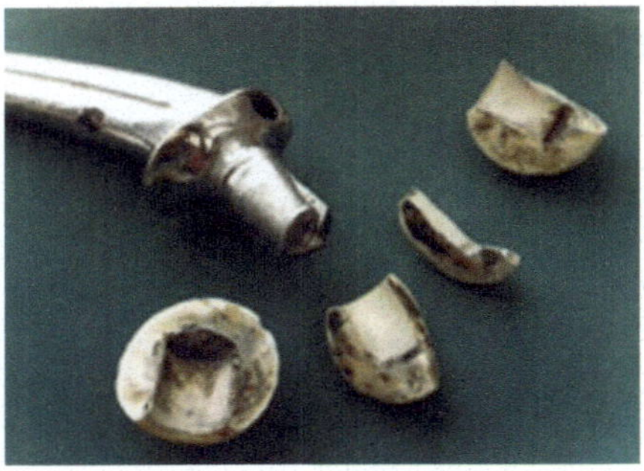

Fig. 1. In vivo destructed head

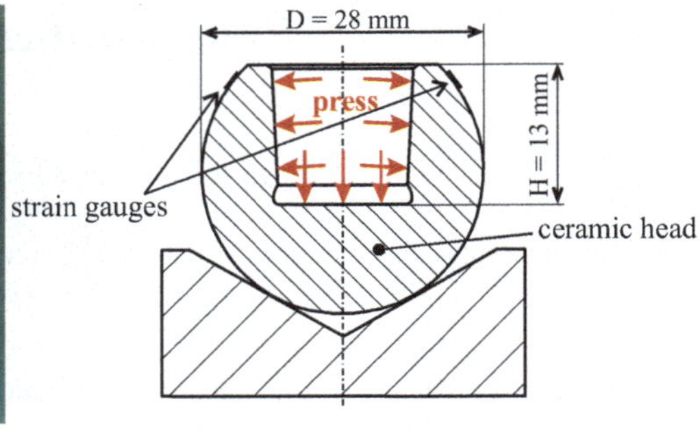

Fig. 2. Scheme of the head's loading

2 Experiments

A special testing jig has been made in which the heads are subjected to compressive load (using the testing device ZWICK Z 020-TND) – Fig. 2 which cause the high tensile circumferential stress in the head. In the course of the test, circumferential strains are measured of the head's external surface – the measured destruction strains are written in the second column in the Tab. 1.

3 Calculation of the material parameters of the bioceramics

At the first the maximal first principal stress (σ_{1max}) in the heads were calculated using FEM. The analysis of the destructed heads shows that the radius in the depth of the cone hole is changed from 0.6 to 0.9 mm (Fig. 3). Therefore the sensitivity analysis of this parameter on the material parameters of the ceramic was realized. The isosurfaces of σ_{1max} for different value of the radius r is shown in Fig. 3 (the circumferential strain in the measured place is for all four shown variants the same – 332 μm/m). The location of the maximum tensile stress is the same for all variants (radius) - in the radius region which is very small. The maximum value of the stress in the head is rapidly increased with the decreasing of the radius value (from 371 MPa for r = 1 mm to 706 MPa for r = 0.25 mm).

Since a linear task is the matter, even the remaining values ε_{dest} can be recounted in a linear way to $\sigma_{1max\text{-}dest}$ for different value of the radius r. Thus we obtain a set of values of destruction stresses that have to be arranged in descending order for further analysis. Each j-th destructed head is assigned the probability of its failure (see the last column in the Tab. 1), e.g. from the relation $P_f(j) = j/(s+1)$, where j is the serial number of the arranged head and s is the total number of destructed heads, in our case s = 40 [5], [14].

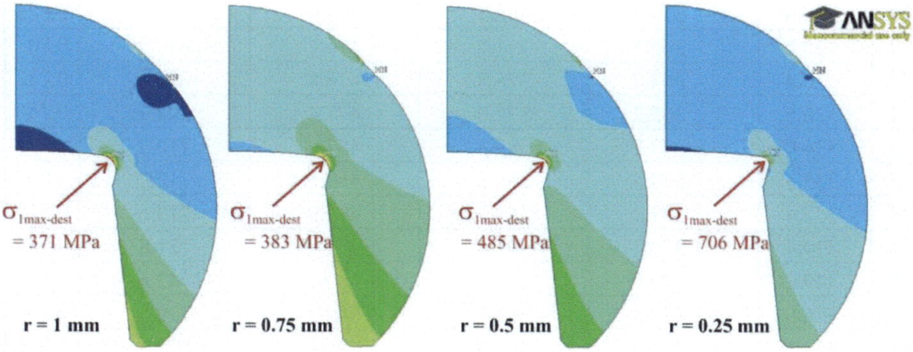

Fig. 3. Isosurfaces of the stress σ_1 in the head for different radius r (loading is pressure p = 100 MPa according Fig. 2)

j	ε_{dest} [μm/m]	$\sigma_{1max\text{-}dest}$ [MPa] r = 1 mm	$\sigma_{1max\text{-}dest}$ [MPa] r = 0.75 mm	$\sigma_{1max\text{-}dest}$ [MPa] r = 0.5 mm	$\sigma_{1max\text{-}dest}$ [MPa] r = 0.25 mm	P_f
1	186.5	208.4	215.1	272.4	396.6	0.024
2	193.7	216.4	223.4	282.9	411.9	0.049
3	194.3	217.1	224.1	283.8	413.2	0.073
4	198.5	221.8	229.0	289.9	422.1	0.098
5	200.9	224.5	231.7	293.5	427.2	0.122
6	224.5	250.8	259.0	327.9	477.4	0.146
7	228.7	255.6	263.9	334.1	486.4	0.171
8	238.6	266.6	275.2	348.5	507.3	0.195
9	242.0	270.4	279.2	353.5	514.6	0.220
10	242.2	270.6	279.4	353.8	515.0	0.244
11	248.0	277.2	286.2	362.4	527.5	0.268
12	253.8	283.6	292.8	370.8	539.7	0.293
13	258.7	289.1	298.4	377.9	550.1	0.317
14	269.6	301.3	311.0	393.9	573.3	0.341
15	271.9	303.9	313.7	397.2	578.2	0.366
16	273.3	305.5	315.3	399.3	581.3	0.390
17	274.2	306.4	316.3	400.6	583.1	0.415
18	275.4	307.7	317.7	402.2	585.5	0.439
19	278.4	311.1	321.2	406.7	592.0	0.463
20	281.4	314.4	324.6	411.1	598.4	0.488
21	281.7	314.8	324.9	411.5	599.0	0.512
22	288.9	322.8	333.2	422.0	614.3	0.537
23	290.4	324.5	335.0	424.2	617.5	0.561
24	295.8	330.5	341.2	432.1	628.9	0.585
25	305,0	340.8	351.8	445.5	648.5	0.610
26	307.0	343.0	354.1	448.4	652.8	0.634
27	309.8	346.2	357.4	452.6	658.9	0.659

28	319.6	357.2	368.7	466.9	679.7	0.683
29	322.2	360.1	371.7	470.7	685.2	0.707
30	326.4	364.7	376.5	476.8	694.1	0.732
31	326.8	365.2	377.0	477.4	694.9	0.756
32	334.0	373.2	385.3	487.9	710.2	0.780
33	335.1	374.5	386.6	489.6	712.7	0.805
34	337.4	377.1	389.3	492.9	717.6	0.829
35	339.4	379.3	391.6	495.9	721.8	0.854
36	349.2	390.2	402.9	510.2	742.6	0.878
37	376.0	420.1	433.7	549.2	799.5	0.902
38	402.1	449.3	463.9	587.4	855.1	0.927
39	423.7	473.4	488.7	618.9	900.9	0.951
40	449.0	501.7	517.9	655.9	954.7	0.976

Tab. 4. Measured destruction strains, computed maximal first principal stress in the head for different radius and failure probability (arranged in descending order)

The first of Weibull parameters is stress σ_u, which must be lower than the minimum values of $\sigma_{1max\text{-}dest}$ (lower than the value for $j = 1$ in the Tab. 1.). In this paper the conservative approach is considered and $\sigma_u = 0$ MPa, then all tensile stresses influence the head's destruction.

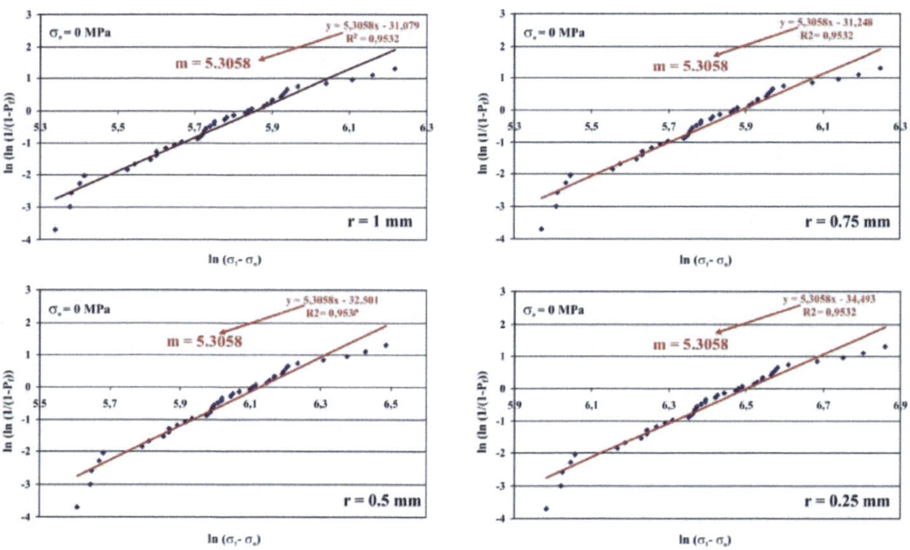

Fig. 4. Weibull plot of the normalize modulus of rupture data for different radius r

The second parameter (Weibull modulus *m*) is connected with the dispersion of experimentally established values and it is determined as a gradient of a line interlaid with logarithmic transformed data from Tab. 1 – see Fig. 4. The value of $m = 5.3058$ and this value is the same for all radius values because the dispersion of the experimental data is the same. Only the values of the stresses $\sigma_{1max\text{-}dest}$ are different. It causes the different values on the horizontal axis, on the vertical axis the values are the same for all variants in Fig. 4.

The last parameter (normalizes volume strength σ_o) is calculated from the following equation:

$$\sigma_o = \sqrt[m]{\sum_{i=1}^{n}(\sigma_i - \sigma_u)^m \Delta V_i}, \ \sigma_i \geq \sigma_u \tag{1}$$

which was derived by modification of basic Weibull equation for calculation of failure probability:

$$P_f = 1 - e^{-\sum_{i=1}^{n}\left(\frac{\sigma_i - \sigma_u}{\sigma_o}\right)^m \Delta V_i}, \qquad \sigma_i \geq \sigma_u, \tag{2}$$

for $P_f = 1\text{-}1/e = 0.63212$. As values σ_i are used the set of the values σ_1 (acted in the volume ΔV_i) calculated by the finite elements method in the all elements in the heads, for which it holds that $\sigma_1 > \sigma_u$. The calculated material parameters for all radius r are shown in the Tab. 2.

The value of the strength σ_o for different radius r is changed from 584 to 668 $MPa.m^{3/5.3058}$ (the increasing is about 14%) but the stress $\sigma_{1max\text{-}dest}$ is changed from 371 to 706 MPa (Fig. 3) – this is increasing 90%. The strength σ_o is influenced not only on the stress σ_1 in the head but on the volume in which it acts too. The volume of the maximal tensile stress in the head is very small (Fig. 3) so the sensitivity of the strength σ_o on the value of the radius r is significantly smaller than on the stress σ_{1max}.

σ_u [MPa]	0 MPa			
m	5.3058			
σ_o [$MPa.m^{3/5.3058}$]	583.7	595.5	614.3	667.7
r [mm]	1	0.75	0.5	0.25

Tab. 2. Calculated material parameters

4 Conclusion

Material parameters of the used bioceramic material (m, σ_u and σ_o) are used by the Weibull weakest link theory from the set of 40 destructed heads. Weibull modulus m is not influenced by the value of the radius in the inner part of the head's hole. Stress σ_u was considered to be equal to zero – conservative approach. The last parameter - strength σ_o is influenced by the radius value, but not as significantly as the value of the maximum principal stress in the head.

Acknowledgement - The research has been supported by the project of the Czech Science Foundation GA CR nr. 13-34632S.

References

1. Basu, B., Tiwari, D., Kundu, D., Prasad, R.: Is distribution the most appropriate statistical strength distribution for brittle materials? Ceramics International 35(1), 237–246 (2009)
2. Fuis, V., Janicek, P.: Stress and reliability analyses of damaged ceramic femoral heads. In: Proc. Conference on Damage and Fracture Mechanics, Structures and Materials, vol. 12, pp. 475–485 (2002)
3. Bush, D.: Designing Ceramic Components for Structural Applications. J. Mater. Eng. Perf. ASM Int. 2, 851–862 (1993)
4. Fuis, V., Koukal, M., Florian, Z.: Shape Deviations of the Contact Areas of the Total Hip Replacement. In: Proc. 9th International Conference on Mechatronics, Mechatronics: Recent Technological and Scientific Advances, pp. 203–212 (2011)
5. Fuis, V., Janicek, P.: Calculation of the Bio-ceramic Material Parameters. In: Proc. 10th International Conference on Mechatronics, Mechatronics, pp. 855-861 (2014)
6. McLean, A. F., Hartsock, D. L.: Engineered materials handbook, Ceramics and Glasses, vol. 4, pp. 676–689. ASM International (1991)
7. Fuis, V., Malek, M., Janicek, P.: Probability of destruction of Ceramics using Weibull's Theory. In: Proc. 17th International Conference on Engineering Mechanics, pp. 155–158 (2011)
8. Fok, S. L., Smart, J.: The Accuracy of Failure Predictions Based on Weibull Statistics. Journal of the European Ceramic Society 15, 905–908 (1995)
9. Fuis, V., Janicek, P., Houfek, L.: Stress and Reliability Analyses of the Hip Joint Endoprosthesis Ceramic Head with Macro and Micro Shape Deviations. In: IFMBE Proceedings of 13th International Conf. on Biomedical Engineering, vol. 23(1-3), pp. 1580–1583 (2008)
10. Fuis, V.: Tensile Stress Analysis of the Ceramic Head with Micro and Macro Shape Deviations of the Contact Areas, Recent Advances in Mechatronics: 2008-2009. In: Proc. International Conference on Mechatronics, pp. 425–430 (2009)
11. Fetta, T., Ernsta E., et al.: Weibull analysis of ceramics under high stress gradients. Journal of the European Ceramic Society 23, 2031–2037 (2003)
12. Fuis, V.: Stress and reliability analyses of ceramic femoral heads with 3D manufacturing inaccuracies. In: Proc. 11th World Congress in Mechanism and Machine Science, pp. 2197–2201 (2004)
13. Jiang, R., Murthy, D. N. P.: A study of Weibull shape parameter: Properties and significance. Reliability Engineering and System Safety 96, pp. 1619–1626 (2011)
14. Fuis, V., Navrat, T., Hlavon, P., et al.: Reliability of the Ceramic Head of the Total Hip Joint Endoprosthesis Using Weibull's Weakest-link Theory. In: IFMBE Proc. World Congress on Medical Physics and Biomedical Engineering, vol. 14, pp. 2941–2944 (2006)
15. Hudak, D., Tiryakioglu, M.: On comparing the shape parameters of two Weibull distributions. Materials Science and Engineering A 528, pp. 8028–8030 (2011)
16. Fuis, V., Navrat, T., Vosynek, P.: Analyses of the Shape Deviations of the Contact Cones of the Total Hip Joint Endoprostheses, In: IFMBE Proceedings of 6th World Congress of Biomechanics, Vol. 31, pp. 1451–1454 (2010)
17. Fuis, V., Varga, J.: Stress Analyses of the Hip Joint Endoprostesis Ceramic Head with Different Shapes of the Cone Opening, In: IFMBE Proceedings of 13th International Conference on Biomedical Engineering, vol. 23, pp. 2012–2015 (2009)
18. Fuis, V., Janicek, P.: Stress and reliability analyses of ceramic femoral heads with axisymmetric production inaccuracies, In: IFMBE Proceedings of 9th MEDICON, pp. 632–635 (2001)
19. Houfek, L., Florian, Z. Brezina, T., et al.: Development of Experimental Devices for Testing of the Biomechanicask Systems, In: IFMBE Proceedings of 13th International Conference on Biomedical Engineering, vol. 13, pp. 2005–2008 (2009)

Multibody model of dynamics and optimization of medical robot to soft tissue surgery

Grzegorz Ilewicz[1],

[1] Department of Mechatronics and Automation, The University of Rzeszów
[1]gilewicz@ur.edu.pl

Abstract. The model of dynamics of the constant point mechanism of Robin heart 1 medical robot to soft tissue surgery that was built in the Matlab and Simulink computer program is presented in the paper. The model describes: kinematics, kinetics and steering function of the mechatronical device. The system of rigid bodies that are connected in rotary joints is steering with use of the PID controllers. Strength optimization with use of the finite element methods and probabilistic metaheuristic algorithm of simulated annealing were performed.

Keywords: medical robot, multibody dynamics, simulated annealing, PID controller

1 Introduction

Medical robots are mechatronical devices used widely in surgery in most developed countries. In Poland, several research centers participate in manufacturing a medical robot for minimally invasive surgery. The most popular is Foundation for Cardiac Surgery Development in Zabrze which produces Robin Heart medical robot. Robin Hearth robot has an effector which is similar to those used in minimally invasive surgery (Fig.1).

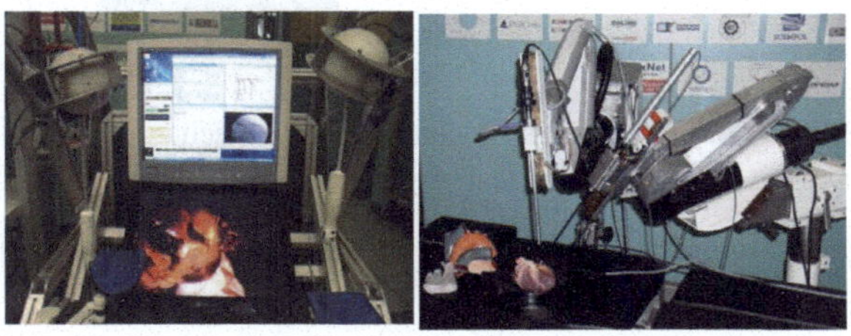

Fig.1. Teleoperator Robin Heart 1 and Robin Heart Vision during experiment electrocoagulation of wall of the heart. The effectors are similar to minimally invasive surgery instrumentation

The works on the development of a medical robot with multibody operating tool are carried out at The University of Rzeszów. In Wroclaw, Da Vinci (Intuitive Surgical Inc., USA) robot is used to operate in soft tissue surgery. Da Vinci and Robin Heart

R. Jabłoński and T. Brezina (eds.), *Advanced Mechatronics Solutions*,
Advances in Intelligent Systems and Computing 393,
DOI 10.1007/978-3-319-23923-1_19

robots have similar structure uses the constant point mechanism. The constant point mechanism of medical robot is illustrated in figure 2.

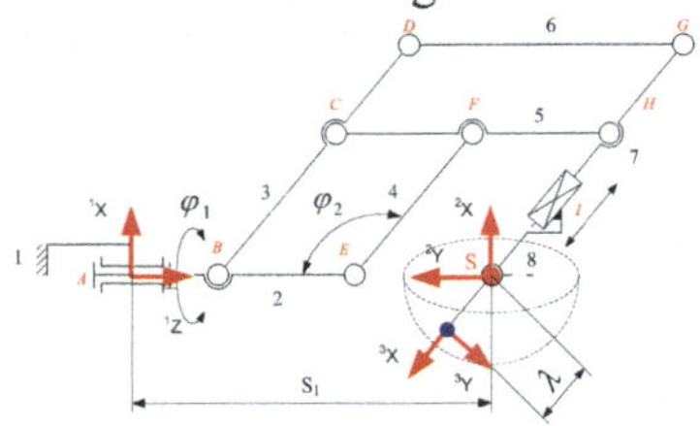

Fig.2. Constant point mechanism of surgical robot with three degrees of freedom and coordinate systems in joints. S_1 – the distance between the first coordinate system and the constant point, λ –distance between the constant point (constant center of rotation) and the end of the effector

Despite the ongoing work, Polish commercial robot to operate of a human body in hospital, has not been built yet.We continue to encounter problems related to the functionality and also problems with dynamics, mechanical vibrations and ergonomy of structures of Polish robots.

The workspace of a medical robot with the constant point mechanism can be obtained making the integration boundaries from configuration variables in the following degrees of freedom of the construction. The workspace is defined as:

$$|V| = 4\int_{0}^{\frac{\varphi_1}{2}} d\varphi_1 \int_{\varphi_2}^{\frac{\pi}{2}} sin(\varphi_2)d\varphi_2 \int_{0}^{\lambda} \lambda^2 d\lambda \tag{1}$$

The configuration variables of the constant point mechanism: φ_1, φ_2 and λ are illustrated in figure 2.The shape of the workspace is illustrated in figure 3.

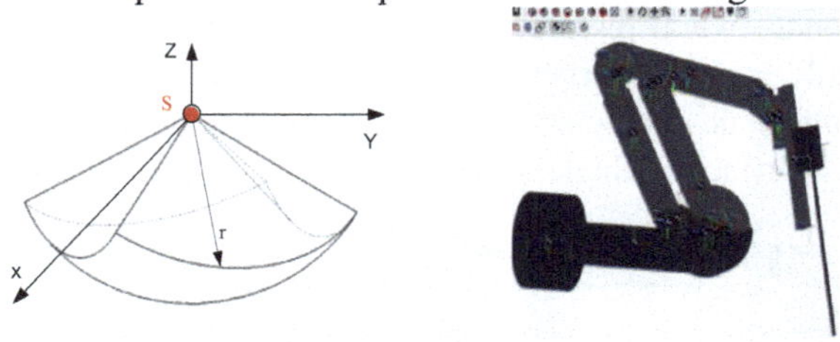

Fig.3. Workspace of the surgical robot with constant point mechanism and model of the constant point mechanism of Robin Heart 1 which was created in Matlab/Simulink

The equations of the inverse kinematics of third degrees of freedom robot model can be written as:

$$\begin{gathered}
\varphi_1 = \arctan2(r_x, r_y), \\
\varphi_2 = arctan2[sin[\arctan2(r_x, r_y)] \cdot r_y + \\
+ cos[\arctan2(r_x, r_y)] \cdot r_x, r_z - s_1], \\
\lambda = (r_z - s_1) \cdot cos^{-1}(arctan2[sin[\arctan2(r_x, r_y)] \cdot r_y]) + \\
+cos[\arctan2(r_x, r_y)] \cdot r_x, r_z - s_1]),
\end{gathered} \tag{2}$$

where: r_x, r_y, r_z are components of the position vector of effector relative to the first coordinate system (Fig.2).

The model of Robin Heart arm (Fig.3) was built with aluminum and steel elements. In fact, the number of the components of the Robin Heart robot is more than 200. The number of the solid elements of the robot arm model was reduced to 20 without affecting functionality and kinetics of the constant point mechanism.

2 Optimization model of Robin Hearth 1 medical robot to soft tissue surgery

An optimization of the medical robot mass is very important because of the dynamics of the robot with the constant point mechanism. The dynamics and strength of the construction of the constant point mechanism deteriorates when mass increasing especially this far distant from the axis of the rotation of a rotational motion which the minimally invasive effector performs [1].
Strength optimization was performed with use of the metheuristic probabilistic iterative algorithm of simulated annealing (SA).
The algorithm allows to resolve the problem of minimize an objective function of several variables on the basis of analogy with the process of change of molecule energy during annealing of a metal. The algorithm is resistant for non-optimal solution.
Acceptance of the value x* which does not improve of the value of the objective function occurs with probability:

$$p = e^{\frac{-|\Delta f|}{T}} \tag{3}$$

Δf - the difference between the value of the objective function in an old and a new point
T – temperature.

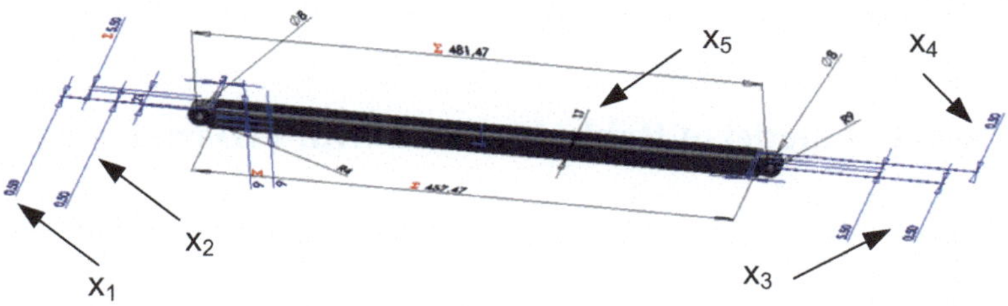

Fig. 4. Steel connector of the kinematic chain of the Robin Heart 1 constant point mechanism

The optimized link of the constant point mechanism of Robin Heart 1 medical robot is illustrated in figure 4.The link is far distant from the axis of rotation of the first degree of freedom of the constant point mechanism.
The decision variables (geometrical dimensions) was designated as: x1, x2, x3, x4, x5.The objective function is defined as:

$$f(x_1, x_2, x_3, x_4, x_5) \rightarrow \min \tag{4}$$

The restrictions of the decision variables and maximum Huber equivalent stress that occurred as a result of the applied forces and adopted supports are defined as:

$$a \leq x_i \leq b \tag{5}$$

$$\sigma_H \leq f[N \cdot m^{-2}] \tag{6}$$

where: f is factor of safety and a, b are geometrical dimensions.
Numerical finite element method (FEM), which gives good approximate solution of strength problems, was applied to define a global value of Huber stress.

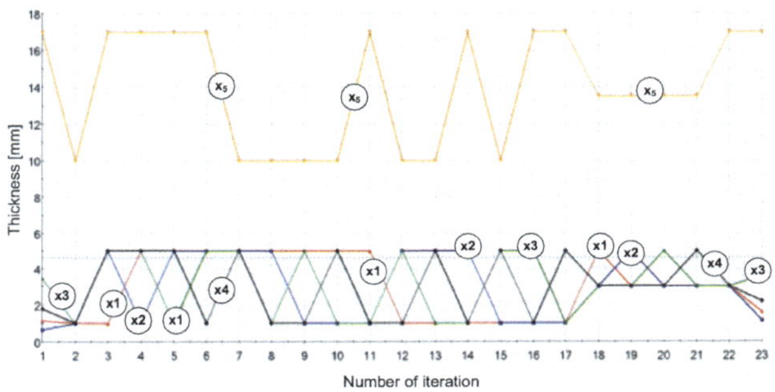

Fig.5. Optimization diagram with marked changes of geometric dimensions during the simulated annealing optimization process

The optimal solution was obtained in 23 iterations. The optimization diagram of the decision variables is illustrated in figure 5.

3 Kinetics model of Robin Hearth 1 with PID controller

The analytical matrix equation which describes the dynamics of movement of the constant point mechanism of a medical robot can be written as:

$$\begin{bmatrix} \boldsymbol{M} & \boldsymbol{\Phi}_q{}^T \\ \boldsymbol{\Phi}_q & \boldsymbol{0} \end{bmatrix} \begin{bmatrix} \ddot{\bar{q}} \\ \bar{\lambda} \end{bmatrix} = \begin{bmatrix} \bar{Q}_e \\ \bar{Q}_d \end{bmatrix} \tag{7}$$

where:

$\boldsymbol{M}$–mass matrix,

$\bar{q}$–vector of generalized coordinates,

$\boldsymbol{\Phi}_q$–matrix of partial derivatives (Jacobian matrix of constraints),

$\bar{\lambda}$ – Lagrange coefficients vector,

$\bar{Q}_d = -\left(\boldsymbol{\Phi}^D{}_q \cdot \dot{q}\right)_q \cdot \dot{q} + 2\boldsymbol{\Phi}^D{}_{tq} \cdot \dot{q} + \boldsymbol{\Phi}^D{}_{tt}$

The dynamics model that allows to solve the equation (7) was created in Matlab programming environment. The model was build based on optimized parametric geometry, which example is shown in Figure 4, with use of the CAD software.
The view of the dynamics model of the Robin Heart constant point mechanism with selected the coordinate systems and the centers of masses of the links is illustrated in figure 3. The model of dynamics was solved with use of the ODE45 solver (ordinary differential equation) of the Matlab software.

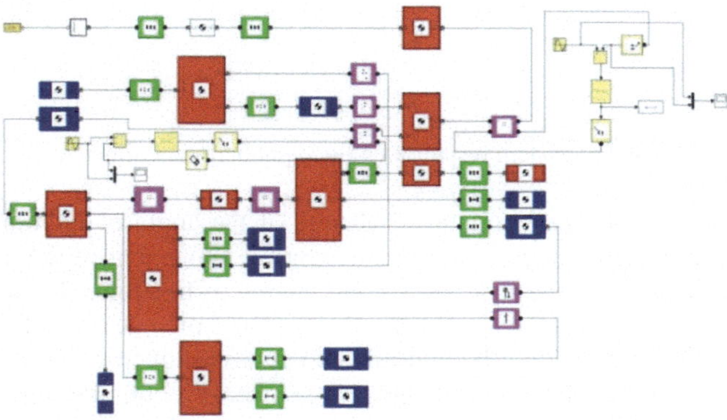

Fig.6. The block diagram of the Robin Heart 1 control system created in the Simulink. Red blocks -links of constant point mechanism, magenta blocks - kinematic pairs, blue blocks– bearings, yellow blocks – sensors, actuators, controllers

The block diagram of the dynamics model was created with use of the blocks and signal transmission lines. It is shown in Figure 6. The PID controllers of continuous time have been added in the following degrees of freedom of the model of the constant point mechanism.

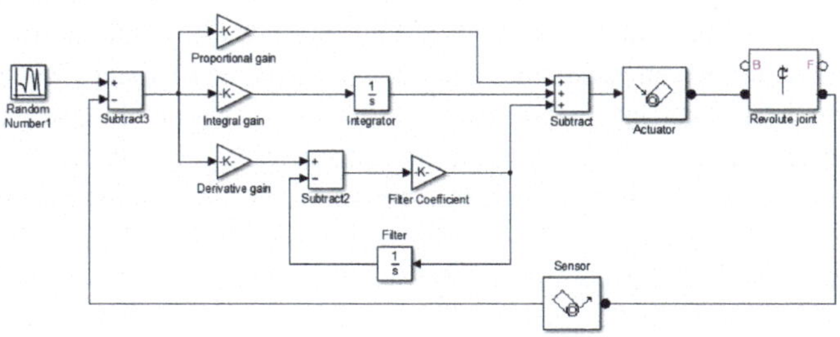

Fig.7. The block diagram of the closed loop control system of the first degree of freedom of the teleoperator Robin Heart 1 with PID controller

The continuous transfer function of PID controller of the Robin Heart 1 first degree of freedom was determined with use of the PID tuner in the Matlab computer program. The continuous transmittance (8) describes the PID controller, when the model of the constant point mechanism is in motion in the first and second degree of freedom.

$$G_1(s) = 20 + 0{,}7 \cdot \frac{1}{s} + 0{,}6 \cdot \frac{122{,}44}{1 + 122{,}44 \cdot \frac{1}{s}} \tag{8}$$

There is no movement in the third degree of freedom of the constant point mechanism during the numerical simulation.

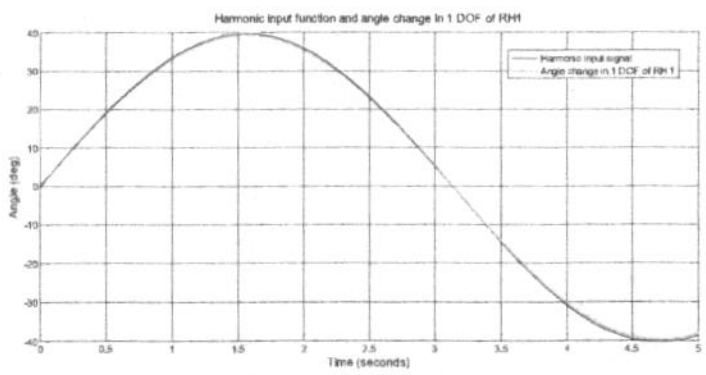

Fig.8. The system response on the input harmonic function

The figure 8 shows the characteristic of change of the angle of rotation in the time (output signal) in the first degree of freedom during movement of the model of constant point mechanism in the first and second degree of freedom. It can be seen that the characteristics of input (reference harmonic signal) coincides with the characteristics of output with use of the PID controller to control movement in the first degree of freedom of the numerical model. It shows the correct functioning of the closed loop control system in first joint of the constant point mechanism. The next stage of the work will be adding models of DC motors with gearboxes to the next degrees of freedom.

4 Conclusions

The model of the dynamics of the constant point mechanism of Robin Heart 1 medical robot which was constructed in the Matlab and Simulink realizes input law of motion with use of the closed loop control systems with PID controllers. The results are similar to those obtained in work [4]. The mass of all elements of the model has been optimized for given boundary conditions with use of the FEM and stimulated annealing method. The optimized structure of constant point mechanism fulfills the strength requirements so the problem of optimization of medical robot mass is solved. The model will allow to test full dynamics of the robot for all movements of cardiosurgeons on the basis of input trajectory which will be obtain with use of the motion capture method during minimally invasive operation.

References

1. Ilewicz, G.: Optymalizacja czynności ruchowych końcówki operacyjnej telemanipulatora kardiochirurgicznego. PhD Dissertation. Gliwice (2011)
2. Nawrat, Z.; Religa, Z.: Stan aktualny i perspektywy stosowania robotów w kardiochirurgii Ekspert Medyczny nr 3(5) s.12-14 (2002)
3. Nikravesh, P.: Computer-Aided Analysis of Mechanical Systems.Prentice Hall (1988)
4. Pérez, V., Escobar, N., Medina, C., Gutiérrez, L., Nawrat, Z., Kostka, P., Bustamante, J.: Simulation of the master slave system Robin Heart: kinematic analysis, dynamic analysis and bilateral control diagrams.pp.116-123Medical Robots. M- Studio Zabrze (2008)

Characterization of low-LET radiation fields for irradiation of biological samples using recombination chamber

Piotr Tulik[1], Sandra Lepak[1], Katarzyna Domańska[1], and Edyta Anna Jakubowska[1]

[1] Institute of Metrology and Biomedical Engineering, Warsaw University of Technology, Warsaw, Poland
{tulik,edyta}@mchtr.pw.edu.pl

Abstract. Irradiation of biological samples for biological dosimetry and radiobiological studies should be performed in well characterized radiation fields. The reference quantity for gamma and X-ray radiations is usually kerma in air, deter-mined with ionization chamber. The paper presents an example of characterization performed for X-rays using a thin-walled, reference ionization chamber. Standard procedure was completed with determination of Recombination Index of Radiation Quality (RIQ), determined with recombination chamber. Irradiations were performed at special stand supporting standard test tubes, which are commonly used in bio-logical dosimetry for blood samples. The influence of all elements of the supporting stand was carefully determined and taken into account in calculations of tissue kerma in blood. The RIQ values reflect microdosimetric parameters of the radiation. The measurements performed for unfiltered X-rays of different energies resulted in markedly higher values of RIQ, comparing to those measured for ^{137}Cs or ^{60}Co gamma radiation fields.

Keywords: radiation quality, low-LET, dosimetry, radiobiology.

1 Introduction

The analysis of dicentric chromosomes in peripheral blood lymphocytes of the exposed person is considered as the most sensitive and specific biological marker of the radiation absorbed dose in the case of accidental overexposure. Since peripheral blood lymphocytes are circulating cells, the radiation-induced damage to their chromosomes reflects the average total-body dose, independent of specific regions of the body that have been exposed. The observed number of the damaged chromosomes is referred to a dose–response calibration relationship obtained from carefully controlled in vitro studies. These curves may differ for different kind and energy of radiations and appropriate calibration curves have to be deter-mined in well characterized radiation fields.

R. Jabłoński and T. Brezina (eds.), *Advanced Mechatronics Solutions*,
Advances in Intelligent Systems and Computing 393,
DOI 10.1007/978-3-319-23923-1_20

For radiobiological studies it is important to correlate the dose-response curves with radiation quality expressed in terms of microdosimetric quantities, which describes the track structure of ionizing particles, i.e. distribution of ionization density along the tracks.

In this paper, the Recombination Index of Radiation Quality (RIQ) was used as a measurable quantity, which values are well correlated with average local ionisation density; therefore they reflect microdosimetric properties of radiation and radiation quality factor [1-4]. The values of RIQ were measured for X-ray voltages from 50 kV to 115 kV. All the measurements were performed using a specially constructed simple stand for blood samples irradiation, mounted at X-ray machine. Influence of all the elements of the stand on the air kerma value has been determined in order to make it possible to monitor air kerma values on-line, during the blood samples irradiations.

RIQ is measured by so called recombination chamber, which is a high-pressure tissue-equivalent ionisation chamber operating not only at saturation but also at lower supplying voltages, under conditions of initial recombination of ions. The last condition is essential, because the initial recombination occurs in single tracks of ionising particles and depends on local density of ions, while does not depend on the dose rate. Most often, RIQ is determined in mixed radiation fields, for purposes of radiation protection. It was also used for characterization of radiation field for irradiation of biological samples at nuclear reactor [5].

Different microdosimetric properties of low- and high-energy photons are not taken into account in radiation protection system, however they can be of interest for radiobiological studies. In our earlier paper [3], the RIQ values were determined for filtered X-ray beams, with X-ray machine working in radiographic mode (several short expositions were used in each series of measurements). In this paper, we used unfiltered beams and fluoroscopy mode of the irradiation. As mentioned above, RIQ depends on local density of ions, which means here the density of ions averaged over a short segment of an ionising particle track, of about 70 nm, or restricted LET with cut-off 500 eV [6]. Therefore, the obtained results might serve as indicators of microdosimetric radiation quality when X-rays of different spectra are compared.

2 Ionization Chambers

The KR-16 chamber was used for determination of kerma in air. It is a planar, tissue-equivalent ionization chamber, with the gas cavity under a thin absorber and backed by a thick layer of tissue-equivalent plastic having similar backscattering characteristics as soft tissue. Almost all the details of the chamber are made of tissue-equivalent material. A thin foil stretched on a solid isolating ring serves as the polarizing electrode. The collecting electrode and the guard ring of the chamber form a common flat surface ensuring the uniform electrical field strength and well defined active volume of the chamber. An additional guarding foil (electrically connected to the guard ring) was mounted at the opposite side of the isolating ring. In effect, the polarizing electrode was placed in the middle between two surfaces being at the same electrical potential. Both, the polarizing electrode and the guarding foil are made of

1.6 mg/cm^2 thick Mylar covered with 0.1 mg/cm^2 of aluminum. A specific property of the KR-16 chamber design is that the voltage insulators are „invisible" from any point of the active volume. Such configuration ensures very good time stability of the electrical field strength in the active volume.

Measurements of RIQ were performed with recombination chamber of F1 type [7]. It is a 3.8 cm^3 high-pressure, in-phantom disc shaped chamber with three parallel plate tissue-equivalent electrodes, 34 mm in diameter (total diameter of the chamber is 62 mm). The distance between electrodes is equal to 1.75 mm. The chamber was filled with ethane up to about 600 kPa. The chambers were calibrated with ^{137}Cs reference radiation source in the Institute for Nuclear Problems, Świerk (Poland) and with reference filtered X-ray beams of different energies, in Central Laboratory for Radiological Protection in Warsaw.

3 Material and methods

The dependence of RIQ on X-ray machine voltage has been investigated by irradiations performed in the Institute of Metrology and Biomedical Engineering of Warsaw University of Technology. All the measurements were performed using a Shimadzu FLEXAVISION HB X-ray machine in fluoroscopic mode of the machine, in order to avoid possible volume recombination during shorter pulses of higher dose rate associated with radiographic mode. Eight beams in the range from 50 kV to 115 kV were used (see Table 2).

A special stand (Fig.1B) was constructed for irradiation of standard test-tubes commonly used for biological dosimetry. All the elements of the stand were made with aluminum in order to minimize radiation scattering. Measurements were performed first without the stand, and then consecutively with all elements of the stand, in order to determine their influence on the kerma values.

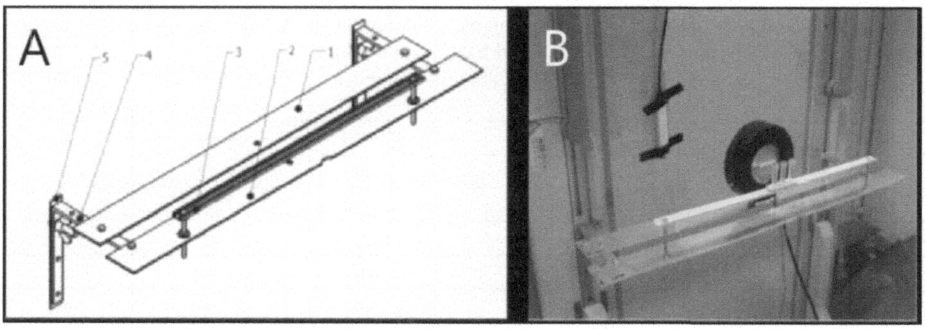

Fig. 1. A - Stand for irradiation of blood samples at X-ray machine, (1 – aluminum bar supporting the ionization chamber, 2 - aluminum bar for test tubes, 3 – elevated support for test tubes, 4 - bracket, 5 – pad); **B** - Stand with test-tubes and KR-16 chamber;

RIQ was determined using the standard procedure [4]. First, the saturation current I_S was determined from the measurements of ionization current at high voltages, close to the maximum voltage allowed for the chamber. Then the ionization current was measured at a specially chosen voltage, denoted as U_R (sometimes it is called "recombination voltage"). U_R had to be determined earlier for each chamber during the calibration in reference field of ^{137}Cs gamma radiation source.

The calibration involves determination of the whole saturation curve. Most often, U_R is chosen as a voltage, which ensures 96% of saturation (i.e. 4% of recombination) for the reference radiation. If the voltages U_R is chosen like described above, the RIQ is denoted as Q_4 and defined as:

$$Q_4 = \frac{(1-f(U_R))}{0.04} \tag{1}$$

where $f(U_R)$ is the ion collection efficiency (the ratio of the ionisation current at voltage U_R to the saturation current) measured in the investigated radiation field.

4 Influence Of The Elements Of The Stand

The stand for test tubes consists of two aluminium bars and an elevated support for test tubes (Fig. 1A) made with propylene. Mounting brackets are outside of radiation beams, so they are made with titanium covered steel. For proper and precise measurements with the presented stand it was necessary to determine the possible influence of each element of the stand on the chamber reading, and to take into account the distance between samples and the KR-16 chamber.

Tab. 1. Influence of elements of the stand on the readings of ionization chamber. All the values are expressed in μGy per 100 ms irradiation pulse. U is voltage of X-ray machine, Ka – kerma in air in the position of test tubes, without the stand. Other values show the influence of different factors on the reading of the chamber, when it is placed in monitoring position, i.e. on the aluminum bar behind the tubes.

U (kV)	K_a	Scattering by the metal bar for the chamber	Scattering by the metal bar for the tubes	Attenuati on by the support	Correction for the distance from the X-ray tube	Scattering by the table
40	34.83	0.3	0.2	-2.989	-5.21	3.81
42	42.41	0.3	0.2	-3.599	-5.82	4.81
44	51.25	0.4	0.2	-4.059	-7.65	5.19
46	60.13	0.3	0.3	-4.691	-8.80	6.03
48	69.02	0.2	0.2	-5.188	-10.01	6.72
50	78.83	0.2	0.3	-5.894	-11.10	7.10
52	88.18	0.3	0.3	-6.572	-12.48	8.30

X-ray machine table and the aluminium bars are the sources of scattered radiation which increases total charge measured by the chamber, while the propylene support attenuates the beam, so it causes a decrease of the reading. The distance between the ionization chamber centre and the X-ray machine table is 100 mm and the distance between the chamber and the test tubes equals to 70 mm. The measurements were performed at vertical position of the table, which was covered with a lead apron in order to limit the scattering.

Comparing chamber readings collected in different configurations with and without the elements of the stand, allows necessary corrections. Results are displayed in Table 1. Influence of the attenuation by the test tubes practically does not depend on the beam energy and constitutes about 4% of the measured kerma for the tube filled with water and 5% for the tube with paraffin.

5 Results of RIQ measurements

Measured values of Q_4 are displayed in Table 2. The results give a clear indication of rise in values of Q_4 for X-rays, comparing with gamma radiation of ^{137}Cs, by a factor of about 1.5. The dependence on X-ray machine voltage is statistically insignificant.

Tab. 2. Values of Recombination Index of Radiation Quality Q_4, measured for Shimadzu FLEXAVISION HB X-ray machine in fluoroscopic mode.

X-ray voltage	RIQ
50 kV	1,55±0,15
60 kV	1,57±0,15
70 kV	1,55±0,15
80 kV	1,52±0,15
90 kV	1,50±0,15
100 kV	1,49±0,15
110 kV	1,49±0,15
115 kV	1,49±0,15

6 Discussion and conclusions

Differences in biological response to different low-LET radiations is in our opinion very important for better understanding the radiation effects in tissue at microscopic levels and then for better estimation of low-dose effects. Dosimetry for such studies has to be performed very carefully, as the use of routine methods may lead to considerable inaccuracy. Measurements performed in this work show that commonly used, and practically always neglected in dosimetric considerations, light supports may cause up to 1.5% increase of the dose absorbed in blood. Attenuation of the beam by the test tubes themselves and by the light support, used in our stand, may decrease the signal of monitoring chamber by more than 10%.

Measurements of RIQ for low-LET radiation provided new experimental data on local ion density in small volumes of tissue, of about 70 nm in diameter. They confirmed qualitatively earlier results [8] obtained for strongly filtered X-ray beams, however the absolute values obtained now are lower by about 12% from the previous ones. The measurements for both studies were performed with the same chamber but filled with different gases - methane at pressure of 1.1 MPa in the earlier paper and ethane at the pressure of 600 kPa now. Comparison of measurements performed with these two gas fillings shows, that RIQ's measured with the chamber containing ethane were always lower by about 7%. With such correction, both series of the measurements are in agreement, within the declared uncertainty.

The mean value of local ion density changes its value from unity to for reference gamma radiation of ^{137}Cs to more than 1.5 for broad range of X-rays spectra used in diagnostic radiology. The results are in qualitative agreement with other microdosimetric measurements and can contribute to the discussion about different biological effectiveness of low and high-energy photons. From practical point of view, this fact can be important e.g. for assessment of radiological risk, associated with mammography screening.

References

1. Golnik, N.; Zielczynski, M.: Determination of quality factor in mixed radiation-fields using a recombination chamber. Radi-at. Prot. Dosim., 44 (1-4),57-60 (1992)
2. Jozefowicz, K.; Golnik, N.; Zielczynski, M.: Dosimetric pa-rameters of simple neutron+ gamma-fields for calibration of radiation protection instruments. Radiat. Prot. Dosim., 44 (1-4), 139-142 (1992)
3. Golnik, N.; Cherevatenko, E.P.; Serov, AY.; Shvidkij, SV; Sychev, BS; Zielczynski, M: Recombination index of radiation quality of medical high energy neutron beams. Radiat. Prot. Dosim., 70 (1-4) 215-218 (1997)
4. Zielczyński M., Golnik N.: Recombination index of radiation quality - measuring and applications. Radiat. Prot. Dosim., 52: 419 422 (1994)
5. Golnik N., Gryziński M.A., Kowalska M., Meronka K., Tulik P.: Characterization of radiation field for irradiation of bio-logical samples at nuclear reactor - comparison of twin detector and recombination methods. Radiat. Prot. Dosim., 161: 196-199 (2014)
6. Golnik, N.: Microdosimetry using a recombination chamber - method and applications. Radiat. Prot. Dosim., 61 (1-3) 125-128 (1995)
7. Golnik N., Zielczyński M.: New applications of the recom-bination method in the dosimetry of mixed radiation fields. Proc. 10th International Congress of the International Radiation Protection Association, IRPA10, Hiroshima, Japan, P-3b-214, www.irpa.net/irpa10/cdrom/00768.pdf (2000)
8. Golnik N., Mayer S., Zielczyński M.: Recombination index of radiation quality of low-LET radiation. Nucl Instr. and Meth. in Phys. Res. B, 213C: 650-653 (2004)
9. Gryziński M.A., Golnik N., Zielczyński M.: Initial recombi-nation of ions in ionization chambers filled with hydrocarbon gases. Nukleonika, 52: 7-12 (2007)
10. Golnik N., Zielczyński M.: New applications of the recom-bination method in the dosimetry of mixed radiation fields. Proc. 10th International Congress of the International Radiation Protection Association, IRPA10, Hiroshima, Japan, P-3b-214, www.irpa.net/irpa10/cdrom/00768.pdf (2000)
11. Golnik N., Zielczyński M., Gryziński M.A., Tulik P.: Recom-bination processes in high-pressure ionization chambers irradiat-ed with low-LET radiation. Nuclear Instruments and Methods in Phys Res. A, 580(1): 25-28 (2007)
12. Zielczyński M., Golnik N., Gryziński M.A.: Determination of initial recombination of ions created in quasi-pulsed radiation fields. Nukleonika, 53 Suppl. 2: 33-36 (2008)

Modular measurement system dedicated as diagnostic equipment in biomedical research studies

Mikołaj Jamroży[1], Miłosz Jamroży[2], and Krzysztof Lewenstein[2]

[1] Warsaw University of Technology, Faculty of Electronics and Information Technology, Warsaw, Poland
m.jamrozy@stud.elka.pw.edu.pl
[2] Warsaw University of Technology, Faculty of Mechatronics, Warsaw, Poland
{k.lewenstein,m.jamrozy}@mchtr.pw.edu.pl

Abstract. The purpose of this study is to develop scalable and modular device dedicated for medical researchers. Experimental medicine often require sophisticated and highly specialized measurement systems. Such devices are developed for specific applications and are difficult to adjust or modify. This causes a problem and generates costs for the research project which require unique diagnostic equipment for i.e. clinical trials. The study aims to provide a solution to those issues by developing low-cost, modular device based on specialized extension measurement cards.

Keywords: Biomedical Engineering·Electronics·Biomedical Research

1 Introduction

Nowadays biomedical researchers and clinical trials MDs require sophisticated and specialized diagnostic equipment. Such devices are not always available, reasons vary from high cost to peculiar demands regarding measurement properties.

Designing dedicated devices for such research is usually difficult due to specific requirements that diagnostic device have to fulfill. Additionally such approach is not always cost-effective, especially if it is not possible to completely specify scope and methodology of the measurements.

In this study, another approach has been presented. The main focus has been put on developing a scalable, modular and flexible platform allowing to conduct wide range of medical measurement with minimum modification and costs. Such result can be accomplished by implemented extension card – carrier scheme used in many electronic standards used by different branches of industry.

This allows to adjust measurement performance or even change measured patient's parameters without the need to modify or redesign the device itself. The only required system adjustment is replacing extension card, which can be tailored to application specific demands.

R. Jabłoński and T. Brezina (eds.), *Advanced Mechatronics Solutions*,
Advances in Intelligent Systems and Computing 393,
DOI 10.1007/978-3-319-23923-1_21

Although device is designed to be easily adjustable to fit needs of medical researchers, carrier with specified set of extension cards can be also used as a standard diagnostic or monitoring equipment in hospitals and clinics. To allow such operation device has been supplied with touch screen interface which allows it to be easily operated by medical personnel. Collected data can be store in the local memory or send via Ethernet connection.
In some experiments [2], medical data has to be processed and interpreted by software algorithms, the device is not designed to directly handle sophisticated digital signal processing. It rather relays on Ethernet connection combined with cloud and cluster computing. Considering Moore's law and rate in which modern devices are becoming obsolete such a computation backbone is fundamental to insure up to date performance. Implementing data processing on external, more powerful machines running on higher abstraction layer reduces time needed for software development.

2 System Architecture

Fig. 1 presents block diagram of the system architecture. Main components are:

- Carrier board,
- LCD touch screen,
- Measurement cards.

Carrier Board (CB) has been designed to minimize amount of necessary electronics circuitry on the extension cards. The system runs on Linux OS hosted by ARM Cortex A8 CPU with Ethernet connectivity and build-in 3D graphics controller. Linux was selected due to high reliability and relatively easy way of creating low level device drivers. For Patient protection purposes, all Measurement Card slots has been featured with power and digital signals isolation.

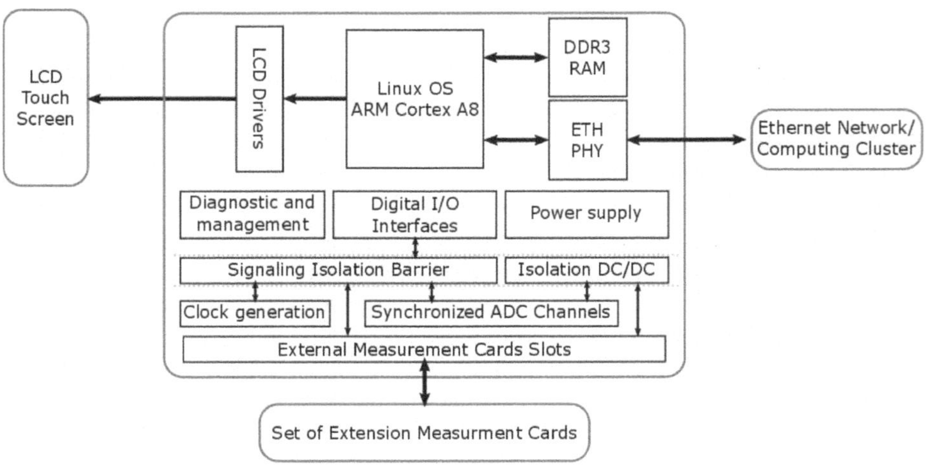

Fig. 1. Block schematic of the system architecture

Safe side of the carrier has been fitted with high precision and resolution ADCs. Used converters have built-in, programed amplifiers and time-synchronized inputs, so multi-card measurements are possible with exact known sample correlations. This solutions allows to reduce some of the measurement cards to analogue front-end. If needed, additional digital circuitry can be placed on the measurement cards, power budget for each slot is around 1W.
Digital signal interfaces include most popular serial data links including:

- Universal Asynchronous Receiver and Transmitter (UART),
- Serial Peripheral Interface (SPI),
- Inter-Integrated Circuit (I2C).

Measurement Extension Cards (MECs) are designed as interchangeable devices, allowing to extend or modify functionality of the measurement system. Modularity allows to experiment with i.e. acquisition methods, sensors, front ends, used electronic chips or even measured patient parameters to achieve optimal results with relative low cost of development. Price of a MEC will vary depending on used electronics, nonetheless production of two layer PCB costs fraction of a multilayer design.

3 Detection of Cardiac Insufficiency

The system is being developed to continue autonomous detection of cardiac insufficiency studies [1]. Data from EKG, accelerometer and infrasound microphone are being collected in order to create classification model which can detect cardiac insufficiency.

Apparatus used previously has proven troublesome due to the nature of the experiment. During the studies a problem arouse in regard of sensors selection and synchronization between them. Latter problem was resolved by implementing software algorithms, but it is expected that synchronous data will be beneficial and will reduce processing time.

Sensor selection in monolithic architectures is a serious issue. Different output signal levels, bandwidths and required front ends makes it almost impossible to change sensor after electronics is developed.

In the case of the detection of cardiac insufficiency study, artificial testes were not sufficient to verify if accelerometer and microphone fit the requirements. The reason is that information is calculated from all the input signals with use of non-linear modeling. It is not possible to directly tie the result to a specific parameter of the sensor such as frequency response or bandwidth. Use of the new, modular system solves most of the problems. Moreover, it gives opportunity to experiment with different kinds of analogue front-ends and sensors. To extend the measurements impedance cardiography will be also introduced as an MEC in course of the study.

Data processing on the old generation of the system was conducted on a host PC computer. Apparatus used small microcontroller with limited calculation possibilities and was used to gather and transfer measurement data. Signal processing rate was strongly dependable on used machine because of selected technologies, such as Lab-

View and Bluetooth. New system can process measurements thanks to used modern microprocessor with high level operation system. Speed of calculation conducted directly on the device will not exceed what was possible on the previous device, but Bluetooth bottleneck is no longer a problem. When Ethernet connection is available system can also benefit from external server services, cloud computing, calculation clusters and other modern internet technologies.

4 Development directions

Subject of the presented study is still in experimental stage. A prototype has been developed with build-in EKG MEC to verify the conception and test selected integrated circuits. Next stage is to prepare modular platform and several test MECs and use it as diagnostic equipment in detection of cardiac insufficiency study.

Usage of FPGA devices is being considered, to handle data and analogue front-ends without overloading system's microprocessor. This solution would allow to aggregate and time stamp data that are not handled by the synchronized ADCs. FPGA could also increase flexibility of the device, thanks to easily implemented low level logic. Linux driver development would be easier, because it would be limited to read/write operation to common memory space. Large FPGA could be even used as a calculation co-processor for the main processor. Such implementation would significantly increase calculation capabilities of the device and made it more powerful when not connected to the Ethernet network.

Another step in the development will be certification and preparing the device to be used in medical studies. Main work that needs to be done to adjust the system to specific diagnostic purposes, is development of MECs and Linux drivers to handle them. If needed, a backbone server applications for data storage and processing will be developed.

5 Conclusions

The paper presents implementation of modular system for measurements in medical diagnostic research. Presented solutions allows the device to be adjusted to meet specific requirements.

Current version of the system is being prepared to continue the study of detection for cardiac insufficiency. ECG, infrasound and accelerometer Measurement Extension Cards are being prepared to assess usability of several sensors in the study, and conduct first measurements.

References

1. Jamroży M., Lewenstein K.: Apparatus System for Detection of Cardiac Insufficiency. In: 9th International Conference Mechatronics 2011, 663-672 (2011).
2. Jamroży M., Leyko T., Lewenstein K.: Early detection of the cardiac insufficiency. Recent Advances In Mechatronics 2008-2009, Springer, 2009 p.407-413.
3. Nguyen C.D., Wilson S.J., Crozier S.: Automated quantification of the synchrogram by recurrence plot analysis. IEEE Trans Biomed, 2012
4. Shukla A., Macchiarulo L.: FPGA based ECG analysis system. BioMED Proceedings of the Sixth IASTED International Conference on Biomedical Engineering, 2008

Conception of Turning Module for Orthotic Robot

Dymitr Osiński, Marcin Zaczyk, Danuta Jasińska-Choromańska

Faculty of Mechatronics, Warsaw University of Technology, Warsaw, Poland
{d.osinski, m.zaczyk, danuta}@mchtr.pw.edu.pl

Abstract. Problems related to designing of orthotic robots (systems for verticalization and aiding motion) are presented. A 'Veni-Prometheus' System for Verticalization and Aiding Motion has been designed and built at the Division of Design of Precision Devices, Faculty of Mechatronics, Warsaw University of Technology. At present, the system can realize five motion functions. Further works aimed at designing and constructing a turning module for the orthotic robot have been carried out. Two concepts of such module are proposed.

Keywords: orthotic robot · exoskeleton · turning module kinematics

1 Introduction

1.1 The 'Veni-Prometheus' System for Verticalization and Aiding Motion

A 'Veni-Prometheus' System for Verticalization and Aiding Motion has been designed and built at the Division of Design of Precision Devices, Faculty of Mechatronics, Warsaw University of Technology (Fig. 1) [1,8,9]. This orthotic robot is designed for the handicapped suffering from paresis or paralysis of the lower limbs. The device clasps the lower limbs of the user, allowing the following motions to be performed within the sagittal plane: straightening and flexion of the hip and knee joints as well as dorsal and plantar flexion of the ankle joints. Hip and knee articulations are driven, whereas articulations of the ankle joints are passive. Actuators *3* are electric DC motors coupled with reducers driving the articulations through flexible connector transmissions (Fig.1). The supply system – a battery of lithium accumulators, and the control system are placed within the backpack *1* of the device. The backpack and the links of the device are connected by means of the hip belt *2*. The orthotic robot is fastened to the user by means of Velcro fasteners and shoulder-straps of the backpack. At present, the system can realize five functions: walking, sitting down, raising, ascending and descending stairs. Further works are carried out in order to enhance capabilities of the device. One of the functions that are to be introduced in the orthotic robot is a capability of realizing a turn. Respective development of the system will probably require not only introducing new control scenarios, but expanding the robot kinematics as well. Besides, it is foreseen that the control unit of the system will realize other tasks, like determining tilt angles by means of a MEMS accelerometer, as suggested in [11], according to a special algorithm, e.g. proposed in [12].

R. Jabłoński and T. Brezina (eds.), *Advanced Mechatronics Solutions*,
Advances in Intelligent Systems and Computing 393,
DOI 10.1007/978-3-319-23923-1_22

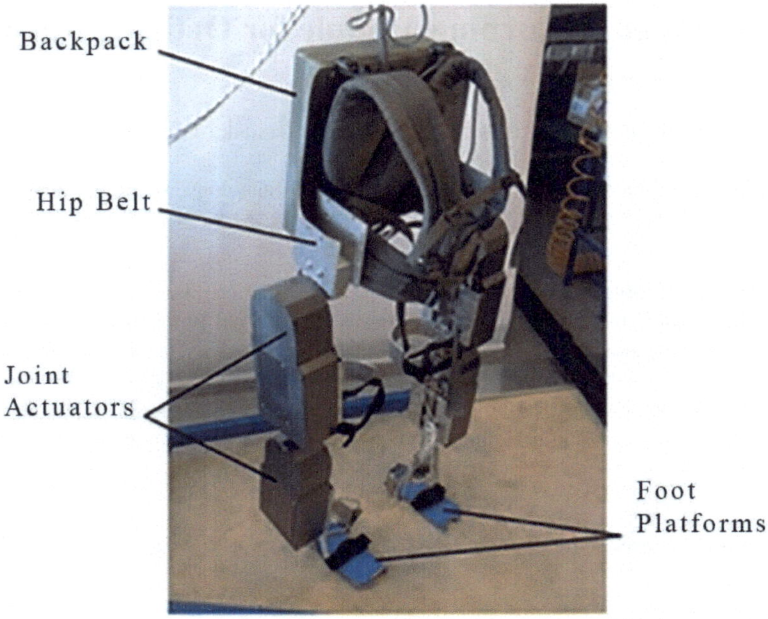

Fig. 1. General view of the 'Veni-Prometheus' System for Verticalization and Aiding Motion

1.2 Turning methods in contemporary orthotic robots and exoskeletons

Among the many orthotic robots and exoskeletons that have been developed so far, only a few are capable of turning, which could be used by an impaired person. In the case of majority of the designs, the number of drives is limited to the corresponding basic motions within the sagittal plane, and the other possible degrees of freedom are not available or the respective articulations are passive. Such designs include, among other devices, ReWalk [5], Ekso [14], Indego [7] or HAL [7]. Realization of turning is possible, however usually a slip of the foot takes place then, and sometimes the motion must be assisted by a therapist. There exist also designs equipped with additional drives making it possible to realize a higher number of movements; these are: XOS2 [6], BLEEX [17], Mindwalker [16] or REX [10]. This group includes a larger number of military exoskeletons. The devices often offer a possibility of adducting and abducting the hip links, as well as a possibility of a hip rotation, yet rather seldom.

Devices designed for the sick usually belong to the first group. As equipped in a lower number of drives, they rarely provide a possibility of realizing a turn without slipping. The most frequent common features of such designs are: electric drive, links located at the outer side of the legs, articulations with one degree of freedom and a necessity of using crutches or a walker in order to provide a support. The turning is often not indicated as a function of a device. Presumably, it is considered an inherent part of walking; however this issue probably has not been developed to a satisfactory extent yet.

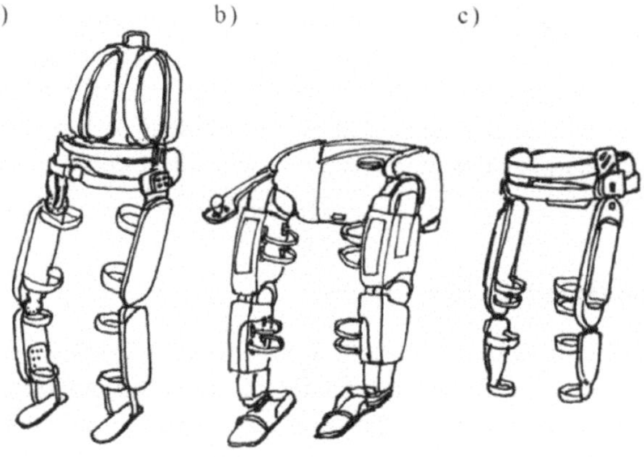

Fig. 2. Illustration of selected exoskeletons and orthotic robots: a) ReWalk; b) REX; c) Indego

2 Conception of the turning module kinematics

2.1 Direct control of the lower limbs

The first of the conceived conceptions (Fig. 3) of kinematics of the turning module for the orthotic robot provides a possibility of realizing both the movement of hip rotation as well as extension and flexion of the hip joint, keeping a small volume. Motion of the lower limb in a transverse plane within a range of few tens of degrees arc would make it possible to perform a turn. Owing to application of a curved guide, the center of the human joint is located at the same spot as the rotation center of the respective device member, and thus there will occur no translations between the system for aiding motion and the human body. Concentration of the drives near the hip belt instead of the leg will not result in an increase of mass and mass moments of inertia of the lower limb, what is important while performing the most frequent movements while walking. Besides, for such configuration, the rotation movement at a bent leg (as in the case of sitting) would result in abduction of the limb – such motion is of an anatomical character and could make it easier to put the orthotic robot on and off.

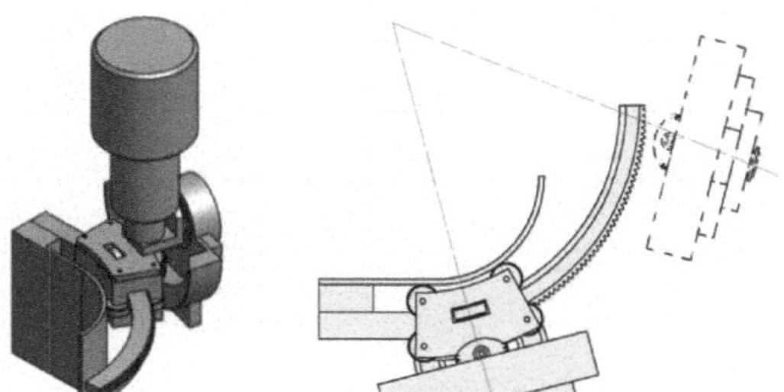

Fig. 3. Three-dimensional model of the device: a) general view (only a part of the hip belt is shown); b) top-down view (extreme positions of hip rotation are shown)

A toothed gear and a mechanical subunit of the hip articulation constitute an integral part. Its frame is the hip belt of the orthotic robot. An electric drive of a bevel gear generates a movement causing hip flexion and extension of the lower limb. The second electric drive with a spur gear realize a movement of hip rotation. A toothed mechanism with rollers provides guidance and makes it possible to transmit the drive at a high efficiency.

2.2 Indirect control of the lower limbs

The second conception can be considered a novel one. A proposal of the kinematics has nothing to do with a common schematic of active articulations equipped with drives – linear or rotary actuators. In the considered case the articulations that are not used during a standard gait cycle (as currently realized by the Veni-Prometheus robot) contain passive articulations equipped only with brakes. It is possible to perform a resultant movement, which depends on trunk movements. The system for aiding motion has been equipped in a positioner with driven degrees of freedom, placed at the back, which controls position of the mass (its role may be played by e.g. the accumulators). Movements of the positioner affect position of the user trunk as well as position of the leg, which at a given moment does not rest on the ground. Such method of indirect control can be compared to a change of position of the support point of a pendulum. Human movements are strictly connected with the concept of balancing, therefore such solution could result in movements resembling the natural ones.

Fig. 4 presents a kinematic diagram of the orthotic robot including the lower limbs, equipped with driven articulations for realization of hip and knee extension and flexion, free articulations for realization of hip adducting and abducting, hip rotation, dorsal and plantar flexion of the ankle articulations, as well as their inversion and eversion. Additionally, the design is equipped with a driven positioner placed at the back, over the rear part of the hip belt.

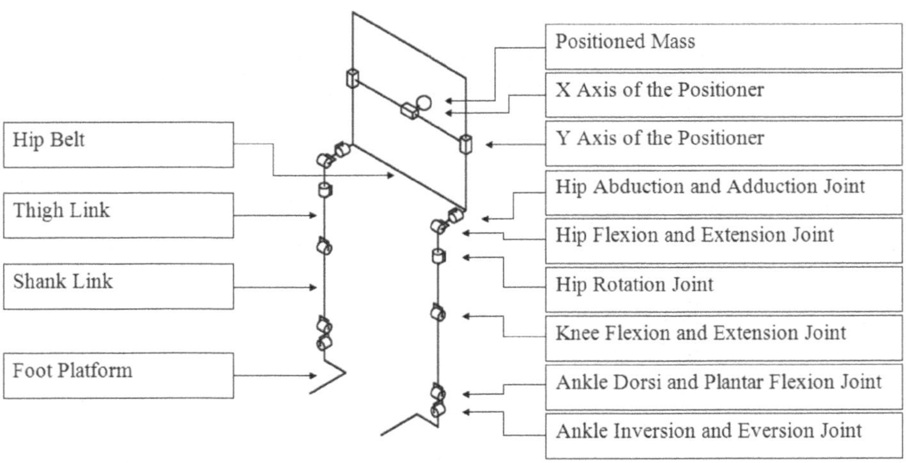

Fig. 4. Kinematic diagram of the orthotic robot employing indirect method of control of the lower limbs by means of a positioner

At such concept, an important issue would be controlling a limb with many degrees of freedom by means of a positioner having only two degrees of freedom – such system might not be deterministic, or it might be chaotic.

3 Discussion

The presented existing exoskeletons and orthotic robots, which offer a capability of realizing a turning motion usually make use of articulations without a drive or special methods of turning that require an assistance of other persons (e.g. a therapist). Sometimes, a turning motion is accompanied by a slip of the foot platform over the ground surface. Such solutions significantly complicate performance of a turn for individuals devoid of a full use of the lower limbs. The existing solutions equipped with driven turning articulations of the lower limb are usually military designs, which rather poorly fit for an application under civil conditions, and even less in the case of the handicapped.

Two presented concepts of the kinematics are connected with new solutions of the hip belt. The first provides a possibility of realizing within a small space two kinds of motion: extension and flexion as well as hip rotations, whereas the rotation centers of the robot articulation and the human joint overlap. Application of passive articulations is characteristic for the second concept. Movement in the articulations is forced indirectly, by balancing the trunk of the user caused by a movement of the positioned mass. It seems that it is worthwhile developing these solutions.

4 Conclusions

One has presented two concepts of developing kinematics of the 'Veni-Prometheus' System for Verticalization and Aiding Motion aimed at realization of a turn. The solutions are to ensure a capability of realizing a turning motion keeping the number of additional drives as low as possible, in order to reduce the mass of the structure, the energy consumption or damageability. At the same time, the drives must fulfill their task in a way that is safe for the user. Further modeling and simulation study as well as experimental works are foreseen in order to validate the analyzed solutions. Experimental works will be realized using dedicated test rigs, presented e.g. in [9], [13], [2,3,4] or [15].

References

1. Bagiński, K., Jasińska-Choromańska, D., Wierciak, J.: Modelling and simulation of a system for verticalization and aiding the motion of individuals suffering from paresis of the

lower limbs. Bulletin of the Polish Academy of Sciences, Technical Sciences, vol. 61, no. 4, pp. 919-928 (2013)
2. Bagiński, K., Wierciak, J.: Effect of Gear Ratio on Energy Consumption of Actuators Used in Orthotic Robot. Mechatronics 2013. In: Jabłoński, R., Brezina, T. (eds.) Recent Technological and Scientific Advances. Springer International Publishing, pp. 511-518 (2014)
3. Bagiński, K., Wierciak, J.: Forming of operational characteristics of an orthotic robot by influencing parameters of its drive systems. In: Szewczyk, R., Zieliński, C., Kaliczyńska, M. (eds.) Progress in Automation, Robotics and Measuring Techniques. Vol. 2 Robotics, Springer International Publishing, pp. 1-9 (2015)
4. Bodnicki, M.: Design of small-outline robot - Simulator of gait of an amphibian. In: Jabłoński, R., Turkowski, M., Szewczyk, R. (eds.) Recent Advances in Mechatronics, pp. 77-81. Springer, Berlin Heidelberg (2007)
5. Esquenazi, A., et al: The ReWalk Powered Exoskeleton to Restore Ambulatory Function to Individuals with Thoracic-Level Motor-Complete Spinal Cord Injury. American Journal of Physical Medicine & Rehabilitation, vo. 91, no. 11, pp. 911-921 (2012)
6. Firestone, J., Maghan, C.: Design and Implementation of the XOS2 Exoskeleton for the United States Military, University of Pittsburgh, Pittsburgh (2012)
7. Herr, H.: Exoskeletons and orthoses: classification, design challenges and future directions. Journal of NeuroEngineering and Rehabilitation, vol. 6, no. 1 (2009)
8. Jasińska-Choromańska, D., et al: Safety module for the system of verticalization and aiding motion for the disabled. In: Brezina, T., Jabloński, R. (eds.) Mechatronics 2013. Recent Technological and Scientific Advances, pp. 79-86. Springer International Publishing (2014)
9. Jasińska-Choromańska, D., et al: Mechatronics System for verticalization and aiding the motion of the disabled. Bulletin of the Polish Academy of Sciences, Technical Sciences, vol. 61, no. 2, pp. 419-431 (2013)
10. Little, R., Irving, R.A.: Self contained powered exoskeleton walker for a disabled user. US patent 20110066088 A1 (2010)
11. Łuczak, S.: Guidelines for Tilt Measurements Realized by MEMS Accelerometers. Int. J. Precis. Eng. Manuf. 15, 489-496 (2014)
12. Łuczak, S.: Advanced Algorithm for Measuring Tilt with MEMS Accelerometers. In: Jabłoński R., Turkowski M., Szewczyk R. (eds.) Recent Advances in Mechatronics, pp. 511-515. Springer, Berlin Heidelberg (2007)
13. Łuczak, S.: Dual-Axis Test Rig for MEMS Tilt Sensors. Metrol. Meas. Syst. 21, 351-362 (2014)
14. Swift, T.: Control and Trajectory Generation of a Wearable Mobility Exoskeleton for Spinal Cord Injury Patients. University of California, Berkeley (2011)
15. Szykiedans, K.: Kinematic interactions between orthotic robot and a human. Progress in Automation, Robotics and Measuring Techniques, vol. 2 Robotics, Springer-Verlag, Berlin Heidelberg, p.265 (2015)
16. Wang, S., et al: Design and Control of the MINDWALKER Exoskeleton. IEEE Transactions on Neural Systems and Rehabilitation Engineering, vol. 23, no. 2, pp. 277 – 286 (2014)
17. Zoss, A., Kazerooni, H., Chu A.: On the Mechanical Design of the Berkeley Lower Extremity Exoskeleton (BLEEX), IEEE/RSJ International Conference on Intelligent Robots and Systems, doi. 10.1109/IROS.2005.1545453 (2005)

Evaluation of cost-efficient auditory MEG stimulation

Anna Jodko-Władzińska[1], Robert Kühler[2], Johannes Hensel[2], Tadeusz Pałko[1], Tilmann Sander[3]

[1] Warsaw University of Technology, Faculty of Mechatronics, Św. Andrzeja Boboli 8, 02-525 Warsaw, Poland
{a.jodko,t.palko}@mchtr.pw.edu.pl
[2] Physikalisch-Technische Bundesanstalt, Department of Acoustics and Dynamics, Bundesallee 100, 38116 Braunschweig, Germany
[3] Physikalisch-Technische Bundesanstalt, Department of Biosignals, Abbestr. 2-12, 10587 Berlin, Germany
{Robert.Kuehler,Johannes.Hensel,Tilmann.Sander-Thoemmes}@ptb.de

Abstract. An acoustic source based on a piezoelectric transducer and commonly available parts was designed to measure auditory evoked MEG. Its magnetic and acoustic properties were tested and MEG measurements were made. The obtained datasets were analyzed and the auditory event related fields were calculated using free software package FieldTrip (www.fieldtriptoolbox.org). The setup successfully delivers sounds at frequencies of interest from 1 kHz to 8 kHz and it is suitable for auditory evoked MEG measurements as data recorded with the MEG showed the auditory M100 response.

Keywords: auditory evoked magnetoencephalography · MEG · auditory event related fields

1 Introduction

Electrophysiological phenomena in the human body lead to the occurrence of magnetic fields. Among others magnetic fields associated with currents arising from electrical activity of the brain (magnetoencephalography) are analysed. Evoked magnetoencephalography is of special interest [1].

Due to the identical source of magnetoencephalographic and electroencephalographic (MEG/EEG) signals, which is the activity of the cerebral cortex neurons, MEG/EEG signals contain fairly similar information and their diagnostic value is comparable. Nevertheless, non-contact measurements of biomagnetic field with MEG avoid difficulties associated with the use of electrodes, such as poor electrode contact with the skin, the presence of insulating layers and varied conductivity of the body tissues. These factors reduce the accuracy of EEG measurements [2].

Measurement of magnetic flux density of magnetic fields generated by living organisms is a challenge due to its extremely low value (10^{-14}-10^{-10} T). Furthermore, magnetic fields caused by technical electromagnetic sources and the Earth's magnetic

R. Jabłoński and T. Brezina (eds.), *Advanced Mechatronics Solutions*,
Advances in Intelligent Systems and Computing 393,
DOI 10.1007/978-3-319-23923-1_23

field ($5x10^{-5}$ T), can significantly influence the measured results [3]. Therefore magnetic shielding (e.g. magnetically shielded room) is necessary for MEG-systems.

Sources for auditory evoked magnetoencephalography have to deliver the sound directly to human's ear and thus have to be placed partly in a magnetically shielded room (MSR). This requires the use of non-magnetic materials for systems delivering stimulus signal to the subject. Commercially available field-free parts are often expensive and some common materials thought to be non-magnetic need to be tested for small remnant fields.

2 Development of auditory source for MEG

A cost-efficient prototype setup based on a piezoelectric transducer (KEMO L010, https://www.luedeke-elektronic.de) and commonly available parts is shown in Fig. 1. Proposed system consists of: piezoelectric transducer (a), funnel (b), polythene tube (c), silicon tube (d) and ear insert (e). Piezoelectric transducer is placed in a funnel, which is connected to a 5 m – long polythene tube. The length of a tube is defined by the magnetically shielded room's dimensions as the sound has to be delivered directly to human's ear and piezoelectric transducer has to be placed outside MSR to avoid a disturbance of the MEG signal from the currents flowing in the transducer. Ear insert is connected with a polythene tube via silicon tube of smaller diameter.

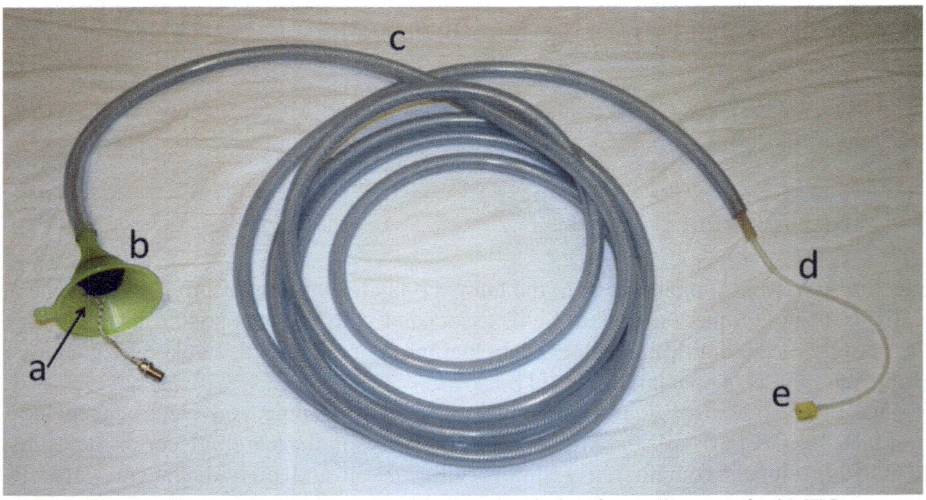

Fig. 1. Designed acoustic source to provide the stimulation for auditory evoked MEG measurements: a) piezoelectric transducer (KEMO L010), b) funnel, c) polythene tube, d) silicon tube, e) ear insert

2.1 Properties

The magnetic and acoustic properties of the acoustic source were tested. The ear insert was placed in the MEG room and the transducer was driven by an electric signal.

No additional signals due to the acoustic source could be detected in the MEG. It can be concluded that the acoustic source will not interfere with MEG recordings of brain magnetic fields.

Sound pressure level (SPL) values for frequencies of interest were measured using Brüel & Kjaer (http://www.bksv.com) type 4157 occluded ear simulator as presented in Fig. 2. For all the frequencies (1 kHz, 2 kHz, 5 kHz and 8 kHz) explicit peaks appeared with SPL values from 64 to 79 dB, audible for measured subjects. Harmonics of the frequencies were observed in the acoustic spectra only for the 5 kHz tone. This indicates, that the transducer does not produce strong harmonics at the applied voltages.

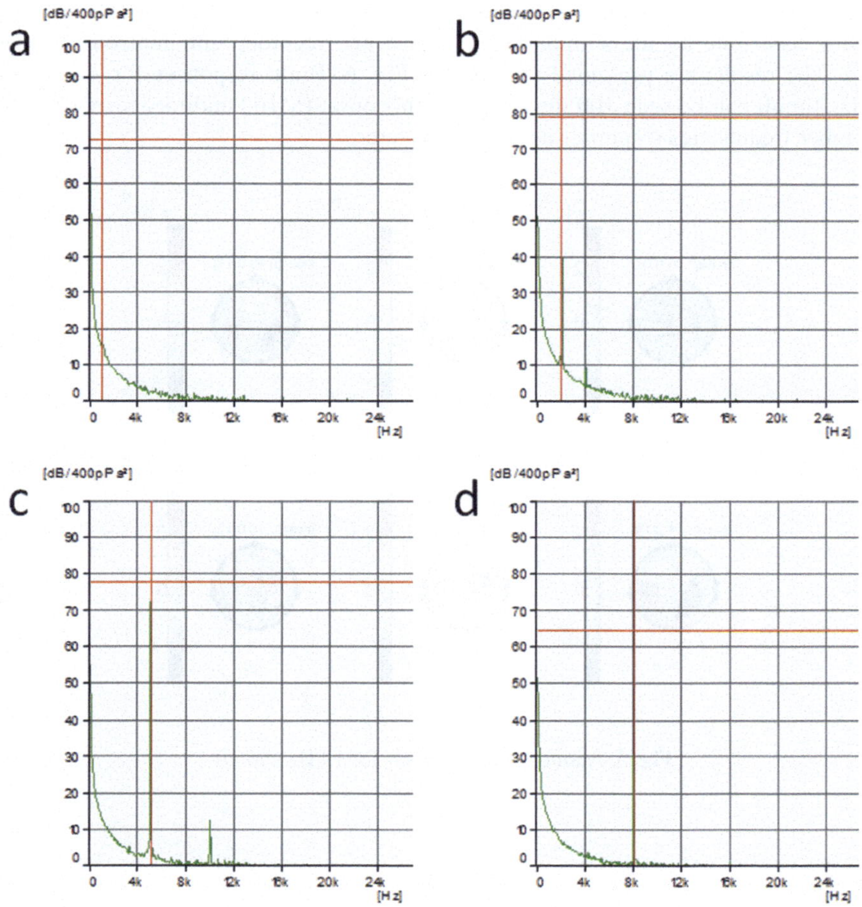

Fig. 2. Sound pressure level of auditory source driven with: a) 1 V at 1 kHz, b) 1 V at 2 kHz, c) 1,2 V at 5 kHz and d) 1,4 V at 8 kHz

3 Results

Auditory evoked MEG measurements were made using the new auditory source. All measurements were made in a magnetically shielded room at PTB Berlin, with a 128-channel gradiometer Yokogawa MEG-system. The stimuli were 500 ms-sine-tones at frequencies of: 1 kHz, 2 kHz, 5 kHz and 8 kHz, presented in random order using MATLAB software (http://www.mathworks.com/matlab). The time interval between sounds was random and had a duration of 2-3 seconds. Tone of each frequency was repeated 60-70 times.

Obtained dataset was analyzed and the auditory event related fields were calculated using free software: FieldTrip [4]. The event related fields were calculated by averaging the trials for each frequency separately using the specific trigger signals associated with each tone by the stimulation sequence and recorded simultaneously to the MEG. The results are presented in Fig. 3 – Fig. 6. Brain responses for 1 kHz and 2 kHz-stimuli can be seen 100 ms after stimulus onset (M100 auditory response), but for higher frequencies responses appear 20 ms later.

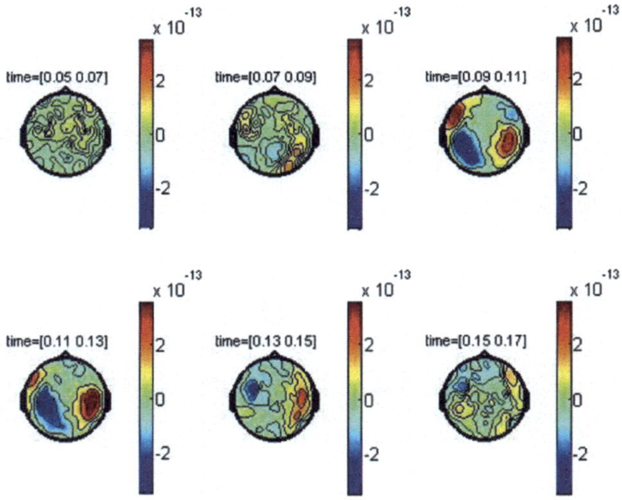

Fig. 3. Auditory evoked fields for 1 kHz stimuli

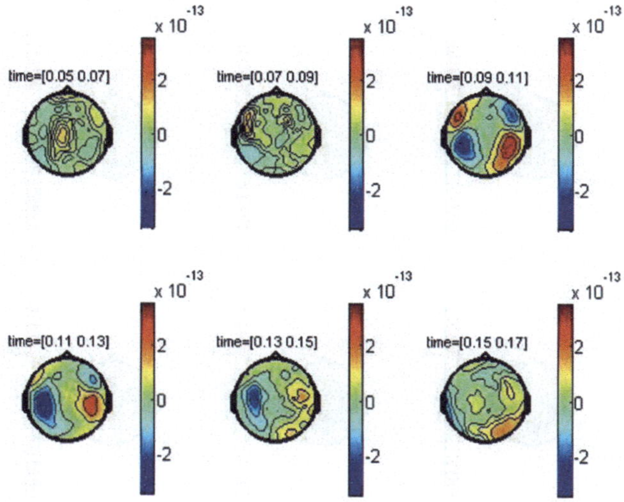

Fig. 4. Auditory evoked fields for 2 kHz stimuli

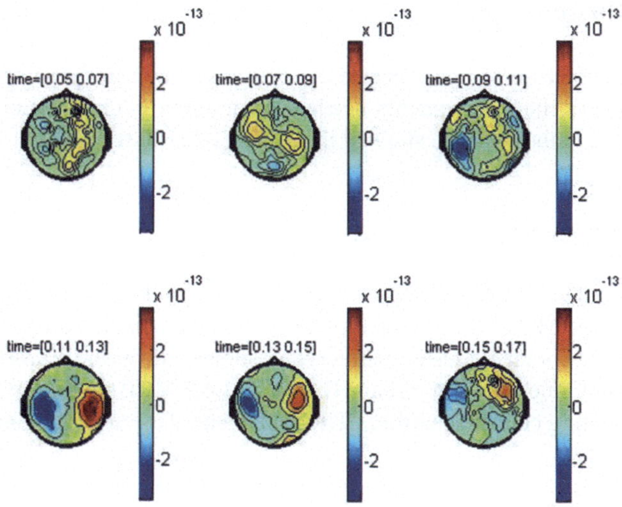

Fig. 5. Auditory evoked fields for 5 kHz stimuli

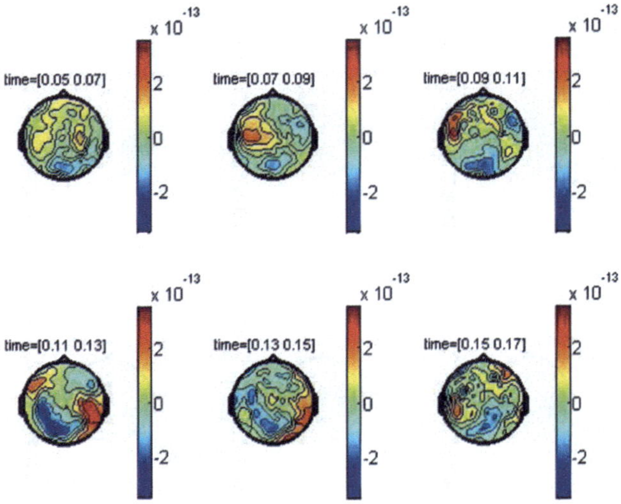

Fig. 6. Auditory evoked fields for 8 kHz stimuli

4 Conclusions

The developed acoustic source successfully delivers sounds at frequencies from 1 kHz to 8 kHz. It meets the requirements needed for auditory evoked MEG measurements as data recorded with the MEG showed the auditory M100 response.

Acknowledgement

The internship of A. Jodko-Władzińska at PTB was conducted within the Foundation for Polish Science SKILLS – Internships Programme co-financed by European Union within European Social Fund. R. Kühler was supported by the European Metrology Research Programme (EMRP, grant HLT01). The EMRP is jointly funded by the EMRP participating countries within EURAMET and the European Union.

References

1. Hansen PC, Kringelbach ML, Salmelin R (editors): MEG: An Introduction to Methods. Oxford University Press, 2010
2. Jodko A, Pałko T: Badania symulacyjne mikroczujnika SQUID pod względem możliwości zastosowań medycznych, Elektronika - konstrukcje, technologie, zastosowania, 2013, 54 (8): 76-79
3. Dunajski Z: Biomagnetyzm, WKŁ, Warszawa 1990
4. Oostenveld R, Fries P, Maris E, Schoffelen JM: FieldTrip: open source software for advanced analysis of MEG, EEG, and invasive electrophysiological data. Comput Intell Neurosci 2011; 2011: 1–9

Detection and evaluation of breast tumors on the basis of microcalcification analysis

Krzysztof Lewenstein[1], Krzysztof Urbaniak[2]

[1]Politechnika Warszawska, Wydział Mechatroniki, Warszawa, Polska
k.lewenstein@mchtr.pw.edu.pl
[2]Politechnika Warszawska, Wydział Administracji i Nauk Społecznych, Warszawa, Polska
k.urbaniak@ans.pw.edu.pl

Abstract. Long-term studies on the possibilities of aiding breast cancer diagnostics with the use of artificial neural networks supported by promising results led the authors of this paper to take up another challenge, the aim of which was localization and classification of microcalcifications. The evaluation of mammographic images by a specialist is not easy, and often ambiguous. Microcalcifications possess 'encoded' significant diagnostic value while having a small size and low contrast. A large number of papers indicates that information included in mammogram can be the basis for extracting the features of microcalcifications and their satisfactory classification with the use of artificial neural networks [9, 10, 12, 13, 14, 15, 16, 19, 20]. Detection of microcalcifications is usually realized in two stages, i.e. by the analysis of mammographic image what results in defining microcalcifications concentration and then by trying to evaluate the degree of their malignance. The paper presents the results of research undertaken in order to evaluate mammograms in terms of detecting places of occurrence and evaluate the degree of their malignance. The evaluation took place on the basis of the analysis of mammographic image with the use of two types of neural networks: feed forward multi-layer MLBP networks and Fahlman networks.
Keywords: microcalcifications, extraction method, segmentation method, binarization method, Fahlman neural network, MLBP neural network

1. Introduction

Studies conducted up to now in the field of supporting breast cancer diagnostics concerned two most significant issues, i.e. detection support and the degree of tumor's malignance, and the classification of microcalcifications with the use of artificial neural networks.

In such studies the most important task to solve was preparing processing methods enabling for obtaining proper feature vectors from mammographic images, constituting coded information describing given medical cases. Principal proceeding methods were taken into consideration, such as image segmentation and binarization [16] and extraction of features from the so-called region of interest [17]. The first step of the examination was implementing image segmentation method, then

R. Jabłoński and T. Brezina (eds.), *Advanced Mechatronics Solutions*,
Advances in Intelligent Systems and Computing 393,
DOI 10.1007/978-3-319-23923-1_24

image binarization and global analysis (line summing algorithm) [22] and finally region analysis (ROI) based on paper [17].
In [22] the described methods, as well as the results of the studies were presented in natural order of their occurrence. At the same time, another modifications of the proceeding methods led to the improvement of the results and achieving better results. Neural networks were decision elements in diagnostic software. For their evaluation we applied the methods of generalization, i.e. cross validation.
Acknowledging numerous studies [1, 2, 3, 4, 6, 7, 8, 16, 17, 21, 23], methods and obtained results [13, 22], another studies were undertaken aiming at trying to evaluate mammograms in terms of microcalcification localization and the degree of their malignance. The evaluation was performed on the basis of mammographic image analysis with the use of two types of neural networks: MLBP and Fahlman networks [5, 11, 18].

2. The algorithm of identification of microcalcification localization.

Difficulties with unambiguous evaluation of microcalcifications provoked dividing works into 2 stages:

- stage I – identifying the localization,
- stage II – evaluation of the degree of microcalcification malignance.

The algorithm of detection and classification of microcalcifications was divided into the following steps:

1. Loading a digital form of mammogram.
2. Putting segments of a given size on mammographic image.
3. Determining features for the image segments.
4. Automatic selection of segments containing microcalcifications.
5. Determining additional features for segments selected in step 4.
6. Automatic classification consisting in the evaluation of the degree of microcalcification's malignance.

Each mammogram was divided into square fields, the size of which depended on the size of the analyzed image and analyzed objects and was accepted by a person conducting the examination. Two sizes of networks were adopted: 32x32 (fig. 1) and 64x64 pixels. Examination was carried out for a greater number of segments, but we intentionally ignored their interpretation due to small differences in their results.
Having known the description of each mammogram it was possible to specify the condition of a particular segment (field). We took up a description in such a way that if a certain field included microcalcification (fig. 1a) [10], the condition of such field was regarded as positive and the value equal to 1 was adopted to describe the features of the vector. When no microcalcifications were found, the condition of the segment received the value equal 0 (fig. 2).

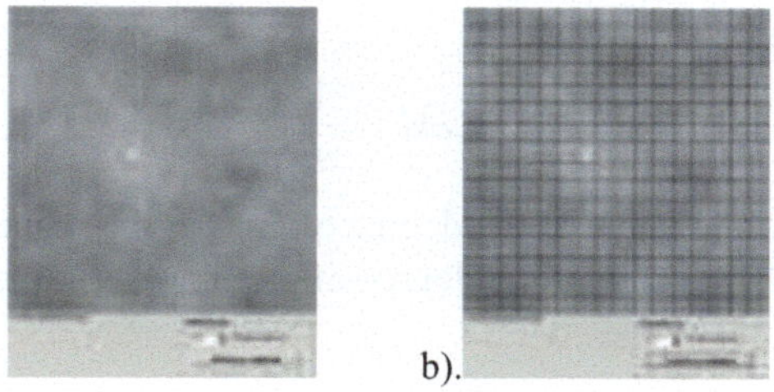

Fig. 1. The field of mammogram of 512x512 pixels size containing microcalcifications a) original image b) image with a chart of fields

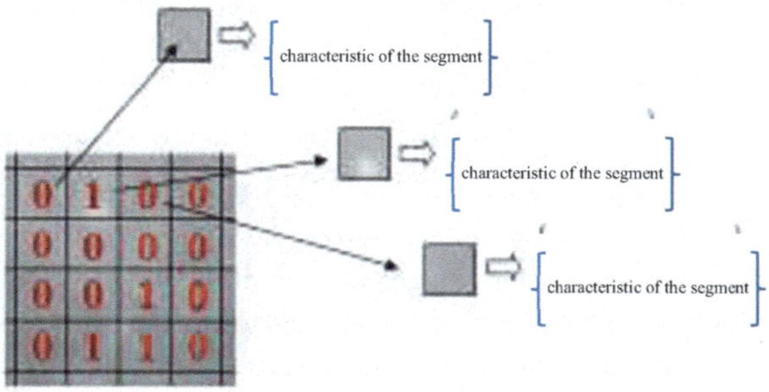

Fig. 2. Segments with the state of activation 0 – negative, 1 – positive

The experience from research on parameterization and evaluation of mammographic images allowed to make use of the previously developed parameters, expanded by new features, for the purposes of identification and evaluation of microcalcifications. For each marked field, the value of the following parameters were calculated (defined in previous research): V – variance, CV – variance coefficient, LFP – longitudinal Fourier power spectrum, AFP – angular Fourier power spectrum [17] and parameters [21, 23] originally proposed for microcalcification detection:

- MG – matrix of the image gradient – describing the difference in the degree of grey of the pairs of pixels adjoining the examinedpixel or located in a distance d from them

$$G_{i,j} = \sqrt{\left(\left(x_{i+d,j} - x_{i-d,j}\right)^2 + \left(x_{i,j+d} - x_{i,j-d}\right)^2\right)} \tag{1}$$

- SG – sum of the image gradients – describing the sum of the differences of the degrees of grey for the pairs of pixels distant for a maximum distance d

$$SG = \sum_{i,j} G_{i,j} \tag{2}$$

- MKG – angular matrix of image gradient – describing the differences of the degrees of grey of the pairs of pixels adjoining the examined pixel or distant from it by a distance d, located on the sections at the angle of: α: 0-180°, 45-225°, 90-270°, 135-315°.

$$G_{i,j(\alpha)} = \sum_{\alpha} \sqrt{\left(\left(x_{i+d,j} - x_{i-d,j}\right)^2 + \left(x_{i,j+d} - x_{i,j-d}\right)^2\right)} \tag{3}$$

- GM – average gradient

$$GM = \frac{1}{m}\sum_{i,j} G_{i,j(\alpha)} \quad (4)$$

- MAX – maximal value of the degree of grey in the segment
- MIN – minimal value of the degree of grey in the segment

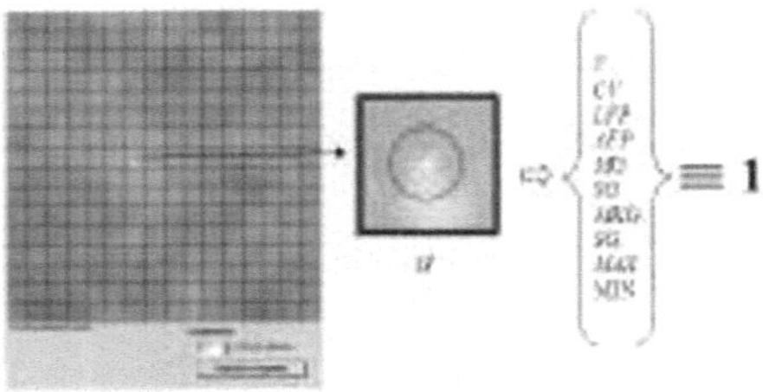

Fig. 3. The pattern of determining the features of mammographic image.

Parameterization of mammographic images carried out this way, together with linking them to medical diagnosis will constitute input sets to the neural networks. During the examination, the evaluation of the size of the feature vector was carried out, as well as its influence on the results obtained with the use of artificial neural networks.
The following sets of features were applied for teaching neural networks [22]:

- V, CV, MAX, MIN
- V, CV, LFP, AFP, MAX, MIN
- V, CV, LFP, AFP, MG, SG, MAX, MIN
- V, CV, LFP, AFP, MG, SG, MKG, GM, MAX, MIN

The best results were achieved for 10-elements vector. The accuracy of localization of microcalcifications was higher for 13% from the weakest result obtained from 4-elements vector of the parameters.

3. The results of identifying the localization of microcalcifications.

The obtained results of the accurate identification of about 92% of microcalcifications and about 90% segments with no microcalcifications, may be considered as very promising. About 5000 segments of different structure were analyzed.
The tools implemented for finding the localization of microcalcifications were Fahlman neural networks [5] and MLBP neural networks [11]. Better results were given by Fahlman neural networks method. It was probably caused by the fact that Fahlman neural network creates architecture individually, what means that during constant updating of the weight, the algorithm automatically analyzes the abilities of generalization of networks and decides on the necessity of adding a hidden neuron. In the case of MLBP network, the architecture is constant. The process of teaching the network consisted in adjusting to patterns the pairs of input vectors containing

features calculated from the segments. In the case of testing the network, for test segments and their features, a result of network was generated, following from the values of test parameters and the importance of networks trained in the process of teaching.

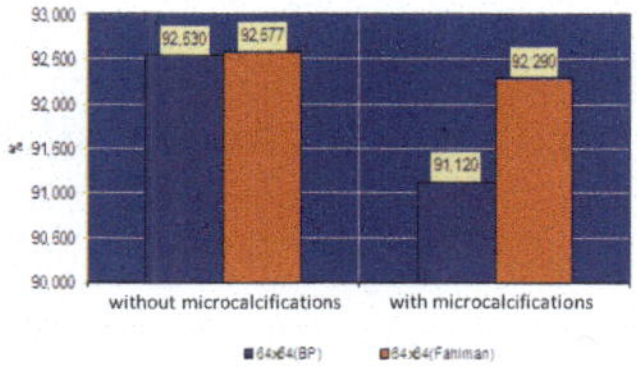

Fig. 4. Comparison of the average results for segments of 64x64 pixels.

4. Application of artificial neural networks in microcalcifications classification.

For the evaluation of the degree of malignance, the same method of research was used due to which the localization of microcalcifications was identified. From among all mammographic images we had to our disposal, 100 segments were chosen with malignant microcalcifications – described in the paper as MZ and 100 segments with benign microcalcifications, called MŁ. The set of 200 segments constituted a training base for neural networks. For testing network and the evaluation of the degree of microcalcifications' malignance, we also chose additional groups of segments of 100 each, containing particular types of microcalcifications. For the evaluation of the degree of malignance we subjected for the analysis of microcalcifications' shape, their number and outlines. As a result of such proceeding method, the feature vector was enriched by additional parameters appointed from binary image.

As a result of preparatory work we obtained a binary image containing microcalcifications (fig. 5.). This form of image allowed for designating additional parameters necessary for the evaluation of the features of microcalcifications' shape.

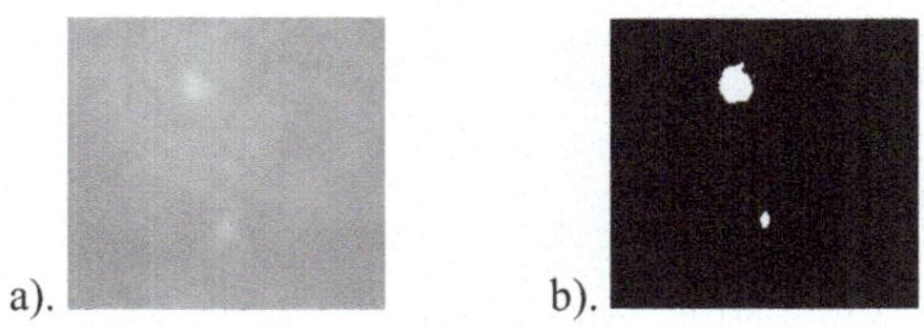

Fig. 5. a) original image and b) its bit map

In the further part of the studies the features were determined which were then introduced into neural network and paired with the appropriate conditions: 0 or 1. Values 0 and 1 corresponded to the conditions of microcalcifications: 0 – MŁ, 1 – MZ. Verification of microcalcifications was carried out by specialists.

For the purposes of describing microcalcifications, each feature vector (V, CV, LFP, AFP, MAX, MIN) described earlier was enriched by the following values:
LK – the number of microcalcifications concentrations
ZM – the outline of microcalcifications (0 – smooth/even, 1 - irregular)
LMK – the number of microcalcifications in a cluster

5. The results of microcalcifications classification.

The results indicate that the number and type of features are of a great significance in the process of microcalcifications evaluation. When classifying such features as: variance, variance coefficient, variance of an average, the degree of minimal and maximal grey – the accuracy of diagnosis is the lowest. Together with the increase of feature vector by angular and linear Fourier power spectrum, the accuracy of BP network grew to over 70%. Whereas a full set of features for network training, increased the accuracy to about 80%. It needs to be emphasized that such parameters as the number of clusters, microcalcification outline, the number of microcalcifications in a cluster, minimal and maximal degree of grey, have a significant impact on the accuracy of microcalcifications' evaluation.
Summarizing, it needs to be stated that as a result of the studies, max. 84% accuracy was achieved for benign microcalcifications and 82% accuracy for malignant microcalcifications. These results were given by Fahlman networks. The studies confirmed previous conclusion indicating that Fahlman neural network is a more effective tool than MLBP.

Feature vector	benign microcalcifications	malignant microcalcifications
V, CV, MAX, MIN.	70%	69%
V, CV, AFP, LFP, MAX, MIN.	75%	76%
V, CV, AFP, LFP, LK, ZM, LMK.	72%	74%
V, CV, AFP, LFP, LK, ZM, LMK, MAX, MIN.	84%	82%

Table 1. The results of the studies for Fahlman networks.

6. Summary.

Studies on the application of artificial neural networks into the analysis of mammographic images in terms of the evaluation of microcalcifications, realized two aims. The first and most important one was preparing a mechanism equipped with artificial neural networks, which was able to locate the places of microcalcifications' occurrence. The second aim was trying to evaluate the places containing microcalcifications in terms of the evaluation of the degree of their malignance.

Mammographic images may be a source of information for diagnostic tools, such as neural networks, through its appropriate parameterization. Due to graphic features of such small objects, the evaluation of the condition of the microcalcifications is a very difficult task. Too little differences in the degrees of grey, together with very small sizes and unclear shapes, are the basic problem in their appropriate evaluation. These problems can be solved by a computer system making use of artificial neural networks. The accuracy of microcalcifications localization of about 90% and their evaluation at the level of 80% absolutely proves it. It has to be added that it was carried out by a system constructed on the basis of MLBP neural network and Fahlman neural network models, the latter demonstrating better abilities of generalization.

References

1. Arbach L., Bennett D.L., Reinhardt J.M., Fallouh G.: Classification of mammographic masses: Comparison between backpropagation neural network (BNN) and human readers. Department of Biomedical Engineering. The University of Iowa. Iowa City. IA 52242 USA. Department of Radiology. The University of Iowa. Iowa City. Department of Biomedical Engineering. Damascus University. Syria (2003)
2. Arodz T., Kurdziel M., Popiela T.J., Sevres E.O.D., Yuen D.A: A 3D Visualization System for Computer-Aided Mammogram Analysis. University of Minnesota, Research Report UMSI 2004/181. Submitted to Computer Methods and Programs in Biomedicine, Elsevier (2004)
3. Burke H.B., Goodman P.H., Rosen D.B., Henson D.E., Weinstein J.N., Harrell F.E., Marks J.R., Winchester D.P., Bostwick D.G: Artificial Neural Networks Improve the Accuracy of Cancer Survival Prediction. American Cancer Society (1997)
4. Collins C.: Breast cancer detection aided by new technology installed at Magee-Womens Hospital of University of Pittsburgh Medical Center. Pittsburgh, Jan. 16, 2002. Magee Womens Hospital (2002)
5. Fahlman S.E., Labiere C.: The cascade-correlation learning architecture Advances in NIPS2, Ed. D. Touretzky, p. 524-532 (1990)
6. Kim J., Park H.: Statistical textural features for detection of mikrocalcifications in digitized mammograms. IEEE Trans. Medical Imag. 18. p.231-238 (1999)
7. Kopans D. (1998) Breast Imaging. Lippincott-Raven.
8. Kuo W., Chang R.F., Moon W.K., Lee C.C., Chen D.R.: Computer-aided Diagnosis of Breast Tumors with Different US Systems. Academic Radiology, Vol 9, No 7, (2002)
9. Lado M.J., Tahoces P.G., Mendez A.J., Souto M., Vidal J.J.: A wavelet based algorithm for detecting clustered nicrocalcifications in digital mammograms. Med. Phys. 26 (7): p. 1294-1305 (1999)
10. Lado M.J., Tahoces P.G., Mendez A.J., Souto M., Vidal J.J.: Evaluation of an automated wavelet-based system dedicated to the detection of clustered microcalcifications in digital mammograms. Departamento de Radiologia Faculted de Medicina, Universidad de de Santiago de Compostela C/San Francisco, 1. 15782 Santiago, Spain (2003)
11. Lewenstein K.: Modele sztucznych sieci neuronowych do oceny testów wysiłkowych stosowanych w diagnostyce choroby wieńcowej. Oficyna Wydawnicza PW. Warszawa (2002)

12. Lewenstein K., Urbaniak K.: Interpretation of mammograms in system with Fahlman's neural network and picture segmentation. Embec VI International Conference on Medical Physics, Monduzzi Editore, p. 181-187 (1999)
13. Lewenstein K., Urbaniak K., Łubkowski P., Pałko T.: The comparison of computer aided discrimination of breast cancer based on the analysis of mammograms transformed digitally in three different ways, in Proc. IX Mediterranean Conf. on Medical and Biological Engineering and Computing – Medicon, Pula, Croatia, June 12-15, 2001, part I pp. 534-538 (2001)
14. Li L., Clark R.A., Thomas J.: Improving Algorithm robustness for mass detection in digital mammography. Department of Radiology, Collage of Medicine, H.Lee Moffit Cancer Center and Research Institute at University of South Florida, Uniformed Services University of the Health Sciences. The U.S. Army Medical Research and Materiel Command under DAMD 17-00-1-0245, (2000)
15. Lundin M., Lundin J., Burke HB., Toikkanen S., Pylkkanen L., Joensuu H.: Artificial neural networks applied to survival prediction in breast cancer. Oncology. 1999 Nov. 57(4):281-6 (1999)
16. Nieniewski M. (2005) Segmentacja obrazów cyfrowych. Metody segmentacji wododziałowej. Akademicka Oficyna Wydawnicza EXIT. Warszawa (2005)
17. Ogawa K., Fukushima M.: Computer-aided Diagnostic for Diffuse Liver Discases with Ultrasonography ny Neural Networks , IEEE Transactions on Nuclear Science, vol. 45, no.6, december (1998)
18. Osowski S.: Sieci neuronowe w ujęciu algorytmicznym. WNT (1996).
19. Pardela M.: Współczesne rozpoznawanie i leczenie guzów sutka u kobiet. Śląska Akademia Medyczna, Katowice, (1997)
20. Ripley B. D., Ripley R. M.: Neural networks as statistical methods in survival analysis. In Dybowski R., Gant V. editors, Artificial Neural Networks: Prospects for Medicine. Landes Biosciences Publishers, (1998)
21. Rudnicki Z.: Wybrane metody przetwarzania i analizy cech obrazów teksturowych. Informatyka w Technologii Materiałów. Nr 1, Tom 2, p. 1-18 (2002)
22. Urbaniak K., Lewenstein K., Łubkowski P., Chojnacki M.: Influence of the digital transformation of the mammograms on computer discrimination of breast cancer 3rd International Conference Mechatronics, Robotics and Biomechanics 2001 Trest, p.181-189 (2001)
23. Wróblewska A., Przelaskowski A., System automatycznej detekcji i klasyfikacji mikrozwapnień w cyfrowej mammografii. Elektronizacja, 3:8–11, Mat. IV Sympozjum 'Techniki Przetwarzania Obrazu', p. 299-305, (2002)

Control Bilateral Teleoperation By Compensatory ANFIS

Rabah Mellah[1] and Redouane Toumi[2]

[1] Design and Control Laboratory Automated Systems, Faculty of Electrical Engineering end Computer Science, UMMTO, PB n° 15000, Tizi-Ouzou, Algeria
mellah.rab@gmail.com
[2] Robotics Laboratory, Faculty of Electrical Engineering, U.S.T.H.B, PB n° 32, El Alia-Bab Ezzouar-16111-Algiers, Algeria
Red_toumi@yahoo.fr

Abstract. This paper presents the force/position architecture of control using ANFIS with compensatory fuzzy in order to get a robust stability and transparency in bilateral teleoperation. The proposed control structure includes both ANFIS controllers with compensatory fuzzy for which one is used to control in force the master robot and the second one for controlling in position the slave robot. The experimental results obtained with this control scheme validate the effectiveness of the proposed teleoperation scheme.

Keywords: Teleoperation; Compensatory Fuzzy; Position/Force Control; ANFIS Control.

1 Introduction

A bilateral teleoperation system is an electromechanical teleoperator composed of master and slave robots, where the signals are exchanged between the two robots through communication channel in order to accomplish tasks in hazardous environment, so that the slave device follows the master motion which is manipulated by human operator [1], [2], [3], [4]. Basically the human operator manipulates the master teleoperator, and drives it with desired trajectories. The computer controlling the robot is responsible for low level activities such as trajectory generation and obstacle avoidance so that the remote slave teleoperator reproduce the task imposed by the master onto the remote environment [5], [6], [7]. One of the major objectives in designing bilateral teleoperation control systems is achieving the fundamental trade-off between performances and stability [8].

Various form of control strategies have emerged the bilateral teleoperation [5]. In contrast, Neural network provides learning ability using nonlinear optimization algorithm, such as back-propagation, thanks their parallel information processing and their inspired mechanism of human neural. In the other hand the strength of fuzzy logic comes from the fact that it can have the ability to make use of knowledge expressed in the form of linguistic rules, thus it offer the possibility of implementing expert human

R. Jabłoński and T. Brezina (eds.), *Advanced Mechatronics Solutions*,
Advances in Intelligent Systems and Computing 393,
DOI 10.1007/978-3-319-23923-1_25

knowledge and experience. The combination of fuzzy logic and neural network is actively study, in which a fuzzy reasoning discussed above is realized within multi-layered hierarchical neural network and the parameters that are represented by connection weights or involved in unit functions can be learned by using the actual data. Unfortunately, conventional neurofuzzy systems can only optimize the fuzzy membership functions under specially defined fuzzy operators which are unchangeable forever, which makes it use the local optimization technique rather than the global optimization technique [9].

The remainder of the paper is organized as follows. Section 2 some fundamentals of ANFIS controller and the learning algorithm for parameter adaption are given. Section 3 describes our teleoperation system. The experiment results are presented in section 4, to illustrate the tracking performance of the proposed control scheme. Finally in section 5, some conclusions are presented.

2 ANFIS Controller With Compensatory Fuzzy

One of the network structures used for neuro-fuzzy control is the adaptive neural fuzzy inference system (ANFIS) introduced by Jang in 1993, which has received the interest of many researchers in various applications. Our contribution is to added a compensatory fuzzy in order to optimize the dynamic of fuzzy rules.

Assume that a fuzzy system consists in L rules, each of which has N input variables $x_1,\ldots\ldots\ldots\ldots x_N$ and one output y. Then the lth rule $\left(R^{(l)}\right)$ is given as:

$$R^{(l)} : If\ x_1\ is\ F_i^l\ and \ldots\ldots x_N\ is\ F_N^l\ Then\ y^l = \sum_{i=1}^{N} a_i^l x_i + a_0^l \tag{1}$$

By using inference product and center of average defuzzification, the fuzzy output can be expressed in this form:

$$u = \sum_{l=1}^{L}\left(\sum_{i=1}^{N} a_i^l x_i + a_0^l\right)\frac{\prod_{i=1}^{N}\mu_{F_i^l}(x_i)}{\sum_{l=1}^{L}\left(\prod_{i=1}^{N}\mu_{F_i^l}(x_i)\right)} = \sum_{l=1}^{L}\left(\sum_{i=1}^{N} a_i^l x_i + a_0^l\right)\overline{w_l} \tag{2}$$

where $\mu_{F_i^l}(x_i)$ is the fuzzy membership value of the ith input for the lth fuzzy rule.

By using the pessimistic operation and optimistic operation, the compensatory form is defined as [7]:

$$C\left(z^k, v^k\right) = (z)^{1-\gamma}\left(v^k\right)^{\gamma} \tag{3}$$

where $\gamma \in [0,1]$ is the compensatory degree

$$z^k = \overline{w_l} \tag{4}$$

$$v^k = \left[\overline{w_l}\right]^{1/N} \tag{5}$$

Finally, the crisp value of the compensatory neural-fuzzy inference is derived as:

$$u = \sum_{l=1}^{L}\left(\sum_{i=1}^{N} a_i^l x_i + a_0^l\right)\left[\overline{w_l}\right]^{(1+1/N)} \tag{6}$$

2.1 Learning Algorithm

In order to train the compensatory ANFIS for the corresponding the optimal vector parameters Φ_l , back propagation method is used. In this method, the following objective function is considered:

$$J = \frac{1}{2}\left[y - y_d\right]^2 \tag{7}$$

Where y and y_d are both actual and desired output respectively. In addition, let be the vector of update parameters. Our objective is to find the vector Φ_l, by using the approach of extended Kalman filter which consist to linearize at each sampling period the control law around the output. This is equivalent to writing:

$$\frac{\partial J}{\partial \Phi_l} = \frac{\partial J}{\partial u}.\frac{\partial u}{\partial \Phi_l} = (y - y_d).\frac{\partial y}{\partial u}.\frac{\partial u}{\partial \Phi_l} = P_l \Psi_l e \tag{8}$$

In, which

$$e = y - y_d;\ \Psi_l = \frac{\partial u}{\partial \Phi_l};\ P_l = \frac{\alpha_1}{\alpha_2 + \Psi_l^T \Psi_l} \tag{9}$$

Where α_1 and α_2 are adaptation gains for varying the convergence rate.
On the other hand, to eliminate the constraint $\gamma \in [0,1]$, we redefine γ as follows:

$$\gamma = \frac{c^2}{c^2 + d^2} \tag{10}$$

Consequently, $$\Phi_l = \left[a_N^l, a_{N-1}^l, \ldots. a_1^l, a_0^l, c_l, d_l\right] \tag{11}$$

Finally, the vector of parameters is adjusted by the following equation:

$$\Phi_l(t+1) = \Phi_l(t) + P_l \Psi_l e \tag{12}$$

Where $$\Psi_l^T = \left[\frac{\partial u}{\partial a_N^l}, \frac{\partial u}{\partial a_{N-1}^l}, \ldots\ldots. \frac{\partial u}{\partial a_1^l}, \frac{\partial u}{\partial a_0^l}, \frac{\partial u}{\partial c_l}, \frac{\partial u}{\partial d_l}\right] \tag{13}$$

3 Description of the Teleoperation System

As illustrated in Fig.2, the teleoperation system is composed of two identical master and slave devices. The slave device is connected to a bilateral spring emulating the environment. The master and the slave consist of one of degree of freedom rotating arms of a length 95 mm. They are actuated by DC motors (Maxon RE 30) via capstan reduction a ration of $\frac{100}{12}$. The arms are connected to the pulley of capstan via torque cells (Futek TFF-325). Positions are measured by 1024 counts per turn encoders connected to the motor shafts. Every DC motor is driven by Maxon ADS 50/5 servo amplifier. The teleoperation controller is implemented using a DSPACE board (DS1103, running at a sampling rate of 10 KHz).

Fig. 1. Master/Slave device

4 Experiment Results

In order to confirm the robustness of methodology for designing controller developed previously, we built two controllers such that the fist is used to control the master device in force by considering the inputs of controller are force error and its derivative, and the second to control the slave device in position by considering the inputs as the position error and its derivative.

Figures 2 and 3 plot the time evolution of master and slave positions using neuro-fuzzy controllers and compensatory neuro-fuzzy controllers respectively. Figures 4 and 5 show master and slave torques in the same conditions as previously. Through these graphics, we can see that, the compensatory Neural-fuzzy controllers provide a good tracking performance, because the slave tracks the master's movement. On the other hand, the force tracking performance of teleoperation system with these compensatory Neuro-Fuzzy controllers is much better.

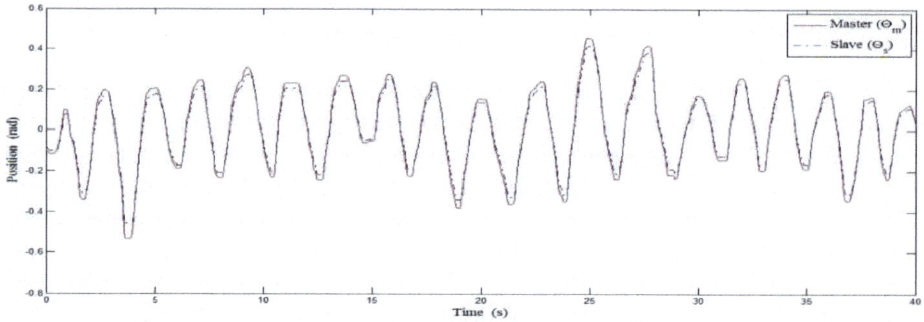

Figure 2. Motion tracking behavior with neuro-fuzzy controllers

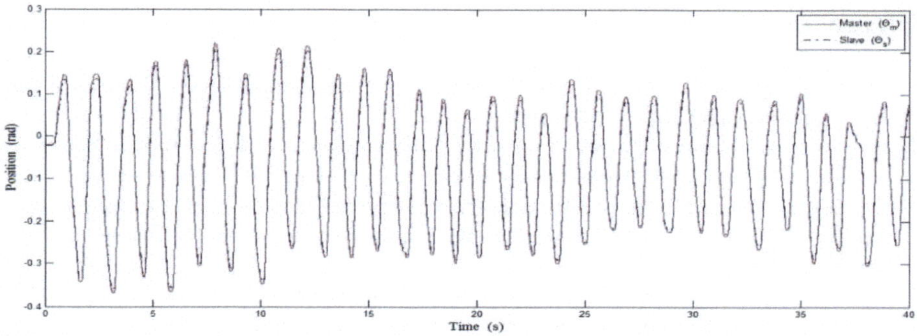

Figure 3. Motion tracking behavior with compensatory neuro-fuzzy controllers

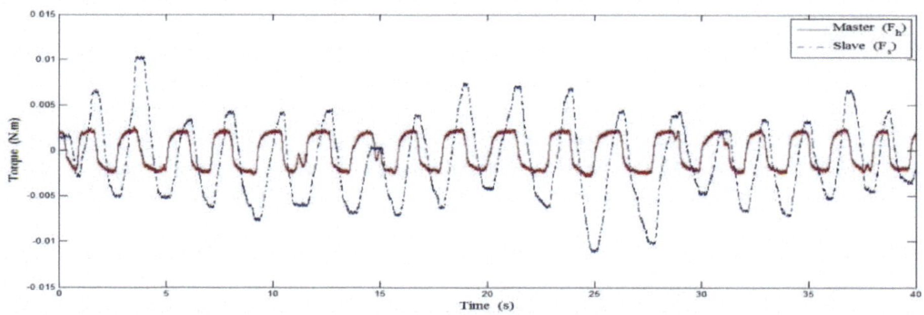

Figure 4. Force tracking behavior with neuro-fuzzy controllers

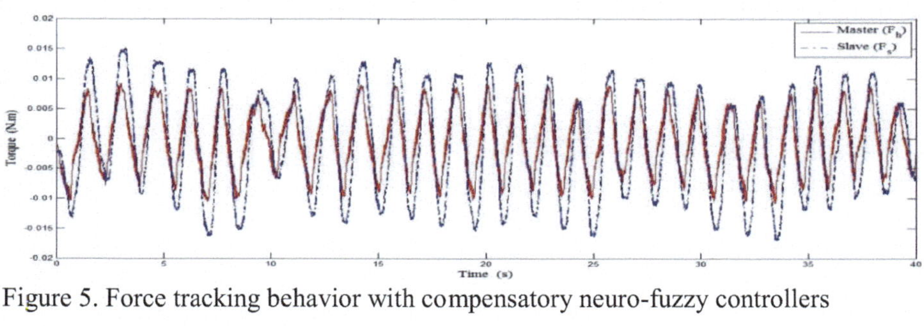

Figure 5. Force tracking behavior with compensatory neuro-fuzzy controllers

5 Conclusion

A bilateral adaptive fuzzy reasoning method using compensatory fuzzy operators has been proposed for the control a one of degree of freedom teleoperation system. The basic idea behind this approach was to make fuzzy logic system more adaptive and more effective by using the compensatory learning algorithm in order to adjust not only parameters of fuzzy rules but also dynamically optimize the adaptive fuzzy reasoning. Through a series of experimental results, it was shown that the proposed controllers were able to reduce the master-slave error. Future work will aim towards to account for the presence varying time delay.

References

1. Liv Y., Chopra N. " Control of Semi-Autonomous Teleoperation System With Time Delays" Automatica.49 (2013) 1553-1565.
2. Haddadi A., and Hashtrudi-Zaad K., "Bounded-Impedance Absolute Stability of Bilateral Teleoperation Control Systems," IEEE Transactions on Haptics, vol. 3, N°.1, pp. 15-27, January-March 2010.
3. Liu X. and Tavakoli M., "Inverse Dynamics-based Adaptive Control of Nonlinear Bilateral Teleoperation Systems," IEEE International Conference on Robotics and Automation, Shanghai, China, May 9-13, 2011.
4. Hashemzadeh F., Hassanzadeh I. and Tavakoli M., " Teleoperation in the Presence of varying Time Delays and Sandwich Linearity in Actuators," Automatica 49 (2013), pp 2813-2821.
5. Veras E., Khorkar K., Alqasemi R. and Dubey R. " Scaled Telerobotic Control of a Man Oulator in Real Time With Faster Assistance for ADL task" Journal of Franklin Institute 349 (2012) 2268-2280.
6. Hua C.C., Yang Y. and Guan X.," Neural Network-Based Adaptive Position Tracking Control for Bilateral Teleoperation Under Constant Time Delay," Neurocomputing 113 (2013), pp 204-212.
7. Zhang Y. Q. and Kandel A., " Compensatory Neurofuzzy Systems With Fast Learning Algorithms, " IEEE Transactions on Neural Networks, Vol. 9, N° 1, January 1998, pp 83-102.
8. Delwiche T., Aberkane S., Catoire L., Torfs S. and Kinnaert M. " Design of a Static Output Feedback Controller for Bilateral Teleoperation" 9 th International Symposium on Robot Control, Gifu, Japan 2009, pp 553-558.
9. Peng L. and Woo P. Y., " Neural-Fuzzy Control System for Robotic Manipulators, " IEEE Control Systems Magazing, February 2002, pp 53-63.
10. Delwiche T., Aberkane S., Catoire L., Torfs S. and Kinnaert M. " Design of a Static Output Feedback Controller for Bilateral Teleoperation" 9 th International Symposium on Robot Control, Gifu, Japan 2009, pp 553-558.

Validation of Finite Element Method Solver for Utilization in Eddy Current Tomography

Paweł Nowak

Institute of Metrology and Biomedical Engineering ,
Warsaw University of Technology, Warsaw, Poland
p.nowak@mchtr.pw.edu.pl

Abstract. Paper presents process of validation of finite element solver utilized for simulations of eddy current tomography forward transformation. Validation of numerical solver is conducted on two models. One test utilizes well known and analytically solved standards of magnetic field: Helmholtz coil and long coil. Second test validates the depth of eddy current inducted in material depending on its conductivity and excitation frequency. Achieved results confirmed proper operation of magnetodynamic solver utilized in Elmer FEM software.

Keywords: FEM, magneto dynamics solver, Helmholtz coil, long coil, eddy current

1 Introduction

Finite element method is a numerical technique for solving partial differential equations which cannot be solved analytically. Complex problem is divided into smaller parts and is solved with utilization of simplified equations. Partial results are summarized and approximated results are achieved.

Finite element method has many applications both in science and in industrial applications. It can be utilized for calculating the stresses in materials [1], distribution of current in ultra-thin layers [2], acoustics [3], magnetovision [4,5] and for many other applications. Currently many finite element programs are available, but most of them has expensive licenses and their code is inaccessible for user. Those drawback do not occur in open-source software in which user does not only have free access to compiled program, but also has possibility for modifying it for their own needs. Elmer FEM is open source software with solvers for multi-physics software [6]. This paper will focus on magneto dynamics solver utilized in forward eddy current tomography transformation [7,8]. During creation of proper inverse tomography transformation software reliability of proper implementation of finite element solver is crucial. Validation of the solver is conducted on two examples. Firstly, the distribution of magnetic field in Helmholtz coil and long coil models is tested and results are compared with

R. Jabłoński and T. Brezina (eds.), *Advanced Mechatronics Solutions*,
Advances in Intelligent Systems and Computing 393,
DOI 10.1007/978-3-319-23923-1_26

analytical solutions. Secondly, tests of eddy current depth in material depending on its conductivity and frequency of excitation are conducted.

2 Validation of magneto dynamics solver on the Helmholtz coil model and on long coil model

Helmholtz coil is standard device utilized for generation of uniform magnetic fields in laboratory conditions [9,10]. The setup consist of two coaxial solenoids with distance between them equal to their radius. Utilization of Helmholtz coil in validation of FEM solvers is caused by the fact, that value of magnetic flux in the center point of setup (on the axis of both coils and halfway between them) is analytically solved [10] and equals:

$$B = \left(\frac{4}{5}\right)^{\frac{3}{2}} \frac{\mu_0 n I}{R} \tag{1}$$

Where B is for the magnetic flux in the center point, μ_0 is free space permeability, n is the number of the turns in wire, I is the current powering the coils and R is their radius and distance.

Model of long coil is approximation of infinite continuous solenoid. This model is based on assumption, that with big enough ratio of solenoid length to its diameter, all fringe effects can be ignored and homogenous magnetic field inside the coil is achieved. Utilization of long coil in validation of FEM solvers is caused by the fact, that value of magnetic flux in the center point of setup (on the axis of both coils and halfway between them) is analytically solved [10] and equals:

$$B = \frac{\mu_0 n I}{R} \tag{2}$$

Where B is for the magnetic flux in the center point, μ_0 is free space permeability, n is the number of the turns in wire, I is the current powering the coils and R is their radius and distance.

2.1 Utilized model of Helmholtz coils

For simulations finite element model was prepared. Model contained two coaxial coils with average 0.9 m radius and with average distance between them 0.9 m. Both coils were placed in a sphere with center point between the coils and with 10 m radius. Big operating range was required for proper application of Dirichlet conditions [6] on far boundaries. Objects were meshed densely – as presented in fig. 1. Both coils were formed by 4482 triangles each and operating sphere contained 9846 triangles. Created model fulfilled all requirements for Helmholtz coil – distance between coils was equal to their radius, coils were coaxial and the cross section of the coil was square. For the simulation purpose coils were formed from one wire turn.

Fig. 1. Meshed model of Helmholtz coils utilized for simulations

2.2 Utilized model of long coil

For simulations finite element model was prepared. Model contained single cylinder representing coil with average 0.9 m radius. Coil was placed in sphere with center point in the middle of the coil and 10 m radius. Big operating range was required for proper application of Dirichlet conditions [6] on far boundaries. Objects were meshed densely – as presented in fig. 2. Coil was formed by 17208 triangles each and operating sphere contained 18090 triangles. Ratio of coil length to its average radius equaled 11, so it can be assumed, that in the coil middle homogenous magnetic field is achieved as for infinite coil.

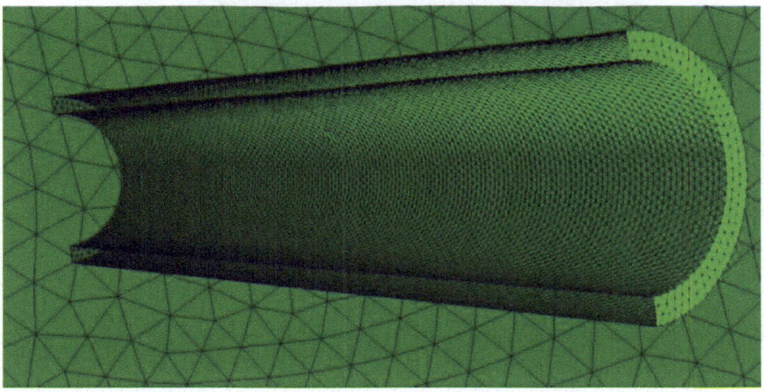

Fig. 2. Meshed model of long coil utilized for simulations

2.3 Simulation results

Graphical representation of simulation results are presented in figures 3 and 4. Results have similar shape to those well known in literature [10,11]. Thus correctness of simulation results was assumed. As presented in figure 4, homogenous magnetic field inside the coil is obtained. Fringe effects on coil edges does not influence distribution of magnetic field in the middle of the coil.

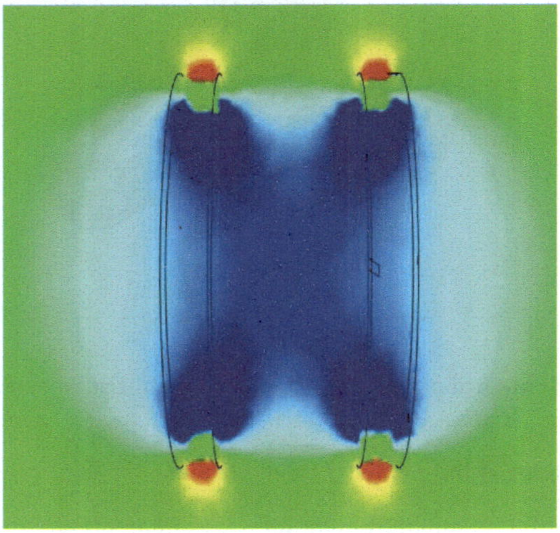

Fig. 3. Graphical representation of magnetic flux in Helmholtz coil setup

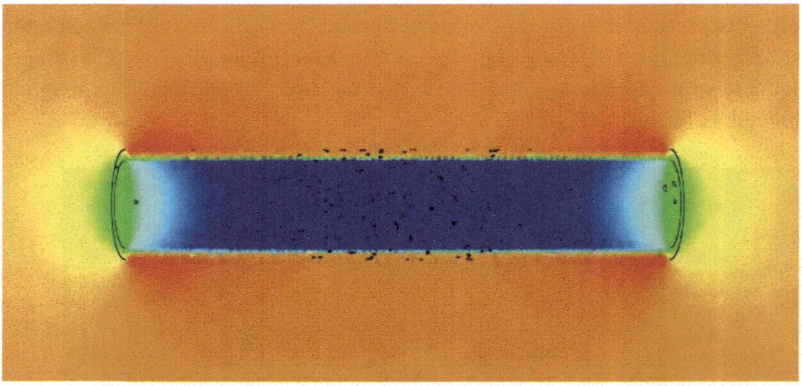

Fig. 4. Graphical representation of magnetic flux in long coil setup

2.4 Results analysis and comparison with analytical values.

Based on (1) and (2) expected magnetic flux value in center point of coils was calculated for the data described above. Also proper values from simulation were obtained. Achieved results, absolute difference between simulation result and analytical solution and relative errors are presented in table 1. Results confirm proper implementation of magnetodynamic FEM solver. Results of simulation are similar to expected analytical values. Relative error is caused by numerical noise and finite dimensions of elements.

Table 1. Comparison and analyses of magnetic flux value in Helmholtz coil center and in long coil center.

Tested model	Analytical value	Simulation result	Absolute error	Relative error
Helmholtz coil	39.96 nT	39.5 nT	0.464 nT	1.16 %
Long coil	0.251 μT	0.246 μT	5.33 nT	2.11 %

3 Tests for eddy current depth in materials

Eddy currents are inducted by variable magnetic fields within conductor and result with skin effect [10] which is major cause of energy losses. On the other hand they are commonly used in many methods of non-destructive testing of magnetic materials. The depth of eddy current penetration inside the material highly depends on the materials parameters- permeability and resistivity [12] as well as on frequency of exciting magnetic field and equals:

$$\delta = \frac{1}{\sqrt{\pi f \mu \sigma}} \tag{3}$$

Where f is the test frequency, μ is magnetic permeability and σ is electrical conductivity. δ represents depth below conductor surface at which eddy current density falls to 1/e of initial density.

3.1 Utilized model

For simulations finite element model was prepared. Model contained single cylinder representing coil with average radius 0.9 m and big block representing tested material. Objects were placed in sphere with center point in the middle of the coil and 10 m radius. Big operating range was required for proper application of Dirichlet conditions [6] on far boundaries. Objects were meshed densely – as presented in fig. 5.

Fig. 5. Meshed model of setup utilized for simulations of eddy current depth of penetration

3.2 Simulation results and comparison with analytical results.

Tests of eddy current depth of penetration were conducted for 3 frequencies and for 3 permeability's values. Based on (3) theoretical values were calculated as well. Comparison of simulation and calculations results is presented in table 2.

Table 2. Comparison of eddy current penetration depth results

	Simulation results			Calculation results		
	Relative permeability			Relative permeability		
Frequency [Hz]	1 000	5 000	10 000	1 000	5 000	10 000
50	70	30	20	71.21	31.84	22.52
100	50	20	10	50.35	22.52	15.92
500	20	10	10	22.52	10.07	7.12

Results confirm proper implementation of FEM solver. Values of penetration depth acquired from simulations are similar to expected. Differences are caused by dimension of finite element in utilized model. Utilization of denser mesh will result with higher convergence of simulation and calculation results.

4 Conclusion

Conducted validation confirmed proper implementation of magnetodynamic solver in Elmer FEM software. Conducted tests verified both proper values of flux density as well as depth of eddy current penetration. Simulation results converge with analytical solutions. Relative errors of magnetic flux density equal up to 2%. Depth of eddy current penetration is similar to expected. Errors are caused by dimension of mesh element in tested model. Thus it can be assumed, that tested solver is suitable for developing software for inverse eddy current tomography transformation, as well as other magnetodynamic applications.

References

1. Jackiewicz, D. et al.: Influence of Stresses on Magnetic BH Characteristics of X30Cr13 Corrosion Resisting Martensitic Steel. In *Recent Advances in Automation, Robotics and Measuring Techniques*, pp. 607-614. Springer International Publishing, (2014).
2. Kozłowska, A et al.: Graphene Joule Heating Measurements in Environmental Chamber. In *Progress in Automation, Robotics and Measuring Techniques*, pp. 129-135. Springer International Publishing, (2015).
3. Malinen, M. et al.: The Parallel Solution of Discrete Linearized Navier–Stokes Equations in the Frequency Domain. *Acta Acust united Ac*, *100*(1), pp. 102-112, (2014).
4. Nowicki, M. et al.: Scanner System Investigation of Magnetic Field Disturbance Sources. In *Mechatronics-Ideas for Industrial Application*, pp. 353-361. Springer International Publishing, (2015).

5. M. Nowicki, R Szewczyk, Modelling of the Magnetovision Image with the Finite Element Method. In *Proceedings of the 20th International conference on Applied Physics Of Condensed Matter*, (2014).
6. Råback, P. et al.: Elmer models manual. CSC–IT Center for Science, (2014)
7. Salach, J. Non-destructive Testing of Cylindrical Ferromagnetic and Non-magnetic Materials Using Eddy Current Tomography. In *Mechatronics-Ideas for Industrial Application*, pp. 373-380, (2015).
8. DeTroye, D. J., & Chase, R. J. The Calculation and Measurement of Helmholtz Coil Fields. ARMY RESEARCH LAB ADELPHI MD. (1994).
9. S. Tumanski, "Handbook of Magnetic Measurements", CRC Press, 2011
10. Cacak, R. K., Craig, J. R. Magnetic Field Uniformity around Near-Helmholtz Coil Configurations. *Review of Scientific Instruments*, *40*(11) ,pp. 1468-1470, (1969).
11. Mook, G., Hesse, O., Uchanin, V. Deep penetrating eddy currents and probes. *Materials Testing*, *49*(5),pp. 258-264, (2007).

Numerical simulation of the effect of supplying arteries occlusion on cerebral blood flow

Adam Piechna[1], Marcin Pieniak[1]

[1] Laboratory of Bioflows, Institute of Automatic Control and Robotics, Warsaw University of Technology, Poland
adam.piechna@gmail.com, pieniak@mchtr.pw.edu.pl

Abstract. As the brain has high metabolic demand, nature created many mechanism of regulating the cerebral blood flow. One of them is a circulatory anastomosis called Circle of Willis (CoW). Using home-made simulation program a blood flow through CoW was modeled. Different scenarios of supplying arteries occlusion were modeled and analyzed. It was shown that even sizable reduction of its diameters do not perturb the cerebral flow rates. However the flow organization of the whole CoW is changed. It is exhibiting especially in the communicating arteries, which compensates the influence of diminished flow in the occluded artery.

Keywords: Circle of Willis, cerebral artery, one dimensional model, blood flow simulation.

1 Introduction

Blood is transported to the brain through a pair of Internal Carotid Arteries (ICA) and a pair of Vertebral Arteries (VA). They originate from aortic arch and lead to the base of the brain, where they interconnect and form cerebral arterial circle, also known as the Circle of Willis (CoW). Six branches of the CoW, namely two Anterior Cerebral Arteries (ACA), two Middle Cerebral Arteries (MCA) and two Posterior Cerebral Arteries (PCA) supply blood to the relevant parts of the brain. Although, the brain is only 2% of body weight, it is supplied by 13-15% of blood pumped by the heart [1]. Maintaining of proper cerebral blood flow is a priority for the human body, so a number of protecting mechanisms has been created by nature. The principal security is given by the circle of Willis redundant connection of blood vessels. In the case of occlusions in one of the main arteries supplying blood to the brain, it is possible, to certain extent, to compensate ischemia of the brain via the bypass blood flow. From the diagnostic point of view it is interesting to assess the impact of any occlusions in the supplying arteries (caused mainly by arteriosclerosis) on blood redistribution within the brain. Mathematical and physical models of cerebral circulation may serve that purpose [2-7]. In this paper we used a home-made software based on the method

R. Jabłoński and T. Brezina (eds.), *Advanced Mechatronics Solutions*,
Advances in Intelligent Systems and Computing 393,
DOI 10.1007/978-3-319-23923-1_27

of characteristics to simulate flows in fully patent (reference state) and partially occluded arterial configuration.

2 Material and methodology

To assess the level of brain ischemia resulting from occlusion in supplying arteries the original program was created to solve flow equations based on the well-known method of characteristics [4,7-12].
The program has been described in details in the Piechna's thesis [9], and was based on equations of conformity on the families of characteristics:

$$du_I = -\frac{dp_I}{\rho_0 c} - \frac{\lambda u^2}{2D} dt_I$$
$$du_{II} = +\frac{dp_{II}}{\rho_0 c} - \frac{\lambda u^2}{2D} dt_{II} \qquad (1, 2)$$

where p is a pressure, u is a blood velocity, λ is a friction coefficient, ρ is a density of blood, D is a vessel diameter and c is a pulse wave velocity defined as:

$$c = \sqrt{\frac{K_Z}{\rho_0}} \qquad (3)$$

where K_Z is bulk modulus complying with arterial wall stiffness and blood compressibility. Characteristics equations has the form:

$$\left(\frac{dx}{dt}\right)_I = u + c \qquad (4)$$

$$\left(\frac{dx}{dt}\right)_{II} = u - c \qquad (5)$$

Where one family (*I* index) of characteristics describes the propagation of disturbance downstream and the second (*II* index) upstream.

Model of the CoW was created based on statistical data provided by prof. Bogdan Ciszek from the Department of Descriptive and Clinical Anatomy Medical University of Warsaw. Dimensions of cerebral segments were averaged basing on 50 samples of actual anatomical specimens. Sketch of the Circle of Willis with major branches is shown in fig. 1.
As the Circle of Willis is only a fragment of the cardiovascular system, it is necessary to define the proper inflow and outflow boundary conditions. Each of the main distributing arteries gives rise of arterial tree which gradually turns into microcirculation. Modeling of all these branches as a structure of wires is impossible due to a lack of data on their topography and geometrical dimensions which, like a finger

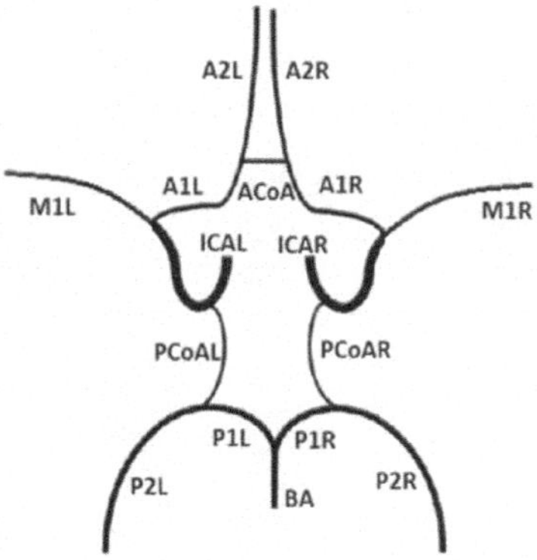

Symbol	Name
A1, A2	Anterior Cerebral Artery
ACoA	Anterior Communicating Artery
ICA	Internal Carotid Artery
PCoA	Posterior Communicating Artery
P1, P2	Posterior Cerebral Artery
BA	Basilar Artery
M1	Middle Cerebral Artery

Fig. 1. Schematic representation of the Circle of Willis (R-right side, L-left side).

print, are an individual feature. Therefore, we replaced peripheral arteries by the integral part reflecting their global resistance. Peripheral resistance can be estimated [6] as inversely proportional to the mass of the supplied brain section.
To properly model the inflow boundary conditions, we have to enlarge the simulated area by including heart with aortic arch. Defining boundary condition by exact value of blood velocity at the inlet of the Basilar Artery (BA) and ICA would prevent simulation of occlusion's impact on cerebral blood flow. Despite resistance increase in one of the arteries supplying blood into the CoW, the flow would remain unchanged. Therefore, the cerebral circulation was supplemented by the rest of the cardiovascular system, which was represented by the global resistance. The whole system was driven by a displacement pump with known flow characteristic, simulating cardiac cycle. A single stroke volume was described by the sum of the first three harmonics [1] having modules equals 225, 65 and 25 [ml / min] successively. The first harmonic has a frequency of 1 Hz.

3 Results

Numerical simulations were carried out using the CoW model. The blood flow in CoW in physiological state and in pathological conditions - with stenosis of arteries - was studied. Flow values in the arteries of A2, M1, P2 for a reference circle were averaged over time and are considered as an optimal for the brain functioning. In the simulations of pathological states these flows were compared and expressed as a percentage of physiological ones. Totally, 21 pathological cases were studied. Obtained results are summarized in Table 1. Whole set of modeled cases was planned in such a way to cover the area of potential relationships between pathologies.

We have presented the impact of the stenosis of supplying arteries on the distribution of blood through the Circle of Willis in the form of radar graphs (fig. 2-3). This presentation will facilitate the analysis of the results. On the fig. 4 we presented occlusion effect on flow in communicating arteries.

Table 1. Summary of obtained results (flow values are time averaged). Bold line denotes reference configuration (no occlusion). Green area denotes flow in the comunicating arteries. Blue area denotes flow distribution in the distributing arteries in % regards to reference state.

Combination of artery occlusion [%]				Flow rate [ml/min]			Flow in distributing arteries [%]					
				AcoA	PCoAL	PCoAR	P2L	P2R	M1L	M1R	A2L	A2R
None	**-**	**None**	**-**	**2.35**	**0.96**	**0.32**	**100**	**100**	**100**	**100**	**100**	**100**
BA	50	None	-	2.38	8.74	4.49	91	90	101	101	101	101
BA	70	None	-	2.46	26.85	24.18	65	65	101	101	101	101
BA	90	None	-	2.44	39.17	35.58	43	42	99	99	99	99
ICAL	50	None	-	47.59	6.14	0.41	101	101	92	101	94	98
ICAL	70	None	-	100.58	14.68	0.54	100	100	80	101	86	94
ICAL	90	None	-	121.04	18.11	0.59	97	98	72	98	80	90
ICAR	50	None	-	41.57	1.02	5.9	101	100	100	92	98	94
ICAR	70	None	-	97.28	1.13	14.01	100	100	100	79	93	85
ICAR	90	None	-	120.82	1.17	17.56	98	98	98	72	89	79
ICAL	50	ICAR	50	3.89	8.80	8.27	101	101	89	89	89	89
ICAL	70	ICAR	70	1.61	32.19	29.83	105	1	58	58	58	58
ICAL	90	ICAR	90	1.65	49.28	45.82	105	105	28	27	28	27
ICAL	90	ICAR	50	92.79	26.40	12.91	99	100	62	81	68	75
ICAL	90	ICAR	70	37.24	40.65	34.04	102	103	42	49	44	47
ICAL	90	ICAR	90	1.65	49.28	45.82	105	105	28	27	28	27
BA	70	PCoAL	70	1.76	0.29	28.16	58	59	102	102	102	102
BA	70	PCoAR	70	3.07	30.74	0.26	60	58	102	102	102	102
ICAL	70	AcoA	70	5.02	30.96	1.26	102	103	57	104	57	104
ICAL	70	PCoAL	70	108.00	0.14	0.38	100	100	77	100	84	93
ICAL	70	AcoA	70	5.02	30.96	1.26	102	103	57	104	57	104
ICAL	70	PCoAL	70	108.00	0.14	0.38	100	100	77	100	84	93

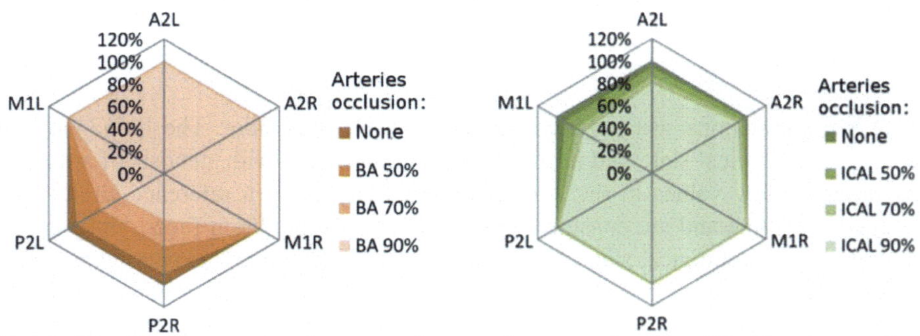

Fig. 2. Distributing arteries flow rate changes due to occlusion of Basilar Artery (left) and left Internal Carotid Artery (right).

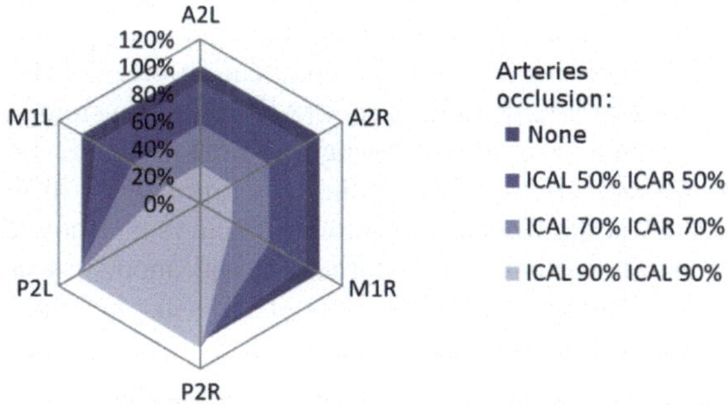

Fig. 3. Distributing arteries flow rate changes due to occlusion of both Internal Carotid Arteries.

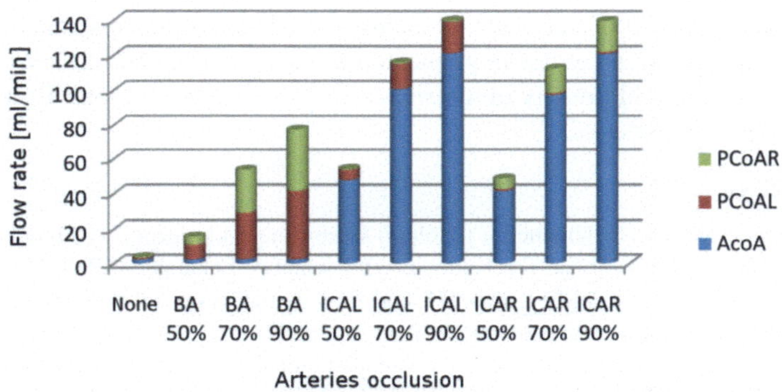

Fig. 4. Flow rates in the communicating arteries caused by different occlusions.

Obtained results will be discussed in the next section.

4 Discussion

Basilar Artery occlusion results in decrease of blood flow in the posterior cerebral arteries (fig. 2). Blood supply to the other brain regions remained unchanged. Interesting fact is that diameter reduction not until greater than 50% results in a significant drop in blood supplies the posterior areas of the brain. This effect is partially compensated by the collateral flow in the communicating arteries. Occlusion in one of the Internal Carotid Arteries causes, compared with the basilar artery occlusion, slightly smaller deficits in brain supply (fig. 3). Middle and front parts of the brain supply was handicapped. Moreover narrowing one of the Internal Carotid Artery induced asymmetric flow in the CoW, which exhibits an increased flow in the Anterior Communicating Artery (fig. 4). Again, reduction greater than just 50% of both internal carotid arteries diameter results in a large impairment of the blood supply to the front and middle parts of the brain. Communicating arteries despite increased flow are not able to compensate deficiencies in the flow in the front of the circle. As the CoW morphometry has high inter-individual variability, and often an asymmetric configuration occurs [13], we have also performed simulations on occluded (at 70% reduction of diameter) communication arteries. Insofar as CoW with single PCoA stenosis still fulfill its role effectively, with reduced diameter of ACoA do not.

As was shown even sizable reduction of diameter do not perturb the cerebral flow rates. However it induces flow in the communicating arteries (fig. 4). Especially interesting is Anterior Communicating Artery (ACoA) as physiologically its blood flow rate should be marginal. Significant change of hemodynamic properties, especially values of shear stresses, could induce pathological processes in the artery wall. As a matter of fact ACoA is the frequent location of aneurysm. Knowledge of the dependency level between hemodynamic disturbances and aneurysm occurrence would be crucial in the medical diagnostic procedures. This is far beyond scope of this paper but was highlighted to show potential relation.

Redundant connections in CoW are not the only one mechanism of providing constant cerebral blood flow. Cerebral autoregulation tends to regulate the blood flow by contraction and dilatation of arterioles due to changes in pressure, CO_2 levels and brain oxygen consumption. It is a complex phenomenon where still is very much to discover. This mechanism was yet not implemented in presented model of CoW. Taking it into account should additionally diminish sensitivity of cerebral blood flow to occlusion of its supplying arteries.

References

1. Fizjologia człowieka z elementami fizjologii stosowanej i klinicznej, pod redakcją Władysława z. Traczyka i Andrzeja Trzebskiego, Wydawnictwa Lekarskie PZWL, 2001
2. Himwich, W.A., Clark, M.E: Cerebral blood flow comparisons between model and prototype. Journal of Applied Physiology 1971, 31, 873–879.
3. Hillen, B., Hoogstraten, H.W., Post, L: A mathematical model of the flow in the circle of Willis. Journal of Biomechanics 1986, 19, 187–194.
4. Cassot F, Zagzoule M, Marc-Vergnes JP: Hemodynamic role of the circle of Willis in stenoses of internal carotid arteries. An analytical solution of a linear model. J Biomech 2000, 33:395-405.
5. Cieslicki K, Ciesla D: Investigations of flow and pressure distributions in physical model of the circle of Willis. J Biomech 2005, 38:2302-2310.
6. Cieslicki K: Experimental and numerical modelling of flow in the human cerebral arteries. Journal of Medical Informatics & Technologies 2004, Vol. 7, IP17-26
7. Alastruey J et al: Modelling the circle of Willis to assess the effects of anatomical variations and occlusions on cerebral flows. J Biomech 2007, 40:1794-1805.
8. Moore SM, Moorhead KT, Chase JG, David T, Fink J: One-dimensional and three-dimensional models of cerebrovascular flow. J Biomech Eng 2005, 127:440-449.
9. Piechna A., Modelowanie rozchodzenia się fali tętna w sieciowej strukturze tętnic zasilających mózgowia. Master Thesis 2008, Warsaw University of Technology.
10. Grinberg L, Cheever E, Anor T, Madsen JR, Karniadakis GE: Modeling blood flow circulation in intracranial arterial networks: a comparative 3D/1D simulation study. Ann Biomed Eng 2011, 39:297-309.
11. Zhang C, Wang L, Li X, Li S, Pu F, Fan Y, Li D: Modeling the circle of Willis to assess the effect of anatomical variations on the development of unilateral internal carotid artery stenosis. Biomed Mater Eng 2014, 24:491-499.
12. Pieniak M. and Cieslicki K: "Properties of Ankle-Brachial-Index (ABI) in the Light of Numerical Simulation of Pulse Wave Propagation," Mechatronics 2013, 809-815 (2014).
13. Cieślicki K: Hydrodynamiczne uwarunkowania krążenia mózgowego, Akademicka Oficyna Wydawnicza EXIT Warszawa 2001

X-ray and neutron radiography studies of archeological objects

Ewelina A. Miśta[1], Jacek J. Milczarek[1], Piotr Tulik[2], and Izabela M. Fijał-Kirejczyk[1]

[1]National Centre for Nuclear Research, Otwock-Świerk, Poland
Ewelina.Mista@ncbj.gov.pl
[2]Institute of Metrology and Biomedical Engineering, Faculty of Mechatronics, Warsaw University of Technology, Warsaw, Poland

Abstract. The aim of the study was application and comparison of neutron (NR) and X-ray radiography (XR) in investigations of archeological and Culture Heritage (CH) objects. The investigation had interdisciplinary character based on physical and archaeological knowledge. The use of two complementary imagining techniques provided data about artefacts structure regarding their composition and shape diversity. The study concerned different objects of metal alloys, corrosion bulk form, ceramics and ancient bone remnants.

Keywords: X-ray radiography, neutron radiography, archaeology

1 Introduction

X-ray and neutron radiography have become important nondestructive tools for studies of archeological and CH objects [1-3] (Deschler-Erb 2004, Lang 1997). X rays have been used for quite long time for imaging of ancient artefacts structure. Replacing X-ray photons with neutrons the NR enables the imaging and determination the elemental composition of the investigated objects [4]. Thermal neutron beams have been used for few decades in autoradiography of paintings as well as for radiography and neutron computed radiography.

In Poland, the neutron techniques are available in the Regional Laboratory of Neutronography (SLN) established at the nuclear research reactor MARIA at the National Centre for Nuclear Research (NCBJ). Collaboration between this Laboratory and Institute of Metrology and Biomedical Engineering of Warsaw University of Technology resulted in work on comparison of neutron and X-ray imaging, provided by standard radiography or computed tomography applied to archaeological objects.

The paper presents the results obtained with both radiographic techniques for variety of archaeological artefacts supplied by few institutions dealing with preservation and conservation of cultural heritage items.

R. Jabłoński and T. Brezina (eds.), *Advanced Mechatronics Solutions*,
Advances in Intelligent Systems and Computing 393,
DOI 10.1007/978-3-319-23923-1_28

2 Methodology

The investigation was based on XR and NR images collected for each object under study. The features concerned not only the object surfaces and were focused on characteristic anomalies revealed by radiography inside the object. For example in case of ceramic vessels it concerned the grains and voids in the wall cross-section, ornament details, surface defects whereas for metal artifacts - their shape and the material composition. The distinction between images of the organic and inorganic parts was studied. For some archeological artifacts only the neutron radiography was applied.

NR images were acquired with a standard dynamic thermal neutron radiography (DNR) station of the Regional Laboratory of Neutronography at NCBJ, Poland. The facility was constructed at the 30 MW nuclear research reactor MARIA [5,6]. The DNR station components include the neutron sensitive AST NDg 6Li:ZnS:Cu,Al,Au screen, converting neutron radiation to visible light, Hamamatsu ORCA-ER camera (1280×1024 pixels, 12 bit dynamic range) and HiPic acquisition software. The neutron beam collimator provided the L/D projection ratio of 165 and the linear resolution was about 153 μm. The white neutron beam of neutrons illuminated the object and the exposition time was 1.6 s. The neutron images were corrected for black current as well for the wide field intensity distribution and white spots were eliminated by appropriate median filter.

XR images were registered at the Institute of Metrology and Biomedical Engineering of Warsaw University of Technology. The X-ray system FLEXAVISION HB package SHIMADZU with Digital Radiography System SDR-100 was used. RTG lamp parameters of focus tube were 0.6 and 1.2 mm; voltage range 40-125 kV, maximum current 630 mA. The automatic exposure control (AEC) system is provided with the CCD camera (1024×1024 pixels). DICOM 3.0 system data recording and the RTG lamp telescopic handle provides a possibility to change focus tube-detector distance in the range of 1.1-1.5 m.

3 Results and discussion

The first objects studied with both techniques were human bones from archaeological site in the Middle East, dated to the 10th century B.C. Radiography analysis of mandible, skull fragment, teeth and long bones of the forearm and metacarpal was performed. The radiographed bone fragments were to some extent corroded due to soil deposition within the crevices. In case of the long bone and skull fragments we found no significant structural differences between images obtained with two techniques. In both NR and XR images clearly outlined bone marrow cavity was revealed. However, in NR images the surface regions with soil contamination as well as resin reconstruction spots were detected (Fig. 1).

The radioscopic images of teeth showed that enamel is less transparent than dentin both for X-rays and neutrons. The fact that dentin absorbed neutrons stronger than X-rays can be attributed to high collagen content in dentin (c. 20%) [7]. One should mention that the collagen main elemental components are nitrogen and carbon, and generally the atoms of light elements absorb neutrons stronger than X-rays [1,8]. Thus, it can be concluded that combined application of both methods permits the detection of preserved collagen.

Burial urn pottery and utility vessels, the clay objects, were studied in order to find some features of ancient technological and ornamentation processes applied in manufacturing processes. The NR yielded some new information about the technology used. The handle of the burial cup (Fig.2) was evidently glued (Fig.3) to the cup body contradicting the wide spread belief that it was attached with a kind of bungs.

Fig. 1. Radioscopic images of the mandible fragment focused on teeth structure.

Fig. 2. A picture and NR-CT reconstruction of outer surface of the burial cup.

There is significant difference between X-ray and neutron images in representation of hydrogen containing regions of the object. Since the X-rays are not absorbed by hydrogen in contrast to strong attenuation of neutron rays passing through the hydrogen rich regions, the grey level of the same regions is different. In effect the regions filled with glue (resin) in the reconstruction of burial vessel are clearly distinguished in neutron radioscopic images from the voids and other air filled crevices (Fig.4).

The heavily corroded metal (iron) objects were also studied before application of usual conservation procedures. About 400 NR images of one of the elongated archeological find (Fig.5), were analyzed with computed tomography treatment and revealed that it is composition of simple sleeve spearheads (Fig.6).

In many cases inner parts of artefacts composed of organic remnants like leather can be visualized with neutron radiography due to their high hydrogen content. For example the results of neutron radioscopic images for an object made of organic (leather belt remnants) and inorganic parts (metal fittings) reveal all components of the artefact (Fig.5). The results of unknown ornamentation procedures with silver left some imprints on the ancient shield grip (Fig.6).

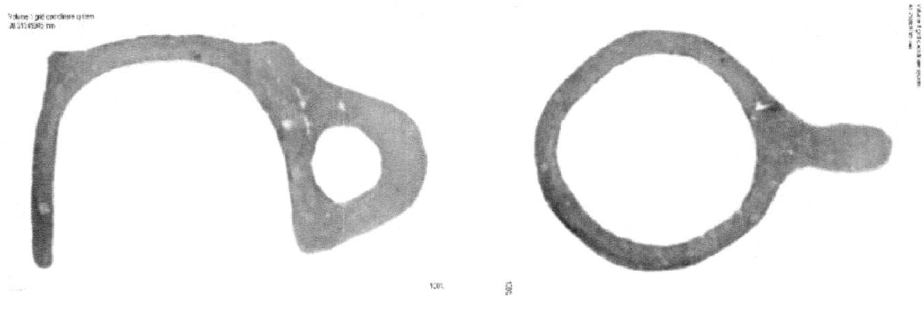

Fig. 3. Cross-sections of the burial cup revealing voids that evidence the gluing of its handle in the manufacturing process.

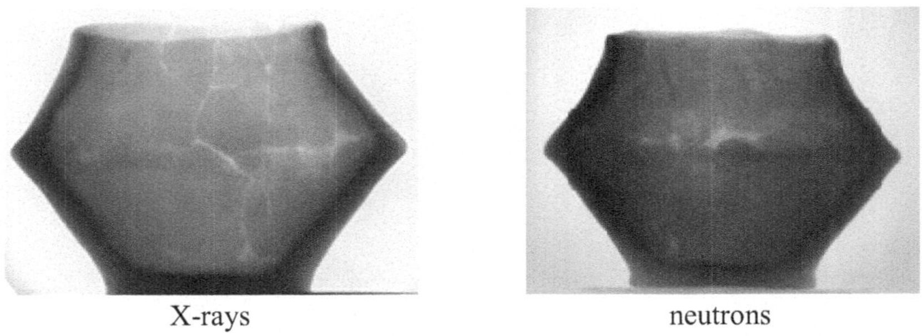

Fig. 4. X-ray and neutron images of the burial vessel. The missing material parts are visible as light regions but X-rays do not distinguish the glue from the empty crevices which are visible as dark lines in neutron image.

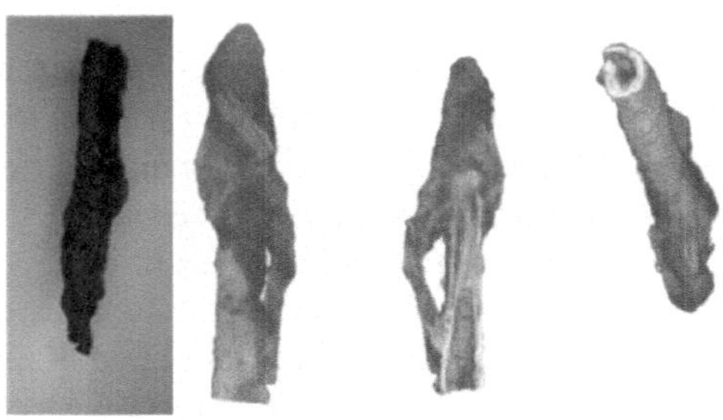

Fig. 5. NR-CT results for outer surface ofan heavily corroded aggregate of iron alloy components (photo image on the left hand side).

Fig. 6. NR-CT cross-sectionsshowing the arrangement of spear heads inside the heavily corroded aggregate presented in Fig.5.

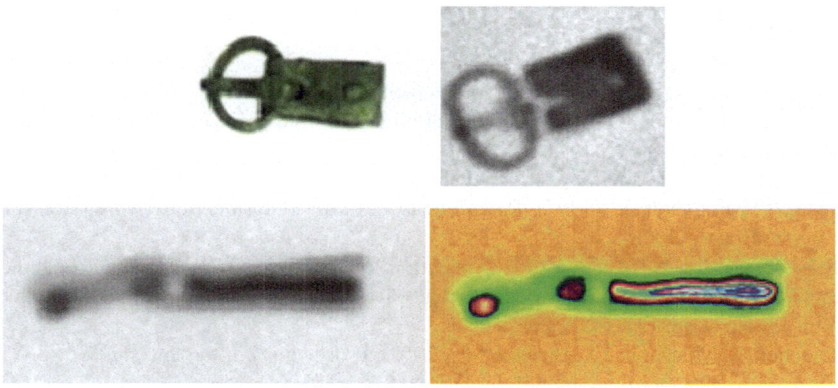

Fig. 7. The small ancient bronze buckle with preserved organic belt elements between metal fittings. Upper row: photo image and NR result, on the left and right side, respectively; lower row: NR image (the image on the right is in false colors) [9].

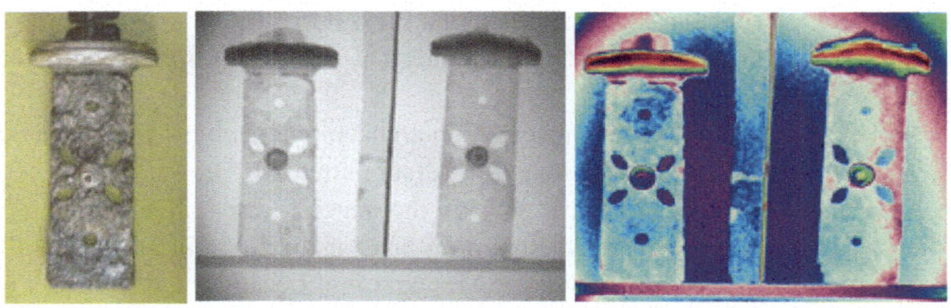

Fig. 8. Ancient shield grip with silvered mounting plates. In NR images there are visible thickness differences due the technology process used in the preservation process [10,11]

4 Conclusions

The radiography imagining is useful in archeology due to its non-destructive nature. The use of X-ray and neutron radiography techniques gave useful information about the structural features of the archeological objects. Due to different interaction of both kind of radiation with matter, the techniques are complementary in estimating the degree of object degradation as well as revealing the shape of components of heavily corroded aggregations of different item. The revealed features can help in studies of ancient technological processes.

Acknowledgements

We would like to thank archaeologists and conservators from Warsaw National Archaeological Museum and Archaeological Institute of Warsaw University for their cooperation and help in supplying us with objects for investigation.

References

1. Deschler-Erb E. et al.: (2004), The complementary use of neutron and X-rays for the non-destructive investigation of archaeological objects from Swiss Collections, Archaeometry 46/4, pp. 647-661.
2. http://penn.museum/sites/applied_science/radiography.html from 01/05/2015.
3. Lang, J., Middleton A., (1997), *Radiology of Cultural Material*, Butterworth-Heinemann, Linacre House, Oxford
4. Stanojev Pereira A. M. et al. (2013), Neutron Imagining techniques applied to studies in archaeolocical and Cultural Heritage fields, Mediterrean Archaeology and Archaeometry, Vol. 13, No 3, pp. 137-143
5. Czachor A., Milczarek J.J., Balasko M., (2002), Proceedings of the Seventh World Conference on Neutron Radiography, Rome, 15–20 September.
6. Czachor A., Abd El-Ghany El Abd, Milczarek J.J., (2002), Acta Phys. Pol. A: 102, pp. 245.
7. Hillson S. (2005), Teeth, IInd ed. NY, University Press.
8. Hutchings M.T., Windsor C. G. (1987), Industrial applications in Neutron Scattering, v.23 of Methods in Experimental Physics, Academic Press, Orlando
9. Miśta A. E. (2015), Study on silvering technology of shield grip from Roman Period, Archaeometry (submitted 2015)
10. Miśta A.E. et al. (2015), Materials research on archaeological object using PIXE and other non-invasive techniques, Acta. Physica. Pol. A (submitted 2015).
11. Miśta A.E., Kalbarczyk P., (2014), Report on the material research concerning cooper alloys item from cemetery in Łężany, Report on analysis of silver items from the cemetery in Łężany, pp. 131-154 in Wiśniewska A Monography: Łężany – the cemetery of the Roman Period and the Migration Period in the Mrągowo Lake District, Research in the season 2013), Warsaw.

Part III
Mechatronics Products: Design, Manufacturing and Testing

Finite Element Analysis of High-Speed Solid Rotor Induction Machine with Copper Cage

Jan Bárta[1], Čestmír Ondrůšek[1]

[1] Brno University of Technology, Faculty of Electrical Engineering and Communication, Technicka 3058/10, 616 69, Brno, Czech Republic
xbarta27@stud.feec.vutbr.cz, ondrusek@feec.vutbr.cz

Abstract. High-speed electric machines are design with respect to mechanical, electromagnetic and thermal requirements. For successful thermal and mechanical design relevant input data from electromagnetic design stage are required. One of the most important data are related to power loss distribution inside the electric machine. That is due to the fact that power losses represent requirement on cooling method and mechanical design. This paper is dealing with electromagnetic calculation of high-speed solid induction machine with copper cage. For this purpose electromagnetic model of this machine has been developed and analyzed. All calculation are performed by using of time-step, 2D finite element analysis. All obtained results are presented and discussed.

Keywords: High-speed solid rotor induction machine with copper cage, eddy current, eddy current losses, finite element analysis

1 Introduction

With modern trends and technologies among high-speed electric machine technology is dealing D. Gerada [1]. In these days there are used mainly four types of high-speed induction machines (Fig. 1). Comparative study on high speed induction machine with different rotor structure was publish by Hao Zhou [2]. This paper is dealing with high-speed solid rotor induction machine with copper cage. Copper bars are inserted in drilled bores. A copper cage produces lower resistance than solid rotor, and hence, a lower slip. On the other hand using a copper cage is mechanically demanding.

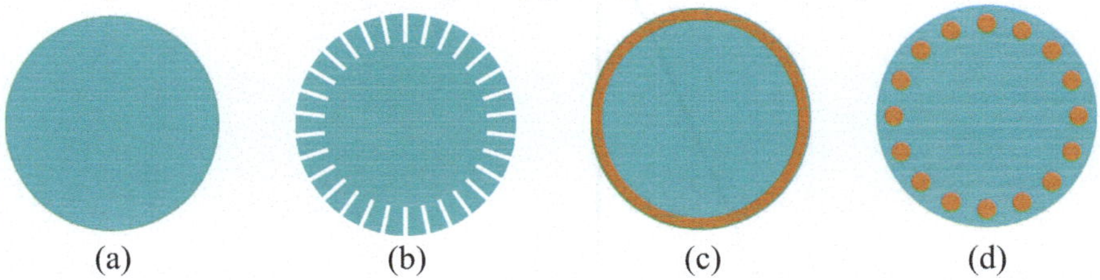

Fig. 1. Topologies of high-speed solid-rotors a) Smooth solid rotor b) Solid rotor with axial slits c) Coated rotor d) Caged solid rotor

R. Jabłoński and T. Brezina (eds.), *Advanced Mechatronics Solutions*,
Advances in Intelligent Systems and Computing 393,
DOI 10.1007/978-3-319-23923-1_29

2 Electromagnetic model description

Analyzed machine parameters are listed in table. 1. Presented high-speed solid rotor induction machine with copper cage have 12 stator and 18 rotor slots. Whole machine design has been made with respect to mechanical and thermal requirements.

Electromagnetic model has been created in Ansys Maxwell according to machine documentation. All calculations are based on time-stepping, 2D finite element analysis [3]. End winding of stator and rotor rings impendances are modeled through external circuits. Only one pole pitch has been modeled and discretized. Periodic boundary have been used at radial boundaries of the problem region. Pole pitch of modeled machine contains total number of 36890 mesh elements. Four layers of finite elements are used in air-gap. Rotor layers element dimensions are smaller than the skin depth associated with the stator slot harmonics and the material characteristics. Used material properties have been measured for higher accuracy of calculated results [4]. With estimation of core losses, field solution is subsequently modified to take core losses effect on field into account. Losses of the machine are estimated in the post-processing phase.

Rotor of the modeled machine has been split out into 7 layers Fig. 2. Each of this layer is 0.1mm depth and layer No. 7 only 0.05 mm depth. These layers allowing more detailed analysis of eddy current losses in rotor surface area.

Table 1 Simulated motor parameters

Parameter name	Unit	Value
Rated power	kW	6
Rated speed	rpm	120 000
Voltage	V	350
Frequency	Hz	2015.488
Number of poles	-	2
Method of cooling	-	water

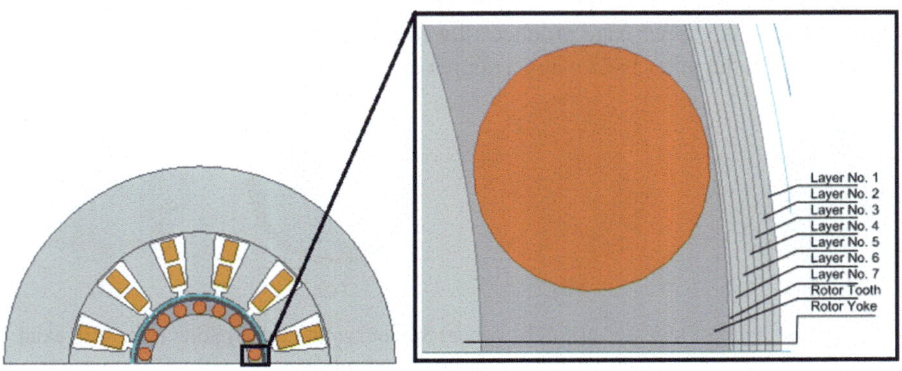

Fig. 2. Electromagnetic model of high-speed solid rotor induction machine with copper cage

3 Calculation results

Proposed electromagnetic model has been used for finite element analysis. Calculated field distributions for rated load conditions are shown in Fig. 3. Penetration depth for higher harmonics is increased thanks to saturation of rotor core. Losses distribution are shown in Fig. 3b,c. Calculated loss distribution can be used for accurate thermal and mechanical machine design.

Results regarding to machine performance are listed in table 2. Power balance represents percentage error between average input and output powers. It is derived by following equation

$$P.B. = \frac{P_{in} - P_{out} - P_{Eloss}}{P_{in}} \cdot 100 \quad (1)$$

where P_{in} is input power, P_{out} is shaft power, P_{Eloss} is sum of all electric losses. Main factors that are having influence on percentage finite element analysis error are following: time step size, steady-state condition, nonlinear residual.

(a)

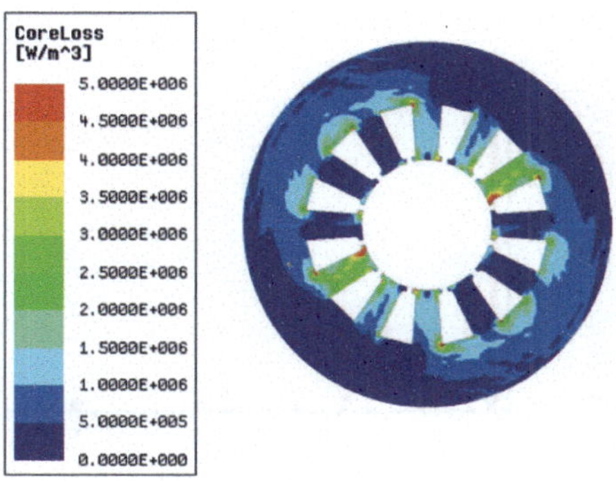

(b)

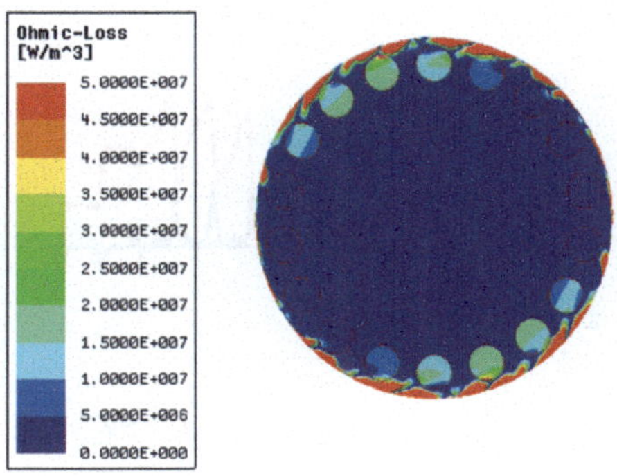

(c)

Fig. 3. a) Flux density distribution in cross-section area b) Core loss distribution in cross-section area c) Eddy current loss distribution in cross-section area

Electric losses P_{Eloss} generated inside of simulated machine are plotted in Fig. 4. It can be seen that dominant component of the losses is located in the rotor. Furthermore it is evident that eddy current losses are decreasing with distance from outer rotor surface.

Fig. 5a shows flux density distribution along the air-gap. The fundamental is the dominant component. Its value is 0.46 T. The next ones in magnitude are the first and second slot harmonic components of the stator and rotor Fig. 5b.

Table 2 Calculated machine performance

Parameter name	Unit	Value
Input power	kW	7.29
Shaft power	kW	6.55
Electromagnetic efficiency	%	90.7
Power factor	-	0.52
Current	A	22
Power balance	%	0.99

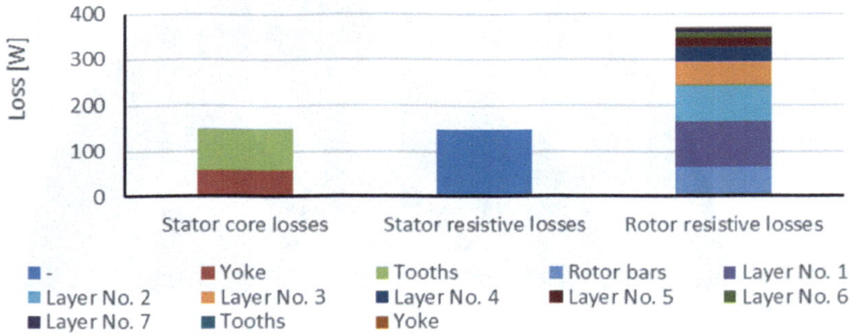

Fig. 4. Distribution of electric losses P_{Eloss}

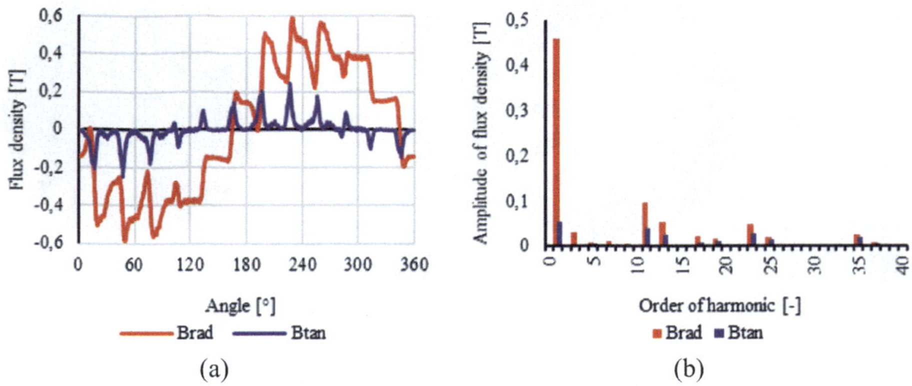

Fig. 5. a) Flux density distribution along air-gap b) Amplitudes of air-gap flux-density harmonics

Fig. 6a shows current time dependency for different rotor layers. It can be seen that highest amplitude of current ripple is closest to the rotor surface. Current is derived by using following equation

$$I = \int_S \vec{J} dS \tag{2}$$

where J is the vector of current density and S is the layer surface. Current spectrum time dependency for different layers is plotted in Fig. 6b. Fundamental amplitude is not included since length of calculated data is not sufficiently long. Dominant are the first and the second slot harmonic components of the stator and rotor. Also sixth harmonic component is significantly included.

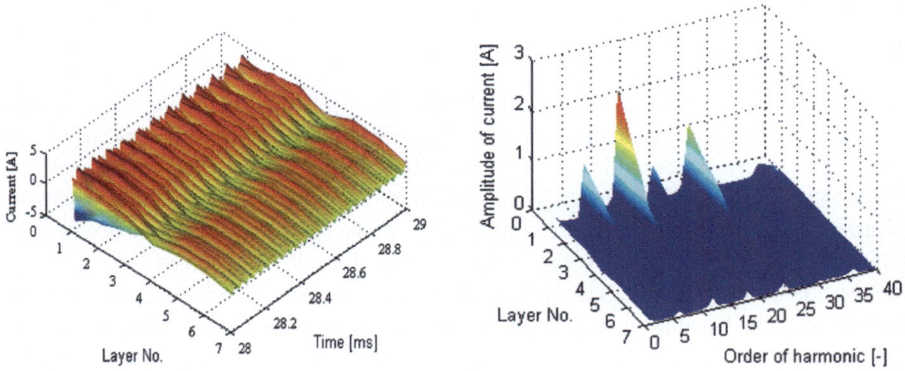

(a) (b)

Fig. 6. a) Current vs. time dependency in different layers b) Amplitudes of current ripple harmonics in different layers

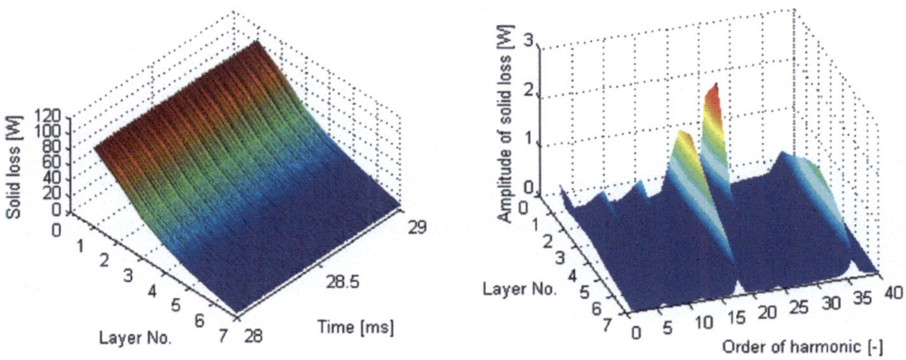

(a) (b)

Fig. 7. a) Eddy current loss vs. time dependency in different layers b) Amplitudes of eddy current losses harmonics in different layers

Fig. 7a shows eddy current losses time dependency for different layers in solid rotor core. It can be seen that eddy current losses are exponentially decreasing with distance from the outer rotor surface. Losses can be considered negligibly under the seventh rotor layer. Eddy current losses are calculated by using following equation

$$P = \int_V \vec{E} \cdot \vec{J} dV = \int_V \frac{\vec{J}}{\sigma} \cdot \vec{J} dV \quad (3)$$

where E is vector of voltage intensity, J is vector of current density and σ is conductivity of rotor material. Eddy current rotor losses spectrum for different layers is plotted in Fig. 7b. Zero order harmonic component value is not included in Fig. 7b. From higher harmonic components are dominant the first and the second slot harmonic components of the stator and rotor.

4 Conclusions

Calculated machine performance and field distribution has been shown. Rotor surface is saturated during rated load condition. Stator core and resistive losses in a stator winding are almost about the same value. Main part of stator core losses arises in stator tooth. Rotor eddy current losses in comparison with other electric losses are dominant. Eddy current losses are concentrated under the rotor surface. Current penetration depth is given by rotor material characteristics. It has been proven that eddy current losses are mainly associated with stator and rotor slot harmonics. Properly calculated loss distribution represents important premise for successful mechanical and thermal design of electric machine. Losses can also be derived by analytical methods. However, due to variable permeability accurate loss derivation is difficult. Compared to that, finite element analysis is relatively fast and relatively accurate method how to derive machine performance. But still for finite element analysis it is necessary to carefully setup calculation and external circuit impendance. Also accurate material data are very important.

Acknowledgments: This research work has been carried out in the Centre for Research and Utilization of Renewable Energy (CVVOZE). Authors gratefully acknowledge financial support from the Ministry of Education, Youth and Sports of the Czech Republic under NPU I programme (project No. LO1210)

References

1. Gerada D. et. al. “High-Speed Electrical Machines: Technologies, Trends, and Developments” IEEE Transaction on Industrial Electronics, Vol. 61, No. 6, June 2014
2. Zhou H., Wang F. “Comparative Study on High speed Induction Machine with Different Rotor Structures” Proceeding of International Conference on Electrical Machines and Systems 2007, Oct. 8-11, Seoul, Korea
3. Arkkio A. “Analysis of Induction Motors Based on Numerical Solution of the Magnetic Field and Circuit Equations“ Acta Polytechnica Scandinavica, Electrical Engineering Series, no. 59, Helsinki 1987
4. Bulin T. et al. “Magnetic and transport properties of 41CrMo4 steel” Proceedings of Abstracts 24th International Conference on Metallurgy and Materials., Brno, Czech Republic,

Initial assessment of multi-mass absorber influence on machine tool vibrations

Tomas Brezina[1], Lukas Brezina[1], Jan Vetiska[1] and Jiri Marek[2]

[1] Brno University of Technology, Faculty Mechanical Engineering, Brno, Czech Republic
{brezina,brezina.l,vetiska}@fme.vutbr.cz
[2] KSK Precise motion, Kurim, Czech Republic
jiri.marek@ks-kurim.cz

Abstract. The paper presents approach to suppress machine tool vibrations which occur during machining process. The approach is based on multi-body model of a machine tool with flexible bodies in linear state-space form with added structure representing the absorber. Such a structure of the model significantly decreases the computational time and makes possible to optimize parameters of the absorber efficiently. There were analyzed two arrangements of multi-mass absorbers – parallel and serial with two types of optimization objective function. The first one that leads to a reduction of the maximum resonant amplitude and the second one reduces the maximum real part of the eigenvalues of the machine.

Keywords: flexible bodies, absorber, machine tool, optimization.

1 Introduction

Finding the weakest points of a machine tool in terms of its dynamic characteristics and assessing the possibilities of their effective compensation plays an important role in the design process.

This is attractive as a design assessment already in pre-production stages. It can mean considerable savings in production costs and improving the useful properties of the final product. The same approach can be applied to a proposal concerning compensations on already finished machine.

The approach has also high usability in the mechatronic design methodology which suggests making the original model more precise by repeated passages of the V-cycle. All of mentioned cases are based on simulation modeling.

The best achievable behavior and the corresponding parameters of the model can be found by optimization methods if following requirements are satisfied:

a) the required behavior of the model is obtained by a simulation fast enough,
b) the model is capable of change of investigated behavior depending on parameters change,
c) there is a suitable objective function of parameters which improves the desired behavior provided by the model the more, the lower value function has.

R. Jabłoński and T. Brezina (eds.), *Advanced Mechatronics Solutions*,
Advances in Intelligent Systems and Computing 393,
DOI 10.1007/978-3-319-23923-1_30

Compensations of the weakest spots of the machine tool can be then formulated as adding a parametric structure which produces locally compensating dynamics and finding the values of its parameters leading to the highest attainable improvements of observed behavior of the machine. The pioneering work in this field was presented in [1] but the tuning methodology for absorber presented in the mentioned article is no longer optimal [2].

In this paper, there are compared solutions found in mentioned way for the serial and parallel arrangement of a multi-mass absorber mounted in the X-Y plane in the center of gravity of the tool holder of machine tool by TOSHULIN, a. s. production. The other possibilities of an absorber attachment are discussed for example in [3], [4] or in [2] and possibilities of using more than one absorber are presented in [5].

There were used two objective functions in the paper - one that leads to a reduction of the maximum resonant amplitude (RAM) and second that reduces the maximum real part of the eigenvalues of the machine (EIR).

2 Model and its structure

2.1 Basic model

The basic model is created according to the simplified CAD model in multi-body simulation software (MBS). The model contains rigid as well as flexible bodies which are imported to MBS from FEM software. All of parts of the model were modeled as flexible except tool holder.

The complete basic model (Fig. 1) is consequently exported from MBS software as a state-space linear time invariant model (LTI), which describes the approximate dynamics of the system with flexible bodies. It is necessary to perform a reduction of the model due to its high order [6].

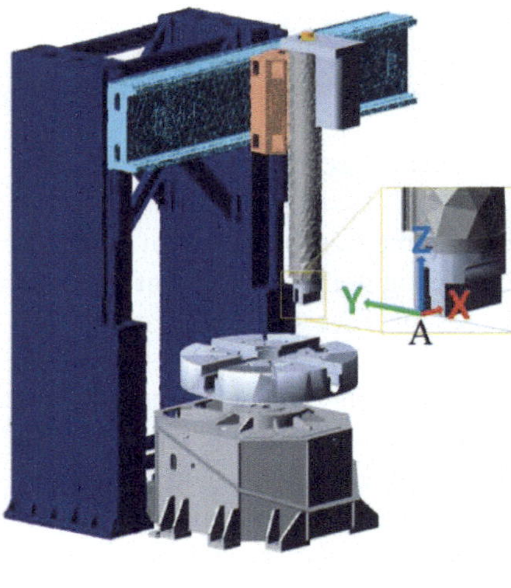

Fig. 1. MBS basic model (machine tool by TOSHULIN, a. s.)

The basic model $\boldsymbol{\alpha}$ (order $n=440$) was exported from MBS as A, B, C, D matrices of the state-space LTI for inputs $\mathbf{u}=[Q_X\ Q_Y\ Q_Z]^T=\mathbf{Q}$, i.e. outer exciting forces acting at *X, Y, Z* of the point *A* and outputs $\mathbf{y}=[\mathbf{q}^T\ \mathbf{q}'^T\ \mathbf{q}''^T]^T$, containing gradually vector of displacements, velocities and accelerations of the point *A* in *X, Y, Z*. Its L-image can be written as

$$\mathbf{y}(s)=\boldsymbol{\alpha}(s)\mathbf{u}(s) \tag{1}$$

and its frequency characteristics of dynamic compliance between axis k and l at the point A can be written as

$$\alpha_{k,l}(j\omega)=\frac{y_k(j\omega)}{u_l(j\omega)}. \tag{2}$$

2.2 Added model

Influence of the configurations of the multi-mass absorbers (r-mass) on the dynamics of the machine tool was investigated. There were considered serial and parallel configurations.

There was used for both of configurations m_i^A as a mass parameter, ideal damping b_i^A and stiffness k_i^A (arranged to the vector of parameters $\mathbf{p}$) and kinematic quantities q_i^A, $q_i'^A$, $q_i''^A$, $i=1,\dots,r$.

a) Serial configuration of the absorber

The added model of the absorber is presented by r masses connected serially via stiffness and ideal damping. Its equations of motion can be written as a system of differential equations

$$m_i^A q_i''^A + b_i^A\left(q_i'^A - q_{i-1}'^A\right) + k_i^A\left(q_i^A - q_{i-1}^A\right) - b_{i+1}^A\left(q_{i+1}'^A - q_i'^A\right) - k_{i+1}^A\left(q_{i+1}^A - q_i^A\right) = Q_i^A,$$
$$i=1,\dots,r \tag{3}$$

for $q_0^A = q_0'^A = q_{p+1}^A = q_{p+1}'^A = k_{p+1}^A = b_{p+1}^A = 0$.

b) Parallel configuration of the absorber

The added model is also presented by r masses. Each of them has own stiffness and ideal damping and oscillates in the same axis. The system of equations of motion is presented by equations of motion for independent single mass absorbers

$$m_i^A q_i''^A + b_i^A q_i'^A + k_i^A q_i^A = Q_i^A,\ i=1,\dots,r. \tag{4}$$

Both of cases of equations describing dynamic compliance constitute L-image of the state-space model $\boldsymbol{\alpha}^A(s,\mathbf{p})$ with known dependency on parameters $\mathbf{p}$ (white box).

2.3 Interface

Interface describes interaction between the basic and added model. It also describes spherical angles presenting the rotation of the coordinate system of the added model with respect to the coordinate system of the basic model (Fig. 2) as $\mathbf{h}=[\cos\theta\cos\varphi \quad \cos\theta\sin\varphi \quad -\sin\theta]^T$. It can also include dead mass of the absorber m_D^A. The model of interface is for $i=1,\dots,r$ defined by $Q_i^A = m_i^A(\mathbf{q}''\mathbf{h})^T$ and $\mathbf{Q}=-\left(m_D^A+\sum_{i=1}^{r} m_i^A\right)\mathbf{q}''+\mathbf{h}\sum_{i=1}^{r} m_i^A q_i''^A$.

High order model of dynamic compliance of the machine tool with multi-mass absorber $\boldsymbol{\alpha}^C(s,\mathbf{p})$ is obtained by the interconnection of the basic and added model via the model of interface and by including of the dead mass m_D^A and spherical angles θ, φ into vector of parameters $\mathbf{p}$. Its dependency on parameters $\mathbf{p}$ still remains known. It is possible to use the model for obtaining the frequency transfers by using of (1), (2).

For the further information on the building of the model suitable for an optimization see [7].

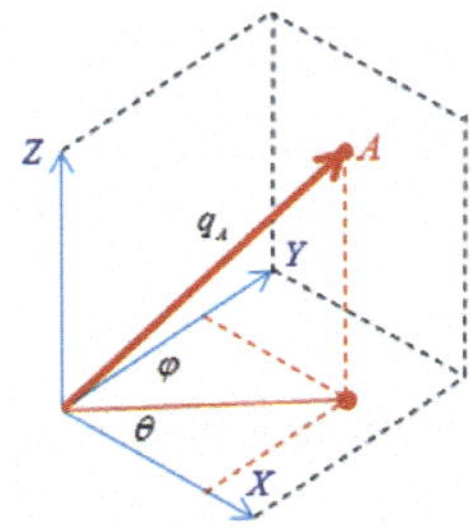

Fig. 2. Spherical angles

3 Optimization

Search for optimal solution was realized by usual way as constrained minimization process of scalar objective function $g(\mathbf{p})$, i.e. $\mathbf{p}_{opt} = \arg\min g(\mathbf{p})$ with such a constraint of the parameters a) $2.5 \le k_i^A \le 30\,\text{N/mm}$ and $3.1 \le b_i^A \le 100\,\text{Ns/mm}$, b) no dead mass was considered $m_D^A = 0\,\text{kg}$, c) $\sum_{i=1}^{r} m_i^A \le 50\,\text{kg}$ and finally d) absorber is expected to be mounted in *x-y* plane ($\varphi = 0^\circ$). The original maximum real part of the eigenvalue of the machine is -3.4974.

There were used two objective functions. The highest amplitude of the dynamic compliance of the machine tool with the multi-mass absorber $\left|\boldsymbol{\alpha}^C(s,\mathbf{p})\right|$ in the investigated frequency range $1*2\pi \le \omega \le 1000*2\pi$ rad/s normed to the highest amplitude of the dynamic compliance of the machine tool without any absorber $\left|\boldsymbol{\alpha}(s,\mathbf{p})\right|$ (RAM)

$$g(\mathbf{p}) = \frac{\max\limits_{k=X,Y}\left(\max\limits_{\omega}\left|\alpha_{k.k}^C(j\omega,\mathbf{p})\right|\right)}{\max\limits_{k=X,Y}\left(\max\limits_{\omega}\left|\alpha_{k.k}(j\omega)\right|\right)} \tag{5}$$

and the highest real part of the eigenvalue from the eigenvalues of the dynamic compliance with absorber $\left|\text{re}\lambda_i\left(\boldsymbol{\alpha}^C(s,\mathbf{p})\right)\right|$ normed again to the same thing without any absorber $\left|\text{re}\lambda_i\left(\boldsymbol{\alpha}(s)\right)\right|$ (EIR)

$$g(\mathbf{p}) = \frac{\max\limits_{i}\left|\text{re}\lambda_i\left(\boldsymbol{\alpha}^C(s,\mathbf{p})\right)\right|}{\max\limits_{i}\left|\text{re}\lambda_i\left(\boldsymbol{\alpha}(s)\right)\right|}. \tag{6}$$

4 Results

The results were analyzed for a) parallel absorbers $r = 1,..,3$ and b) one and two in parallel connected serial absorbers with $r = 2$. There was no significant difference between using one or three absorbers in the parallel configuration or one or two absorbers for the serial configuration. Additional absorbers were attached to the same point of the basic body as the first absorber in both of cases. The presented results describe behavior only for single parallel and single serial absorber from the reasons mentioned above for RAM (Fig. 3.) and EIR (Fig. 4.) optimization.

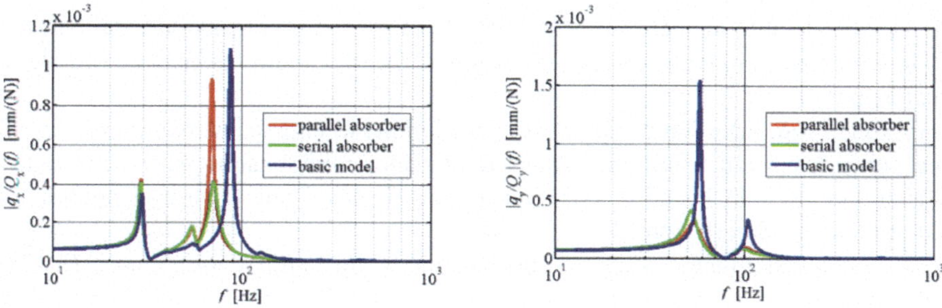

Fig. 3. Amplitude characteristics for best parallel and serial absorber – RAM optimization; transfer q_x/Q_x (left), transfer q_y/Q_y (right)

Values of the parameters obtained by the RAM optimization are following for the a) parallel absorber configuration $b_1^A = 12.82 N.s/mm$, $k_1^A = 30 N/mm$, $m_1^A = 49.20 kg$, $\theta = 90°$ and maximum real part of the eigenvalue is -2.36, b) serial absorber configuration $b_1^A = 17.40 N.s/mm$, $b_2^A = 100 N.s/mm$, $k_1^A = 30 N/mm$, $k_2^A = 30 N/mm$, $m_1^A = 24.70 kg$, $m_2^A = 24.60 kg$, $\theta = 67.65°$ and maximum real part of the eigenvalue is -0.30.

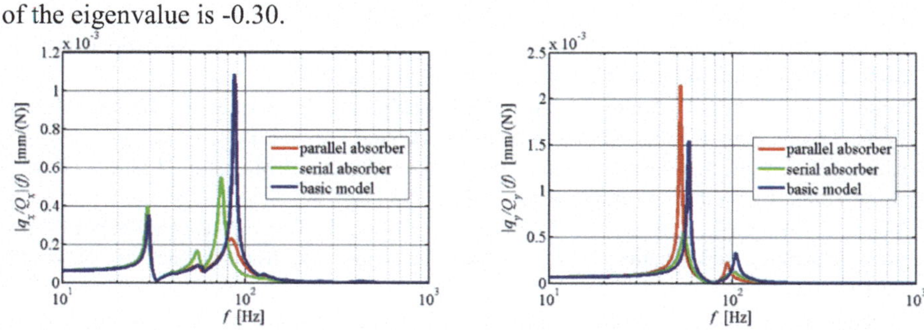

Fig. 4. Amplitude characteristics for best parallel and serial absorber – EIR optimization; transfer q_x/Q_x (left), transfer q_y/Q_y (right)

Values of the parameters obtained by the EIR optimization are following for the a) parallel absorber configuration $b_1^A = 7.66 N.s/mm$, $k_1^A = 30 N/mm$, $m_1^A = 35.30 kg$,

$\theta = 0°$ and maximum real part of the eigenvalue is -3.99, b) serial absorber configuration $b_1^A = 8.43 N.s/mm$, $b_2^A = 8.53 N.s/mm$, $k_1^A = 30 N/mm$, $k_2^A = 30 N/mm$, $m_1^A = 25 kg$, $m_2^A = 15.30 kg$, $\theta = 67°$ and maximum real part of the eigenvalue is -3.53.

5 Conclusions

a) Compared to a single absorber, a configuration of two or three absorbers acting at the same direction does not bring significant improvement in the behavior either for parallel or for serial absorber configuration.

b) A considerable increase in the magnitude of maximum eigenvalue real part occurs in the case of simultaneous minimization of the maximum resonant amplitude in X and Y axis.

c) Explicit minimization of the maximum (dominant) eigenvalue real part does not bring better results in amplitudes than simultaneous minimization of the maximum resonant amplitude in X and Y axis. But it leads on small reduction in magnitude of maximum eigenvalue real part.

Let's note that higher values of ideal damping for serial configuration (RAM) would have to be implemented as an active damping. Presented results are reached exactly for studied machine tool. It may differ for other machine.

Acknowledgement

This work is supported by project NETME Centre PLUS. The results of this project NETME CENTRE PLUS (LO1202) were co-funded by the Ministry of Education, Youth and Sports within the support programme „National Sustainability Programme I".

References

1. Den Hartog, J.P.: Mechanical Vibrations. McGraw-Hill, New York (1956)
2. Sims, N. D.: Vibration Absorbers for Chatter Suppression: A New Analytical Tuning Methodology. J. of Sound and Vibration. 301, 592–607 (2007)
3. Leigh, E.P., Tlusty, J., Smith, S.: Advanced Machining Techniques on Titanium Rotor Parts. Article, American Helicopter Society 56th Annual Forum (2000)
4. Chung, B., Smith, S., Tlusty J.: Active Damping of Structural Modes in High Speed Machine Tools. J. of Vibration and Control. 3, 279–295 (1997)
5. Yang, Y., Muñoa J., Altintas Y.: Optimization of Multiple Tuned Mass Dampers to Suppress Machine Tool Chatter. Int. J. of Machine Tools and Manufacture. 50, 834–842 (2010)
6. Brezina, T., Hadas, Z., Vetiska J.: Simulation Behavior of Machine Tool on the Base of Structural Analysis in Multi-Body System. In: Proc. 15th Int. Conf. on Mechatronics - Mechatronika 2012, pp. 347-350. Czech Technical University, Prague (2012)
7. Brezina, T., Brezina, L., Stetina, J., Marek J.: Design of Optimal Parameter Values of Mechatronic System with Flexible Bodies Using a Block Model. In: Proc. 16th Int. Conf. on Mechatronics - Mechatronika 2014, pp. 301-307. Brno University of Technology, Brno (2014)

Dynamic Simulation of Progressive Crank Train

Lubomír Drápal[1], Pavel Novotný[1], Václav Píštěk[1]

[1] Brno University of Technology, Institute of Automotive Engineering, Brno, Czech Republic
{drapal,novotny.pa,pistek.v}@fme.vutbr.cz

Abstract. This paper presents computational approaches for verification of conceptual study of low-friction-losses crank train design and its vibration analysis. The decrease of friction losses is realized by reduction of crankshaft main bearings number while the influence of a crankshaft pulley torsional vibration and main bearings load are investigated.
The computational model is assembled as well as numerically solved in a Multi-Body System whereas the modally reduced bodies are incorporated into it and hydrodynamic problem is also taken into account.

Keywords: crank train · friction losses · main bearings · Multi-body System · torsional vibration

1 Introduction

A measure of engine design excellence is described by the overall engine efficiency. This parameter is defined as a product of several partial efficiencies. One of them is mechanical efficiency which is influenced by ventilation losses, friction losses and auxiliaries power requirement which are altogether known as mechanical losses. Friction losses originate in the contact of parts where relative movement occurs, e.g. journal bearings, piston assemblies and cylinder liners, cam-tappet contacts, etc.

Contemporary research shows relatively big potential for engine fuel saving by means of friction losses reduction. This potential can be exploited by changes in material, geometry, surface treatment, and design modifications of appropriate engine parts. One of the design modification methods rests in reduction of the crankshaft main bearings number because the share of crankshaft main bearings is considerable in engine friction losses.

In case of an in-line four-cylinder internal combustion engine, the changes are realized by engine block and crankshaft modification in order to decrease the main bearing number from 5 to 3.

The above mentioned modifications are implemented at the new, naturally aspirated spark-ignition engine having 1.6 l displacement. By using state-of-the-art computational methods the influence over crank train dynamics and power losses is evaluated.

R. Jabłoński and T. Brezina (eds.), *Advanced Mechatronics Solutions*,
Advances in Intelligent Systems and Computing 393,
DOI 10.1007/978-3-319-23923-1_31

2 Crankshaft design

The new 3-main-bearing crankshaft is based on a turbocharged version of the standard engine. Missing main pins are replaced by sheet-metal webs due to mass and inertia moments reduction. A connection between the web and the crankshaft is realized by laser welding for low thermal load of the weld.

3 Crank train dynamics simulation

In order to investigate a dynamic response of excited crank train, the state-of-the-art computational approaches are used.

A complex computational model of an engine (i.e. virtual engine) is solved in the time domain. This enables different physical problems, including various non-linearities to be incorporated. The virtual engine is assembled as well as numerically solved in MBS (Multi-Body System) ADAMS. ADAMS is a general code and enables integration of user-defined models to be made directly using ADAMS commands or user-written FORTRAN or C++ subroutines. [1]

In general, the virtual engine includes all significant components necessary for dynamics analyses. The included module is a crank train, a valvetrain, a timing drive, and a rubber damper. Following analyses just deal with the crank train as a main module of the virtual engine.

The crank train module consists of solid model bodies, linearly elastic model bodies and constraints between them.

Solid model bodies are piston assembly, connecting rod assembly, and dynamometer rotor.

The linearly elastic model bodies are modally reduced Finite Element (FE) models suitable for dynamic simulation. These are crankshaft, crankshaft pulley, flywheel, engine block, cylinder head, crank train sump, and gear case.

A dumb-bell shaft connecting a flywheel with a dynamometer rotor is represented by a body with defined torsional stiffness and damping. These characteristics are adjusted on account of torsional vibration measurement.

The interaction between the crankshaft and the engine block is ensured via a non-linear hydrodynamic journal bearing model, where pre-calculated force databases obtained when solving separate hydrodynamic problem are used.

Virtual engine is excited by means of cylinder pressure, defined by high-pressure measurement, and via inertial forces from moving parts. Simulation starts from 1000 rpm and is carried out to 6000 rpm.

4 Results of Multi-body dynamics solution

4.1 Main bearing friction losses

In general, friction losses solution of relatively low loaded powertrain slide bearings such as main and crank pin bearings of a naturally aspirated engine, camshaft bearings and eventually balancing shaft bearings can neglect the elastic deformations. The solution is based on the Reynolds equation:

$$\frac{\partial}{\partial x}\left(\frac{\rho h^3}{12\eta}\frac{\partial p}{\partial x}\right)+\frac{\partial}{\partial y}\left(\frac{\rho h^3}{12\eta}\frac{\partial p}{\partial y}\right)-U\frac{\partial(\rho h)}{\partial x}-\frac{\partial(\rho h)}{\partial t}=0, \tag{1}$$

where x and y are coordinates, t is time, h is oil film thickness, ρ is oil density, η is dynamic oil viscosity and U is relative velocity.

The dependency of oil viscosity on pressure can be expressed with the use of the Roelands relation.

The solution of a hydrodynamic problem requires reaction forces (F_x and F_y), and angular velocity (ω). These values are obtained from the virtual engine. Subsequently, the force equilibrium condition is solved.

The slide bearing model is realized in Matlab. Gauss-Seidel method together with SOR (Successive Over-Relaxation) techniques is used to solve the hydrodynamic problem.

If the pressure, p, distribution in oil is known, the friction torque can be calculated as follows:

$$M_t = R\iint_S\left(\frac{\eta\omega R}{h}-\frac{h}{2}\frac{\partial p}{\partial x}\right)\mathrm{d}S, \tag{2}$$

where R denotes the bearing radius and S is bearing surface.

Crank train main bearing power losses can be obtained as a sum of all main bearings if friction torque and instantaneous pin angular velocity are considered.

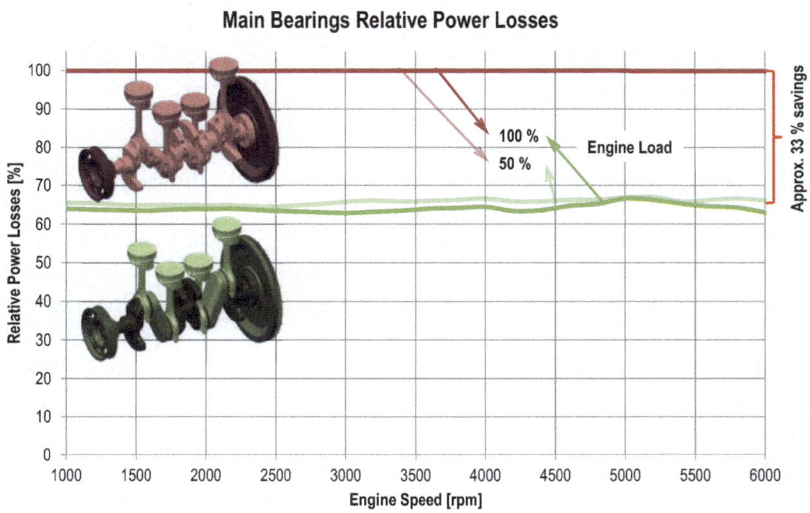

Fig. 1. Main bearing relative power losses

The main bearing power losses analysis brings very interesting results concerning the mechanical efficiency while using only three main bearings for an in-line four-cylinder spark-ignition naturally aspirated engine shows out the reduction of power losses approximately by third for whole speed operating range and even for the full and the partial engine load compared to the standard crank train design (Figure 1).

4.2 Torsional vibration

The results obtained from harmonic analysis of the crankshaft pulley angular displacement of the standard crank train are illustrated to the left in Figure 2. It is obvious that there is a strong impact of the sixth harmonic component of the torsional vibration that is reaching resonance under the high speed. At 4000 rpm the resonance of the eighth harmonic component is evident.

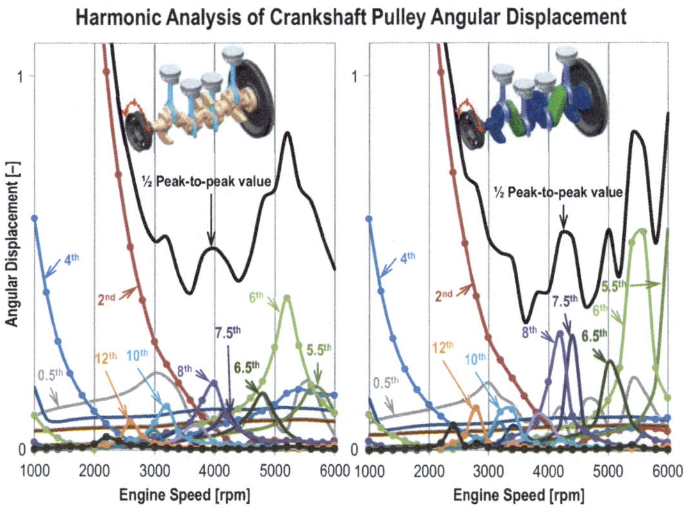

Fig. 2. Harmonic analysis of crankshaft pulley angular displacement

The occurrence of half-harmonic components (6.5th, 7.5th etc.) is caused by the fact that whole working cycle of a four-stroke internal-combustion engine takes two crankshaft revolutions while the fundamental frequency for engine dynamics is crankshaft rotational frequency.

Harmonic analysis of the torsional vibration proved the theoretical assumptions – the fact that the major orders take the highest part in vibrations. For the given engine concept these orders are integer multiples of two. The synthesis is equal to the half-size of peak-to-peak value from the periodic torsional oscillation.

Figure 2 also presents the results of the modified crank train harmonic analysis (to the right). Static stiffness analyses, as well as the virtual engine simulation, have proved that the 3-main-bearing crankshaft is torsionally stiffer than the standard one. Higher torsional stiffness is evident, for example, at the resonance from the eighth harmonic component which occurs at 4200 rpm (plus 200 rpm in comparison with the standard one).

Nevertheless, the small number of the crankshaft pivot points brings specific changes in crank train dynamics behaviour. The resonance amplitudes, not only of the major harmonic orders, but also the other harmonic orders are increased. "The sharpness" of resonance curves is caused in particular by lower damping of crank train torsional vibration since the dominant sources of damping are crankshaft main bearings.

4.3 Main bearings load

Main bearing load can be obtained by a numeric integration of the bearing pressure field. The load is described by resulting reaction force and moment. An important parameter of main bearing load is the maximum of bearing reaction force for one engine operating cycle.

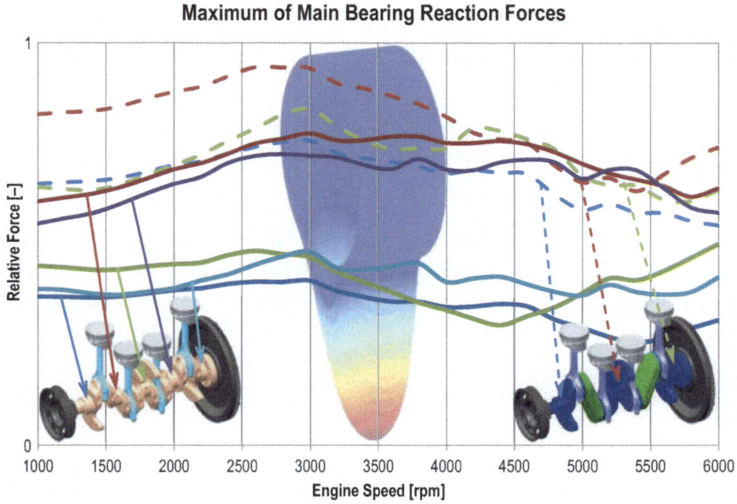

Fig. 3. Maximum of main bearing reaction forces

An analysis of the maximum of main bearing reaction forces during the whole engine operating speed for the standard and the modified crank train is performed (Figure 1). The growth of main bearing load at the 3-main-bearing crankshaft is detected, however, it is not expected to be critical. Nevertheless, the new design of bearing shell should be considered.

5 Conclusion

In general, the virtual engine results can help to understand complex dynamic behaviour of a new mechanically efficient powertrain and enable to speed-up the development process together with reductions of expensive prototypes. Therefore, the computational tools based on FEM and MBS principles play an important role in the modern powertrain research and development.

Dynamics analysis results demonstrate the potential of crankshaft main bearing reduction for decrease of crankshaft power losses expressively. If a naturally aspirated spark-ignition in-line four-cylinder engine is considered, the power loss savings reach around 33 % in comparison to the standard configuration having 5-main-bearing crankshaft.

The less main bearings, the more torsional vibration of crank train is excited though, however, this effect can be successfully suppressed by using a torsional damper whose parameters and impact can also be verified by using the virtual engine.

The 3-main-bearing crankshaft conception seems to be an advance way to increase the mechanical efficiency of a crank train. Therefore, it will be further evolved in terms of fatigue life computational prediction, detail design modifications, and subsequent prototype manufacturing.

Acknowledgement

The research leading to these results has received funding from the MEYS under National Sustainability Programme I (Project LO1202) and Specific research project of the Faculty of Mechanical Engineering, Brno University of Technology (FSI-S-14-2334).

References

1. Novotný P.. Virtual Engine – A Tool for Powertrain Development, Inaugural Dissertation, Brno University of Technology, Czech Republic, 2009.
2. Novotný P., Píštěk V.. New Efficient Methods for Powertrain Vibration Analysis, Proceedings of the Institution of Mechanical Engineers, Part D, Journal of Automobile Engineering, 2010, pp. 611–629. ISSN 0954-4070.
3. Butenschön H. J.. Das hydrodynamische, zylindrische Gleitlager endlicher Breite unter instationärer Belastung. PhD dissertation, Universität Karlsruhe, Germany, 1976.
4. Rebbert M. Simulation der Kurbewellendynamik unter Berücksichtigung der hydrodynamischen Lagerung zur Lösung motorakustischer Fragen. Ph.D. dissertation, Rheinisch-Westfälischen Technischen Hochschule, Aachen, Germany, 2003.
5. Roelands C. J. A.. Correlational Aspects of the Viscosity-Temperature-Pressure Relationship of Lubricating Oils. Ph.D. Thesis. Delf: Technical University Delft, The Netherlands, 1966.
6. Craig R. R.. Structural Dynamics. John Willey & Sons, 1981. ISBN 0-471-04499-7.

Design of a Special Purpose Permanent Magnet Motor

Karel Hruska[1], Vladimir Kindl[1], Roman Pechanek[1] and Lukas Skrbek[2]

[1] University of West Bohemia, Faculty of Electrical Engineering, RICE, Pilsen, Czech Republic
{khruska, vkindl, rpechane}@rice.zcu.cz
[2] E-Glider.com, Lucany nad Nisou, Czech Republic
info@e-glider.com

Abstract. The paper deals with the complete on-demand design of a permanent magnet electric machine for special purposes. It includes information about the whole design procedure – winding design, electromagnetic design, thermal and ventilation calculations leading to manufacture of first prototype of the machine. Further it includes information about re-design used for the second prototype based on measurements performed on the first prototype. Additionally the whole control system of the machine is described in the article.

Keywords: permanent magnets, delta connection, BLDC machine, design, analysis, propeller, controller

1 Introduction

The development on permanent magnets led to possibility of design of high power density machines – machines with high output power and low mass. The total mass of the machine may be lowered by further adjustments of the winding of the machine and its chassis. As a good solution used for permanent magnet machines has evolved the tooth winding [1, 2, 3], which is nowadays commonly used in many designs of permanent magnet synchronous machines (PMSM) and especially brushless direct current machines (BLDC) [4, 5].

Both mentioned designs of permanent magnet machines may be evaluated from many points of view – design of the machine, necessity of sophisticated power electronic converter, presence of some kind of feedback sensor etc. After considering all pros and cons there are many mid-power applications, where the usage of a BLDC machine is the most suitable solution. The main reasons are:

- slightly lower mass of the machine
- simple power electronic converter
- simple feedback sensor

These reasons led to decision of design of a BLDC machine with delta connected winding, which would be equipped by a simple converter, which would also have low mass [5]. As a power source a direct current source (batteries) would be used.

R. Jabłoński and T. Brezina (eds.), *Advanced Mechatronics Solutions*,
Advances in Intelligent Systems and Computing 393,
DOI 10.1007/978-3-319-23923-1_32

2 Design requirements

The main target was the design of a light yet powerful machine with possibility of simple control of its speed and output power. Although it could be possibly used in many application, the fundamental purpose of its usage was the aviation – as an electric drive for a propeller of an electric paraglide.

For the purpose of recognition of proper working point the HELIX H25F propeller with diameter 1.30 *m* has been chosen. The characteristics of the propeller are shown in Figure 1. Lines drawn in Figure 1 correspond with propeller output power (blue line) and its loading torque (red line). Both characteristics are measured at 15 degrees Celsius and atmospheric pressure 1013 *hPa*. Considering aerodynamic behavior of the air flowing around the propeller a working point at 2300 rpm has been considered as reference point for the design of the machine.

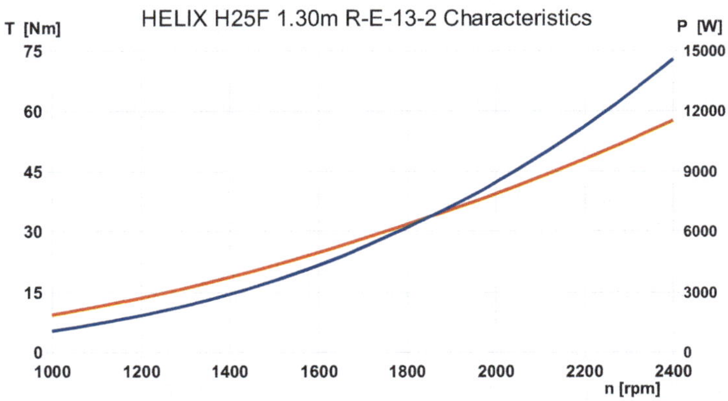

Fig. 1. HELIX H25F Characteristics

Since the machine is considered to be supplied from batteries from the very beginning of the design process, the discharge process of the batteries must be considered in the design. Therefore the machine is not designed for the exact working point resulting from characteristics of the propeller, but from a range nearby this working point. This results in design of a 15 *kW* BLDC machine, which will force the propeller to work at slightly higher speed at fully charged batteries and with lowering of the batteries voltage its speed and output power will slightly drop according to the DC bus voltage. The battery pack supplying the machine provides voltage 62 *V* at no-load state and while loading it drops from 58 *V* to lower values according to discharge of the battery pack itself. The resulting rated parameters of the machine are therefore:

- output power: $P = 15\ kW$
- terminal voltage: $U = 58\ V$
- speed: $n = 2300\ rpm$
- torque: $T = 62\ Nm$
- DC bus current: $I = 280\ A$

3 Design of the machine

The basic electromagnetic design of the BLDC machines is very similar to fundamental design of permanent magnet synchronous machines with main significant difference in considering currents and voltage on terminals of the machine and within the winding of the machine [3, 4, 5]. In this case the delta connection of the winding has been used for the design, which results in winding phase current formed by steps (see Fig. 2, denoted as hatched element, value marked as I is the current in DC bus). Each phase is then shifted by 120 degrees electrical to create a rotating magnetic field.

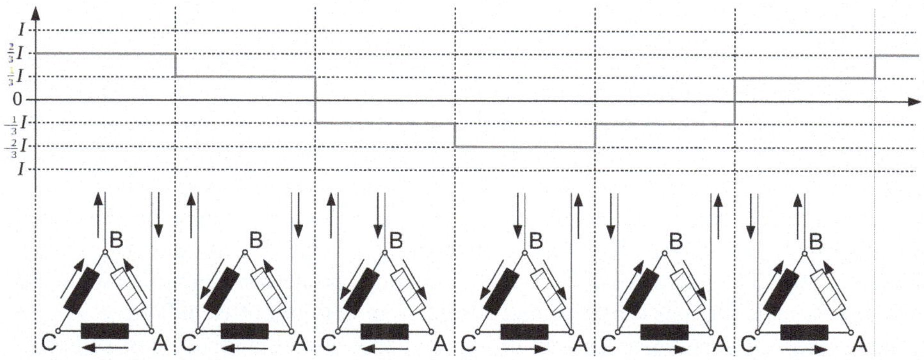

Fig. 2. Current flowing through winding phase between terminals A and B (hatched)

The design of the machine is based on the outer rotor topology, where the rotor with permanent magnets is mounted in the outside of the stator. The stator utilizes appropriate topology of tooth winding for creation of rotary magnetic field. Three teeth of the machine are equipped with Hall sensors for the purpose of detection of rotor position for appropriate switching of the converter [6].

The presumed discharge time of the batteries also allows further optimization of the machine from thermal point of view – the expected load of the machine should not exceed 15 minutes, therefore some weight may be saved by appropriate choice of saturation of the machine and the cross-section of the winding. Further savings of the weight of the machine offers a completely opened construction of the chassis of the machine (see Fig. 3).

Fig. 3. Detail of manufactured machine (left) and machine with mounted propeller (right)

As seen from Fig. 3, the machine is designed as fully opened, therefore the cooling air gets directly into contact with the winding of the machine. This effects leads to fundamental improvement of cooling of the machine and helps in reduction of dimensions of the machine. Using mentioned facts and optimization using latest computing techniques like finite element method (FEM) for analysis of the electromagnetic design [7, 8, 9] and losses [10] and computation fluid dynamics (CFD) the whole design resulted in a 15 *kW* BLDC machine with total weight only 5,4 *kg*.

4 Controller of the machine

The controller of the machine integrates three main functions – switching of the DC bus onto three terminals of the machine (including recuperation), the control of the speed of the machine and monitoring of status of the batteries. The whole controller is embedded into a box with dimensions 100 x 100 x 66 mm and it includes a high efficiency rectifier (approximately 98 per cent) controlled by a 32-bit ARM microprocessor. The electronics is cooled by two fans mounted on top of the box.

The controller may transfer all measured data into the computer or show them through small telemetry display. The connection may be performed through CAN bus, RS-485, RS-232 and USB port. The measured properties include monitoring of all phases and DC bus (voltages and currents), monitoring of each battery cell, unmatched protection and management of lithium batteries and monitoring of controller, motor and battery temperatures.

The battery pack is charged via separate charger equipped with battery management system (BMS) to keep the whole battery pack in balance and optimal conditions for use. The designed BLDC motor with all supporting devices is shown in Figure 4.

Fig. 4. BLDC motor and the whole control system

5 Measurement of the first prototype and re-design

After the manufacture of the prototype it was tested by its producer with mixed results. The prototype showed the possibility of obtaining high torque and output power, but it was not able to reach its working point with desired load type, since its speed rapidly dropped according to the load of the machine. The measured load characteristic in form of dependency of speed depending on provided output power is shown in Figure 5.

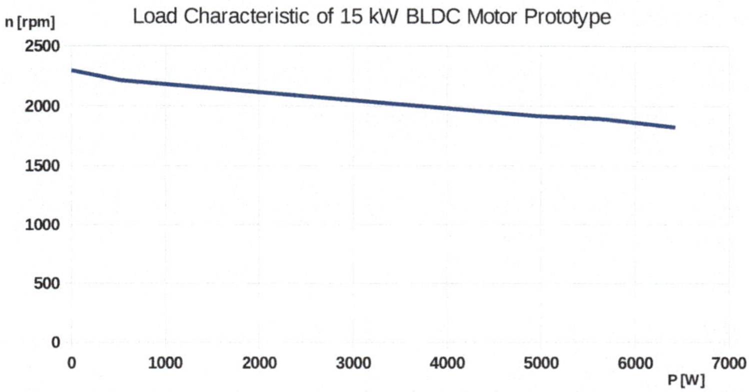

Fig. 5. Measured load characteristic of the first prototype

From the measured characteristic it is clear that the rated speed of the machine is reached only in no-load state while its rated power would be produced at less than 1500 *rpm*. This effect is caused by a high value of induced-back electromotive force (EMF) induced in the winding. The high value prevents the current from flowing through the winding, therefore the rated output power may be never reached with intended load. In this case there are two options possible to compensate this effect – increase of the terminal and DC bus voltages or lowering of induced-back EMF.

Since any increase of the terminal voltage is not possible for safety reasons, the only remaining possibility is removing of one coil turn from each coil of the winding to lower the induced-back EMF. The performed change in design of the machine results in increase of no-load speed (2790 *rpm* at terminal voltage 57,5 *V*) but the nominal working point of the machine corresponds with the demands of the load. Further space obtained by removal of the winding turn further improves cooling conditions of the machine allowing usage of class B insulation materials.

6 Conclusion

Although the initial design of the machine was marked with design inaccuracies, a slight redesign of the topology of the winding compensated this mistake and resulted in a successful design of a BLDC machine with efficiency approximately 93 percents. The total efficiency of the whole control system with the motor is then approximately 88 percents at nominal working point while keeping small dimensions and low weight of the drive.

The current development is focused on design and manufacture derivations of the initial 15 *kW* BLDC machine including derivatives with different rated speeds and output powers. The application of these versions also includes the electric paragliding, but they are also intended for use in different areas like automotive industry and pilotless aviation.

Acknowledgement

This research has been supported by the European Regional Development Fund and the Ministry of Education, Youth and Sports of the Czech Republic under the Regional Innovation Centre for Electrical Engineering (RICE), project No. CZ.1.05/2.1.00/03.0094, project TA ČR number TA002000103, Centre for Intelligent Drives and Machines (CIDAM) and funding program of the University of West Bohemia number SGS-2015-038.

References

1. Salminen, P. "Fractional slot permanent magnet synchronous motors for low speed applications" Lappeenranta University of Technology, 2004. ISBN 951-764-983-5
2. Hruska, K.; Sokol, M. "Advanced numeric tooth winding analyses," MECHATRONIKA, 2012 15th International Symposium, pp. 1, 7, 5-7 Dec. 2012
3. Müller, G.; Vogt, K.; Ponick, B. "Calculations of electric machinery (Berechnung elektrischer Maschinen)" Weinhein, Germany: Wiley-Vch, 2007. ISBN 978-3527405251
4. Hanselman, D. C. "Brushless Permanent-Magnet Motor Design" Mcgraw-Hill, USA, 1994. ISBN 978-0070260252
5. Dorrell, D.G.; Staton, D.A.; McGilp, M.I. "Design of Brushless Permanent Magnet Motors - A Combined Electromagnetic and Thermal Approach to High Performance Specification," IEEE Industrial Electronics, IECON 2006 - 32nd Annual Conference on, pp. 4853, 4858, 6-10 Nov. 2006
6. Simpkins, A.; Todorov, E. "Position estimation and control of compact BLDC motors based on analog linear Hall effect sensors," American Control Conference (ACC), 2010, pp. 1948, 1955, June 30 2010-July 2 2010
7. Bianchi, N. "Electrical machina analysis using finite elements" Boca Raton, USA, Taylor&Francis, 2005. ISBN 0-8493-3399-7
8. Reece, A. B. J.; Preston, T. W. "Finite Element Method in Electrical Power Engineering" Oxford University Press, UK, 2000. ISBN 978-0-19-856504-8
9. Schmidt, E.; Susic, M.; Eilenberger, A., "Finite element analysis of an external rotor permanent magnet synchronous machine with star- and delta-connected tooth coil windings," Electrical Machines (ICEM), 2010 XIX International Conference on, pp. 1, 6, 6-8 Sept. 2010
10. Markovic, M.; Perriard, Y., "A simplified determination of the permanent magnet (PM) eddy current losses due to slotting in a PM rotating motor," Electrical Machines and Systems, 2008. ICEMS 2008. International Conference on, pp. 309, 313, 17-20 Oct. 2008

Transport Duty Cycle Measurement of Hybrid Drive Unit for Mixing Drum

Lenka Jakubovičová[1], Peter Zavadinka[2], Ján Jakubovič[3]

[1] Department of Applied Mechanics, Faculty of Mechanical Engineering, University of Zilina,010 26 Zilina, Slovak Republic
lenka.jakubovicova@fstroj.uniza.sk
[2] Institute of Solid Mechanics, Mechatronics and Biomechanics Faculty of Mechanical Engineering, Brno University of Technology, Technicka 2896/2, 616 69 Brno, Czech Republic
[3] Jura Hronca 3415/4, 010 15 Zilina, Slovak Republic

Abstract: The article show some information and chosen measurement results of a hybrid drive unit for Concrete Transit Mixer. The unit was developed based on patent DE102010047314B4 and tested in the field with the intention prove the technology and promote potential advantages for future solutions. The hybrid drive unit consists from electric and hydrostatic motor connected through addition mechanical transmission. Motors are working in different working conditions and by this way optimizing efficiency of the power flow. The results show fuel saving potential, better controllability and performance through the concrete transport. In addition showing potential for next benefits for further utilizations and applications.
Keywords: hybrid drive unit; concrete transit mixer

1 Introduction

Concrete Transit Mixer Application consists of chassis, drum and a transmission which is transferring power from engine (diesel) to the drum. The drum and chassis technology did big steps from evolution point of view but the transmission is nearly same more than 40 years.

The transmission which is transferring power from chassis diesel to the drum has physical limitations and these limitations are a barrier for next better and more efficient application performance. However units of the transmission and regulation were improved through last years the system concept (hydrostatic pump, motor and gearbox) is the same more than 40 years and is only one existing solution today. As the next step in this evolution is utilization of two parallel drives, where electric drive will be used for low power working condition and hydrostatic drive for high power working conditions what will improve today total system efficiency and will provide new dimension of total performance.

The prototype hybrid drive unit for a mixing drum that is disposed on a motor vehicle which, as compared with the prior art, exhibits a further improvement to the efficiency of the drive system of the mixing drum and offering possibility eliminate unneces-

R. Jabłoński and T. Brezina (eds.), *Advanced Mechatronics Solutions*,
Advances in Intelligent Systems and Computing 393,
DOI 10.1007/978-3-319-23923-1_33

sary exhaust gas and noise emissions in a particular working arrears as underground and tunnels. The drive engine, in particular the internal combustion engine of the truck mixer, is capable of being shut down while the rotation of the mixing drum is maintained. The drive system will be capable of implementation inexpensively with little effort and robust and reliable in its realization.
This object is achieved as a result of the fact that in addition to driving the mixing drum via a hydraulic drive train with displacement pump and hydraulic motor, the electric motor, with a relatively low output as compared with the hydraulic drive, is or can be permanently connected to the truck mixer gearbox in such a way that the mixing drum can be driven solely by the electric motor.

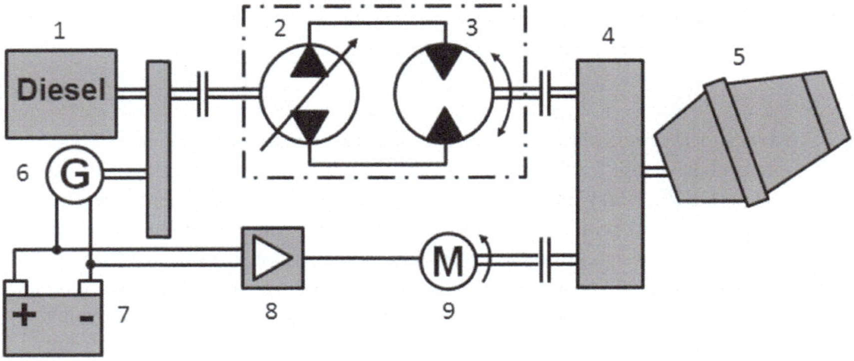

Fig. 1. The main components of a drive unit in a schematic illustration
1- Diesel engine, 2- Hydraulic pump, 3- Hydraulic motor, 4- Gearbox, 5- Drum, 6- Alternator, 7- Electric battery, 8- Inverter, 9- Electric motor

2 Measurement

2.1 System Solution

System was installing on the 4 years standard transit mixer with 9 m^3 drum where the drum drive was updated from hydrostatic to hydroelectric.

Fig. 2. Concrete Transit Mixer Application and prototype of hybrid drive unit

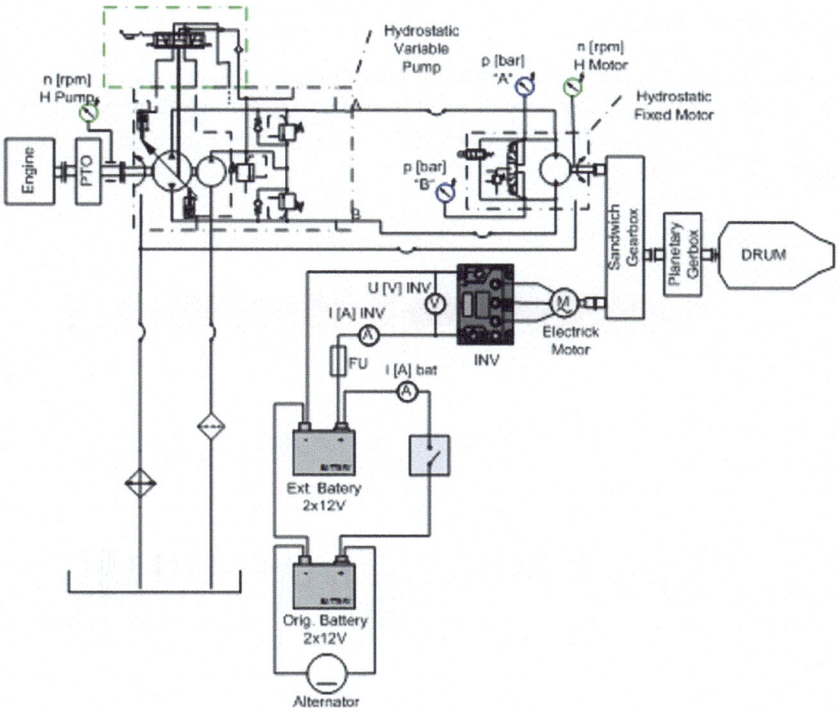

Fig. 3. Schematic diagram with measure points

Fig. 4. Visualization of particular components where are A- Alternator, B- Inverter, C- External Batteries, D- Electric Motor, E- Hydrostatic Motor, F- Sandwich Gearbox, G- Transit Mixer Planetary Gearbox, H- Power Torque Output, I- Hydrostatic Pump

2.2 System Performance

There was simulated transport of concrete with slump S3 by using dry gravel. The average loads of S3 slump concrete through transport was known from past measurements. Transport was simulated in town traffic conditions in a circuit 24 km with average transport speed 40 km.

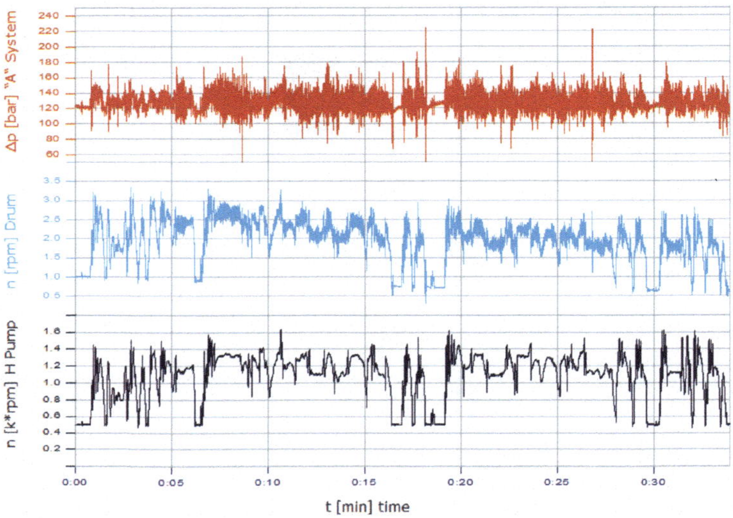

Fig. 5. Hydrostatic drive with manually actuated hydrostatic pump where drum speed was adjusted 1 rpm at 500 rpm engine speed

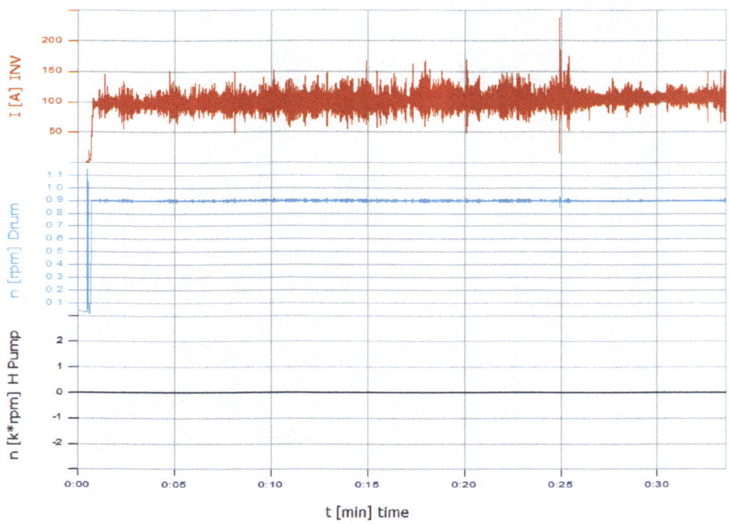

Fig. 6. Electric drive with electric motor with constant speed drum 0.9 rpm where the Power Output Torque (PTO) was switch off

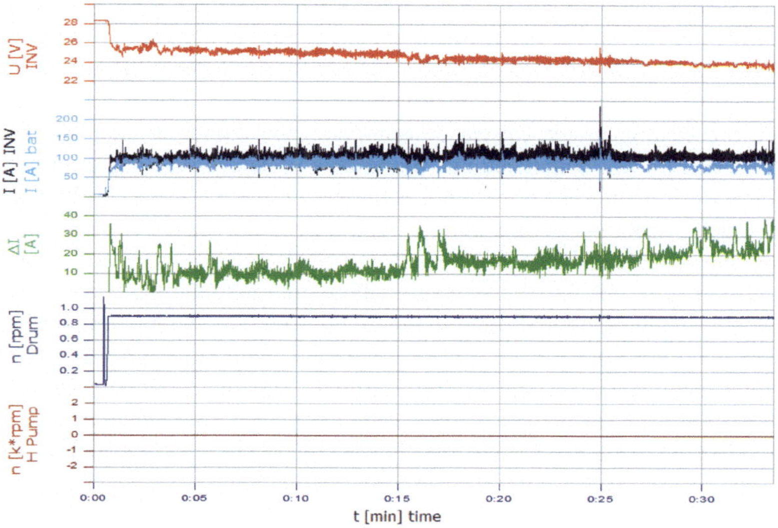

Fig. 7. Electric drive with electric motor with constant speed drum 0.9 rpm where the Power Output Torque (PTO) was switch off. Here was measured current consumption on electric motor side and capability of standard truck alternator supply

3 Results

Tab. 1. Measured fuel saving through simulated transport

	Record through driving	Load	Drive	Drum Speed	Time [min]	Distance [km]	Consumption [l/100km]
1	DCA-TN-DCA Hydrostatic	120-180 bar	Hydraulic	1 rpm at low idle	34	24	41.9
2	DCA-TN-DCA Electric	80-150 A	Electric	0.9 rpm constant	33	24	36.8
						Delta	5.01
3	DCA-TN-DCA Hydrostatic	120-180 bar	Hydraulic	1 rpm at low idle	37	24	37.3
4	DCA-TN-DCA Electric	80-150 A	Electric	0.9 rpm constant	37	24	36.4
						Delta	0.9
5	DCA-TN-DCA Hydrostatic	Empty drum	Hydraulic	1 rpm at low idle	33	24	28.8
6	DCA-TN-DCA Electric	Empty drum	Electric	0.9 rpm constant	32	24	28.1
						Delta	0.7

Cycles 1&2 was measured through standard traffic where driver was simulating aggressive driving according possibilities. Cycles 3&4 was measured through relatively high traffic where driver was driving relatively slowly without any aggressive accelerations. Cycles 5&6 was measured through standard traffic where driver was driving according his needs but with empty drum.

4 Conclusions

Character of driving and temperature of the diesel engine has big influence to the consumption. Due limited possibilities simulate the same and repeatable conditions, results have only informative character. Even through this the saving potential and the

system performance where stable constant speed of drum has big influence to concrete quality and life of the drum, was proved. In order to achieve more precise data would be necessary to drive much longer distances, eliminate traffic influences and achieve average drive speed at least 60 km/hour.
Based on the electric motor current consumption and alternator supply capabilities comparison (Fig 7) was concluded that the truck will need to use bigger or additional alternator and different kind of external batteries for electric drive. From this reason was concluded to continue with additional optimization of sizing through one dimension simulation model and lab tests.
The main motivation for future customers will be not really the fuel saving but possibility used this technology in closed arrears in construction as underground buildings or tunnels where the trucks must switch off diesels engine through waiting time but on the other hand cannot switch off drum turning when the concrete is inside. There are lots of additional advantages of this separated power flow which can be used as well for other applications. At the first the marked needs must be investigated and then optimize design and system solution for particular application needs.

References

1. Carter, D. E., Alleyne, A. G.: Earthmoving vehicle powertrain controller design and evaluation. In: American Control Conference, June 30 2004-July 2 2004. Proceedings of the 2004, vol.5, pp. 4455-4460 (2004)
2. Kiencke, U., Nielsen L.: Automotive Control Systems for Engine, Driveline and Vehicle. 2nd ed., XVIII, p. 512. ISBN 978-3-540-23139-4. (2005)
3. Kriššák, P.: Popis systému merania a riadenia hydrostatického prevodu. In: Vedecká konferencia „Nekonvenčné technológie“, pp. 146-163. ISBN 80-8070-373-6. Žilina (2005)
4. Kriššák, P., Jakubovič, J., Zavadinka, P.: Transport Duty Cycle Simulation of Electro-hydro-mechanical Drive Unit for mixing Drum. In: Mechatronics 2013 – Recent technological and scientific advaces, Springer International Publishing Switzerland, pp. 235-241. ISBN 978-3-319-02293-2. (2014)
5. Kriššák, P., Kučík, P.: Computer aided measurement of hydrostatic transmission characteristics. In: Hydraulika i Pneumatyka, 4/2005, pp. 22-25. INDEKS 37726, ISSN 1505-3954.
6. Kroschel, B., Hergt, J., Jakubovic, J., Krissak, P.: DE102010047314B4
7. Zavadinka, P., Kriššák, P.: Simulation of vehicle working conditions with hydrostatic pump and motor control algorithm. In: JAMRIS, 3/2012, pp. 40-46. ISSN 1897-8649.
8. Zavadinka, P., Kriššák, P.: Dynamický model vozidla pri prejazde zákrutou. In: Hydraulika a pneumatika, 1-2/2010, pp. 53-57. ISSN 1335-5171. (2010)
9. Zavadinka, P., Kriššák, P.: Simulačný model dieselového motora: In: Acta Hydraulica et Pneumatica, 1/2009, pp. 34-38. ISSN 1336-7536. (2009)
10. Zavadinka, P., Kriššák, P.: Modeling and simulation of mobile working machine powertrain. In: Technical Computing Prague 2009, 19.11.2009, Praha, Česká republika, 2009, p. 118. ISBN 978-80-7080-733-0. (2009)
11. Zavadinka, P., Kriššák, P.: Modeling and simulation of diesel engine for mobile working machine powertrain. In: Hydraulics and pneumatics 2009, 7-9.10.2009, Wroclaw, Poľsko, pp. 339-348. ISBN 978-83-87982-34-8. (2009)
12. Sauer-Danfoss: Series TM Transit Mixer Drive System, Catalogue 520L0482, Rev BC, Dec 2010.

Temperature Field Optimization on the Mould Surface

Jaroslav Mlynek[1], Roman Knobloch[1], Radek Srb[2]

[1] Department of Mathematics, FP
jaroslav.mlynek@tul.cz, roman.knobloch@tul.cz
[2] Institute of Mechatronics and Computer Engineering
radek.srb@tul.cz
Technical University of Liberec, Studentska 2, 461 17 Liberec, Czech Republic

Abstract. This article focuses on issues of uniform heating of the shell metal moulds. Infrared heaters located above the aluminium mould surface heat the mould. This is one of the economical ways of artificial leathers production in the automotive industry (e.g. the artificial leathers for car interiors). The described mathematical model allows us to specify the locations of infrared heaters over the mould to obtain approximately a uniform temperature field on the mould surface. In this way we can obtain a uniform material structure and colour shade of the artificial leather and thereby prevent the scrap production. We used a differential evolution algorithm during the optimization process. The optimization procedure was programmed in the Matlab development environment. The software package ANSYS was used for temperature calculations. A practical example of optimization of heaters locations over the mould and calculation of the temperature across the mould surface is included at the end of the article.

Keywords: intensity of heat radiation; temperature field; mathematical model; differential evolution algorithm; software implementation.

1 Introduction

This article focuses on issues concerning the technology used for the production of artificial leathers. The leathers are used as the final surface for some parts of car interior equipment (e.g. car doors, dashboards). The product technology consists in sprinkling PVC powder onto a hot inner aluminium mould surface. The mould is heated by infrared heaters located above the outer mould surface. The mould is heated for a few minutes. Uniform temperature field across the whole mould surface during heating of the mould is required to produce leathers with proper surface structure. The infrared heaters are of tubular shape, the tube length is from 15 to 25 cm. The directional heat radiation characteristic is not usually supplied by heater manufactures. Therefore, the heat radiation intensity around the heater was determined experimentally by means of a sensor (see [1]).
The question is how to locate a relatively high number of infrared heaters above the mould surface, so that the generated temperature field on the working surface of the mould is as uniform as possible.

R. Jabłoński and T. Brezina (eds.), *Advanced Mechatronics Solutions*,
Advances in Intelligent Systems and Computing 393,
DOI 10.1007/978-3-319-23923-1_34

2 Heat Radiation Model

In this chapter we will briefly describe the mathematical model of the heat radiation generated by infrared heaters on the heated mould surface. The heaters and the mould are located in 3-dimensional Euclidian space E_3 described by the Cartesian coordinate system (O, x_1, x_2, x_3) with orthogonal base vectors $e_1 = (1, 0, 0)$, $e_2 = (0, 1, 0)$, $e_3 = (0, 0, 1)$. The model is described in more detail in [1, 2].

2.1 Infrared Heater Representation

The heaters used for a specific mould are all of the same type (i.e. they have the same geometric shape and heating power). The individual heater can be modelled by a straight line segment with length d (see Figure 1). For the subsequent use of a differential evolution algorithm for optimization of heat radiation intensity it is necessary to keep the minimal number of the parameters that unambiguously define the location of heater. The position of every heater H can be defined by 6 parameters

$$H\colon\ (s_1, s_2, s_3, u_1, u_2, \varphi), \tag{1}$$

where the first three parameters are the coordinates of the heater centre S, the following two parameters are the first two coordinates of the unit vector **u** of the heat radiation direction (the third coordinate is negative, i.e. the heater radiates downward) and the last parameter is the angle φ between the vertical projection of unit vector **r** of the heater axis onto the x_1x_2 − plane and the positive part of the axis x_1 (the vectors **u** and **r** are orthogonal).

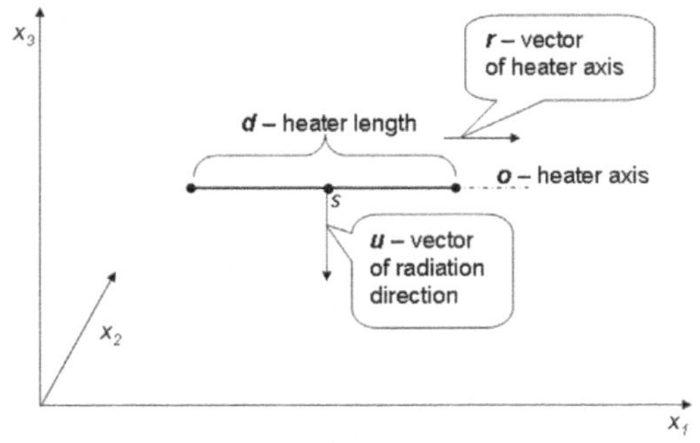

Fig. 1. Representation of the heater in the model.

2.2 Mould Representation

The outer mould surface P is described by elementary surfaces p_j, where $1 \le j \le N$. It holds that $P = \cup p_j$, where $1 \le j \le N$ and $\operatorname{int} p_i \cap \operatorname{int} p_j = \emptyset$ for $i \ne j$, $1 \le i, j \le N$. Each elementary surface is described by the centroid $T_j = [t_1^j, t_2^j, t_3^j]$, by the unit normal vector $\mathbf{v}_j = (v_1^j, v_2^j, v_3^j)$ at the point T_j (we suppose $\mathbf{v}_j$ faces upwards and is therefore defined through the first two components v_1^j and v_2^j) and by the area of elementary surface w_j. Each elementary surface can thus be defined by the following 6 parameters $p_j : (t_1^j, t_2^j, t_3^j, v_1^j, v_2^j, w_j)$.

2.3 Calculation and Optimization of Heat Radiation Intensity

We describe the numerical computation procedure for the total heat radiation intensity on the mould surface. We denote L_j as the set of all heaters radiating on the *j*th elementary surface p_j $(1 \le j \le N)$ for the fixed position of heaters, and I_{jl} the heat radiation intensity of the *l*th heater on the p_j elementary surface (I_{jl} is a constant value on the whole p_j in our model). Then the total radiation intensity I_j on the elementary surface p_j is given by the following relation (see [3] for more details)

$$I_j = \sum_{j \in L_j} I_{jl} \,. \tag{2}$$

The producer of artificial leathers recommends the constant value of heat radiation intensity I_{rec} on the whole outer mould surface. Then we can define the deviation function *F* by the relation

$$F = \left(\frac{1}{W} \cdot \sum_{j=1}^{N} (I_j - I_{rec})^2 w_j \right)^{½} \text{, where } W = \sum_{j=1}^{N} w_j \tag{3}$$

and I_j is given by relation (2). We look for the minimum of function F.

Function *F* has many local extremes. Using gradient methods for finding the minimum of the function *F* is not appropriate. If we use a gradient method, there is a high probability that we find only a local minimum of this function. Therefore, we use a *differential evolution algorithm* named *DE/rand/1/bin* (for more details see [4]). The optimization procedure was programmed by the authors in the software Matlab and is described in detail in [1]. A certain disadvantage of evolution optimization algorithms is their computational demandingness and slow convergence.

The location of each heater is defined in accordance with relation (1) by 6 parameters. Therefore, 6*M* parameters are necessary to define the locations of all *M* heaters. One individual *y* of differential evolution algorithm represents one possible location of all *M* heaters. We seek the individual $y_{\min} \in C$ satisfying the condition

$$F(y_{\min}) = \min\{F(y)\,;\, y \in C\}, \tag{4}$$

where $C \subset E_{6M}$ is the searched set. The identification of the individual $y_{\min}$ defined by (4) is not realistic in practice. But we are able to find the optimized solution y_{opt}.

We describe the procedure of calculation of the temperature field in the mould, which corresponds to the location y_{opt} of heaters in following chapter.

3 Mathematical Model of the Heat Conduction

In this part we describe briefly a mathematical model of the temperature field in the mould body. We will solve the parabolic evolutionary equation of heat conduction

$$c\rho \frac{\partial T(x,t)}{\partial t} = \lambda \, \Delta T(x,t) + Q \; , \; \text{ where } \Delta T(x,t) = \sum_{i=1}^{3} \frac{\partial^2 T(x,t)}{\partial x_i^2} \; , \tag{5}$$

on the domain $\Omega \subset E_3$, where Ω represents the mould. Function $T(x,t)$ denotes a temperature field in relation (5), point $x = (x_1, x_2, x_3) \in \Omega$, time $t \in (0, \tau)$, where τ is the duration of the heat radiation. The values c and ρ stand for specific heat and mass density of the mould material. The value λ denotes the heat conductivity of the mould material. We assume a homogeneous material of the mould. Symbol Q represents volume density of heat sources ($Q = 0$ in our model).

We consider the initial condition $T(x,0) = T_0 \quad \forall x \in \Omega$. We choose the Newton boundary condition which suits best the situation when the hot body is surrounded by an environment (air) and heat transfer between the body and environment is possible. The simple Newton boundary condition is usually expressed as $\lambda \frac{\partial T}{\partial \nu} = -\alpha (T - T_{air})$, where α is the coefficient of the heat transfer between the mould material and air temperature T_{air}, ν stands for the unit vector of the outer normal. Nevertheless this simple linear formulation of the boundary condition is not sufficient for our application. There are two principal reasons: 1/ the own heat radiation of the mould according to Stefann-Boltzmann law cannot be neglected; this means that the simple Newton boundary condition has to be supplemented by the term representing own radiation; 2/ there are no volume heat sources Q in the body of the mould; the heat is supplied exclusively by infrared heaters through the upper part of the surface of the mould.

When these two facts are considered we have the following versions for the boundary condition (see e.g. [5]):

$$\lambda \frac{\partial T}{\partial \nu} = -\alpha \,(T - T_{air}) - \varepsilon \sigma \left(T^4 - T_{air}^4\right) + I \tag{6}$$

for the upper part of the surface of Ω and

$$\lambda \frac{\partial T}{\partial \nu} = -\alpha \,(T - T_{air}) - \varepsilon \sigma \left(T^4 - T_{air}^4\right) \tag{7}$$

for all other parts of the surface of Ω. Here value ε denotes emissivity of the mould and σ denotes Stefann-Boltzmann constant, $\sigma = 5{,}775 \cdot 10^{-8}$ Wm^2K^{-4}. Symbol I stands for the heat radiation intensity generated by infrared heaters (heat flux) incident onto the upper surface of the mould. Both boundary conditions (6) and (7) are not linear (which is apparent by comparison with simple Newton boundary condition) which makes the process of searching for a solution more complicated. In fact the temperature field has to be calculated numerically.

4 Example of the Calculation of the Temperature Field

In this part we describe the numerical calculation of the temperature field by means of the optimization computational procedure programmed by the authors in the system Matlab and software package ANSYS 15.0.7. The used finite elements according to the ANSYS specification were SOLID90 and SURF152. The corresponding base functions were quadratic.

The size of the mould is 0,6x0,4x0,15m^3, the mould surface is described by 2187 triangular elementary surfaces. Mould material parameters are as follows: aluminium alloy, $\rho = 2770 \text{kg/m}^3$, $\lambda = 160\ \text{W/mK}$, c = 875J/kgK, $\alpha = 20\ \text{W/m}^2\text{K}$, $\varepsilon = 1$, $T_0 = T_{air} = 22$ °C, thickness of the mould is 0,008 m. The recommended radiation intensity by the producer of leathers is $I_{rec} = 47\ \text{kW/m}^2$. We used 16 heaters of the same type (producer Philips, power 1000W; length 0,15m, width 0,04m).

The heaters in the initial positions lie in the plane parallel with the plane given by axes x_1 and x_2 and in distance 10cm over the mould and with heater axis **r** parallel to axis x_1. The aberration $F(y_0)$ for the initial heaters locations defined by relation (3) is $F(y_0) = 20{,}3435$. We received y_{opt} by the use of *differential evolution algorithm* after calculation of 20000 generations. The aberration decreased to $F(y_{opt}) = 2{,}6258$. The optimized heaters locations y_{opt} and the corresponding heat radiation intensity on the upper surface of the mould are represented in Figure 2.

The temperature field on the lower surface of the mould for the optimized locations y_{opt} for heating time t =180s is shown in Figure 3. The temperature difference between minimum and maximum temperature attains 202,67°C for the initial heaters locations y_0 and 14,39°C for the optimized locations y_{opt}. To attain an even more uniform temperature field on the mould surface it would be necessary to use more heaters. However, this solution is not feasible because of higher energy consumption and in the practical production it is achieved by switching some heaters off and on.

5 Conclusion

Manual positioning of heaters above the mould based only on experience of competent technicians that was used before was inexact, vague and besides time consuming. We are able to determine the heaters positions on the basis of the presented model and differential evolution algorithm in a way that is theoretically justified. Besides we can calculate the temperature field corresponding to the optimized positioning of the heaters on the lower mould surface by means of the ANSYS system. In this way the approximately uniform temperature of the mould surface is ensured and so the same material structure and colour shade of artificial leather is guaranteed on its whole surface.

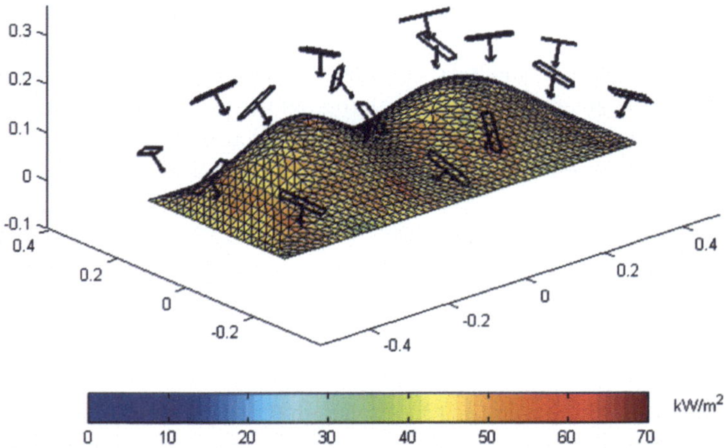

Fig. 2 The optimized heaters locations y_{opt} and the corresponding heat radiation intensity.

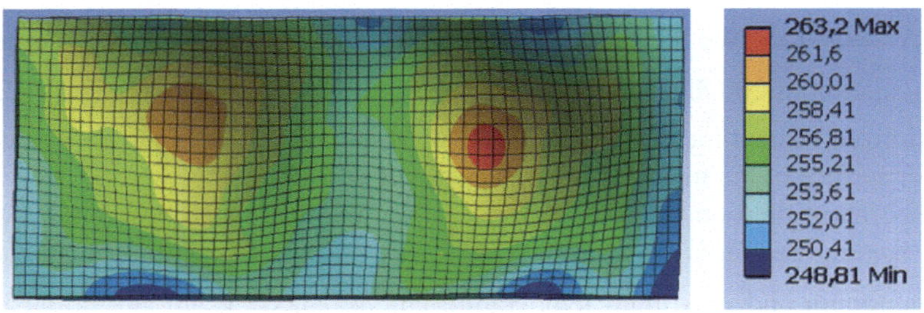

Fig. 3 The temperature field at the time of heating t =180s on the lower part of the mould for the case optimized heaters locations y_{opt} .

This work was supported by grant SGS-FP-TUL 21049 and by grant SGS-FM-TUL.

References

1. Mlynek, J., Srb, R.: Differential Evolution and Heat Radiation Intensity Optimization, Proc. of Conf. on Mathematics and Computers in Sciences and Industry, Varna, Bulgaria, IEE Computer Society, 2014, pp. 135-138. DOI 10.1109/MCSI.2014.11
2. Mlynek, J., Srb, R.: The Process of an Optimized Heat Radiation Intensity Calculation on a Mould Surface, Proc. of the 26th European Conference on Modelling and Simulation, Koblenz, Germany, 2012, pp. 461-467. DOI: 10.7148/2012-0461-0467
3. Cengel, Y.A.: Heat and MassTransfer, McGraw-Hill, New York, 2007
4. Price, K. V., Storn, R. M., Lampien, J. A.: Differential Evolution, Springer-Verlag, Heidelberg, 2005
5. Incropera, F.P., DeWitt, D.P.: Fundamentals of Heat and Mass Transfer, John Wiley & Sons, Inc., 2002

21st Century Lecturer – from University to Secondary and Primary Schools

Dusan Maga, Boris Simak

Czech Technical University in Prague
{dusan.maga,simak}@fel.cvut.cz

Abstract. Cedupoint™ (Continuing Education Point) is an integral part of the Department of Telecommunication Engineering at the Czech Technical University in Prague, Faculty of Electrical Engineering. Due to its mission, it provides a lot of activities focused on promotion of university education techniques and technologies to primary and secondary schools. The activity presented in this paper is based on introduction of touchscreen devices (tablets) into education at small- and medium-sized primary and secondary schools. The paper presents a summary of all related activities and provides an overview of relevant technologies and technical solutions as well as fundamental ideas of the designed education program; nevertheless, it is focused on the major activity related to classroom management systems designed for teachers of these schools.

Keywords: Education • ICT • Collaboration between universities and primary/secondary schools.

1 Introduction

Cedupoint™ (Continuing Education Point) at the Department of Telecommunication Engineering of the Czech Technical University in Prague (CTU), Faculty of Electrical Engineering (FEE), has been a leader of many different educational projects focused on students and adults education. Collaboration between the university and primary and secondary schools has a long tradition. These activities were described e.g. in [1-4]. Following these projects, a new one has been realized, entitled "Modern Teacher of the 21st Century" (supported by the Education for Competitiveness Operational Programm – OP VK, No. CZ.1.07/1.3.00/51.0019). This project is carried out by Cedupoint, together with 40 partners (32 primary and 8 secondary schools) having independent financial budgets. The main goal of the project is to equip all the 40 partners with touchscreen devices (tablets) and to educate the teachers in modern technologies related to the obtained hardware. However, the final vision of future project calls is to equip the complete classrooms with tablets; the present project is focused "only" on teachers – they will receive the hardware as well as the necessary knowledge. The aim is to prepare the teachers to be able to operate the hardware and to use the appropriate software for managing the classroom and the education process.

R. Jabłoński and T. Brezina (eds.), *Advanced Mechatronics Solutions*,
Advances in Intelligent Systems and Computing 393,
DOI 10.1007/978-3-319-23923-1_35

The presented ideas are based on the authors' previous experience with education in the field of power electronics supported by modern and IT-based learning technologies [5-8].

An overview describing the project activities will be presented in the following chapters, as well as the basic architecture of the used classroom management system. The most important parts of the project activities will be described in chapter 2 and 4; the presented study programs were specifically designed for the purposes of this project and they have been successfully accredited by the Ministry of Education, Youth and Sports of the Czech Republic as further education and lifelong learning study programs for pedagogical staff.

2 Structure of the project

The project is based on seven key activities. The initial three of them are focused on individual participants (teachers or managers of the schools) while the fourth, fifth and sixth one target the complete group of the teaching staff. The last activity is dedicated to electronic support of the above-mentioned activities, so it will not be discussed separately; however, the inputs and outputs of this activity will be included in the description of the first six ones.

It has to be stated here that the activities No. 4, 5 and 6 are still in progress, while the first three ones have already been finished.

2.1 Key Activity No. 1

The first key activity is focused on managers (directors) of 40 partner schools. The aim of the activity is to present the basic economical rules as well as the aims and timelines of the project to school managers.

2.2 Key Activity No. 2

The second key activity is based on support of individuals belonging to the local teaching staff. These people are educated in fundamentals of e-learning systems and electronic learning/teaching materials preparation, techniques and technologies. The aim of this activity is to prepare an expert (or two) at every local partner school to be able to help the other local teachers with basic tasks of electronic education and preparation of electronic educative materials when necessary.

2.3 Key Activity No. 3

This is the last activity focused on individuals. Unlike the previous two ones, it is focused on local IT experts. The aim of this activity is to present and explain the classroom management system to local IT experts so that they are able to provide the necessary support and assistance to their colleagues (the architecture of the classroom management system itself will be briefly presented in one of the following chapters).

The local IT experts will become familiar with the used system, they will be able to administer user management of the system, to set and control the hardware and software requirements; eventually, they will be able to define and formalize the necessary troubleshooting topics to the university or to another team of experts.

2.4 Key Activity No. 4

The scope of this activity is different as it is focused on the complete group of teachers (groups of less than 20 people per partner). Its content, however, is similar to the key activity No. 2 – the local teachers will obtain the knowledge on electronic education bases, their history and possibilities, together with their positive and negative features. Also, the necessary legislative background is presented (including the description of copyright rules and rules for downloading and reuse of any third-party content – especially images and plain text).

2.5 Key Activity No. 5

The content of this activity is also focused on the complete group of local teachers. The teachers will obtain basic knowledge on how to operate the received hardware. They will practice the basic skills and basic functions with the received tablets, including the protection against cybercrime threats and other possible risks.

2.6 Key Activity No. 6

This is the most important activity (from the viewpoint of this paper). The activity is based on classroom management system, generally on the use of the received hardware in education process. More detailed description will be presented below.

3 ICT in the education process

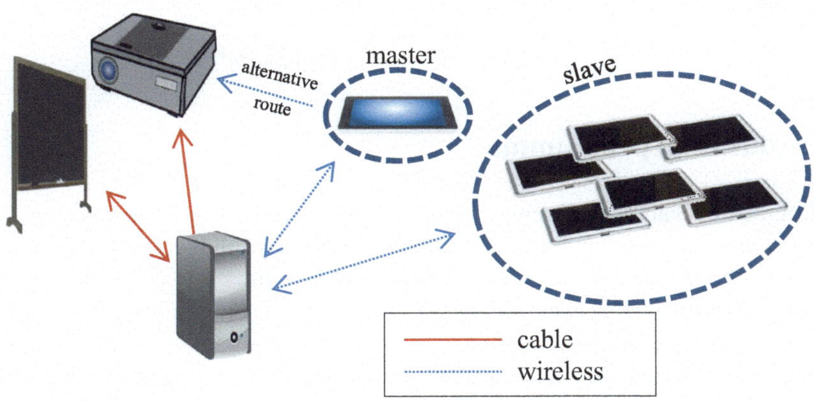

Fig. 1. Hardware solution used in the project

The possible practical use of the received ICT hardware in the education process is the most important part of the described project. The idea of the hardware architecture is based on [9]. The hardware solution (Fig. 1) consists of students' tablets (slave), teachers' tablets (master), PC server based on MS Windows system, and a data projector or an interactive whiteboard (depending on the equipment already existing in the classrooms). The software solution [9] allows the teacher to control the slave tablets (including the restrictions concerning specific application), screen sharing, screen monitoring on the whiteboard, testing, group reporting, etc.
The complete solution requires a classroom management system [9] running on MS Windows platform; nevertheless, an alternative route directly from the teacher's (master) tablet to the projector already exists and has been introduced. On the one hand this allows the partners to use their existing equipment; on the other hand, this solution does not allow full controlling of the slave tablets.

Fig. 2. Classroom setup at Cedupoint™

4 Education programme

The designed education programme is titled *"Use of ICT in the education process – classroom management"* and it is based on three mutually interconnected blocks:

- Introduction to classroom management system,
- Professional/specialized part,
- Applications for education.

The content of the first block is based on comparison of different and commonly used mobile operating systems; however, it has to be mentioned here that the presented

solution is based on Android devices. The basic ideas on managing the classroom are presented here, including hardware requirements and the installation of required software both on server and master/slave tablets. A brief overview of system configuration and initiation is also presented.
The most important part deals with administration of data necessary for classroom realization – technical part of the course preparation, management of teachers' accounts and management of students' accounts. Based on generally included rules, the teachers are able to run and control the classroom from theirs master tablets. They are able to monitor the individual slave devices – which means that they can allow/deny running of an application on slave tablets, as well as remotely run or stop an application on slave tablets. It is also possible to share the screen of a slave device on the whiteboard or to use other public/private communication channels from the master to the selected slave device(s).
The users (teachers) are able to prepare separate lectures, and to lead and manage them both from the master tablets and from the PC server.
The second block of the education program is focused on a specialized part of the education, depending on teachers' professional qualification. Despite the many different professional interests, the team is focused on mathematics, physics and languages (English and French). However, the professional orientation is not the bottleneck of the problem at this moment; the physics, mathematics (etc.) is used to demonstrate specific applications and the use of the above-presented classroom management system directly in the classrooms. The used examples are easy to extend into a lot of other different branches, including technical, social or any other ones.
The short third block is based on introducing of the third-party applications suitable for education process, depending on the required level given by the skills of the target audience (students).

5 Conclusions

The basic ideas of the current activity (realized by Cedupoint – part of the Department of Telecommunication Engineering at the Czech Technical University in Prague, Faculty of Electrical Engineering) has been described in the this paper. The educational activity is focused on primary and secondary school teachers, and its aim is to support these people with necessary touchscreen hardware and related knowledge in using the received equipment in the education process. All important activities of the project have been briefly described:

- The hardware architecture of the system has been presented,
- The new technologies in the hardware architecture are included; however, an alternative route has also been designed. This allows the partner schools to use their existing equipment more efficiently (with inclusion of touchscreen master devices),
- The major educational activity (called *Use of ICT in the education process – classroom management*) has been described in details.

Also the response of the target group is very important for the authors. The key activities No. 1–3 are already in the final phase, while No. 4–6 are running fully right now; therefore, the feedback cannot be presented in the paper and will be introduced later.

Acknowledgement

The presented activities are supported by the Education for Competitiveness Operational Programme (OP VK) No. CZ.1.07/1.3.00/51.0019 “Modern Teacher of the 21st Century” (Moderni ucitel 21. stoleti).

References

1. Maga, D., Simak, B.: Cooperation between Universities and Secondary Schools in the Czech Republic. In: Crisis and Progress, pp. 43 – 52, ISBN 978-963-8445-34-6. Edutus, Tatabanya (2012)
2. Maga, D., Dudak, J., Pavlikova, S., Hajek, J., Simak, B.: Support of Technical Education at Primary and Secondary Level. In: Proceedings of 15th Mechatronika, pp. 195-198. CVUT, Praha (2012)
3. Hrad, J.: Motivating Youngsters for Engineering Study, In: Proceedings of the 24th Annual Conference of European Association for Education in Electrical and Information Engineering. Heraklion: Technological Educational Institute of Crete, pp. 245-246. ISBN 978-960-99889-5-7. (2013)
4. Zeman, T., Hrad, J.: Benefits of collaboration between a university and a vocational college, In: Proceedings of the 25th International Conference on European Association for Education in Electrical and Information Engineering, EAEEIE 2014, pp. 37-40, ISBN 978-147994205-3. IEEE Press (2014)
5. Bauer, P., Dudak, J., Maga, D., Hajek, V.: Distance Practical Education for Power Electronics. International Journal of Engineering Education. vol 23, Nr. 6. pp. 1210-1218. ISSN 0949-149X. (2007)
6. Maga, D., Sitar, J., Bauer, P.: Automatic Control, Design and Results of Distance Power Electric Laboratories. In: Recent Advances in Mechatronics 2008-2009, pp. 281 – 286, ISBN 978-3-642-05021-3. Springer Verlag, Berlin-Heidelberg (2009)
7. Bauer, P., Staudt, V.: Remote controlled practical education for Power Electronics. In European Conference on Power Electronics and Applications, EPE 2007, 10 pages. ISBN 978-92-75815-10-8. EPE Association, Brussels (2007)
8. Hrad, J., Zeman, T., Hajek, J.: Multilanguage e-Learning Course for Industrial Automation. In: IFIP 20th World Computer Congress, IFIP TC 3 ED-L2L Conference, pp. 371-372. ISSN 1571-5736. ISBN 978-0-387-09728-2. Springer, Boston (2008)
9. Zeman, T., Hajek, J., Hrad, Progressive Approaches to Education of Modern Modulation Methods. In: 16th International Workshop on Systems, Signals and Image Processing, IWSSIP 2009; Chalkida. pp. 674542 - 674542, ISBN 978-1-4244-4530-1. (2009)
10. Samsung School | Educational Technology. *Http://www.samsung.com/us/business/by-industry/education-solutions/samsung-school* [online]. © 1995-2015 [ref. 2015-04-15]. Available from: http://www.samsung.com/us/business/by-industry/education-solutions/samsung-school

Ultra-precision machine system feedback-controlled with hexapod-type measurement device for six-degree-of-freedom relative motions between tool and workpiece

Takaaki Oiwa[1], Kazuki Kobayashi[1], Kenji Terabayashi[1], and Junichi Asama[1]

[1] Faculty of Engineering, Shizuoka University, Hamamatsu, Japan
tmtooiw@ipc.shizuoka.ac.jp

Abstract. This study deals with a feedback-controlled precision machine system based on a hexapod-type measurement device for six-degree-of-freedom (6 DOF) relative motions between the tool and workpiece. The hexapod mechanism installed between the tool spindle and surface plate measures the relative motions because the hexapod consists of six passive extensible struts equipped with linear length measurement instruments. This paper describes improvements of the test apparatus for length measurement accuracy of the strut when it is expanded and contracted in the linear motion stage. We show that there is little effect from either room temperature fluctuation or angular motion error of the stage on apparatus accuracy because of the small dead path of the laser interferometer system and small Abbe offset between the strut and the laser path.

Keywords: precision machine system• machine tool• parallel kinematics• Abbe's principle• six-degree-of-freedom motions

1 Introduction

Recently, to realize submicron-order volumetric accuracy of spatially moving mechanisms for precise machining or coordinate measurement, there has been serious interest in a machine system able to generate accurate relative motion between the tool and workpiece, as well as accuracy improvements for each element of the machine [1]. In general orthogonal coordinate machine structure, feedback control is implemented for each feed drive mechanism, which is based on one axis displacement measurement or a one degree-of-freedom (DOF) motion measurement, even though the machine consists of three feed axes. In each feed mechanism, 5-DOF motion errors, including three angular motion errors, are usually not measured to compensate for the Abbe errors caused by the angular motion errors and the Abbe offsets. On the other hand, prediction methods have been investigated to compensate for the thermal deformation of the machine structure based on temperature sensors and thermal deformation analysis. However, thermal deformation is very difficult to predict precisely under rapid temperature change [2].

R. Jabłoński and T. Brezina (eds.), *Advanced Mechatronics Solutions*,
Advances in Intelligent Systems and Computing 393,
DOI 10.1007/978-3-319-23923-1_36

This study proposes a feedback control system based on a hexapod-type measurement device for 6-DOF relative motions between the tool and workpiece [3]. We first describe the system concept. We then compare displacement measured by a linear scale unit installed in the strut with a laser interferometer length measurement system on a test bed when a strut of the hexapod-type measurement device is expanded and contracted by a linear motion stage. In a previous report, the 400-mm dead path of a laser interferometer system and the 57-mm Abbe offset between the laser path and the strut spoiled the performance of the test bed. In this report, the dead-path and Abbe offset have been decreased to mitigate the influence of room temperature fluctuation and angular motion error of the linear stage on the measurement accuracy

2 Machine system based on 6-DOF motion feedback control

Figure 1 shows the components of the proposed machine system based on a measurement device for the 6-DOF motions between the tool and the workpiece. This system consists of a hexapod-type parallel kinematic mechanism (PKM), a conventional machine structure, and a controller. To measure the relative 6-DOF motions, the base platform and moving platform of the PKM are mounted on the surface plate and the machine spindle, respectively. Both platforms are connected through six extensible struts with prismatic joints. Since the struts do not have actuators, the moving platform of the PKM is passively moved in three-dimensional space by the conventional machine. As a displacement measurement unit measures the change in the length of each strut, the 6-DOF motions can be calculated by the forward kinematics of the hexapod-type PKM. Consequently the controller compensates for the motion errors and accurately actuates the tool. Because the measurement device measures the 6-DOF motions, including angular motions, no Abbe error is produced even if the orthogonal coordinate machine structure is employed. Moreover, this measurement device is independent of any elastic and thermal effects and motion errors of the machine structure because the device is separated from the machine's main structure. In other words, the system can not only compensate for the systematic motion error caused by geometrical deviations of the machine elements, but non-repetitive motion errors caused by the elastic and thermal

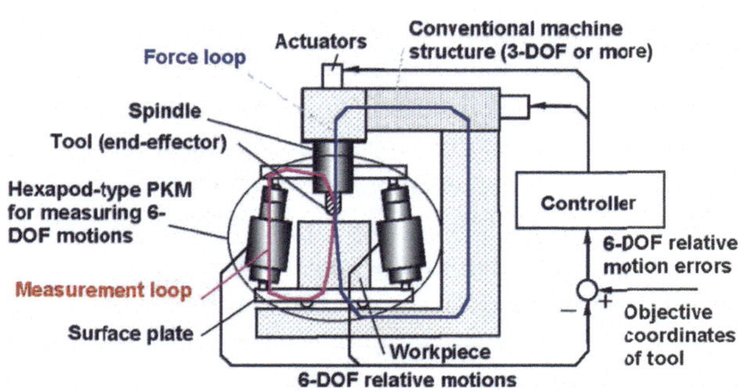

Fig. 1 Fundamentals of proposed machine system based on 6-DOF measurement device between tool and workpiece

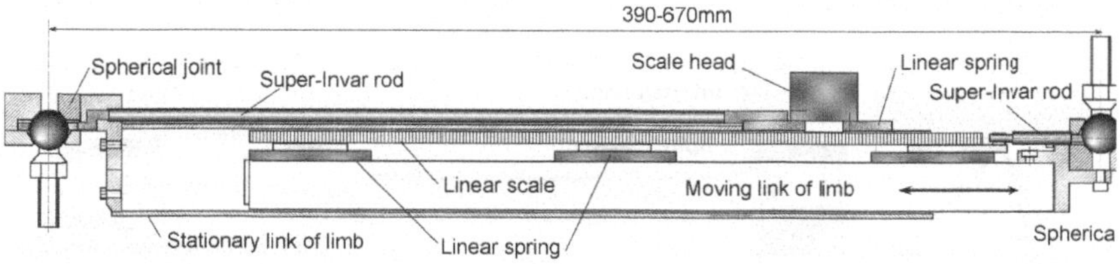

Fig. 2 Passively-extensible strut with compensation device

deformations if the 6-DOF motion can also be precisely measured. If this hexapod is used for the coordinate measuring machine, the coordinates of the probe tip are directly measured by the PKM.

3 Extensible strut with compensation device

Figure 2 is a sectional view of the extensible strut. Four linear-motion roller bearings mounted on the stationary link allow the moving link to move only in the longitudinal direction. The spherical joints are fixed at each end of the stationary link and moving link. The scale unit (Heidenhain LIP481R, with a measuring length of 270 mm and a resolution of 2 nm) is installed on a straight line between the spherical joints so that the Abbe offset is minimized. Moreover, each strut has a mechanical compensation device for both elastic and thermal deformations of joints and links. A rod connects the scale unit head with a spherical joint, while another rod connects the linear scale with another spherical joint. The head and the scale mounted on plates are guided by simple notch-type linear springs that allow them to move only in the longitudinal direction. Thus, the scale unit can measure not only the displacement change of the prismatic joint, but also the spherical joint errors and the link deformation in the strut axis direction. Furthermore, because these rods are made of Super-Invar (thermal expansivity of approximately 0.5 ppm/K), temperature changes have little effect on distances between the scale unit and the spherical joints. Additionally, these distances are not influenced by external forces because no external force is applied to the rods and the scale unit. To put it briefly, even if the strut is thermally or elastically deformed, the scale unit can accurately measure the length change of the strut.

4 Improvement of test bed

To ensure the spatial measurement accuracy of the Hexapod-type 6-DOF measurement device, it is very important to evaluate the length measurement accuracy of each strut. In a previous paper [4], the test bed shown in Figure 3 was used to evaluate the accuracy and the temperature dependency of the struts. A laser-interferometer length measurement system (Renishaw ML-10) was employed to measure the displacement of the moving stage. Basically, the difference between readings of these two displacement measurement systems represented the measurement error of the passive strut. From the results of repeated experiments, we found that when the length of the strut changed by 270 mm, the displacement measurement accuracy was approximate-

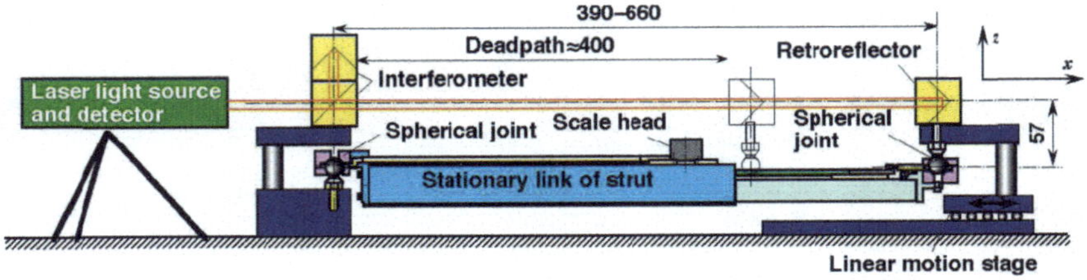

Fig. 3 Experimental setup for measuring linear stage displacement in previous study [4]

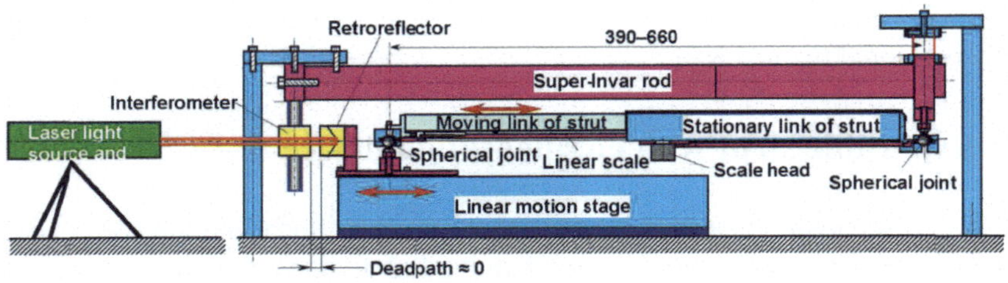

Fig. 4 Improved experimental setup in this report

ly 1 $\mu m_{p\text{-}p}$ in a thermostatic room. However, it was also revealed that a 57-mm Abbe's offset and 20-μrad angular motion of the stage produced an Abbe error of 1.14 μm. In addition, the 400-mm-long dead path of the laser interferometer system definitely caused length measurement errors associated with the change in environmental parameters during a measurement. According to the Edlen's equation on the refractive index of air, a one-degree change in room temperature varies the refractive index by approximately –1 ppm, even if humidity and air pressure are constant. Thus, a 400-mm dead path has the potential to produce a length measurement error of 0.4 μm.

Figure 4 illustrates the test bed improved to reduce the Abbe offset and dead path length. The laser path of the interferometer system is arranged coaxially with the strut axis to decrease the Abbe offset. In addition, the interferometer is mounted as close to the retroreflector as possible to reduce the dead path. In this configuration, the test bed frame from which the interferometer and spherical joint of the stationary link hang is made of Super-Invar to prevent thermal expansion

5 Experimental result

5.1 Constant temperature test

First, displacement measurement was carried out in a thermostatic room. Figure 5 shows measurement errors measured on the previous and present test beds. In this experiment, the maximum error reached about 2.5 $\mu m_{p\text{-}p}$, 1 μm larger than in the previous experiment (i.e., approximately 1.5 μm). The former error is obviously larger than nominal scale accuracy of the linear scale unit (i.e., ±0.5 μm). The pitching motion error of the strut's prismatic joint was approximately 0.8 $mrad_{p\text{-}p}$ as measured

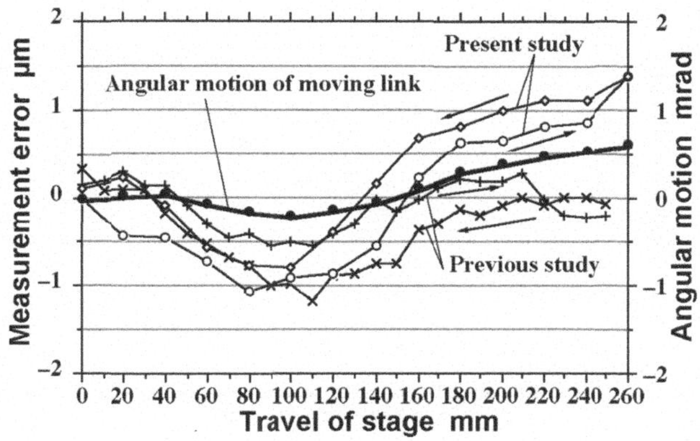

Fig. 5 Length measurement error measured on two test beds and angular motion of strut

by an inclinometer (Jewell LCF-100-1, range: ±1°). Measured angular motion is also plotted in Fig. 5. This chart indicates that variation of the angular motion with the stage travel very much resembles that of the length measurement error of the strut. This angular motion error certainly caused the Abbe error of 0.96 µm, because the linear scale unit has a 1.2-mm offset between the head and the linear scale. From now on, improvement of the angular motion error caused by the prismatic joint will be needed to reduce the length measurement error of the strut.

5.2 Fluctuated temperature test

Next, to verify the improvement in temperature dependency of the test bed, the room temperature was changed from 20 °C to 28 °C during the measurement. Figure 6 shows the difference between the strut's length measurement errors in both constant and fluctuating temperature environments. Although the error difference measured on the present test bed was ±1 µm or almost constant, the error difference measured on the previous test bed increased up to 6 µm in proportion to the stage travel. This confirms that the 400-mm dead path caused such measurement error. Moreover, Fig. 7 shows the relationship between the room temperature and error differences during measurement. This chart indicates that error in the previous test bed increased with the temperature. In other words, the previous test bed had a strong temperature dependency of 0.766 µm/K. On the other hand, the error measured in the present test bed showed almost no change with room temperature (i.e., approximately ±1 µm), demonstrating that it is little affected by temperature.

6 Conclusion

We described an ultra-precision machine system based on 6-DOF motion feedback control between the tool and workpiece. A passive hexapod-type PKM consisting of six extensible struts with linear scale units was used as a feedback sensor to measure the relative motions and orientations between the tool spindle and the surface plate

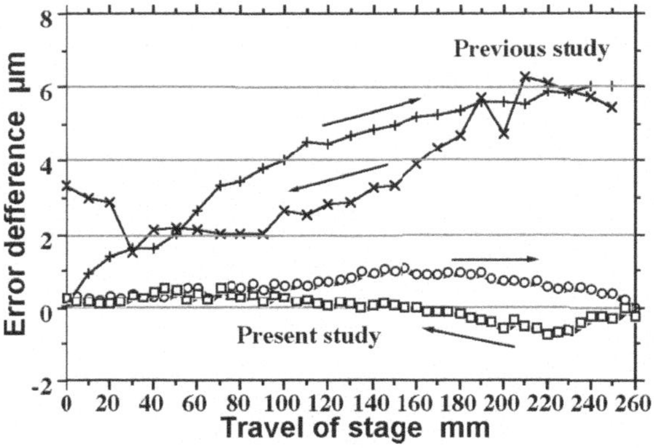

Fig. 6 Effect of temperature disturbance on length measurement error measured on two test beds

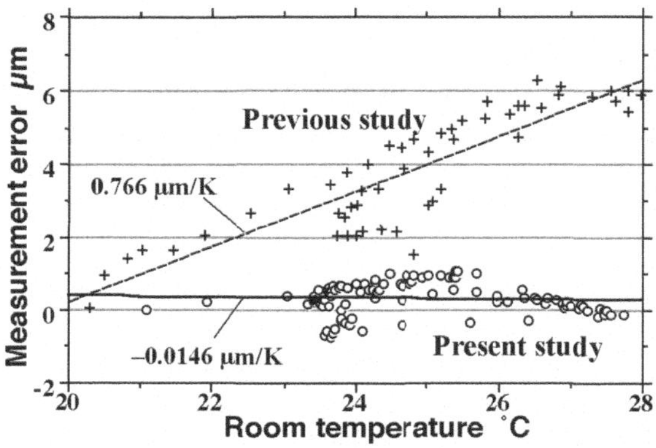

Fig. 7 Temperature dependency of two test beds

regardless of temperature change and external forces. Two kinds of test beds have been investigated to evaluate the length measurement accuracy of each strut. Shortening the dead path of the laser interferometer system successfully improved temperature dependency of the length measurement.

References

1. Ramesh, R., Mannan, M.A., Poo, A.N.: Int. J. Machine Tools & Manufacture, Vol. 40, (2000) pp. 1235-1256.
2. Ramesh, R., Mannan, M.A., Poo, A.N.: Int. J. Machine Tools & Manufacture, Vol. 40, (2000) pp. 1257-1284.
3. Oiwa, T.: An ultraprecise machining system with a hexapod device to measure six-degree-of-freedom relative motions between the tool and workpiece, Int. J. Precision engineering and manufacturing, Vol. 8, No. 1 (2007) pp.3-8.
4. Oiwa, T., Yao, W., Asama, J.: 10th Int. Conf. and Exhibition on Laser Metrology, Machine Tool, CMM & Robotics Performance (Lamdamap 2013), pp. 69-78.

Analysis of the Influence of Input Function Contact Parameters of the Impact Force Process in the MSC. ADAMS

Alžbeta Sapietová[1], Lukáš Gajdoš[2], Vladimír Dekýš[1], Milan Sapieta[1]

[1] Department of Applied Mechanics, Faculty of Mechanical Engineering,
University of Zilina, 010 26 Zilina, Slovak Republic
{alzbeta.sapietova, valadimir.dekys, milan.sapieta}@fstroj.uniza.sk,
[2] Schaffler Slovakia, Dr. G. Schaefflera 1, 02401 Kysucké nové mesto, Slovak Republic
gajdolka@schaeffler.com

Abstract: Integration of differential algebraic equations system in MSC. ADAMS is a process which could not be done without the active participation of the user. The process of integration is necessary to interactively directing with entering appropriate input parameters. This article presents a sensitivity analysis of the input parameters for the function of IMPACT in the MSC. ADAMS. The solution will be presented on simulation of the impact force of the tensioning mechanism. Impact effects will be solved with flexibility of chosen member of the mechanism in mind. The solution will be realized in the environment of software MSC.ADAMS and ANSYS.

Keywords: mechanism, dynamic analysis, impact, ADAMS, tensioning mechanism, rigid body, flexible body, sensitivity analysis

1 Introduction

Analysis of a wide range of problems in technical practice uses mathematical representation of the future product in a virtual computerised form. Such modelling of the problem requires that the parameters input to the solution, and also the outputs, be consistent with reality [1].

Keeping this objective in mind, this paper deals with setting parameters of impact solutions function of load by forces – in this case the so-called impact forces – that cause destruction of the device. If there is interaction between two bodies and the bodies do not move relative to each other, we talk about their contact. Interaction of bodies that move relative to each other is called impact. Impact is characterised by short-term significant forces occurring in the contact area. These impact forces act for a very short time – in the order of milliseconds. Despite the above, these forces cause significant changes in the velocity and thus in the momentum of the participating bodies. The change is by a final value. Therefore, each body must receive a final impulse equivalent to the change in its momentum. However, the impact forces must be

R. Jabłoński and T. Brezina (eds.), *Advanced Mechatronics Solutions*,
Advances in Intelligent Systems and Computing 393,
DOI 10.1007/978-3-319-23923-1_37

very large in order to generate the final impulse in a very short time. This is then followed by the above-mentioned destruction of the material when the forces act [12]. A better understanding of the interdependence between the impact force and the mechanism parameters allows engineers to find more efficient and innovative solutions [2,3,4]. Therefore we carried out analysis of the impact force in the tensioning mechanism using the IMPACT function in the MSC.ADAMS software.

2 Theoretical background of contact forces in the MSC.ADAMS software environment

When the bodies in the virtual prototype models are in contact with each other, the CONTACT function allows making or modifying the contact in the ADAMS/View software environment. Bodies in the contact models can touch each other at points, touch points can travel along pre-described curves, or the curves of one body can be in contact with the curves upon the other body. We can also define the contact between non-deformable bodies at a standstill and in motion. We can model the contact of bodies with stiff (rigid) as well as flexible elements [3].

Solver algorithms detect the contacts of two geometric objects, determine the touch point position, calculate the position of the common normal at the touch point, and calculate the normal and slipping relative velocity of the point in contact.

In terms of dynamics, it is necessary to distinguish:

- discontinuous (pulse) contacts occurring in short periods of time – impacts, for which ADAMS/Solver creates an estimate of the contact force by modeling the course of local deformation. Loss of energy during the collision is represented by the damping force with the corresponding damping coefficient or restitution coefficient. In the first phase, compression occurs during which kinetic energy is converted to potential and dissipative energy of the compressed material. In the second phase, the accumulated potential energy causes reverse motion accompanied by the transformation of potential energy into kinetic and again dissipative energy [3,4,5].

- continuous contacts represented by a model in the form of a non-linear spring-damper system in which stiffness takes into account elasticity of contacting surfaces and damping, in turn, takes into account energy dissipation, and the bodies are not separated after the collision. Calculation of contact forces is carried out separately for each point, and the resulting effect is their sum.

The force acting in the contact point is calculated either by the restitution method (POISSON model) or by the IMPACT function [2,3,4]. Both contact force models result from the regulation of normal contact constraint penalty, where the constraint in this modelling technique is secured mathematically by applying forces according to the constraint gradient, and the magnitude of forces is a function of the constraint breach level [5].

Theoretically, there is no mutual penetration of surfaces at the contact of non-deformable bodies, which can be expressed in the form of inequalities for unilateral constrain. Additional constrain conditions can be expressed also by introducing Lagrange coefficients, but the penalty regulation is advantageous due to its simplicity since no additional equations or variables are needed. The advantage is the ability to apply algorithms of active and inactive conditions for unidirectional constraint as well as for clear interpretation from the physical point of view. The disadvantage of the penalty regulation is that the user is responsible for setting a suitable penalty parameter in the form of material stiffness, a high value of which may cause difficulties in the work of solver integration algorithms. Mutual penetration of the body surfaces is monitored by the gap function g, whose positive value indicates penetration. A positive value of the normal contact force in turn means that it is trying to separate the bodies, and its value is non-zero while the surfaces are in contact [5].

The impact force model in MSC.ADAMS is implemented using the IMPACT function. An example of modelling using the IMPACT function is a ball falling towards the ground, plotted at the beginning of the investigated motion and at impact (Fig.1). If the distance between the I and J markers reaches x_1, the impact force F_{IMPACT} turns on. This occurs when two bodies collide. As long as the distance between the I and J markers is greater than x_1; the force is zero.

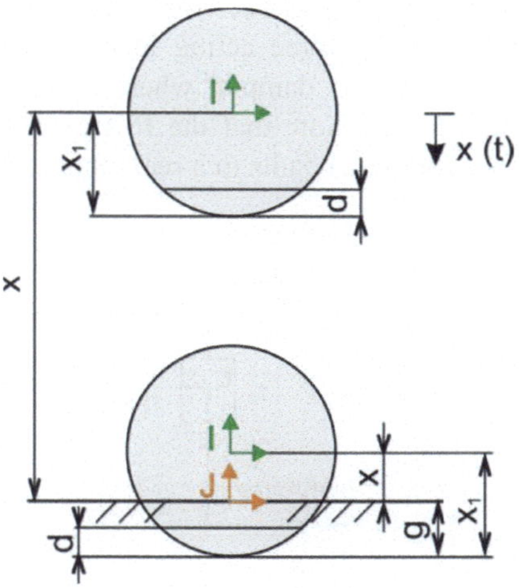

Fig. 1. Example illustrating the IMPACT function

The impact force has two components, a spring or stiffness component, and a damping or viscous component. The stiffness component is proportional to the stiffness coefficient *k*, that is load per unit length, i.e. it is a function of changes in the distance between the I a J markers. The stiffness component opposes the mutual penetration of the body surfaces. The damping coefficient *c* belonging to the damping component of the force is a function of the velocity of mutual penetration of the body surfaces. The damping opposes the direction of relative motion. The damping coefficient achieves a maximum value c_{max}, at a user-defined penetration *d* (Fig. 2b).

Mathematical model of IMPACT force correlates stiffness *k* as a scalar penalty parameter, gap function $g=x_1-x$, and the exponent *e* for modelling a stiffening (e >1) or softening (e <1) spring characteristic of the force (Obr.2a) [4].

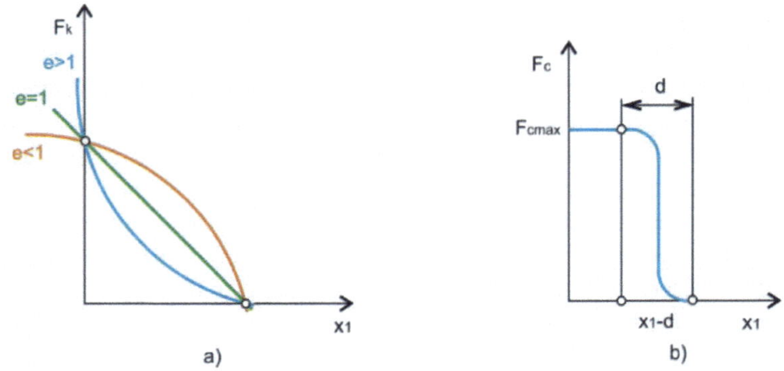

Fig. 2. a) Exponent e of the force in a non-linear spring, **b)** Depth d of the penetration to the maximum damping force

Figure 3 shows a local coordinate frame of the I marker that represents the first body and a local coordinate frame of the J marker that represents the second body. The impact force model consists of the force acting in the compression spring, where $F_k=kx$, and of the force acting in the damper, where $F_c=c\dot{x}$. Then the impact force magnitude is $F_{impact}=F_k+F_c=kx\text{-}c\dot{x}$. Note that the IMPACT function solution in the MSC.ADAMS software environment results in a real number [5].

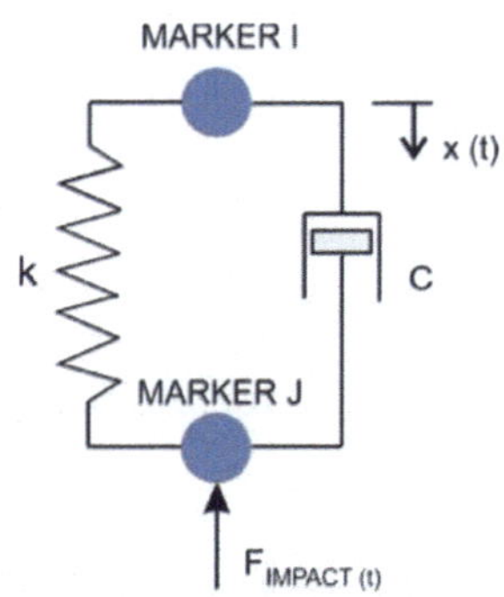

Fig. 3. Model of the unilateral impact force by the IMPACT function

Given the solver problems with high stiffness values and the need to include critical material properties into the impact model, the solver algorithm works with an extended mathematical model that includes non-linear viscous damping dependent on the magnitude and velocity of mutual penetration of surfaces.

The equation defining IMPACT is:

$$IMPACT = \begin{Bmatrix} Max\ (0, k(x_1 - x)^e - STEP(x, x_1 - d, c_{max}, x_1, 0).\dot{x}) & : x < x_1 \\ 0 & : x \geq x_1 \end{Bmatrix},$$

Based on the presented impact model, the following applies (Fig. 1):
$x > x_1$, no mutual surface penetration occurs and the force is zero (penetration $g = 0$)
$x < x_1$, penetration g occurs at the end closer to the J marker, and the force is > 0.
Also note that when $g < d$, the instantaneous damping coefficient is a cubic step function of the penetration g.
When $g > d$, the instantaneous damping coefficient is c_{max}.
The MSC.ADAMS (C ++) software solver never returns a negative force for impact. If the above expression is negative, the solver returns a force value of zero.
It must be said that in terms of theory there has been a failure so far to mathematically describe the phenomena during impact, although the impact models based on energy balance before and after the impact correlate the corresponding parameters with sufficient precision [5,6].

3 Introduction to the solutions of impact forces in a tensioning mechanism

To solve the contact effects using the IMPACT function we selected a wedge belt tensioning mechanism in an automobile engine. Purpose of the mechanism is to generate a sufficient contact force on the wedge belt so that the wedge belt tensioning ensures correct functioning of the devices depending on the wedge belt movement. Another purpose of the mechanism is to regulate the tension force depending on the transient phenomena (change in engine revolutions, load changes in individual devices) [13].
Figure 4a) shows a kinematic diagram of the tensioning mechanism that has two degrees of freedom ($n = 2$). To maintain the mechanism in equilibrium, one degree of freedom is taken by a pre-stressed spring positioned in cylinder 2 and the other degree in the belt girded on pulley 5. When the belt is girded on pulley 5, the girding force acts through a lever mechanism 4 on the spring that is hereby compressed. Thus, its stiffness allows eliminating the undesirable effects of transients acting on the mecha nical system of the car. Practical experience shows that no impact occurs under normal operating conditions. Problems occur at the moment when the wedge belt breaks or suddenly loosens as a result of assembly error. This rapidly releases the spring from its compressed state and the structural elements of its containment in the cylinder become subjected to such impact force F_{IMPACT}, which can cause destruction of the tensioning mechanism.

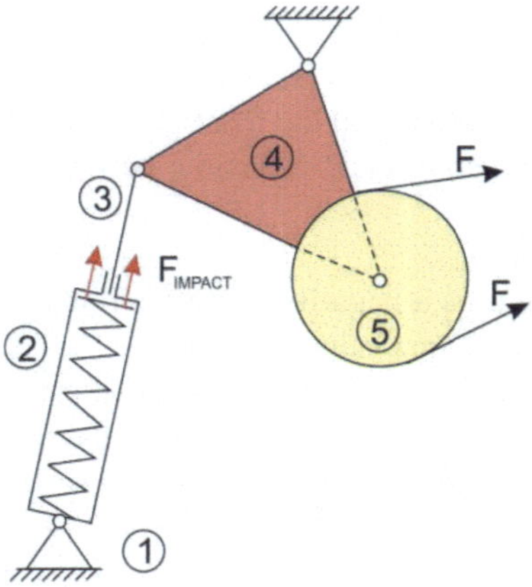

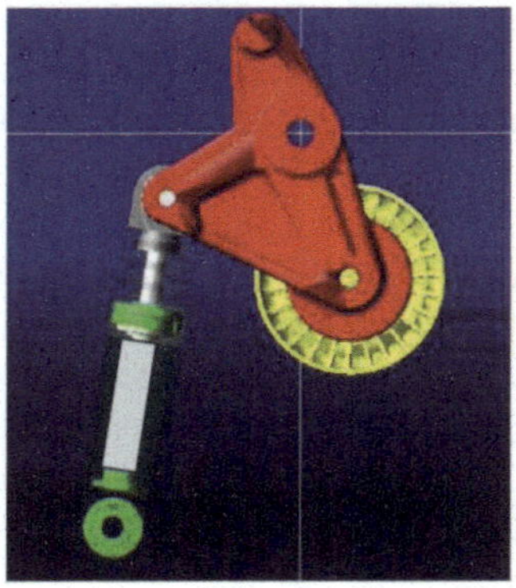

Fig. 4. a) Kinematic scheme of a tensioner, **b)** Virtual prototype of a tensioner in MSC.ADAMS

4 Creation of a virtual prototype in the MSC.ADAMS environment

Individual members of the mechanical tensioner system were imported from the ProE 4.0 software to the MSC.ADAMS/View software. In addition to solid (rigid) members, the system also included one flexible member modelled in the MKP environment of the ANSYS software [7,8,9]. The flexible member was a snap ring, which rammed into the guide ring during destruction (Fig.5). This MKP model attempted to ensure that the addressed parameters be in conformity with reality. After defining the dynamic parameters, geometric and kinematic constraints in the MSC.ADAMS/View environment, the system was prepared for analysis [10,11].
After defining the spring parameters that were obtained from real measurements (prestress, spring stiffness and damping effects of oil through the damping coefficient), calculation was conducted.

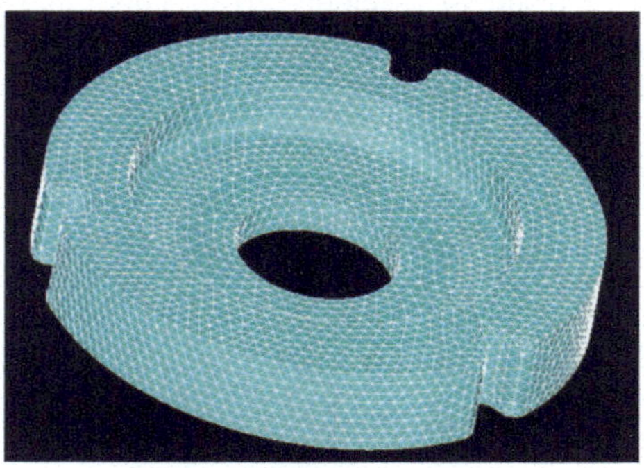

Fig. 5. FEM mesh of guide ring.

5 Sensitivity analysis of the contact parameters for a flexible body impacting a solid one (flex to solid)

A series of sensitivity analyses was performed to define contact parameters between a solid body and a flexible one. Sensitivity analyses were carried out using the Design Study method, which is a tool for studying the influence of one parameter on the objective function in a specified range of its changes. First, we performed sensitivity analysis of the influence of the *calculation step size*. The smaller the time step, the longer the computation time. A too large time step would increase the inaccuracy of numerical computation, and the impact force might not even be recorded.

Simulation was before the impact performed with a large time step and time in which it is assumed the impact has been set very small time step because of faster computation progress. The following simulation script was created to adjust the step size during simulation:

SIMULATE/TRANSIENT, END=0.0055, DTOUT= 0.00002

SIMULATE/TRANSIENT, END=0.006, DTOUT= 0.00000005

Figure 6 shows the waveforms of the impact force at each time step sizes. The analysis shows that the impact force computation is stable as soon as the time step is 50*ns* (Fig.6).

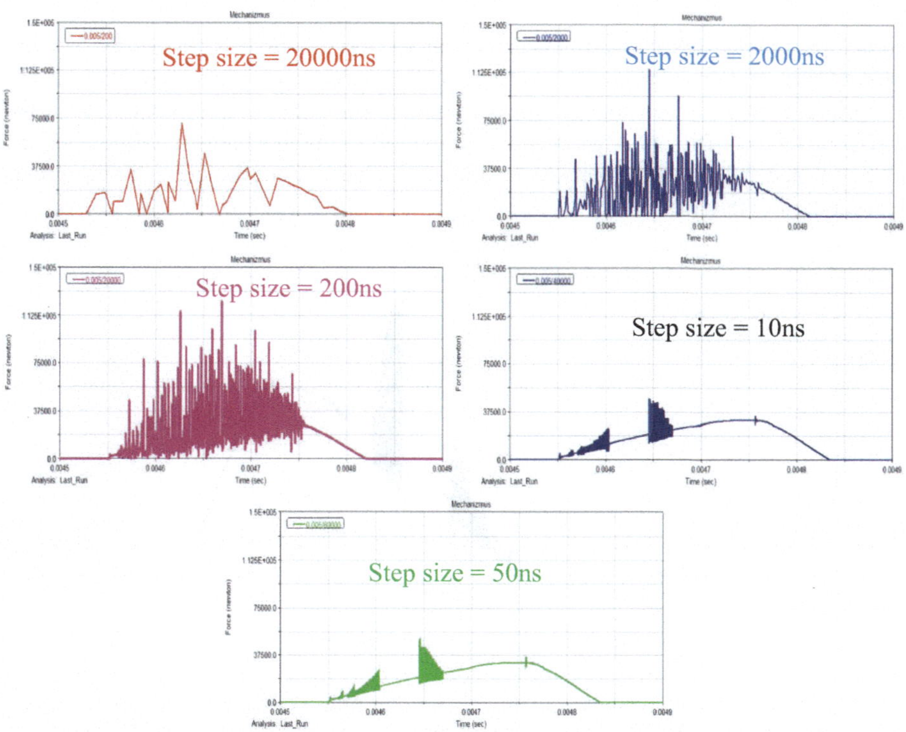

Fig. 6. Sensitivity analysis of the influence of the calculation step size

Next, we investigated the influence of the *surface interaction stiffness coefficient k* on the impact force computation for a flexible body. Input stiffness parameter values were chosen between $1 \cdot 10^4$ and $1 \cdot 10^6$ N/mm. The analysis shows that the given parameters of contact stiffness in impact affect the course of the impact force but do not affect the impact force magnitude. The $1 \cdot 10^5$ N/mm stiffness value can be considered stable (Fig.7).

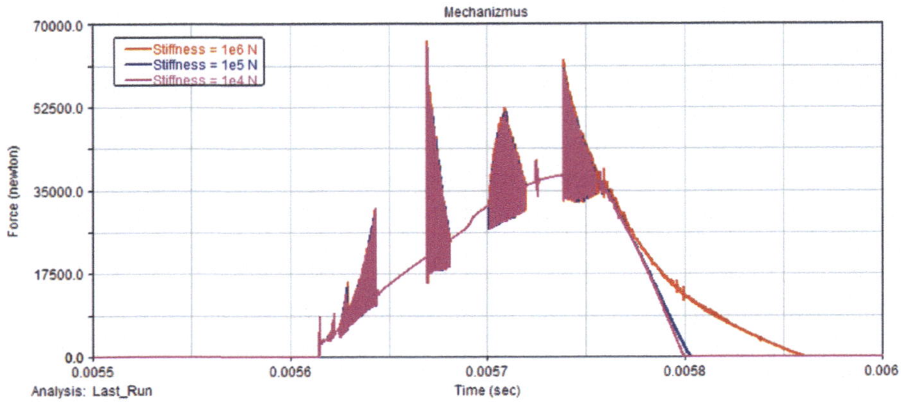

Fig. 7 The influence of the surface interaction stiffness coefficient *k*

Another parameter that can influence the impact force course is the *stiffness force exponent* (Fig.8). For soft materials such as rubber we choose values close to 1.1. It is appropriate to choose the value 1.5 for soft metals, and 2.2 for hard metals. To verify the effect of the coefficient *e*, we made a simulation of the mechanism with the values of 1.5, 1.7 and 2.2 [4]. The analysis results show that the coefficient *e* had minimal effect on the flexible model.

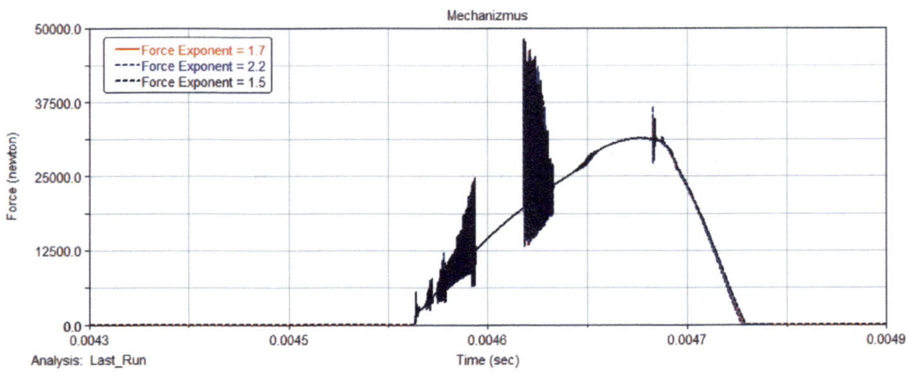

Fig. 8. the stiffness force exponent *e*

In another sensitivity analysis we changed the damping coefficient size. The values varied in the range from 1 to 1000 Nsmm^{-1}. For the damping value of 1 Nsmm^{-1} the force pulse of the first phase (from contact to maximum load) is identical to the force pulse of the second phase (from maximum load to the end of contact), which corresponds to the impact of bodies without damping. At higher damping values the force pulse of the second phase of impact is reduced, thus corresponding more credibly to real deformable bodies. However, larger computation instabilities occur at higher damping values, which are manifested by peaks in the course (Fig.9). For further calculations we chose 100 $Nsmm^{-1}$ (blue course) as a suitable value of the damping parameter.

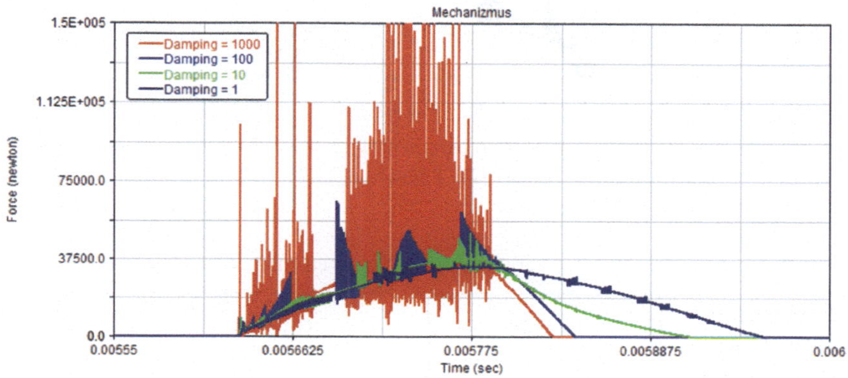

Fig. 9. Influence of the contact damping parameter *c* on the impact computation using the IMPACT function

The last parameter defining the CONTACT function in MSC.ADAMS is *penetration depth d* of the contact surfaces (Fig.10). The material reaches the maximum damping force in the overlap value. The recommended value is 0.01 [4]. During the sensitivity analysis we changed the overlap in values 0.01, 0.1, 1, and additionally 0.5 mm. The results show that this parameter has a great influence on the calculation stability and the impact course because it cooperates closely with the damping coefficient. As long as the penetration depth value is too high there is no full damping, thereby the damping course approaches elastic impact. Lowering the penetration value brings a more realistic impact course, but with certain instability that is manifested by high peaks. These peaks can be reduced using increased calculation accuracy by solver parameter adjustment. However, increased calculation accuracy increases severalfold the time required for calculation.

At first sight, the most stable impact course can be observed in penetration 0.01 mm. However, after displaying the course of the guide ring deformation we concluded that the calculation is not stable in the first phase of the impact. At the penetration value of 0.1mm, the guide ring shape was taken into account from the beginning, and the ring deformation was uniform in the entire impact range. Therefore, we considered the penetration value of 0.1mm for further analysis. Table 1 shows an overview of parameter influence of values for the computation of the maximum overlap.

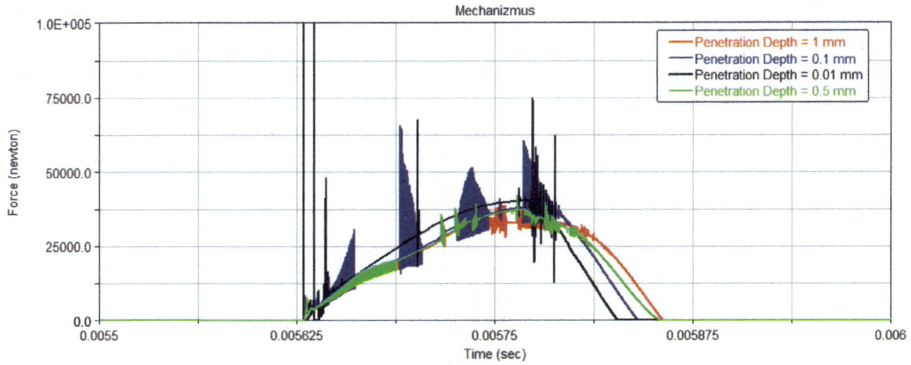

Fig. 10. Effect of the maximum overlap parameter *d* on the impact calculation using the IMPACT function

Tab. 1. Summary of influence parameter of values to calculate the maximum overlap

Penetration depth (mm)	0.01	0.1	1
Max. impact force (kN)	-	31.8	28.5
End of impact (s)	0.00481	0.00483	0.00486

Building on the experience gained in sensitivity analyses we chose parameters shown in table 2 to analyse the impact force of a flexible body.

Tab. 2. Parameters of the IMPACT function for a flexible body to a solid one

IMPACT	
Stiffnes [N/mm]	1.0E+005
Force exp.	1.7
Damping [$Nsmm^{-1}$]	100
Penetration depth [mm]	0.1

6 Summary

In previous sensitivity analysis we can see the effect of input contact parameters of impact force process. Significant influence on the accuracy of the impact force solutions are the choice of time steps. The time step is smaller the computation is longer. A too large time step would increase the inaccuracy of numerical computation, and the impact force might not even be recorded. Generally speaking, the input parameters for the function Impact have no significant effect on the maximum size of the impact force of flexible body. The influence of these parameters has significant effect on timing of the impact mainly in the second phase of the contact force, causing a change of restitution coefficient in the solution.

Acknowledgements

This work was supported by the Slovak Research and Development Agency under the contract No. APVV-0736-12.

References

1. Grega, R.: The Chance for Application of the Pneumatic Dual: Mass Flywheel in Drive Combustion Engine. In: 54th International Conference of Machine-Design-Departments, MODERN METHODS OF CONSTRUCTION DESIGN , Book Series: Lecture Notes in Mechanical Engineering, pp. 43-47, (2014)
2. Stronge, W. J.: Impact Mechanics, Cambridge University Press, (2004)
3. Johnson, K. L.: Contact Mechanics. Cambridge University Press. Cambridge, (2013)
4. Giesbers, J.: Contact Mechanics in MSC Adams – A technical evaluation of the contact models in multibody Dynamics software MSC Adams: Bachelor Thesis. University of Twente. 42 pages. Information on http://purl.utwente.nl/essays/62109 (2012)
5. Information on http://atc.sjf.stuba.sk/files/tm2/8_cvic_TM_II_naraz_sustav.pdf
6. Dekys, V., Kopas, P., Sapieta, M., Stevka, O.: A detection of deformation mechanisms using infrared thermography and acoustic emission. In: NOVEL TRENDS IN PRODUCTION DEVICES AND SYSTEMS Book Series: Applied Mechanics and Materials, 474, pp. 315-320, (2014)
7. Kopas, P., Vaško, M., Handrik, M.: Computational modeling of the microplasticization state in the nodular cast iron. In: Applied Mechanics and Materials, Vol. 474, pp. 285-290,(2014)
8. Shimanovsky, A. O., Kuznyatsova, M. G., Pleskachevskii, Yu. M.: The strength analysis of the partitions in road tank reservoirs. In Procedia Engineering, vol. 48, pp. 607-612, (2012)
9. Vasko, M., Guran, A., Jakubovičová, L., Kopas, P.: Determination the contact stress depending on the load rate of the NU220 roller bearin. In Komunikacie, Vol. 15, No 2, pp. 88-94. (2013)
10. Vavro, J., Vavro, jr. J., Vavrova, A., Kovacikova, P.: Dynamic Analysis of Winding Mechanisms for Manufacturing of Raw Tyres. In: Hutnické listy. 64, pp. 189-192, (2011)
11. Zmindak, M., Saga, M.: Structural Optimization of Trusss Structures for Deterministic and Nondeterministic Lodas Using MATLAB. In: Proceedings of the VII th International Conference on Numerical Methods in Continuum Mechanics. High Tatras, pp. 368-373, (1998)
12. Stancekova, D., Czan, A., Derbas, M.: Investigation of defects in forging tools by nondestructive detection method. In: 22nd International Conference on Metallurgy and Materials, pp. 485-490, (2013)
13. Bąkowski, A., Jaroszewicz, J. Radziszewski, L.: Assessment of the Mechanical Properties of a Diesel Engine Injector. In: Applied Mechanics and Materials, 630, pp. 334-340. (2014)

Experimental measurements of levitation force generated by high-temperature superconductors in magnetic field

Anna Sibilska-Mroziewicz[1], Edyta Ładyżyńska-Kozdraś[1], Krzysztof Falkowski[2], Krzysztof Wolski[1], Wojciech Credo[1], Andrzej Skalski[1]

[1] Politechnika Warszawska Wydział Mechatroniki, {A.Sibilska, E.Ladyzynska, K.Wolski, W.Credo, A.Skalski}@mchtr.pw.edu.pl
[2] Wojskowa Akademia Techniczna kfalkowski@wat.edu.pl

Abstract. This paper describes the measuring procedure of the levitation force, as a function of levitation gap. Levitation force is a consequence of the Meissner effect occurring between superconductor and the magnetic field. Those experiments are part of studies on magnetic Unmanned Air Vehicles Catapult dynamics. Described experiments used height quality testing machine, box with four height-temperature superconductors YBCO and part of magnetic catapult rail system made of neodymium magnets.

Keywords: magnetic levitation, high-temperature superconductors, Meissner effect

1 Introduction

The levitation term comes from the Latin levitas - lightness and is a physical phenomenon, in which the body is at rest, without any physical contact with the substrate material. Uses of the levitation phenomenon in mechanical systems eliminate the friction between the body and the substrate, which significantly improve the energy efficiency of the system. At the present technology development the phenomenon of levitation may occur mostly by magnetic and electromagnetic interactions.

Fig. 1. Commercial Mag-Lev solutions: Transrapid (EMS suspension), Japan Railways Company (EDS suspension)

R. Jabłoński and T. Brezina (eds.), *Advanced Mechatronics Solutions*,
Advances in Intelligent Systems and Computing 393,
DOI 10.1007/978-3-319-23923-1_38

To date, magnetic levitation had the strongest impact on the rail transport and magnetic bearings industry. There are currently two commercial solutions of magnetic transport suspensions (Mag-Lev)[1]. Electromagnetic suspension (EMS) developed by the German Transrapid system, based on the strength attraction force between the metal rails and mounted on the underside of the train electromagnets. The second solution is electrodynamics suspension (EDS) used by Central Japan Railway Company , that use strong magnetic fields generated by superconducting electromagnets. An attractive alternative for those solutions is to use levitating diamagnets, in particular high-temperature superconductors. Described in paper research is part of an innovative solution of using magnetic suspension system, based on high-temperature superconductors, in the UAV (Unmanned Air Vehicle) catapult.

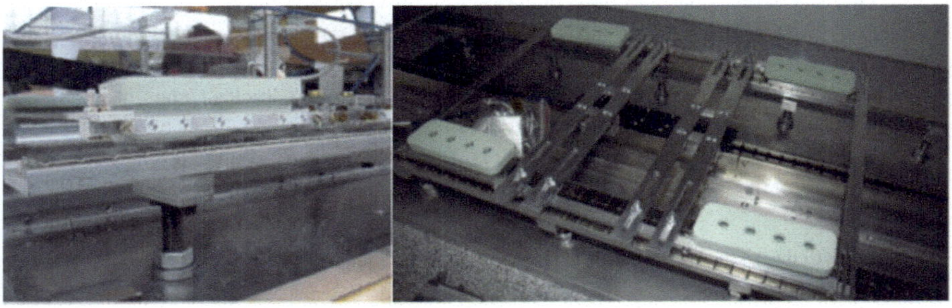

Fig. 2. Magnetic UAV catapult based on Meissner effect

1.1 Superconductivity

The superconductivity phenomena dates back to the first decade of the twentieth century. Liquefying helium, boiling at 4.2 K, allowed the Dutch physicist Heike Kamerlingh-Onnes, study the properties of metals at very low temperatures [2]. Scientist discovered that the electrical resistance of mercury is not decreasing gradually, but at a certain temperature suddenly fades to zero. The same properties were found for others non-magnetic metals, such as lead and tin. Physics works are currently regarded as one of the greatest discoveries in the history of physics. Superconductivity is a separate thermodynamic state of matter and occurs in very specific, for a given material, external conditions. The critical temperature Tc, intensity of the magnetic field Bc, and the critical current Jc characterize superconducting state Fig. 2. The best theory describing superconductivity is BCS theory [3], saying that lossless current flowing inside the superconductor is the result of so-called Cooper pairs movement, rather than single electrons. Superconductors exhibit many interesting physical and chemical properties, but from the magnetic suspension standpoint the most interesting is the Meissner effect and flux pinning.

Meissner effect exhibit in pushing off the magnetic field, outside the superconductor [4]. According to Lentz rule, if the conductor is placed in the magnetic field, then, to neutralize external field, the current flow inside the conductor material is induced. In the case of superconductors, in which the current flows without any resistance, field shielding is complete. Shielding currents flow in a thin layer near the surface of the superconductor and change superconductor into magnet. Superconductor magnetization vector is always facing away from the external field (owned by diamagnetic) what results in stable levitation.

Many metallic elements, alloys, intermetallic compounds and doped semiconductor, characterized by different crystallographic structure, exhibit superconductivity. The critical temperature range extends from 0.001 K for rhodium to 165 K for $HgBa_2Ca_2Cu_3O_{8+\delta}$ at a pressure of 20 GPa. The adjective "high-temperature" is therefore a relative term and refers to a superconductor with a critical temperature above 30K. The significant milestone in superconductivity was discovering of YBCO, which can be cooled down below critical temperature with liquid nitrogen. not dangerous and expensive liquid helium.

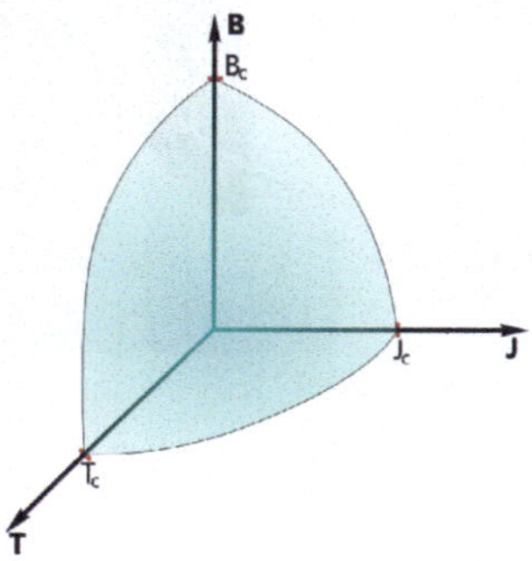

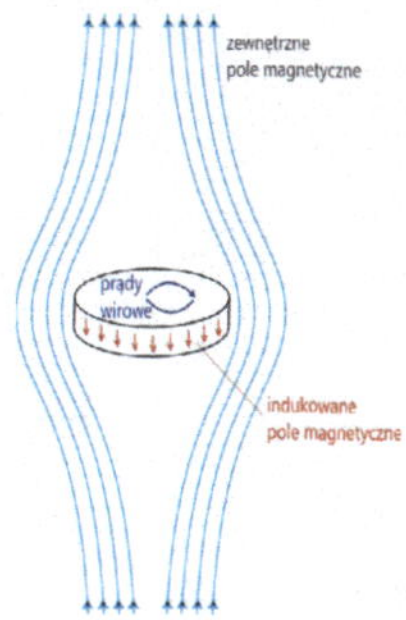

Fig. 3. Superconducting state as a separate thermodynamic state and Meissner effect

2 Research post

The study used the MTS Bionics 793.00 testing machine. The machine lower jaws are stationary. The force sensor, model 661.18H.01, 1kN scope is mounted below the lower jaws. The force sensor measure the levitation force generated by high temperature superconductors placed in static magnetic field. The lower jaws hold part of magnetic catapult rail system. Rails are made of steel beams with three rows of neodymium magnets. During experiments two kinds of magnets configurations showed in fig.5 were tested. The upper jaws of testing machine move vertically, with different velocities. The upper jaw position is measured by laser extensometer. The upper jaws hold fixing device designed and manufactured at Warsaw University of Technology, Faculty of Mechatronics. The device is made of non-magnetic materials and allows to safety flooding of superconductors with liquid nitrogen, in order to cool YBCO below a critical temperature. Four high-temperature superconductors were placed in box made of material with high thermal insulation. Box with superconductors is attached to fixing device hold by upper jaws.

During experiments box with superconductors is moving vertically in stationary magnetic field generated by neodymium magnets. The levitation force resulting from Meissner effect is measured by testing machine force sensor. Experiments results show the levitation force characteristics as a function of distance between magnetic rails and container with superconductors.

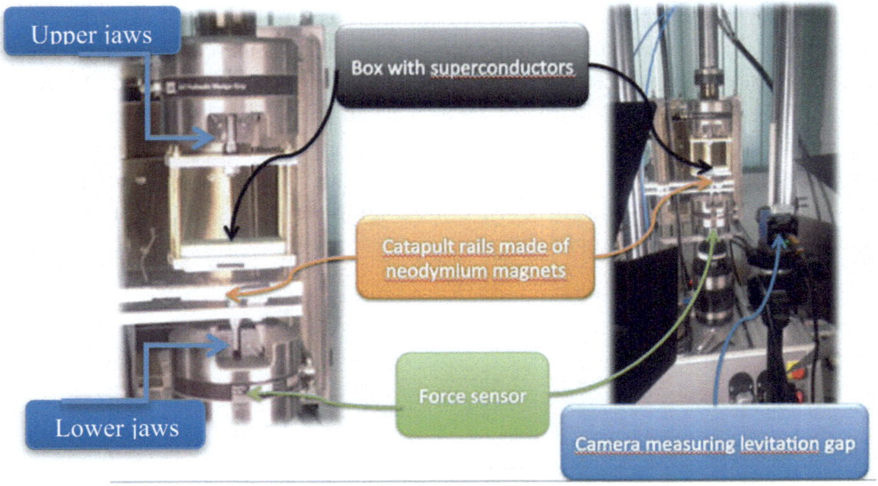

Fig. 4. Research post

3 Test results

Below the measurements results of two different experiments are shown. Each result consists of three charts. Top chart shows changes over time the levitation force recorded by testing machine force transceiver. The middle chart shows changes over time of the relative distance between the box with superconductors and generating magnetic field catapult rail. The last chart shows hysteresis loops. In the first experiment the value of levitation force, depending on magnetic field shape was examined. Catapult rails, fixed to the bottom jaws of testing machine, are made of three rows of neodymium magnets.

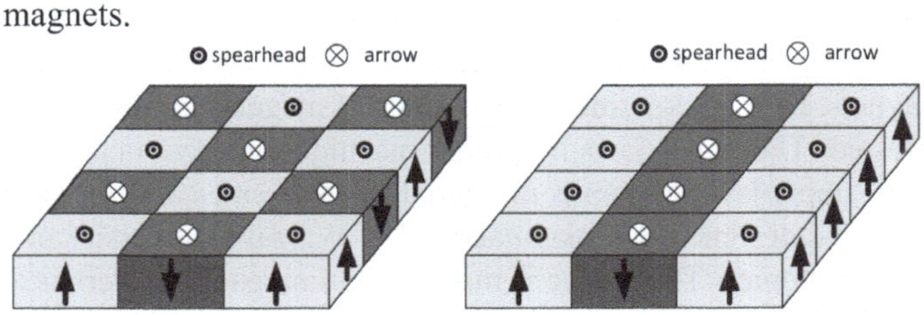

Fig. 4. Magnets configurations of the magnetic UAV catapult rail system

Two configurations of magnets polarization were examined. The first one is "chest" configuration (Magnets Configuration A, left in fig. 5). All magnets edges touch edges of opposite polarization magnets. The second one is "gutter " (Magnets Configuration B, right in fig. 5). Magnets edges along the rails touch magnets with the same configuration, while edges across the rails touch magnets with opposite configuration.

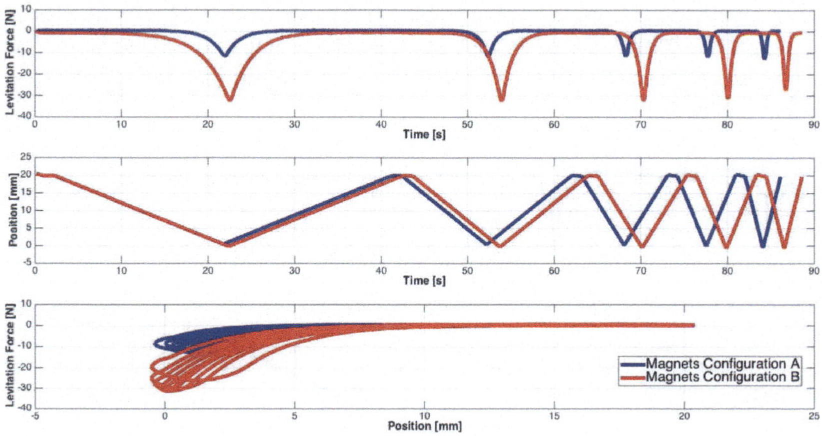

Fig. 6. First experiment for different magnets configuration

The experiment results are shown in Fig. 6. The box consisting of superconductors was moving continuously up and down with rising velocities. It can be observed that levitation force does not depend much on velocity, so we can assume that dissipation coefficient is quite low. For the first magnets configuration ("chest") the maximum force for 1mm levitation gap is 11N. For the same levitation gap, but different magnets configuration ("gutter") the levitation force is 34 N. So we observe that the levitation force value is about three times higher for the second magnets configuration.

The second experiment shows extremely interesting and surprising results. They were obtained for change height of flooding superconductors with liquid nitrogen. At distance of 20 mm it can be assumed that magnetic field generated by catapult tracks have zero value. At the superconductor flooded with liquid nitrogen in zero magnetic field acts only levitation force resulting from Meissner effect. In case when superconductor is flooded in lower height, part of magnetic field lines penetrates into a superconductor. During transition YBCO into superconducting state, penetrating a superconductor field lines are "trapped" inside. This phenomenon is known as flux pinning effect [10]. Movement of trapped inside of a superconductor vortices is a source of energy dissipation. Superconductor position, coinciding with the stored arrangement of magnetic field lines, is an attractor position. In other words, it appears additional force attracting superconductor to a position where the external magnetic field configuration coincides with the configuration "remember" during the transition in the superconducting state.

Fig. 7 shows a comparison of dynamical measurement of levitation force depending on height of flooding at 20mm and 3mm. In the case of flooding superconductor in a non-zero magnetic field (blue line) recorded maximum value levitation force is much lower than in the case of flooding superconductor at zero magnetic field (red line). For levitation gap of 3mm levitation force is zero, while above that value force takes positive values. It means that on the superconductor acts two forces: levitation force repulsing superconductor from magnetic rails and additional force attracting superconductor to place of flooding.

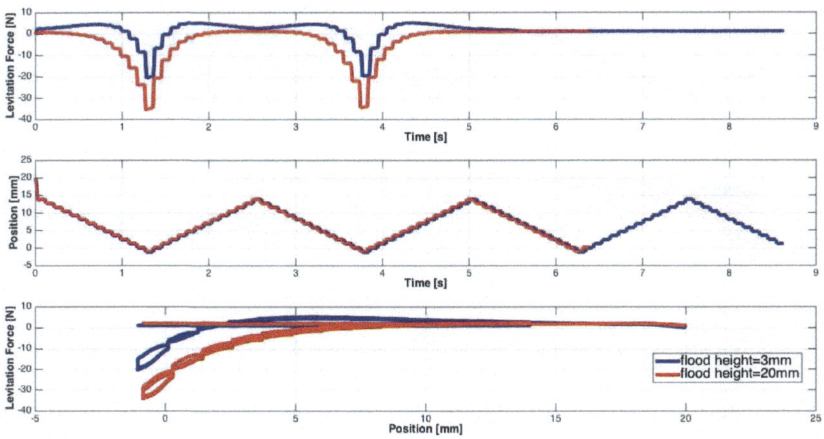

Fig. 7. Second experiment for different height of flooding superconductor

4 Conclusions

Presented in the paper research method provides promising results. This method allows to experimental determination of the stiffness coefficient of levitation force. In future we plan to repeat the measurements for different number and type of superconductors. The disadvantage of this method is height inertia of testing machine, which can be seen in dynamic measurement, especially at high velocities.

References

1. Francis C. Moon.: Superconducting Levitation: Applications to Bearings & Magnetic Transportation. ISBN 0-471-55925-3. Wiley-VCH, August 1994
2. Karol Izydor Wysokiński :Nadprzewodnictwo – pierwsze 100 lat. Postępu Fizyki, 62:222, 2011
3. Robert Schrieffer John Bardeen, Leon Cooper .: Theory of superconductivity. *Phys. Rev.*, 108:1175, 1957
4. A. Szewczyk, A. Wiśniewski, R. Puźniak, H. Szymczak : Magnetyzm i nadprzewodnictwo PWN, 2012
5. R. Feynman: Wykłady z Fizyki

Accuracy of the Parts from Iron Powder Manufactured by Injection Moulding

Andrzej Skalski[1], Dionizy Bialo[2]

[1] Institute of Metrology and Biomedical Engineering,
Warsaw University of Technology, Warsaw , Poland
askalski@mchtr.pw.edu.pl
[2] University of Ecology and Management, Warsaw , Poland
dbialo@mchtr.pw.edu.pl

Abstract. This paper presents the results of investigation on the influence of injection moulding conditions on the challenge of the element diameters after injection and after sintering. The feedstock composed of iron powders with granularity of 1 and 4 μm and special thermoplastic binder based on paraffin, wax, polyethylene and stearic acid were used. The paper looks at the influence of powder content V_p and powder granularity on shrinkage and spread of the element diameters. As the indicating parameter of element accuracy standard deviation was selected.

Keywords: injection moulding, iron powder , shrinkage, accuracy

1 Introduction

Several methods of micro-machining are applied to make of micro parts and micro structures. The most important of these include micro-milling, etching, LIGA technology, erosive and laser micro-machining, plastic micro–moulding, etc. [1, 2, 3].
One of the more promising methods is forming of micro-elements through injection. Such a method of fabrication makes possible the making of micro-elements with complex forms at high replication accuracies in the form of large-scale series at high efficiency that are competitive with respect to other fabrication methods.
The essence of the injection process for forming elements using powders – Powder Injection Moulding (PIM) – involves the preparing of feedstock composed of special thermoplastic binder and powder, preparing the granulate, and forming the parts by way of injection using plastic injection moulding machines [4,5]. Subsequently, the part is subjected to a debinding process, i.e. the removal of binder, and sintering. The process of micro part injection moulding – μPIM and μMIM (Micro Powder Injection Moulding, Micro Metal Injection Moulding) – is significantly more difficult than any macro injection moulding of products due to the small mass of the micro parts (<0.01 g) and different thermal conditions of the process. It is necessary to use injection

R. Jabłoński and T. Brezina (eds.), *Advanced Mechatronics Solutions*,
Advances in Intelligent Systems and Computing 393,
DOI 10.1007/978-3-319-23923-1_39

moulding machines of special design and operation (different plastifying and feedstock feeding systems).
Schematic diagram of the process of powder injection moulding is presented on Fig.1.

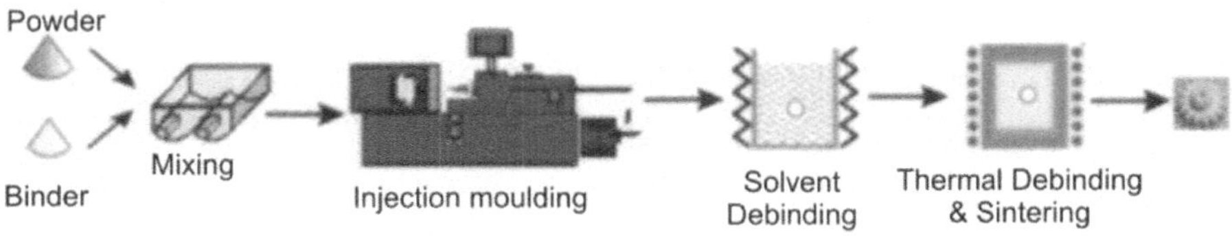

Fig. 1. Schematic diagram of the process of powder injection moulding

2 Material

The feedstocks for injection included the iron powder and special thermoplastic binder were prepared by hot mixing. Iron powders with granularity of 1and 4 μm were used in this study. The view of the powders is shown on Fig.2.

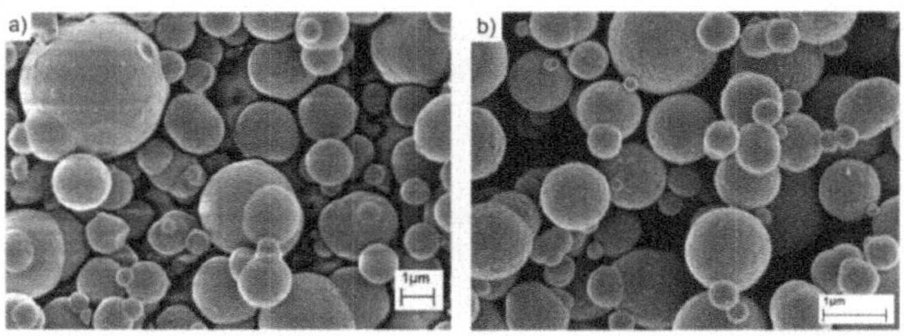

Fig. 2. Iron powders: a) OM – 4 μm, b) HQ – 1 μm

The complex binder system based on paraffin, low density polyethylene, wax and stearic acid has been applied here. Binder composition is presented on Table 1.

Table 1. Binder composition

Material	Percentage share
Paraffin	64%
LD polyethylene	21%
Wax	10%
Stearic acid	5%

Volume fraction of powder in feedstocks V_p was in the range of 45 do 60% vol.
Feedstocks were injected to the mould cavity with folowing dimentions:
width b = 1,209 mm, high h = 1,175 mm and length l_w = 12,978 mm (Fig. 3).
The range of parameters of injection moulding has been established:

- pressure of injection p = 80 MPa
- feedstock temperature T_w =125°C
- mould temperature T_f = 75°C

3 Tests

Shrinkage of the samples after injection (s_w) was defined:

$$s_w = \frac{l_w - l_f}{l_w} * 100\% \quad (1)$$

where : l_w – length of mould cavity
l_f– length of the sample after injection

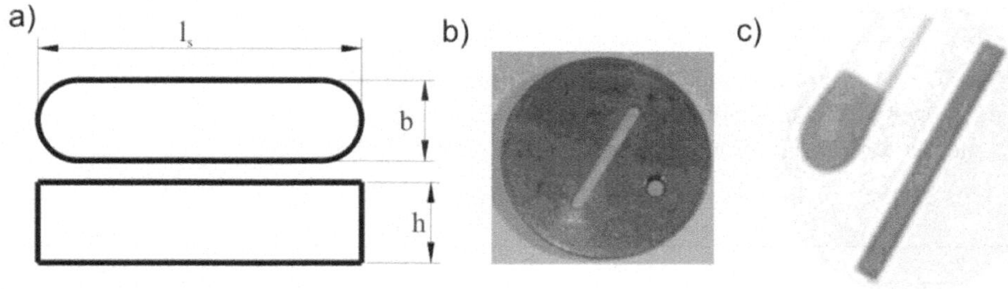

Fig. 3. Shape of the sample a), mould b), sample c)

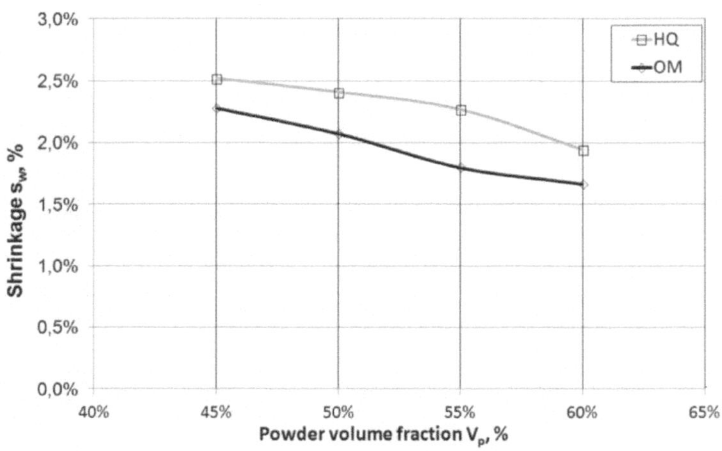

Fig. 4. Shrinkage sample S_w vs volume fraction of the powder in the feedstock; at: T_f = 70°C and T_w = 125°C

Shrinkage of the sample vs volume fraction of powder in the feedstock is shown on Fig.4. Value of shrinkage was in the range of 1,8 – 2,5%.

As can be seen shrinkage with the increase of V_p decreases. This dependence occur for both type of iron powder. It is worth to note that granularity of powder in the feedstock also affect shrinkage. The shrinkage was higher for smaller powder particle in the samples. This dependence was also note by Autors [10] for another powders i.e. stainless steel or ceramic.

After injection the samples were subjected to debinding and sintering process. The debinding (removal of binder) was currying out in the form of solvent debinding using tetrachloroethylene. Sintering took place at temperature of 1100°C in protective atmosphere $80H_2/20N_2$, for 1 hour.

Value of dimensions and their standard deviation for samples prepared from 1 μm iron powder (HQ) after sintering is shown on Table 2. Similar results for both powder used is presented on Fig.5.

Dependence between sample length l_s, powder volume fraction V_p and powder granularity is evidence. The finer powder used and smaller V_p, the greater shrinkage of the finished product and as a result smaller is l_s with respect to the mould cavity.

Above was mentioned that standard deviation of sample dimension may be used as indicator of sample accuracy. From presented results we can note that this parameter is greater for smaller V_p and changed for the l_s from 0,087 to 0,024.

Table 2. Dimensions of the samples made of iron powder HQ (1μm) after sintering

V_p, %	45%	50%	55%	60%
with b, mm	0.984 ± 0.048	1.008 ± 0.022	1.037 ± 0.015	1.076 ± 0.017
hight h, mm	0.972 ± 0.058	0.990 ± 0.027	1.016 ± 0.023	1.028 ± 0.018
length l_s, mm	10.784 ±0.087	11.070 ±0.055	11.245 ±0.023	11.445 ±0.024

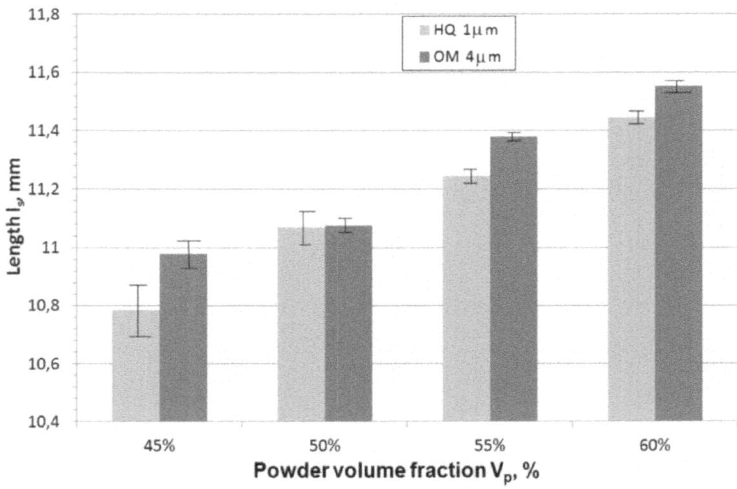

Fig. 5. Influence of powder volume fraction V_p on length and standard deviation of sample after sintering

Sample dimension after sintering dependence on granularity of powder [8]. Powder with small particle easier undergo sintering and sample prepared from this kind of powder shrink more. Shrinkage of parts during sintering must be considerate at mold design, dimensions of the mould cavity are greater than part dimensions.

Shrinkage of the samples after sintering (s) was defined [9]:

$$s = \frac{l_w - l_s}{l_w} * 100\% \quad (2)$$

where: l_w – length of mould cavity

l_s – length of the sample after sintering

Results of measurement of length and shrinkage after sintering the samples from iron powder HQ is shown on Tab.3. Increasing of V_p effect decreasing of shrinkage. This dependence is nearly linear (Fig.6).

Table 3. Length and shrinkage of the sample from iron powder HQ (1 μm).

V_p, %	l_s, mm	s, %
45%	10.784 ± 0.087	15.87 ± 1.15
50%	11.070 ± 0.055	14.08 ± 0.78
55%	11.245 ± 0.023	12.86 ± 0.64
60%	11.445 ± 0.024	11.69 ± 0.43

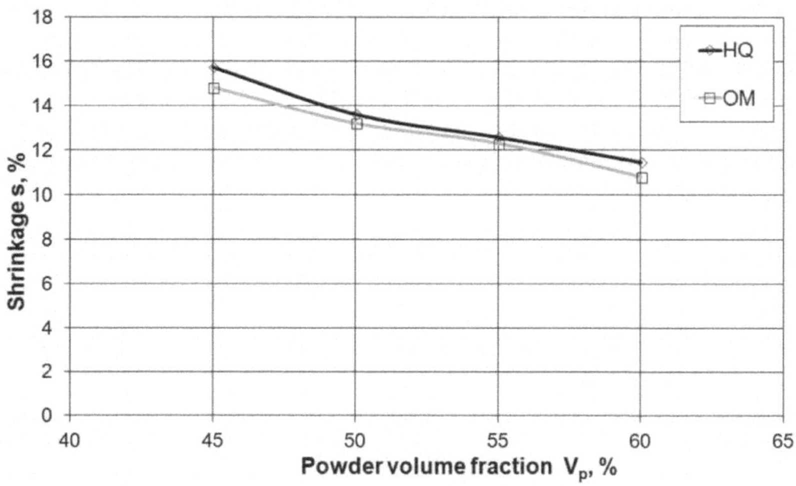

Fig. 6. Influence of powder volume fraction V_p and powder granularity on shrinkage of the samples from HQ (1 μm) and OM (4 μm) powders

4 Conclusion

The following conclusion may be drawn on the basis of conducted experiments:

- Dimension accuracy of elements manufacturing by injection moulding depends on powder volume fraction in feedstock and granularity of powder
- The greater the volume fraction and greater powder granularity, the smaller shrinkage of parts is observed

References

1. Piotter, V., et al.: Micro Powder Injection Moulding, pp. 259-264 EURO PM2000, Oct. 18-20. 2000, Munich, Germany, vol. PIM, (2000)
2. Oczoś, K.E.,: Micro Parts Forming: Characteristics of Micro Machining Methods, [Mechanic],. pp. 309-327Mechanik, 5-6, (1999), (in Polish)
3. Hasselbach, J., et al.: Investigation on the International State of Art of Micro Production Technology. pp.11-18, Euspen Int. Topical Conf., Aachen, Germany, May 19-20, (2003)
4. Piotter, V., et al.: Micro Powder Injection Moulding – New Developments and Results; pp. 331-335Proc. of the Euro PM 2008 – Powder Injection Moulding, vol 2, (2008)
5. Piotter, V., Gietzelt T., Merz, L.,; Micro Powder – Injection Moulding of Metals and Ceramics, pp. 299-306, Sadhan, vol 28, part 1, (Febr/Apr. 2003),
6. Ph. Imgrund, Haack, J., Rota, A.,: Current Developments in Micro Moulding of Functional Materials, pp. 115-118, Proc. of the 7-th Int. Conf. Bremen, (May 2007)
7. Biało, D., Skalski, A.,: Powder Injection Moulding of Micro Parts. Journal of Automation, pp. 87-90, Mobile Robotics & Intellient Systems, No.4, (2009)
8. Missol, W.,: Sintered machine parts, pp.218, Wyd. Śląsk, , (1978), (in Polish)
9. Billiet, R.,: The challenge of tolerance In P/M Injection molding, pp. 91 – 109, Progress in Powder Metallurgy, vol.4, (1986)
10. Paszkowski, L., Biało, D., Wiśniewski, W.,: Injection moulding of the samples from composites with a polymer matrix and metal powder, pp.35-40, Kompozyty (Composites), nr 1 (10), (2010), (in Polish)

New mechatronic stabilographic device – design and software

Anna Sokół[1], Monika Kwacz[2]

[1] Institute of Radioelectronics (MSc Student), Warsaw University of Technology, ul. Nowowiejska 15/19, 00-665 Warsaw, Poland

[2] Institute of Micromechanics and Photonics, Warsaw University of Technology, ul. św. A. Boboli 8, 02-525 Warsaw, Poland

Abstract. This work presents design of a new stabilographic force platform and software dedicated to this device written in Java. The software provides a graphical user interface and easy analysis of the posturograph. The algorithm calculates the following parameters: total length of stabilogram, total and average distance from main position, average speed of center of pressure, percentage of measurements beyond the scope of stable sways, maximum sway in antero-posterior and medio-lateral surface, and total length of stabilogram in both planes.

Keywords: stabilography, static posturography, postural stability, balance

1 Introduction

Postural stability is an ability to restore the balance lost due to sways and any another external or inner perturbation. The postural stability is often impaired by aging process, vestibular pathologies, stroke, head injuries, orthopedic disorders, and neurological diseases [1,2]. This can lead to frequent slips and dangerous falls.

The maintenance of balance is possible by postural reactions, which can be analyzed using stabilographic devices, e.g. a stabilographic platform. The platform allows to record the time-varying coordinates of the center of pressure (COP) in static or dynamic trials. The diagnostic process is based on measuring the ground reaction force. Determination of the force components allows to quantify postural stability and somatosensory function. In contrast, determination of the COP movement range allows to assess the patient's vestibular system. Therefore, the stabilographic platforms are commonly used in the diagnosis of neurological disorders (e.g. patients after stroke, Parkinson's patients) [1], the pathology of the vestibular [3] and musculoskeletal [4] system. Additionally, aircraft pilots, gymnasts, people working at heights, whose work requires an efficient balance, represent also a large group of examined people.

The aim of this work is to develop a design and software of the new mechatronic stabilographic testing device.

R. Jabłoński and T. Brezina (eds.), *Advanced Mechatronics Solutions*,
Advances in Intelligent Systems and Computing 393,
DOI 10.1007/978-3-319-23923-1_40

2 Design of the stabilographic device

The new stabilographic device is designed to measure the ground reaction force, convert the measured analog signal into digital signal, process the signals, calculate the parameters, visualize, and archive the resulting posturographic data. The design has been described in detail in [5]. Briefly, the platform is composed of a circular plate with a diameter of 40 cm and a thickness of 1 cm. The plate is made of construction glass (Float ESG, Press Glass SA, Poland) produced by the float method. Three capacitive force sensors (iLoad Mini^{TM} Stainless Steel Miniature Load Cell, Loadstar Sensors Ltd, Fremont CA, USA; Fig. 1a) are placed below the plate surface at the 17.5-cm distance from the plate center. The angular distance between sensors is 120°. In the glass plate, three holes with a diameter of 6.35 mm were drilled. A bushing with internal thread was glued into each hole. Then, a screw pin of the iLoad Mini sensor was screwed into the bushing. The scheme of attachment of the sensor to the plate is shown in Fig. 1b.

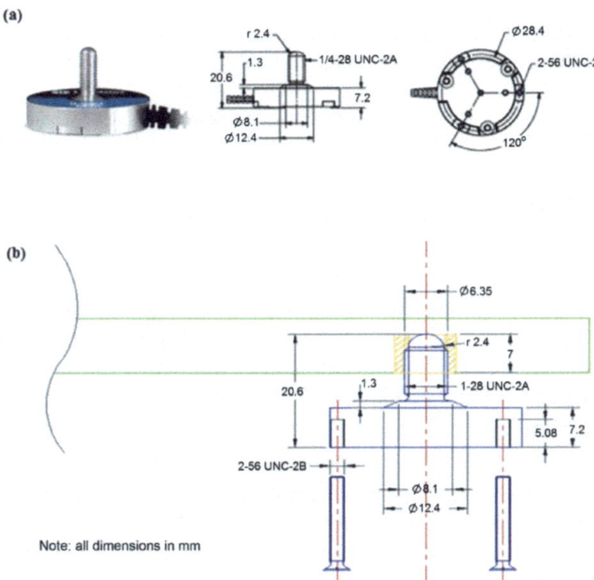

Fig.1. (a) Force sensors iLoadMini^{TM} and its dimensions (www.loadstarsensors.com), (b) scheme of attachment of the sensor to the plate [5]

The four channel frequency interface DQ-4000U (Loadstar Sensors Ltd, Fremont CA, USA) was used to convert an analog signal from the sensors to a USB digital output. The DQ-4000U interface was connected to the PC via the USB port.

3 Software for the stabilographic device

Our software provides a graphical user interface (GUI), easy analysis of the posturograph and archiving of the results. The software has been written in Java. The algorithm determines the following parameters: total path length of posturographic plot, total and average distance from main position, average speed of COP, percentage of measurements beyond the scope of stable sways, maximum sway in anteroposterior and medio-lateral surface, and total path length in both surfaces.

The software calculates the coordinates (COP_x, COP_y) of the COP based on the following relationships (derived from the equilibrium equations):

$$COP_x = \frac{M_y}{F_{z1}+F_{z2}+F_{z3}} \quad (1)$$

$$COP_y = \frac{M_x}{F_{z1}+F_{z2}+F_{z3}} \quad (2)$$

where F_{z1}, F_{z2}, and F_{z3} are the forces along the z axis measured by the sensor 1, 2, and 3, respectively (Fig. 2), and

$$M_x = 0.5a \cdot (2F_{z1} - F_{z2} - F_{z3}) - 0.5t \cdot (F_{y1} + F_{y2} + F_{y3}) \quad (3)$$

$$M_y = \frac{\sqrt{3}}{2} a \cdot (-F_{z2} + F_{z3}) + 0.5t \cdot (F_{x1} + F_{x2} + F_{x3}) \quad (4)$$

where $a = 17.5$ cm is the distance between the sensor and the plate center, t = 1.0 cm is the plate thickness, and F_{x1}, F_{y1}, F_{x2}, F_{y2}, and F_{x3}, F_{y3} are the forces along the x and y axis measured by the sensor 1, 2, and 3, respectively.

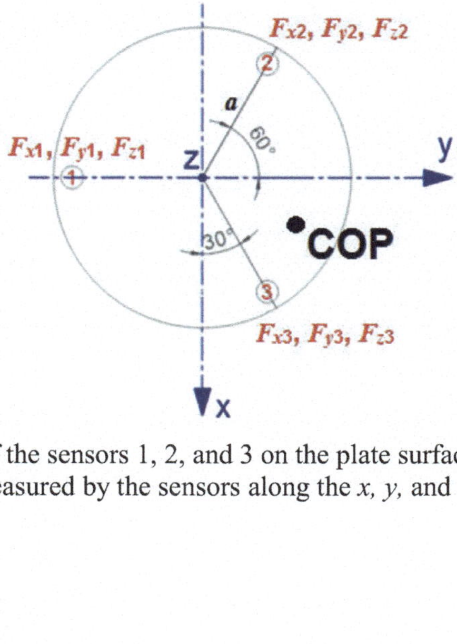

Fig. 2. Location of the sensors 1, 2, and 3 on the plate surface. *Fx*, *Fy* i *Fz* – forces measured by the sensors along the *x*, *y*, and *z* axes.

Figure 3 shows the GUI of our software with an exemplary stabilogram. The software allows user to observe drawing stabilogram and changing parameters during the procedure. The user can also control time of the patient trial and export the results to a 'Patient Card' saved in xls format. Figure 4 shows an example of the 'Patient Card' created by our software. For exporting the xls file, Java Excel API and Java Excel Apache libraries have been used.

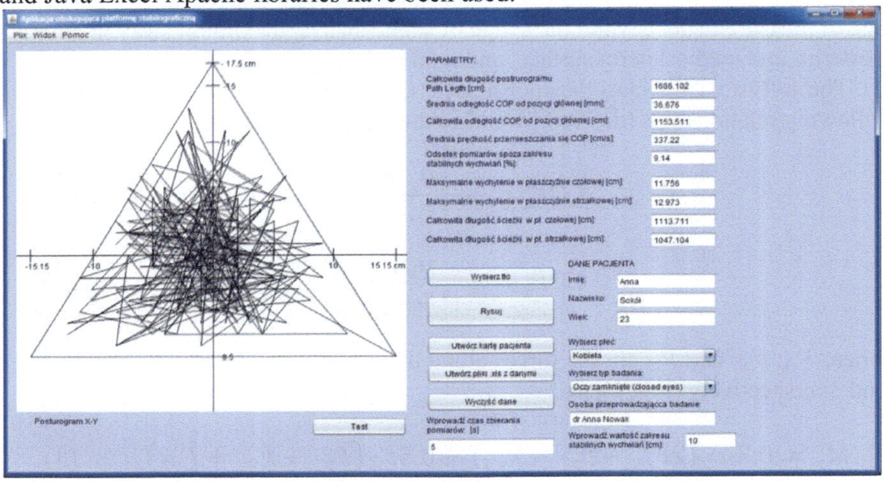

Fig. 3. The GUI of the software with an exemplary stabilogram.

PATIENT CARD		
Name:	Anna	Person performing procedure:
Surname:	Nowak	dr Janusz Kowalski
Age:	23	Date of procedure:
Gender:	Kobieta	2015-22-04

Procedure type:	Oczy zamknięte (closed eyes)

Values of parameters:	
1. Total path length [cm]:	1990.4
2. The average distance COP of the main position [cm]:	27.5
3. The total distance COP of the main position [cm]:	1381.7
4. The average speed of COP [cm/s]:	331.7
5. Percentage of measurements beyond the scope of stable sways	8.3
6. Maximum sway in antero-posterior surface [cm]:	13.5
7. Maximum sway in medio-lateral surface [cm]:	12.9
8. Total path length in antero-posterior surface [cm]:	1342.4
9. Total path length in medio-lateral surface [cm]:	1216

Fig. 4. Example of the 'Patient Card' created by the software.

4 Discussion

In this work, we describe design and software of a new stabilographic device. Our stabilographic platform is light, stiff, and its estimated cost is about $1200. The platform stiffness has been numerically confirmed using 'PlaTo 4.0' software (INTERsoft Ltd., Poland, www.intersoft.pl). The simulation results showed that the maximal deflection of the glass plate is about 0.074 mm (Fig. 5). This result was obtained for the glass Young's modulus of 90 GPa, the glass Poisson's ratio of 0.24, and the patient weight of 100 kg.

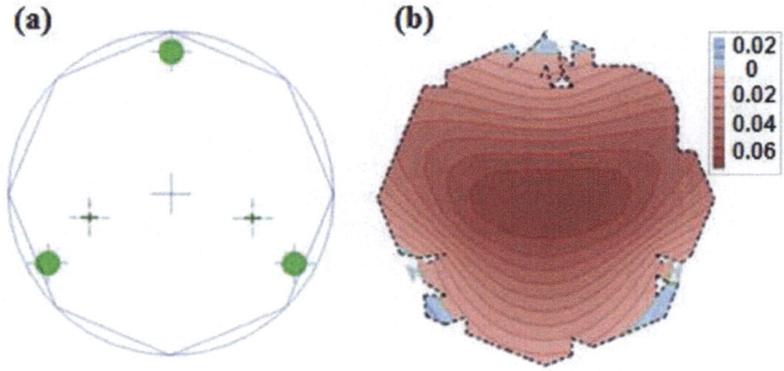

Fig. 5. (a) Plate with sensors (green circle) and points of load (green crosses), (b) visualization of the plate deflection (in millimeters).

The proposed device allows to analyze the balance in quiet standing by determine the COP displacement, as in many commercial solutions. However, the commercial platform are more costly (~ $6500) compared with our device. Our platform is provided with the intuitive GUI and the utility software. This software allows to analyze the data and to quantify the postural stability. Further work is needed to develop an accurate calibration procedure.

Several studies have shown that the static posturography test is unable to identify the etiology of every balance disorders. The sensitivity and specificity of balance testing can be increased complementing the static posturography by dynamic tests, vestibular caloric and rotational tests and analysis of gait. However, the COP path is accurate in discriminating patients with multiple sclerosis and assessing the risk for falling among elderly population [6].

5 Conclusion

The new stabilographic device might be useful for clinical use for screening analysis of balance disorders and assessing the risk for falling among elderly population. Balance can be assessed in undisrupted stance with closed or opened eyes using the standard Romberg test.

Acknowledgements

The prototype is currently investigated at the Student Research Group of Medical Devices at the Institute of Micromechanics and Photonics Warsaw University of Technology. The work is supported by Robert Bosch Ltd. (Warsaw, Poland).

References

1. Visser J.E., Carpenter M.G, van der Kooij H., Bloem B.R.: The clinical utility of posturography. Clinical Neurophysiology 119, 2424–36 (2008)
2. Nardone A., Schieppati M.: The role of instrumental assessment of balance in clinical decision making. Eur. J. Phys. Rehabil. Med. 46, 221–37 (2010)
3. Di Fabio R.P.: Sensitivity and specificity of platform posturography for identifying patients with vestibular dysfunction. Phys. Ther. 75(4), 290–305 (1995)
4. Boucher P., Descarreaux M., Norman M.C.: Postural control in people with osteoarthritis of the cervical spine. J. Manipul. Physiol. Therap. 31(3), 184–190 (2008)
5. Mach E.: Force platform for measuring ground reaction forces during biomechanical tests, Master Thesis, Warsaw University of Technology (2012), (in polish)
6. Piirtola M., Era P.: Force platform measurements as predictors of falls among older people – a review. Gerontology 52, 1–16 (2006)

E-vehicle energy consumption optimization based on fleet and infrastructure information

Pavel Steinbauer, Petr Denk[1], Jan Macek[1], Josef Morkus[1], Zbyněk Šika[1]

[1] Czech Technical University in Prague, Faculty of Mechanical Engineering, Prague, Czech Republic
{Pavel.Steinbauer}@fs.cvut.cz

Abstract. Nowadays vehicles, especially the electric ones, are complex mechatronic devices. The pickup vehicles of small sizes are currently used in transport considerably. They often operate within a repeating scheme of a limited variety of tracks and bigger fleets. Thanks to mechatronic design of the vehicles and their components, there are many means of optimizing their performance. The paper shows an approach developed to increase the range of e-vehicle operation substantially. It is based on prior information about the route profile, traffic density, road conditions, past behaviour, mathematical models of the route and vehicle and dynamic optimization. The a-priori knowledge is taken into account.

Keywords: E-Vehicle. Dynamic optimization. Range extension. Model Based Predictive Control. Fleet optimization.

1 Introduction

The current transport is heavily using small size pick up vehicles, so optimization of energy consumption based on current ICT means is of great interest ([1]). Furthermore, they often operate within repeating scheme on a limited variety of tracks and bigger fleets. Thanks to mechatronic design of the vehicles and their mechatronic components, there are many means of optimizing their performance. These calculations could be done prior to journey using powerful computers residing in the cloud.
The prior information about the route profile, traffic density, road conditions, history behaviour etc. is thus available and can be used to optimize the vehicle route profile to achieve minimal energy consumption. This concept is even more important for e-vehicles which have quite limited energy storage (comparing to cars with regular combustion engines).
The means for determination of the optimal solution are formally developed, e.g. by dynamic programming or Pontryagin principle of minima. However, the application for a complex mechatronic system is usually not possible, as the mathematical model is highly nonlinear, partially given by measured look-up table or hidden in the code of a simulation tool. The search for the optimal solution is thus difficult or even impossible analytically.

R. Jabłoński and T. Brezina (eds.), *Advanced Mechatronics Solutions*,
Advances in Intelligent Systems and Computing 393,
DOI 10.1007/978-3-319-23923-1_41

The paper introduces an approach of the optimal velocity route profile design, based on dynamic optimization for pre-trip calculation. The goal is to find such velocity profile along selected route, which minimizes energy consumption. The velocity is thus parametrized by position of the vehicle on the route.
Realistic path discretization is most important part of the approach. It is based on physical inside into driving mode. The route is divided into sections, where important properties of the section remain constant, e.g. maximum achievable velocity or slope. The maximum achievable velocity is either given by the legal or physical limit (whichever is lower). The physical limit is calculated from the route curvature and elaborated from the fleet historical measurements. Using maximum velocity and available acceleration and deceleration in each section, minimum achievable time is calculated. This provides information about maximum energy consumption. It is optimized in next optimization based on route sections.
The main goal of the optimization process is to find the optimal velocity profile with minimal energy consumption for required travelled time limit.
The most accurate model with differential equations of motion, measured maps of batteries and motor characteristics has to be numerically integrated and thus requires a lot of computational power and time. That`s why the initial calculation and optimization is based on a simplified algebraic model. It enables to employ the multi-parameter optimization along the whole route at once. To further improve optimization convergence, the sub-optimal velocity profile is calculated. The calculations are based on physical insight into the vehicle mechanics and the driving strategy performed by experienced drivers, as well as on the backward principal known from dynamic programming.
The above described procedure is computationally costly and cannot be performed by on-board CPU power. So the calculations are performed on powerful servers placed in cloud service. The optimal velocity route profile is fetched via cloud services into vehicle ECU. ECU local energy manager running in real-time, which uses the vehicle sensor measurements and the pre-optimized velocity route profile, adapts the route profile for actual traffic and route situation.
The vehicle also collects data from its sensors during the path for post-trip analyses in the cloud. These historical data are used to improve future route planning and optimization.

2 Problem statement

The e-vehicle is a mechatronic system which can be generally described by non-linear state space model

$$\dot{x} = f(x, u, t) \quad (1)$$
$$y = g(x, u)$$

It can be optimized by dynamic system optimization methods, which were developed for optimization of complex and nonlinear dynamic processes ([2],[3]). They are based on discretization of the state trajectory and replacement of highly non-linear dynamic model by algebraic functions.

However a-priori knowledge can be utilized. The reduced state trajectory is given by the position and velocity of the vehicle. The objective is to fulfil given route $y = y_g(t)$ with restricted time of travel T while minimizing energy consumption.

The optimal control is described by a function describing the optimal velocity along the given path

$$v = v_{opt}(s) \tag{2}$$

The obvious discretization is based on the concept of sections which are further discretized into natural driving modes shown on the **Fig. 2**.

3 Vehicle model

The vehicle model used for optimization is based on the analyses of energy flow and losses. On the other hand, the simple algebraic equations must be derived to achieve rapid evaluation. So it is necessary to include analyses of vehicle components into the calculations and optimization of control strategy.In principal, the most important losses are depicted on the **Fig. 1**.

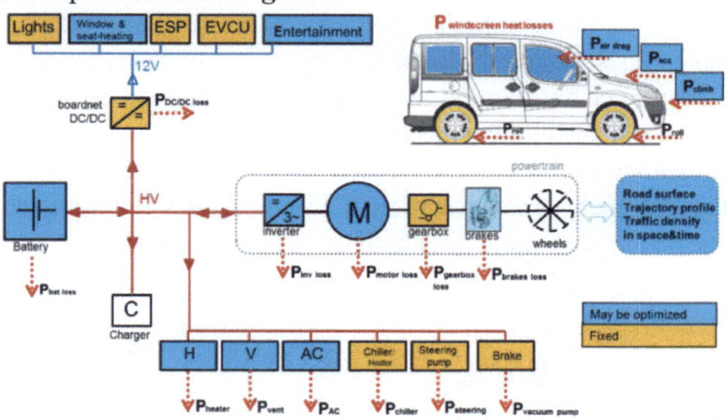

Fig. 1. Vehicle conceptual model

3.1 Control structure

The optimal velocity route profile is determined prior to the ride on powerful CPU located in the cloud service. It is based on the knowledge of the expected route (its height and allowed speed profile), vehicle parameters, the expected level of traffic density and driving style of the driver.

The feedback controller is introduced at ECU to manage the difference between predicted and measured values – the distance travelled, the time and the consumed energy during the ride. The controller output cannot be fed directly into the vehicle or engine, but mainly uses the driver as an intelligent actuator. It means that the calculated control actions are displayed as the recommended action for the driver who is motivated to follow the instructions by gamification approaches.

4 Route definition

The route is defined in a way suitable for dynamic optimization. The approach is based on identification of sections, which have a constant upper velocity limit and slope.

The route-trip planning data generated by route planning algorithm or history fleet data from previous trips include the route definition including longitude, latitude, altitude, legal speed limits, traffic density, weather conditions, etc.

4.1 Route parametrization

For the purpose of energy optimization procedure the route is divided into particular sections. In general each section represents a part of the route in which all the relevant properties such as the legal speed limit, the safe and comfortable speed limit, the slope, the weather conditions and the traffic density are constant.

However, the velocity profile optimization requires much more subtle parametrization. Thus the section is further divided into logical phases. In each section four phases of vehicle driving mode may occur (**Fig. 2**): The phase of acceleration, the phase with constant velocity of the vehicle, the phase of coasting, and the phase of deceleration. The order of phases cannot be changed, but some of them are skipped in particular cases,.i.e they have zero length. E.g. the coasting and deceleration will not occur if consecutive section has higher constant velocity.

The section velocity profile is fully described by quintuplet $[v_1, a_a, s_a, s_d, a_d]$. The main advantage of such parametrization is that it allows the zero length of section phases without getting into numerical and mathematical problems.

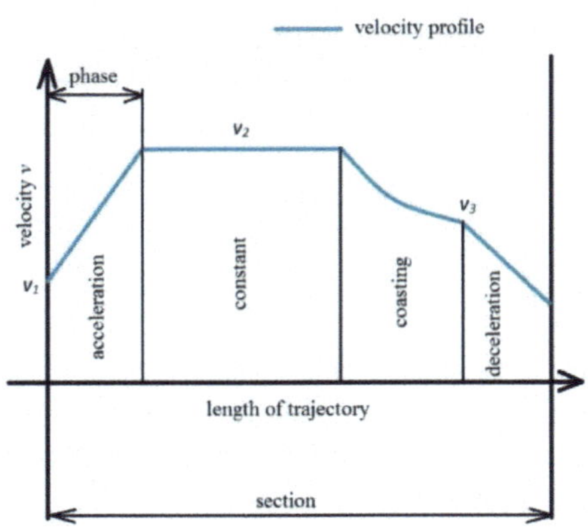

Fig. 2. The section discretization into phases

The set of optimization parameters creates the trajectory description as follows.

$$v_2 = \sqrt{2a_a s_a + v_1^2}, \qquad v_3 = \sqrt{2a_d\, s_d + v_4^2}$$
$$s_{coa} = f\,(v_2, v_3), \quad s_{con} = s - s_a - s_d - s_{coa}(x, u, t) \tag{3}$$

4.2 Route decomposition

The information about the roads, or streets along the specific route can be obtained from several sources. The route description coming from route planning algorithm accompanied with fleet history data is in a raw GPS coordinates format. The decomposition strategy is based on the evaluation of legal restrictions (maximum velocity limits), the analyses of route curvature and by defining the maximum acceptable velocity from the comfort point of view. The planar coordinates of the route points are fitted with polynomial functions and the curvatures over the whole route are calculated. The knowledge of the curve radius and static vehicle parameters makes it possible to calculate the maximum velocity from the fraction of adhesion limit to achieve a comfortable speed profile.

5 Optimization procedure

The optimal velocity route profile is calculated using a hierarchical approach. After the route is planned and decomposed into sections, the fastest velocity profile is found using the maximum legal or physical limits in each section as well as maximum achievable acceleration and deceleration. It shows whether overall time limit requirement can be achieved at all.

Then pre-optimization procedure calculates suitable velocity parameters. It is based on empiric rules extracted from driver experience. These rules are formed into simple algebraic equations and applied in reversed time using dynamic programming principal.

Finally, numerical optimization of the whole parametrized velocity route profile is performed to fine tune the results. The fast numerical vehicle model based on mainly algebraic equations of energy flows is used.

6 Results

The optimization results for reference route show that most influencing factor is the time restriction. If riding time is tightly restricted, the optimum velocity route profile uses the maximum allowed velocities in each section. Thus a lot of energy demanding acceleration and deceleration is necessary. However, if the time is not an issue, quite interesting savings can be achieved.

The following table shows preliminary result for reference route (**Fig. 3**). It is based on a general mathematical model of the e-vehicle, however with realistic coefficients. The comparison is made with respect to the fastest achievable velocity profile (respecting legal limits).

Total ride time [s]	1300	1500	2500
Energy savings [%]	7,65	11,7	14,82

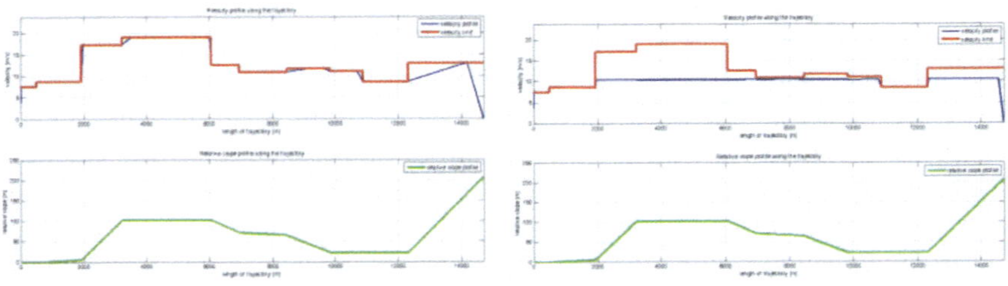

Fig. 3. Optimization result for maximum time 1300s and 1500s on reference route

7 Conclusions

The procedure to determine the optimal behaviour of the driver from the energy consumption point of view was described. The considerable energy savings can be achieved by pre-trip optimization. However, the driver must be motivated to follow pre-optimized suggestions. It can be done using gamification methods. Also changing traffic and weather situation does not allow the driver to fulfil given velocity profile exactly. Deviations always occur, so the described prior dynamic optimization approach and the open-loop control is supplemented by the local closed loop controller based e.g. on the model based predictive control approaches.

Nevertheless, it is expected that the volume of production and sales of electronic equipment for the "connected cars" will soon be comparable with the production of mobile phones. This power can be used not only for entertainment, but truly connect the vehicle with others via cloud and share the "experience" to achieve better energy efficiency, better traffic flow. And this can already be applied nowadays for cars operating in fleets, where there are no legal or organizational obstacles for sharing data.

References

1. PAPAJEWSKI, J. Predictive operating strategy – A possibility to optimize customer wish and energy efficiency, Tage des Hybrids 2012, Aachen
2. BIEGLER, L. T. An overview of simultaneous strategies for dynamic optimization. Chemical Engineering and Processing: Process Intensification, 2007, 46.11: 1043-1053.
3. LOGSDON, J. S.; BIEGLER, L. T. Decomposition strategies for large-scale dynamic optimization problems. Chemical engineering science, 1992, 47.4: 851-864.

Acknowledgments

This research has been realized using the support of EU FP 7 Project No. 608756, Integration and Management of Performance and Road Efficiency of Electric Vehicle Electronics and using the support of The Ministry of Education, Youth and Sports program NPU I (LO), project # LO1311 Development of Vehicle Centre of Sustainable Mobility. This support is gratefully acknowledged.

The effect of the Halbach array on FSPM machine magnetic field

Pavel Svetlik[1], Karel Hruska[1]

[1] University of West Bohemia, Faculty of Electrical Engineering, Pilsen, Czech Republic
{psvetlik, khruska}@kev.zcu.cz

Abstract. This paper deals with improvement of behavior of the flux-switching permanent magnet machine through adding the Halbach array on the outer diameter of the stator. Different variants of Halbach arrays were modeled and compared. A comparison of obtained results shows an interesting way of flux-switching permanent magnet machine magnetic field shielding.

Keywords: permanent magnets, Halbach array, FSPM machine, magnetic flux density, analysis

1 Introduction

The flux-switching permanent magnet machine (FSPM) belongs to stepper motor class. Against a classical stepper motor, it is enhanced with permanent magnets on the stator which excites a permanent magnetic field. This gives this machine a stand-still torque.

The Halbach array is a group of permanent magnets with special arrangement. It was designed by Klaus Halbach in 1985. This group of magnets creates strong magnetic field on one side and significantly weaker on the other side of the group. In this time, the Halbach array is used in magnetic bearings, maglev trains, aerospace applications and many others. [1,2,3]

The models were created on behalf of basic 12-10 poles FSPM machine design without any additional design changes. [4,5] The improvement was enhancing the stator with different variants of the Halbach array only.

R. Jabłoński and T. Brezina (eds.), *Advanced Mechatronics Solutions*,
Advances in Intelligent Systems and Computing 393,
DOI 10.1007/978-3-319-23923-1_42

2 FSPM machine analyses

2.1 FSPM Design

The design of the FSPM machine considers the easiest way of construction. The stator consists of twelve C-shaped metal pieces. For prototype construction, M350-50A steel type was used. For this reason this steel was used in models too. Between these profiles are N35H type neodymium (NdFeB) permanent magnets. The stator outer diameter is 110 mm. Around the stator a Halbach array is placed (see Fig.1.) in several configurations also created of the same type of neodymium magnets. As last possibility the ferrite permanent magnets were considered. Thickness of magnets is 1mm as the ring between Halbach arrays in double Halbach array configurations.

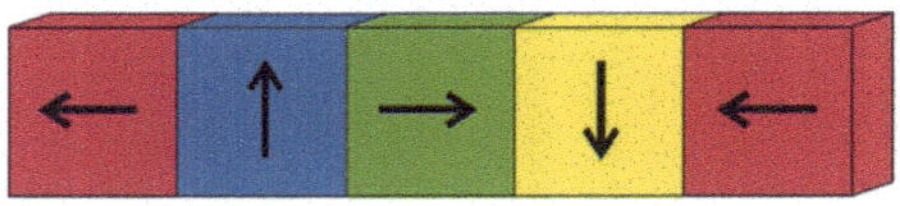

Fig. 1. Halbach group of permanent magnets [1]

The rotor is constructed of same steel as stator and it has ten cylindrical sallies. The cross section through the machine is on Fig. 2.

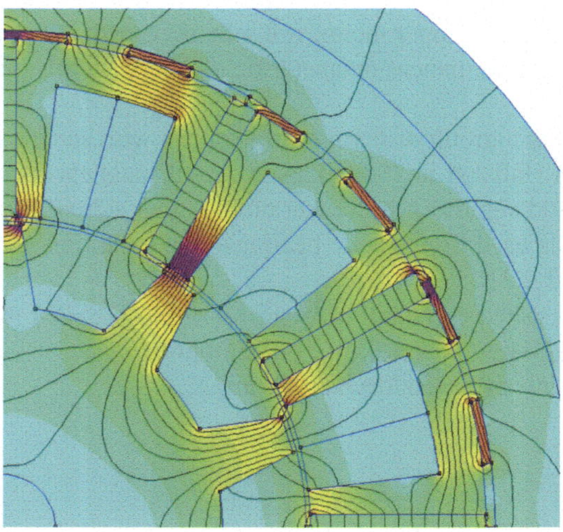

Fig. 2. Cross section through the machine with Halbach array

2.2 Magnetic field analyses

All of the modifications of the FSPM machines with Halbach array have been modeled in finite element method software FEMM 4.2 [6,7]. In every case the normal component of magnetic induction had been analyzed in two places:

- In the center of the air gap
- 1 cm distance from the outer diameter of the stator

2.3 Halbach arrays modification and modeling

Modifications of the Halbach arrays were focused on several parameters:

- Density of the array (length of the magnets)
- Number of the Halbach arrays
- Orientation of magnets

These modifications give a lot different models for further examination and comparison. There are eight variants:

- FSPM without Halbach array - no improvements.
- FSPM machine with total of 48 magnets in Halbach array
- FSPM machine with total of 96 magnets in Halbach array
- FSPM machine with total of 192 magnets in Halbach array
- FSPM machine with total of 192 magnets in each of two circles of Halbach array (opposite orientation)
- FSPM machine with total of 192 magnets in each of two circles of Halbach array (identical orientation)
- FSPM machine with total of 192 magnets in each of two circles of Halbach array (identical orientation) – ferrite magnets used
- FSPM machine with total of 192 magnets in each of two circles of Halbach array (rearranged magnetization direction) [5]

All of them are shown on Fig.3 (from top left corner to bottom right corner).

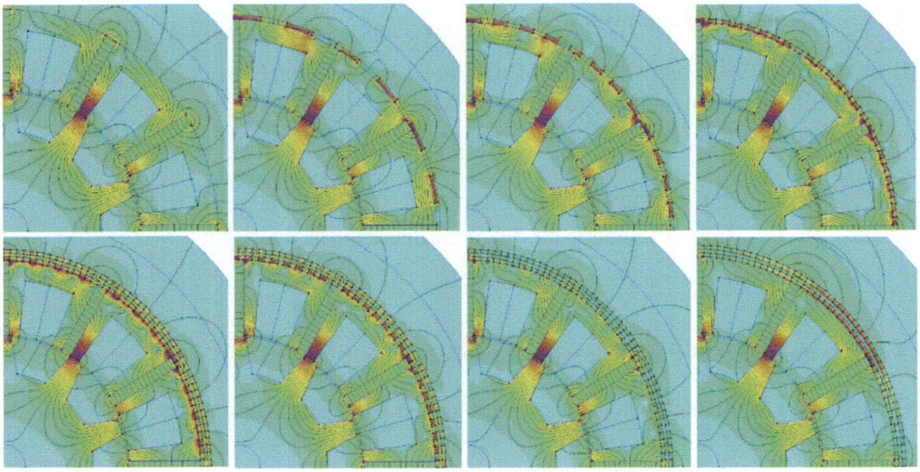

Fig. 3. Cross sections through analyzed machines

As seen in Fig.4, The Halbach array application has almost none effect on magnetic flux density in the air gap. The small difference among variants is only in peaks of curves.

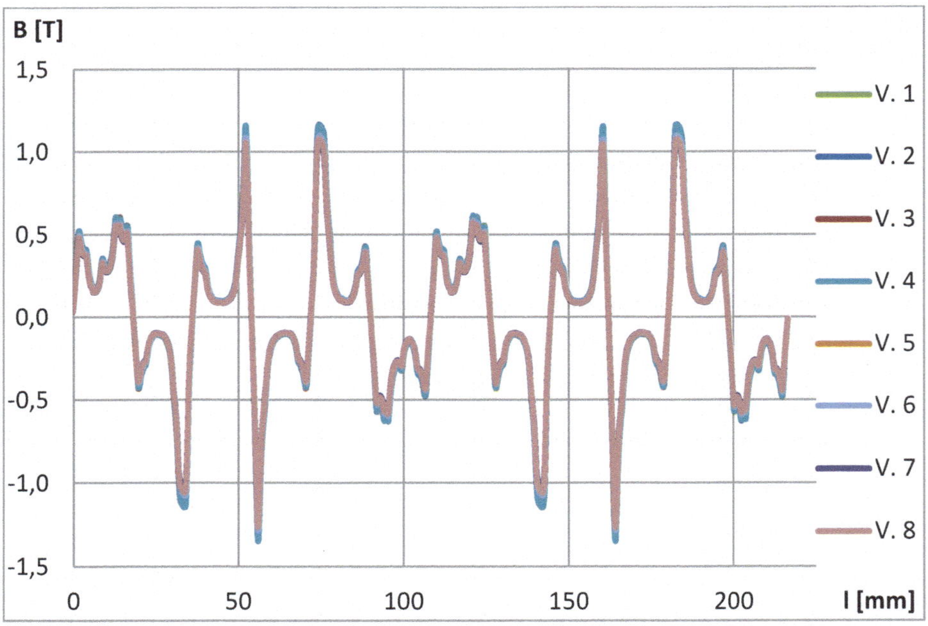

Fig. 4. Magnetic flux density in the air gap of analyzed machines

On the other hand, magnetic field outside the stator shows bigger differences (see Fig.5).

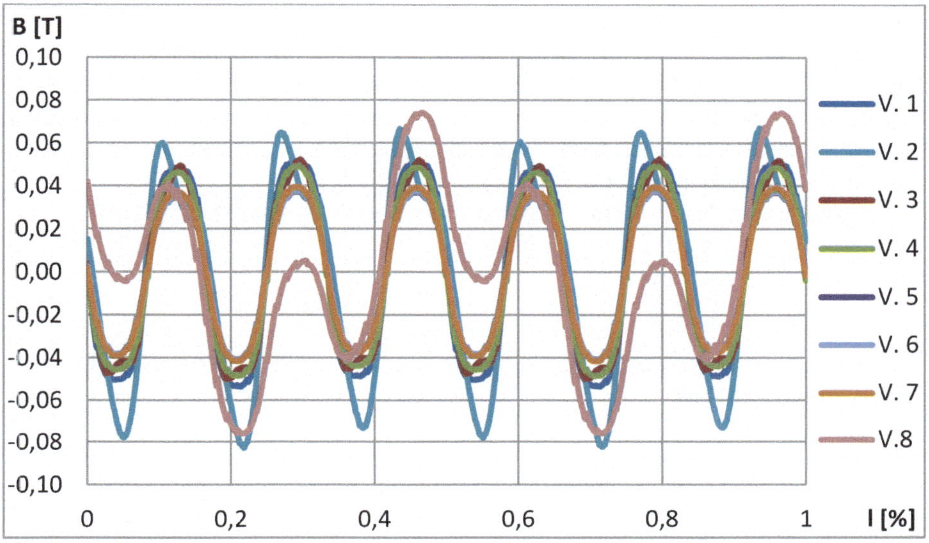

Fig. 5. Magnetic flux density in 10 mm distance from the stator of analyzed machines

Some of the variants exhibit increase of magnetic induction around the stator and also they changed a shape to non-sinusoidal. But some of them decrease the magnetic flux density of value 12mT in peaks. This means almost 25% decrease of magnetic flux density at 10mm distance out of stator in comparison with FSPM machine without any Halbach array.
12mT is not very significant value but 25% definitely is. Moreover modeled machine is relatively small and its dispersion of magnetic field is small too but for bigger machines this could mean an interesting way how to reduce magnetic field emissions into surroundings.

3 Conclusion

The best result mentioned before has been achieved with variant 6 (two Halbach arrays with identical orientation). Also variant 7 shows very similar results. This is because of the same configuration of the Halbach array but cheaper ferrite magnets have been used. The third best result achieves variant 5 with opposite configuration of the Halbach array. All of these three configurations look like suitable to use as magnetic field shielding.

The disadvantage of these configurations is bigger stator diameter. This could be troubling especially for small machines. For bigger machines this is just small complication.
The single Halbach array could not be recommended for this. Variant number 3 shows some improvement but this is just a few percent. Other variants exhibit more impairment than improvement. Moreover the shape of some curves exhibits some higher harmonics distortion (like variant 8) which is not suitable. This is probably because of insufficient shielding effect and this leads to superposition of magnetic fields of stator and Halbach array itself.
The magnetic flux density in the air gap decreases with stronger shielding effect. This is caused by magnetic flux closing more in stator yoke in compare with FSPM machine with no Halbach array.

Acknowledgement

This paper has been supported by funding program of the UWB number SGS-2015-038

References

1. Mayer, D.,Ulrych, B. "Halbachova soustava magnetů a její modifikace," Diagnostika '07, Pilsen: University of West Bohemia, 2007, ISBN: 978-80-7043-557-1
2. Halbach, K. "Application of permanent magnets in accelerators and electron storage rings," Journal of Applied Physics, Vol. 57 (1985)
3. Sanabria-Walter, C., "Design of a 600kW ring-type direct-drive Flux-Switching Permanent Magnet machine for aerospace main propulsion," *Power Electronics and Applications (EPE'14-ECCE Europe), 2014 16th European Conference on* , vol., no., pp.1,10, 26-28 Aug. 2014
4. Hruska, K.; Svetlik, P., "Improvement of FPSM's behavior through geometry adjustments," *Power Electronics and Applications (EPE'14-ECCE Europe), 2014 16th European Conference on* , vol., no., pp.1,9, 26-28 Aug. 2014
5. Zhu, Z.Q., "Recent Development of Halbach Permanent Magnet Machines and Applications," *Power Conversion Conference - Nagoya, 2007. PCC '07* , vol., no., pp.K-9,K-16, 2-5 April 2007
6. Bianchi, N. "Electrical machine analysis using finite elements," Boca Raton, USA, Taylor&Francis, 2005, ISBN 0-8493-3399-7
7. Reece, A. B. J., Preston, T. W. "Finite element method in electrical power engineering," Oxford university press, UK, 2000, ISBN 978-0-19-856504-8

Limits of spatial sensitivity in eddy current tomography of spindle-shaped elements

Roman Szewczyk

Institute of Metrology and Biomedical Engineering, Warsaw University of Technology, Poland
szewczyk@mchtr.pw.edu.pl

Abstract. Paper presents the analysis of spatial sensitivity of eddy current tomography of spindle-shaped elements made of constructional steel. Analysis was based on finite element magnetodynamic simulations utilizing open source software: NETGEN, ELMER FEM and OCTAVE. During the simulations, both location and diameter of the test cylindrical hole was changed to detect the edge of sensitivity of eddy current tomography. The results of simulations indicate, that for spindle-shaped elements with 30 mm diameter, cylindrical holes may be detected at the depth of about 4 mm. Moreover, changes of phase of signal from eddy current tomography devices are more significant than changes of amplitude.

Keywords: eddy current tomography, finite element method, magnetodynamic simulations.

1 Introduction

Eddy current tomography (ETC) is one of the most promising methods of non-destructive testing [1, 2], which may be introduced to the industry [3] during the next couple of years. In the opposite to X-ray tomography, the eddy current tomography is safe and user-friendly [4]. On the other hand, signal processing required for eddy current tomography is extremely sophisticated [5, 6] and calculation power consuming. However, due to the Moore's law [7], it is expected, that calculation power required for inverse tomographic transformation will be soon accessible in the industrial conditions.

Recently both hardware and processing methods required for eddy current tomography are intensively developed [8, 9, 10]. However, methods of optimisation of measuring conditions as well as efficiency of ETC in the case of different elements are still not sufficiently described. Method of analysis and determination of optimal frequency from the point of sensitivity in ETC of spindle-shaped elements [11] was presented previously. On the base of this optimisation, for 1000 Hz driving current frequency, the spatial sensitivity of eddy current tomography of steel spindle was identified during the finite element method analysis.

R. Jabłoński and T. Brezina (eds.), *Advanced Mechatronics Solutions*,
Advances in Intelligent Systems and Computing 393,
DOI 10.1007/978-3-319-23923-1_43

2 Measuring setup and method of investigation

The idea of measuring setup for eddy current tomography is presented in figure 1. Tested spindle-shaped element (1) is placed between driving coil (2) and sensing coil (3). During the measurement, the spindle is rotated by the step motor (4) as well as moved by linear actuator (5).

When driving coil is driven by the sine wave current, the sensing coil detects sine wave voltage. This confirms, that system is nearly linear. Information about the discontinuities and other properties of the spindle is carried out by the changes of both amplitude and phase of signal on sensing coil. On the base of measurements of these changes, the plots of amplitude $a(\phi,x)$ and phase $f(\phi,x)$ can be presented [12] as a function of linear movement x of the spindle and its rotation ϕ.

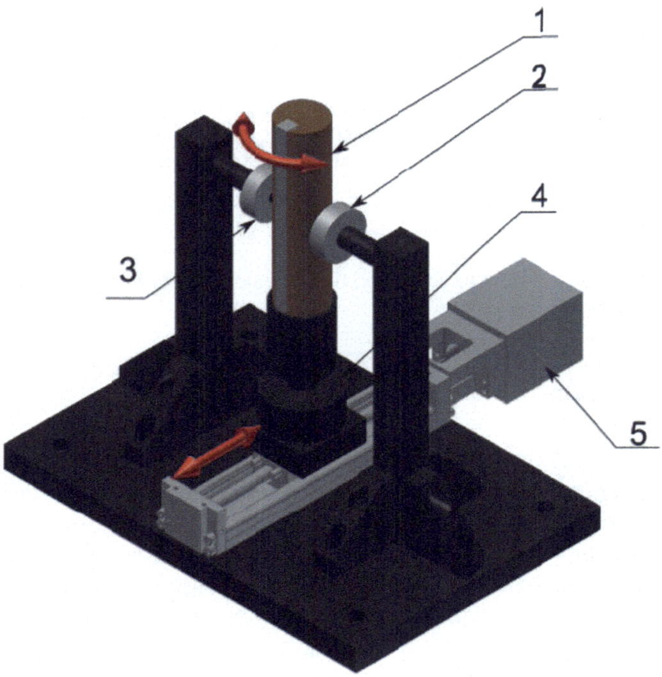

Fig. 1. Measuring setup for eddy current tomography: 1 – tested spindle-shaped object, 2 – driving coil, 3 – sensing coil, 4 – stepper motor (rotation), 5 – linear actuator

For assessment of the sensitivity of eddy current tomography setup, the finite element method was used. All simulations were carried out on the base of open-source software to guarantee the possibility of validation of the results.

Each step of simulation generated one set of information about the signal on the sensing coil (amplitude and phase shift of output signal). Such step consists two parts:

1. Generation of tetrahedral mesh (first order elements) for modelling the system. This mesh was created by NETGEN 5.3 software [13] developed as an open-source by team located at the Johannes Kepler University Linz.
2. Definition of magnetic properties and solving Maxwell equations for magnetodynamic system. This part was done with use of ELMER FEM [14] software developed by CSC, Finland. Moreover, ELMER FEM calculated the value of magnetic flux on the sensing coil.

As it was proved previously [15], due to the application of Whitney-based method for solving Maxwell equations in ELMER FEM, high level of conformity was achieved between simulation and experimental results. For this reason, it is possible to use simulation based on finite element method for assessment of sensitivity during testing elements using eddy current tomography.
The cross-section of modelling test spindle made of constructional steel with cylindrical discontinuity is presented in figure 2. Spindle diameter D was 30 mm, whereas its average permeability μ equals 1000 and conductance σ equals 1.45 S/m. Diameter of discontinuity $2r$ was changed from 1 to 7 mm. The deepness of discontinuity location k was varying from 0 to 4 mm.

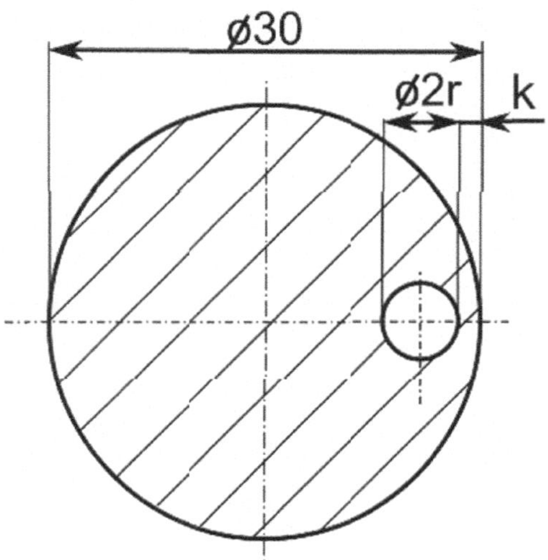

Fig. 2. Cross-section of test spindle for modelling

Sensitivity of eddy current tomography in the case of spindle with discontinuity presented in figure 2 was estimated separately from the point of view of amplitude of the signal on the sensing coil as well as its phase shift versus driving coil current. The assessment of sensitivity was made on the base of special indicator equalling the sum of square differences between signals for spindle without discontinuity and spindle with discontinuity with diameter $2r$ located at the deep k. Differences for sum operation were calculated for each position of spindle described by its linear movement and rotation.

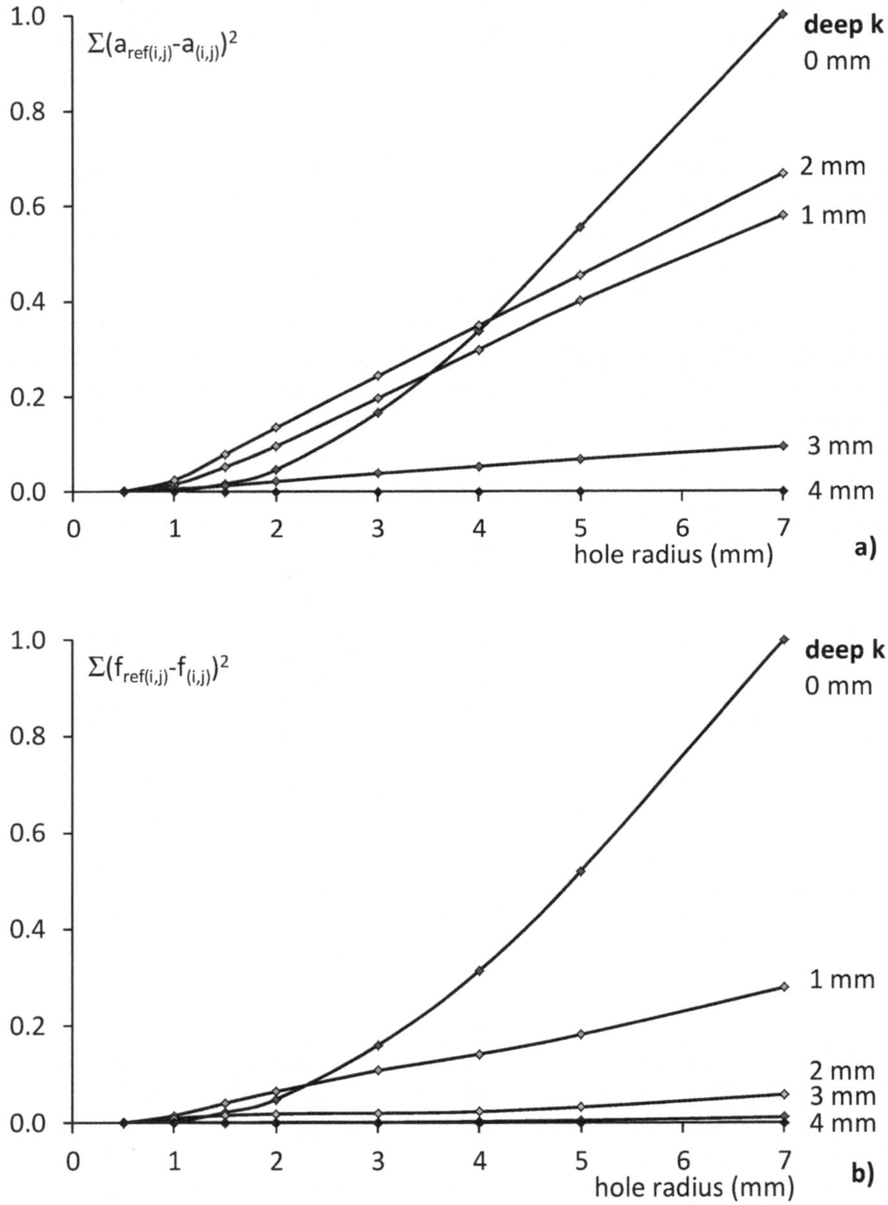

Fig. 3. Indicators of sensitivity for cylindrical discontinuity detection as a function of hole radius *r* and location deepness *k*: a) amplitude of signal on the sensing coil, b) phase shift of signal on the sensing coil

3 Results of investigation

Figure 3 presents the results of simulation utilizing both NETGEN and ELMER FEM open source software. As it is expected, output signal change increases monotonously for increase of hole radius *r*. However, for lower values of hole radius *r* (up to about 4 mm from the point of view of changes of amplitude and up to about 2 mm from the point of view of phase shift) holes located at some depth (about 1 – 2 mm) can be better detected, than holes located on the surface of the spindle.

On the other hand, sensitivity of eddy current tomography reduces significantly for discontinuities located at the depth of over 4 mm. In such a case observation of changes of phase shift of output signal is more effective than observation of its amplitude.

Presented results clearly indicates, that eddy current tomography is not suitable for detection of discontinuities located over 5 mm below the surface. This observation is very important from practical point of view and should be considered during the development of eddy current tomography systems for industrial applications.

4 Conclusion

Presented in the paper, method of simulation of signal from eddy current tomography system enables calculation of sensitivity for given shape of tested element, considering both its permeability and conductivity. As a result, sensitivity can be assessed and optimised from the point of view of parameters of driving coils and driving current.

In presented case, for spindle with 30 mm diameter, relative permeability $\mu = 1000$ and conductivity σ equal 1.45 S/m, cylindrical holes may be detected only if located at the deep of up to 4 mm. Moreover, for discontinuities located at about 4 mm, output signal phase shift is more sensitive on discontinuities, than changes of amplitude of output signal.

References

1. Soleimani M., Lionheart W. R. B., Peyton A. J., Xiandong M., Stuart H.R.: A three-dimensional inverse finite-element method applied to experimental eddy-current imaging data. IEEE Transactions on Magnetics 42, 1560 (2006)
2. Ioan D., Rebican M.: Numerical model for eddy-current testing of ferromagnetic steel parts. IEEE Transactions on Magnetics, 38, 629 (2002)
3. Soleimani M., Lionheart W. R. B., Peyton A. J.: Image reconstruction for high-contrast conductivity imaging in mutual induction tomography for industrial applications. IEEE Transactions on Instrumentation and Measurement, 56, 2024 (2007)
4. Soleimani M.: Improving the temporal resolution of magnetic induction tomography for molten metal flow visualization. IEEE Transactions on Instrumentation and Measurement, 59, 553 (2010)

5. Soleimani M.: Simultaneous reconstruction of permeability and conductivity in magnetic induction tomography. J. of Electromagn. Waves and Appl. 23, 785 (2009)
6. Ioan D., Rebican M.: Numerical model for eddy-current testing of ferromagnetic steel Parts. IEEE Transactions on Magnetics 38, 629 (2002)
7. Keyes R.W.: The Impact of Moore's Law. IEEE Solid-State Circuits Society Newsletter 11, 25 (2006)
8. Tamburrino A., Rubinacci G.: Fast methods for quantitative eddy-current tomography of conductive materials. IEEE Transactions on Magnetics 42, 2017 (2006)
9. Prkmel D., Mohammad-Djafari A.: Eddy current tomography in cylindrical geometry. IEEE Transactions on Magnetics 31, 2000 (1995)
10. Salach J., Szewczyk R.: High resolution eddy current tomography setup and possibilities of its signal filtering. Pomiary Automatyka Robotyka 17, 112 (2013)
11. Szewczyk R.: Frequency dependence of the sensitivity of an eddy current tomography system. Proc. Int. Conf. "Problems and Progress in Metrology PPM'15", 07-10.06.2015, Koscielisko, Poland
12. Salach J., Szewczyk R., Nowicki M.: Eddy current tomography for testing of ferromagnetic and non-magnetic materials. Meas. Sci. Technol. 25, 025902 (2014)
13. http://www.hpfem.jku.at/netgen/
14. http://www.csc.fi/english/pages/elmer
15. Szewczyk R., Salach J., Ruokolainen J., Råback P., Stefko K., Nowicki M. : Noise assessment in Whitney elements based forward transformation for high resolution eddy current tomography. Advances in Intelligent Systems and Computing 352, 219 (2015)

The Analysis of Bennett's linkage as a part of deployable mechanism

Ksawery Szykiedans[1], Grzegorz Baska[1], Paweł Nowakowski[1]

[1] Warsaw University of Technology, Faculty of Mechatronics, Warsaw, Poland
k.szykiedans@mchtr.pw.edu.pl, grzegorz.baska@gmail.com,
p.nowakowski@ovi.com

Abstract. The main goal of presented work was to analyse, design and model a 1-dof deployable mechanism with circular shape. The Bennett's linkage and scissor linkage have been used as the unit mechanisms. Kinematics of the Bennett's linkage has been analysed based on symmetry. The parameters of the Bennett's linkage have been calculated through the inverse kinematic calculation. Then the principle of connecting different Bennett's linkages has been proposed. The mechanism can be folded into a bundle of links and deployed into a circular surface.

Keywords: deployable mechanism, Bennett's linkage, scissor linkage, shape approximation.

1 Introduction

1.1 Mechatronic design

Designing of mechatronic devices challenges engineers to fulfil very sophisticated requirements. One of them is a capability of self-optimization. This applies as much to a small MEMS devices, such as position sensors [1, 2] and complex robots. These robots are both inspection drones ever more sophisticated [3, 4] as well as wearable robots [5, 6]. Among this new group, the most spectacular are exoskeletons and orthotic robots. All mentioned devices have to be able to perform self-optimizing procedures. In most cases self-optimizing is aimed on power consumption or security [7] but there also some devices where self-optimization is based on its geometrical features. There are deployable mechanisms that optimize their shape depending on their actual function.

1.2 Deployable mechanisms

This paper presents some works of two students of WUT Faculty of Mechatronics – Grzegorz Baska and Piotr Nowakowski. Under supervision of Ksawery Szykiedans they designed deployable mechanism intended to be used as a mobile shelter.

R. Jabłoński and T. Brezina (eds.), *Advanced Mechatronics Solutions*,
Advances in Intelligent Systems and Computing 393,
DOI 10.1007/978-3-319-23923-1_44

The term Deployable Structure/Mechanism [8] is used to describe a device which has at least two distinct states. A simple example is an umbrella. An umbrella has two distinct states (open and closed), and displays one of the most common design requirements of deployable structures in general, which is that in its inoperative state it is much more compact. A number of typical requirements of a deployable structure are illustrated in this simple example. These include:

- Structure has at least two distinct states: stowed and deployed
- The deployed state is stable (meaning the deployed shape does not need to be actively maintained)
- Structure is typically more compact and robust in the stowed state
- Structure occupies a larger volume in the deployed state. This is usually the state in which the deployable structure is operational, although it might be more fragile:
- A smooth deployment path exists between the two states:
- The structure's deployment is reliable and, sometimes, repeatable. This points to a need for simplicity in design.

Deployable mechanisms (DMs) have their ability to alter their shape to meet different practical requirements. DMs based on linkages have been studied by a number of researchers over the past decades. Several types have been proposed such as Wohlharts polyhedral star-transformer [9], Hoberman's polyhedral mechanism [10], and the cubic/octahedral linkages which are described by Kipper [11]. Deployable mechanisms can have different types of expansion patterns, e.g., one-dimensional, planar, polyhedral, and cambered. Deployable/foldable prisms represent an important example of one-dimensional DMs. They are widely used in large space structures and can conceivably find applications as fixture mechanisms in manufacturing, as well as in civil engineering applications.

1.3 Bennett's linkage

Common 4R mobile loops can normally be classified into two types: the axes of rotation are all parallel to one another, or they are concurrent, i.e. they intersect at a point, leading to 2D 4R or spherical 4R linkages, respectively. Any disposition of the axes different from these two special arrangements is known usually to be a chain of four pieces which is, in general, completely rigid and so furnishes no mechanism at all. But there is an exception, which is the Bennett's linkage [12].

The Bennett's linkage is a skewed linkage of four bars having the axes of revolute joints neither parallel nor concurrent. This linkage was also found independently by Borel [13]. The four links are connected by revolute joints, each of which has axis perpendicular to the two adjacent links connected by it. The lengths of the links are given alongside the links and the twists are indicated at each joint.

2 Bennett's linkage geometry and computations

The kinematics of the Bennett's linkage has been analysed in this project. The bar in the completely foldable Bennett's linkage is not the common normal of the two rotating axes. Two rotation axes of each bar are described as the link in Fig. 1. The two axes are placed on two perpendicular planes, the bar is the intersection of the two plane. The twist of the Bennett's linkage is defined by the angle t_1 and t_2 in Fig. 1.

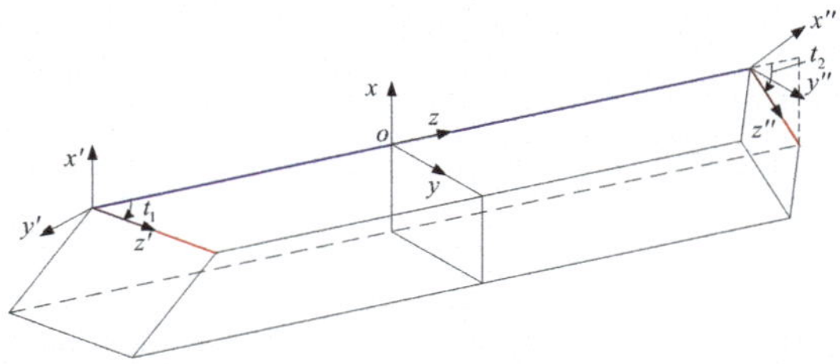

Fig. 1. Angle Definition of the two revolute joints of a bar

There are two coordinate systems have been built on the Bennett's linkage. The first one is established as shown in Fig. 2, z axis is along the length direction of the bar, y axis is placed on the plane generated by s_1 and s_3. The original point is located on s_1.

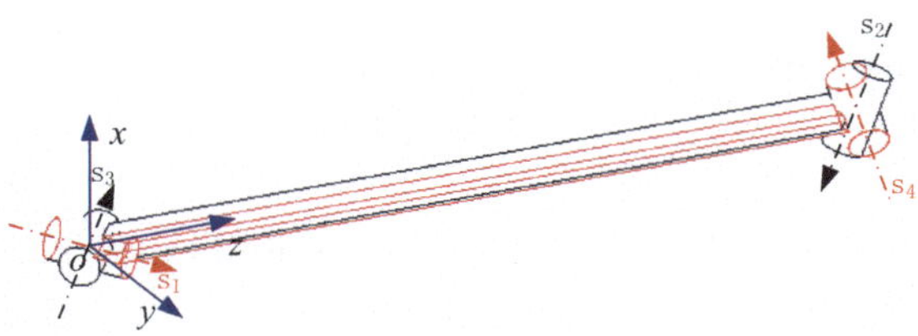

Fig. 2. The first coordinate system of the Bennett's linkage in the original configuration

Knowing the normalized twists of the four revolute joints ζ_i and the length of a bar l, position and orientation of each twist can be calculated through the exponential formula

$$g_{st}(\theta_2) = exp(\theta_1\hat{\zeta}_1)g_{st}(0_2) \quad (1)$$

$$g_{st}(\theta_3) = exp(\theta_1\hat{\zeta}_1)exp(\theta_2\hat{\zeta}_2)g_{st}(0_3) = exp(\theta_4\hat{\zeta}_4)g_{st}(0_3) \quad (2)$$

where,

$$exp(\theta\hat{\zeta}) = \begin{bmatrix} e^{\theta\widehat{\omega}} & (I - e^{\theta\widehat{\omega}})p \\ 0 & 1 \end{bmatrix} \quad (3)$$

The second coordinate system of the Bennett's linkage is established as, y axis is along the rotation axis s_1, z axis is on the place generated by s_1 and s_2, the original point is located on s_1, Fig. 3.

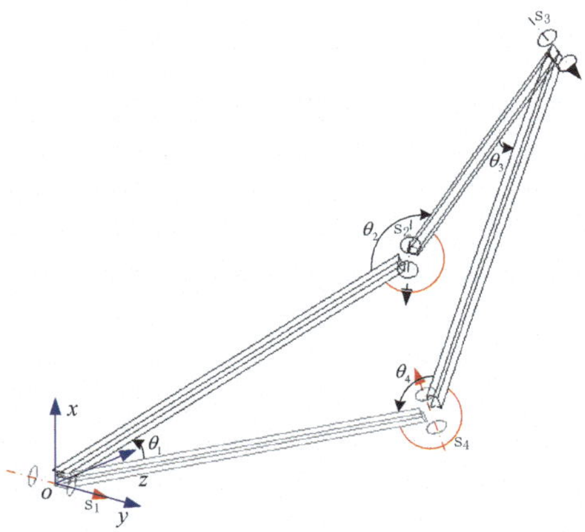

Fig. 3. The second coordinate system of the Bennett's linkage

The relation between the two coordinate systems is

$$R\left(x,-\frac{\pi}{2}+t_1\right)R\left(y,-\frac{\theta}{2}\right)=\begin{bmatrix}1 & 0 & 0\\ 0 & \cos\left(\frac{\pi}{2}-t_1\right) & \sin\left(\frac{\pi}{2}-t_1\right)\\ 0 & -\sin\left(\frac{\pi}{2}-t_1\right) & \cos\left(\frac{\pi}{2}-t_1\right)\end{bmatrix}\begin{bmatrix}\cos\frac{\pi}{2} & 0 & -\sin\frac{\pi}{2}\\ 0 & 1 & 0\\ \sin\frac{\pi}{2} & 0 & \cos\frac{\pi}{2}\end{bmatrix} \quad (4)$$

Therefore the position and orientation of each rotation joint is

$$g_{st}(\theta)=R_yR_x exp(\theta\hat{\zeta})g_{st}(0_2) \quad (5)$$

For the Bennett's linkage with same twists, planes formed by two alternating joints are perpendicular, furthermore, the linkage is symmetric with the two planes. Based on symmetry, once the t_l is fixed, position of point *2* and point *4* which are the intersecting points of two bars are

$$P_2=\left[l\sin t_1\sin\frac{\theta_1}{2},l\cos t_1,l\sin t_1\cos\frac{\theta_1}{2}\right] \quad (6)$$

$$P_4=\left[-l\sin t_1\sin\frac{\theta_1}{2},l\cos t_1,l\sin t_1\cos\frac{\theta_1}{2}\right] \quad (7)$$

where θ_1 is the rotating angle of the first revolute joint.
The normal vector of the plane generated by s_2 and s_4 is

$$ss=s_2\times s_4=[a \quad b \quad c] \quad (8)$$

and the equation of the plane is

$$ax+by+cz+d=0 \quad (9)$$

Due to symmetry, point 3 and point 1 are symmetric with the plane, position of point 3 is

$$P_3 = \left[\frac{-ad}{a^2+b^2+c^2}, \frac{-bd}{a^2+b^2+c^2}, \frac{-cd}{a^2+b^2+c^2}\right] \quad (10)$$

If t_1 is known, (3) is a function of t_2, by substitute it into (11) t_2 can be solved.

A circular surface has been designed in presented project. There are analysed properties of the two component linkages Bennett's linkage and scissor ones. In the design, the deployable assembly is composed of two types of Bennett's linkage. Their sizes are considerably different. All the intersection points of the Bennett's linkage are located on a circle, while only two points of the small Bennett's are on the circle. In order to solve the kinematic calculation of these two types of Bennett's linkage, the circle is divided by several features, each point indicate one or two intersections of the Bennett's linkage. Location of feature points are shown in Fig. 4.

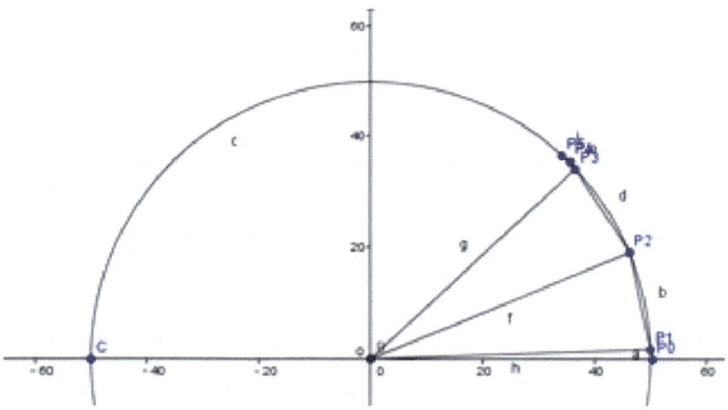

Fig. 4. Feature points of a desired circle

If we shift the original from point O to point P_1, for the big Bennett's linkage, the projection of intersecting points on the circle are $P_1' = (0; 0)$, $P_2' = (-3.78; 17.39)$, $P3' = (-13.4; 32.36)$. We choose $t_1 = 40°$, so other parameters can be calculated through the above equations. The following results were obtained for the big linkage

$$l = 18.78,\ \theta = 1.0403 \text{ rad},\ t_2 = 68.291°.$$

These parameters of the small Bennett's are:

$$l = 1.8794,\ \theta = 1.0403 \text{ rad},\ t_2 = 78°.$$

The big Bennett's and small Bennett's linkage can be connected through scissor linkage. It is found that, the big and small Bennett's have same t_1, but t_2 differ. During the assembly, the Bennett's linkages should be connected along the diagonal that connected the two t_1 axes. Therefore, the movement can be transferred along this direction to the end Bennett's linkage. Also, this is a necessary condition for completely folding.

3 Conclusions

In this project, we presented a deployable mechanism that can be deployed into a circular surface and folded completely. The CAD model of the circular Bennett's assembly has been built in PTC Pro Engineer 4.0 Wildfire software. All the intersecting points of the big Bennett's linkage are located on the curved surface which satisfies the design requirements. Scissor linkage has been used as a connection between the small Bennett's and the bigger one. Kinematics of the Bennett's linkage has been analysed by using the exponential formula and its symmetric feature. It has been found that by connecting the Bennett's linkage through scissor linkage, the degree of freedom of the mechanism does not increase, which guarantees the mechanism can be assembled with indefinitely many unit mechanisms, retaining the same mobility.

References

1. Łuczak, S.: Experimental Studies of Hysteresis in MEMS Accelerometers: a Commentary. IEEE Sensors J. 15, 3492-3499 (2015)
2. Łuczak, S.: Guidelines for Tilt Measurements Realized by MEMS Accelerometers. Int. J. Precis. Eng. Manuf. 15, 489-496 (2014)
3. Bodnicki, M., & Kamiński, D. (2014). In-pipe Microrobot Driven by SMA Elements. In Mechatronics 2013 (pp. 527-533). Springer International Publishing.
4. Osiński, D., & Szykiedans, K. (2015). Small Remotely Operated Screw-Propelled Vehicle. In Progress in Automation, Robotics and Measuring Techniques (pp. 191-200). Springer International Publishing.
5. Jasińska-Choromańska, D., Szykiedans, K., Wierciak, J., Kołodziej, D., Zaczyk, M., Bagiński, K., & Kabziński, B. (2013). Mechatronic system for verticalization and aiding the motion of the disabled. Bulletin of the Polish Academy of Sciences: Technical Sciences, 61(2), 419-431.
6. Wierciak, J., Bagiński, K., Jasińska-Choromańska, D., & Strojnowski, T. (2014). Orthotic Robot as a Self Optimizing System. In Mechatronics 2013 (pp. 607-614). Springer International Publishing.
7. Bagiński, K., Jasińska-Choromańska, D., & Wierciak, J. (2013). Modelling and simulation of a system for verticalization and aiding the motion of individuals suffering from paresis of the lower limbs. Bulletin of the Polish Academy of Sciences: Technical Sciences, 61(4), 919-928.
8. Gan, W. W., & Pellegrino, S. (2003, April). Closed-loop deployable structures. In Proceeding of the 44th AIAA/ASME/ASCE/AHS/ASC Structures, Structural Dynamics, and Materials Conference, Norfork, VA.
9. Wei, G., Ding, X., & Dai, J. S. (2010). Mobility and geometric analysis of the Hoberman switch-pitch ball and its variant. Journal of Mechanisms and Robotics, 2(3), 031010.
10. Hoberman, C. (1991). U.S. Patent No. 5,024,031. Washington, DC: U.S. Patent and Trademark Office.
11. Kiper, G., Söylemez, E., & Kişisel, A. Ö. (2008). A family of deployable polygons and polyhedra. Mechanism and Machine Theory, 43(5), 627-640.
12. Bennett's, G T. A new mechanism, Engineering (1903). 76. 777-778.
13. Bennett's, G T. The parallel motion of Sarrus and some allied mechanisms. Philosophy Magazine (1905), 803-810.

Selection of electric driving modules for orthotic robot

J. Wierciak[1], W. Credo[1] and K. Bagiński[1]

[1] Warsaw University of Technology, Faculty of Mechatronics,
8 Św. A. Boboli St., 02-525 Warsaw, Poland
{j.wierciak, w.credo, k.baginski}@mchtr.pw.edu.pl

Abstract. Driving modules of orthotic robots are employed for moving paralyzed limbs of the disabled. They operate under strongly variable loads disturbing realization of the desired motion profiles to be reconstructed at the driven human joints. A way of selecting an electric driving module for an orthotic robot, designed and built at the Warsaw University of Technology, is discussed. The presented works suggest that selection of driving modules for orthotic robots, which are to meet rigorous functional and operational requirements, is not possible unless a simulation model study is carried out.

Keywords: electric drives, gear ratio, orthotic robots

1 Introduction

Orthotic robots (Fig. 1a), substituting motor functions of human lower limbs [1], should make it possible to realize the basic motor activities, including, first of all, a gait over a flat surface, then sitting and raising as well as climbing stairs [2]. Realization of these activities essentially boils down to a program reconstruction of certain motion profiles by the articulations of the robot, i.e. time courses of the bending angle at particular joints (Fig. 1b). These profiles are usually a modification of known, experimentally determined courses characterizing movements of healthy individuals, gait in particular [3].

Beside realization of typical motion functions, the robot actuators are also used by some safety systems. Their task is to prevent from occurrence of hazardous events – falls of the user especially, and in the case of such event, to minimize the resultant negative effects [4, 5]. The safety systems are quite sophisticated and usually employ various kinds of sensors, e.g. miniature displacement sensors [6] for detection of contact between feet and the ground as well as a tilt sensor based on an accelerometer or a gyroscope [7] as detector of a hazard. Micromachined accelerometers seem preferred in such applications, as they feature many advantages [8].

2 Structure of the driving module

Designing of a driving module for an orthotic robot is a complex process, with respect to both: complicated courses of the related mechanical quantities as well as a need of

R. Jabłoński and T. Brezina (eds.), *Advanced Mechatronics Solutions*,
Advances in Intelligent Systems and Computing 393,
DOI 10.1007/978-3-319-23923-1_45

appropriate spatial configuration of its mechanical structure. The above needs can be met while employing a structure presented in Fig. 2. Such diagram has been accepted for a robot designed at the Warsaw University of Technology [9].

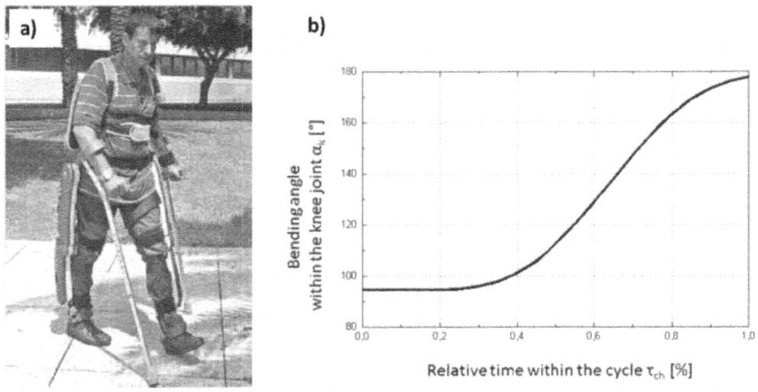

Fig. 1. Orthotic robot (a) [10] and exemplary motion profile at a joint (b) [11]

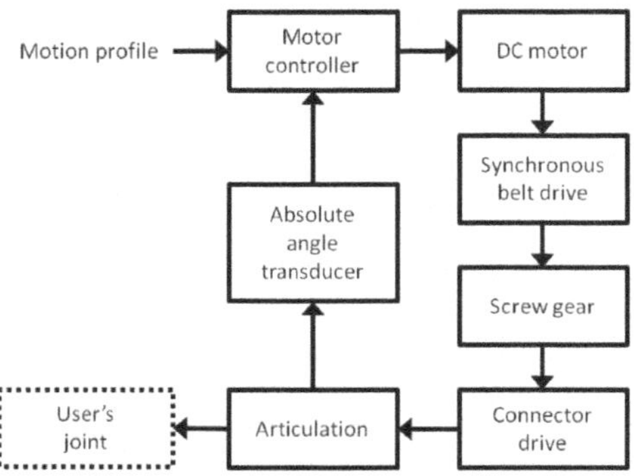

Fig. 2. Block diagram of the driving module

3 Ways of selecting the driving motors

Well-known algorithms of selecting DC motors and dedicated gears, recommend by the manufacturers of electric machines [12, 13], are related to relatively simple cases of their applications. Therefore, the relevant notes issued by the manufacturers present methods of selecting motors for static and dynamic applications. Algorithms related to the first group usually apply to systems that are to develop a given velocity, especially a constant velocity. Algorithms related to the second group apply, first of all, to design of positioning systems. It is noteworthy that in both cases the motor is selected on the basis of its capability of developing a given maximum continuous torque. The torque loading the motor results from reduction of torques acting in the mechanism and being transmitted by a gear that has been selected from a catalog beforehand. An essential difference between static and dynamic systems is the source of the loading torque. In the case of static systems, there occur active and frictional loads, which are accepted as constant, as far as the calculations are concerned. In the case of positioning systems, dominating loads result from accelerating and decelerating rotating elements having particular mass moments of inertia.

The applied algorithms of motor selection, both for static and dynamic applications, assume that the motor is constantly loaded with a computed reduced torque, even if in fact its operation is of a bit different nature.

A continuous torque results in the fact that the motor consumes a constant current, which causes thermal losses at the winding resistance, called also winding losses. While knowing thermal parameters of a motor it is possible to check if during its continuous operation the permissible temperature of the windings would be exceeded. Only this calculation finishes the selection or is a basis for repeating it again.

So, the aforementioned typical algorithms of selecting motors cannot be directly applied while designing driving modules for an orthotic robot, due to a strong variability of the loading torque. Therefore, designers of the *BLEEX* exoskeleton applied their own complicated algorithm, which is minutely described in [14]. Thus, while designing a drive for actuators of an orthotic robot, we have also used an original method [15], in which the sequence of selecting a motor and a gear is reversed with respect to the standard one proposed in related catalogs.

4 Computations of the driving module

As far as the accepted method is concerned, the key point in selecting a motor with a gear is to evaluate power demand of the driven mechanism and to initially select from related catalogs such motors that can develop this power without unnecessary surplus. In the case of driving modules of an orthotic robot, the demand for mechanical power is variable and depends on the realized activity as well as its phase. It was accepted that a relevant power demand occurs while the patient gets up from a chair. The maximal torque loading the driven knee articulation of the robot was assumed to be $M_j = 225$ Nm, and its rotational speed of about $n_j = 0.125$ rev/s. Taking into account a safety factor of 1.2, the following value of power was obtained,

$$P_{max} = 200 \text{ W} .$$

On this basis, motors whose maximal power equals the calculated demand or exceeds it by no more than 20% were found in catalogs. Further proceeding is presented for an exemplary Dunkermotoren GR63Sx55 motor, which has been selected for application in the design because of its characteristics as well as appropriate design features, especially the overall dimensions. The applied algorithm determines such gear ratios i, for which the assumed motion parameters, i.e. the torque and the speed will be developed at a standard supply voltage specified in the related catalog.

$$n_s = n_j \cdot i, \quad M_i = \frac{M_j}{i \cdot \eta}, \tag{1}$$

where: M_l – loading torque reduced the motor; n_s – rotational speed of motor under load; η – efficiency of the unit transmitting the drive.

In order to calculate these ratios, a mechanical characteristic of the motor

$$n_s = -\frac{n_0}{M_r} \cdot M_s + n_0, \tag{2}$$

is to be compared with a mechanical characteristic of the reduced load

$$n_s = \frac{M_j \cdot n_j}{\eta} \cdot \frac{1}{M_l}, \tag{3}$$

where: M_r – starting torque of the motor; M_s – torque developed by the motor under the load; n_0 – no-load speed of the motor.

The characteristic of the reduced load illustrates a dependency of the required speed on the reduced torque for a given range of the ratios. If the motor is capable of developing the required power, the aforementioned characteristics have usually two (at least one) common points (Fig. 3). A specific value of the gear ratio i corresponds to each of the two intersection points. These values were calculated as follows:

$$i_1 = 375, \quad i_2 = 98.$$

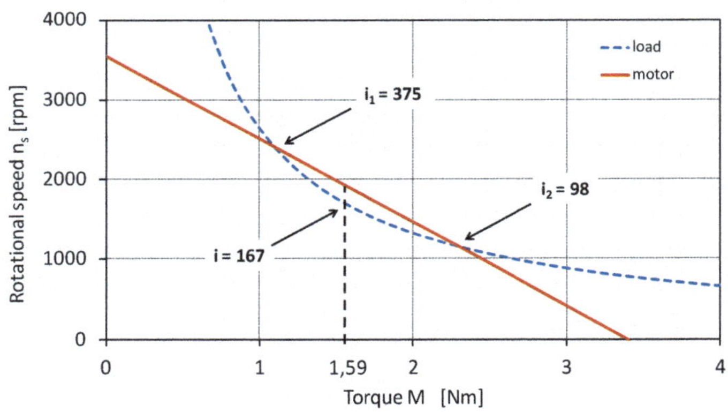

Fig. 3. Determination of the range of the gear ratios

The motor will satisfy the accepted assumptions, provided the gear ratio can be found within the determined range. Due to design reasons, the designed driving module realizes the total ratio of

$$i = 167.$$

A view of the designed driving module is presented in Fig. 4. The driving motor drives the ball screw through the toothed belt gear. The ball nut displaces along the guideway, which prevents from its rotation. A cable, which actuates the wheel of the driven articulation, is attached to the nut, which is subjected to displacements. The total efficiency of the unit transmitting the drive is of about *85%*.

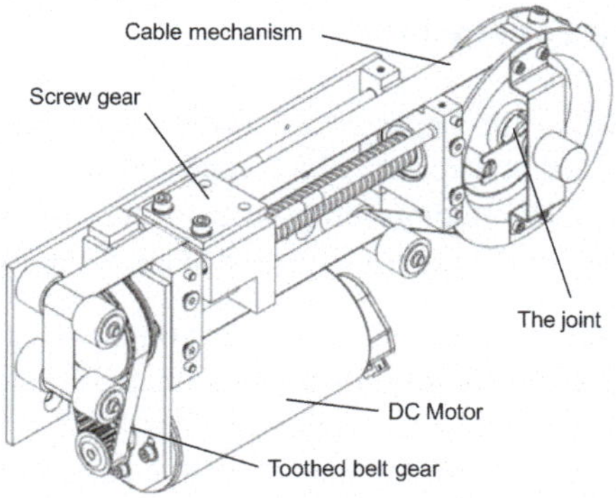

Fig. 4. Structure of the driving module with the ball screw; view without the plate composing the casing of the mechanism

5 Model study of the actuators

Evaluation criteria of operation of the robot driving modules go considerably beyond its capability of realizing the assumed motor activities. Designers are not only to ensure a correct reconstruction of the accepted motion profiles at particular articulations of the robot, but also, among other things, to take care of its energetic effectiveness. Minimization of consumption of the energy supplied from a battery, which is carried by the user, extends the time of autonomic operation of the robot between charging operations. Optimization of the design of the driving modules aimed at achieving their satisfactory characteristics is possible owing to results of simulation studies realized with application of a dedicated model [16] created according to the same principles, which are commonly used in the case of modeling of walking robots, presented in e.g. [17, 18]. In Fig. 5, exemplary results of a simulation study are presented, which reveal dependencies of the energetic demand of the driving module on the gait velocity, for two different driving motors.

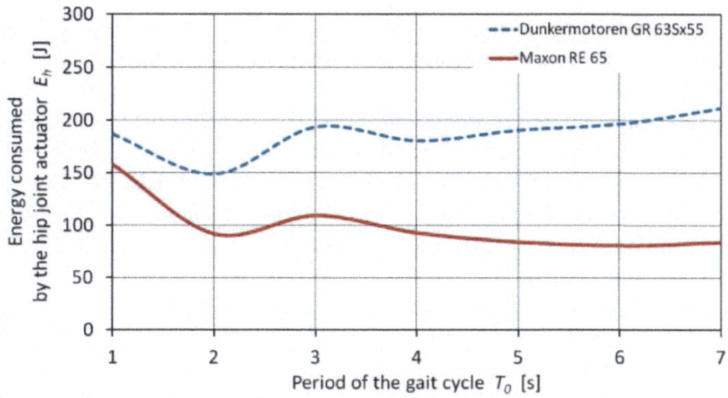

Fig. 5. Amount of energy consumed by the drive of the hip joint within a single gait cycle in function of period of the cycle – exemplary results of a simulation study

6 Summary and conclusions

Considerations presented in the paper reveal difficulties occurring while designing driving modules for an orthotic robot, which are related to techniques of selecting components of the drive. The authors propose to overcome them by using modern computation tools. Results of the described works suggest the following conclusions:

- according to our knowledge, there is a lack of commonly available algorithms of selecting driving gears and motors for systems that are loaded with considerably variable torques, realizing nonlinear motion profiles;
- while selecting elements of a driving module on the basis of static data, it is advantageous to reverse, with respect to traditional algorithms, the sequence of actions by computing ratio of a gear for a motor that has been selected beforehand;
- in order to optimize a drive with respect to advanced criteria related to its operation, e.g. energetic effectiveness, it is necessary to carry out a study using simulation models, which allow for both: complicated description of the loads, as well as the phenomena taking place in the module itself.

References

1. Pons, J. L.: Wearable Robots: Biomechatronic Exoskeletons. John Wiley & Sons, Chichester (2008)
2. Wierciak, J., Jasińska-Choromańska, D., Szykiedans, K.: Orthotic Robot as a Mechatronic System. In: Jabłoński, R., Březina, T. (Eds.) Mechatronics. Recent Technological and Scientific Advances, pp. 579-588. Springer, Heidelberg (2011)
3. Winter, D. A.: The biomechanics and motor control of human gait: Normal, elderly and pathological. Univeristy of Waterloo Press (1991)
4. Jasińska-Choromańska, D., Kabziński, B., Matyjewicz-Maciejewicz, M., Kołodziej, D.: Safety Module for the System of Verticalization and Aiding Motion of the Disabled. In: Březina, T., Jabłoński, R. (Eds.) Mechatronics 2013. Recent Technological and Scientific Advances, pp. 79-85. Springer International Publishing (2014)
5. Jasińska-Choromańska, D., Matyjewicz Maciejewicz, M.: Safety Module Based on Gyroscopie in the System for Verticalization and Aiding Motion of the Disabled. In: Advances in Intelligent Systems and Computing Vol. 351, Progress in Automation, Robotics and Measuring Techniques, Vol. 2 Robotics, pp. 119-128. Springer (2015)
6. Bodnicki, M., Grzybowski, J.: Miniature transducer of linear displacement based on miniature Hall Effect sensors. In: Szewczyk, R., Zieliński, C., Kaliczyńska, M. (Eds.) Advances in Intelligent Systems and Computing vol. 352, Progress in Automation, Robotics and Measuring Techniques vol. 3, Measuring Techniques and Systems, pp. 21-26. Springer International Publishing (2015)
7. Łuczak, S.: Guidelines for Tilt Measurements Realized by MEMS Accelerometers. Int. J. Precis. Eng. Manuf. 15, 489-496 (2014)
8. Łuczak, S.: Effects of Misalignments of MEMS Accelerometers in Tilt Measurements. In: Březina, T., Jabłoński, R. (Eds.) Mechatronics 2013. Recent Technological and Scientific Advances, pp. 393-400. Springer International Publishing (2014)

9. Jasińska-Choromańska, D., et al.: Mechatronic system for verticalization and aiding the motion of the disabled. Bulletin of the Polish Academy of Sciences: Technical Sciences 61(2), pp. 419–431 (2013)
10. Bogue, R.: Exoskeletons and robotic prosthetics: a review of recent developments. Industrial Robot: An International Journal 36(5), pp. 421-427 (2009)
11. Yoshioka, S., Nagano, A., Hay, D. C., Fukashiro, S.: Biomechanical analysis of the relation between movement time and joint moment development during a sit-to-stand task. BioMedical Engineering OnLine (2009)
12. MAXON. High Precision Drives and Systems, Catalogue (2012)
13. PORTESCAP. A Danaher Motion Company: Motion Solutions that Move Life Forward, Catalogue
14. Zoss, A., Kazerooni, H.: Design of an electrically actuated lower extremity exoskeleton. Advanced Robotics 20(9), pp. 967-988 (2006)
15. Wierciak, J., Oleksiuk, W.: Selection of DC motor as well as gear ratio for reliable operation of mechatronic devices. 6th International Conference "Dynamics of Gear Drives", June 19-22, 2002, Zavazna Poruba, Slovak Republic (2002)
16. Bagiński, K., Jasińska-Choromańska, D., Wierciak, J.: Modelling and simulation of a system for verticalization and aiding the motion of individuals suffering from paresis of the lower limbs. Bulletin of the Polish Academy of Sciences: Technical Sciences 61(4), pp. 919–928 (2013)
17. Grepl, R.: Extended kinematics for control of quadruped robot. In: Jabłoński R., Turkowski, M., Szewczyk, R. (Eds.) Recent Advances in Mechatronics, pp. 126-130. Springer, Heidelberg (2007)
18. Zezula, P., Vlachy, D., Grepl, R.: Simulation modeling, optimalization and stabilisation of biped robot. In: Jabłoński R., Turkowski, M., Szewczyk, R. (Eds.) Recent Advances in Mechatronics, pp. 120-125. Springer, Heidelberg (2007)

Modelling and Verification of Piezoelectric Vibration Energy Harvester

Zdenek Hadas[1] and Radek Lan[1]

[1] Brno University of Technology, Czech Republic
hadas@fme.vutbr.cz

Abstract. This paper deals with a modelling and verification of the commercial piezoelectric vibration energy harvester Volture. The model is created as an electromechanical energy transducer with 1 DOF and an equivalent electrical circuit is used. On the base of experiments the verified model could be used for a design of a complex vibration energy harvesting system. The design of a vibration energy harvester with 2 DOF is proposed in this paper. This vibration energy harvester operates with an extended bandwidth and it could be used as autonomous power supply for wireless sensors in vibratory environments. The tune up of the operation frequency bandwidth and output power depends on a piezoelectric generator choice, used seismic tip mass and design of a mutual stiffness.

Keywords: Energy harvesting, piezoelectricity, vibration, model, verification.

1 Introduction

This paper deals with a modelling and verification of a piezoelectric vibration energy harvesting system. The commercial piezoelectric generator Volture is manufactured by MIDE company and it can provide a useful piezoelectric element for customize energy harvesting applications. Piezoelectric vibration energy harvesting systems could be used as autonomous power supplies for several wireless applications, which operate in vibratory environment [1]. The maximal power is harvested only in a resonance operation and it can be limited for several engineering applications [2]. The proposal and modelling of vibration energy harvesters with extended an operation bandwidth is presented. Two piezoelectric generators are used for extending of operation bandwidth and the design proposal based on piezoelectric model is presented in this paper. The operation with common power management electronics is expected and it can provide an alternative power supply for some wire-less applications.

2 Model of Piezoelectric Vibration Energy Harvester

The piezoelectric vibration energy harvester is designed as a flexible cantilever with a tip seismic mass. This design provides mechanical a resonance system and it is excited by mechanical vibrations and kinetic energy of the seismic mass movement is har-

R. Jabłoński and T. Brezina (eds.), *Advanced Mechatronics Solutions*,
Advances in Intelligent Systems and Computing 393,
DOI 10.1007/978-3-319-23923-1_46

vested to electricity using piezoelectric elements. Piezoelectric elements are fixed on both sides of the flexible cantilever and cantilever deformations during mass movement provide mechanical strain of piezoelectric elements. The CAD model and the tested Volture product are shown in Fig. 1. The model of piezoelectric cantilever is presented in a form of equivalent circuit model [3]. This model is shown in Fig. 2 and the electromechanical transducer is described by equation (1).

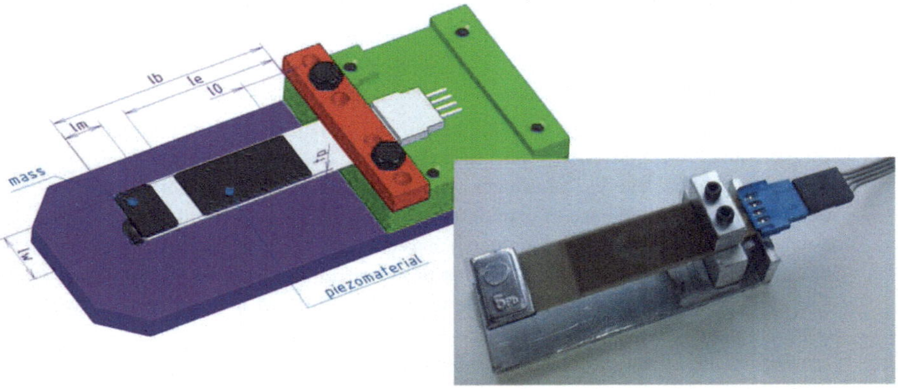

Fig. 1. Model of piezoelectric cantilever and tested commercial Volture product

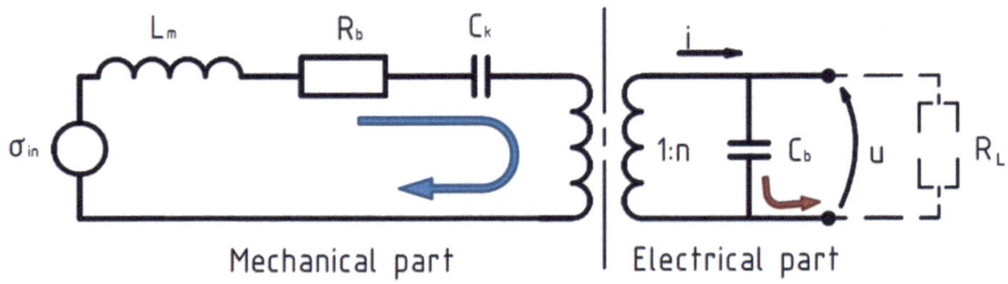

Fig. 2. Equivalent circuit model of piezoelectric cantilever

$$\sigma_{in} = L_m \ddot{S}_1 + R_b \dot{S}_1 + \frac{S_1}{C_k} + nu$$
$$u = \frac{1}{C_b}\int i\mathrm{d}t \tag{1}$$

where S_1 is the mechanical strain in poling direction, L_m is the equivalent inductor which is proportional to the tip seismic mass m at the cantilever end, R_b is the equivalent resistor which presents harvester mechanical damping, C_k is the compliance of the cantilever and i is the current flows by piezo-elements. Transformer n represents ratio between mechanical strain and the electrical voltage. The input mechanical strain is represented by σ_{in} and it is proportional with input vibrations, C_b is the capacitance of the piezoelectric material and u is the output voltage.

The equation (1) can be derived in a form of the cantilever and piezoelectric parameters and these differential equations (2) are used for simulation modelling of this energy harvester.

$$\begin{aligned} m\ddot{y}k_1 &= mk_1k_2\ddot{S}_1 + b_m k_1 k_2 \dot{S}_1 + YS_1 - \frac{a}{2}\cdot\frac{d_{31}Y}{t_p}u \\ \dot{u} &= \frac{2t_p d_{31}Y}{a\varepsilon_{33}}\dot{S}_1 - \frac{2t_p}{R\varepsilon_{33}a^2 w(l_e - l_0)}u \end{aligned} \tag{2}$$

where $\ddot{y}$ is the acceleration of the vibration excited mass movement, b_m is the mechanical damping, Y is Young's modulus, $a/2$ covers influence of the series/parallel connection of the piezo elements wires ($a = 1$ for parallel, $a = 2$ for series), ε_{33} is absolute permittivity, d_{31} is piezoelectric strain coefficient in the poling direction 31, t_p is the thickness of one single layer of the piezoelectric material. Parameters k_1 and k_2 are functions of dimensions and they are presented by relations (3), which are used for a definition of a relation between a force transfer to the mechanical strain and in the second case relation between strain and displacement. I is the cantilever's area moment of inertia, dimension b is approximately $t_p/2$.

$$\begin{aligned} k_1 &= \frac{b}{2I}\cdot(2l_b - l_m - l_e) \\ k_2 &= \frac{(2l_b - l_m)^3}{12b\cdot(2l_b - l_e - l_m)} \end{aligned} \tag{3}$$

3 Measurement and Model Verification

The verified model of the piezoelectric generator as energy harvesting device is very important for design of customize applications and prediction of harvested power [4].

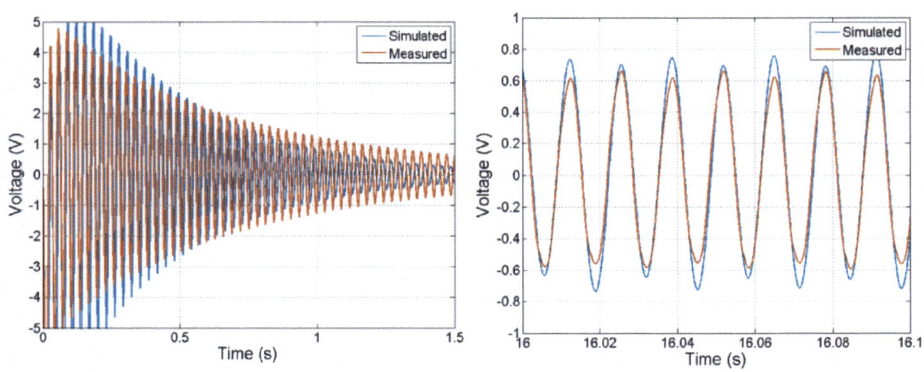

Fig. 3. Model verification: step response and harmonic excitation

Measurements of the commercial Volture product and results of the piezoelectric simulation model were compared. Mechanical and piezoelectric parameters were verified with measurements and both results are shown in Fig. 3; for a step response and harmonic oscillations (response of vibration excitation).

The harvested output voltage depends on a connected electrical load. Measurements and simulation results were compared with pure resistance loads and the comparison of measurement and simulation results of verified model are shown in Fig. 4. This model will be used for design of more complex vibration energy harvesting system and mainly for a prediction of harvested electrical power.

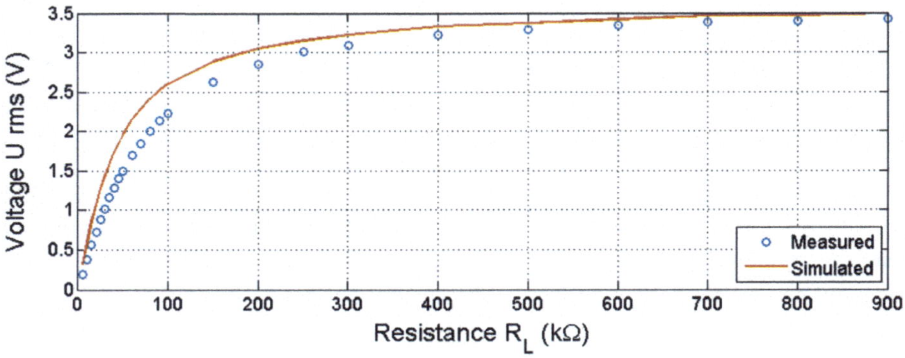

Fig. 4. Output voltage vs. resistance load

4 Proposal of 2 DOF Piezoelectric Vibration Energy Harvester

The verified model is used for the design of a complex piezoelectric vibration energy harvester with 2 DOF. This 2DOF design proposal is shown in Fig. 5.

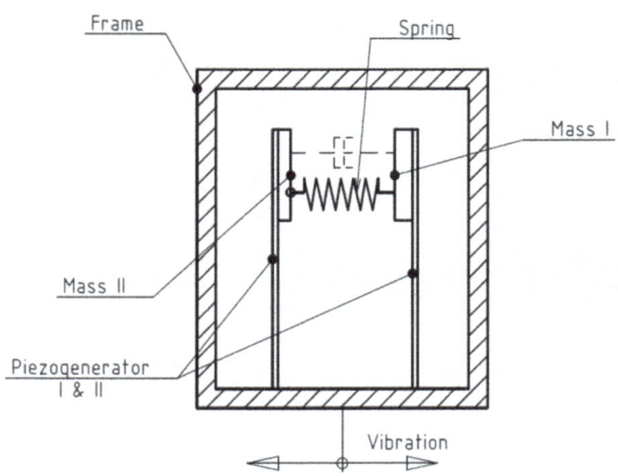

Fig. 5. Proposal of piezoelectric vibration energy harvester with 2 DOF

The 2 DOF design were several times published for an energy harvesting operation improvement [5], [6]. This proposal consists of two Volture piezoelectric generators and mutual spring, which is in a form of an elastic cantilevers linkage.

Our aim is to use the design with 2 DOF for extending of operation bandwidth and the two verified models of piezoelectric generators with the mutual spring are used. The model of this energy harvesting device is used as the equivalent circuit model, Fig. 6, and this model is excited by ambient mechanical vibrations.

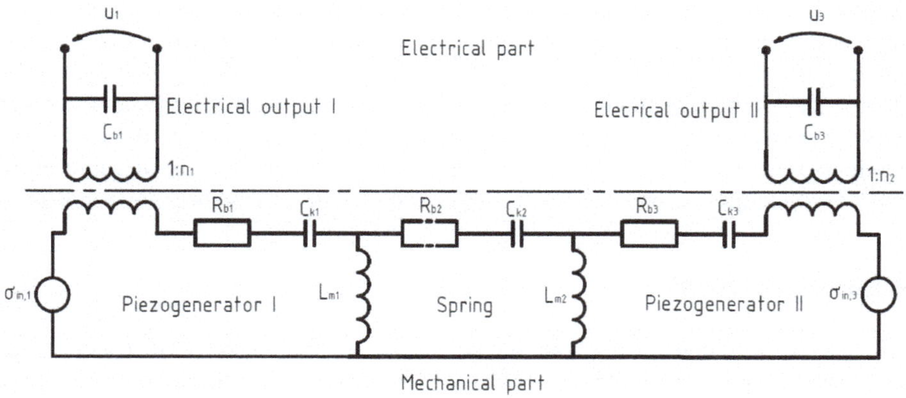

Fig. 6. Equivalent circuit model of two connected piezoelectric beams

Used piezoelectric generators have the different tip mass and the frequency response of this design is shown in Fig. 7. The amplitude of output voltage is proportional with excited vibration level and the operation frequency depends on the used tip mass. The presented vibration energy harvesting model is linear, however, the design of the mutual elastic spring or a mutual springy element can be in form of nonlinear permanent magnet forces [7] or other nonlinear stiffness [8].

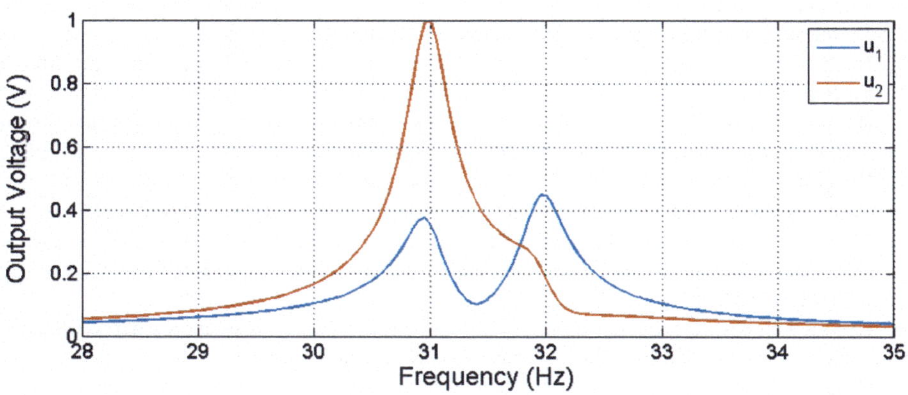

Fig. 7. Frequency response of both piezoelectric energy harvesters

5 Conclusions

The model-based design of the piezoelectric vibration energy harvesting system is presented in this paper. The verified model of Volture piezoelectric generator is used for the proposal of the complex harvester design with 2 DOF and this system can provide the extension of the operation frequency bandwidth.

The main aim of this paper is the presenting of the methodology for the development of complex vibration piezoelectric energy harvesting systems. The proposed design has to be optimized for the sufficient energy harvesting from mechanical vibrations and the operation frequency range and output power will be optimized on the base of customer requirements. Due to the complexity of this device some artificial intelligence optimize methods could be used [9]. There is a space for parameters and shape optimization studies. The optimal harvester design can provide the useful autonomous source of energy for several wireless applications, mainly in engineering branches of monitoring or autonomous diagnostic systems in vibratory environment.

Acknowledgement

This work is an output of research and scientific activities of NETME Centre, supported through project NETME CENTRE PLUS (LO1202) by financial means from the Ministry of Education, Youth and Sports „National Sustainability Programme I".

References

1. Anton SR, Sodano HA (2007) A review of power harvesting using piezoelectric materials (2003–2006). Smart Mater Struct 16:R1–R21.
2. Hadas Z, Ondrusek C, Singule V, Kluge M (2008) Vibration power generator for aeronautics applications. Proc. 10th Anniv.Int.Conf.Eur.Soc.Precis.Eng. Nanotechnology, EUSPEN 2008. euspen, pp 46–50.
3. Khaligh A (2010) Kinetic Energy Harvesting Using Piezoelectric and Electromagnetic Technologies—State of the Art. IEEE Trans Ind Electron 57:850–860.
4. Hadas Z, Singule V, Ondrusek C (2010) Verification of Vibration Power Generator Model for Prediction of Harvested Power. Solid State Phenom 164:291–296.
5. Halim MA, Park JY (2014) Performance enhancement of a low frequency vibration driven 2-DOF piezoelectric energy harvester by mechanical impact. J Phys Conf Ser 557:012093.
6. Kucab K, Górski G, Mizia J (2013) Energy harvesting in the nonlinear two-masses piezoelastic system driven by harmonic excitations. Eur Phys J Spec Top 222:1607–1616.
7. Martens W, von Wagner U, Litak G (2013) Stationary response of nonlinear magneto-piezoelectric energy harvester systems under stochastic excitation. Eur Phys J Spec Top 222:1665–1673.
8. Kitio Kwuimy CA, Litak G, Nataraj C (2015) Nonlinear analysis of energy harvesting systems with fractional order physical properties. Nonlinear Dyn 80:491–501.
9. Hadaš Z, Ondrůšek Č, Kurfürst J (2009) Optimization of vibration power generator parameters using self-organizing migrating algorithm. Recent Adv. Mechatronics 2008-2009. pp 245–250

Part IV
Metrology and Nanometrology

Miniature displacement sensor

Maciej Bodnicki[1], Paulina Pakuła[2], Martyna Zowade[2]

[1] Institute of Micromechanics and Photonics, Warsaw University of Technology, Warsaw, Poland
m.bodnicki@mchtr.pw.edu.pl
[2] Graduate of Faculty of Mechatronics, Warsaw University of Technology, Warsaw, Poland
{p.pakula, m.zowade}@gmail.com

Abstract. Nd-Fe-B micromagnets and miniature Hall-effect sensors enable to build small measuring systems for micro-displacements. They can be applied as individual sensors, elements of sensor arrays or sources of position signals in miniature servodrives. A prototype miniature displacement transducer, with Hall-effect devices operating as a differential system, was elaborated and manufactured using 3D printing technology. A computer aided measurement system was launched allowing the prototype to be tested. The text presents the obtained basic metrological characteristics of the converter. Conclusions are drawn and future works are specified. It is possible to propose configuration ensuring a linear signal with resolution of at least 0.3 micrometer at a displacement range of 1 millimeter.

Keywords: linear displacement · measurement · Hall-effect sensor · micro sensor

1 Introduction

Various miniature measuring devices have been built over the recent years at the Institute of Micromechanics and Photonics, Warsaw University of Technology (WUT). An example can be here tilt sensors based on MEMS accelerometers [1] minutely described in [2] or devices for measuring micro-displacements, of which few prototypes were built.

One of directions of the undertaken research was development of a technique of contactless measurements with application of optoelectronic technology, being useful especially in the case of studying or operating miniature devices (with dimensions within a range of few to few tens of millimeters). The related solutions are based on open optoelectronic connections or two-dimensional CCD arrays. The solutions allow many problems of a technical nature to be solved – e.g. problems related to type of the surface of the studied object. Methods based on light reflection (e.g. laser interferometry), featuring the highest accuracy, can be applied only in the case of surfaces with a high reflectivity, what essentially limits functionalities of the elaborated apparatuses. However, in many cases, a contact of the studied element (via e.g. a contact tip in the case of a linear microdrive) with the transducer can be allowed.

R. Jabłoński and T. Brezina (eds.), *Advanced Mechatronics Solutions*,
Advances in Intelligent Systems and Computing 393,
DOI 10.1007/978-3-319-23923-1_47

Other solutions of contact transducers of displacement/position are based on the following ideas: inductance displacements sensor (including LVDT), mechanical dial gauges (currently equipped with electric outputs), capacitive displacement differential systems; strain gauges, incremental sensors (e.g. optoelectronic encoders) [3,4].

The presented works concern a prototype of an inexpensive contact sensor built on a basis of miniature permanent Nd-Fe-B magnets and Hall-effect devices fabricated in SMD technology. Tests of the prototype had been preceded by a theoretical analysis presented in [5], which indicated a possibility of achieving a satisfactory resolution at a measuring range of ca. 1 mm.

2 Demonstrator of the technology

At the first stage, a study of a dependency of the signal on the position of a micro-magnet with respect to the sensor was carried out. Then, it was possible to create a mathematic model of the dependence of the signal generated by a HW 108A miniature Hall-effect sensor in magnetic field generated by a Nd-Fe-B micro-magnet (cylinder, volume of about 1 mm^3). According to the results presented in paper [5], it is possible to obtain for a ±0.5 mm (1 mm) displacement transducer nonlinearity less than 0.1 % F.S.

The configuration of two Hall-sensors and one micro-magnet system generating signals for differential amplifying used in the demonstrator is presented in Fig. 1.

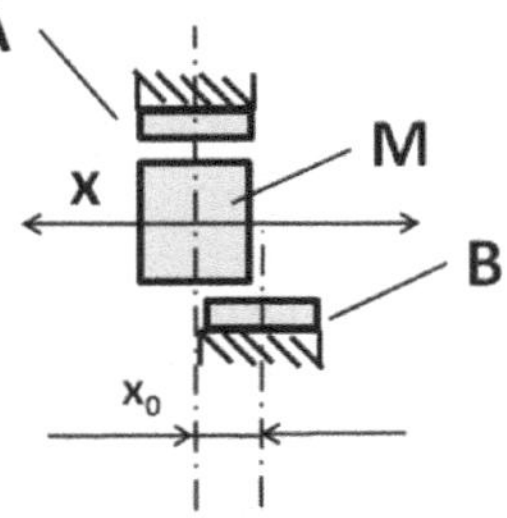

Fig. 1. Configuration of the differential transducer with Hall-effect sensors - one magnet system applied in the technology demonstrator:
M – micromagnet, A and B – sensors, x – linear displacement, x_0 – phase shift at maximum sensitivity of Hall sensors

All the constituent parts of the demonstrator were manufactured employing 3D printing technique , except for a reverse spring of the measurement tip made of a bronze strip. The demonstrator is presented in Fig. 2.

Differential signal is performed from signals of sensors A and B by INA 105 KP precision differential amplifier. Supply voltages of the amplifier and HW 108A sensors are ±15 V.

Fig. 2. Technology demonstrator: 3D visualization and design

3 Test stand

A computer supported test stand, presented in Fig. 3, was development for testing accuracy and linearity of the aforementioned prototype. The proposed method is based on using a signal of high-resolution linear displacement transducer as the reference.

The main components are:

- mechanical part integrating the prototype, actuator positioning the tip (linear screw gear) and the reference linear displacement meter (optoelectronic digital length meter Heidenhain MT 25: range – 25 mm, accuracy - ±0.2 μm);
- electronic unit of the stand connected with a PC-type microcomputer via a Data Acquisition System (DAS) card: NI USB-6259 DAS. The general parameters of the card, important from the point of view of the discussed application are: 32 programmable analog inputs (16-bit); 1.25 MS/s single-channel (1 MS/s all channels) 4 analog outputs (16-bit, 2.8 MS/s) and software support by NI LabView tools for Windows: icon-based intuitive graphical programming software using OPC standard driver interface.

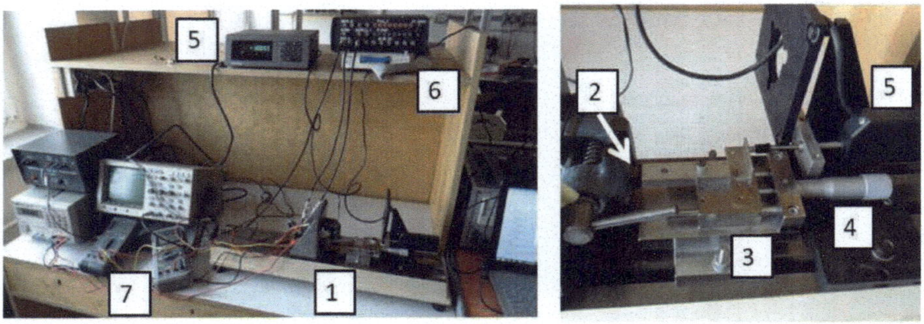

Fig. 3. View of the test stand for determination of the demonstrator characteristics

1 – mechanical body of the stand, 2 – the tested demonstrator in a special holder, 3 – linear guide system with a micrometer screw mechanism for positioning of the measuring tip, 4 – micrometer screw, 5 – reference linear displacement meter (MT 25), 6 – NI USB-6259 DAS, 7 – additional meters (multimeter, digital oscilloscope) and supply units

4 Testing of the demonstrator

A number of experiments have been performed on the technology demonstrator using the test stand . The tests included:

— checking linearity of the transducer
— measuring resolutions of the transducer
— determining real detection range,
— estimating impact of disturbances.

Differential signal obtained during the tests is presented in Fig. 4.

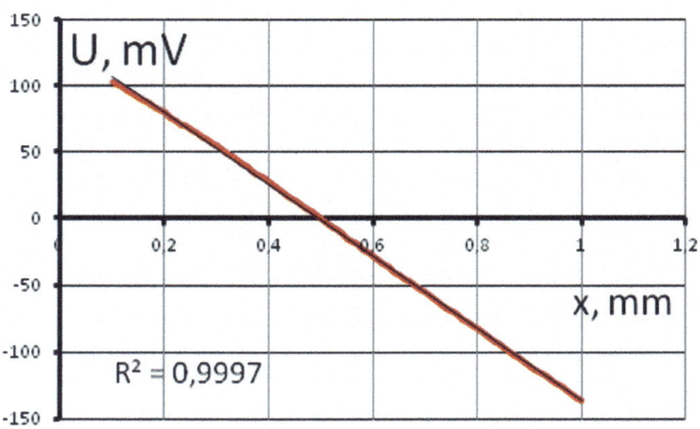

Fig. 4. Physical experiments – experimental characteristics of the transducer U – differential voltage, x – signal from the MT 25 reference meter, red thick line – experimental data, black thin line – trend line

Linearity of the presented output signal is satisfactory. They were also carried out research repeatability of measurements. The results obtained for two test positions are shown in Table 1.

Table 1. Results of the repeatability tests

Position	Mean (n=10)	Standard deviation	Standard deviation	Standard deviation
mm	mV	mV	%	µm
0.2	69.9	0.316	0.45	1.2
0.6	-130.5	0.527	0.4	1.9

Some technical problems have been encountered when installing the cable connection between the Hall sensors and the amplifier. This clearly indicates a need of placing an integrated electronics within the body of the transducer.

5 Summary and conclusions

The presented results of the experimental tests confirm that it is possible to build miniature converters of displacements featuring satisfactory metrological parameters. So, it would be possible to design linear servo drives, whose micromotor, coupled with a system of converting motion into a linear one, is integrated with a transducer of the pusher displacement, based on the presented concept, and thus composes a small actuator. Such units enable to build miniature electromagnetic systems for spatial positioning.

It should be emphasized that these possibilities do not refer to measurements of linear displacements only. They can be adapted to measurements of angular displacements, and also measurements of such mechanical quantities, in which a measurement of displacement of the element subjected to some other interaction occurs (e.g. measurement of forces, torques, pressure). Systems of such sensors will enable to build measuring arrays or matrixes of relatively high density of information. That will create e.g. a possibility of building low-cost pressure mats for measuring parameters related to gait in medicine, or motion in bioengineering – especially foot sensors for self optimizing orthotic robots, which have been designed at WUT [6,7].

References

1. Łuczak, S.: Effects of Misalignments of MEMS Accelerometers in Tilt Measurements. In: Březina T., Jabloński R. (eds.) Mechatronics 2013 Recent Technological and Scientific Advances, pp. 393-400, Springer, Heidelberg (2014)
2. Łuczak, S.: Guidelines for Tilt Measurements Realized by MEMS Accelerometers. Int. J. Precis. Eng. Manuf. 15, 489-496 (2014)
3. Wilson J. S. (ed.), Sensor Technology Handbook, Elsevier, Oxford, (2005)

4. Bishop R. H. (ed), Mechatronic Systems, Sensors, and Actuators. Fundamentals and Modeling, CRC Press, Boca Raton (2008)
5. Bodnicki M., Grzybowski J.: Miniature transducer of linear displacement based on miniature Hall Effect sensors. In Szewczyk R., Zieliński C. and Kaliczyńska M. (eds), Progress in Automation, Robotics and Measuring Techniques. Vol. 3 Measuring Techniques and Systems, pp. 21-26, Springer, Heidelberg (2015)
6. Wierciak J., Bagiński K., Jasińska-Choromańska D., Strojnowski T.: Orthotic Robot as a Self Optimizing System. In: Březina T., Jabloński R. (eds.) Mechatronics 2013 Recent Technological and Scientific Advances, pp. 607-614, Springer, Heidelberg (2014)
7. Szykiedans K.: Kinematic interactions between orthotic robot and a human. In: Szewczyk R., Zieliński C., Kaliczyńska M. (eds.) Progress in Automation, Robotics and Measuring Techniques Vol. 2 Robotics, pp. 265-274. Springer International Publishing (2015)

Liquid-filled adjustable optical wedge applied for deflection of the laser beam

Olga Iwasińska-Kowalska

Warsaw University of Technology, Faculty of Mechatronics, Institute of Metrology and Biomedical Engineering, Warsaw, Poland
O.Iwasinska@mchtr.pw.edu.pl

Abstract. Studies on the application of a liquid-filled variable angle prism as precise beam deflector are presented. The system operates on the basis of a liquid-filled variable angle prism. The basic idea is to use two flat optical plates separated by optical fluid. To avoid distortion of wavefront the liquid layer is very thin. The optical fluid has a refractive index matching the refractive index of plates. Plates are adjusted at an angle in order to obtain the optical fluid wedge between them. Theoretical basis, experimental set up and experimental results are presented. As a result a range of deflection $5 \cdot 10^{-6}$ radians were obtained.

Keywords: laser beam angular drifts, beam deflector, liquid filled variable angle, prism wedge prism, carrier fringes

1 Introduction

Practical usage of optical methods such as interference distance and straightness measurement [1], may need performing very small angle deviation of a light beam. The angular stability of a laser system and the active beam displacement compensation is often the main factor limiting application precision [2]. The axis instability of an angular laser beam can reach up to 20 μrad per hour in laser interferometer systems. A number of concepts of beam deflection measurement have been developed, examples are listed [3,4,5]. The information about measured displacement may be used to compensate the angular displacement of the beam [6]. Only the method, that does not cause distortion of the beam wavefront [5,7,8], can be applied in interference displacement measuring systems.

1.1 Principle of operation

Developed systems are based on two flat plates P_1,P_2 (fig. 1) and an optical fluid that has a refractive index matching the refractive index of plates. Plates are adjusted at an angle in order to obtain the optical liquid wedge. The deflection δ of the transmitted

R. Jabłoński and T. Brezina (eds.), *Advanced Mechatronics Solutions*,
Advances in Intelligent Systems and Computing 393,
DOI 10.1007/978-3-319-23923-1_48

optical beam is caused by a change of the wedge angle. The deflection can be described as:

$$\delta = \arcsin\,(n_l \cdot \sin(\theta)) - \theta \qquad (1)$$

where n_l is the refractive index of glasses plates and optical fluid, θ is angle of the optical liquid wedge. Angle of incidence is 0.
For small angle θ (1) can be simplified to form:

$$\delta = (n_l - 1) \cdot \theta \qquad (2)$$

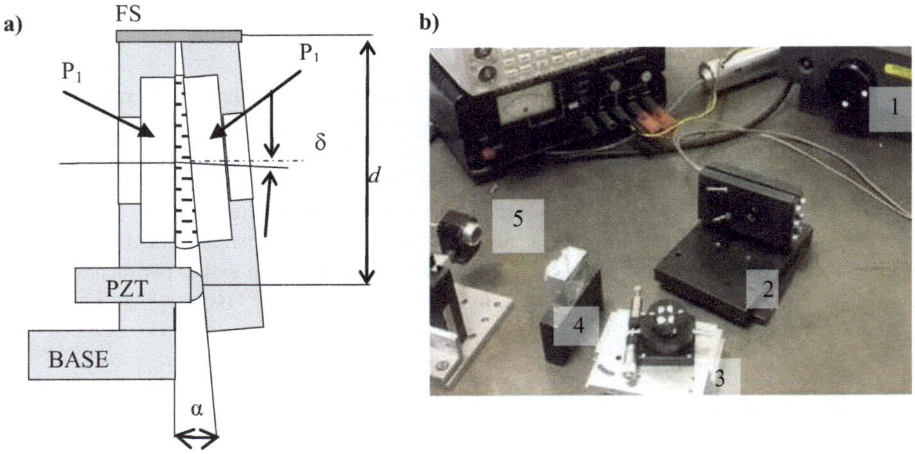

Fig. 1. a) Optical set up of liquid filled variable angle prism b) View of the experimental set-up

In Fig. 1, a) an outline of an experimental system for steering micro-deflections of the laser beam is represented. The plate P_1 in the casing is fixed to the base. The plate P_2 is subjected to an angular deflection. The element steering the deflection is a piezoelectric transducer PZT. The casing of plate P_1 is suspended on a flat spring FS.
View of the experimental set-up is shown in fig. 1. b). The beam of metrological laser 1 passes through the liquid filled variable angle prism 2. The deflection of the laser beam is analyzed in specialized interferometer 3 [6]. Interfering beams are transmitted by a polarizer 4 to photodetectors 5.

1.2 Experimental results

To evaluate the metrological proprieties of the described deflector, an analysis based on the measurements of angular microdeflection was performed. During the experiment, a voltage of given range was used to control the piezoelectric element. The change in the deflection angle was registered. An interferential method, described in the [6], was used for measuring the angular drift of the beam.
In fig. 2, examples of the signals registered on the oscilloscope are represented: the piezoelectric controlling signal (channel 2) and the output beam deflection measuring device signal (channel 1). The value of angle drift is given by the following equation:

$$\delta = (n_l - 1) \cdot \mathrm{arctg}\,(\frac{d}{\Delta x}) \quad (3)$$

where d – the length of lever arm, Δx – displacement of piezotransducer, n_l – the refractive index of the optical fluid.

The exemplary result is shown in fig. 3. The range of the interferometer signal is 1,25 V. This corresponds to the beam drift 50 μrad. The value was calculated on the basis of control voltage (5 V), that caused displacement of piezoelectric actuator Δx=5μm and length of the lever arm d = 75 mm.

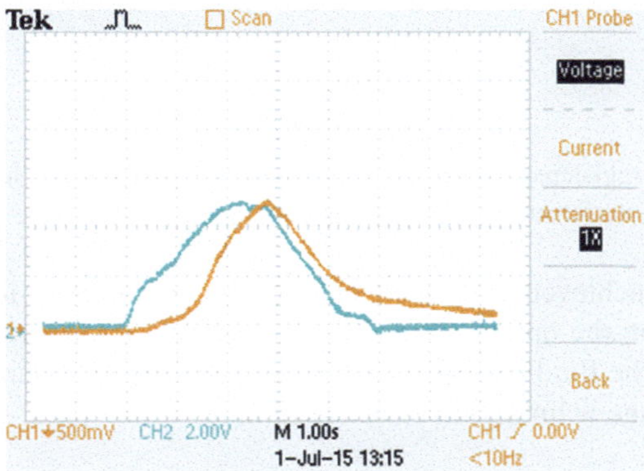

Fig. 2. Examples of registered signals: the device-controlling signal channel CH2 and the interferometer s measuring signal [6] channel CH1.

Since capillary forces counteract the angle change, the deflection response to an applied voltage is not linear. The delay of the response depends on the displacement direction. This gives rise to hysteresis (fig. 4).

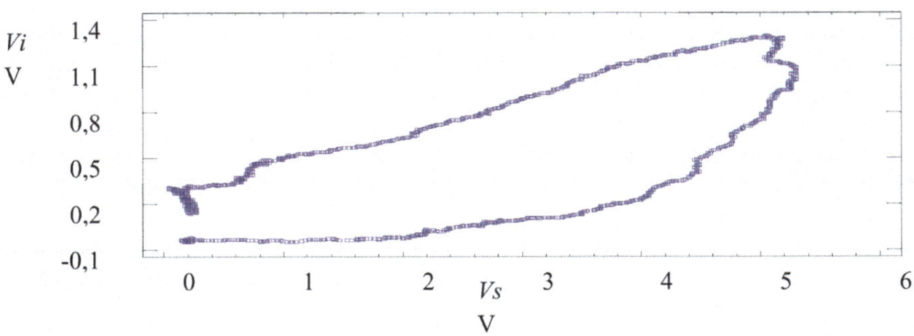

Fig. 3. Hysteresis of *Vi* - signal of the deflection measurement versus *Vs* - the steering voltage of PZT

The disadvantage of the liquid optical wedge is limited frequency response. In fig. 5 the time of manual change of the supply is 5 s. The ratio control to response signals was 4. In fig. 3 the frequency is about 2Hz. The ratio of control to response signals was 10.

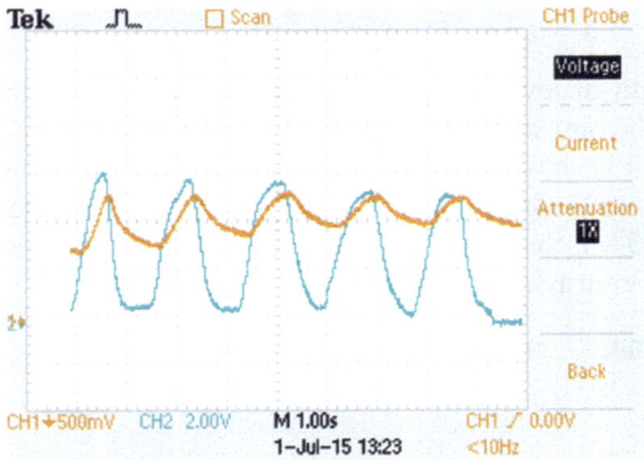

Fig. 4. Examples of registered sinusoidal signals: controlling signal channel CH3 and the interferometer s measuring signal [6] channel CH1.

Precision cannot be achieved with this technique due to capillary forces, pressure and temperature effects in the optical fluid that can lead to a variation of the index of refraction throughout the fluid volume affecting both beam quality and deflection angle. The main disadvantage is limitation of bandwidth.

Acknowledgements

The research described in the article have been financed with funds from the National Research and Development Centre. Grant NCBiR no. PBS3/A6/21/2015.

References

1. Dobosz M.: Laser diode distance measuring interferometer - metrological properties. Metrol. Meas. Syst.,Vol. XIX, No. 3, pp. 553-564, (2012)
2. Zamiela G., Dobosz M.: Corner cube reflector lateral displacement evaluation simultaneously with interferometer length measurement. Optics and Laser Technology, vol. 50, pp. 118-124 (2013)
3. International Standard EN ISO 11670. Lasers and laser-related equipment Test methods for laser beam parameters Beam positional stability. (2003)
4. Garc a-Valenzuela A., Sandoval-Romero G. E., S nchez-P rez C., High-resolution optical angle sensors: approaching the diffraction limit to the sensitivity, Appl. Opt. 43, 4311 4321 (2004).
5. Sender kov , D., trba, A.: Analysis of a wedge prism to perform small-angle beam deviation. Proceedings of SPIE, 5036, pp 148 151. (2003).
6. Dobosz M., Iwasińska-Kowalska O.: Interference method for ultra-precision measurement and compensation of laser beam angular deflection. Applied optics, Vol. 53, No. 1, pp 111-122 (2014)
7. Roland E., Dub J., Dub G.: .Refractive Beam Steering. Proceedings of SPIE Vol. 5528, Bellingham, WA, (2004) \
8. Iwasińska-Kowalska O.: A system for precise laser beam angular steering. Metrol. Meas. Syst., Vol. XXI, No. 1, pp. 27 36, (2014)

Master artifacts for testing the performance of probes for CNC machine tools

Michał Jankowski[1], Adam Woźniak[1]

[1] Institute of Metrology and Biomedical Engineering, Warsaw University of Technology, Poland
{m.jankowski,a.wozniak}@mchtr.pw.edu.pl

Abstract. Methods of testing the performance of probes for CNC machine tools, the same as methods of testing the performance of probes for coordinate measuring machines (CMMs) can be divided into indirect methods and direct methods. Indirect methods give information about performance of the whole on-machine measuring system, including the probe and the machine tool. Direct methods enable separation of the probe's errors and the machine tool's errors, but they need an additional equipment. One of the direct methods of probe testing is a method with moving master artifacts. Performance of this method depends on the master artifact used. In the paper various master artifacts are presented, as far as results of probe testing with use of them.

Keywords: on-machine measurement · probes · machine tools · direct methods.

1 Introduction

Probes for CNC machine tools are widely used for workpiece setup and its dimensional control after machining or in-process. They can be also used to determine the CNC machine tool's volumetric errors by measurement of the master artifact [1]. In most of these applications, errors of the probes influence machining errors. That's why probe's user needs information about its performance, especially about its repeatability in a single probing direction and about its spatial errors.

The most popular methods of probes' metrological parameters determination base on on-machine measurement of master artifact, e.g. ring gauge or test sphere [2 - 7]. This measurement is conducted the same way as workpiece measurements on the machine. These methods, called indirect methods, are simple, fast and cheap, because no additional equipment except the master artifact is needed. Unfortunately, they do not enable to separate probe's errors and machine tool's errors, while this separation is necessary in many applications, e.g. during determination of machine tool's volumetric errors with use of the probe.

That's why direct methods, basing on use of specialized test setups, are developed. They were successfully used to test the probes for coordinate measuring machines (CMMs) [8 – 10], which are somewhat similar to the probes for CNC machine tools.

R. Jabłoński and T. Brezina (eds.), *Advanced Mechatronics Solutions*,
Advances in Intelligent Systems and Computing 393,
DOI 10.1007/978-3-319-23923-1_49

In direct methods probe is tested out of the machine tool or on the machine tool which is not moving. These methods enables to determine metrological parameters of the probe only – machine tool's errors have no influence on the results.
One of the direct methods is method with the moving master artifact developed by authors [11]. This method enables probe testing both in 2D and 3D, depending on used master artifact. In the paper various master artifacts for testing the probes for CNC machine tools are presented and results obtained them are shown.

2 Probe testing method with moving master artifact

In method with moving master artifact stylus tip (1) of tested probe (2), which is fixed, is placed in a centre of master artifact (3), which is mounted on three-axial translator (4), as it's shown in the fig. 1.

Fig. 1. Mechanical part of setup for probe testing

To test the probe's operation in direction i, the master artifact is moved by the positioner in this direction. When the probe triggers, the coordinates X, Y and Z of the master artifact's position $TG_{i,j}$ are saved, where j is a number of measurement in the given direction.
To test the probe's operation in a plane or space, measurements are done for various directions. Than the best-fitted element (circle or sphere) is determined. The distance from this element's centre O_S and the saved position $TG_{i,j}$ is a measured triggering radius $r_{i,j}$. For every direction i, the mean measured triggering radius in this direction

$\bar{r}_i$ is calculated. To characterize systematic errors of the probe the triggering radius variation parameter V_r is used:

$$V_r = \max(\bar{r}_i) - \min(\bar{r}_i). \quad (1)$$

Tests can be carried out in laboratory or on machine tool, in workshop conditions – because the measurement is very quick, there is no need for temperature stabilization.

3 3D inner hemisphere, araldite master artifact

To test the spatial performance of the CNC machine tool probe's, the inner hemisphere master artifact shown in the fig. 2 is used.

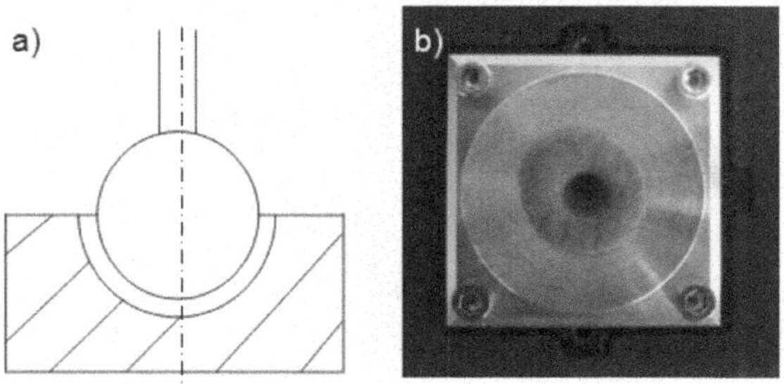

Fig. 2. Inner hemisphere master artifact: a) scheme of the master artifact working with the stylus tip; b) view of the artifact

This master artifact was manufactured by pressing a precise ball into the araldite and its diameter is 6.0208 mm. Because form error of this master artifact is about 2.8 μm, its error map was created using the Renishaw TP200 precise, strain-gauge probe. After implementation of numeric compensation of master artifact errors, which bases on the created error map, the uncertainty (k = 2) of determination of triggering radius variation using this master artifact is 0.6 μm [11].
Exemplary results obtained for Renishaw OMP40-2 probe with 5.9989 mm stylus, using described master artifact are shown in the fig. 3, where β is an angle between the probe's operation direction and the plane XY, i.e. plane perpendicular to probe's axis. The test speed was 0.5 mm/s. Because the tested Renishaw OMP40-2 probe is a 3-point kinematic probe, 3-lobed characteristics of triggering radius are consistent with the theoretical predictions. The value of triggering radius variation in 3D for this probe V_r = 10.7 μm.

4 2D steel ring gauge

To test the performance of the probes in a plane perpendicular to their axes and to verify results obtained using 3D inner hemisphere master artifact, the steel ring gauge is used. The ring diameter is 6.0181 mm. Because its form error does not exceed 0.2 μm, no numeric compensation is needed. In the fig. 4 the comparison of Renishaw

OMP40-2 probe's triggering radii in XY plane measured using 2D and 3D gauges is shown. The test speed was 0.5 mm/s. As it can be seen, the obtained characteristics are very similar and the 3-lobed shape is preserved. The obtained values of triggering radius variation in XY plane are, respectively for 3D and 2D gauges, $V_r = 8.5$ μm and $V_r = 9.4$ μm.

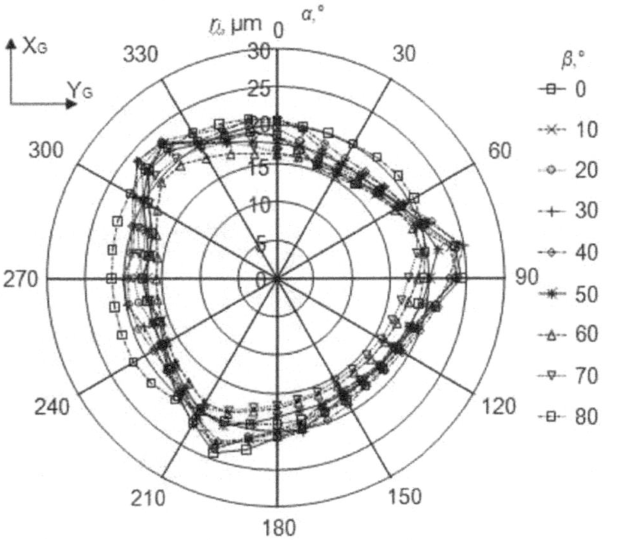

Fig. 3. Triggering radius characteristics of Renishaw OP40-2 3-point kinematic probe

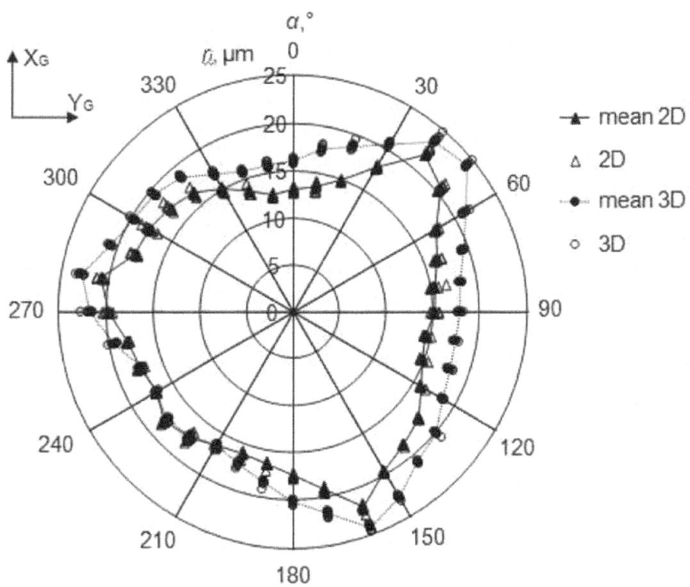

Fig. 4. Comparison of triggering radius values of Renishaw OP40-2 3-point kinematic probe obtained using 2D and 3D gauges

5 Ring gauge with a cut

Ring gauge enables testing the probes' performance in the plane perpendicular to their axes, but does not enables testing probe's performance in planes containing their axes. To overcome this limit, the ring gauge with a cut was proposed. In the fig. 5 the view and the scheme of this master artifact's usage with a probe tip are shown. As it can be seen in the figure, the stylus tip can be placed inside the gauge thanks to the cut, in which the stylus can be placed.

The form error of the proposed gauge does not exceed 0.2 μm. Thanks to that there is no need to use the numeric error compensation. The diameter of this master artifact is 6.0193 mm.

In the fig. 6 triggering radius characteristic of Renishaw TP200 precise, strain-gauge probe for CMMs in plane containing this probe's axis is shown. The test speed was 0.5 mm/s. The obtained value of triggering radius variation in XZ plane is V_r = 0.8 μm.

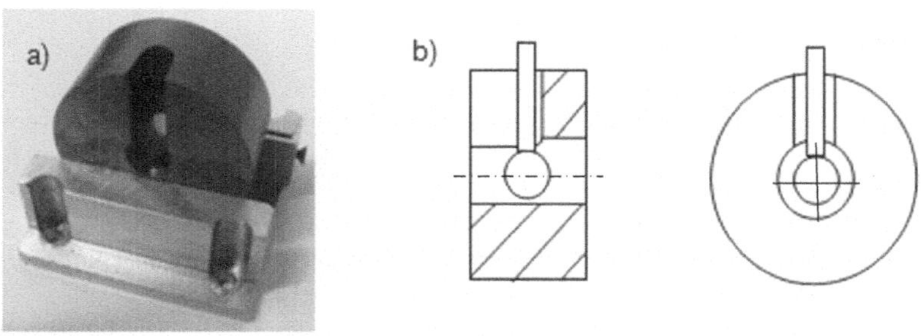

Fig. 5. Ring gauge with a cut: a) view; b) scheme of the master artifact working with a stylus tip

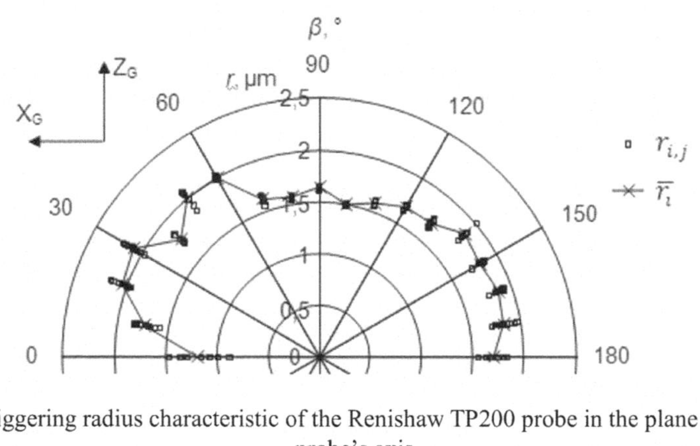

Fig. 6. Triggering radius characteristic of the Renishaw TP200 probe in the plane containing probe's axis

6 Conclusions

The method with moving master artifact enables testing the performance of the probes for CNC machine tools both in 2D and in 3D. The performance of the test setup using this method depends on the used master artifact. To determine probes' triggering radius characteristics in the plane perpendicular to their axes, the ring gauge is used. To determine characteristics in 3D space, the inner hemisphere master artifact made of araldite is used. As it was shown, in the plane perpendicular to the probe's axis, both master artifacts gave a similar results.

To enable probe testing in planes containing probes' axes, but without using the numeric error compensation necessary in case of the 3D master artifact, the ring gauge with a cut was proposed. As it was shown, this master artifact enables not only testing the probes for CNC machine tools, but also probes for CMMs.

Because the used master artifact is a main source of errors of the test setup using moving master artifact method, further development of master artifacts would enable to improve setup's performance.

References

1. Mayer, J.R.R.: Five-axis machine tool calibration by probing a scale enriched reconfigurable uncalibrated master balls artifact. CIRP Annals – Manufacturing Technology 61, 515–518 (2012)
2. ISO 230-10:2011 standard: Test code for machine tools -- Part 10: Determination of the measuring performance of probing systems of numerically controlled machine tools. International Organization for Standardization, Geneva (2011)
3. Cho, M.W., Seo T.I.: Machining error compensation using radial basis function network based on CAD/CAM/CAI integration concept. Int. Prod. Res. 40, 2159-2174 (2002)
4. Cho, M.W., Seo T.I.: Inspection planning strategy for the on-machine measurement process based on CAD/CAM/CAI integration. Int. J. Adv. Manuf. Tech. 19, 607-617 (2002)
5. Cho, M.W., Seo T.I., Kwon H.D.: Integrated error compensation method using OMM system for profile milling operation. J. Mater. Process. Tech. 136, 88-99 (2003)
6. Choi J.P., Min B.K., Lee S.J.: Reduction of machining errors of a three-axis machine tool by on-machine measurement and error compensation system. J. Mater. Process. Tech. 155-156, 2056-2064 (2004)
7. Zeleny, J., Janda M.: Automatic on-machine measurement of complex parts. Modern Machinery Science Journal 2, 92-95 (2009)
8. Cauchick-Miguel, P.A., King, T., Abackerli, A.J.: CMM touch trigger performance verification using a probe test apparatus. J. Braz. Soc. Mech. Sci. & Eng. 25, 147-153 (2003)
9. Dobosz, M., Woźniak, A.: Metrological feasibilities of CMM touch trigger probes: Part II: Experimental verification of the 3D theoretical model of probe pretravel. Measurement 34, 287-299 (2003)
10. Dobosz, M., Woźniak, A.: CMM touch trigger probes testing using a reference axis. Precis. Eng. 29, 281-289 (2005)
11. Jankowski, M., Woźniak, A., Byszewski M.: Machine tool probes testing using a moving inner hemispherical master artifact. Precis. Eng. 38, 421-427 (2014)

Acoustic detector of argon content in air-argon mixture

Artur Jedrusyna, Andrzej Noga

Wroclaw University of technology, Faculty of Mechanical and Power Engineering
artur.jedrusyna@pwr.edu.pl, andrzej.noga@pwr.edu.pl

Abstract. In the paper a prototype of ultrasonic argon detector that can be used as a warning device in safety systems for cryogenic installations has been described. The sensor is operating at the frequency of 40kHz and it has a response time below 1 second. The device after minor modifications could be also used for detection of other gases. A brief information about the construction of the sensor together with its preliminary test results were presented.

Keywords: ultrasonic gas detector, argon content measurement

1 Introduction

One of the gases most often used both in modern industry and science is argon. Among its most important applications cryogenic systems should be mentioned. Argon is also used in many other high temperature processes like arc welding (especially TIG technology). It is also a component of air - argon molar content of argon in air in normal conditions is (0.9332 ± 0.0006) % according to the new measurements performed in 2004 [1,2].

Argon as a inert gas is not toxic, but should be considered as an asphyxiant - a release of large amount of argon inside a confined space may lead to the situation described as Oxygen Deficiency Hazard (ODH) where a major danger of suffocation occurs for a personnel present in the location of the gas release. After such an accident an immediate evacuation action should be started. ODH is a typical hazard for large cryogenic installations and should be taken into consideration during the safety analysis for such a system.

It is obvious that a fast detection method of argon detection would a vital diagnostic tool for every security system installed at the workplace. However, most of the currently used argon detection methods are inherently slow. In this paper we would like to present a prototype of an acoustic argon detector characterized by a sub-second response time.

R. Jabłoński and T. Brezina (eds.), *Advanced Mechatronics Solutions*,
Advances in Intelligent Systems and Computing 393,
DOI 10.1007/978-3-319-23923-1_50

2 Methods of argon detection

Argon content in air could be detected in an indirect way – by measurement of oxygen content in air. If the oxygen content is reduced and the only additional gas present in the installation was argon then an electrochemical oxygen sensor can be utilized. Standard commercially available ODH detectors usually operate on the galvanic cell principle and they measure directly the oxygen concentration. Their drawback is a long time constant (usually exceeding 20s).
A much faster response of the sensor could be achieved by utilization of acoustic method, where the relation between gas concentration and sound velocity is used.

3 Acoustic argon detection

Another possible technique for argon detection is an acoustic method. In this approach, oxygen deficiency detection is based on measurements of a sound velocity. The derivation of the argon content is based on the changes in sound velocity in argon-gas mixture that are related to argon content (the composition of the mixture).

The oxygen concentration is derived on the assumption that the gas added to the air is argon alone.
In general case, the sound velocity a in a perfect gas mixture is given by equation:

$$a = \sqrt{\kappa RT} = \sqrt{\left(\frac{1}{\sum \frac{z_i}{\kappa_i - 1}} + 1\right) \cdot \frac{\bar{R}T}{\sum z_i M_i}} \tag{1}$$

Where κ denotes specific heats ratio, T - temperature , $\bar{R}$ is universal gas constant, R - gas constant of the mixture and z_i and M_i are respectively molar concentration and molar weight of the i-th component.

One of the possible practical techniques used here could be the TOF (Time of Flight) method, where the propagation time of an ultrasonic pulse between the transmitter and the receiver is measured. If the distance between the transmitter and receiver is known, then sound velocity could be derived.
The TOF technique was already utilized for helium content measurement in air-helium mixture. It should be noted that the sound velocity in air-helium mixture at 293K varies between 343 m/s (0% of He content,) and 1007 m/s (100 He % content). In the case of argon the changes in velocity are less prominent – from 343 m/s in pure air to 319 m/s in pure argon. Therefore the change in sound velocity over

the measurement range is smaller and in this case additional effects changing the sound velocity cannot be neglected. Especially during the acoustic sensing of the argon content in air the influence of temperature changes have to be taken into account.

4 A prototype of an acoustic sensor for argon detection

A prototype of acoustic sensor for argon detection was constructed at Wroclaw University of Technology. The instrument was used for verification of relation between argon content and the sound velocity in argon-air mixture.

The principal aim was to build a small, portable and reproducible acoustic argon detector which can be used in confined spaces like corridors or test rooms at atmospheric pressure. The device operation should not be harmful to the human personnel.

The block diagram of the prototype system is shown in Fig.1 and the examples of waveforms present in the system during normal operation were shown in Fig.2.

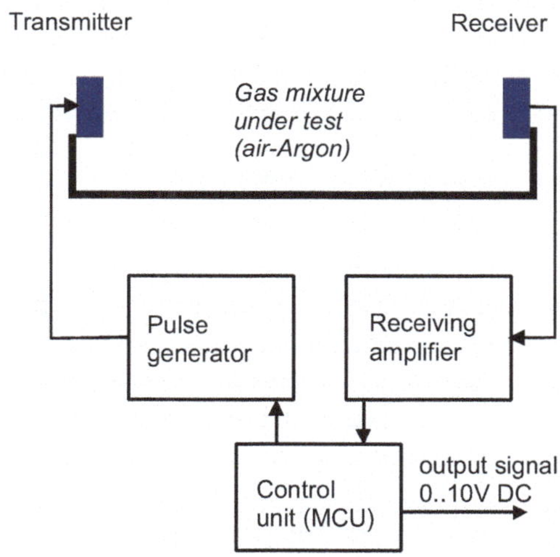

Fig. 1. Ultrasonic argon sensor – internal structure

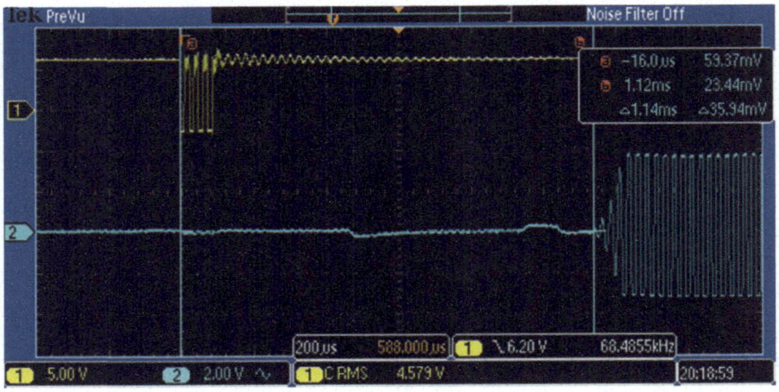

Fig. 2. The time relation between the transmitted (upper line) and received ultrasonic signal.

The sensor was operating in pulse mode. An ultrasonic burst (base frequency 40 kHz) was sent from a transmitter to the receiver which was located in a known distance. The time-of-flight method was used for sound velocity calculation. Burst repetition rate was set to 10Hz.

The sound velocity c was evaluated from the classic equation:

$$c = x/\Delta t \tag{2}$$

Where x is the spacing of the sensors and Δt is the time of propagation of ultrasonic burst between the transmitter and receiver.

The thermal correction was introduced by an additional Pt100 thin-film temperature sensor located in the space between the transmitter and the receiver. The readout from the sensor was sent to external devices as an analog 0-10V DC signal. Two different prototypes with x=340mm and x=605 mm were tested.

The content of argon in air-argon mixture was evaluated from an inversed formula [1].

The sensor was initially calibrated using a test chamber with the volume of about 1m^3, into which argon was injected from the high-pressure cylinder. The experimental setup was shown in Fig. 3. Additional internal fan was used for mixing of the gas in order to ensure a uniform distribution of argon in the whole volume of the test chamber. After each series of measurements the chamber was flushed with clean air.

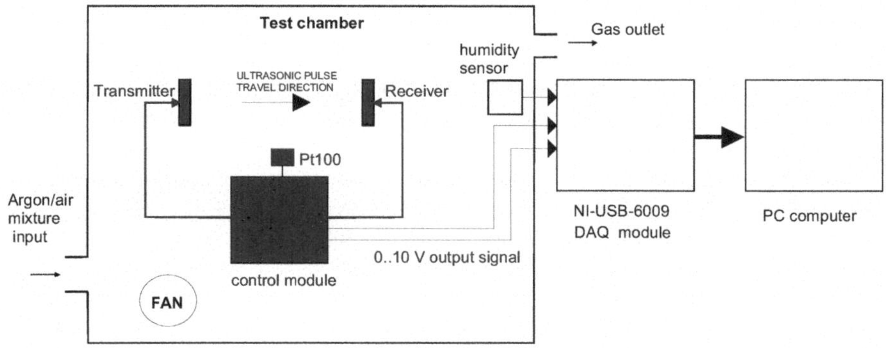

Fig. 3. Experimental setup for ultrasonic argon detector testing and initial calibration.

The chamber was equipped with additional Pt100 temperature sensor with resistance-current transducer, capacitive-type humidity sensor with 0-10V output signal and electrochemical oxygen sensor which was used as a reference device for determination of oxygen content in air-argon mixture. A multi channel USB-controlled data measurement module was used for acquisition of all signals.

After cross-checking of acoustic sensor readings with electrochemical oxygen sensor signal, an error below 5% of Ar content was achieved using the present version of software. The time constant of the ultrasonic sensor was below 1s which is about an order of magnitude faster than the electrochemical sensor. Such a fast response is a very useful parameter in safety applications where gas sensors are to be used as warning devices.

The acoustic sensor's accuracy could be most likely improved by introduction of more advanced signal processing algorithm. More precise calibration could be commenced using gas chromatography and with the help of set of reference gas mixtures of precisely known composition.

5 Final remarks

The prototype sensor required a correct positioning for proper operation. The best location should be in at least 1m from the wall. Such a position reduced the parasitic reflections of ultrasonic signals from the wall (this effect may lead to generation of false echo signals and degradation of measurement quality). It should be noted that the presented device could be easily used for detection of other gases (especially helium). The only necessary changes are the minor modification of hardware and control software. Please note that helium detectors should be located in the area of ceiling of the test room, while the argon detector require positioning near the floor for optimal performance.

References:

1. Park S.Y., Kim J.S., Lee J.B., Esler M.B., Davis R.S., Wielgosz R.I., A redetermination of the argon content of air for buoyancy corrections in mass standard comparisons. Metrologia, 2004, 41(6), 387-395
2. Picard A., Fang H., Gläser M.: Discrepancies in air density determination between the thermodynamic formula and a gravimetric method: evidence for a new value of the mole fraction of argon in air. Metrologia, 2004, 41(6), 396-400.
3. Zuckerwar J.: Handbook of the Speed of Sound in Real Gases A. Academic Press, 2002.
4. Kortbeek P. J., Muringer M. J. P., Trappeniers N. J. ,Biswas S. N.: Apparatus for sound velocity measurements in gases up to 10 kbar: Experimental data for argon. Rev. Sci. Instrum. 56, 1269 (1985).
5. Boyes S.J. : The Speed of Sound in Gases with Application to Equations of State and Sonic Nozzles : PhD Thesis, University of London, 1992.
6. A.F. Estrada-Alexanders, J.P.M. Trusler : The speed of sound in gaseous argon at temperatures between 110 K and 450 K and at pressures up to 19 MPa. The Journal of Chemical Thermodynamics, Volume 27, Issue 10, October 1995, Pages 1075–1089

Precision increase in automated digital image measurement systems of geometric values

Igor Korobiichuk[1], Yurij Podchashinskiy[2], Oksana Shapovalova[2], Valentina Shadura[2], Michał Nowicki[3], and Roman Szewczyk[3]

[1]Industrial Research Institute for Automation and Measurements, Warsaw, Poland
ikorobiichuk@piap.pl
[2]Zhytomyr State Technological University, Zhytomyr, Ukraine
ju.podchashinskiy@gmail.com, ksyxon@gmail.com, valshad@mail.ru
[3]Institute of Metrology and Biomedical Engineering, Warsaw University of Technology, Warsaw, Poland
nowicki@mchtr.pw.edu.pl, r.szewczyk@mchtr.pw.edu.pl

Abstract. The article is devoted to the development of the algorithmic method of video image processing of natural stone products. These video images contain the measurement information about geometric parameters of products, and the parameters of their movement during production at stone processing plants. The developed methods provide the algorithmic compensation of random, dynamic and geometric errors of video images which occur at the forming stage. The measurement results of the above mentioned values are used to control the technological process in the stone processing plants, and to improve the quality of natural stone products.

Keywords: geometric parameters, motion parameters, algorithmic error compensation, video image.

1. Introduction

The enterprises of stone mining are the important component of the industry of Ukraine. Unfortunately, the effective realization of their capabilities is obtained by the outdated equipment and the means of mechanic values do not possess high precision and fast-acting. That is why, the task of development of new precision means and the increase of their fast-acting features in order to control the mechanic values at natural stone mines is becoming more and more urgent.

The modern visualization methods are suggested to be applied to increase the precision and fast-acting of measurements the geometrical parameters of product movement. These methods are aimed to form the product video image and to perform their algorithmic processing [1, 2]. The geometrical parameters of the natural stone product outer contour, along with the geometrical parameters of the processed surface structure of these products are measured on the basis of the captured on-line video

R. Jabłoński and T. Brezina (eds.), *Advanced Mechatronics Solutions*,
Advances in Intelligent Systems and Computing 393,
DOI 10.1007/978-3-319-23923-1_51

images. The modern means of mechanic values include computer or microcontroller. As a result, these means are capable to process the measuring signals to obtain the mechanic values measurement and to increase their precision [3-5]. The algorithmic methods of video image processing have been considered in many papers. They are: the investigation of machine vision [6-8], the automated control systems with video sensors [9, 10], the television measuring systems [11, 12]. But these papers do not fully focus on the requirements of the metrological characteristics of the video images measurement information. .

2. The formulation of the task of the algorithmic error compensation of the geometric values measurement information

The existing methods of algorithmic processing of measuring information about mechanic values are mostly used to process one-dimension signals [13-15]. The existing solutions for two-dimension signals do not apply all possibilities of modern information computer technologies concerning video image processing [13-15].

There are also transient and adverse factors which influence the measurements of the mentioned mechanic values. The solution of this task is performed in the following sequence: the identification of transfer function of video image forming device, the parameters of geometric errors, the parameters of video image correlation function errors and random errors; the calculation of video image spectrum density and random errors by means of two-dimensional Fourier transformation of their correlation functions; the performance of algorithmic compensation of random, dynamic and geometric errors based on the results of their parameters identification; the estimation of geometric parameters precision measurements. The problem of video image forming device transfer function, the error parameters, and correlation functions is investigated in [16].

Therefore, the purpose of the given article is to develop new methods of algorithmic processing of video images with the measurement information about the geometric parameters and parameters of products movement, which provide significant precision increase of these measurements

3. Algorithmic compensation of random errors of video images

The method of algorithmic compensation of random errors [17] is developed to increase the precision mechanic value measurement. According to this algorithm a sample of natural stone from mineral deposit is fixedly placed at the digital camera; a set of video images is formed and input into computer; the video image averaging is performed and the estimation of measuring information about mechanic values is obtained.

The known criteria of video image fidelity reproduction are aimed to estimate the amplitude errors and visual quality of video images. So, the following indexes are suggested to estimate the precision of geometrical parameters in automated systems: the coordinates of the determination of the coordinates of the center of mass, the linear dimensions and area of a product and the structural elements of its surface, and also the average value of the error of a product coordinate contour measurement:

$$\Delta_{\text{cont}} = \frac{\sum_{n,m} \hat{f}_{0\text{c}}(n,m) \oplus f_{0\text{c}}(n,m)}{l_{\text{к}}}, \tag{1}$$

where

$$f_{0\text{c}}(n,m) = \begin{cases} 1, & \text{if} \quad f_0(n,m) \in Q_{\text{OB}}, \\ 0, & \text{if} \quad f_0(n,m) \notin Q_{\text{OB}}, \end{cases} \tag{2}$$

$$\hat{f}_{0\text{c}}(n,m) = \begin{cases} 1, & \text{if} \quad \hat{f}_0(n,m) \in \hat{Q}_{\text{OB}}, \\ 0, & \text{if} \quad \hat{f}_0(n,m) \notin \hat{Q}_{\text{OB}}, \end{cases} \tag{3}$$

where: $\oplus$ – is the logical operation of the sum determination by the module 2; $l_{\text{к}}$ – is the contour length of a product in discrete points; Q_{OB} i $\hat{Q}_{\text{OB}}$ – is the set of points which belong to a product at the initial and the processed video image after their product/background segmentation, correspondingly.

This method provides the precision of the geometrical parameters determination of the fixed products made of natural stone (in 2,0…9,5 times for the correlation signal/noise $\psi_{\text{c/h}}$ = 55 dB on the video image and in 3,0…12,5 times for $\psi_{\text{c/h}}$ = 40 dB).

4. Algorithmic compensation of dynamic errors of video images

The method of reproduction of video images with dynamic errors for the measurement of geometric parameters and movement parameters under the conditions of non-stationary factor influence [18] is developed:

I. It is recommended to form the video image of a product $f_{\text{H}}(n,m)$, and to compensate the errors in accordance with the mentioned above recommendations. As a result, the estimation of the image with dynamic and geometric errors $\hat{f}_{\text{d}}(n,m)$ is obtained.

II. The image amplitude in relation to maximum value $\hat{f}_{\text{d}\max}$ is normalized and the normalized amplitude derivative of the video image is calculated.

III. The maximum value $f'_{\max}$ of the normalized amplitude derivative of the video image between the values is found.

IV. The discrete readouts transfer frequency function is calculated for the device which forms video images $H_{\text{H}1}(u,v)$ on the basis of [15], taking into account the correlations $\omega_1 = 2\pi u/(N\delta_x)$, $\omega_2 = 2\pi v/(M\delta_y)$, $u \in \overline{0, N-1}$, $v \in \overline{0, M-1}$:

$$H_{\text{H}1}(u,v) = \exp\left(-\frac{\pi\left(u^2/N^2 + v^2/M^2\right)}{\Delta^2_{f\max}} \right) \tag{4}$$

where ω_1, ω_2 are the space frequencies in the video image spectrum.

V. The estimation $\hat{f}_0(n,m)$ is calculated by means of $\hat{f}_{\text{Д}}(n,m)$ usage with the digital fitter with frequency characteristics $H_{\text{H1}}^{-1}(u,v)$, obtained from the formula (4), and the geometric errors compensation according to the instructions [11, 12].

The developed method of reproduction uses the frequency transfer function of the device for video image forming. It absolutely corresponds to the conditions of the measurements at the current moment of time in contrast to the known methods which use the prior information of the average values for measurement conditions having been found before. It provides the precision increase of the contour points coordinate of products of 1,5…6,3 times.

5. The algorithmic processing of the video image sequence with the measuring information about geometric values

The time sequence of video images $f_i(x,y)$, $i \in \overline{1,K}$ characterizes the current position of the products and the industrial equipment at the moments of time $t_i = i \cdot \delta_{\text{d}}$. The i-video image has to be determined by the coordinate of the mass center and equipment x_{ci}, y_{ci}, and its angle position α_i. The indicated parameters characterize the two-dimensional movement of these objects in the image area. Thus area consists of the mass center onward movement, and the rotational movement around it.

The movement parameters of the product and equipment are determined in every coordinate x_{ci}, y_{ci}, α_i resulting in the obtained transition vectors, velocity and acceleration (the variable x_i can be represented by any coordinates among x_{ci}, y_{ci}, α_i).

The following mathematical model is used to estimate the movement parameters:

$$x_i = \sum_{l=1}^{N_{\Pi}} \theta_{il}\beta_l\,,\; X = (x_1 \;\; \ldots \;\; x_K)^T = \Theta\beta\,, \tag{5}$$

where $\Theta = [\theta_{il}]$ – is the value matrix prior to the known functions at the moments of time $i\delta_{\text{д}}$; $\theta_{il} = \dfrac{(i\delta_{\text{д}})^{l-1}}{(l-1)!}$; $\beta = (\beta_1 \;\; \ldots \;\; \beta_{N_{\Pi}})^T$ – is the vector of movement parameters which describe the process of movement.

The result of movement parameters estimation based on the maximum likelihood method equals:

$$\hat{\beta} = B^{-1}A\,,\; B = \Theta^T R_{\Delta x}^{-1}\Theta\,,\; A = \Theta^T R_{\Delta x}^{-1} X^*\,,\; \Psi_{\hat{\beta}} = B^{-1} = \left(\Theta^T R_{\Delta x}^{-1}\Theta\right)^{-1}, \tag{6}$$

where $R_{\Delta x} = \sigma_{\Delta x}^2 \cdot I_k$ – is the correlation matrix of measurement errors Δ_{xi}; $\sigma_{\Delta x}^2$ – is the error dispersion of the x_i coordinate measurement; I_k – is the unitary matrix of the $K \times K$ size; $\Psi_{\hat{\beta}}$ – is the correlation matrix of the movement parameter vector estimation.

That is why; the formulas (5) and (6) are used to determine the estimation of movement parameters of products and equipment in real time. The calculation of the movement measurement errors on the basis of errors correlation matrix and their comparison to the known methods is shown in Fig. 1. The estimations are calculated for time moments $j\delta_{д}$, $j = N_{ос},...,K$, each of them is preceded by the estimation time $T_{ос} = (N_{ос} - 1)\delta_{д}$. It is found by means of the numerical modeling and the experimental research, that the measurement precision of the current coordinates and product movement parameters for the accumulated sequence of video images increases in 3, 7…6, 7 times, for the measurements in real time in 2,1…3,7 times .

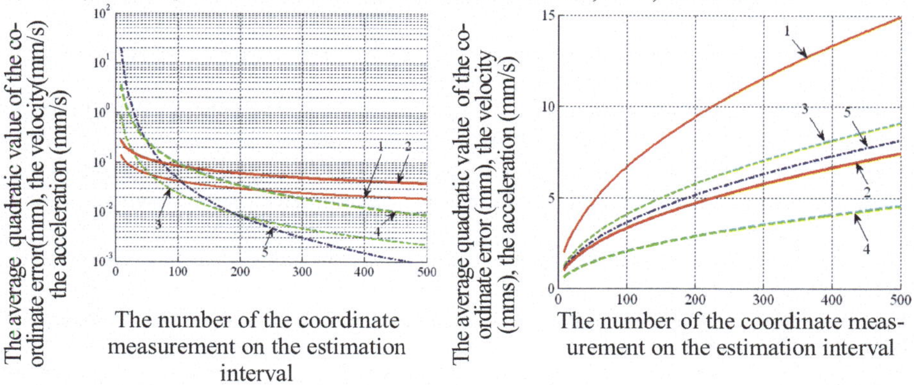

a) b)

Fig. 1. The theoretical precision calculation of the current coordinates determination and object movement parameters in accordance with the developed methods (a) and their comparison (b) to the method of numerical differentiation; 1- is the current coordinate, the numerical differentiation: 2- is the current coordinate, the developed method; 3- is the velocity, the numerical differentiation; 4- is the velocity, the developed method; 5- is the acceleration, the developed method

The results of parameter movement determination are used in to control the technological standards when producing the products made of natural stone. These results are also used for the compensation of video image dynamic errors which are caused by the product movement at the digital camera. It allows increasing the precision and velocity of the equipment which measures the products directly at the production process. As a result, the quality of products made of natural stone is increased.

Conclusions

The new methods of algorithmic error compensation of video image with the measurement information about geometric values are developed. These methods provide the precision increase of the determination of geometric parameters and movement of products made of natural stone of 1,5…6,3 times compared to the existing means of measurements.The measurement results of the mentioned values can be used to control the technological process of mining enterprises, and to increase the quality of products made of natural stone.

References

1. Pat. 82859 Ukraine, MPK (2013.01) G 06 B 7/00.Equipment for the product geometric parameter measurement /Podchashinskiy Y.O.; the applicant and the owner of the patent

Zhitomyr State Technological University. – № u2012 08341 ; appl. 07.07.12; publ. 27.08.13, Biyl. № 16.
2. Bezvesilna O.M.. The principles of the design of the system for mechanic values measurement on the basis of two-dimensional information / O. M. Bezvesilna, Y.O.Podchashinskiy // Visnyk of Engineering Academy of Ukraine. – 2009. – № 4. – P. 52–55.
3. Gutnikov V.S. The filtration of measuring signals / V.S.GutnikovB. . – L. : Energoatomizdat , 1990 . – 192 p.
4. Granovskiy V.A. The method of data processing at measurements / V.A.Granovskyi, T.N.Siraya. – L. : Energatomizdat , 1990. – 288 p.
5. Yatsuk V. The methods of measurement precision increase : text-book / V.O.Yatsuk, P.S.Malachivskiy. – Lviv : Beskyd Bit, 2008. – 368 p.
6. Taylor G. Visual Perception and Robotic Manipulation. 3D Object Recognition, Tracking and Hand-Eye Coordination / G. Taylor, L. Kleeman. – Berlin, Heidelberg : Springer-Verlag, 2006. – 230 p.
7. Pauli J. Learning-Based Robot Vision. Principles and Applications / J. Pauli. – Berlin, Heidelberg : Springer-Verlag, 2001. – 288 p.
8. Vision Systems Applications / Edited by G. Obinata and A. Dutta. – I-Tech Education and Publishing, 2007. – 616 p.
9. The basics of contour analysis:; application to image and signal processing / Y.A.Furman, A.V.Kravetskiy, A.K.Peredreyev and oth.; edited byY.A.Furman . – 2-nd ed . – M. : Phismatlit, 2003. – 592 p.
10. Kriksunov L.Z. Tracking systems with optic and electron coordinate / L.Z/Krikunov . – K.: Technique, 1991. – 156 p.
11. Gorelik S.L.Television measuring systems /S.L.Gorelik,B.M.Kats,V.I.Kivrin. – M.: Communication , 1980. – 168 p.
12. GryazinG.N. The system of applied television: manual for universities /G.N.Gryazin – Politechnica , 2000. – 277 p.
13. Forsight D The computer eyesight. The modern approach / D.Forsight, J.Pons. – M. : Williams, 2004. – 928 p.
14. H. R. Sheikh and A. C. Bovik, "An evaluation of recent full reference image quality assessment algorithms," IEEE Trans. Image Process., vol. 15, no. 11, pp. 3440-3451, November 2006.
15. Shliht G.Y. Digital processing of coloured images. – M. : ECOM, 1997. – 336 p.
16. Podchashinskiy Y.O. Identification of dynamic characteristics of devices forming two-dimensional measuring information about mechanic values / Y.O.Podchashinskiy //Autmation:problems, ideas,solutions : ISC , 8-12September 2008 г. : abstr,rep / SevNTU. – Sevastopol, 2008. – P. 175 – 178.
17. Pat. 91174 Ukraine , МПК (2009) G 06 K 9/40, G 06 K 9/00. The way of filtration of video images which contain measuring information about geometric parameters of measurement objects /Podchashinskiy Y.O. the applicant and the owner of the patent . – № a2009 13355 ; appl. 22.12.09 ; publ. 25.06.10, Vol. № 12.
18. Pat . 98035 Ukraine, МПК (2006.01) G 06 F 7/00, G 06 K 9/42. The way of reproduction of the images which contain measuring information about geometric parameters of measurement objects / Podchashinskiy Y.O.; The applicant and the owner of the patent Zhitomyr State Technological University. – № a2010 09758 ; appl.з 05.08.10; publ. 10.04.12, Vol. № 7.

Methods for verifying the dimensional and material properties on industrial CT scanners according to VDI / VDE 2630 Blatt 1.3.

Tomasz Kowaluk, Adam Woźniak

Warsaw University of Technology, Faculty of Mechatronics, Institute of Metrology and Biomedical Engineering, Warsaw, Poland
{t.kowaluk, a.wozniak}@mchtr.pw.edu.pl

Abstract. In the article methods for verifying the dimensional and material properties like GS, GF i GG on industrial CT scanners according to VDI / VDE 2630 Blatt 1.3. „*Computertomografie in der dimensionellen Messtechnik*“ guidelines were described. The proposed gauges in the form of a stepped cylinder were presented. In studies two types of different materials of gauges with different X-ray absorption coefficients were used. The study on METROTOM CT 800 manufactured by C. Zeiss with the X-ray tube voltage of 130 kV was performed. As the material for the gauges aluminum (aluminum alloy PA6) and ertacetal was selected. Both gauges in an accredited laboratory were calibrated.

Keywords: computed tomography, dimensional and material parameters, gauges for testing of industrial CT scanners,

1 Introduction

The idea of computer tomography measurements consists of assembling three-dimensional spatial object from multiple flat images resulting from the scanning of measured element in the selected angular position. While the measured object is illuminated by an X-ray beam generated in the tube, the image detection is done at certain angular positions of rotary table on the detector array. Acquired images are processed by specialized software, producing a three-dimensional model of the measured element. The change of the intensity of the parallel beam of radiation of the same energy, during the transition through the measuring object is described by the dependency:

$$I = I_0 e^{-\mu g}, \tag{1}$$

where:

I – intensity of radiation after passing through the object,
I_0 – initial intensity of radiation,
μ – linear absorption coefficient of radiation,
g – thickness of the material.

R. Jabłoński and T. Brezina (eds.), *Advanced Mechatronics Solutions*,
Advances in Intelligent Systems and Computing 393,
DOI 10.1007/978-3-319-23923-1_52

X-ray absorption by the measurement object is proportional to the density and atomic structure of the material and wall thickness of the element [1-2]. The main factors influencing on the weakening of X-rays are the type and thickness of the material. Therefore, it was decided to check the dimensional and material properties on industrial CT scanners using appropriate gauges made of two different materials: aluminum (aluminum alloy PA6) and plastic material – ertacetal [2-4].

2 Accuracy tests of the computer tomographs according to VDI/VDE guidelines.

Computed tomography for industrial applications are a relatively new area of research and have not been developed the ISO standard for testing the accuracy. Currently produced computer tomographs can be tested according to VDI/VDE 2630. Blatt 1.3. „Computertomografie in der dimensionellen Messtechnik“ [5]. The guidelines included criteria for evaluating computer tomography for the four groups of parameters:

- parameters associated with the scan error (PF, PS),
- parameters associated with the length measurement error (E),
- parameters defining the resolution (Dg),
- parameters defining the depending dimensions - material **(GS, GF, GG),**

The most important element in dimensional and material properties research are applicable gauges. The material from which it is made gauges should be as similar as possible to the elements measured on computer tomography. For checking the influence of dimensions-material dependence is recommended as the gauge stepped cylinder with a minimum of five and maximum of ten greatly differing diameters with a coaxial straight through hole. The diameter of the largest step cannot be less than five times the diameter of the hole. The height of each step must be the same, and their diameters should be graded linearly. In addition, the ratio of height of gauges of the largest diameter should be in the range of <0.8; 1.2>. According to the guidelines VDI/VDE was designed and manufactured gauges visible on Fig. 1.

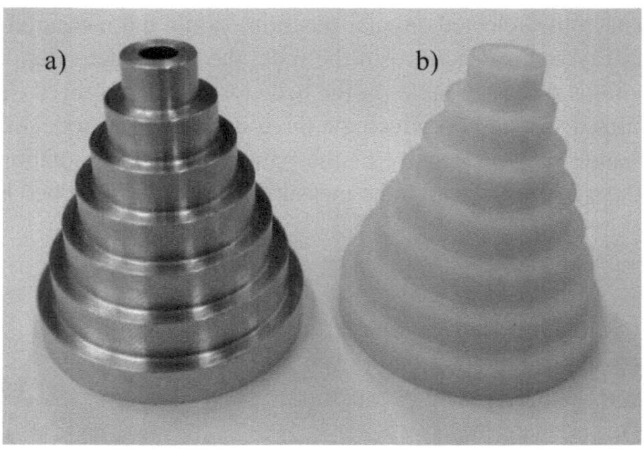

Fig. 1. View of the gauges. A- made of aluminum alloy PA6, b – made of ertacetal.

On the left (a) gauges made of aluminum on the right (b) made of ertacetal. Gauges made in the form of stepped cylinder with a seven steps (height of a step – 6 mm) with greatly differing diameters (from 10 to 40 mm) with a coaxial straight through hole (5 mm). Both gauges were calibrated in an accredited laboratory.

3 Proposal of study procedure with results

Studies were performed using computer tomograph METROTOM 800 Carl Zeiss company which is in equipment of the laboratory of the Institute of Metrology and Biomedical Engineering of Faculty of Mechatronics of Warsaw University of Technology. To the main assembly of computer tomography included the X-ray tube, the radiation detector and a system for positioning the measuring element consisting of a rotary table and the linear guideways. The obtained data are processed by a separate computer. Computer tomograph METROTOM 800 is equipped with a constant X-ray tube with a power of 39 W (at maximum parameters – 130 kV voltage and 300 μA current) and a detector with a resolution of 1536x1920 pixels. Each of the gauges was measured ten times with identical settings for a given material. Depending on the X-ray absorption coefficient appropriate measured parameters for the gauges material was set. Summary of measurement parameters are presented in Table 1. The measurement results were determined with the use CALYPSO software version 5.2. The nominal values were taken from certificates of gauges from an accredited laboratory.

Table1. Summary of measurement parameters.

Parameter	Aluminum	Ertacetal
X-ray tube voltage	130 kV	85 kV
X-ray tube current	118 μA	110 μA
Integration time	1000 ms	800 ms
Detector gain	8,0 x	8,0 x
Number of projections	1500	
Voxel size	30,09 μm	
Mechanical filters	Cu – 0,5 mm	NO
Focal spot control	YES – Frame interval 64	
Noise reduction filter	Shepp Logan	

3.1 Determination of the parameter GS

GS parameter is given for each step internal and external diameter. ***GS*** parameter value is a measure of the deviation of the measured inner diameter, or the outer diameter D_a of the reference diameter D_r.

On the basis of 10 times of the tomographic measurement for both types of gauges determined average values of internal and external diameters. Exterior and interior diameters was determined with numbers from 1 to 7 (nominal values outside diameters from 10 to 40 mm in steps of 5 mm). Numbers of internal diameters with nominal value of 5 mm were corresponded to appropriate numbers of external diameters. Table 2 shows the results of measurements of the GS parameter.

Table 2. Summary of measurements results of GS parameter. Values is [mm]

Diameter	1	2	3	4	5	6	7
Aluminum external							
Average deviation	-0,0020	-0,0009	-0,0005	-0,0004	0,0004	-0,0001	-0,0001
Standard deviation	0,0003	0,0004	0,0007	0,0007	0,0010	0,0008	0,0008
Range	0,0010	0,0013	0,0021	0,0024	0,0030	0,0025	0,0024
Aluminum internal							
Average deviation	-0,0021	-0,0017	-0,0020	-0,0018	-0,0021	-0,0009	-0,0001
Standard deviation	0,0006	0,0004	0,0008	0,0015	0,0014	0,0010	0,0015
Range	0,0016	0,0013	0,0023	0,0043	0,0048	0,0040	0,0045
Ertacetal external							
Average deviation	0,0090	0,0118	0,0135	0,0152	0,0168	0,0143	0,0124
Standard deviation	0,0010	0,0011	0,0014	0,0016	0,0012	0,0010	0,0013
Range	0,0032	0,0038	0,0040	0,0039	0,0038	0,0032	0,0039
Ertacetal internal							
Average deviation	-0,0047	-0,0044	-0,0033	-0,0031	-0,0027	-0,0021	-0,0028
Standard deviation	0,0008	0,0004	0,0004	0,0009	0,0006	0,0011	0,0011
Range	0,0025	0,0012	0,0012	0,0025	0,0019	0,0031	0,0037

In the case of a gauges made of aluminum alloy PA6, a maximum deviation value (GS parameter) is 2,1 μm. The maximum standard deviation value is 1,5 μm and maximum range value is 4,8 μm. In the case of a gauges made of ertacetal GS parameter for internal diameter is 4,7 while for external diameter is 16,8 μm. The maximum standard deviation value is 1,6 μm and maximum range value is 4,0 μm.
Summary of the GS parameter measurement results shows a graph in Figure 2.

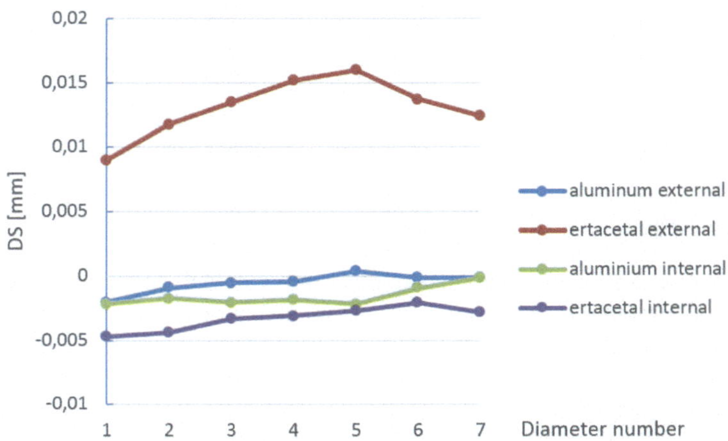

Fig. 2. The results of measurement of the GS parameter

3.2 Determination of the parameter GG

GG - is a parameter defining the internal hole axis straightness. When determining the uncertainty should take into account the actual deviation of the hole straightness and the shape of the cross section.
The reference value of straightness of aluminum alloy PA6 gauges is 0,6 μm and of ertacetal gauges is 1,2 μm. The calculated value of the GG parameter are: for aluminum gauges 3,9 μm and for ertacetal gauges 6,9 μm.

3.3 Determination of the parameter GF

GF parameter of measured arc is defined as the radial extent of the deviation of matching shape. The calculated value of the GF parameter are dependent on the reconstruction of the measured object. Measurement of shape deviation is carried out on a reconstructed surface which consisting of a triangle mesh. View of reconstructed surface was presented on Fig. 3 (a). In view (b) is presented a graph of shape deviations obtained from the CMM, in view (c) from the CT. Both graphs were obtained for the same diameter with the same measurement parameters and magnifications. The differences in the results are related to the surface reconstruction method (triangle mesh). The calculated value of the GF parameter (maximum value) are: for aluminum alloy PA6 gauges 20,2 μm and for ertacetal gauges 22,9 μm.

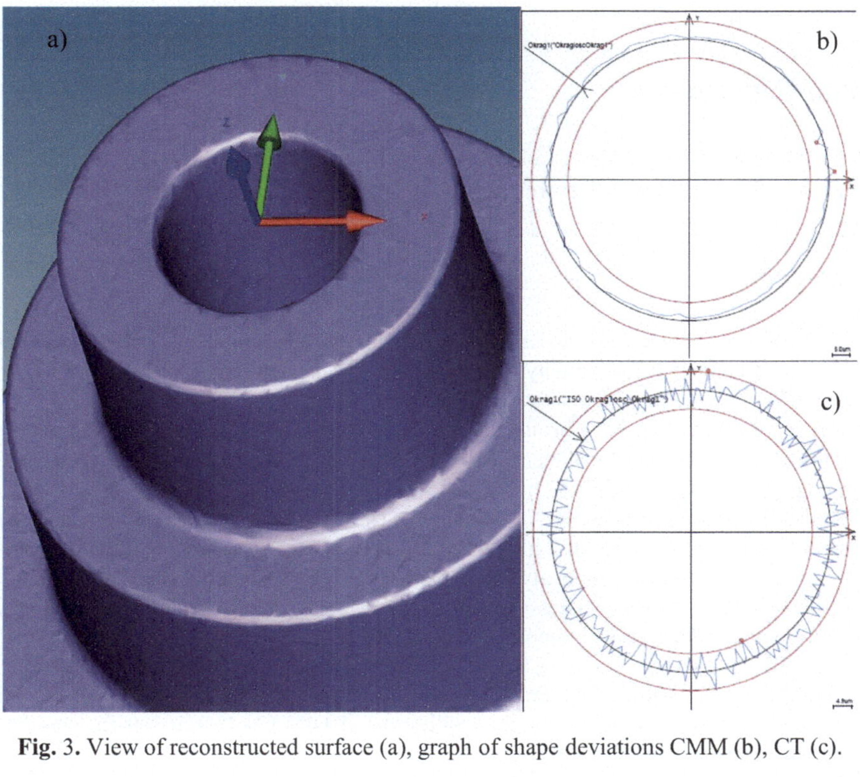

Fig. 3. View of reconstructed surface (a), graph of shape deviations CMM (b), CT (c).

4 Conclusions

Due to the nature of the tomographic measurements a very important issue is verifying the dimensional and material properties like GS, GF i GG. During the analysis of GS parameter expected greater influence of material and geometric depending for gauges made of aluminum alloy PA6 due to a higher X-ray absorption coefficient in relation to the plastic material.
The maximum value of the GS parameter obtained for aluminum gauges is equal 2,1 μm. Only one of the average value of deviations was greater than reference value from calibration certificate. In the case of a gauges made from ertacetal all external diameters have larger deviations than the reference value while the internal diameters have lower values. This results in the reconstruction of a larger volume of material than in reality. The border value (ISO surface) between the material and the background is not correctly defined. The maximum value of the GS parameter obtained for ertacetal gauges is equal 16,8 μm.
The GG parameter is defined as an internal hole axis straightness. For both types of gauges defined parameters are at a similar level. The calculated value of the GG parameter are: for aluminum gauges 3,9 μm and for ertacetal gauges 6,9 μm. The differences in obtained values are related to the accuracy of the gauges.
The calculated value of the GF parameter are dependent on the reconstruction of the measured object. The calculated value of the GF parameter (maximum value) are: for aluminum gauges 20,2 μm and for ertacetal gauges 22,9 μm.

Acknowledgements

The research described in the article have been financed with funds from the National Research and Development Centre. Grant NCBiR no. PBS2/B6/16/2013.

References

1. De Chiffre, L., Carmignato, S., Kruth, J.P., Schmitt, R., Weckenmann, A.: Industrial applications of computed tomography. CIRP Annals - Manufacturing Technology. 63 (2014) 655–677
2. Kruth, J.P., Bartscher, M., Carmignato, S., Schmitt, R., De Chiffre, L., Weckenmann A.: Computed tomography for dimensional metrology. CIRP Annals - Manufacturing Technology. 60 (2011), pp. 821–842
3. Müller P., Hiller J., Cantatore A., Tosello G., De Chiffre L.: „New reference object for metrological performance testing of industrial CT systems". Departement of Mechanical Engineering, Technical University of Denmark (DTU), Kgs. Lyngby (Dania), 2012
4. Ryniewicz A.: „Accuracy assessment of shape mapping using Computer Tomography". Metrology and Measuring Systems. Quarterly of Polish Academy of Sciences. Vol.XVII, No.3/2010.482-491
5. VDI/VDE 2617 Blatt 13: Accuracy of coordinate measuring machines - Characteristics and their testing - Guideline for the application of DIN EN ISO 10360 for coordinate measuring machines with CT-sensors.

AFM studies of stiction properties of ultrathin polysilicon films

Marcin Michałowski[1], ZygmuntRymuza[1], Rodica Voicu[2], Cosmin Obreja[2], Angela Baracu[2,3], Raluca Muller[2]

[1]Warsaw University of Technology, Department of Mechatronics, Institute of Micromechanics and Photonics, Warsaw, Poland
mm.michalowski@gmail.com, z.rymuza@mchtr.pw.edu.pl
[2]National Institute for Research and Development in Microtechnologies (IMT), Bucharest, Romania
[3]University Politehnica of Bucharest, Romania Splaiul Independentei 313, 060042, Bucharest, Romania

Abstract. The paper presents the analysis of the sticking properties of ultrathin polysilicon films and their correlation with the temperature (and resulted roughness) at which the film has been deposited. The force distance curves (FDC) has been done with an atomic force microscope (AFM). Temperature impacts roughness of the surface, which in turn influences adhesion of the sample. Study has been done for various humidities to find optimal parameters for deposition of the films. Additional properties of the samples such as wettability, roughness and mechanical parameters have been taken into consideration in the stiction studies.

Keywords: stiction·MEMS·polysilicon films

1 Introduction

Radio frequency micro electromechanical systems (RF MEMS) switches are considered one of the most promising technologies to improve the capabilities of wireless and telecommunication systems [1]. The MEMS reliability is still an open challenge [2]. One of the most important and unresolved problems in RF MEMS switches is the narrowing of the actuation and release voltages during long period of continuous actuation. This behavior is due to both electrical and mechanical phenomena which alter the characteristics of the device eventually resulting in its failure. The relative smoothness of surfaces, the small interfacial gap, and the small restoring forces of the compliant structures, the contact between their components can lead to the permanent adhesion of their moving parts called stiction [3]. The stiction is the combination of stick and friction which is due to the adhesive force. The stiction strongly depends on the surfaces topographies of the contacting surfaces and their mechanical stiffness in particular of the thin surface layers of the contacting MEMS components.

R. Jabłoński and T. Brezina (eds.), *Advanced Mechatronics Solutions*,
Advances in Intelligent Systems and Computing 393,
DOI 10.1007/978-3-319-23923-1_53

We used atomic force microscopy (AFM) to investigate the sticking properties of the polysilicon films fabricated at various conditions in order to have the most background to choose the best materials for the use in the fabrication of devices such as RF MEMS. AFM measurements are frequently used for acquiring surface parameters of samples in recent years [4-5].

2 Experimental details

During the studies the AFM NT-206 system (Microtestmachines Ltd, Gomel, Belarus) has been used. For measurements of adhesion (pull-off force) a beryllium bronze cantilever with a 0.35 mm diameter SiO_2 ball has been used. A force distance curve (FDC) [6] (Fig. 1) has been acquired by coming into contact and after a short hold cantilever was retracted from the sample. During the unloading cantilever sticks to the sample and at the last moment a deflection in the opposite direction than previous displacement occurs. By the measurement of that deflection, which after calibration can be transformed into pull-off force we can state the adhesive force between the sample and the cantilever tip (ball). Each sample at varied humidity has been measured 10 times over 10 μm x 10 μm areas, in three different locations on the sample for total of 30 measurements. The FDC curves have been performed with a rate of 100 nm per second during both loading and unloading.

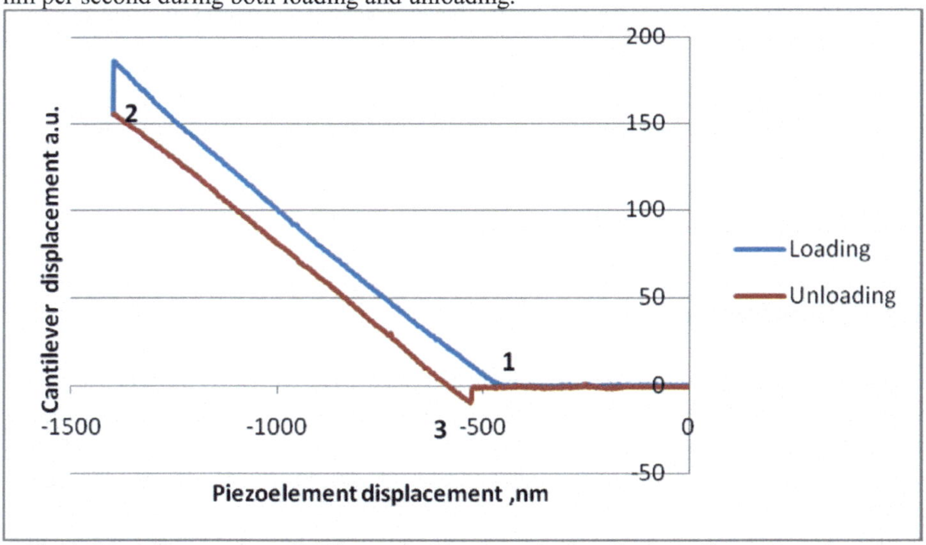

Fig. 1. Force distance curve (FDC) on polysilicon sample. At point 1 cantilever comes into contact with sample, deflection of the cantilever rises up to point 2 where the load is at maximum. Afterwards unloading starts and at the end of it we can see pull-off force at point 3.

In order to receive pull-off force values in nano Newtons calibration of AFM has been performed before testing. This involved usage of a stiff sample such as silicon wafer to calibrate the photodiode signal into nano meters of cantilever deflection. Afterwards testing on a membrane with a known stiffness such as Nanoidea`s membrane calibration structure can be used to identify the actual cantilever stiffness, which allows us to acquire the final FDC curve in nano Newtons to nano meters of piezoelement displacement.

The same AFM has been used to measure the roughness of the samples. Those measurements were done by contact mode scanning over 5μm x 5 μm with 512 x 512 measurement points. A sharp tip with radius of under 8 nm has been used. Total of 10 scans for each sample have been made to avoid significant impact of surface pollution and to decrease measurement error.

Additionally wettability studies of the polysilicon films were carried out. Those tests were done by a droplet method, which means placing a small droplet of fluid on the sample surface and measuring the angle the droplet forms with the surface. The measuring device has been developed and constructed at Warsaw University of Technology and is located in a clean-room laboratory in the Institute of Micromechanics and Photonics.

The samples were fabricated starting from a 4" silicon wafer with a thermally oxidized layer of 1.7 μm thickness. The polysilicon film was deposited after that by LPCVD process in the horizontal furnace reactor chamber (Annealsys LC100) at three different temperatures: 580°C, 610°C, and 630°C. The pressure was 0.2mbar. For each run, deposition time and silane flow were adjusted for a polysilicon layer with thickness of 50 nm.

3 Results and discussion

The roughness (Fig. 2) increases with increasing temperature of deposition. The differences are not big, however are still statistically significant.

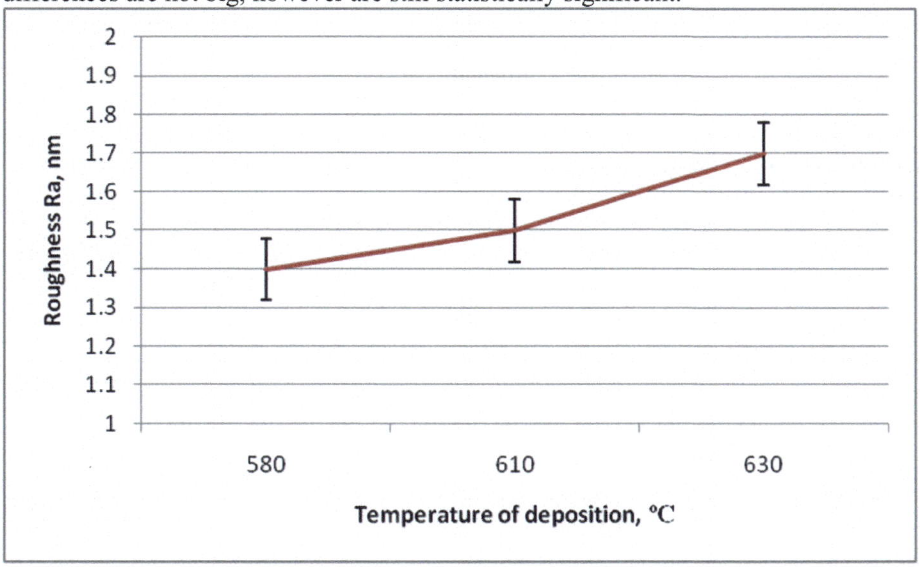

Fig. 2. Roughness vs. deposition temperature

The pull-off force multiple measurements show the standard deviation of about 10% of the average value. It is slightly higher for lower results and lower for higher values. Pull-off force measurements show that in this range of roughness local maximum of adhesion can be found at about 40% humidity (Fig. 3).The local maximum of adhesion at average humidity is presumably caused by creation of a thin film of water between sample and the tip. The difference between each level of humidity is greatest for the most smooth sample, which is the sample deposited in the lowest temperature. Additionally we can see that for lower humidities the adhesion is lower the higher the roughness and the temperature of production of the polysilicon layers (Fig. 4). For higher humidity this effect was not observed.

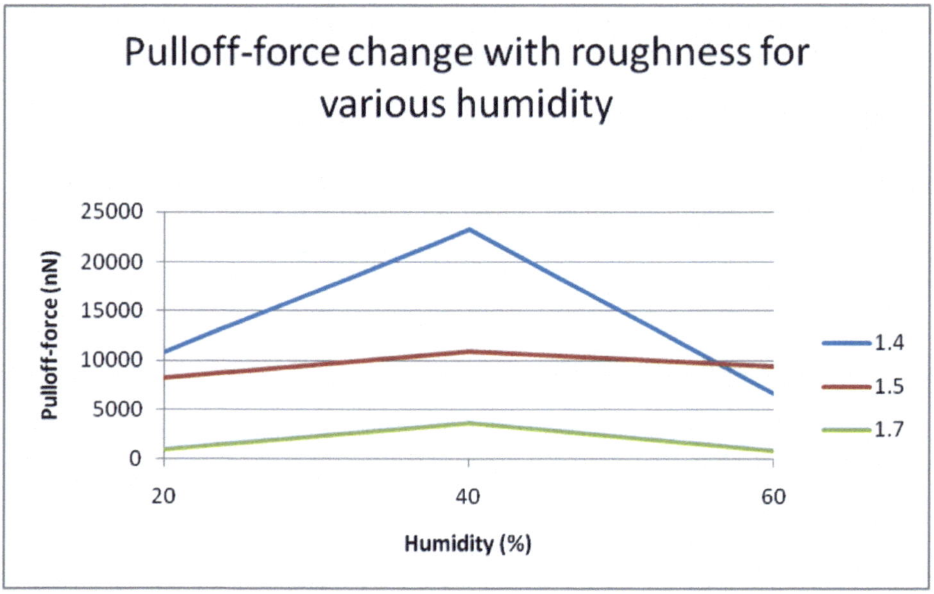

Fig. 3 Results of pull-off force measurements vs. humidity and roughness (R_a 1.4 , 1.5, and 1.7 nm)

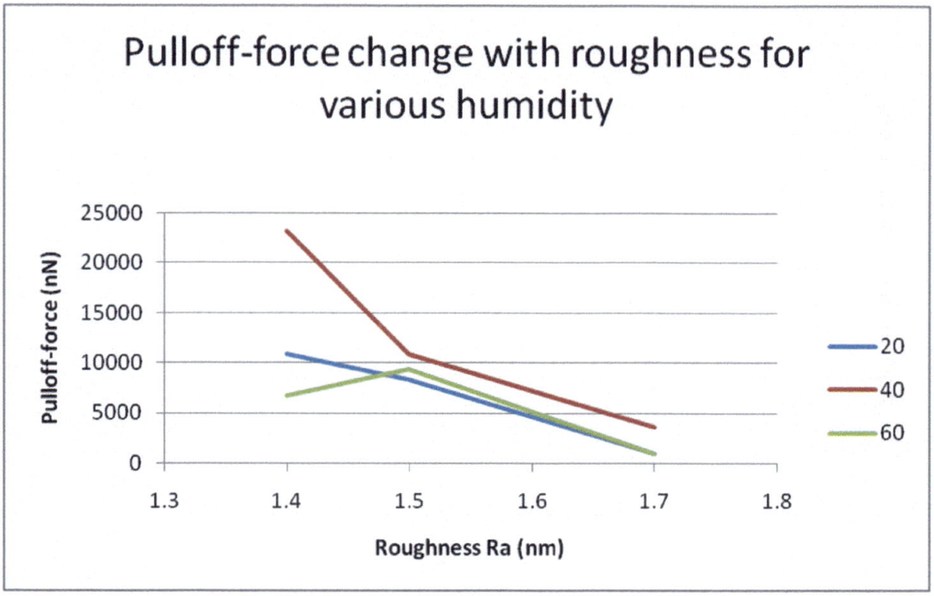

Fig. 4. Results of pull-off force measurements vs. roughness and humidity (20, 40 and 60 %)

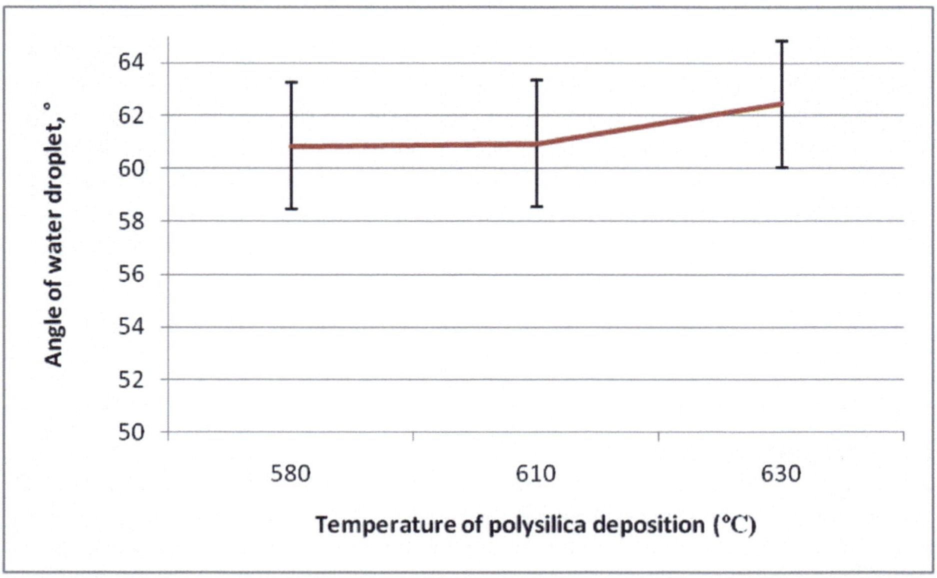

Fig. 5. Contact angle of water droplet vs. temperature of deposition

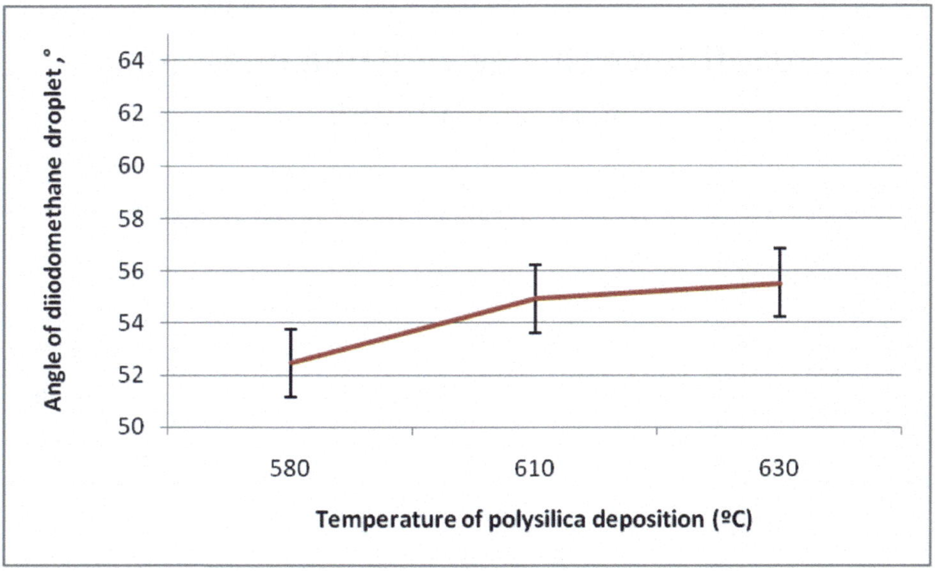

Fig. 6. Contact angle of diiodomethane droplet vs. deposition temperature

4 Conclusions

In summary, this paper presents properties of polysilicon layers on silicon. Those properties vary based on the deposition temperature during production. With increasing deposition temperature roughness of the surface increases as well. From the measurements we can observe some trends in adhesion while humidity of the environment and roughness of the sample change. Local maximum of adhesion has been found for humidity of 40%, for lower and average humidities the adhesion increases with decreasing roughness. There seems to be no significant impact of temperature of deposition on wettability (contact angles) in the tested range.

5 Acknowledgements

We acknowledge the support from the project ERANET 3SMVIB project IMT contract 7-063/2012 as well as in the case of IMT the support from the Sectoral Operational Program Human Resources Development 2007-2013 of the Ministry of European Funds through the Financial Agreement POSDRU/159/1.5/S/132397.

References

1. Rebeiz G.M., "RF MEMS: Theory, Design and Technology", Wiley, Hoboken, (2004).
2. **Hartzell** A.L., **da Silva** M.G., **Shea** H., "MEMS Reliability", Springer, Berlin, (2011).
3. Zhang Y.,.Zhao Y.P, Microsyst. Technol. 21 (2015).
4. Voicu R., Rymuza Z., Michałowski M.*et al.* in : CAS 2013 Proceedings , IMT Bucharest, (2013).
5. Jarzabek D. M., Kaufmann A. N., Schift H., Z Rymuza , Jung T. A., Nanotechnology 25 (2014).
6. Cappella B., Dietler G., Surface Science Reports 34 (1999).

Systematic errors of measurements on a measuring arm equipped with a laser scanner on the results of optical measurements

Michał Rak[1,2], Adam Woźniak[1]

[1] Warsaw University of Technology, Faculty of Mechatronics, św. Andrzeja Boboli 8 Street, 02-525 Warsaw, Poland
{m.rak,wozniaka}@mchtr.pw.edu.pl
[2] École Polytechnique Montréal, Department of Mechanical Engineering, P.O. Box 6079, Station "Centre-Ville", Montréal, Québec, Canada H3C 3A7
michal-bartosz.rak@polymtl.ca

Abstract. Laser triangulation is an optical measuring method which characterizes with relatively high precision. There are many factors which can affect this type of measurement. They depend on a nature of optical measurements. Equally important from the viewpoint of accuracy is how precisely the position of the scanner in measuring volume is determined. In this paper we present tests performed using measuring arm equipped with a laser scanner. As in measuring arms position of the probe depends on the accuracy of angular encoders their influence on the results was tested. Five different strategies were chosen to show the significance of employment of additional encoders. Results showed that if less encoders are used, better results are obtained. Also overlapping of individual scans affects precision of results. Therefore one should pay attention on choosing proper measuring strategy, to not change the scanning direction excessively and to not scan the same area many times.

Keywords: measuring arm · laser scanning · angular encoders · coordinate measuring techniques

1 Introduction

The beginnings of coordinate measurements date back to the seventies of the 20th century. The development of this measurement technique was conditioned by progress of computer technology [1]. Coordinate measuring techniques rely on computer processing of information from a measurement. This information is represented by coordinates of points gathered from the surface. Points can be collected either in contact or in non-contact way. Contact measurements are accurate but number of collected points is rather small because of the time needed to measure every single one. On the other hand non-contact method provide many points from the entire object in a

R. Jabłoński and T. Brezina (eds.), *Advanced Mechatronics Solutions*,
Advances in Intelligent Systems and Computing 393,
DOI 10.1007/978-3-319-23923-1_54

short time. The non-contact methods most common in metrology [2]: triangulation, ranging, interferometry, structured lighting and image analysis.

In this paper we focused on laser triangulation as this method has higher precision and lower costs comparing to other optical measuring methods [3]. In triangulation a laser beam is aimed in the direction of the measured surface, producing a spot, a stripe or a cross, depending on the used device. Then it is imaged on a detector at a position dependent on the distance of the measured area from the source of the laser. There are some factors affecting the accuracy of this optical measurement and they stem from the nature of this type measurements.

These factors are [4]: laser scanning depth, projected angle, environment effects or operation error. Nevertheless we should always remember that the positioning of the scanner has an utmost importance on accuracy of the measurement. There are three technical solutions which enable determination of the position of a laser scanner. First is a self-positioning laser scanner [5]. Since this type of scanner has to find its position into space by itself, it must determine an origin for each scan. The position of the scanner when the acquisition starts is set as an origin of the reference system and all subsequent acquisitions are performed in this fixed coordinate system. Attaching of targets to the measured part, or around it in case of small parts, enables positioning of the scanner in the measuring space.

The other two solutions are attachment the laser scanner to the devices which enable determination of its location: coordinate measuring machine (CMM) and coordinate measuring arm. As spatial accuracy of measuring arms is of the same order as accuracy of laser scanners we found this combination of devices interesting for the research.

2 Investigation setup

The tests were performed using a coordinate measuring arm with the accuracy [6] of ± 0.028 mm according to the B test (point repeatability) and ± 0.040 mm according to the C test, which stands for spatial accuracy of the arm . The arm was equipped with a laser scanner head working on a principle of triangulation with 2sigma accuracy of ± 0.032 mm.

A measuring arm consists of three tubes connected with joints inside which there are angular encoders. At the end of the arm there is a measuring probe which can be either tactile or optical. Position of the probe in Cartesian system is determined based on angular positions of encoders and lengths of the tubes. Scheme of the arm used for testing is presented in Fig. 1

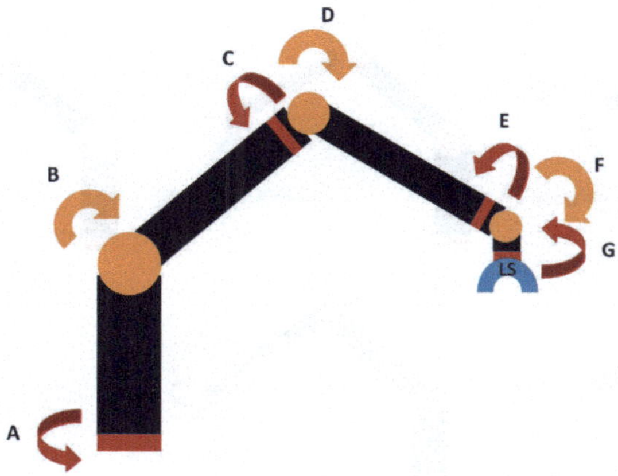

Fig. 1. Scheme of a measuring arm

As it can be seen from the above scheme this 7-axis arm has four rotary axes marked red (A, C, E, G) and 3 tiltable axes marked orange (B, D, F), which can rotate only to a certain extent depending on construction of the arm. Angular coordinates of these seven axes are used to calculate the position of the measuring probe in the Cartesian system. In this paper we present the influence of use these axes on the final result of the measurement.

3 Description of the method

As it was described in the previous section there are angular encoders in each of 7 rotation axes. These encoders have limited accuracy and error from each one influences on the error in the positioning of the measuring probe at the end of the arm, in our case the probe is a laser scanner.

Using the measuring arm equipped with the laser scanner we measured the planar surface of a marble block. The block was of the length of 600 mm and of width of 100 mm which enabled measurement of the whole surface with a single pass of the scanner.

Five different measuring strategies were chosen:

1) The rotary axes were immobile, only tiltable axes were used in the process of data acquisition.

2) Axis “A” was rotated by 180°, during the measurement only tiltable axes were moved.

3) Axes “A” and “G” were rotated by 180°, during the measurement only tiltable axes were moved.

4) The rotary axes were immobile, only tiltable axes were moved but the surface was measured three times in the row.

5) Movement in all seven axes.

These strategies are presented in Fig. 2

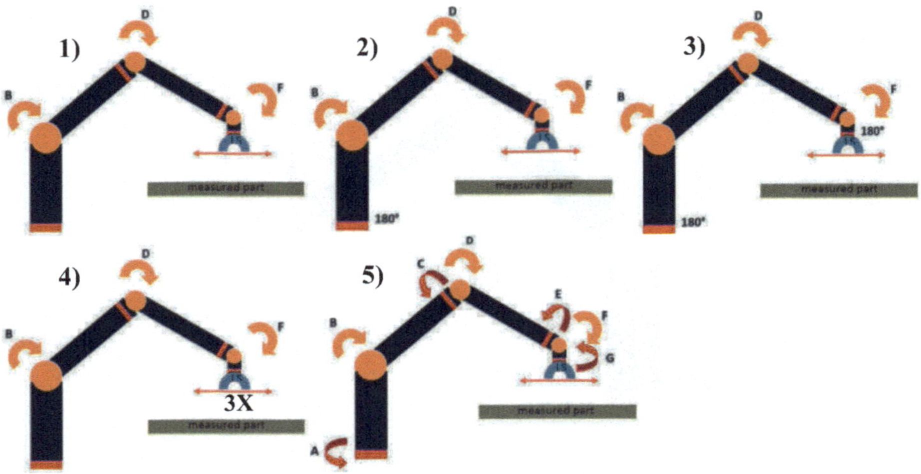

Fig. 2. Measuring strategies

We took care to maintain the same conditions for all strategies: temperature, scanning depth, incident angle, lightening conditions so only the influence of the encoders was tested.
First strategy is the simplest one and can be used in a very limited number of cases, where a part can be measured with a single scanning path without employment of rotary axes. In second and third strategies before the measuring process one or two axes respectively were rotated by 180° but during the measurement their positions were not changed. In the fourth strategy the same surface was measured three times so individual scans overlapped. In many situations in real measurements it is impossible to avoid overlapping of scans, which might be caused by complex shape of the surface or unsuitable properties for optical measurements . The last strategy represents the situation in which big parts with quite complex shape are measured. Then all joints are rotated to cover the part with a laser beam.
From the measurements with different strategies point clouds were obtained. Outlying points were removed but any filtration method was applied. A single cross-section from each point cloud was determined and compared with the corresponding line measured on a CMM. Accuracy of the CMM is one order of magnitude higher so this data could be treated as reference. The straightness of the line measured on the CMM with a maximum permissible error for length measurement (E_{MPE}) of (1,7 + L/333) μm was of 10 μm.

4 Results

To calculate the position of the laser scanner, angular coordinates in joints are used. Observing the cross-section of the point cloud from optical measurement and comparing with the reference data (CMM) it can be seen that this cross-section is waved. This is caused by inaccuracies of the encoders of the arm not by actual distortion of the measure surface. A sample comparison on two lines measured with different methods for the first strategy is presented in Fig. 3.

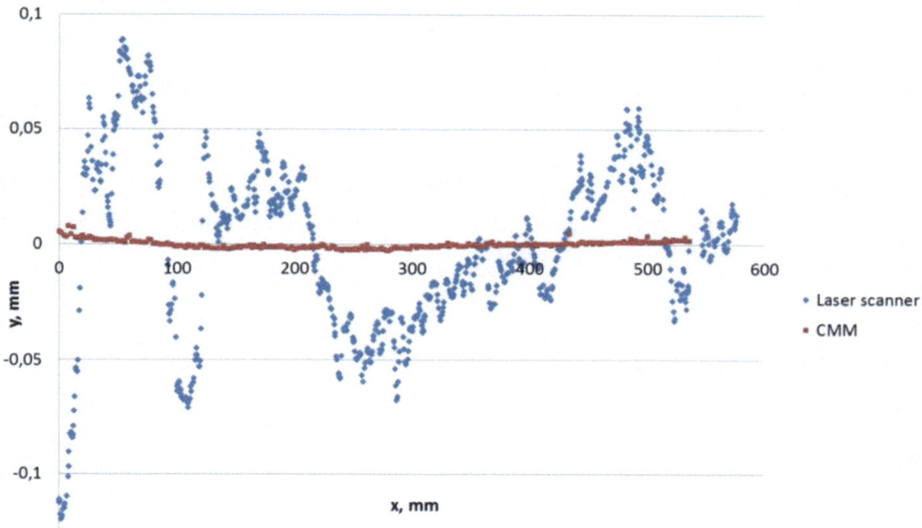

Fig. 3. Comparison of a cross-section from optical measurement with reference data

The amplitude of the presented waviness shows the actual inaccuracy of measuring system. It is different for different strategies.
As the evaluation parameters range of result and number of points were chosen. Measurements for all strategies were repeated ten times and average values were calculated. The results are presented in Table 1.

Table 1. Results for all strategies

	strategy 1	strategy 2	strategy 3	strategy 4	strategy 5
range, μm	168.8	179.6	184.2	213.7	188.9
st. deviation, μm	30.5	23.5	28.2	28.3	22.7
no of points	2297	2342	2206	4630	2070

5 Discussion of results

During all tests the best conditions for optical measurements were obtained. Therefore the differences in results for all strategies are caused only by different use of arm encoders. Considering the number of gathered points it can be noted that the rotations in joints do not influence on the number of points. More points are gathered if the laser scanner is led over the surface for a longer time, for example if measurements of the surface are repeated (fourth strategy).
What is more important is how the employment of axes of the arm influences on accuracy of performed measurement. The best results were obtained when the use of

encoders was minimized. Results from the second strategy show that only by adding constant rotation in one axis provide worsening of around 10 μm. Employment of another encoder increase the error. In fifth strategy any constant rotation was added but during data acquisition all encoders were moving. The worst results were obtained from fourth strategy. This error has two components. One comes from rotation in "B", "D" and "F" axes and the other comes from overlapping of individual scans.
The results showed how big is impact of the operator on obtained results. Neither frequent changes of scanning direction nor overlapping of scans influence positively on the measurement accuracy. Comparing strategies first and forth it can be noted that lack of experience in performing measurements on arm equipped with a laser scanner might introduce about 25% bigger error

6 Conclusions

In this paper we touched on the important topic of influence of the operator and selected measuring strategy on the results of measurements performed using manual devices. As an example of a manual device we used a measuring arm equipped with a laser scanner. Spatial accuracy of the arm is of 40 μm and accuracy of the laser scanner is of 32 μm. Assuming equality of impact factors the accuracy of the whole system is of 51 μm, while obtained errors are much higher. Part of this error is due to mentioned factors affecting accuracy of optical measurements. Nevertheless significant differences between strategies proves the influence of the manner in which the surface was measured. The conclusion from our research is that, when using manual devices such as measuring arm, scanning directions should not be changed frequently and overlapping of individual scans should be avoided.

References

1. Ratajczyk E. „Coordinate measuring technique" OWPW, Warsaw 2005 (in Polish)
2. Várady T., Martin R., Coxt J., Reverse engineering of geometric models – an introduction. Computer-Aided Design, Vol.29, No.4 pp,255-268, 1997
3. Martínez S., Cuesta E., Barreiro J., Álvarez B., Analysis of laser scanning and strategies for dimensional and geometrical control. The International Journal of Advanced Manufacturing Technology (2010) 46:621-629
4. Wang L., Ding H., Wang S.: Measurement Error Compensation Using Data Fusion Technique for Laser Scanner on AACMMs. ICIRA 2010, Part II, LNAI6425, pp.576-586, 2010
5. http://www.creaform3d.com/en/metrology-solutions/portable-3d-scanner-handyscan-3d
6. ASME B89.4.22-2004 „Methods for Performance Evaluation of Articulated Arm Coordinate Measuring Machines".

Compensation of influence of element's eccentric positioning on the result of roundness deviation measurement of discontinuous sections by radial method

Maciej Sieniło[1]

[1] Warsaw University of Technology, Faculty of Mechatronics, św. Andrzeja Boboli 8, 02-525, Warsaw, Poland
m.sienilo@mchtr.pw.edu.pl

Abstract. This article proposes a way of compensating the influence of element's positioning in relation to the FMM (Form Measuring Machine) table's axis of rotation on the result of roundness deviation measurement. The research concerned the elements with discontinuous surfaces e.g. with holes. Using the devised software simulation study was conducted and verified experimentally. The experiments were conducted with Taylor-Hobson Talyrond 365 machine. The obtained results confirmed far greater resistance of proposed calculating method to the positioning of element in regard to the axis of rotation.

Keywords: roundness deviation, radial method, eccentricity

1 Introduction

Roundness deviation measurements of rotatory elements are mostly performed with either universal coordinate machines (CMM - Coordinate Measuring Machines) or specialized FMM stations (Form Measuring Machines), which perform measurements using radial method.

The first method is more universal, allowing measurements of both continuous and discontinuous surfaces of various shapes and dimensions. The time of measurement is relatively short and attained accuracy reaches values even below 1 μm.

FMMs are not as universal. However, among their assets one should first of all mention greater accuracy of measurement, reaching even the tenths of micrometer. The condition of obtaining such high accuracy is meeting adequate requirements [1, 2, 3] namely that the centre of studied profile is placed as close as possible to the axis of rotation, i.e. the eccentricity should be few orders lower than the profile's radius.

To attain such positioning of element, it is necessary to exercise adequate preparatory actions. In case of elements of simple geometry, such actions are not a considerable problem. In case of more complicated elements, consisting of several non-coaxial surfaces or elements of discontinuous surface e.g. with holes, their proper positioning on the machine's table is very complex, time-consuming and sometimes even impossible. In that case, the occurrence of errors that increase measurement uncertainty has to be taken into account. Such problem does not occur when measuring with coordinate machines.

R. Jabłoński and T. Brezina (eds.), *Advanced Mechatronics Solutions*,
Advances in Intelligent Systems and Computing 393,
DOI 10.1007/978-3-319-23923-1_55

2 The influence of eccentric positioning of element's section on the result of roundness deviation measurement.

Radial method relies on measuring the increments of element's radius r_i as a function of angle of rotation φ. This method is sensitive to element's positioning in relation to the table's axis of rotation. Mapping of element ideally round (Fig.1a) positioned eccentrically results in the profile that is not a circle. Fig. 1b and 1c demonstrate that eccentric positioning of element caused (according to the theory) profile's elongation in the direction of given eccentricity and flattening in perpendicular direction [1]. To indicate profile's distortion the sum of the highest rise RONp and the deepest valley RONv, i.e. roundness deviation RONt determined in relation to mean circle was used.

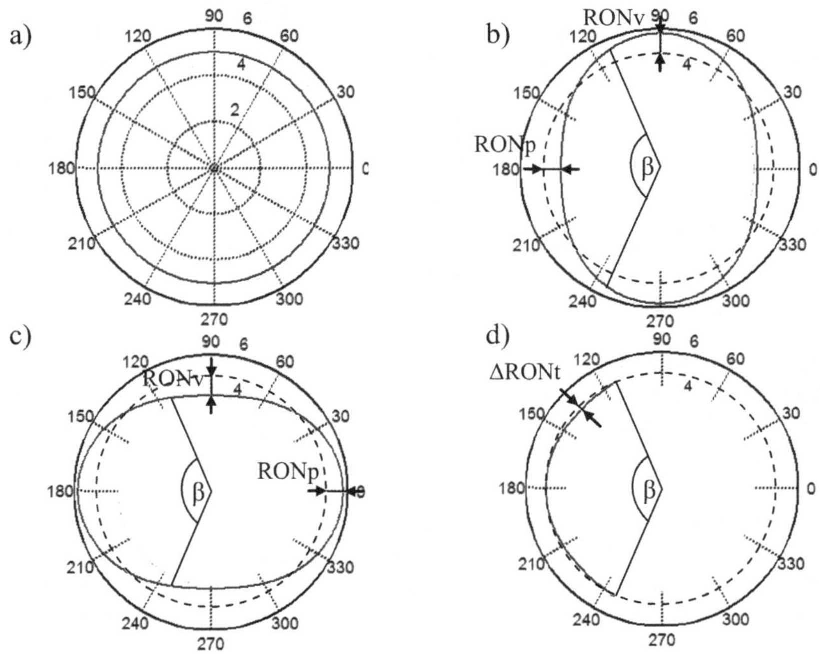

Fig. 1. Exemplary shape of section's segment a) axial positioning, b) eccentric positioning, y direction, c) and d) eccentric positioning, x direction

Fig. 1d presents the result of roundness deviation measurement of chosen profile's fragment of arc's length determined by angle β. The element was positioned in the same way as in Fig. 1c.

In general, it can be observed that the shape of studied section's segment depends on 3 parameters: *β*, *θ*, *e*. Geometric dependencies between these parameters are presented in Fig. 2.

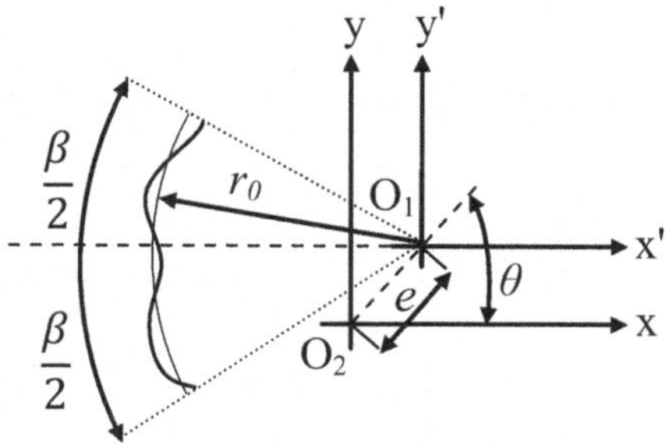

Fig. 2. Geometric dependencies between the parameters of measured section

The measurement with CMMs relies on collecting the contact points between the probe and the surface of element. The registered data allow determining local radii of measured section. In case of measurement with FMMs local radii are not determined and the registered signal allows determining only the increments of radii.

The concept of compensating the influence of eccentricity *e* on the result of measurement RONt relies on replacing the algorithm of determining mean circle used in FMM with the algorithm used in CMMs. The essential part of this concept is adding to FMM measuring signal a mean value of element's radius, determined e.g. on the basis of documentation of studied section. This results in transforming the set of radii increments expressed as a function of angle of rotation to the set of local radii. Conversion of coordinates of measurement points from polar system to Cartesian system constitutes the next step. As a result of this operation a data set similar to the set derived from measurement with coordinate machine is obtained, which allows using CMM algorithm, later termed the improved algorithm. Detailed overview of this concept was presented in [4].

3 Simulation study

Simulation study was aimed at investigating the results of employing the improved algorithm. The segments of sections of various lengths were analysed, while value *e* and the direction of eccentricity vector θ were altered. For clarity of argumentation, the percentile share of continuous part in regard to overall perimeter of section was marked as *p*. Roundness deviation was determined with two methods: traditional (RONt) and improved (with additional index $RONt_u$). Simulation study neglected shape deviation of element.

Simulation program devised in Matlab environment allows introducing various values of element's radius and measuring tip, and of eccentricity. For illustrative purposes very large eccentricity e=450 μm was assumed. The illustrative material placed in the further part of this article pertains to simulation study for element of 30 mm radius and stylus with round measuring tip of 2 mm radius.

3.1 Simulation study – traditional algorithm

The results of study on the influence of length of measured arc represented by parameter p on the roundness deviation measurement error is presented in Fig. 3.

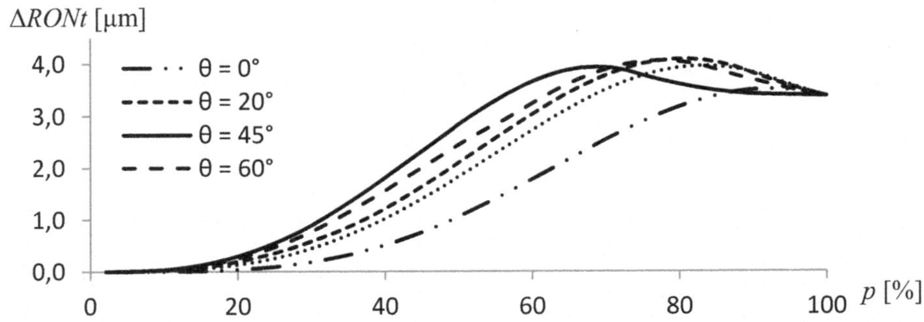

Fig. 3. The influence of parameter p on the value of roundness deviation determined by traditional method

Measurement error $\Delta RONt$ reaches maximum value exceeding 4 μm for p between 60% and 90% depending on the direction of eccentricity vector θ.

The results of study on the influence of direction of eccentricity vector θ on the roundness deviation measurement error $\Delta RONt$ is presented in Fig. 4.

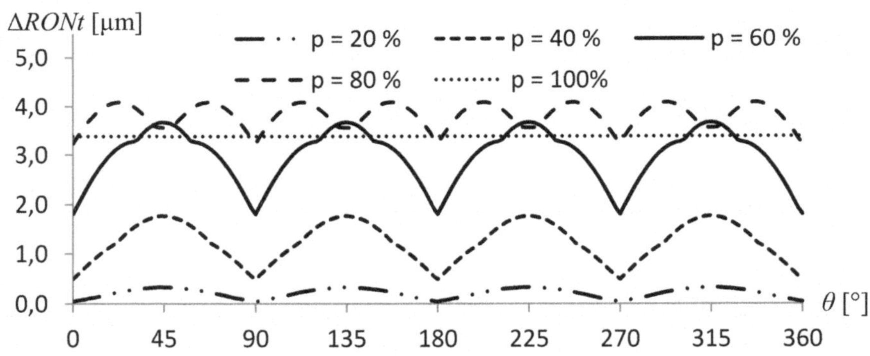

Fig. 4. The influence of direction of eccentricity vector on the value of roundness deviation determined by traditional method

Characteristic oscillation is observed (for p between 40% and 80%). Maximum value of this oscillation exceeds 1,5 μm. As observed in Fig. 4 for continuous section, ideally round element (p=100%) the direction of eccentricity vector has no influence on the measurement error ΔRONt. In case of discontinuous sections the directions either less or more favourable in regard to the obtained results may be distinguished. The smallest errors occur for directions overlapping with the directions of axes ($\theta = 0 \pm k\frac{\pi}{2}$). One should add that the operator cannot decide about the direction of eccentricity when measuring discontinuous elements.

3.2 Simulation study - improved algorithm

The results of analogous simulation study conducted by using the improved algorithm are shown in Fig. 5.

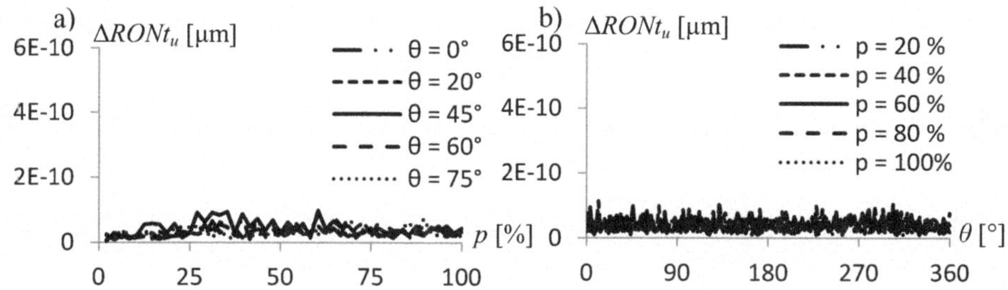

Fig. 5. a) The influence of parameter p on the value of roundness deviation determined by improved method, b) The influence of direction of eccentricity vector on the value of roundness deviation determined by improved method

Lack of influence of both the arc's length and the direction of eccentricity vector θ on the roundness deviation measurement error $\Delta RONt_u$ was observed. The observed errors are negligible and random. In conclusion, based on simulation study high resistance of improved algorithm to eccentricity can be expected.

4 Experimental studies

Experimental study was performed with Taylor Hobson Talyrond 365 machine equipped with round tip Ø4. Cross sections of cylindrical roundness master Ø60, characterized by roundness deviation RONt=0,2 μm, and sections of combustion engine piston Ø64 of roundness deviation RONt=15 μm were studied. Roundness deviation of sections of each element positioned at various distances from the axis of rotation were measured. Continuous sections were measured and then the fragments of profiles were numerically removed, obtaining segments of assumed angular length. A part of study presented below pertains to the segments of proportion of continuous part of section to whole perimeter p=80%.

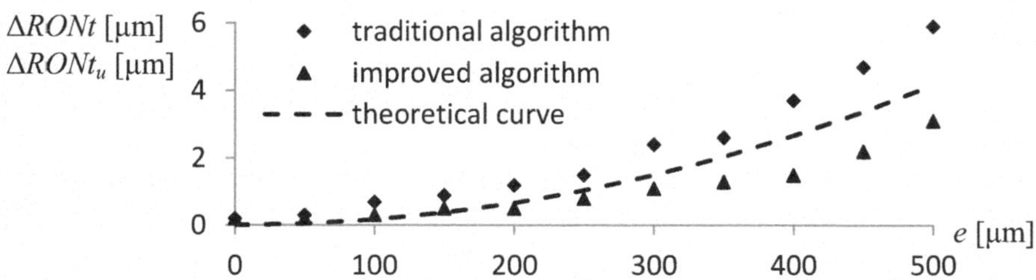

Fig. 6. The influence of eccentricity e on the roundness deviation measurement result for section of master Ø60 by employing traditional and improved algorithms (p=80%, θ=0).

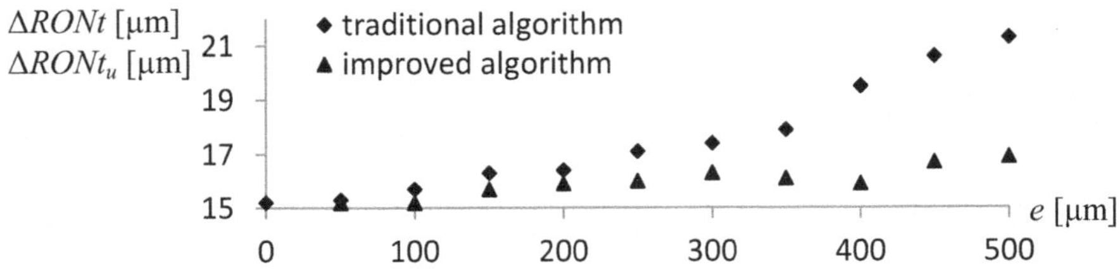

Fig. 7. The influence of eccentricity e on the roundness deviation measurement result for section of a piston Ø64 by employing traditional and improved algorithms (p=80%, θ=0).

Figures 6 and 7 present cumulative charts of roundness deviation measurement results for master Ø60 and piston Ø64 obtained by using both algorithms. The observable difference between the theoretical curve and the charts obtained by traditional method attests to the influence of other factors (other than eccentricity) on the measurement result in real conditions. Among these one could mention: non-linearity of probe characteristic, shape errors of round measuring tip, discretization errors, other than radial direction of displacement, friction, oscillation. The above factors are the source of errors in both methods. They also influence the increase in value of roundness deviation obtained by using improved algorithm. Despite this, both in case of master and piston sections the results obtained by using the proposed method are burdened with error more than two times lower.

5 Conclusions

The article presents the results of simulation and experimental studies on roundness deviation measurements in elements of discontinuous surface obtained by using two algorithms: traditional and improved. Based on the results of simulation study it could be concluded that using improved algorithm allows fully eliminating the errors resulting from eccentric positioning of element in relation to the axis of rotation. However, experimental study revealed the influence of other factors associated with the construction of measuring device. Nevertheless, significant improvement of measurement results were observed when employing improved algorithm.

References

1. Żebrowska–Łucyk S., Bezodniesieniowa metoda badania makrogeometrii powierzchni elementów obrotowych. Oficyna Wydawnicza Politechniki Warszawskiej, Warszawa 2001
2. Adamczak S., Janecki D., Problems cencerning the measurement of form profiles of non-closed cylindrical surfaces of machine parts, Measurement Science Review, no. 1 (2001).
3. Zakharov O. V., Brzhozovskii B. M., "Accurancy of centering during measurement by roundness gauges", *Measurement Techniques*, vol. 49, 2006.
4. Rudziński R., Tomasik J.: Możliwości wykorzystania rozszerzonego zakresu pomiarowego profilometrów, VII Konf. Metrologia w Technikach Wytwarzania Maszyn.

Proposal of a wireless measurement system for temperature monitoring of biological active materials

Martin Skovajsa[1], Peter Fabo[1], Ľubomír Pepucha[1], and Ivan Sládek[1]

[1] Research Centre University of Žilina, Žilina, Slovakia
{martin.skovajsa, peter.fabo, lubomir.pepucha, ivan.sladek}@rc.uniza.sk

Abstract. Industrial systems for the collection of environmental data play an important part in the production processes and also in storage facilities. When storing large volumes of bio-materials such as wood chips (wood pellets) or different silage, problems arise with dangerous self-heating of the stored material. This paper deals with the design of an autonomous system for collecting such data. In the view of the needed measurements, there were designed and developed measuring probes with wireless data transmission. The proposal includes an information system for archiving and evaluation of measured data.

Keywords: measurement system, transmitter, receiver, communication

1 Introduction

Transition from conventional to renewable sources of energy is currently preferred in the field of energy sources. One that is already quite widespread, is the use of biomass in the form of wood chips or wood pellets as a source for heat production. Quite often it comes hand-in-hand as a combination with CHP units to generate electricity.

Wood chips, as the most commonly used wood material have properties of firewood. The chips are made of wood waste from branches and raw materials acquired in the thinnings of the forests. Typically, the length of the chips is in the range of 5 to 50 mm, width of 5 to 30 mm and a thickness of 5 to 15 mm. Wood chips are biologically active material. By activities of the living parenchymal cells, chemical oxygenation, hydrolysis of the cellulose components in an acidic environment and the biological activity of bacteria and fungi, the wood chips begin to decompose relatively quickly. This will reduce the volume and increases the humidity of the material. At the same time the temperature of the chips is growing to 50°C - 70°C. In case the temperature rises over 100°C, self-ignition of the wood chips may also happen. If the water content of the wood chips is greater than 25% to 30%, which is the optimal moisture content for the combustion, after a certain time (depending on temperature) the wood chips start degrading and are affected by mildew.

The Slovak Republic stipulates the storage of wood chips decree [1], which defines the need for temperature measurement in chips at a depth of 1.5 m at a maximum

R. Jabłoński and T. Brezina (eds.), *Advanced Mechatronics Solutions*,
Advances in Intelligent Systems and Computing 393,
DOI 10.1007/978-3-319-23923-1_56

distance of measuring points 10 m from each other every day. If the temperature of the chips in the first week does not exceed 35°C, the sample period may be extended to every 3 days. After three weeks of storage, the measurements interval can be extended to once a week. If the temperature in the pile reaches 50°C or if the temperature rises by more than 3°C in 24 hours, the wood chips must be scuffed using appropriate mechanisms. The records of temperature measurements of wood chips are in accordance with the provisions of Section 19 of the decree [1]. In addition of fire risk in chips storage, other hazards are associated, i.e. the occurrence of organic dust contaminated with microorganisms - bacteria and fungal spores.

2 Measuring system requirements

The primary objective in the design of the entire measuring system was the elimination of manual temperature measurements of biomass and complete automation of the process of measuring and evaluating temperature of the material being measured. The entire system consists of the measurement hardware, storage and archiving of measured data, visualization and application software. Following requirements were specified for individual system parts:

Measuring probe

- measurement of temperature from -10°C to 90°C with an accuracy of +/- 0.5°C and a resolution of 0.1°C
- autonomous operation for at least three months
- robust construction which withstands the pressure of the stored wood chips and resistant to mechanical stress caused by mechanisms in the processing and handling of wood chips, probe transmits in free band 433.92MHz,
- one-way communication protocol enabling collision detection of simulcasting of several probes on the receiver side,
- frequency of data transmission is parameterized by the temperature of chips,
- probe dimensions at the upper limit of typical size of wood chips

Receiver

- operation in the industrial environment,
- operating multiple receivers improves the system parameters to capture the signal from probes in deep layers

Software

- The data model of the system - internal model of measurement system holding information about the hardware configuration, active sensors, their physical positions, measured data and other additional information,
- Communication via the network - secure communication channel using SSL encryption,
- Presentation software - platform independent, user-friendly client software. The possibility of extending the web version of the software with the same or similar functionality as the desktop application.

3 Concept and implementation of hardware parts

The proposed measuring system consists of two basic parts: measuring probes that transmit the measured data and receiver modules, whose role is to intercept transmitted data. Hardware configuration of both will be described in the next section.

3.1 Measuring probe

Measuring probe in block diagram in Fig.1a is used to measure environmental parameters such as temperature, humidity, atmospheric pressure, solar irradiation and others. In the current embodiment, a temperature probe is used to measure the ambient temperature. The role of the test probe is processing the measured data, preparation of telemetry data, format the data according to designed format and transmit data via RF interface at 433MHz.

Following requirements were identified in the design of the probe:

- autonomous operation over an extended period, typically several months,
- minimal power requirements,
- minimal mechanical dimensions - in its placement in the bio-materials, the probe shall not be an obstacle,
- stiffness and tightness of the outer enclosure - version for environments with high humidity and temperature,
- modular design - option to add, respectively replace the sensor module for another module to measure other physical quantities.

MCU: Microprocessor with control software. It is used 32-bit microprocessor ST STM32F030F4T6 [4]. Selected microcontroller has 16kB Flash memory for program and 4KB RAM for data. The internal A/D converter connected via a resistive divider measures the battery voltage during the measurement interval. This value is used for telemetry to monitor the battery status, a status review is provided by the control software.

Sensor: Represents the sensor of environmental parameters such as temperature, humidity, pressure, light intensity, and others. The current version uses a temperature sensor LM75AD with resolution of 0.125°C. The working range for higher absolute accuracy is set in range of -25°C to 100°C. The sensor uses 11-bit A/D converter for temperature conversion. This type's advantage is also in a minimal consumption in inactive mode (3.5 uA). To the MCU it is connected via I2C bus.

Power supply: Power supply block with a stabilized power output for the module circuits. The basic requirement was to create a design with as small energy consumption of the entire solution as possible. As an energy source is used one AA battery with increasing converter. StepUp converter ON Semiconductor NCP1402 used in the designs has efficiency reaching 85% and passive current consumption 30uA.

Another variant of the power module allows drawing energy from renewable sources such as thermocouple, solar cell, wind turbine and others.

Transmitter: To transmit data was chosen transmitter module for the band 433MHz with FSK modulation.

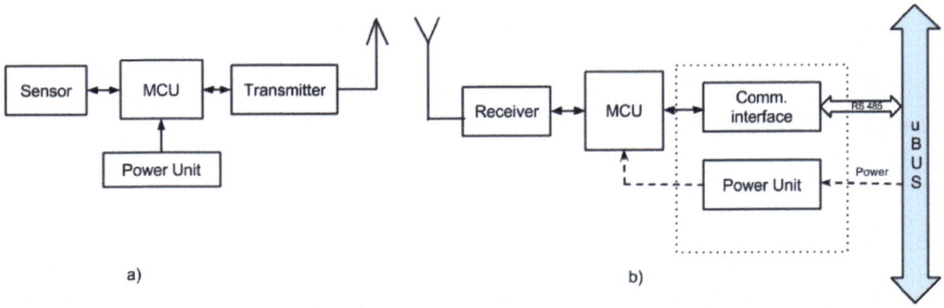

Fig. 1 Block diagram of the a) measuring probe, b) receiver module

3.2 Receiver module

Module for wireless reception of data transmitted from measuring probes in block diagram in Fig. 1b. Basic processing of the data - checking consistency of the received block of data, preparation of data for transmission to the control center.

Receiver: wireless reception of broadcast data from the measuring probes. Superreactive receiver module is used with a demodulator and an output stage to connect to the digital input of the control microprocessor.

MCU: microprocessor with control software. The same microprocessor as in a measuring probe is used, thus STM32F030.

Power supply: power supply block with a stabilized power output for the module circuits. Linear regulators with low voltage drop are used. Input part with a surge protection, overcurrent protection and a broadband filter.

Interface: RS485 bus interface module.

For connection to RS485 bus is used Texas Instruments SN65HVD10 which includes circuits for ESD protection. The overvoltage protection input circuit is complemented by a specialized Bourns circuit "High-Speed Protector". Solution using RS485 bus allows to include more transceiver modules on a single cable system. Using multiple transceiver modules allow better coverage of the area with the location of measuring probes and also increase resistance to interference.

4 Design and implementation of software

The measuring system consists of a number of probes and a number of receivers. The number of measuring probes is in tens and the number of receivers in the order of units. The final number of the receivers depends on several factors such as the actual area of monitored site, its segmentation and the presence of interference sources such as radio signals, metal structures, power lines, distribution of technical media, etc. In the design of firmware, respectively the application logic of transmitter and receiver, was needed to ensure reliable data transfer.

4.1 Firmware

Measuring probe

As was explained in the previous section, the probe is powered by one AA battery. Therefore, the firmware design dealt with the important aspect of energy saving and energy management. A microcontroller has the possibility of transition to standby mode in which current consumption drops to 20uA. In standard mode, the current consumption at the maximum frequency of 48MHz rises near 30 mA. To reduce the power consumption following settings were applied:

- as the clock source internal RC oscillator was used,
- the frequency of the clock signal is set to the lowest value, that is f=8MHz,
- during the standby period the microcontroller is put into standby mode (lowest power mode, where most parts of the microcontroller are disconnected from the power),
- measurement time during which the probe collect the data is reduced to the minimum necessary - 100 milliseconds.

As the communication between the probe and receiver module is one-way, it was needed to include information that probe sends and state information of the voltage level of the battery. This information is needed to assess the battery life.

The data that the probe sends are: probe address, the voltage level of the battery, measured temperature (humidity, pressure, etc.) and an 8-bit CRC sum.

Period during which the probe is in standby mode is designed to be 5 minutes. To avoid interference of transmission data from various probes, probe sleep time is parameterized using the measured temperature. By increasing the temperature by 10°C the sleep time is reduced by 60s and the minimum sleep time is set to 3 minutes. Reducing the temperature by 10°C the sleep duration is increased by 60s.

Receiver module

Receiver module is built with the same microcontroller as measuring probes, thus STM32F030 [4]. These modules are powered by an external supply voltage and therefore there was no need to address the energy-saving modes. To achieve the required performance and processing speed following settings were applied:

- the frequency of the clock signal (internal RC oscillator) is set to the maximum value, that is f=48MHz,
- communication with host component (measuring computer) via the RS485 serial interface is set at speed 19200 baud/s

Other features of the receiver module are as follows: includes internal memory with the option to store data from 250 different probes, detects incorrectly received packets and supports data encryption of measured data towards measuring computer.

4.2 Middleware

An important part of the whole measurement system is a system of data collection, data storage and communication interface to the end application. To access the measurement module and the measured data Java and Python library was developed allowing communication with the connected hardware. In implementation of the RF measurement system it was necessary to consider the typical scenarios, such as data from some sensors may not be available for a long time. This can occur while transmission of several probes at the same time, where their RF signals interferes with each other. Another example is that the measurement data of the measuring probe are received by several receiver modules, where is need to eliminate such redundant data.

5 Conclusions

Low power consumption of the probe itself is a key parameter. Verification of the module functionality proved that in standby mode the current drawn from the AA battery is 0.43mA and in the active mode is 20 mA. The measuring probe wakes from standby in average every 300s and subsequent data measurement and transmission takes 0.4s max. This gives an average consumption of 0.463 mA. Using AA batteries with a capacity of 2000 mAh, expect life of the probe circa 180 days. Practical tests of the proposed system in the state of a working prototype demonstrated the correctness of the solution concept. Probe systems are sufficiently mechanically resistant to ambient pressure. The concept of multiple receiving antennas eliminates the loss of the carrier signal at 433 MHz in passing the tests with a five meters tall layer of wood chips.The whole concept can be used for monitoring the status of other commodities for storage and processing such as food products, in agriculture for the quality control of the fermentation process and the temperature conditions in the layer of snow with hills exposed to avalanche breakdown. Interesting is also the possibility of monitoring environmental parameters with a probe with autonomous power photovoltaic cells and storage of energy in the supercapacitor - for example, in agriculture for monitoring surface and subsurface temperature and humidity.

Acknowledgement
The research is supported by the European Regional Development Fund and the Slovak state budget by the projects "Research Centre of the University of Žilina" - ITMS 26220220183.

References

1. Vyhláška MV SR č. 258/2007 o požiadavkách na protipožiarnu bezpečnosť pri skladovaní ukladaní a manipulácii s tuhými horľavými látkami, §19-21
2. Chaczko, Z., Ahmad, F. Wireless Sensor Network Based System for Fire Endangered Areas, In: Information Technology and Applications, 2005. ICITA 2005. Third International Conference, Volume: 2, ISBN 0-7695-2316-1, pp 203-207
3. Ďuďák, J., Skovajsa, M., Sládek, I.: Proposal of a Communication Protocol for Smart Sensory Systém, In: Proceedings of the 16th International Conference on Mechatronics-Mechatronika 2014, Brno, 3.-5. December, 2014. Editors: Brezina, T., Maga, D., Stefek, A., ISBN 978-80-214-4817-9. S. 107-112
4. STMicroelectronics. *STM32F030x4: datasheet -production data* [online]. 2015 [cit. 2015-06-29]. Available online: http://www.st.com/st-web-ui/static/active/en/resource/technical/document/datasheet/DM00088500.pdf

Reconstruction of dopant vertical position from Kelvin probe force microscope images

Krzysztof Tyszka[1, 2], Ryszard Jabłoński[2]

[1] Research Institute of Electronics, Shizuoka University, Japan
{tyszka,yabu}@mchtr.pw.edu.pl
[2] Institute of Metrology and Biomedical Engineering, Warsaw University of Technology, Poland
tyszka@rie.shizuoka.ac.jp

Abstract. In novel nano-scale electronic devices the number and location of individual dopant atoms within a device determine its characteristics. Therefore, precise control over these parameters is required. In this paper we describe Kelvin Probe Force Microscope (KPFM) designed for dopant detection. We propose a method for reconstruction of dopant position by using comparison between KPFM images with surface potential simulation based on Thomas-Fermi approximation. This is the first step to allow 3D reconstruction of many-dopant arrangements from 2D KPFM images.

Keywords: KPMF · AFM · dopant · detection · transistor

1 Introduction

In the era of continuous miniaturization of electronic devices, silicon technology is approaching the stage where further development of conventional devices, like field-effect transistors (FET), is significantly hindered by random dopant fluctuation [1]. To reach beyond the limitations of present devices, new concept using dopant itself as a building block was recently introduced, namely the single-dopant transistor [2]. This approach is based on single-electron tunneling through individual dopants working as Quantum Dots (QDs) in the Field-Effect Transistor (FET) channel. For such devices number and position of dopants in the channel is essential. However, most single-dopant transistors characterized so far were doped in purely random manner by diffusion [2, 3]. Only few reports introduced devices with more controlled dopant arrangement (by single-ion implantation [4] or STM manipulation [5]). However, the number and position of embedded dopants in the final devices is not fully confirmed by these methods. Moreover, the commonly used electrical characterization (I-V), in fact, describes only the electron transport properties and does not allow clear resolving of internal structure of a transistor channel. Therefore, full characterization of devices based on dopants requires new and more direct investigation methods.

R. Jabłoński and T. Brezina (eds.), *Advanced Mechatronics Solutions*,
Advances in Intelligent Systems and Computing 393,
DOI 10.1007/978-3-319-23923-1_57

The Kelvin Probe Force Microscope (KPFM) was proposed for dopant detection in [6, 7, 8] as a tool capable of resolving individual donors embedded in Si FET channel and later deep-level dopants in 2D *pn* junctions [9]. Although KPFM allows detection of individual dopants embedded in the Si channel (up to 5-10nm deep, it provides only information about lateral position. The vertical position (Z coordinate) can be extracted from the potential depth measured by KPFM. However, metrological standard as a reference is needed. Until now, there is no measurement method allowing high precision detection of vertical dopant position in materials thicker than few atomic layers [10]. Therefore theoretical predictions have to be used as a reference for vertical position extraction from KPFM potentials. For this purpose we propose surface potential simulation using Thomas-Fermi approximation allowing calculation of potential induced by donor's distributions known *a priori*. Similar approach was used to simulate potentials in delta-doped devices [11].

In this work we describe the KPFM setup designed for detection of dopants in FET channel. We propose a method for defining the dopant position in depth by using comparison between KPFM images with surface potential simulation - simply because there is no any other standard which can be used as reference. This is an important step to allow full 3D reconstruction of many-dopant arrangements from 2D KPFM images.

2 KPFM setup

Originally the KPFM was proposed by M. Nonnenmacher [12] as Atomic Force Microscope (AFM) employing the principal of Kelvin probe. Therefore, KPFM allows also determination of surface topography. The schematic illustration of KPFM

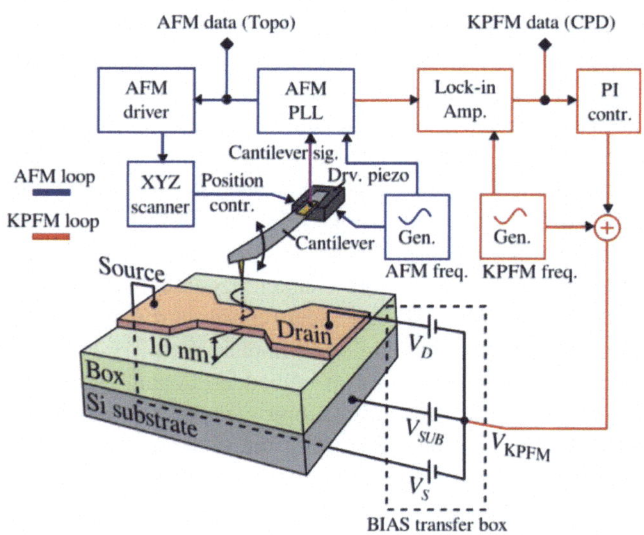

Fig. 1. The schematic illustration of KPFM setup and additional circuitry allowing depletion of FET channel (bias transfer box); FET structure is also schematically shown

setup is shown in Fig.1. Measurements are done in high vacuum (~ 4×10^{-9} Torr) and room temperature (296 K). The KPFM working principle is based on detection of long range electrostatic forces by the piezo-resistive cantilever tip scanning in proximity of the sample surface. The constant distance between the cantilever and the sample is maintained by AFM (AFM topography detection loop is marked blue in Fig.1). The cantilever oscillations are first excited by driving the piezo with specific AFM frequency matching the mechanical resonance frequency of a cantilever. In proximity of sample surface the attractive forces are acting on a cantilever. Due to tip-sample interactions the cantilever bends and its resonant frequency is changing. This change is detected by Phase-Locked Loop (PLL) and recorded as a topography data. Simultaneously the "topo" signal is given as a feedback to AFM driver and cantilever's Z position is readjusted to keep the same tip-sample distance. While the

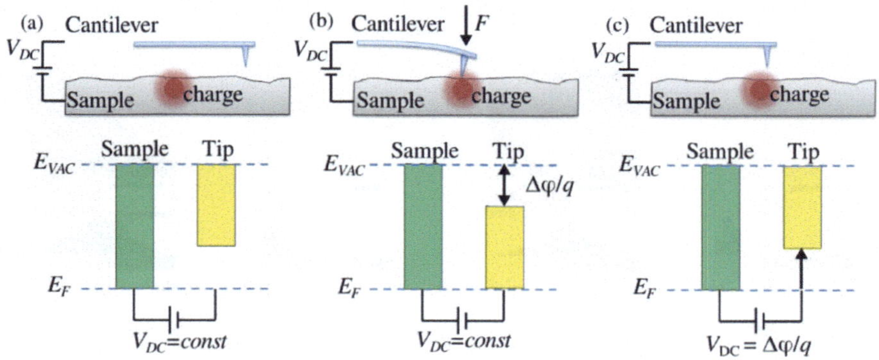

Fig.2. Schematic explanation of KPFM working principle; (a) System in equilibrium; (b) Electrostatic force (F) builds up; (c) Cancelation of F by V_{DS} bias application.

cantilever is scanning the surface in above manner it also detects the electrostatic forces which arise due to work function difference between tip and sample, as shown in Fig.2. When the cantilever and the sample are in equilibrium, the V_{DC} is applied between them to compensate the initial work function difference between tip and sample. The attractive electrostatic force (F) builds up if some local change in work function occurs, as in Fig.2 (b). This force is cancelled by readjustment of V_{DC} which corresponds to a local Contact Potential Difference (CDP). To allow continuous detection of CPD, V_{DC} is modulated by specific KPFM bias frequency resulting in total tip-sample bias (V_{KPMF}). Electrostatic forces due to V_{KPMF} and local CPD cause deflection of cantilever which is detected first by AFM PLL, then directed to KPFM loop (marked red in Fig.1) and demodulated in Lock-in amplifier. This way KPFM data is obtained. The important difference between conventional KPFM setup and one presented here is the possibility of additional biasing of the sample (in this case the FET transistor), realized by bias transfer box. For transistor V_{KPMF} works as a virtual ground. The FET is realized as a thin constriction (10 nm thick channel) between two wider pads of silicon - working as source (S) and drain (D) electrodes. The channel is doped with phosphorus in concentration of $N_D \approx 1\times10^{18}$ cm^{-3}. The S/D electrodes are grounded (virtually) while the back gate bias (V_{SUB}) is applied to the substrate (also in reference to V_{KPMF}). Such circuitry allows application of negative biases and depletion of free electrons from the channel leaving mostly ionized P-donors. Potential image induced by these P-donors is then detected by KPFM.

3 Donor's vertical position reconstruction

Figure 3(a) shows exemplary potential image measured for different V_{SUB}. It is clearly visible that when negative bias is applied the distinct circular potential dips appear (darker potential areas in average size of ~10-20 nm and depth of ~20-40 mV). These can be ascribed to individual donors [9] and their lateral position can be easily resolved. The vertical (Z coordinate) position of P-donors embedded in the channel can be extracted from the potential depth measured by KPFM. For this purpose, we refer to surface potential simulation as a theoretical metrological standard.

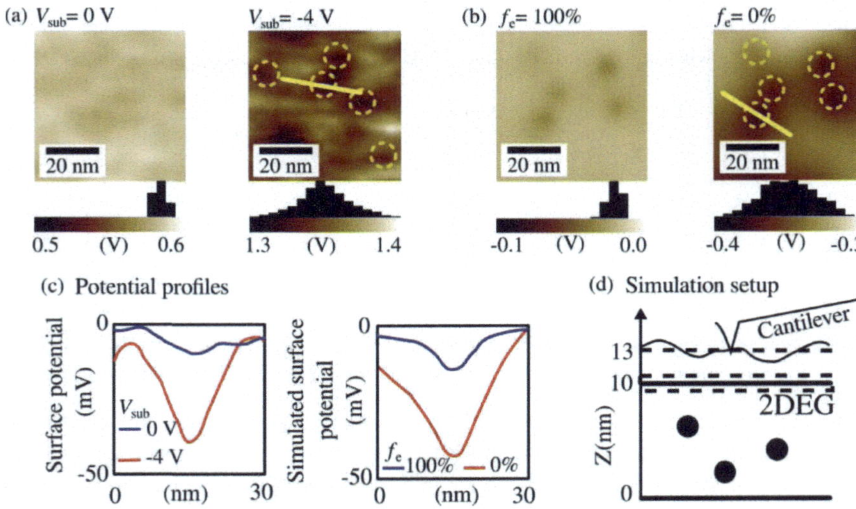

Fig.3. (a) Example of KPFM images acquired under different biases V_{SUB}. (b) Exemplary results of simulation for different electron coverage fraction f_e (c) Comparison between measured and simulated potential profiles [taken along solid yellow line from (a) and (b)] for single P-donor. (d) Schematic illustration of simulation setup. The surface potential scale bars along with histograms are shown below every potential map.

The schematic simulation setup is shown in Fig. 3(d). In our model, donors are randomly distributed in a 3D Si sheet. We first calculate bare donor potential by summation of Coulomb potentials from each donor, assuming periodic boundaries. The electrons are confined in two dimensional electron gas (2DEG) on the surface of the Si sheet. Density of 2DEG is calculated by solving 2D Thomas-Fermi equation in low temperature. The total potential is obtained by solving Fourier-transformed Poisson equation. The potential sensed by cantilever is estimated at level of 3 nm from the sample surface (it includes ~2 nm protective SiO_2 layer and ~1nm tip-sample separation). The system is solved self-consistently by relaxation assuming 0.2 nm grid. The electron coverage fraction parameter (f_e) is equivalent to experimental V_{SUB}, therefore effect of screening electrons can be included.

The simulated potential images exhibit similar features as measured ones - marked by yellow circles in Fig. 3(a) and Fig. 3(b). The comparison between potential profiles induced by individual donors is shown in Fig. 3(c). It can be seen that simulation is in good agreement with experimental results. The experimental results, Fig. 3(a), show features with different intensities. These intensities are related to vertical position of a donor in most cases, although other factors like donor-donor

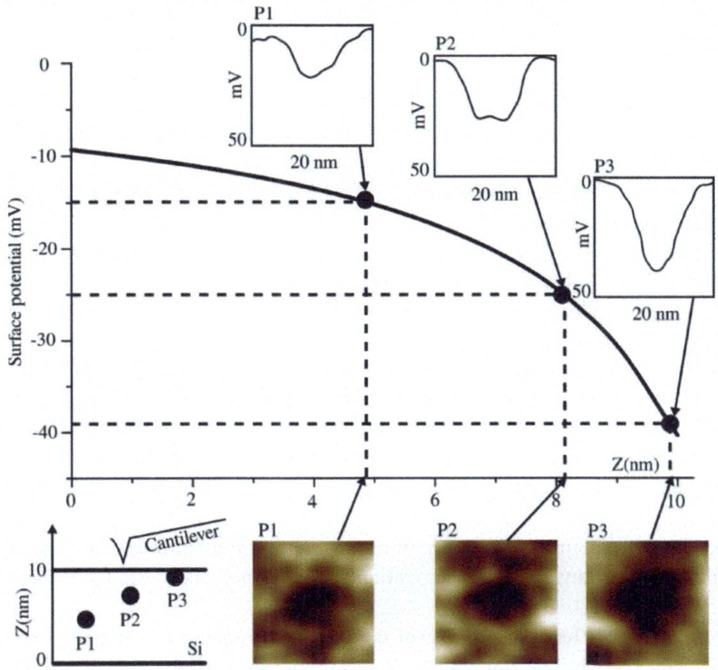

Fig.4. Simulated dependence between surface potential and vertical donor position, correlated with experimentally measured features (ascribed to individual donors) showing different intensities. Potential landscapes and corresponding profiles are shown for three different donors (P1, P2, P3). The inset shows schematic arrangement of the donors.

interaction or measurement noise cannot be fully excluded. This relation can be theoretically resolved by simulation of surface potential vs. vertical donor position (Z) dependence (as shown in Fig.4). Features measured in experiment for three different donors (P1, P2, P3) are assigned, as an example, to the points on the curve in Fig.4. This way vertical donor position can be resolved.

4 Conclusions

We presented the method for reconstruction of vertical position of donors embedded in randomly doped FET. By using KPFM with modified feedback circuit we measured potentials induced by ionized P-donors. Experimental results were referred to surface potential simulation used as a theoretical metrological reference standard. This is first approach in such reconstruction and it still requires deep analysis of influential factors. However, the most important factors, i.e. free electrons screening and tip-sample distance are included allowing successful correlation between simulation and experiment. This is important first step towards full 3D reconstruction of many-dopant arrangements from 2D KPFM images.

References

1. Asenov, A.: Random dopant induced threshold voltage lowering and fluctuations in Sub-0.1 m MOSFET's: a 3-D "atomistic", simulation study. IEEE Trans. Electron Devices 45, 2505 (1998)..
2. Sellier, H. et al.: Transport Spectroscopy of a Single Dopant in a Gated Silicon Nanowire. Phys. Rev. Lett. 97, 206805 (2006)
3. Tabe, M. et al.: Single-electron transport through single dopants in a dopant-rich environment. Phys. Rev. Lett. 105, 0160803 (2010)
4. Prati, E. et al.: Anderson-Mott transition in arrays of a few dopant atoms in a silicon transistor. Nature Nanotechnol. 7, 443 (2012)
5. Fuechsle, M. et al.: A single-atom transistor. Nature Nanotechnol. 7, 242 (2012)
6. Ligowski, M. et al.: Observation of individual dopants in a thin silicon layer by low temperature Kelvin Probe Force Microscope Appl. Phys. Lett. 93, 142101 (2008)
7. Anwar, M., et al.: Effect of electron injection into phosphorus donors in silicon-on-insulator channel observed by Kelvin probe force microscopy Appl. Phys. Lett. 99, 213101 (2011)
8. Tyszka, K. et al.: Comparative study of donor-induced quantum dots in Si nano-channels by single-electron transport characterization and Kelvin probe force microscopy. J. Appl. Phys. 117, 244307 (2015)
9. Nowak, R. et al.: Effects of deep-level dopants on the electronic potential of thin Si pn junctions observed by Kelvin probe force microscope. Appl. Phys. Lett. 102, 083109 (2013)
10. Hwang, J. et al.: Three-Dimensional Imaging of Individual Dopant Atoms in $SrTiO_3$, Phys. Rev. Lett. 111, 266101 (2013)
11. Nixon, J.A., Davies, J.H.: Potential fluctuations in heterostructure device. Phys. Rev. B 41, 7929 (1990)
12. Nonnenmacher, M. et al.: Kelvin probe force microscopy Appl. Phys. Lett. 58, 2921 (1991)

Acknowledgments

We thank Prof. Michiharu Tabe (Research Institute of Electronics, Shizuoka University) who provided insight and expertise that greatly assisted the research.

Study of the clock signal quality impact on the PWM signal generated in FPGA

Robert Ugodziński[1], Michał Nowicki[1], Roman Szewczyk[2]

[1)] Industrial Research Institute for Automation and Measurements,
Al. Jerozolimskie 202, 02-486 Warsaw, Poland
rugodzinski@piap.pl, mnowicki@piap.pl
[2)] Institute of Metrology end Biomedical Engineering, Warsaw University of Technology,
A. Boboli 8, 02-525 Warsaw, Poland
szewczyk@mchtr.pw.edu.pl

Abstract. In the paper the clock signal quality impact on the PWM signal generated in FPGA is investigated. The study used a Spartan-6 system from Xilinx. In the first case the clock signal was generated by an external PLL, and then raised to a high frequency through an internal PLL. As an accurate source the 'Fox XPRESSO Crystal Oscillator' precision oscillator was used. Additionally, the PWM signal generated in FPGA was compared with the signal generated by the Atmega8 microcontroller.

Keywords: FPGA, PWM, PLL, precision crystal oscillator, Xilinx, Spartan-6

1 Introduction

The PWM signal is commonly used as a source of variable power [1,2,3]. This enables us as a simple way to control the actuator, providing him the required amount of power. Traditionally, this is done through a digital-to-analog circuits and power amplifiers [4]. However, the analog signal generation is highly problematic, and there is movement away from these solutions towards a fully digital signal generation.

However, conventional solutions in terms of generation of the PWM signal are often not precise enough. And while in the LED control circuits it does not really matter, in the precise systems signal accuracy on the order of picoseconds is needed.

This article presents a comparison of the PWM signal generated by the FPGA depending on the source of the clock signal. Additionally, the PWM signal generated by the popular Atmega8 microcontroller was tested for comparative purposes.

R. Jabłoński and T. Brezina (eds.), *Advanced Mechatronics Solutions*,
Advances in Intelligent Systems and Computing 393,
DOI 10.1007/978-3-319-23923-1_58

2 Test equipment

2.1 FPGA with development board

The Spartan-6 FPGA with XC6SLX9-CSG324 catalog number was used during the research [5], mount on the 'Spartan-6 FPGA LX9 MicroBoard' development board.

This board is factory-equipped with a 27MHz quartz and an external PLL, which is factory programmed for the synthesis of 100 MHz frequency signal, which is then raised to a of 0.5 GHz by the internal PLL.

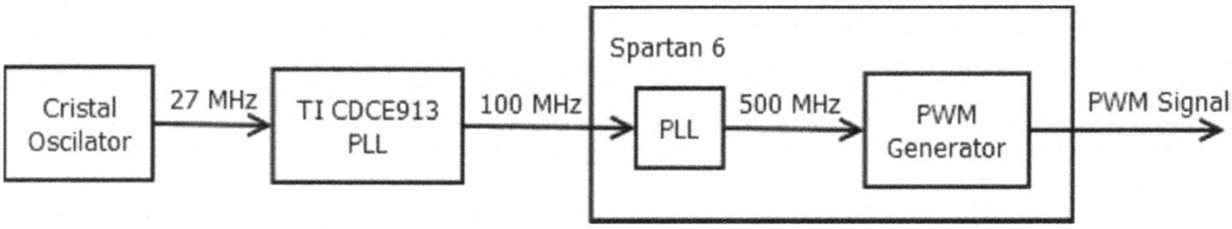

Fig. 1. Poor conditioned PWM signal generation.

Importantly, on the board a place for the reserve oscillator was allocated, in place of which oscillator model produced by Fox Electronics part number FXO-HC736R-125 has been soldered [6]. It generates a clock signal with a frequency of 125 MHz with an accuracy of ± 25 ppm. Then, as in the previous case it will be raised to a frequency of 0.5GHz.

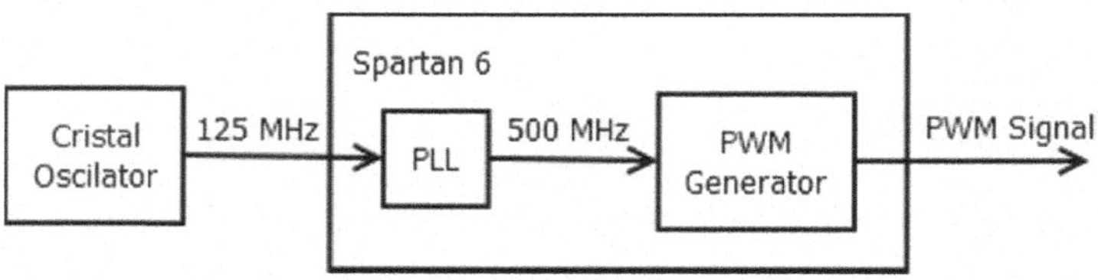

Fig. 2. Good conditioned PWM signal generation.

2.2 Atmega8

The Atmega8 microcontroller was utilized for the comparative purposes [8]. It has been configured to use the maximum available clock frequency, ie. 8 MHz. This allows us to examine the quality of the signal using all of the systems possibilities.

The signal is software generated in the „Fast PWM" mode, with a resolution of 8 bits. This configuration allows the generation of 31,25kHz frequency signal with a scale interval equal to 125ns.

2.3 Oscilloscope

Since the work involves the study of signals with very high resolution, adequately accurate oscilloscope is required. The Tektronix TDS5104B oscilloscope was used for this purpose [7]. It is characterized by a 1 GHz bandwidth and 5GS/s maximum sample rate.

It has also extensive software to help analyze the signals, among which there is FastAcq mode which allows to average waveforms, or Jitting Analyzer allows for automatic and accurate measurements of clock signal errors.

3 Results

Circuit	Atmega8	FPGA with non-precise clock	FPGA with precise clock
Nominal values			
Reference clock frequency	8 MHz	0,5 GHz	
PWM signal frequency	31,25 kHz	~ 30,518 kHz	
PWM signal period	32 μs	32,768 μs	
Scale interval (period)	125 ns	2 ns	
Resolution	8 bit	14 bit	
Scale interval (in %)	390 m%	6,10 m%	
The experimental values - from the primary application			
The ambiguity of the slope pulse	16 ns	300 ns	> 150 ps
Pulse rise time	3.4 ns	~ 3 ns	~ 3 ns
Pulse fall time	3.6 ns	~ 3 ns	~ 3 ns
The experimental values – from the Tektronix Jitter Analysis software			
PWM signal period	31,149 μs	32,771 μs	32,768 μs
Absolute error of PWM signal period	851 ns	2,611 ns	62 ps
The standard deviation of the PWM signal period	13,450 ns	50,065 ns	41 ps
Maximum PWM signal period	31,196 μs	32,848 μs	32,7681 μs
Minimum PWM signal period	31,112 μs	32,691 μs	32,7678 μs
Jitter peak-to-peak	64,379 ns	156,333 ns	264 ps
The calculable values[1]			
Maximum possible resolution	8.9 bitów	7.7 bitów	16.9 bitów

Table 1. Summary of the results

The table above presents results obtained from the basic oscilloscope application and advanced Tektronix Jitter Analysis tool, to examine the quality of the generated clock signals. The research was carried out by setting the duty cycle for 50%.

[1] According to Tektronix Jitter Analysis.

As it was expected the biggest period stability error could be observed in the case of the PWM signal generation by the FPGAs using less accurate clock. The standard deviation amounted up to 50 ns. This is more than thousand times more than using version with accurate clock. It also means that despite the fact that the scale interval is 2 ns, we are not able to provide such a resolution. Atmega8 system errors are smaller compared with the FPGA version with the inaccurate clock, but much larger than the version with accurate clock. The standard deviation of the period in this case is 13.5 ns. But from the available interval point of view (125 ns), this error is not as significant.

'Peak-to-peak' errors are also an important parameter. This allows us to observe what is the biggest discrepancy of the signal period in the registered PWM frames population. And as in the previous discussed parameters the widest period spread was recorded for the inaccurate FPGA clock. It amounted to up to 156 ns, which compared with the version with accurate clock is 500 times greater than 264 ps. The Atmega8 error was 64 ns.

Based on this data, the maximum possible resolution for each individual case can be calculated. The formula for obtaining the maximum possible resolution for given errors is as follows:

$$Max.\,avaible\ resolution = log_2\left(\frac{Period\ of\ PWM\ signal}{Peak{-}to{-}peak\ period\ error}\right) \quad (1)$$

Using this relationship it can be seen that the Atmega8, if only had a higher clock timing, could generate a signal nearly one bit more accurate. On the other hand, for the Spartan 6 system the calculated resolution is nearly two times less than nominal one (7,7 bit vs 14 bit nominal). For a system with a precise clock applied the maximum resolution that could be obtained is 16,9 bit.

One can see that it's a lot more than the control clock allows. In this way, if the same FPGA can work at higher than 500 MHz frequencies, the reference clock used would still be a proper reference signal source for the internal PLL circuit, enabling the generation of PWM signal with higher resolution for a given period.

Another investigated parameter specifying the quality of the generated PWM signal is the falling edge ambiguity. With this parameter, we can see the scattering between different implementations, in regards of the stability of the slope in time, and thus also the accuracy of the energy transfer in the a single frame period of the PWM. The slope ambiguity was equal to 16 ns for the Atmega8 system. This is a good result for this system, given that the scale interval is 125 ns. The biggest error can be seen in the case of the use of inaccurate clock in FPGA. The ambiguity in this case is as high as 300 ns, which is equal to 150 scale intervals. The smallest error was observed for the precise clock FPGA version. Then the error is almost negligible and within the limits of accuracy of the oscilloscope. It is less than 150 ps.

4 Conclusions

Summarizing the results of the research, a big advantage of the FPGA systems over other solutions can be seen. As a result, the user has the possibility to generate a PWM signal with very high resolution. But an appropriate quality of the clock signal controlling the system is crucial, because as it has been shown it has a huge impact on the accuracy of the generated signal.

The developed solution has been tested, inter alia, to control the electromagnetic coils in ultra-high resolution analytical balances [9].

RADWAG Company is kindly acknowledged for their help.

5 Acknowledgements

The research presented in this work has been supported by the European Union within the European Regional Development Fund program no. POIG.01.03.01-14-086/12.

6 References

1. Barr, Michael.: Pulse width modulation. Embedded Systems Programming, 2001, 14.10: 103-104.
2. Holmes, D. Grahame; LIPO, Thomas A.: Pulse width modulation for power converters: principles and practice. John Wiley & Sons, 2003.
3. Arrillaga, J.; Liu, Y. H.; Watson, N. R.: Pulse Width Modulation. Flexible Power Transmission: The HVDC Options, 127-140.
4. Magrath, A. J.; Clark, I. G.; Sandler, M. B.: Design and implementation of a FPGA sigma-delta power DAC. In: Signal Processing Systems, 1997. SIPS 97-Design and Implementation., 1997 IEEE Workshop on. IEEE, 1997. p. 511-521.
5. Xilinx, „Spartan-6 Family Overview," [Online].
6. Fox Electronics, „HCMOS 7 x 5mm 3.3V Oscillator Datasheet" [Online]. Available: http://www.foxonline.com/pdfs/FXO_HC73.pdf.
7. Tektronix, „Digital Phosphor Oscilloscope TDS5034B, TDS5054B, TDS5104B Data Sheet".
8. Atmel, „ATmega8 Datasheet" [Online]. Available: http://www.atmel.com/.
9. Ugodziński, R.; Szewczyk, R.: Determination of Parameters of Second Order Integration Model for Weighing Scales. In: Mechatronics 2013. Springer International Publishing, 2014. p. 161-167.

Correction of the influence of not ideal geometric profile on the conductivity of reference cell

Aleksander A. Mikhal[1], Zygmunt L. Warsza[2], Vladimir. G. Gavrylkin[3]

[1)] Institute of Electrodynamics, National Academy of Science of Ukraine (NANU), Peremogy av. 56, 03680 Kyiv-57
a_mikhal@ukr.net

[2)] Industrial Research Institute of Automation and Measurements, PIAP, Al. Jerozolimskie 202, 02-486 Warszawa, Poland
zlw@op.pl

[3)] Ukrmetrteststandard, Metrologichna Str. 4, 03143 Kiev, Ukraine
vgavrilkin@ukrcsm.kiev.ua

Abstract. The principle of operation and metrological model of the Ukrainian primary standard of electrolytic conductivity (EC) is presented. The equations for calculating the cell constant and the budget of unknown errors for estimation the uncertainty B-type are given. Strategies: how to minimize components of this uncertainty are proposed and obtained results, verified by Key Comparisons, are given.

Keywords: conductivity, primary cell, geometric errors, uncertainty.

1 Introduction

The electrical conductivity of solutions is measured in chemistry, biology, medicine, environment and see monitoring, electrotechnology industry and many other fields. Reliable and comparable results of these measurements are obtained due creation the national metrological system with the traceability, calibration and international comparison regulations. Several countries, including Ukraine, have established national standards of the electrolytic conductivity (EC). In the world practice a different ways to implement this standard are used, however, the principle of its operation is almost the same - the “absolute", i.e. direct method of reproduction of a unit of physical quantity [1-3]. This “absolute” method is based on the measurement of the resistance of liquid column in the conductivity cell and on the calculation of the EC from the known length and cross-sectional area of this column.

The EC primary standard of Ukraine as basic components includes the four-electrode cell with calculated constant K, special high precision conductivity AC bridge, thermostat with temperature control at 25°C and precision digital temperature meter. Value k of EC is determined as

$$k = GK\left(1 + \alpha t_{25}\right) \tag{1}$$

where: G is the conductance of liquid column; K is the cell constant calculated from

R. Jabłoński and T. Brezina (eds.), *Advanced Mechatronics Solutions*,
Advances in Intelligent Systems and Computing 393,
DOI 10.1007/978-3-319-23923-1_59

its geometric dimensions; α is the temperature coefficient for conductivity of the solution; t_{25} is the temperature deviation from 25°C.

On the example of this EC standard the method of estimation and minimizing the uncertainty of type B of the measurement of the electric conductivity EC of electrolytes is presented.

2 Measurement system for reproduction of the EC unit

The instrumental part of EC standard is a information-measuring system and it consists of several subsystems: SSR EC – subsystem of EC reproduction, SST EC – subsystem of EC transmission, SST MC – subsystem of thermo stabilization, subsystem of the preparation of solutions, subsystem of control and processing of the results of measurement.

The main element of the EC standard is the subsystem SSR EC (ВЭП). Their simplified functional diagram is shown in Figure 1.

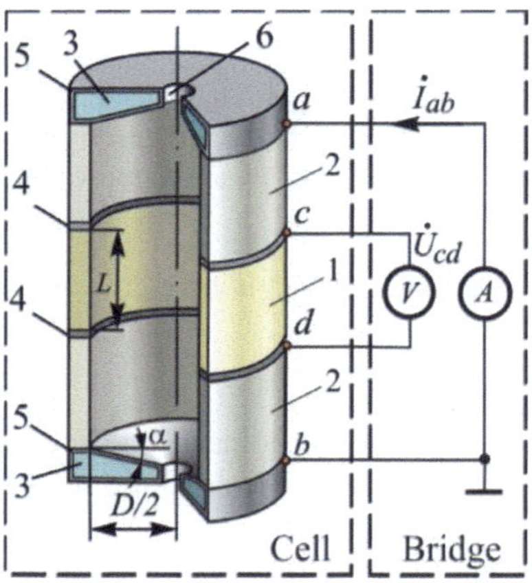

Fig. 1. Sketch of the primary conductivity cell and its connections

The main element of the conductivity sensor is the tube with internal diameter D, which is filled with electrolyte solution. Typically, this is a solution of potassium chloride. The tube is used to fix the geometry of the liquid conductor and it consists of three parts. The central part of the tube 1 has length L and two side portions 2 have the same length l. The ends of the central portion of the tube 1 are coated with circular potential electrodes 4. Their width corresponds to the tube wall thickness. At the edges of the tube two discs 3 are fixed. Inner surface of the discs is coated with metallic film 5 and are current electrodes. Discs 3 have central holes 6 of the diameter d, which serve for filling liquid. Inner disc surface has the form of cone with an angle α. Such configuration is intended to facilitate the removal of air bubbles when filling cell with the liquid. The tube and the discs are made of quartz glass which has good insulating properties, temporal stability and minimum coefficient of thermal expansion. The cell electrodes are made from platinum which has a minimum polarization effect. These electrodes, in its four points a, b, c and d, are connected to the AC bridge (left side of Fig. 1), which measures the conductance G of liquid column. This is not a standard automatic AC bridge, but it is designed specifically for the EC standard. The admittance components of the parallel two-element equivalent circuit of the conductivity cell are measured by this bridge. The voltage comparison circuit is also used in it. That allows to eliminate the influence of impedances which occur on the border of electrolyte with current and potential electrodes of the conductivity cell.

3 Budget of systematic errors

One of the major metrological parameters of this standard is the uncertainty u_k of the EC unit reproduction (here referred only as type B standard uncertainty). Upon absence of correlation between the parameters of the equation (1), the combined standard uncertainty u_k is calculated in accordance with Guide GUM [11]:

$$u_k = \sqrt{\left(\frac{\partial k}{\partial G}u_G\right)^2 + \left(\frac{\partial k}{\partial K}u_K\right)^2 + \left(\frac{\partial k}{\partial t}u_t\right)^2} \tag{2}$$

where: u_G, u_t are standard uncertainties of measurement the conductance G and temperature t correspondingly and u_K uncertainty of the constant K calculation.

Uncertainties u_G, u_K, and u_t will depend on errors from sets of various factors $x \in \{x_i, x_j, x_n\}$. The major factors x are metrological characteristics of measuring devices and instruments, errors of calibration standards, methodological errors of calculation models, manufacturing accuracy of calculated elements, environmental conditions, parameters of power supply, etc. Usage of relative values on the stage of preliminary evaluation gives us possibility to compare contribution of each error component and to estimate a reasonable amount of errors upon budget formation. Thus, to estimate the uncertainty u_k^* from the budget of the limited relative errors $\delta_{iG} = \Delta G_i / G$ of conductance G measurement, limited relative errors $\delta_{jK} = \Delta K_i / K$ of the constant K calculations and δ_{nt} of the maximum temperature increment $\Delta t\alpha$, we obtain the following formula:

$$u_k^* = C_P \sqrt{\sum_{i=1}^{N} (A_{iG}\delta_{iG})^2 + \sum_{j=1}^{M} (A_{jK}\delta_{jK})^2 + \sum_{n=1}^{H} (A_{nt}\delta_{nt})^2} \tag{3}$$

where: the limited errors δ_{iG}, δ_{jK}, δ_{nt} influence coefficients

$$C_{iG}\frac{\partial G(x_i)}{\partial x_i} \equiv A_{iG},\quad C_{jK}\frac{\partial K(x_j)}{\partial x_j} \equiv A_{jK},\quad C_{nt}\frac{\partial t(x_n)}{\partial x_n} \equiv A_{nt} \tag{4}$$

Coefficients C_{iG}, C_{jK}, C_{nt} depend on the probability distributions of the relevant errors.

Type B uncertainty of the measurement system including several instruments, almost entirely dependent on estimation of unknown values of systematic errors. Therefore, a complete set or budget of systematic errors of the EC standard is basic to assess the quality and adequacy of the mathematical model describing the accuracy of any standard. The budget "tree" of Ukrainian EC standard is shown in Figure 2.

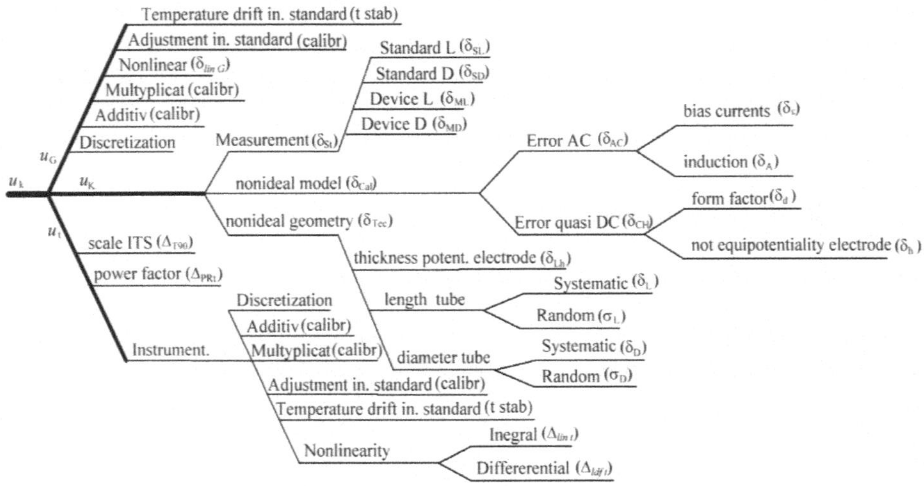

Fig. 2. Tree of the uncertainty u_k budget of the electrolytic conductivity k

In the equation (1), two parameters only (conductance G and temperature increment nt) are measured directly with instruments. In this case the main methods of reducing uncertainty is well-known technique of calibration and temperature stabilization of these instruments. The third parameter (cell conductivity constant K) is calculated. Therefore, the estimation of its uncertainty needs the increased attention. The cell constant is determined by calculating the ratio of the length of the tube 1 to the cross-section area. However, this definition is true for an idealized object of measurements with uniform distribution of current flow lines. Distortion of such lines will be due to: the presence of air holes in the solution filling; the form of current electrodes and the presence of potential electrodes; non-ideal profile of the inner tube 1 in Fig. 1. Therefore, the calculations constant K will have errors.

The error budget for the constant calculation of the cell, shown in Fig. 1, can be written as an set of unknown systematic relative errors:

$$U_K \in \{\delta_{St}, \delta_{Cal}, \delta_{Geom}(\delta_{Tec}, \delta_{PE})\} \tag{5}$$

where: δ_{St} is an error due to the accuracy of measurement standards and measuring instruments to determine the length and diameter of the tube; δ_{Cal} is an error due to the deviation of the calculation model for the cell constant in real conditions relative to the idealized model; δ_{Geom} is an error in assessment of geometrical dimensions.

Each argument of set (5) has several components. Let us consider them in detail.

Minimizing of error δ_{St} is limited by the level of metrological assurance for measurements of tube length and diameter. It is defined by the metrological parameters of the standards and instruments for length measurement.

An error δ_{Cal} has two components: an error δ_{Tec} due to alternating current measurement and an error δ_{PE} due to discontinuity of an electric field in the cell because of its finite dimensions and design features. Analysis of all δ_{Cal} components is described in detail in [5-7].

Geometric error δ_{Geom} also has two components: δ_{Tec} is an error due to manufacturing technology for the tube sections and their assemblage; δ_{PE} is an error due to the presence of potential electrodes. The latter component of the error depends on the finite thickness of potential electrodes and on changing position of a singular point of potential electrodes upon assemblage. This component is related to the calculation of the electric field inside the cell. It will be considered in other papers. In this paper, we examine the component of an error δ_{Tec}. This error is due to the deviation of actual profile of the inner surface of the tube 1 (Fig. 1) from the ideal profile of the tube. The latter one is presented as rectangle along the longitudinal section and as circle in cross-section.

It should be mentioned that the cost of tube production from a monolithic quartz crystal is extremely high. As a rule, tubes are manufactured from work pieces (preform) which undergo precision machining. If precision machining of the inner surface is too deep, the mechanical resistance of the tube will reduce significantly. Tubes of less than 1mm in thickness will crack (fracture) under elastic forces (adhesive polymerization, temperature differences). Therefore, grinding of the work piece inner profile should be of minimum depth. On the other side, the work piece inner surface can have wedge-like cracks which are parallel to the axis of the work piece. These cracks are due to manufacturing techniques of work piece production and depend on the quality of nozzles through which the work piece is pulled itself. Therefore, due to the lack of deep machining of tubes, we can observe deviation from circle in cross-section along the entire profile. The second reason of non-ideal profile may be the precession of the grinding tool. During processing, quality control of the tube is practically impossible. After final grinding, the tube profile may differ from the ideal rectangle. As a result there is a systematic error that cannot be completely excluded. Let us consider in detail this component of error.

4 Manufacturing errors

To determine the actual profile of the tube, its diameter and length should be measured according to the following algorithm. Uniformly by circle, p measurements of tube length are made in different directions L. Conventionally, uniformly along the length, the tube is divided into m sections. To define the diameter, its n measurements are made in the cross-section of each part of the tube in different directions. As a result, we obtain $n \times m$ measurements of the tube diameter and p values of the tube length. The constant can be determined from the results of measurements through average values of diameter D_{av} and tube length L_{av}.

$$K = \frac{4L_{av}}{\pi D_{av}^2} \tag{6}$$

Modern technologies for processing of quartz glass are those that it is much easier to manufacture a tube with stable length value than a tube with stable internal diameter. From the experimental data we observe distortions of the internal profile of two types. The first one is a deviation of the tube profile from rectangle along the longitudinal section due to precession of the grinding tool. The second one is a deviation of the tube profile from circle in cross-section due to presence of the wedge-like cracks on the inner surface of work pieces.

The geometric dimensions of the actually manufactured tube can be measured much more accurately than the error of profile. Measurement of the parameters is generated with a device with LSB 0.1 μm. A random component of the device has not exceeded 0.5 LSB, while deviations of the actual profile from the ideal one have exceeded LSB of the device tenfold. Therefore, the following expression can be taken as a metrological model for the constant calculation:

$$K = K_0(1+\delta_K), \qquad \sigma_K = \sqrt{2\sigma_D^2 + \sigma_L^2} \tag{7a,b}$$

where: K_0 is the true value of the constant which is determined by the actual profile of the tube inner surface; δ_K is a systematic relative error of the constant calculation; σ_K, σ_D and σ_L are standard deviations of the mean cell constant, diameter and length of the tube correspondingly. Then, an error due to manufacturing δ_{Tec} has two components: systematic δ_K and random σ_K. Then the mean value of tube diameter has the standard deviation (SD). The diameter and its SD for each i-th section of the tube are determined from the expressions:

$$D_{av} = \frac{\sum_{i=1}^{m} D_{i,av}}{m} = \frac{\sum_{i=1}^{m}\sum_{j=1}^{n} D_{ij}}{m \cdot n} \tag{8}$$

$$\sigma_{iD} = \sqrt{\frac{1}{n(n-1)}\sum_{j}^{n}\left(D_{ij} - D_{i,av}\right)^2} \tag{9}$$

where: $D_{i,av}$ - mean diameter in the i-th section of the tube, D_{ij} is j-th mean diameter in the i-th section of the tube.

The value SD from m section can be expressed as:

$$\sigma_D = \sqrt{\frac{1}{m(m-1)}\sum_{i}^{m}\sigma_{iD}^2} \tag{10}$$

4.1. *Error in the longitudinal section*

The measuring results for the mean diameter of one of the sections of the tubes upon n=8 and m=10 are shown in Fig. 3.

Represented data indicate that the average diameters in sections 1 and 6 differ almost by 20 μm. In general, the profile of the tube internal section can be expressed through arbitrary function $D(x)$. The measuring results for average diameters along the length of the tube (Fig. 3) show that this dependence has the clearly determinate character.

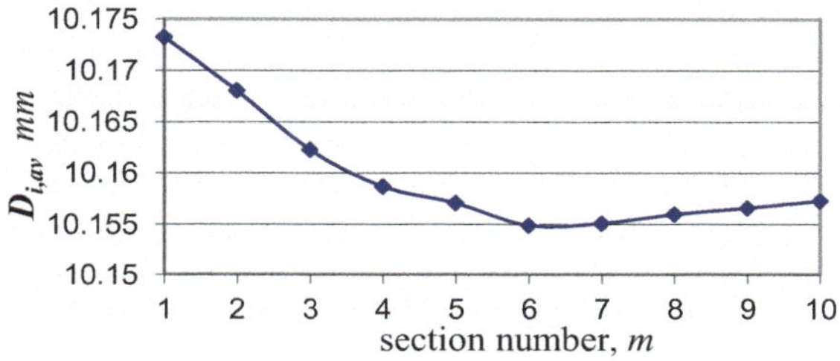

Fig. 3. Profile along the axis of the tube

The discrete nature of the data allows us to use linear interpolation for the function $D(x)$. The results are shown in the following formula:

$$K = \frac{4}{\pi}\int_0^L \frac{dx}{D(x)^2} = \frac{4}{\pi}\sum_{i=0}^{m}\int_0^{\Delta X_i} \frac{dx}{(ax+b)^2} \tag{11}$$

where: a and b are linear interpolation coefficients; ΔX_i is the length of the area between i-th and $i+1$ section.

Polynomial coefficients are expressible as:

$$a = \frac{D_{i+1} - D_i}{\Delta X_i}, \qquad b = D_i \tag{12a,b}$$

After simple transformations, the expression for calculating the constant with the proposed correction takes the form:

$$K = \frac{4}{\pi}\sum \frac{\Delta X_i}{D_{i,av} D_{i+1,av}} \tag{13}$$

If we use procedure for the classical averaging of the diameter measuring results, eq. (14) -(16), we receive significant rises in the standard deviation. Let us show on one figure the graphs for SDM of diameter measuring results without deterministic component and with deterministic component by using linear interpolation (Fig. 4).

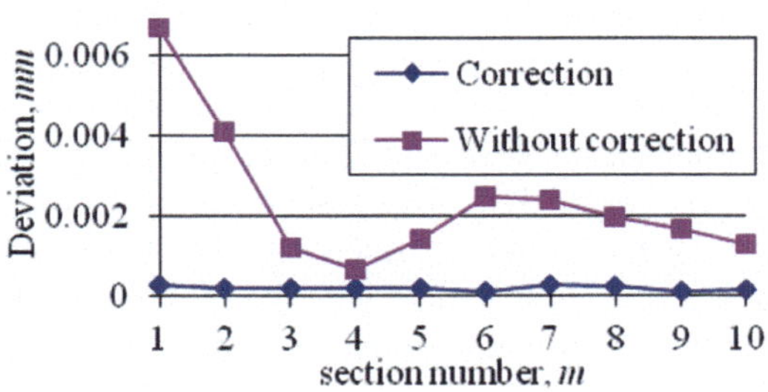

Fig. 4. Influence of linear interpolation on the level of SDM (σ_{iD})

As it can be seen from Fig. 4, SDM of the diameter measuring results, taking into account deterministic component, is close to almost one value and has a very small scatter of results compared with the case where the deterministic component is ignored and the average diameter value is calculated according to equation (8). As a result of correction, change of SDM along the cross-section is reduced by 10 -15 times. SDM of diameter with correction, calculated according to formula (10) is less than 0.0004 mm. This parameter without correction is 3 times larger. For the eq. (13), we obtain an error reduction by two times.

It should be noted that such correction method shifts the average value of the diameter. Thus, the constants calculated by the formulas (6) and (13) for the tube 1 in Fig.1, differ by 0.027%. This is the rate by which systematic relative errors δ_K (7a) differ when calculating the cell constant with and without correction.

4.2. *Error in the cross-section*

Let us consider another tube for which the values of mean radius in each of the sections are grouped along the virtually horizontal line. However, in each individual section the surface profile differs from the circle. Example of the function for the deviation from the mean D_{iav} (Fig. represented by the circle $0.5D_{3,av}$=4.569 mm) in one of the sections with m=3 is shown in Fig. 5.

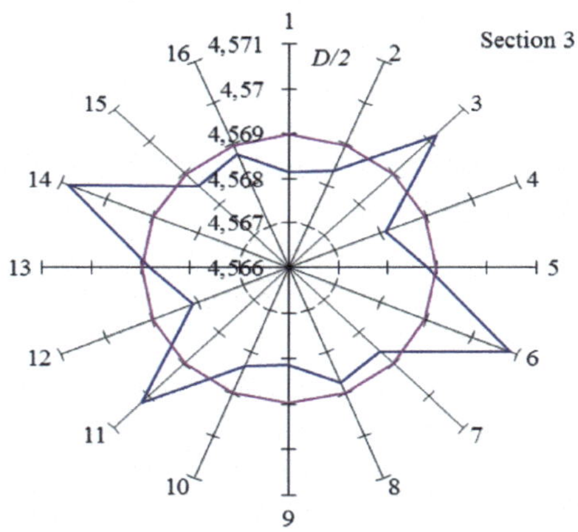

Fig. 5. Real profile - broken line (blue) across the axis of the tube for section m=3 and its equivalent polygon (red) of the average circle diameter $0{,}5D_{3,av}$=4.569 mm.

In all ten sections of m = (1 –10) we observe triangular run outs along the lines 3 -11 and 6 – 14 (Fig. 5). Such character of profile distortions allows using a method of the equivalent triangles. This method involves assessment of the effective area S_{ief} of the tube section and subsequent diameter corrections. The algorithm for calculation is to replace diameter values in section 3 and 6 with the values of mean diameter D_{av}, then to use standard formula for calculation of the basic mean diameter D_{bias} and basic area of the tube section S_{bias}.

$$S_{bias} = \pi(D_{bicr})^2/4 \tag{14}$$

Next, we calculate the correction value in each section separately. It is represented as areas of triangles:

$$S_i = c_i h_i/2 \tag{15}$$

where: c_i is the base of assumed triangle in direction 3, 11 or 6, 14 (Fig. 5); and h_i is the height of assumed triangle in direction 3, 11 or 6, 14 (Fig. 5).

We take into account the influence of the deterministic component by forming the effective area of each section. It is expressed as:

$$S_{ief} = S_{bicr} + 2\sum_{i=3,6} S_i = S_{bicr} + 2\sum_{i=3,6} c_i h_i \tag{16}$$

According to the following formula, we calculate the corrected D_{icor} which is put in the equation (13) instead of D_{iav}.

$$D_{icor} = \sqrt{4S_{ief}/\pi} \tag{17}$$

The difference between SDM values of the measured result of diameter D_{iav} before correction and of the mean diameter D_{icor} after correction is shown in Fig. 6:

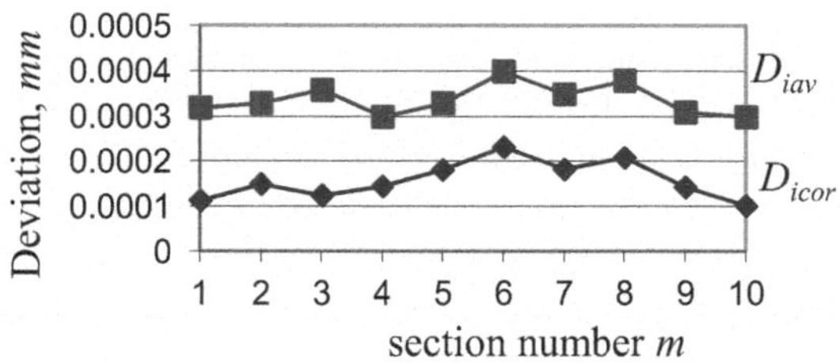

Fig. 6. SDM (σ_{iD}) without correction D_{iav} and with correction D_{icor}

As it can be seen from Figure 6, that SDM with the correction of deterministic component is by 2.5-3 times less than one without correction. The use of an algorithm of effective areas shifts the mean diameter value.

The constants calculated by formulae (11, 13) and (11, 17) differ by 0.015%. Just as is in the previous case, this value represents the difference of systematic errors (13) in calculations of the cell constant K.

5 Experimental results

Both above methods were used to calculate the corrections to the cell constant K [8, 9]. For the primary EC standard of Ukraine, the several cell designs were made. The appearance view of the primary cell is shown in Fig. 7.

Fig. 7. The appearance of cells for primary EC standard of Ukraine

Correctness, sufficiency and adequacy of the selected models of correction of the cell non-ideal profile are confirmed by international comparisons (P22, P47, K36), which involved the primary standard of Ukraine (laboratory UkrCSM). The best results were obtained in international comparisons CCQM-K36 [4].

Errors described in 4.1 and 4.2 sections are used for corrections δ_C of the equation (9). After accounting these and some other corrections, obtained are parameter values listed in Table 1.

Table 1. Final values of parameters of the Ukrainian primary EC standard

Source of uncertainty	Sensitivity coefficient, (3)	Standard uncertainty	Contribution to standard uncertainty
Parameter	$\partial k/\partial i$	U_i	$u_i(k)$
Conductivity G	297 1/m	2,2 E-7, S	6,5 E-5
Constant cell *K*	1,7 E-3 S	4,0 E-2, 1/m	6,8 E-5
Temperature *t*,	1,01 E-2 S/m°C	0,005 °C	5,1 E-5

Hence, the value of expanded relative uncertainty U_k^* from eq. (5) is equal to

$$U_k^* = 2\sqrt{(6{,}5E-5)^2 + (6{,}8E-5)^2 + (5{,}1E-5)^2} \approx 2{,}2E-4$$

This result is put as final in the statement of international comparisons of CCQM-K36 [4], presented in Figure 8. According to the results of comparisons the CMC (calibration and measurement capabilities) of Ukraine in the field of EC measurements are added to the BIPM database.

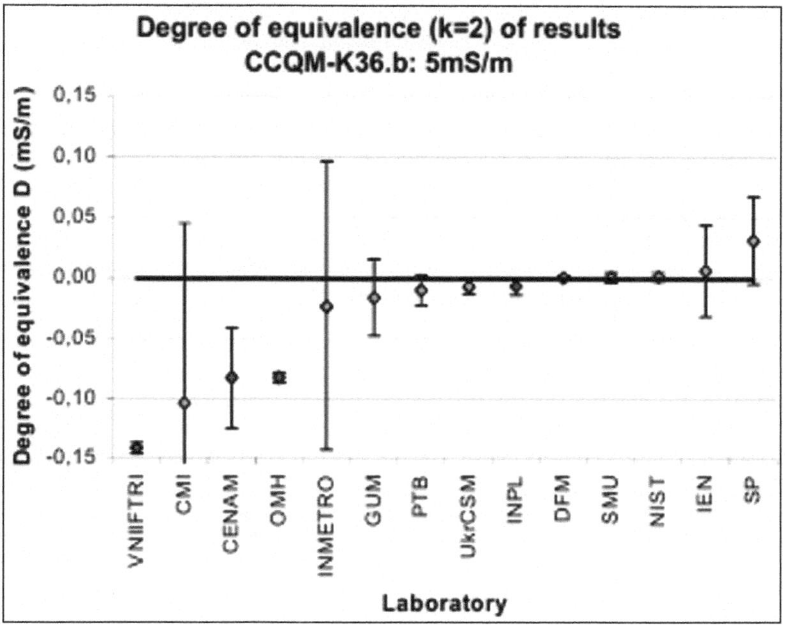

Fig. 8. Results of Key Comparisons K36

6 Conclusions

Described methods for the correction of cell constant and other methods (for example, calibration of all devices) allowed to solve two main problems.

Firstly, the value obtained for laboratory UkrCSM k_{lab}, practically coincides with the value k_{ref}. This is the value of electrolytic conductivity which was reproduced by the leading countries of the world. Results were achieved by minimizing the systematic errors.

Secondly, by minimizing the random component of the error, we have obtained the minimum value of uncertainty $u(k_{lab})$.

References

1. Shreiner, R.H., Pratt, K.W.: Standard Reference Materials: Primary Standards and Standard Reference Materials for Electrolytic Conductivity. NIST Special Publication 260 -142, (2004)
2. Mariassy, M., Pratt, K.W., Spitzer, P.: Major applications of electro-chemical techniques at National Metrology Institutes. Metrologia 46, 199 –213 (2009)
3. Brinkmann, F., at all.: Primary methods for the measurement of electrolytic conductivity. Accred Qual. Assur, 346 – 353 (2003), doi: 10.1007/s00769-003-0645-5
4. Jensen, H. D.: Final report of Key Comparison CCQM-K36 (2006), http://kcdb.bipm.org/AppendixB/appbresults/ccqm-k36/ccqm-k36_final_report.pdf
5. Hlukhenky, A.I., Mikhal, A.A.: Analysis of circular conductor impedance component within AC measurement. Tehnichna elektrodynamika 1, 15-22 (2010), http://www.techned.org.ua (in Russian)
6. Hlukhenky, A.I., Mikhal, A.A.: Error estimation of standard conductometric cell conditional on non-equipotentiality of electrode fase. Tehnichna elektrodynamika 1, 76-82 (2011), http://www.techned.org.ua (in Russian)
7. Gavrylkin, V.G., Glukhenkiy, A.I., Mikhal, A.A.: An Analysis of the error when determining the constant of the primary standard conductometric cell. Measurement Techniques, 56 (8), 935-941 (2013)
8. Mikhal, A.A., Warsza, Z.L.: Influence of geometric uncertainties on the accuracy of calculated constant of the primary conductivity cell. Proceedings of 11th ISMQC on Measurement and Quality Control, TC-14 IMEKO International Symposium, Cracow-Kielce, Poland, 85 (2013)
9. Mikhal, A.A., Warsza, Z.L.: Geometric part of uncertainties in the calculation constant of the primary four electrode conductivity cell. ACTA IMEKO 4 (2), (2015)

Errors and uncertainties of imbalanced bridge-circuits as primary converters for RTD sensors

Zygmunt L. Warsza[1] and Adam Idzkowski[2]

[1] Industrial Research Institute of Automation and Measurement (PIAP), Warsaw, Poland
zlw@op.pl
[2] Bialystok University of Technology, Faculty of Electrical Engineering, Bialystok, Poland
a.idzkowski@pb.edu.pl

Abstract. The original way of evaluating accuracy of the resistance-to-voltage converter (imbalanced bridge-circuit) is presented. It is important because this converter is used as the input stage of industrial and laboratory temperature measurement systems. From circuit theory point of view it is the two-port circuit of type X with four variable resistances and it is supplied by a current (or voltage) source. Formulas of transfer coefficients are presented in general forms and in relative units. Moreover, two particular cases of this converter used in temperature measurements, i.e. with one or two RTD elements, are considered in detail. Error propagation functions are determined in two-component form, i.e. separately for both initial value and relative increment of a transfer coefficient (dependent on temperature). Values of limited errors and expanded uncertainties are calculated for a few variants of this converter with the Class A and B industrial Pt100 sensors. Formulas and example results allow to determine accuracy measures for the selected configuration of bridge-circuit and to compare with other configurations. Null adjustment, inside and outside the bridge, is also considered.

Keywords: temperature sensors · imbalanced bridge-circuit · limited error · uncertainty.

1 Introduction

Resistance temperature detectors (RTDs) are commonly used in temperature measurement instruments and systems. A signal converter is used at the input stage of these systems, e.g. [1]. This circuit transforms the temperature dependent changes in resistance into an analogue or digital signal, which can be useful for further processing. The 2-wire RTD sensor connected to the imbalanced bridge is applied in control systems, e.g. to produce a corrective signal which is sent to the final control element to bring the controlled variable to the proper value [2]. Special resistance temperature detectors (SRTD) are used in the highest accuracy temperature measurements in metrological laboratories, also in the cryogenic temperature range [3]. In this case measurements base mainly on the AC automatic bridge techniques [4].

R. Jabłoński and T. Brezina (eds.), *Advanced Mechatronics Solutions*,
Advances in Intelligent Systems and Computing 393,
DOI 10.1007/978-3-319-23923-1_60

The imbalanced resistance bridge is one of the popular analogue circuits which converts the combination of changes in resistance ΔR_i into a voltage (current, frequency, time period) variation. Output signal can also be digital. The circuit should be supplied by a stable current or voltage source (Fig. 1). Either from one to four single resistance sensors or one or two differential sensors may be applied in such bridge-converter. The independent settings of zero and sensitivity of output voltage is easy to determine. This bridge-circuit can work both in static and dynamic measurements without going into saturation and that is a common disadvantage of circuits based on differential operational amplifiers, e.g. Anderson loop circuit. The imbalanced four-resistor bridge-circuit (4R) is well-known for many years. However, the error (or uncertainty) propagation formulas in general form (assuming the arbitrary increments of resistances) were not published in the literature. The probable reason is that the complicated propagation formulas contain components of all the errors (or uncertainties) that result from constant and variable bridge parameters (Fig. 1), i.e. current (or voltage) and internal resistance of a supply source, nominal values of four arm resistances (in a state of balance), increments of arm resistances and load resistance RL.

In some publications, e.g. [5], accuracy was only considered in particular cases, i.e. when a bridge-circuit approached a balance condition or when the fractional changes in resistance of sensors were low. This problem is especially important in these cases of temperature measurements when the RTD element has a large fractional change in resistance, even several times higher than nominal resistance.

The electric circuit theory considers an imbalanced bridge as a passive two-port circuit (type X) of variable internal parameters (Fig. 1). A bridge-circuit is characterized by variable terminal functions (transfer voltage or impedance, input and output impedances) because it is connected with other circuit building blocks [6]. These terminal functions are dependent on varying resistances of sensors and they can be defined using the relative values of circuit and sensor parameters [7-8].

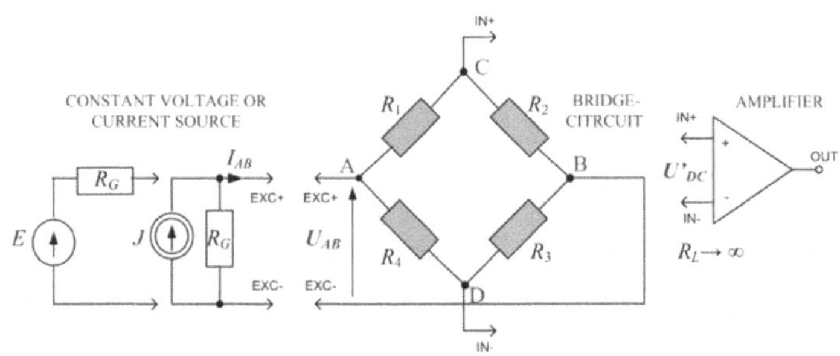

Fig. 1. Four-resistor bridge-circuit working as a two-port circuit (of type X) powered by a voltage or current source

The aim of this paper is to conform the general analysis of transfer coefficients, limited errors and uncertainties to the two most common variants, i.e. with one or two platinum RTDs in the 4R bridge-circuit. It is considered as resistance-to-voltage converter. This theory can be useful in industrial and laboratory experiments as well as in temperature measurements of medium accuracy.

2 Transfer coefficients of the bridge-circuit as R/U converter

The four-resistance (4R) bridge, presented as a two-port circuit (type X) with two pairs of terminals AB and CD, is shown in Fig. 1. If some of its branch resistances R_i are variable, the output voltage U'_{DC} depends on R_i and may be equal to zero for some values of these resistances. Then the bridge is balanced. When the bridge has to work as the imbalanced measurement circuit it should be specially designed. It is preferred to use the ideal supplying source: a current source $I_{AB} \to J = const.$ and $R_G \to \infty$ or a voltage source: $E = U_{AB} = const.$, $R_G = 0$ and unloaded output $R_L \to \infty$, i.e.: $U'_{DC} \to U_{DC}^{\infty}$. If only one variable, e.g. temperature, is measured then it is enough to know the change of one parameter on bridge terminals, e.g. the bridge output voltage U'_{DC}. If the circuit is supplied by the current source J, the output voltage is a product of current I_{AB} and r_{21} (transfer impedance of the unloaded circuit)

$$U'_{DC} \to U_{DC}^{\infty} = I_{AB}\, r_{21}. \tag{1}$$

However, if the circuit is supplied by a voltage source E, the output voltage is a product of voltage U_{AB} and k_{21} as transfer voltage of the unloaded circuit

$$U'_{DC} \to U_{DC}^{\infty} = U_{AB}\, k_{21}. \tag{2}$$

The transfer coefficients r_{21}, k_{21} depend on the bridge arm resistances R_i, as it is given in (3) and (4). Their generalized forms can also be introduced

$$r_{21} \equiv \frac{U_{DC}^{\infty}}{I_{AB}} = \frac{R_1 R_3 - R_2 R_4}{\sum R_i} \equiv t_0 f(\varepsilon_i), \tag{3}$$

$$k_{21} \equiv \frac{U_{DC}^{\infty}}{U_{AB}} = \frac{R_1 R_3 - R_2 R_4}{(R_1 + R_2)(R_3 + R_4)} \equiv k_0 f_E(\varepsilon_i). \tag{4}$$

If the transfer coefficient is zero ($r_{21}=0$ or $k_{21}=0$) then the bridge is in a state of balance. The balance equation: $R_1R_3=R_2R_4$ is the same for E or J supply. It can occur for many different combinations of resistances R_i. Each variable resistance R_i of sensor can be expressed as

$$R_i \equiv R_{i0}(1 + \varepsilon_i), \qquad i = 1,2,3,4. \tag{5}$$

In this analysis the lead resistances of any 2-wire, 3-wire or 4-wire sensor are included in R_i and the influence of their changes is negligible. If all relative increments ε_i are zero and the initial values R_{i0} of all resistances R_i fulfil condition

$$R_{10}R_{30} = R_{20}R_{40}, \tag{6}$$

then the bridge is in a state of balance. Then transfer coefficients r_{21}, k_{21} can be simplified to products of their initial sensitivities t_0, k_0 (in a initial state of balance) and normalized unbalance functions $f(\varepsilon_i)$, $f_E(\varepsilon_i)$ -see the right sides of formulas (3) and (4).

Furthermore resistances R_{i0} can be referred to resistance R_{10} (of the first arm), i.e.: $R_{20} \equiv mR_{10}$, $R_{40} \equiv nR_{10}$ and from (6): $R_{30}=mnR_{10}$. Thus the initial sensitivities t_0, k_0 of transfer coefficients (3), (4) can be expressed as

$$t_0 \equiv \frac{R_{10}R_{30}}{\sum R_{i0}} = R_{10}\frac{mn}{(1+m)(1+n)}, \tag{7}$$

$$k_0 \equiv \frac{R_{10}R_{30}}{(R_{10}+R_{20})(R_{30}+R_{40})} = \frac{m}{(1+m)^2}. \tag{8}$$

Value of $t_0 \rightarrow R_{10}$ when $m \rightarrow \infty$, $n \rightarrow \infty$. Parameter m or n is limited by maximum permissible power of the resistance connected in series with the smallest one. If all initial resistances $R_{i0}=R_{10}$ ($m=n$) are equal then $t_0 = 0.25\ R_{10}$. The k_0 value is maximal for $m=1$ and n is arbitrary. Normalized imbalance functions $f(\varepsilon_i)$, $f_E(\varepsilon_i)$ in equations (3) and (4) can be expressed as relative parameters

$$f(\varepsilon_i) = \frac{\Delta L(\varepsilon_i)}{1+\varepsilon_{\Sigma R}}, \tag{9}$$

where $\Delta L(\varepsilon_i) = \varepsilon_1 - \varepsilon_2 + \varepsilon_3 - \varepsilon_4 + \varepsilon_1\varepsilon_3 - \varepsilon_2\varepsilon_4$, $\quad \varepsilon_{\Sigma R} = \dfrac{\varepsilon_1 + m\varepsilon_2 + n(\varepsilon_4 + m\varepsilon_3)}{(1+m)(1+n)}$,

$$f_E(\varepsilon_i) = \frac{\Delta L(\varepsilon_i)}{(1+\varepsilon_{12})(1+\varepsilon_{34})}, \tag{10}$$

where $\quad \varepsilon_{12} = \dfrac{\varepsilon_1 + m\varepsilon_2}{1+m}, \quad \varepsilon_{43} = \dfrac{\varepsilon_4 + m\varepsilon_3}{1+m}.$

Dependences of the transfer coefficients r_{21}, k_{21} on the relative increment of sensor resistances ($\varepsilon_i = \pm\ \varepsilon$) are presented as ‘1’, ‘2’, ‘4’ particular cases in Fig. 2 and Fig. 3. It is assumed that all initial resistances R_{i0} are equal to R_{10} ($m=n=1$). Other particular cases are analyzed in paper [8]. Transfer impedance r_{21} of the bridge-circuit, supplied by current source (J), is linear in cases ‘2’ and ’4’ (Fig. 2). If the circuit are powered by voltage (E) then the transfer voltage k_{21} is linear in case ‘4’ only (Fig. 3). The normalized imbalance function (9) is less nonlinear than (10), if the value of m is the same. Then for the bridge-circuit in case ‘1’ the nonlinearity of r_{21} is lower than nonlinearity of k_{21}.

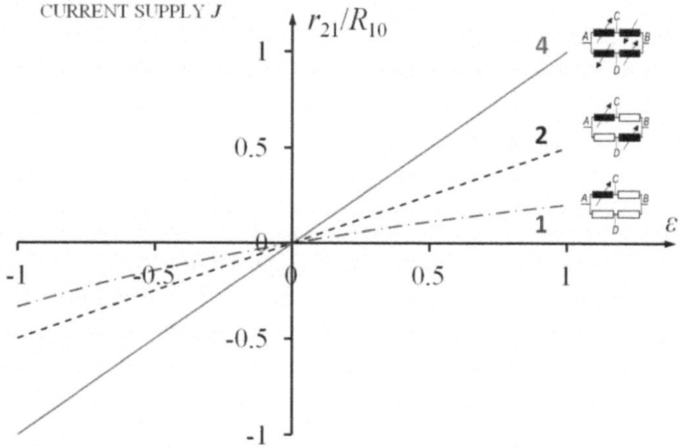

Fig. 2. Dependence of the transfer impedance r_{21} on the relative resistance increment ε (sensor configuration in the bridge-circuit in three cases)

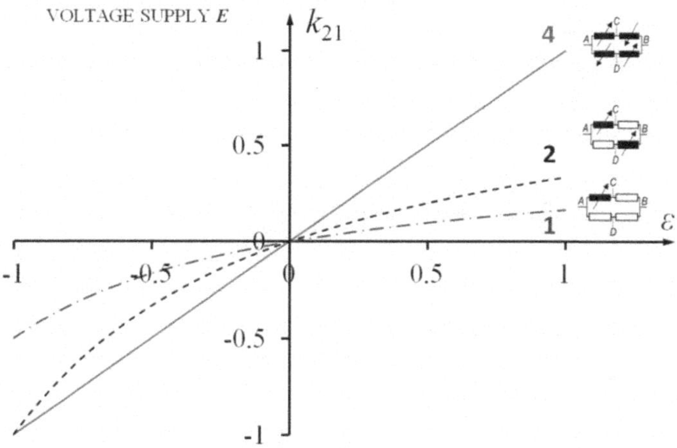

Fig. 3. Dependence of the transfer voltage k_{21} on the relative resistance increment ε (sensor configuration in the bridge-circuit in three cases)

3 Accuracy Analysis of Bridge Transfer Coefficients

Sensor accuracy and bridge accuracy in general forms and in particular cases can be specified with the use of two methods: single component and two-component [8-9]. Basing on the second method, the formulas of accuracy measures were depicted in a graphical form (Fig. 4 and 5). The two-component method specifies accuracy as it is usually stated for many measuring instruments, e.g. digital voltmeters.

Sensor accuracy can be specified as follows. For any variable resistance R_i given in (5) the absolute error is $\Delta_i = \Delta_{i0}(1+\varepsilon_i)+\Delta_{\varepsilon i} R_{i0}$. Hence the relative error δ_{Ri} can be expressed by two relative errors: initial error δ_{i0} and error of increment $\delta_{\varepsilon i}$

$$\delta_{Ri} \equiv \frac{\Delta_i}{R_i} = \frac{\Delta_{i0}}{R_{i0}} + \frac{\Delta_{\varepsilon i}}{1+\varepsilon_i} = \delta_{i0} + \frac{\varepsilon_i}{1+\varepsilon_i}\delta_{\varepsilon i} \,. \tag{11}$$

Bridge accuracy can be represented by the instantaneous (actual) values of measurement errors. They can be created for two transfer coefficients - r_{21} and k_{21}. They result from the total differential of analytical equations (3), (4) and they are known as the error propagation functions [10]. After ordering all components of δ_{Ri}, the absolute error of transfer impedance r_{21} is

$$\Delta_{r21} = R_1 \frac{R_3 - r_{21}}{\Sigma R_i}\delta_{R1} - R_2 \frac{R_4 + r_{21}}{\Sigma R_i}\delta_{R2} + R_1 \frac{R_1 - r_{21}}{\Sigma R_i}\delta_{R3} - R_4 \frac{R_2 + r_{21}}{\Sigma R_i}\delta_{R4} \equiv \sum_{i=1}^{4} w_{Ri}\,\delta_{Ri}, \tag{12}$$

where: $w_{Ri} \equiv R_i \dfrac{(-1)^{i+1} R_j - r_{21}}{\sum R_i}$ - weight coefficient of the error δ_{Ri} component; symbols: i=1, 2, 3, 4 and j=3, 4, 1, 2 (in order), multiplier $(-1)^{i+1}$=+1 (positive) if i is 1, 3 or –1 (negative) if i is 2, 4.

Equation (12) can be transformed to the following (generalized) form

$$\Delta_{r21} = \sum w_{Ri}\,(\delta_{i0} + \frac{\varepsilon_i}{1+\varepsilon_i}\,\delta_{\varepsilon i}), \tag{13}$$

where a weight coefficient equals

$$w_{Ri} = \frac{t_0}{1+\varepsilon_{\Sigma R}}[(-1)^{i+1}(1+\varepsilon_j) - \frac{R_{i0}}{R_{j0}}\Delta L(\varepsilon_i)](1+\varepsilon_i). \tag{14}$$

If resistance R_i is constant, i.e. $\varepsilon_i = 0$, then from (11) it follows that $\delta_{Ri} = \delta_{i0}$. However, the weight coefficient w_{Ri} multiplied by δ_{i0} in formula (13) causes that Δ_{r21} still depends on the other increments ε_j. This subject has been taken up in any literature except of author's papers [7-8]. Similarly, after derivation of (4), absolute error Δ_{k21} can be obtained.

Bridge is in the initial state of balance when all increments ε_j=0. Thus the nominal transfer coefficients should be $r_{21}(0)\equiv r_{210}=0$ and $k_{21}(0)\equiv k_{210}=0$, according to (3), (4) and (6). Nevertheless, the real resistances R_i have some initial errors δ_{i0} and usually the initial values of absolute error of r_{21} and k_{21} are not equal zero ($\Delta_{r210} \neq 0$, $\Delta_{k210} \neq 0$). From (13) and (13a) they can be expressed as

$$\Delta_{r210} = t_0\,\delta_{210}, \tag{15}$$

$$\Delta_{k210} = k_0\,\delta_{210}, \tag{16}$$

where: $\delta_{210} = \delta_{10} - \delta_{20} + \delta_{30} - \delta_{40}$.

In general case, i.e. for the 4R bridge-circuit with four variable resistances, the formulas of instantaneous errors, limited errors and random measures, e.g. type B uncertainties introduced in GUM [10], are presented in [9]. It is assumed that all components of random measures are not correlated, i.e. all correlation coefficients $corr_{ij}$=0.

Two-component absolute error of the transfer impedance r_{21} (11), after subtracting its initial value, is expressed as

$$\Delta_{r21} - \Delta_{r21\,0} = \sum_{i=1}^{4} \left[w_{Ri} - (-1)^{i-1} \right] \delta_{i0} + \sum_{i=1}^{4} w_{Ri} \frac{\varepsilon_i}{1+\varepsilon_i} \delta_{\varepsilon i} . \tag{17}$$

After referencing (16) to r_{21} and substitution w_{Ri} (13a) the relative error of difference r_{21}-r_{210} is

$$\delta_{r21r} \equiv \frac{\Delta_{r21} - \Delta_{r210}}{r_{21}} = \frac{\delta_{r21} - \delta_{210}}{f(\varepsilon_i)} = \sum_{i=1}^{4} w'_{ri0}\, \delta_{i0} + \sum_{i=1}^{4} w'_{r\varepsilon i}\, \delta_{\varepsilon i} , \tag{18}$$

where:

$$w'_{ri0} = (-1)^{i-1} \frac{\varepsilon_i + \varepsilon_j + \varepsilon_i \varepsilon_j - \varepsilon_{\Sigma R}}{\Delta L(\varepsilon_i)} - t_0 \frac{1+\varepsilon_i}{R_{j0}(1+\varepsilon_{\Sigma R})}, \quad w'_{r\varepsilon i} = \left[\frac{(-1)^{i-1}(1+\varepsilon_j)}{\Delta L(\varepsilon_i)} - \frac{t_0}{R_{j0}(1+\varepsilon_{\Sigma R})} \right] \varepsilon_i \tag{19}$$

Weight coefficients (17a,b) are finite for any value of r_{21}, including r_{21}= 0, because if all $\varepsilon_i \rightarrow 0$ also $\Delta L \rightarrow 0$. Error δ_{r21r} represents adequately as the error $\delta_{\varepsilon i}$ (of increment ε_i) in formula (11).

From (13a) and (17) the absolute error of transfer impedance r_{21} and this error related to initial sensitivity t_0 is

$$\Delta_{r21} = t_0\, \delta_{210} + r_{21} \delta_{r21r} , \tag{18}$$

$$\delta_{r21} = \delta_{210} + f_{21} \delta_{r21r} , \tag{18a}$$

Accordingly, it is defined for the transfer voltage k_{21}

$$\Delta_{k21} = k_0\, \delta_{210} + k_{21} \delta_{k21k} , \tag{19}$$

$$\delta_{k21} = \delta_{210} + f_{21E} \delta_{k21k} , \tag{19a}$$

where: $t_0\,\delta_{210} = \Delta_{r210}$, $k_0\,\delta_{210} = \Delta_{k210}$ – absolute errors of the initial values of r_{21} or k_{21}, e.g. r_{210} = 0 or k_{210}= 0; δ_{r21r} and δ_{k21k} – relative errors of increments r_{21} - r_{210} or k_{21} - k_{210}.

Table 1. Accuracy measures of the 4R bridge-circuits (R_L=∞) with equal initial resistances ($4R_{10}$)

Relative accuracy measures of the transfer impedance r_{21}	
Absolute accuracy measures related to initial sensitivity t_0 and their initial values for bridge zero (r_{210}=0)	Relative accuracy measures of difference r_{21} - r_{210} (related to r_{21})
Case '1' variable $R_1=R_{10}(1+\varepsilon_1)$, $R_2=R_3=R_4=R_{10}$, $\varepsilon_1 \geq -1$ $r_{21} = \frac{R_{10}}{4}\frac{\varepsilon_1}{1+0.25\,\varepsilon_1}\quad r_0 = \frac{R_{10}}{4}$	
$\delta_{r21} = \frac{(1+\varepsilon_1)\delta_{10} + \varepsilon\delta_{\varepsilon 1} + (1+0.5\,\varepsilon_1)^2\delta_{30} - (1+0.5\,\varepsilon_1)(\delta_{20}+\delta_{40})}{(1+0.25\,\varepsilon_1)^2}$ $\delta_{210} = \delta_{10} - \delta_{20} + \delta_{30} - \delta_{40}$	$\delta_{r21r} = \frac{(\frac{1}{2}-\frac{1}{16}\varepsilon_1)\delta_{10} + (\frac{1}{2}-\frac{3}{16}\varepsilon_1)\delta_{30} + \frac{1}{16}\varepsilon_1(\delta_{20}+\delta_{40}) + \delta_{\varepsilon 1}}{1+0.25\,\varepsilon_1}$
$\lvert\delta_{10}\rvert, \lvert\delta_{\varepsilon 1}\rvert, \lvert\delta_{20}\rvert = \lvert\delta_{30}\rvert = \lvert\delta_{40}\rvert \equiv \lvert\delta_0\rvert$ $\lvert\delta_{r21}\rvert = \frac{(1+\varepsilon_1)\lvert\delta_{10}\rvert + \lvert\varepsilon_1\rvert\,\lvert\delta_{\varepsilon 1}\rvert + (3+2\,\varepsilon_1+0.25\,\varepsilon_1^2)\,\lvert\delta_0\rvert}{(1+0.25\,\varepsilon_1)^2}$ $\lvert\delta_{210}\rvert = \lvert\delta_{10}\rvert + 3\lvert\delta_0\rvert$	$\lvert\delta_{10}\rvert, \lvert\delta_{\varepsilon 1}\rvert, \lvert\delta_{20}\rvert = \lvert\delta_{30}\rvert = \lvert\delta_{40}\rvert \equiv \lvert\delta_0\rvert$ $\lvert\delta_{r21r}\rvert = \frac{\frac{1}{2}\,(1-\frac{1}{8}\varepsilon_1)\,\lvert\delta_{10}\rvert + \frac{1}{2}\,(1-\frac{5}{8}\,\varepsilon_1)\,\lvert\delta_0\rvert + \lvert\delta_{\varepsilon 1}\rvert}{1+0.25\,\varepsilon_1}$
$\bar{\delta}_{10}, \bar{\delta}_{\varepsilon 1}, \bar{\delta}_{20} = \bar{\delta}_{30} = \bar{\delta}_{40} \equiv \bar{\delta}_0,$ $\bar{\delta}_{r21} = \frac{\sqrt{(1+\varepsilon_1)^2\bar{\delta}_{10}^2 + \varepsilon_1^2\bar{\delta}_{\varepsilon 1}^2 + (1+0.5\,\varepsilon_1)^2[(1+0.5\,\varepsilon_1)^2+2]\bar{\delta}_0^2}}{(1+0.25\,\varepsilon_1)^2}$ $\bar{\delta}_{210} = \sqrt{\bar{\delta}_{10}^2 + 3\bar{\delta}_0^2}\quad corr_{ij} = 0$	$\bar{\delta}_{10}, \bar{\delta}_{\varepsilon 1}, \bar{\delta}_{20} = \bar{\delta}_{30} = \bar{\delta}_{40} \equiv \bar{\delta}_0,$ $\bar{\delta}_{r21r} = \frac{\sqrt{(1-\frac{1}{8}\varepsilon_1)^2\bar{\delta}_{10}^2 + (1-\frac{3}{4}\varepsilon_1+\frac{11}{64}\varepsilon_1^2)\,\bar{\delta}_0^2 + 4\bar{\delta}_{\varepsilon 1}^2}}{2(1+0.25\,\varepsilon_1)}$ $corr_{ij} = 0$
Case '2' variable $R_1=R_3=R_{10}(1+\varepsilon)$, $R_2=R_4=R_{10}$, $\varepsilon \geq -1$ $r_{21} = \frac{R_{10}}{4}2\varepsilon$	
$\delta_{r21} = (1+\varepsilon)(\delta_{10}+\delta_{30}) - \delta_{20} - \delta_{40} + \varepsilon(\delta_{\varepsilon 1}+\delta_{\varepsilon 3})$ $\lvert\delta_{r21}\rvert = (1+\varepsilon)(\lvert\delta_{10}\rvert + \lvert\delta_{30}\rvert) + \lvert\delta_{20}\rvert + \lvert\delta_{40}\rvert + \lvert\varepsilon\rvert(\lvert\delta_{\varepsilon 1}\rvert + \lvert\delta_{\varepsilon 3}\rvert)$ $\bar{\delta}_{r21} = \sqrt{(1+\varepsilon)^2(\bar{\delta}_{10}^2+\bar{\delta}_{30}^2) + \bar{\delta}_{20}^2 + \bar{\delta}_{40}^2 + \varepsilon^2(\bar{\delta}_{\varepsilon 1}^2+\bar{\delta}_{\varepsilon 3}^2)}$ $corr_{ij} = 0$	$\delta_{r21r} = 0.5\,(\delta_{10}+\delta_{30}+\delta_{\varepsilon 1}+\delta_{\varepsilon 3})$ $\lvert\delta_{r21r}\rvert = 0.5(\lvert\delta_{10}\rvert + \lvert\delta_{30}\rvert + \lvert\delta_{\varepsilon 1}\rvert + \lvert\delta_{\varepsilon 3}\rvert)$ $\bar{\delta}_{r21r} = \frac{1}{\sqrt{2}}\sqrt{\bar{\delta}_{10}^2+\bar{\delta}_{30}^2+\bar{\delta}_{\varepsilon 1}^2+\bar{\delta}_{\varepsilon 3}^2}$ $corr_{ij} = 0$
$\lvert\delta_{10}\rvert = \lvert\delta_{30}\rvert, \lvert\delta_{20}\rvert = \lvert\delta_{40}\rvert \equiv \lvert\delta_0\rvert, \lvert\delta_{\varepsilon 1}\rvert = \lvert\delta_{\varepsilon 3}\rvert \equiv \lvert\delta_\varepsilon\rvert$ $\lvert\delta_{r21}\rvert = 2\,(1+\varepsilon)\,\lvert\delta_{10}\rvert + 2\lvert\delta_0\rvert + 2\lvert\varepsilon\rvert\,\lvert\delta_\varepsilon\rvert$ $\lvert\delta_{210}\rvert = 2\lvert\delta_{10}\rvert + 2\lvert\delta_0\rvert$	$\lvert\delta_{10}\rvert = \lvert\delta_{30}\rvert, \lvert\delta_{20}\rvert = \lvert\delta_{40}\rvert \equiv \lvert\delta_0\rvert, \lvert\delta_{\varepsilon 1}\rvert = \lvert\delta_{\varepsilon 3}\rvert \equiv \lvert\delta_\varepsilon\rvert$ $\lvert\delta_{r21r}\rvert = \lvert\delta_{10}\rvert + \lvert\delta_\varepsilon\rvert$
$\bar{\delta}_{10} = \bar{\delta}_{30}, \bar{\delta}_{\varepsilon 1} = \bar{\delta}_{\varepsilon 3} = \bar{\delta}_\varepsilon, \bar{\delta}_{20} = \bar{\delta}_{40} \equiv \bar{\delta}_0$ $\bar{\delta}_{r21} = \sqrt{2((1+\varepsilon)^2\bar{\delta}_{10}^2 + \bar{\delta}_0^2 + \varepsilon^2\bar{\delta}_\varepsilon^2)}$ $\bar{\delta}_{210} = \sqrt{2\bar{\delta}_{10}^2 + 2\bar{\delta}_0^2}$ $corr_{ij} = 0$	$\bar{\delta}_{10} = \bar{\delta}_{30}, \bar{\delta}_{\varepsilon 1} = \bar{\delta}_{\varepsilon 3} = \bar{\delta}_\varepsilon, \bar{\delta}_{20} = \bar{\delta}_{40} \equiv \bar{\delta}_0$ $\bar{\delta}_{r21r} = \sqrt{\bar{\delta}_{10}^2 + \bar{\delta}_0^2 + 2\bar{\delta}_\varepsilon^2}$ $corr_{ij} = 0$

Accuracy measures of the 4R bridge in general case (four variable resistances) are presented in [9]. Limited errors $|\delta_{r21}|$ and random (standard) measures (of random errors or type B uncertainties) $\bar{\delta}_{r21}$ in two particular cases, used with Pt100 sensors, '1' – only R_1 variable or '2' – variable resistances R_1 and R_3 and their relative increments equal to ε, are shown in Table 1. Limited errors and uncertainties in the function of resistance increment ε, for two kinds of the bridge supply, are presented in Fig. 4 and Fig. 5. All accuracy measures of the coefficient r_{21} rise from ε=0 until 2.5 (Fig. 4). In the same range of ε the accuracy measures of coefficient k_{21} rise and diminish. They are less than the limited error $4|\delta_0|$ and the mean square measure (uncertainty) $2\bar{\delta}_0$ (Fig. 5).

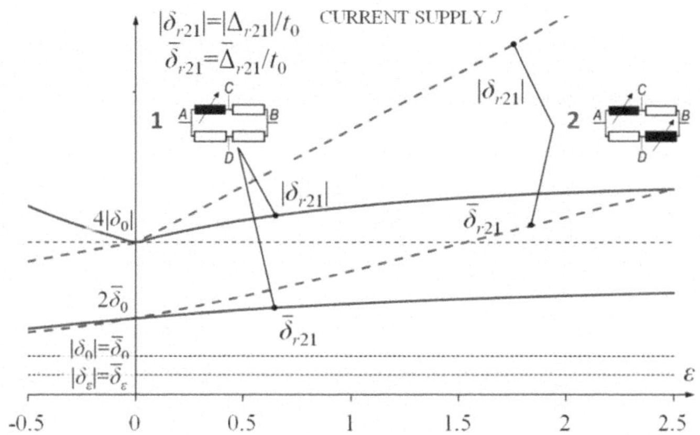

Fig. 4. Accuracy measures of the transfer impedance r_{21} (bridge-circuit with all initial resistances equal R_{10} supplied by a current source)

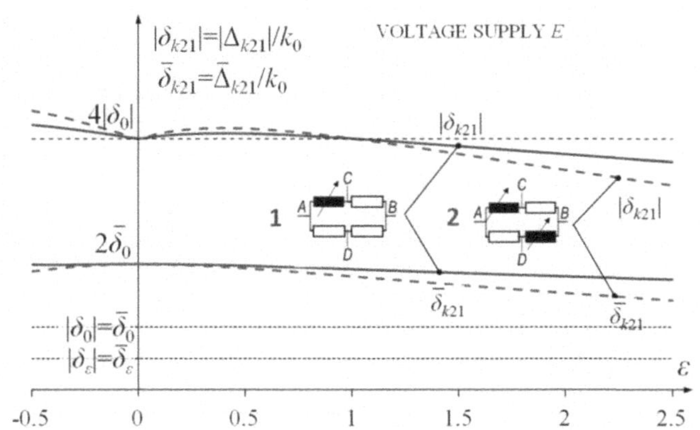

Fig. 5. Accuracy measures of the transfer voltage k_{21} (bridge-circuit with all initial resistances equal R_{10} supplied by a voltage source)

Table 2. Formulas and values of limited errors and uncertainties of the industrial Class A or B Pt 100 sensor in a few variants of the 4R bridge-circuit supplied by a current source

No	Limited errors $\lvert\delta_{r21}\rvert$ or $\lvert\delta_{r21\varepsilon}\rvert$ if $R_{i0}=R_{10}$, $\lvert\delta_{20}\rvert=\lvert\delta_{30}\rvert=\lvert\delta_{40}\rvert\equiv\lvert\delta_0\rvert$ Combined standard uncertainties $\bar{\delta}_{r21}$, $\bar{\delta}_{r21\varepsilon}$ if $\bar{\delta}_{10}=\frac{\lvert\delta_{10}\rvert}{2\sqrt{3}}$, $\bar{\delta}_{\varepsilon 1}=\frac{\lvert\delta_{\varepsilon 1}\rvert}{2\sqrt{3}}$, $\bar{\delta}_{20}=\bar{\delta}_{30}=\bar{\delta}_{40}\equiv\bar{\delta}_0$	Cl.	Limited errors and expanded uncertainties in % Standard Pt100 sensors, T=650°C, ε_1= 2.296 $\lvert\delta_0\rvert=\lvert\delta_{10}\rvert$=0.06% (Class A) or 0.12% (Class B) $\lvert\delta_{\varepsilon 1}\rvert$=0.11% (Class A) or 0.32% (Class B)			
			$\lvert\delta_{r21}\rvert$	$\lvert\delta_{r21\varepsilon}\rvert$	$U_{r21}=2\bar{\delta}_{r21}$ ($p=0.95$)	$U_{r21\varepsilon}=2\bar{\delta}_{r21\varepsilon}$ ($p=0.95$)
1	Single sensor (without bridge circuit) $\lvert\delta_{R1}\rvert=\frac{\lvert\Delta_{R1}\rvert}{R_1}=\lvert\delta_{10}\rvert+\frac{\lvert\varepsilon_1\rvert}{1+\varepsilon_1}\lvert\delta_{\varepsilon 1}\rvert$	A	0.14	0.08	0.06	-
	$\bar{\delta}_{R1}=\sqrt{\bar{\delta}_{10}^{\,2}+\frac{\varepsilon_1^{\,2}}{(1+\varepsilon_1)^2}\bar{\delta}_{\varepsilon 1}^{\,2}}$	B	0.34	0.24	0.14	-
2	One-sensor bridge without null adjustment $\lvert\delta_{r21}\rvert=\frac{(1+\varepsilon_1)\lvert\delta_{10}\rvert+\lvert\varepsilon_1\rvert\,\lvert\delta_{\varepsilon 1}\rvert+(3+2\varepsilon_1+0.25\varepsilon_1^2)\,\lvert\delta_0\rvert}{(1+0.25\,\varepsilon_1)^2}$	A	0.40	-	0.10	-
	$\bar{\delta}_{r21}=\frac{\sqrt{(1+\varepsilon_1)^2\bar{\delta}_{10}^2+\varepsilon_1^2\bar{\delta}_{\varepsilon 1}^2+(1+0.5\,\varepsilon_1)^2[(1+0.5\,\varepsilon_1)^2+2]\bar{\delta}_0^2}}{(1+0.25\,\varepsilon_1)^2}$	B	0.89	-	0.24	-
3	One-sensor bridge, external null correction $\lvert\delta_{r21\varepsilon}\rvert=\lvert\delta_{r21}-\delta_{210}\rvert=$ $=\frac{\lvert\varepsilon_1\rvert\,(\frac{1}{2}(1-\frac{1}{8}\varepsilon_1)\,\lvert\delta_{10}\rvert+\frac{1}{2}(1+\frac{5}{8}\varepsilon_1)\,\lvert\delta_0\rvert+\lvert\delta_{\varepsilon 1}\rvert)}{(1+0.25\varepsilon_1)^2}$	A	-	0.19	-	0.07
	$\bar{\delta}_{r21\varepsilon}=\overline{\delta_{r21}-\delta_{r210}}$ $=\frac{0.5\,\lvert\varepsilon_1\rvert\sqrt{(1-\frac{1}{8}\varepsilon_1)^2\,\bar{\delta}_{10}^2+(1+\frac{3}{4}\varepsilon_1+\frac{13}{16}\varepsilon_1^2)\,\bar{\delta}_0^2+4\bar{\delta}_{\varepsilon 1}^2}}{(1+0.25\varepsilon_1)^2}$	B	-	0.47	-	0.19
4	One-sensor bridge, internal null correction $\lvert\delta_{210}\rvert=0$ $\lvert\delta_{r21\varepsilon}\rvert=\frac{\lvert\varepsilon_1\rvert[\frac{1}{2}\lvert\delta_{10}\rvert+(\frac{1}{2}+\frac{\varepsilon_1}{4})\lvert\delta_0\rvert+\lvert\delta_{\varepsilon 1}\rvert]}{(1+0.25\varepsilon_1)^2}$	A	-	0.19	-	0.07
	$\bar{\delta}_{r21\varepsilon}=\frac{0.5\lvert\varepsilon_1\rvert\sqrt{\bar{\delta}_{10}^{\,2}+(1+\frac{1}{2}\varepsilon_1)^2\,\bar{\delta}_0^{\,2}+4\bar{\delta}_{\varepsilon 1}^{\,2}}}{(1+0.25\varepsilon_1)^2}$	B	-	0.47	-	0.19
5	Two-sensor bridge without null adjustment ($\varepsilon_1=\varepsilon_3=\varepsilon$, r_{21} related to $2t_0$) $\lvert\delta_{r21}\rvert=(1+\varepsilon)\lvert\delta_{10}\rvert+\lvert\delta_0\rvert+\lvert\varepsilon\rvert\lvert\delta_\varepsilon\rvert$	A	0.51	-	0.17	-
	$\bar{\delta}_{r21}=\sqrt{0.5((1+\varepsilon)^2\bar{\delta}_{10}^{\,2}+\bar{\delta}_0^{\,2}+\varepsilon^2\,\bar{\delta}_\varepsilon^{\,2})}$ where $\bar{\delta}_{30}=\bar{\delta}_{10}$, $\bar{\delta}_{20}=\bar{\delta}_{40}\equiv\bar{\delta}_0$ and $\bar{\delta}_{\varepsilon 1}=\bar{\delta}_{\varepsilon 3}\equiv\bar{\delta}_\varepsilon$	B	1.25	-	0.46	-

4 Tolerances of Industrial Pt100 Sensors

Industrial Class A and B Pt100 RTDs are a good example of large variable resistance sensors. They are commonly used in temperature measurements. Tolerances of these sensors are listed according to the IEC751 norm. They are expressed in °C or, as values of permissible resistance changes, in Ω. The $R=f(T)$ characteristic curve of Class A Pt100 sensors is determined up to 650°C and in case of less accurate Class B sensors – up to 850°C. Limited errors of nominal resistance $|\delta_{10}|$ are equal 0.06% (Class A) and 0.12% (Class B) respectively.

5 Uncertainty of transfer impedance r_{21} for a few variants of the bridge-circuit with Pt100 sensors

The transfer impedance uncertainties $\bar{\delta}_{r21}$, $\bar{\delta}_{r21\varepsilon}$ of the circuit with a single Class A or B Pt100 sensor were estimated on the basis of formulas quoted in Table 2. It was assumed that the limited errors $|\delta_{i0}|$ of constant resistances in the bridge were equal and not higher than the sensor initial error $|\delta_{10}|$. The bridge was in a state of balance at temperature 0°C and the current supplying source was stable enough. The maximum value of range (0 – 650°C) was considered in calculations. Within this range the relative increment of sensor resistance was: $\varepsilon_1=\varepsilon_{max}=2.296$.

5.1 Example calculations: case '1'- one sensor, 4R$_{10}$ bridge with external null correction, current supply

- Determination of the limited error of r_{21}:

According to IEC751 norm, the limited error of nominal resistance Class B Pt100 sensor equals $|\delta_{10}|=0.12\%$ and the limited error of increment $|\delta_{\varepsilon 1}|= 0.32\%$. It is assumed that the limited errors of nominal resistances R_2, R_3, R_4 are equal $|\delta_{20}|=|\delta_{30}|=|\delta_{40}|\equiv|\delta_0|=|\delta_{10}|=0.12\%$. The limited relative zero error $|\delta_{210}|$ of the bridge (if $\varepsilon_1=0$) comes to 0.48%. The initial sensitivity t_0 equals 25 Ω and therefore the limited absolute zero error $|\Delta_{r210}|=0.12$ Ω. On the basis of equation in Table 2, the limited relative error $|\delta_{r21\varepsilon}|$ (related to $t_0= 0.25\ R_{10}$) of the one-sensor double-symmetric bridge ($R_{i0}=R_{10}$) equals

$$\left|\delta_{r21\varepsilon}\right| = \frac{\left|\varepsilon_1\right|\,(0.5\,(1-0.125\left|\varepsilon_1\right|)\left|\delta_{10}\right| + 0.5\,(1+0.625\left|\varepsilon_1\right|)\left|\delta_0\right| + \left|\delta_{\varepsilon 1}\right|)}{(1+0.25\,\varepsilon_1)^2} = \frac{0.44\left|\varepsilon_1\right| + 0.03\,\varepsilon_1^2}{(1+0.25\,\varepsilon_1)^2} \tag{20}$$

For the full range of the converter, i.e. $\varepsilon_{1max}=2.296$ ($T=650$°C) determined values are equal to $|\delta_{r21\varepsilon}|_{max}=0.47\%$ and $|\Delta_{r21\varepsilon}|_{max} = 0.12$ Ω. These absolute errors, related to the full-range transfer impedance of the bridge

$$r_{21\max} = \frac{R_{10}}{4}\frac{\varepsilon_{1\max}}{1+0.25\,\varepsilon_{1\max}} = 25\,\frac{2.296}{1+0.918} = 19.67\ \Omega \tag{21}$$

are equal to 0.6% ≈4°C (zero error) and 0.6% ≈4°C (full-range error), adequately.

The minimal errors of this R(T)/U converter with a platinum sensor can be obtained if $|\delta_J|= 0$ and the errors of resistors R_2, R_3, R_4 are negligible in comparison with sensor errors, i.e. $|\delta_{20}|=|\delta_{30}|=|\delta_{40}|=0$. Then relative error of the bridge with single Class B Pt100 sensor is expressed as

$$|\delta_{r21\varepsilon}| = |\delta_{r21} - \delta_{210}| = \frac{|\varepsilon_1|\,(0.5(1-0.125\,|\varepsilon_1|)\,|\delta_{10}| + |\delta_{\varepsilon 1}|)}{(1+0.25\varepsilon_1)^2} = \frac{0.38\,|\varepsilon_1| - 0.0075\,\varepsilon_1^2}{(1+0.25\,\varepsilon_1)^2} \tag{22}$$

From (24) the limited zero error values are: $|\delta_{210}|$=0.12%, $|\Delta_{r210}|$=0.03 Ω. And for the full range $\varepsilon_{1\max}$ the values of limited error are $|\delta_{r21\varepsilon}|_{\max} = 0.34\%$, $|\Delta_{r21\varepsilon}|_{\max} = 0.084$ Ω), respectively. Absolute errors, related to the full range $r_{21\max}$, are equal to 0.15% ≈1°C (zero error) and 0.43% ≈2.8°C (full-range error), adequately.

- Determination of the expanded uncertainty of r_{21}:

Let us determine the expanded uncertainty when the parameters of particular sensor and of the bridge-circuit have unknown values. However these parameters have uniform distributions of the same range as above mentioned limited errors.

Type B standard uncertainties of sensor R_1 and other resistances R_2, R_3, R_4 in the bridge are as follows

$$\bar{\delta}_{10} = \frac{|\delta_{10}|}{2\sqrt{3}},\ \bar{\delta}_{\varepsilon 1} = \frac{|\delta_{\varepsilon 1}|}{2\sqrt{3}},\ \bar{\delta}_{20} = \bar{\delta}_{30} = \bar{\delta}_{40} \equiv \bar{\delta}_0 . \tag{23}$$

They are components of the combined standard uncertainty of transfer impedance r_{21}. According to Table 2, this is expressed as

$$\bar{\delta}_{r21\varepsilon} = \frac{0.5\,|\varepsilon_1|\sqrt{(1-0.125\varepsilon_1)^2\,\bar{\delta}_{10}^2 + (1+0.75\varepsilon_1 + 0.812\varepsilon_1^2)\,\bar{\delta}_0^2 + 4\bar{\delta}_{\varepsilon 1}^2}}{(1+0.25\varepsilon_1)^2} = \frac{|\varepsilon_1|\sqrt{0.438 + 0.007|\varepsilon_1| + 0.012\varepsilon_1^2}}{4\sqrt{3}(1+0.25\varepsilon_1)^2} \tag{24}$$

If $\varepsilon_{1\max}$ is equal to 2.296 (T=650°C) then $\bar{\delta}_{r21\varepsilon}$=0.096%. The standard uncertainty of zero $\bar{\delta}_{210}$ equals 0.07%.

In this way the expanded uncertainty $U_{r21} = 2\bar{\delta}_{r21\varepsilon}$ is determined. The result is 0.19% (coverage factor k=2 based on Gauss distribution, level of confidence p=0.95 [10]). It is assumed that all correlation coefficients are equal to 0.

5.2 Summary - limited error and uncertainty formulas and values

The influence of lead resistances of RTD elements was assumed in calculations as negligible. Numerical formulas of limited errors and uncertainties were presented in Table 2. Three different variants of the one-sensor bridge were considered, i.e. without any adjustment of zero error, with external null correction and with internal null correction. If zero is adjusted outside the bridge then the limited error is a little higher (Table 2, line 3) than one of with internal null correction (Table 2, line 4). The values of errors in lines 2, 3 and 4 in Table 2 (sensor in circuit) are higher than the values from line 1 (single sensor).

The values in line 5 are not more than two times greater in comparison to values in line 2. However, the two-sensor bridge output voltage depends linearly on the resistance increment ε (Fig. 2). Both the sensors have equal increments $\varepsilon_1 = \varepsilon_3$. The expanded uncertainty (k=2, p=0.95) of the bridge with two Class B Pt100 sensors, calculated for the same temperature range, does not exceed 0.46% (Table 2, line 5) - without null correction. The initial sensitivity of output voltage in the two-sensor bridge is two times higher in comparison to the bridge with single sensor. In this case all accuracy measures related to t_0 or k_0 are divided by 2 (Table 2, line 5).

6 Conclusions

The formulas of r_{21} or k_{21} accuracy measures were obtained after transforming the error propagation formulas. All these accuracy measures were presented in the two-component form. The two-component method relies on presenting accuracy of transfer coefficients separately for the initial value (e.g. equal to zero) and for the increment dependent on values of the measured variable. This method is similar to accuracy specifications of digital instruments and sensor transmitters.

Main conclusions of this paper are:

- The curves of r_{21} and k_{21} in the function of ε (Fig. 2, Fig. 3) show that the current supply is preferable one. The bridge-circuit with the single variable resistance sensor supplied by a current source has smaller nonlinearity. The larger values of sensitivity to the sensor changes in resistance are also gained in case of the current supply in comparison to the same circuit supplied by a voltage source. If two similar RTD sensors are used and placed in the same temperature then the transfer impedance r_{21} of the bridge-circuit depends linearly on the relative resistance increment ε and sensitivity to temperature is two times higher.
- The two-component formulas of accuracy measures are also simpler for the two-sensor bridge with linear output (Table 1).
- In case of the two-sensor bridge supplied by a voltage source the values of limited errors and uncertainties (Fig. 4, Fig. 5) are the lowest, however bridge output voltage is nonlinear.
- The way of null correction is important to diminish a value of expanded uncertainty (Table 2), the values are different, e.g. 0.24% (without adjustment), 0.19% (zero adjustment outside the bridge-circuit), 0.19% (zero adjustment in the bridge-

circuit). The results were listed in case of the bridge-circuit supplied by a current source and cooperated with a single Class B Pt100 sensor. It is assumed that all limited errors $|\delta_0|$, $|\delta_{10}|$, $|\delta_{\varepsilon 1}|$ have uniform distributions. Expanded uncertainties can be calculated if results in Table 2 are multiplied by coverage factor k=2, having a level of confidence p= 95% [10].

More information on the accuracy measures in other cases of the 4R bridge-circuit are presented in [7-8]. The two-component formulas of accuracy measures can also be used for other electrical circuits, e.g. unconventional double-current circuit in simultaneous measurement of strain and temperature [9].

References

1. Yurish, S.Y.: Universal Interfacing Circuit for Resistive-Bridge Sensors. IEEE Conference publications - 2010 First International Conference on Sensor Device Technologies and Application, 211-216 (2010), doi:10.1109/SENSORDEVICES.2010.46
2. He, H., Feng, J., Li, K.: Development of a Temperature Control System Based on DSP for a Real-time PCR Instrument. Progress in Mechatronics and Information Technology, Applied Mechanics and Materials 462-463, 549-552 (2014), doi: 10.4028/www.scientific.net/AMM.462-463.549
3. Kołodziej, B., Manuszkiewicz, H., Szmyrka-Grzebyk, A., Lipiński, L., Kowal, A., Steur, P.P.M., Pavese, F.: Argon Triple-Point Device for Calibration of SPRTs. Int. J. Thermophys. 36 (2-3), 229-239 (2015), doi:10.1007/s10765-014-1768-0
4. Mikhal, A.A., Warsza, Z.L.: Simple Methods to Measure the Additive Error and Integral Nonlinearity of Precision Thermometric Bridges. Advances in Intelligent Systems and Computing 352, (2015) 157-170, doi:10.1007/978-3-319-15835-8_18
5. Kreuzer, M.: Linearity and Sensitivity Error in the Use of Single Strain Gages with Voltage-Fed and Current-Fed Circuits. Experimental Techniques 8 (10), 30-35 (1984), doi: 10.1111/j.1747-1567.1984.tb01835.x
6. Sydenham, P.H., Thorn, R.: Handbook of Measuring System Design. Wiley (2005)
7. Warsza, Z.L.: Accuracy Measures of the Four Arm Bridge of Broadly Variable Resistances. Proceedings of 15th IMEKO TC4 International Symposium on Novelties in Electrical Measurement and Instrumentations, 17-28 (2007)
8. Warsza, Z.L.: Four Terminal 4T Immittance Circuits in Multivariable Measurements - Immitancyjne układy czterobiegunowe 4T w pomiarach wieloparametrowych. Przemysłowy Instytut Automatyki i Pomiarów PIAP, Warszawa (2004)
9. Warsza, Z. L., Idzkowski, A., Walendziuk, W.: The parameters of unconventional double-current circuit, their accuracy measures and measurement of strain and temperature. International Journal of Electronics and Telecommunication 60 (4), 327-330 (2014), doi: 10.2478/eletel-2014-0043
10. Guide to the expression of uncertainty in measurement JCGM 100: 2008, http://www.bipm.org/utils/common/documents/jcgm/JCGM_100_2008_E.pdf

Evaluation of damping characteristics of a damper with magneto-rheological fluid

Vytautas Bucinskas[1], Andrius Klevinskis[1], Nikolaj Sesok[1], Igor Iljin[1] and Zygmunt Lech Warsza[2]

[1] Vilnius Gediminas Technical University, Vilnius, Lithuania
{vytautas.bucinskas,andrius.klevinskis,nikolaj.sesok, igor.iljin}@vgtu.lt
[2] Industrial Research Institute for Automation and Measurements, Warsaw, Poland
zlw@op.pl

Abstract. There are many applications of magneto-rheological (MR) damping devices, utilizing their properties of simple damping control under the application of a magnetic field. Nevertheless, damping properties of the known MR dampers remain non-linear; therefore, implementation of such devices within a wide range of vibrations requires further research. This paper is one of the attempts aimed at investigating damping characteristics of an MR damper through experimental research and further processing of experimental data according to the proposed methodology. The experimentally obtained data processed to a form of hysteresis curve; its numerical integration gave the amount of energy loss during one cycle of vibration. After mathematical transformations, in a result of calculus, the coefficient of friction obtained. Finally, the MR damper characteristics, as dependency of a damping coefficient on frequency of vibration under the application of different voltage to an MR damper solenoid, obtained. The paper concludes by providing the summary of the results and recommendations.

Keywords: Nonlinear damping; rheological liquid; damping control; experimental research

1 Introduction

Magneto-rheological (MR) fluids are suspensions of magnetically responsive particles in a liquid carrier [1, 2 and 3]. The usage of such liquids improves technical characteristics of shock absorbers because of ability to change its rheological properties in the presence of a magnetic field [4]. Magnetic flux density B depends on magnetic field intensity H in a non-linear manner. The B–H curve for MR fluids throughout their useful range is non-linear and of such a magnitude placing them into a very unique category of intermediate materials between low susceptibility materials such as aluminum, and ferrous materials such as steel [1]. In general, two methods for modelling MR shock absorbers can be distinguished. One is a non-parametric modelling tech-

R. Jabłoński and T. Brezina (eds.), *Advanced Mechatronics Solutions*,
Advances in Intelligent Systems and Computing 393,
DOI 10.1007/978-3-319-23923-1_61

nique that expresses analytical equations to describe the characteristics of the MR shock absorber. There are several different models of non-linear models used for the characterization of MR dampers including Bingham plastic [5], Stanway et al. [6], Herschel–Bulkley [7], Gamota and Filisko [8]. The second technique to model MR shock absorbers is a parametric modeling technique, which characterizes the device as a collection of springs, dampers and other physical elements. In this study, the parametric modeling technique for MR shock absorber analysis will be used.

There are three main (basic) types of linear MR shock absorbers: the single tube, the twin tube and the double ended [9, 10]. Previous studies [11] have shown that, taking into account a number of components of a construction, size, ease of use, ease of fabrication (simplicity) and ease of assembly/disassembly, a single cylinder and a slide valve - shock absorbers offer the highest performance. Therefore, for the purpose of this study a shock absorber of this particular structure analyzed in this paper. Moreover, a double-ended MR shock absorber (Fig. 1) is investigated since there is no change in volume as the piston rod moves relative to the damper body, and the double-ended MR shock absorber does not require an accumulator.

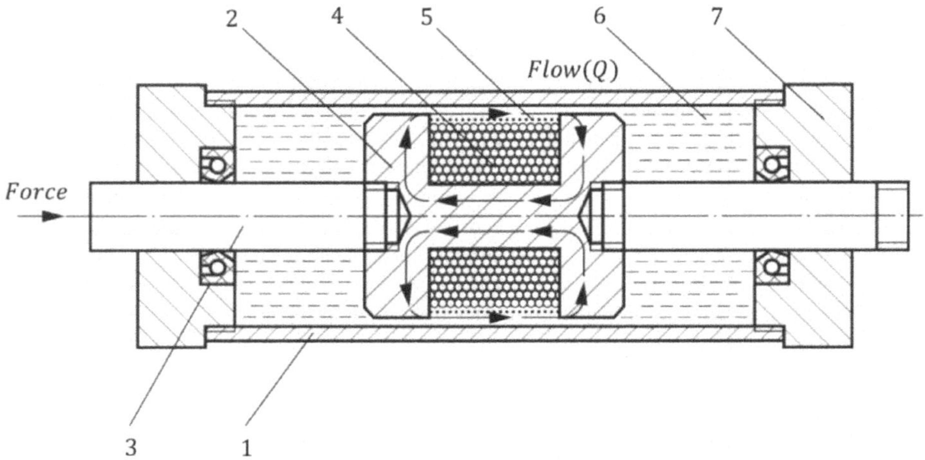

Fig. 1. Principle scheme of MR shock absorber: 1 – cylinder; 2 – piston; 3 – rod; 4 – wires; 5 – silicone; 6 – MRF–122EG; 7 – core

The objective of this study is to investigate the non-linear damping characteristics of a magnetic damper, find their dependencies from the frequency of vibrations, and control current in the damper magnetic coil.

2 Experimental research

In order to conduct an experiment on MR damper properties, an original test bench, presented in Fig. 2, designed and built. A special test rig consists of an MR shock absorber 1, rigidly attached to the table 3 through the frame 2. The electro-dynamic vibrator 4, mounted on the basement 3 and an active platform of the electro-dynamic vibrator part tightly connected to a damper rod through a spring 5. Both ends of spring 5 are equipped with accelerometers 6 and 7. The accelerometers register vibrations acceleration and transfer signals to data acquisition system for data storage and further processing.

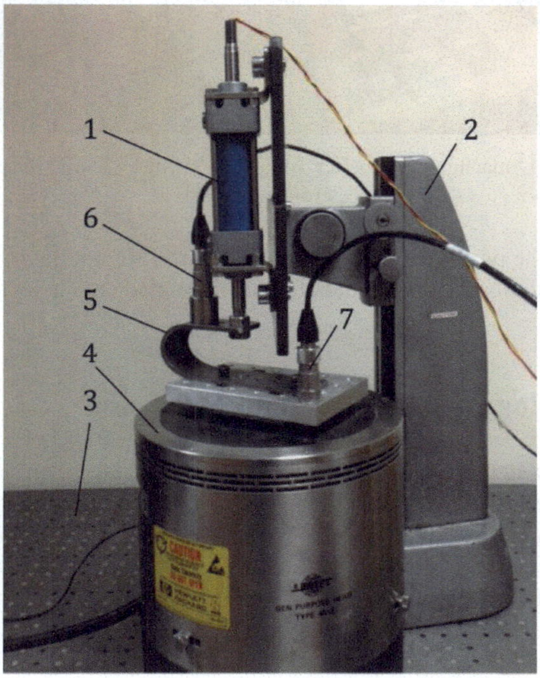

Fig. 2. MR shock absorber test bench: 1 – MR shock absorber, 2 – stand, 3 – table, 4 – electro-dynamic vibrator, 5 – spring, 6, 7 – sensor

The body of MR shock absorber rigidly attached to the table 3 through frame 2, so the electro-dynamic vibrator, by moving the spring, moves along with the damper rod. In this case, the test rig installation has only one active degree of freedom. The idea of this installation is in small mass spring 5 placed between the vibrator 4 and MR damper 1; then damping force of the damper 1 will bend the spring 5. Knowing stiffness of the spring 5, the damping force here estimated as the difference of displacements on both ends of the spring 5. These displacements, obtained from values of accelerations from sensors 6 and 7, after numerical integration two times. Signal of exciting force recorded along the accelerator signals; this allows keeping the correct phase during integration procedure. Used experimental data acquisition system and the sensors presented in Fig. 3.

A possible use of sensors of a different type (contactless Eddie current or capacitance) here is very inconvenient due to moving of both ends of spring 5. Data, collected from accelerometers, is stored in registering computer and processed afterwards. The data processing methodology described below.

Fig. 3. „Brüel&Kjær“ machine diagnostics Toolbox – “Type 9727“; a) diagnostics toolbox, b) accelerometer “8341”

The results of the performed experimental research are not directly discussed here; the detailed presentation (and interpretation) of generated data is included in the result section.

3 Methodology of analytical data processing

A methodology to define the damping force based on the main assumption that it represented by the spring mounted between the vibrator and the damper, and strain value. Some methodologies, proposed in [9, 18] required special means while this method allows to achieve the same results using a special spring in the workbench dynamic system. For further dynamic system analysis, a dynamic model of experimental test bench was built. In this single DOF model, the damping element is nonlinear and depends on voltage on the solenoid coil and the frequency of vibration. Also, it depends on velocity of system (including vibrator) movement, but in terms of vibration frequency at the same amplitude, it is sufficient [13]. A similar experimental evaluation of an MR damper is done in the papers [15, 16, and 18].

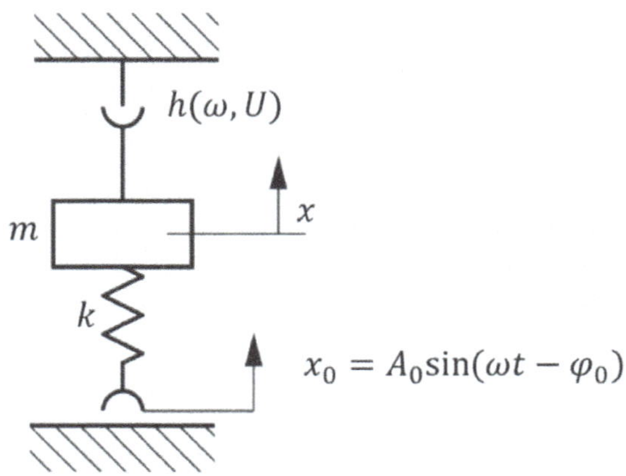

Fig. 4. Dynamic model of the MR shock absorber test bench

A dynamical model of the workbench contains the three components: moving mass m, stiffness of the spring k, and damping coefficient $h(\omega,U)$ of a damper (Fig. 4). Additionally, the experiment performed under the following assumptions: spring damping and mass are not considered, because they are much smaller than other parts of the workbench. Mass m in the model is the sum of vibrator and beam mass.

The excitation system is applied cinematically to coordinate x_0. The vibrator creates certain force with given amplitude, but in this case, much more convenient is the use of kinematic parameters, while the force from the vibrator in this case cannot be defined directly. The characteristics of the vibrator as a mechanical system available from its technical documentation, its influence added into dynamic model. The system was excited harmonically:

$$x_0 = A_0 \sin(\omega t - \varphi_0), \tag{1}$$

where x_0 – initial coordinate of system, A_0 – initial amplitude of vibration, ω – angular frequency of excitation, φ_0 – phase angle.

Kinetic energy expressed as:

$$T = \frac{1}{2} m \dot{x}^2, \tag{2}$$

where T – kinetic energy, m – mass, $\dot{x}^2$ – derivative of coordinate in respect to time, velocity.

Dissipation function:

$$\Phi = \frac{1}{2} h(\omega,U)\, \dot{x}^2, \tag{3}$$

where Φ – dissipation function, $h(\omega,U)$ – coefficient of friction, which is a function of angular frequency and voltage on the damper solenoid coil.

Potential energy:

$$\Pi = \frac{1}{2} k(x_0 - x)^2, \tag{4}$$

where Π – potential energy, k – coefficient of stiffness, x – active coordinate.

Using the Lagrange equation of the second kind [13]:

$$\frac{d}{dt}\left(\frac{\partial T}{\partial \dot{x}}\right)-\frac{\partial T}{\partial x}+\frac{\partial \phi}{\partial \dot{x}}+\frac{\partial \Pi}{\partial x}=F(t), \tag{5}$$

where $F(t)$ – force of excitation.

We obtain the vibration equation of the system in terms:

$$m\ddot{x}+h(\omega,U)\dot{x}+kx=kx_0\,. \tag{6}$$

In the second stage of this study, the MR shock absorber is excited at different frequencies and solenoid acting at various voltages. The experiment was performed at 0V, 5V, 10V and 15V of electrical excitation voltage, and a frequency of 80 Hz, 120 Hz, 200 Hz, 240 Hz and 280 Hz.

The achieved experimental research results for the test bench suggest that sinusoidal excitation gives sinusoidal response, therefore it can be stated that:

$$\ddot{x}=-A\omega^2\sin(\omega t+\varphi), \tag{7}$$

$$\dot{x}=-A\omega^2\cos(\omega t+\varphi), \tag{8}$$

$$x=A\omega\sin(\omega t=\varphi), \tag{9}$$

$$\ddot{x}_0=-A\omega_0^2\sin(\omega t+\varphi_0), \tag{10}$$

$$\dot{x}_0=-A\omega^2\cos(\omega t+\varphi_0), \tag{11}$$

$$x_0=A_0\omega\sin(\omega t+\varphi_0). \tag{12}$$

In this way, with the accelerometers data $\ddot{x}$ and $\ddot{x}_0$, the incoming displacements x and x_0 for Eq. (6) could be obtained. Next, the acceleration data stored on computer disk and filtered as illustrated in Fig. 5.

Knowing $\ddot{x}$, x and x_0, from Eq. (5), the damping force can be expressed as:

$$F_{DM}=h(\omega,U)\dot{x}=k(x_0-x)-m\ddot{x}\,. \tag{13}$$

The damping force values, obtained from the experimental research are presented in Fig. 6:

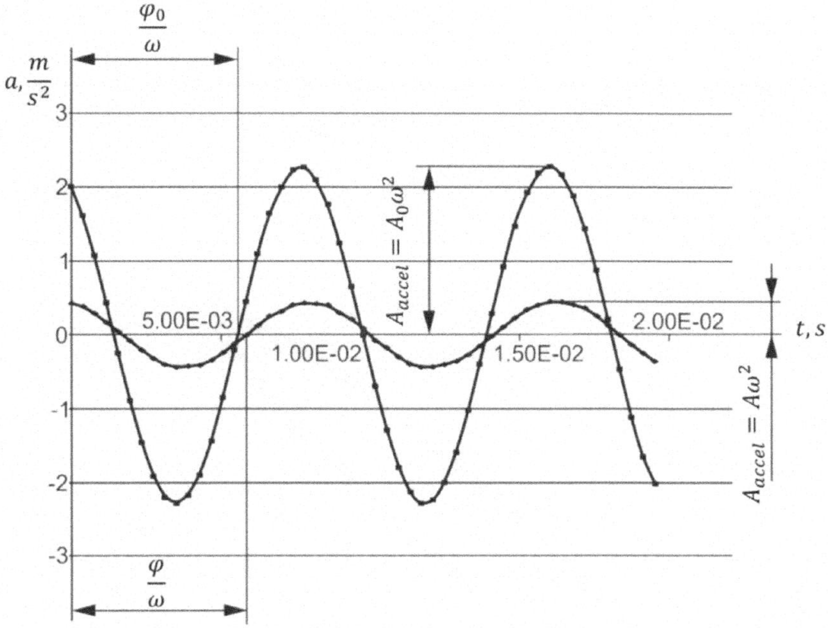

Fig. 5. Experimentally obtained oscillation amplitude and phase identification

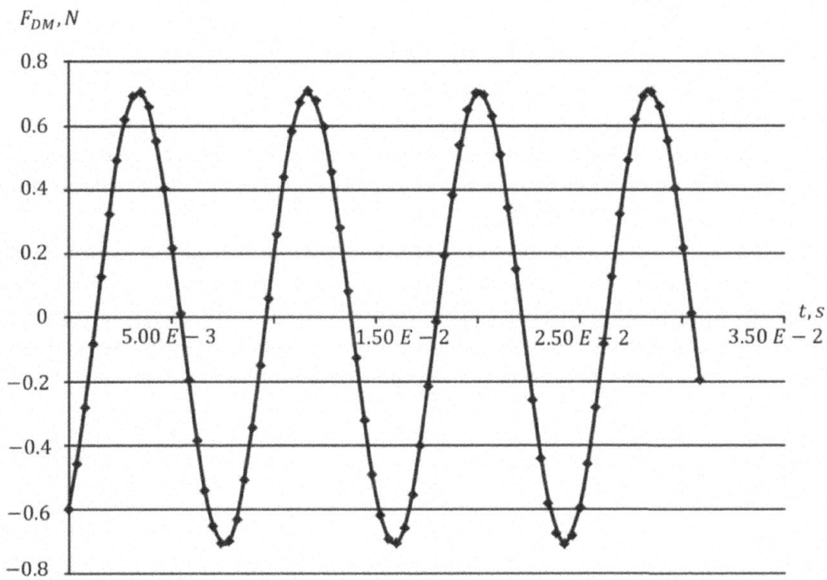

Fig. 6. The damping force change F_{DM} when $\nu = 120$ Hz and $U = 5$ V

The energy loss per cycle equals:

$$\Delta E = \pi \omega h A^2 . \tag{14}$$

Since ΔE is equal to the hysteresis loop area, which defined numerically, when having values F_{DM} and x, we obtaining wanted damping coefficient from the following equation:

$$h = \frac{\Delta E}{\pi \omega A^2} \tag{15}$$

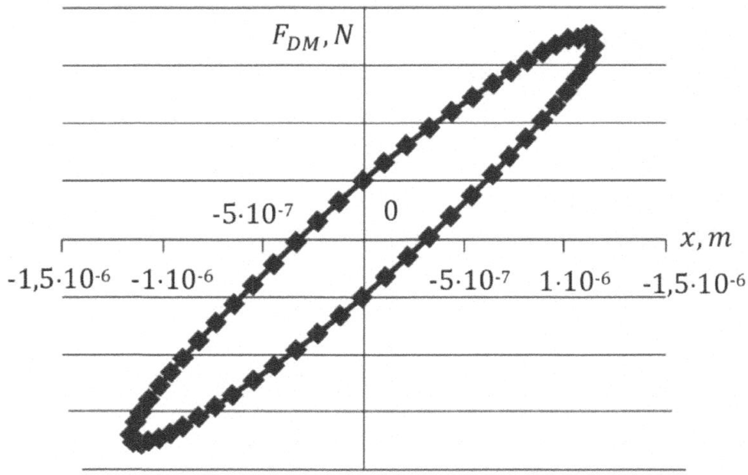

Fig. 7. Damping process hysteresis loop, when $\nu = 120$ Hz and $U = 5$ V

The gained values from this research presented in the result section as a graph of dependency of damping coefficient h values to excitation frequency and voltage, applied to the solenoid coil in the MR damper.

The current flow resistance through a coil and friction forces increasing temperature of the device and causes decreasing of the fluid‘s viscosity, which, in consequences, causes decreasing of a dissipative properties of a MR shock absorber [22]. Within the research process, the temperature the MR shock absorber remained low, therefore the changes of dissipative properties of a magneto-rheological fluid not taken into account.

4 Results

The graphs of damping coefficient change of the investigated MR damper are presented in Fig. 8. They not only indicate sophisticated processes within the damper but also (clearly) illustrate the effects of different physical phenomena.

A practical value of such a dependency consists in the range of frequency for this shock absorber, when magnetic field values have no significant influence on damper efficiency; the frequency range from 80 Hz to 240 Hz is operational for such a damper. In case of a damping control within one vibration period cycle, the frequency range will be significantly lower, as concluded from [18, 19 and 20].

5 Conclusions

The performed analysis on damping characteristics of a magneto-rheological damper has brought interesting results.

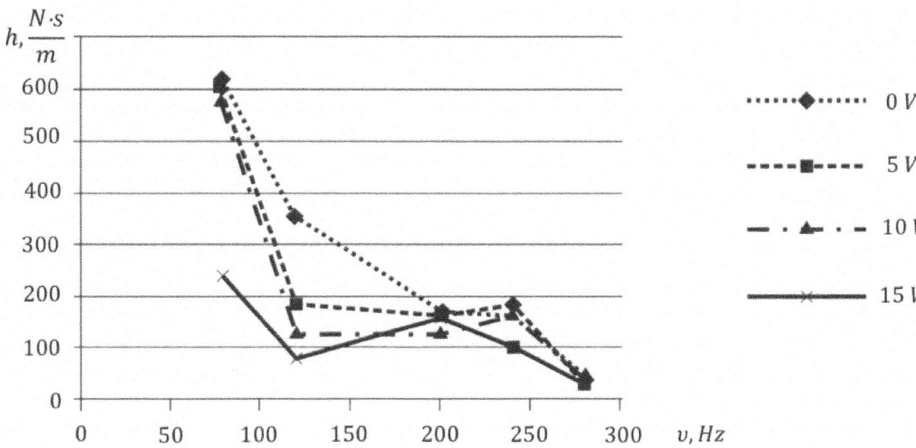

Fig. 8. Diagram of MR shock absorber damping coefficient variation

Firstly, it has shown that the proposed experimental-analytical methodology for detection of damping properties is applicable for various applications.

Secondly, it has revealed this methodology to be particularly useful for practical installations because of its simplicity – experimental research requires only two accelerometers, applied to two sides of the MR damper. Taken together, in the result of the experiment, the following conclusions can be drawn:

- The coefficient of damping generally decreases with an increase of damping frequency;
- An MR fluid in the frequency range higher than 240 Hz is not applicable;
- In higher end of frequency range, an MR fluid behaves like a hydraulic fluid rather than fluid with MR properties;
- The behaviour of MR damper influenced by several physical phenomena and experimental research delivers interference of them.

References

1. Carlson, D.: Magnetorheological Fluids, Smart Materials. Sud. CRC Press (2008)
2. Berger, J., Rubinstein, J.: Connectivity and superconductivity, 62. Springer (2001)

3. Bajkowski, J., Skalski, P.: Analysis of Viscoplastic Properties of a Magnetorherological Fluid in a Damper. J. Acta Mechanica et Automatica, 6 (3), 5–10 (2012)
4. Karimi, H. R.: A Semiactive Vibration Control Design for Suspension Systems with MR Dampers (1998)
5. Phillips, R. W.: Engineering applications of fluids with a variable yield stress (1969)
6. Stanway, J. L. S. R.: Non-linear modelling of an electro-rheological vibration damper, J. Electrost, 2, 167–184 (1987)
7. Ashour, O., Rogers, C. A., Kordonsky, W.: Magnetorheological Fluids: Materials, Characterization, and Devices, J. Intell. Mater. Syst. Struct, 7 (2), 123–130 (1996)
8. Gamota, D. R., Filisko, F. E.: Dynamic mechanical studies of electrorheological materials: Moderate frequencies, J. Rheol. 1978-Present, 35 (3), 399–425 (1991)
9. Ashfak, A., Saheed, A., K. A. Rasheed, A., Jaleel, J. A.: Design, Fabrication and Evaluation of MR Damper, Int. J. Aerosp. Mech. Eng., 1, 27–33 (2011)
10. Norris, J. A.: Behavior of magneto-rheological fluids subject to impact and shock loading, Virginia Polytechnic Institute and State University (2003)
11. Li, W. H., Du, H., Guo, N. Q.: Design and testing of an MR steering damper for motorcycles, Int. J. Adv. Manuf. Technol., 22 (3–4), 288–294 (2003)
12. Catania, G., Sorrentino, S.: Experimental evaluation of the damping properties of beams and thin-walled structures made of polymeric materials, In the IMAC-XXVII, Orlando, Florida USA, Society for Experimental Mechanics Inc. (2009)
13. Tse, F. S., Morse, R., T. Hinkle, F. S.: Solutions manual to accompany Mechanical vibrations: theory and applications, second edition. Boston: Allyn and Bacon (1978)
14. Bert, C. W.: Material damping: An introductory review of mathematic measures and experimental technique, J. of Sound and Vibration, 29 (2), 129–153 (1973)
15. Mucchi, E.: Experimental evaluation of modal damping in automotive components with different constraint conditions, J. Meccanica, 47 (4), 1035-1041 (2012)
16. Biswal, D. K., Bandopadhya, D., Dwivedy, S. K.: Evaluation of Electromechanical, Damping and Dynamic Mechanical Properties of Silver Electrode IPMC Actuator, Adv. Nanomaterials and Nanotechnology, Springer Proc. in Physics, 143, 321-331 (2013)
17. Regazzo, P., Ghiotti, A.: Development of innovative systems based on smart fluids technology for the improvement of the dimensional accuracy in sheet metal forming operations, Int. J. of Material Forming, 2 (1), 841-844 (2009)
18. Mcmanus, S. J., St. Clair, K. A., Boileau, P.É., Boutin,J., Rakheja, S.: Evaluation of vibration and shock attenuation performance of a suspension seat with a semi-active magnetorheological fluid damper, J. of Sound and Vibration, 253 (1), 313–327 (2002)
19. Metered, H., Bonello, P., Oyadiji, S. O.: The experimental identification of magnetorheological dampers and evaluation of their controllers, Mech. Syst. and Sig. Processing, 24 (4), 976–994 (2010)
20. Koo, J. H., Goncalves, F. D., Ahmadian, M.: Investigation of the response time of magnetorheological fluid dampers, Proc. SPIE 5386, Smart Structures and Materials: Damping and Isolation, 63 (2004)
21. Wereley, N. M., Pang, Li, Kamath, G. M.: Idealized Hysteresis Modeling of Electrorheological and Magnetorheological Dampers, J. of Intelligent Material Systems and Structures, 9 (8), 642-649 (1998)
22. Bajkowski, J., Bajkowski, M., Grzesikiewicz, W., Sofonea, M., Shillor, M., Zalewski, R.: Analysis of the dependence between a temperature and working parameters of the MR damper, J. Mechanics, 26 (4), 149–155 (2007)

Method of evaluation the measurement uncertainty of the minimal value of observations and its application in testing of plastic products

Mykhaylo Dorozhovets[1], Ivanna Popovych[2] and Zygmunt L. Warsza[3]

[1] Rzeszow University of Technology, Rzeszow, Poland
michdor@prz.edu.pl
[2] National University - Lviv Politechnic, Lviv, Ukraine
popovych.i@ukr.net
[3] Industrial Research Institute of Automation and Measurement (PIAP), Warsaw, Poland
zlw@op.pl

Abstract. In this paper the non-standard statistical method for evaluating uncertainty, when a minimal observation is the measurement result, is proposed. This method is based on properties of minimal order statistic. As example, in measurement practice this method is used to evaluate the uncertainty of a percent elongation and tensile strength in testing plastic products. The accuracy of obtained experimental results is also investigated by Monte Carlo simulation. Proposed method can be used to uncertainty evaluation of another quantities, when the minimal (or maximal) value of several normally distributed random observations is the result of measurement.

Keywords: quality testing of plastic tubes · uncertainty of minimal observation.

1 Introduction

The products' quality control and testing is the necessary element of any industrial production. In this article the example of measurements for testing the quality control of plastic tubes is considered. A general approach of processing of these measurements is given. During this test two parameters are measured - percent elongation and tensile strength of the plastic tube in the process of their break. The uncertainty of the minimal value of observations should be found. Methods of its estimation has to be found, developed and tested.

In accordance with regulatory documents [1, 2, 3], the test procedure of plastic tube has to be carried for not less than 24 h after manufacturing the tubes. The experimental research with quality control of plastic tubes used for gas and water networks or other needs is performed in the laboratory of "Elplast-Lviv" Ltd. [4]. The test specimens of shapes, shown in Fig. 1, are prepared by cutting from segments tube, with length not less than 150 mm. From each control segment at least five specimens should be cut uniformly around the tube perimeter.

R. Jabłoński and T. Brezina (eds.), *Advanced Mechatronics Solutions*,
Advances in Intelligent Systems and Computing 393,
DOI 10.1007/978-3-319-23923-1_62

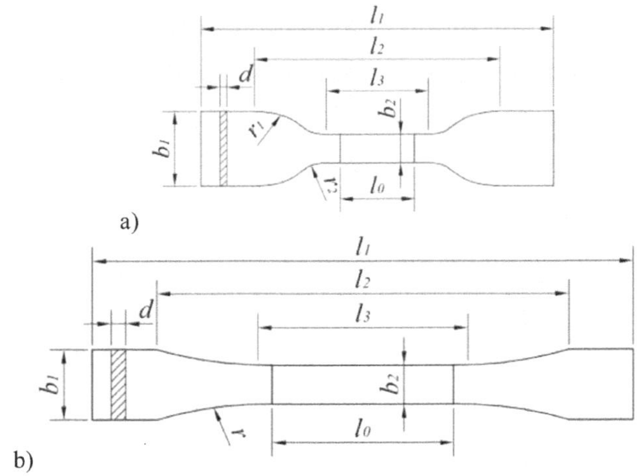

Fig. 1. Shape and dimensions of the test specimens: a) Type 1, b) Type 2

Before the beginning of test procedure, samples are conditioned for minimum 2 h at temperature (23 ± 2) °C. For the cold water tubes testing is made on type 2 test specimens. It is carried with the speed tensile of the grips of testing machine (100 ± 10) mm/min according to the nominal wall thickness. It is up to 5 mm for test specimens of type 1 and more than 5 mm for test specimens of type 2 [6]. Tubes for combustible gases are also tested by type 2 specimens [1, 2]. The tensile tests are carried on testing machine which provides measurement accuracy of load (from 5000 to 10000 N) with permissible error less than ±1% of measuring value. In addition, the speed of testing may be controlled in a wide range. Calculation of the percent elongation ε_p at break, given in (1), should be rounded to two significant figures [1, 2]:

$$\varepsilon_p = \frac{\Delta l_p}{L_0} \cdot 100\%, \tag{1}$$

where: Δl_P - elongation of the test specimen at the moment of break (in mm), as the measurement result of its smallest value, L_0 - initial length of test specimen (in mm).

For determination the tensile strength at yield, the thickness D and width b_2 in working part is measured with a micrometer. It is made minimum in three cross-sections with permissible deviations of the width up to ±0.05 mm and of the thickness - up to ± 0.01 mm. Based on the measurement results with an accuracy up to ± 0.001 mm². The initial cross-sectional area of each test specimen is calculated as

$$A_0 = D \cdot b_2. \tag{2}$$

In calculating of the tensile strength at yield σ_{PT}, the minimum of cross-sectional area of the test specimen A_0 (in mm^2) and load F_{PT} (in N) are used. Then the tensile strength at yield is achieved:

$$\sigma_{PT} = \frac{F_{PT}}{A_0}. \tag{3}$$

2 Measurement results

For each tested specimen the testing machine records the elongation Δl_P (in mm) at the moment of the specimen break. Results of these measurements are given in the third column of Table 1 [4], Results of load F_{PT} (in N), at which the tensile strength is achieved at yield, are in the fifth column of Tab. 1 [4].

Table 1. Results of measurement obtained from testing machine and calculated parameters of test specimens [4]

Number of specimen	A_0, mm^2	Δl_P, mm	ε_P, %	F_{PT}, N	σ_{PT}, N/mm^2
The test specimen of type 1					
1	20.030	145	583.50	453	22.62
2	19.405	140	**563.38**	440	22.68
3	18.942	147	591.55	426	**22.49**
4	19.250	146	587.53	433	22.49
5	19.034	145	583.50	430	22.59
The test specimen of type 2					
1	93.425	286	575.45	2090	22.37
2	93.845	287	577.47	2064	21.99
3	93.085	280	**563.38**	2060	22.13
4	93.859	285	573.44	2051	**21.85**
5	94.276	287	577.47	2078	22.04

A_0 = initial cross-sectional area; Δl_P = valid elongation of test specimen at the moment of break; ε_P = percent elongation of the test specimen at the moment of break; F_{PT} = load of the tensile strength at yield; σ_{PT} = tensile strength at yield.

Obtained results of measurements are used for calculation the percent elongation at break ε_P and tensile strength at yield σ_{PT} for each of five test specimens. According to the test requirements, as the result the minimum values of the tensile strength at yield (3) and tensile strength at yield are calculated, and can be rounded to two significant digits.

The test specimen of type 1 (for water pipes norm of $\varepsilon_{P,\,per} = 350$ %) has the percent elongation at break 563.38 % > 350 % and the test specimen of type 2 (for gas pipe norm of $\varepsilon_{P,\,per} = 500$ %) has 563.38 % > 500 %. Then, due to the percent elongation at break, these products meet the requirements of the Ukraine standards, [3], which are in practice very near to EC and ASTM standards [1], [2] (cold water pipe norm: $\sigma_{PT,\,per} = 21.0$ N/mm^2; gas pipe norm: $\sigma_{PT,\,per} = 20.0$ N/mm^2). The actual value of the tensile strength at yield of type 1 is 22.5 N/mm^2 > 21.0 N/mm^2 and for the test specimen of a type 2 is 21.9 N/mm^2 > 20.0 N/mm^2. Then the tensile strength at yield products meet the requirements of Ukraine standard documents [3].

3 Theory of minimal value uncertainties

The uncertainty of the minimum value of percent elongation and tensile strength has two components:

- instrumental component caused by uncertainty indications of the measuring machine, additional equipment (such as caliper and micrometer) and measurement conditions different from normal,

- statistical component caused by the dispersion of observations.

Therefore the processing uncertainty of obtained results is estimated by two GUM methods [5]:

- Method type B is based on analysis and calculation of the uncertainty of metrological properties of used measuring instruments, measurement conditions and equations describing the dependence of desired parameters from directly obtained measurements results. This method is used to evaluate the standard uncertainty $u_{cB}(\varepsilon_{p1})$.
- Method type A - component $u_A(\varepsilon_{p1})$ of uncertainty of a minimal value is evaluated for n independent observations taken from the normal distribution.

The component type B of uncertainty is not analyzed below in detail, the previously calculated in [4] these components are presented in Table 1. There are:

- the relative combined standard uncertainty of relative elongation calculated $u_{cB,\,rel}(\varepsilon_{pl}) = 0.41\%$ for the test specimen of type 1 and $u_{cB,\,rel}(\varepsilon_{pl}) = 0.20\%$ for the test specimen type 2 [4];

- the relative combined standard uncertainty of the tensile strength $u_{cB,\,rel}(\sigma_{PT}) = 0.614$ % for the test specimen type 1, and $u_{cB,\,rel}(\sigma_{PT}) = 0.584$ % for the test specimen of type 2 [4].

The next stage is the estimation of uncertainty component of the minimal value by statistical method (type A). Problem of the evaluation type A uncertainty in testing of the percent elongation and tensile strength is, as noted above, that have to be find the *minimum values* of parameters of the test specimens. Therefore, it is impossible to apply directly the standard GUM method of uncertainty evaluation of measurements with multiple observations [5].

In this work estimation of the uncertainty of minimum values of controlled parameters from the sample of five elements is performed. The minimal observation $x_{\min} = x_{(1)} = \min(x_1, x_2, \ldots, x_n)$ is the first one from the set of ordered observations: $x_{(1)} \le x_{(2)} \le x_{(3)} \le \ldots \le x_{(n)}$. The result of a test measurement is not as usual the arithmetic mean $(\bar{x})$ but the minimal (or maximal) value of observations. Then, the standard and expanded uncertainties of test results cannot be calculated according to standard GUM procedures [5]. Another procedure should be used.

It is obvious that minimal value is a random value, however its probability density function (PDF) is not equal to the PDF $p(x)$ of population.

In the next sections of this paper the minimal observation $x_{(1)}$ is denoted by x_1. Theoretical distribution $p(x_1)$ of minimal value x_1 for the normally distributed observations ($m=0, \sigma=1$, $p(x_1)=\frac{1}{\sqrt{2\pi}}\exp(-x^2/2)$, $F(x_1)=\frac{1}{\sqrt{2\pi}}\int\limits_{-\infty}^{x_1}\exp(-x^2/2)dx$) can be described [6] by formula:

$$p(x_1)=n\cdot\frac{1}{\sqrt{2\pi}}\exp(-x^2/2)\cdot[1-F(x_1)]^{n-1}. \tag{4}$$

This distribution for $n=5$ is presented on Fig. 2a. From (4) the expected value m_{01} of x_1 can be calculated as:

$$m_{0,1}=\int\limits_{-\infty}^{\infty}x_1 p(x_1)dx_1 \tag{5}$$

and $\sigma_{0.1}$ standard deviation of the minimal observation:

$$\sigma_{0,1}=\int\limits_{-\infty}^{\infty}x_1^2 p(x_1)dx_1-m_{0,1}^2. \tag{6}$$

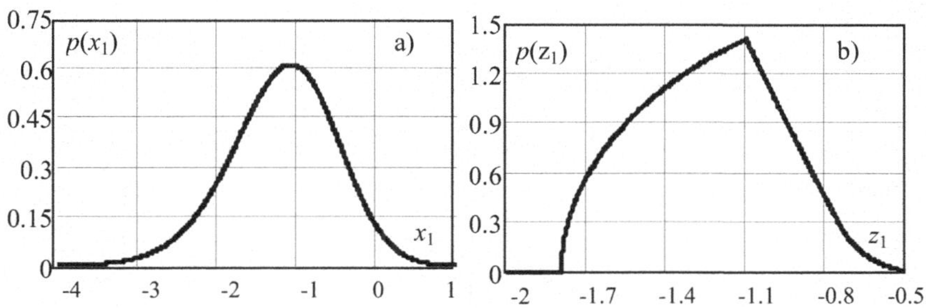

Fig. 2. Distribution of the minimal observation x_1 a) and normalized deviation z_1 ($n=5$) b)

For $n=5$ observations from (5) the expected value of normalized $m_{0.1}=-1.16296$ and from (6) the standard deviation $\sigma_{0.1}=0.66898$. If $m\neq 0$ and $\sigma\neq 1$ then expected value m_1 and standard deviation σ_1 of the minimal observation of x_1 are:

$$m_1=m+m_{0.1}\cdot\sigma;\qquad \sigma_1=\sigma_{0.1}\cdot\sigma. \tag{7}$$

In practice the expected value m_1 of minimal observation of x_1 is unknown, but after (7) the estimate $\hat{x}_1$ of m_1 can be calculated as

$$\hat{x}_1 = \bar{x} + m_{0,1} \cdot s_x \ , \tag{8}$$

where arithmetical mean and experimental standard deviation of observations are

$$\bar{x} = \frac{1}{n}\sum_{i=1}^{n} x_i \ , \ s_x = \sqrt{\frac{1}{n-1}\sum_{i=1}^{n}\left(x_i - \bar{x}\right)^2} \ . \tag{9}$$

From (7) the experimental standard uncertainty of minimal value is calculated as

$$u_A(x_1) = \sigma_{0.1} \cdot s_x \ . \tag{10}$$

The distribution $p_{z1}(z_1)$ of the minimal observation x_1 deviation from mean value $\bar{x}$, normalized to s_x is

$$z_1 = \frac{x_1 - \bar{x}}{s_x} \ . \tag{11}$$

This distribution does not depend on $\bar{x}$ and s_x. It depends only on population distribution $p(x)$ and number of observation n. It can be shown that the range of random value z_1 is independent from population PDF and equals to

$$(n-1)/\sqrt{n} \le z_1 \le 1/\sqrt{n} \ . \tag{12}$$

The distribution $p_{z1}(z_1)$ consists of n - 1 sections, with bounds $z_{b,i}$ (i = 1, 2,.. n - 1) that are determined by the formula:

$$z_{b,i} = -\sqrt{(n-1)(n-i)/(n \cdot i)}, \quad i = 1, 2, \ldots, n-1 \ . \tag{13}$$

In test procedure the minimal observation x_1 is compared with the critical value x_{critic}, then after determination x_1, the left side of expanded uncertainty $U_{p,low}(x_1)$ must be calculated as

$$x_1 - U_{p,low}(x_1) \ge x_{critic} \ . \tag{14}$$

For the very small number of observations (for example n = 5) the most important is the first part (left section) with bounds

$$z_{b,1} = -(n-1)/\sqrt{n}, \quad z_{b,2} = -\sqrt{(n-1)(n-2)/2n} \ . \tag{15}$$

If $n = 5$ then $z_{b,1} = -4/\sqrt{5} \approx -1.7889$, $z_{b,2} = -\sqrt{6/5} \approx -1.0954$, because at the end of the first part the cumulative function is

$$F_{z1}(z_1) = \int_{-(n-1)/\sqrt{n}}^{-\sqrt{(n-1)(n-2)/2n}} p_{z1}(z_1) dz_1 > 0.10 \ . \tag{16}$$

For normally distributed n =5 observations, the theoretical distribution $p_{z1}(z_1)$ at the left side part can be described as

$$p_{z1}(z_1)=\frac{5\sqrt{5}}{2\pi}\sqrt{1-\frac{5}{16}z_1^2}\,, \qquad -\frac{4}{\sqrt{5}}\le z_1\le -\sqrt{\frac{6}{5}}\,. \tag{17}$$

From (17) cumulative function in this part is:

$$F_{z1}(z_1)=\int_{-\sqrt{2}}^{z_1} p_{z1}(z_1)dz_1=\frac{5}{2}\left[\frac{1}{2\pi}\cdot z_1\sqrt{5-\left(\frac{5}{4}z_1\right)^2}+\frac{2}{\pi}\arcsin\left(\frac{\sqrt{5}}{4}z_1\right)+1\right]. \tag{18}$$

For $z_1=-\sqrt{6/5}$ the cumulative function is $F_{z1}\left(-\sqrt{6/5}\right)=0.6806$. Total distribution $p_{z1}(z_1)$ of z_1 is shown in Fig. 2b. The lower $k_{low}(p)$ coverage factor for the confidence level p can be calculated from equation

$$\int_{-2}^{k_{low}(n,p)} p_{z1}(z_1)dz_1=F_{z1}(z_1)=1-p\,. \tag{19}$$

The values of $k_{low}(5,p)$ for p = 0.90; 0.95; 0.975 and 0.995 and for n =5 are presented in Tab. 2.

Table 2. The numeric values of coverage factors

p	0.90	0.95	0.975	0.99	0.995
$k_{low}(5,p)$	-1.6016	-1.6714	-1.7156	-1.7489	-1.7637

From (11) and Tab. 2 the lower limit $U_{1-p,low}(x_1)$ of expanded uncertainties of minimal value is

$$x_{1,p}=\bar{x}+k_{low}(n,p)\cdot s_x\,. \tag{20}$$

4 Evaluation of uncertainties from experimental results

Below the uncertainty $u_A(x_1)$ is evaluated and analyzed. Using results of a measurement presented in Tab. 1, the following parameters of minimal value of percent elongation and tensile straight of each specimen were calculated (results of calculations are shown in Tab. 3):

- minimal value x_{min}, arithmetic mean $\bar{x}$ and standard deviation s_x (9);
- left limit $x_{1,\min}$ (11), (12) and the estimate $\hat{x}_1$ of expected value (8);
- estimated standard uncertainty $u_A(x_1)$ (10);
- relative standard uncertainty $u_{A,rel}(x_1)$;

- lower $x_{1,1-p}$ limit of expanded uncertainty ($p = 0.95$) (20),
- relative combined standard uncertainty $u_{c,rel}(x_1)$ (using previously relative standard uncertainty $u_{cBrel}(x_1)$, Tab. 3);
- combined standard uncertainty $u_c(x_1)$.

From the last line of Tab. 3 one can see that these lower limits of expanded uncertainties ($p = 0.95$) are very closed to experimental values (Tab. 1), i.e. for percent elongation calculated minimal values are: 563.72 % (specimen 1), 563.84 % (specimen 2) and both experimental minimal values are equal to 563.38% (Tab. 1), for tensile strength calculated values for specimen 1 is 22.44 N/mm^2 - experimental values is 22.49 N/mm^2 and for specimen 2 calculated value is 21.76 N/mm^2, and experimental values is 21.85 N/mm^2.

Table 3. Results of uncertainty calculations of percent elongation and tensile strength

Parameters	Percent elongation		Tensile strength	
	Specimen 1	Specimen 2	Specimen 1	Specimen 2
Minimal value: ε_{p1}, $\sigma_{PT,1}$	563.38 %	563.38 %	22.49 N/mm^2	21.85 N/mm^2
Arithmetic mean; $\bar{\varepsilon}_p$, $\bar{\sigma}_{PT}$	581.89 %	573.44 %	22.57 N/mm^2	22.08 N/mm^2
Standard deviation: s_{ε_p} , $s_{\sigma_{PT}}$	10.87 %	5.87 %	0.081 N/mm^2	1.19 N/mm^2
Relative combined uncertainties $u_{cB,rel}(\varepsilon_{p1})$, $u_{cB,rel}(\varepsilon_{p1})$ calculated by method type B:	0.41 %	0.20 %	0.61 %	0.58 %
Relative standard uncertainty $u_A(\varepsilon_{p1})$, $u_{c,rel}(\sigma_{PT,1})$ calculated by method type A	1.29 %	0.70 %	0.24 %	0.59 %
Relative combined standard uncertainty of a minimal value $u_{c,rel}(\varepsilon_{p1})$, $u_{c,rel}(\sigma_{PT,1})$	1.36 %	0.73 %	0.66 %	0.83 %
Combined standard uncertainty of the minimal value: $u_c(\varepsilon_{p1})$, $u_c(\sigma_{PT,1})$	7.6 %	3.9 %	0.15 N/mm^2	0.18 N/mm^2
The lower limit of expanded uncertainties ($p = 0.95$) of minimal % elongation: $\varepsilon_{p1,0.05}$, $\sigma_{PT,1.0.05}$	563.72 % ≈ 563.7 %	563.84 % ≈ 563.8 %	22.44 N/mm^2	21.76 N/mm^2

5 Simulations by Monte Carlo method

To verify the quality of determination of statistical parameters and type A uncertainty, the Monte-Carlo method was used [7]. The number of simulations is $M = 10^5$; number of observations $n = 5$; parameters of a test specimen 1 are from the population of normal distribution, the expected value $m = \bar{\varepsilon}_p = 581.891\%$ and standard deviation $\sigma = s_{\varepsilon_p} = 10.873\%$. The histogram ($w_{\varepsilon p1}$) of the minimal value ε_{p1} of the percent elongation is presented on Fig. 3a and histogram of standardized deviation the minimal value from mean: $z_1 = \frac{\varepsilon_{p1} - \overline{\varepsilon_p}}{s_{\varepsilon_p}}$ of the percent elongation is shown in Fig. 3b.

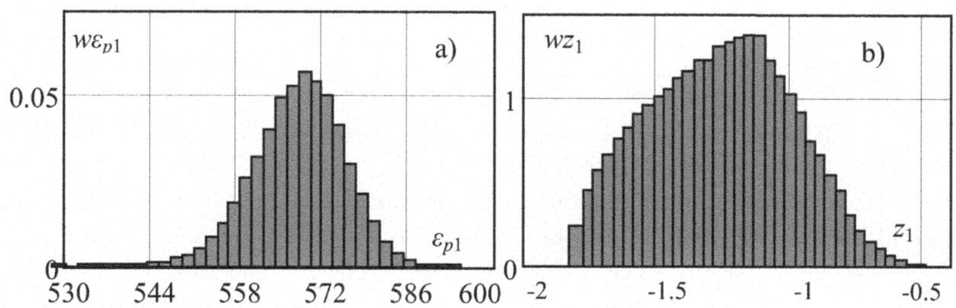

Fig. 3. Histograms of minimal value ε_{pl} (a) and normalized deviation z_1 (b)

Three main statistical parameters of the minimal value of relative elongation ε_{pl} of the test specimen 1, obtained by Monte-Carlo simulations, and corresponding to them experimental values are presented in Table 4.

Table 4. The experimental results and results from Monte-Carlo simulation of parameters of minimal value of relative elongation ε_{pl} of a test specimen 1

Parameters	Experimental results	Monte-Carlo simulation
Expected minimum value	569.26 %	569.24 %
Standard deviation (uncertainty) of minimum value	7.274 %	7.275 %
Left limit of minimal value, $p = 0.99$	563.72 %	563.30 %

From comparison of histogram shapes (Fig. 2a, b) and theoretical distribution (Fig. 3a, b) and also from comparison of experimental and Monte-Carlo simulation results (in Tab. 4) we can see good convergence of experimental results with results of simulation.

6 Conclusions

A method of the uncertainty evaluation, which is based on properties of minimal order statistic, is proposed.

The standard uncertainties of the percent elongation and of the tensile strength of plastic tubes calculated from experimental measurements are very close to the results of simulations by the Monte-Carlo method.

The PDF of maximal value is symmetrical to the PDF of minimal value. Then parameters of uncertainty of maximal value can be calculated in the same way as for uncertainty of minimal value [9]. Only the opposite sign of the deviation of maximal value from the expected value should be taken into account.

Proposed method can be used for evaluating uncertainties of results for the test measurements of other quantities, where informative parameter is the minimal (or maximal) value of the sample of several observations.

References

1. Davis, J.R.: Tensile Testing, Second Edition. ASM International (2004)
2. ASTM International, Standard Test Method for Tensile Properties of Plastics D638-14, http:// http://www.astm.org/Standards/D638.htm
3. Ukraine standards of testing methods and conditions of plastic materials and products, GOST 11262-80, GOST 26277-84 ГОСТ 12423-66
4. Avramenko, O.V., Dorozhovets, M.M., Popovych, I.V.: Evaluation of uncertainty of measurement results in testing of percent elongation and tensile strength of plastic products (in Ukrainian). Automation, Measurement and Control, Lviv Polytechnic National University (2014)
5. ISO, Guide to the Expression of Uncertainty in Measurement JCGM 100:2008, http:// www.bipm.org/utils/common/documents/jcgm/JCGM_100_2008_E.pdf
6. Fisz, M.: Probability Theory and Mathematical Statistics. John Willey & Sons, London (1963)
7. BIPM, IEC, IFCC, ILAC, ISO, IUPAC, IUPAP and OIML, Evaluation of measurement data – Supplement 1 to the Guide to the Expression of Uncertainty in Measurement – Propagation of distributions using a Monte-Carlo method
8. Cox, M.G., Siebert, B.R.L.: The use of a Monte Carlo method for evaluating uncertainty and expanded uncertainty. Metrologia 43, 178-188 (2006)
9. Botsiura, O. A, Zharko, Yu.G., Zakharov, I. P.: Measurement uncertainty evaluation of the maximum observed value of the test parameter (in Russian). Information Processing Systems (COI) 2 (127), pp. 21-23, Kharkiv, (2015)

Method of upgrading the reliability of measurement inspection

Eugenij Volodarsky[1], Zygmunt Warsza[2], Larysa Kosheva[3], and Adam Idzkowski[4]

[1] Science National Technical University of Ukraine "MCI", Department of Automation of Experimental Studies, Kiev, Ukraine
vet-1@ukr.net

[2] Industrial Research Institute of Automation and Measurement (PIAP), Warsaw, Poland
zlw@op.pl

[3] Department of Biocybernetics and Aerospace Medicine, Kiev, Ukraine
l.kosh@ukr.net

[4] Bialystok University of Technology, Faculty of Electrical Engineering, Bialystok, Poland
a.idzkowski@pb.edu.pl

Abstract. Applying an auxiliary quantity, homogeneous with the measured one, is proposed in the quality control of an object made with the use of measuring channel and decision-making module. The result of processing this quantity in the measurement channel is applied for the additive and multiplicative correction of decision limits. This allows to reduce the impact of imperfect processing characteristics of measuring channel. The significant effectiveness of such correction, at different ratio of the components of processing error, is presented. The influence of errors of forming auxiliary quantity is also estimated.

Keywords: quality control · reliability assessment · measurement inspection.

1 Introduction

The reliability of measurement inspection is a confidence measure in decision-making. The inspection is characterized by its uncertainty [1], [4]. If the inspection is based on a measurement then quantity X is measured and it contains information about the specific properties of a controlled object. The comparison procedure of the value x with limits x_l , x_h of the tolerance interval is preceded by the sequence of "measurement/processing" operations of this quantity in a measuring channel (MC). The real characteristics $\varphi(x)$ of measuring channel differs from its nominal characteristics $\varphi_0(x) = x$. The decision of the inspection can be taken only after processing x by the channel. The decision is described by inequality $\varphi_0(x_l)<\varphi(x)<\varphi(x_h)$, which does not coincide with the required decision-making rule $x_l \leq x \leq x_h$.

If the channel output quantity is calibrated in units of the measured quantity then $\varphi_0(x) = x$ and $\varphi(x) \approx (x+\Delta)(1+\gamma)$ will be obtained. The function $\varphi(x)$ can be expressed as a series expansion

R. Jabłoński and T. Brezina (eds.), *Advanced Mechatronics Solutions*,
Advances in Intelligent Systems and Computing 393,
DOI 10.1007/978-3-319-23923-1_63

$$\varphi(x)-\varphi_0(x)=a+(b-1)x+\sum_{i=2}^{n}c_i x^i \; . \tag{1}$$

The main systematic error, resulting from the non-ideal (real) MC characteristics, commonly contains in practice two components: additive $\Delta = a/b$ and multiplicative coefficient $\gamma=(b-1)$, because a third term describing the nonlinearity of MC characteristic is usually negligible. Combinations of these components, both as absolute values and signs, result in the correctness of inspection to a different extent. This effect was analyzed in [2]. There were shown that the effect of non-ideal characteristics of the measurement channel might be included with the use of additional segments θ_l and θ_h. Their lengths are dependent on the combination of two error components. These segments cause the shift of x interval. If a decision-making rule is transformed to $x_l+\theta_l \le x \le x_h+\theta_h$ then probability of the incorrect decision is defined by area under the curve of the probability distribution of controlled quantity. Type of wrong decisions, false refuse or unfounded refuse, depends on the position of segments θ_l and θ_h in relation to the lower x_l and upper x_h limits of the tolerance interval of x values.

The traditional way to improve the reliability of measurement inspection involves the use of more accurate measuring instruments in the channel MC. Typically, this reduces the speed and increases the costs of control. Also not always, there is a suitable measuring equipment available. An influence of deviation, between the real processing characteristics and the nominal one, on the correctness of inspection can be reduced by other means. One of the methods is described below.

2 Theoretical background of the method

The purpose of this paper is to present the ways to increase reliability of decisions taken in an inspection process and to reduce incorrect decisions [3].

It is proposed to achieve this without employment of more precise measuring equipment. This will be done by reducing the impact of errors in measuring channel MC. In the inspection, in contrast to the measurements, one cannot determine the precise value of a measured physical quantity x of controlled object. It is sufficient to determine whether it is within the tolerance interval $x_l \le x \le x_h$. Therefore, the obtained result does not need to be corrected for reducing the resultant error of sequence operation "measurement/processing" in the measuring channel. Instead of that, a number of other equations, obtained in a structural or algorithmic way which contain information about the real characteristics $\varphi(x)$ of the measuring channel, is possible to use. The calculations based on the results obtained from these equations are used for adjusting the limits of the acceptance interval of x values. This adjustment is made to bring the relationship between result of processing $\varphi(x)$ and adjusted settings $\varphi_0(x_l)$, $\varphi_0(x_h)$ into conformity with the relationship between x and predefined limit values x_l and x_h of the tolerance interval. If the correction associated with calculating the correction d, which is added to settings, then it is additive correction. And if it is associated with calculating the multiplier c, which is multiplied by settings, then it is multiplicative correction.

An auxiliary quantity x_0, homogeneous with the measured one, should be applied to implement the method for increasing the reliability of measurement inspection. If the value x_0 is known then it will be also known the result of its ideal processing $\varphi_0(x_0)$. Divergences between real $\varphi(x_0)$ and ideal result $\varphi_0(x_0)$ are considered in calculations of correction a, or in adjustment of b parameter. Thus, before the actual control procedure, auxiliary quantity x_0 should be given to input of the MC measuring channel. The result of its processing $\varphi(x_0)$ contains information about the actual channel characteristics (Fig. 1).

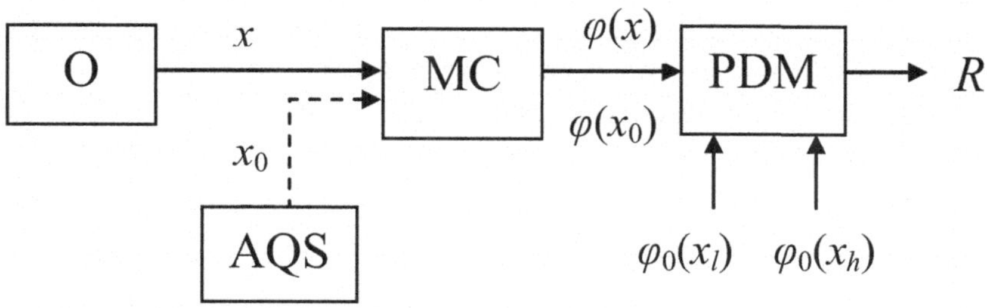

Fig. 1. Method for increasing the reliability of measurement inspection: O – controlled object; AQS - auxiliary quantity x_0 source; MC – measuring channel; PDM – processing and decision making module, R- result of inspection

2.1 Additive correction

As the result $\varphi_0(x_0)$ of ideal processing of x_0 is known, on the basis of $\varphi(x_0)$ and $\varphi_0(x_0)$ one can determine the proportional factor $d= \varphi(x_0) - \varphi_0(x_0)$ of the multiplicative component of measuring channel error. Value d is used for additive correction i.e. to shift the initial settings

$$\varphi_a(x_l) = \varphi_0(x_l) + d \, , \; \varphi_a(x_h) = \varphi_0(x_h) + d \, . \tag{2a,b}$$

In this way the additive correction was done. Without loss of generality it can be assumed that the characteristics, real and nominal, are as follows

$$\varphi(x) = (x + \Delta)(1 + \gamma), \; \varphi_0(x_0) = x \, , \tag{3a,b}$$

where: Δ – additive component and γ – multiplicative component of absolute processing error.

Decision rule "Update" for the input of measurement channel is as follows

$$\frac{x_l}{1+\gamma} - \Delta < x < \frac{x_h}{1+\gamma} - \Delta \, . \tag{4}$$

After adjusting the settings for value $d = x_0\gamma + \Delta(1+\gamma)$ one receives

$$\frac{x_l}{1+\gamma} + x_0 \frac{\gamma}{1+\gamma} < x < \frac{x_h}{1+\gamma} + x_0 \frac{\gamma}{1+\gamma}. \tag{5}$$

It results from (5) that the additive error does not affect the test results. To separate the ideal decision, the rule expression (5) is converted to the following

$$x_l + \alpha(x_0 - x_l) < x < x_h + \alpha(x_0 - x_h), \tag{6}$$

where $\alpha \equiv \frac{\gamma}{1+\gamma}$.

The right parts from both sides of the equation (6) include, remaining after correction, partial impact of measuring channel errors. Substituting different values x_0 in the expression (6), the tolerance interval can be changed. From the expression (6) it results that probability and type of incorrect decisions depend on the values x_0, x_l, x_h. There are three cases of selecting the value x_0: within the tolerance interval, and two out of its limits, i.e.

$$x_0 < x_l,\; x_l < x < x_h,\; x_0 > x_h.$$

From the detailed analysis in [2] it results that the impact of measuring channel error on the reliability of measurement inspection will be minimal when x_0 will be selected within the tolerance interval. Then the decision rule after correction of limits will be

$$x_l + \alpha(x_0 - x_l) < x < x_h - \alpha(x_h - x_0) \tag{7}$$

or

$$x_l + \theta_l \le x \le x_h - \theta_h. \tag{8}$$

Equal intervals θ_l and θ_h capture the impact of measuring channel errors on accuracy of inspection.

The area under the probability density function (pdf) for the value of controlled quantity is proportional to the probability of incorrect decisions (Fig. 2).

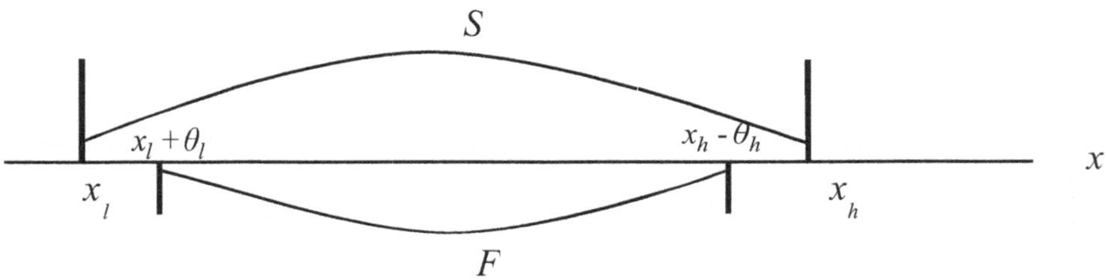

Fig. 2. Location of the intervals of parameter x for the objects which met efficiency standards S (tolerance interval) and found as fit F (acceptance interval)

The length of intervals θ_l and θ_h has the positive sign. A sign in (8) gives information about a type of incorrect decision. Thus, in the case of controlled objects whose value of information parameter is within the range θ_l or θ_h, it can only take place a *false refuse* due to their uselessness although in fact they are efficient. The probability of a false refuse determines the expression

$$P_{err} = P_f = f(x_l)\,(\theta_l + \theta_h) + \tfrac{1}{2} f(x_l)\,(\theta_l^2 + \theta_h^2) \tag{9}$$

where: $f(x_l)$ - density distribution of the possible values of controlled quantity for x_l.

For uniform distribution the second component on the right side of formula (9) disappears. If the multiplicative component of measuring channel error is negative, there is only a decision: *unfounded refuse.*

Let us consider the impact of error x_0 on the effectiveness of additive correction setting limits decision. In this case the value $\tilde{x}_0 = x_0 + \Delta_0$ is given to the input of measuring channel, with an error Δ_0. The result of ideal processing of this quantity will not be equal to $\varphi_0(x_0)$. Then, according to the algorithm $\tilde{b}$ is calculated instead of b. After the additive correction one receives

$$x_l + \alpha(x_0 + x_l) - \Delta_0 < x < x_h - \alpha(x_h - x_0) - \Delta_0\,. \tag{10}$$

It results from equation (10) that on the one hand an error of auxiliary quantity increases the replacement interval – see right side of the equation and on the other hand it reduces the interval – see on the left side of the equation. The only one requirement is that the value $\tilde{x}_0$ should be within the initial tolerance interval. To fulfill this condition, assuming an uniform distribution of controlled size, the auxiliary quantity error will not affect the conformity of inspection. The optimal selection of auxiliary quantity value is $(x_h+x_l)/2$ assuming the normal distribution of a controlled variable value and the positive multiplicative component of measuring channel error. However, for the negative multiplicative component of error, a reduction of the impact of remnant errors (remained after correction) occurs for $x_0=x_l$ or $x_0=x_h$. In setting a final probability of incorrect decisions there is need to consider the second component of right side of equation (6). It depends on the ratio of the standard deviation of the dispersion values in controlled amounts σ and the interval duration (x_h-x_l) of limit values of controlled variable x. In this case, as opposed to a dependence for uniform distribution, the auxiliary quantity error increases the probability of incorrect decision. However, an additional increase in the probability of incorrect decisions will be the second order of smallness.

2.2 Multiplicative correction

As it is apparent from the expressions (6) and (9), the probability of incorrect decisions depends on the multiplicative component of MC error. If preliminary analysis shows that the effect is significant, there is a need to perform multiplicative correction. With this correction the factor c of both real and ideal results of auxiliary quantity processing x_0 is equal to $c = \frac{\varphi(x_0)}{\varphi_0(x_0)}$. This factor is used to correct the limits of tolerance interval. If the real characteristic of the channel is $\varphi(x) = (x+\Delta)(1+\gamma)$, then $c = \frac{(x_0+\Delta)(1+\gamma)}{x_0}$. After correction a decision rule is as follows

$$\frac{x_h}{x_0}(x_0+\Delta) < (x+\Delta) < \frac{x_h}{x_0}(x_0+\Delta). \tag{11}$$

From the last expression it results that the multiplicative error component does not affect the test result. Similarly, as in the previously discussed case, one stands out ideal (nominal) decision rule $x_l < x < x_h$. Then one goes to inequality in which additional restrictions, on the right side and the left side, take into account the impact of non-ideal processing characteristics of the MC

$$x_l + \frac{\Delta}{x_0}(x_l - x_0) < x + \Delta < x_h + \frac{\Delta}{x_0}(x_h - x_0). \tag{12}$$

Thus, the influence of multiplicative component of the measurement error on the reliability of measurement inspection was eliminated.

To minimize the remnant impact of constant value a of error, the auxiliary value x_0 within the permissible range should be also chosen. However, using a multiplicative correction, regardless of the probability distribution of a controlled quantity value x, the selected value x_0 should be near to x_h. Inequality (13) (decision rule) determines the reliability of decisions

$$x_l - \frac{\Delta}{x_0}(x_0 - x_h) < x < x_h + \frac{\Delta}{x_0}(x_h - x_0). \tag{13}$$

From a comparison of expressions (6) and (13) it results that a type of incorrect decisions changes to opposite then for the additive correction (in this case a failure will be detected).

Let us also consider the impact of error of creating the auxiliary quantity x_0 on the effectiveness of the multiplicative correction. As a result of inaccuracy during the creation of this quantity, i.e. when $\tilde{x}_0 = x_0 + \Delta_0$ then the real value of the correction factor is

$$\tilde{c} = \frac{(x_0 + \Delta + \Delta_0)(1+\gamma)}{x_0}. \tag{14}$$

If the decision rule is selected taking into account the error of creation x_0 then

$$x_l - \frac{\Delta}{x_h}(x_h - x_l) + \frac{\Delta_0}{x_h} x_l < x < x_h + \Delta_0, \; \frac{\Delta_0}{x_h} < 1. \tag{15}$$

From the expression (15), the effect of errors during creation the quantity x_0 can be expressed as a relatively small reduction in the length of the smaller equivalent interval, and equivalent increase the length of the upper replacement interval, proportional to the error of forming the auxiliary quantity. This may increase the probability of incorrect decisions and significantly reduces the effectiveness of the multiplicative correction.

3 Numerical example

The functionality of a product is characterized by a certain output quantity x. According to the requirements of production technology this quantity can occur with equal probability in the range from 0 to 20 mV. These objects should be selected whose x value is within the interval (10 -14) mV, thus $\varphi_0(x_l) = 10$, and $\varphi_0(x_h) = 14$. The real characteristics of measuring channel is

$$\varphi(x) = (x + 0.1)(1 + 0.1).$$

According to the equation (9) the probability of incorrect decisions with uniform probability density distribution of controlled quantity $f(x) = 0.05$, will be $P_{err} = 0.13$, i.e. the conformity of inspection is 0.87. Auxiliary quantity x_0 was used, located within the tolerance interval (10 - 14) mV. After appropriate corrections, the probability of incorrect decisions decreased to $P_{err} = 0.02$. So the conformity of inspection increased to 0.98. For $x_0 = 12$ mV permissible error of auxiliary quantity cannot exceed ±2 mV (±17%). Error of auxiliary quantity creation does not affect the conformity of this inspection.

When choosing $x_0 = 14$ mV and an additive error $\Delta = 0.1$ mV, with a multiplicative correction, the probability of incorrect decisions will be $P_{err} = 0.002$. Thus, the probability is decreased by an order of magnitude. The impact of error in creating the auxiliary quantity x_0 will be also considered. When $\Delta_0 = \pm 2$ mV, an additional component of probability of incorrect decisions is 0.025. This corresponds to the reliability of inspection $D = 0.975$ – by an order of magnitude in relation to the additive correction of settings. In Table 1 the reliability of inspection is shown. It is calculated for the case where the processing characteristics of the measuring channel MC has the form of $\varphi(x) = (x + 0.1)(1 + 0.1)$.

Table 1. The reliability of inspection and the types of incorrect decisions for different ways of limits correction

Way of limits correction	Reliability of inspection		Types of incorrect decisions
	Error of x_0		
	negligible	10%	
Without correction	0.87	0.87	for x_l – unfounded refuse for x_h – false refuse
Additive correction	0.98	0.98	for x_l and x_h –false refuse
Multiplicative correction	0.998	0.895	for x_l and x_h – unfounded refuse

From Table 1 it results that by reducing the impact of measuring channel error, the correction allows to increase the reliability of inspection. In the presented example

$$x_i=10\text{ mV},\ \alpha=\frac{\gamma}{1+\gamma}=0.999\ \text{ and } \Delta=0.1\text{ mV, i.e. } \alpha x_l>\Delta.$$

Therefore, more appropriate is the use of a multiplicative correction. The data in Table 1 confirm that the reliability of inspection will be greater then. It also results that the reliability of the multiplicative correction decreases as the auxiliary quantity x_0 is formed with an error. This should be taken into account when choosing the type of correction. A type of incorrect decision is also significant.

4 Summary

Appropriate correction of the tolerance interval of measured quantity reduces the impact of the error components of the measurement channel on the reliability of inspection. An impact of additive error component is completely eliminated after applying the additive correction. At the same time accuracy requirements decreases concerning the auxiliary quantity. However, if the multiplicative component of processing error in measuring channel is dominant, the effectiveness of additive correction decreases.

Multiplicative correction of the tolerance interval of measured quantity eliminates the influence of multiplicative error component. At the same time the auxiliary quantity error causes a proportional increase in the probability of incorrect decisions. Thus, the choice of correction depends on the ratio of both components: the measuring channel error and the accuracy in creating the auxiliary quantity.

Taking into account the probability distribution of controlled quantity and properly selecting the value of auxiliary quantity, the impact of a multiplicative component of error in measuring channel can be minimized. Depending on the sign of this component, the results of inspection may include falsely detected or undetected malfunctions.

References

1. Grous, A.,: Analysis of Reliability and Quality Control: Fracture Mechanics 1. ISTE Ltd and J Wiley & Sons Inc (2013)
2. Volodarsky, E.T., Koshevaya, L.A., Warsza Z.L.: Uncertainty of the decision about the compliance of the measured object to given requirements. Measurement Automation and Monitoring (PAK) 58 (4), 391-395 (2012)
3. International Organization for Standardization, Statistical methods - Guidelines for the evaluation of conformity with specified requirements - Part 1: General principles ISO 10576-1:2003
4. Joint Committee for Guides in Metrology, Evaluation of measurement data — The role of measurement uncertainty in conformity assessment JCGM 106:2012

Part V
Nanotechnology, MEMS, New Materials

Magnetic measurement of lamination punched with the press

Tomáš Bulín[1], Čestmír Ondrůšek[2]

[1] FEEC, BUT, Brno, Czech Republic
xbulin01@stud.feec.vutbr.cz
[2] FEEC, BUT, Brno, Czech Republic
ondrusek@feec.vutbr.cz

Abstract. This article deals with a problem of processing material for electrical machines. Magnetic properties of every material can be changed relatively easily for example after mechanical or thermal stress. Then it is suitable to measure material after processing and consider to provide measurements on samples with dimensions similar to real magnetic circuits. Edges of laminations, which are machined, affect the size of magnetic losses and when these edges are closer together then the value of losses is increasing. Value of losses given by manufacturer is often different from real application, because they use only standardized method, which ignore the change of dimensions during the machining.

Keywords: RemaCOMP; magnetic properties; losses; electrical steel

1 Introduction

It is difficult to design electrical machine according to parameters from manufacturer. Properties of magnetic material are changing almost every time, when the material is processed. Magnetic domains can be influenced by heating close to Currie temperature or it can be changed by mechanical stress. Stator or rotor laminations are usually punched by press and the shapes of this lamination are very various. There are some wide and some very narrow places, so the mechanical stress is not the same in all part of lamination. It is important to know the change of magnetic parameters after material processing. The knowledge of this change is an advantage and designer can calculate or simulate machine with greater accuracy. Also designer has to remember, that different methods for magnetic measurements do not give the same results.

Manufacturer is usually measuring magnetic properties with device called Epstein frame. This device is easy to use but the shape of the measured sample is not perfect. Much more convenient is toroidal shape, field lines of magnetic flux pass uniformly through the whole cross section. When the Epstein frame is used, there are some places, where magnetic flux is not spread uniformly, so there can be some deviations in the measured magnetic properties.

R. Jabłoński and T. Brezina (eds.), *Advanced Mechatronics Solutions*,
Advances in Intelligent Systems and Computing 393,
DOI 10.1007/978-3-319-23923-1_64

2 Equations and figures

2.1 Theory of standardized device for magnetic measurement

Epstein frame is a standardized device for reproducible magnetic measurements of soft magnetic materials. It uses the square assembly of steel strips, which must overlapped in the corners. Number of strips have to be four or multiples of four. Each corner of overlapping sheets must be pressed by force of 1 N. The width of a strip is 30 mm and length is variable from 280 mm to 305 mm. There are solenoids with defined number of primary and secondary turns around each side of the square, which is made from sheets. Total number of turns is 700. Solenoids are connected in series and guarantee the same number and uniformly distributed turns. Coils are associated as non-detachable and connected with measuring devices. [1] In this work magnetic properties, which was measured by Epstein frame, are taken from catalogue of manufacturer.

2.2 Dimensions of measured samples

The test material was selected from company called Acroni. The chosen type of material is a common type of non-oriented cold rolled fully-processed electrical steel called M470-50A. The thickness of this material is 0.5 mm and it is used in many types of electrical machines. [2]

The shape of measured samples is toroid and the test samples are assembled from single laminations, which are in shape of annulus. Three samples with different dimensions have been prepared for measurement. All three samples have outer diameter 120 mm, but there are difference between inner diameter and size of thickness.

Table 1 Parameters of the samples and dimensions of toroid

◇ Sample 1: bundle of sheets with outer diameter 120 mm, inner diameter 96 mm and overall thickness 13 mm (26 laminations); Winding $N_1 = 227$; $N_2 = 47$ turns.
+ Sample 2: bundle of sheets with outer diameter 120 mm, inner diameter 106 mm and overall thickness 7 mm (14 laminations); Winding $N_1 = 251$; $N_2 = 52$ turns.
× Sample 3: sheet with outer diameter 120 mm, inner diameter 106 mm and thickness 0.5 mm (one lamination); Winding $N_1 = 268$; $N_2 = 50$ turns.

2.3 Measured data

Magnetic measurements were performed on devices RemaCOMP and RemaGRAPH, which can be used for AC and DC measurements. These devices are based on using inductive law. There were three samples with different cross-sections. These samples were measured at rated excitation 1.5 T and 1 T. Amount of losses in the first sample was initially smaller, than amount of losses, which are specified by manufacturer. This was probably caused by larger cross section than in rest of samples. The first sample was the biggest one and had the smallest amount of losses at every excitation. The worst sample with the biggest losses is Sample 3, because there was measured only one sheet. This sample probably is burdened by measuring error, because the cross section is very thin and it is not square or rectangle at all. Amount of losses is bigger than from manufacturer in the last case.

Table 2: Comparison of measurement with catalogue data

M470-50A	Acroni [2]	Sample 1	Sample 2	Sample 3	
P_s at 1.5T	4.7	3.948	4.152	5.286	W/kg
P_s at 1T	2	1.778	1.933	2.077	W/kg
B_{min} at 2500A/m	1.54	1.513	1.512	1.463	T

Very important input parameter for simulation is commutation curve. Even in this type of measurement it can be seen, that mechanical stress from processing this lamination was significant and influenced value permeability and magnetic field strength in commutation curve. This measurement was provided with excitation 1.5 T and permeability is decreasing at samples with processed edge due to punching.

Table 3 Parameters of commutation curve with excitation $B = 1.5T$

	Sample 1	Sample 2	Sample 3	
μ_{max}	5114	4059	3380	
$H(\mu_{max})$	85.1	97.6	114	A/m
H_{max}	2191	2285	2360	A/m

The effect of the mechanical stress can be seen even in measurement with zero frequency. Magnetic field strength and loss of energy are increasing with greater damage of sheets. This effect appears and is rapidly increasing when the sample is near to saturation.

The accuracy of the measurement was ensured by repeated measurement (six times) with the same excitation. Average value has been obtain from the repeated measurements and then relative deviation was calculated for each measurement. There can be seen that the biggest relative deviation appears in small or high excitation in Table 3, where the deviation reached almost 5%. Uncertainty of measurements was not calculated and instead of it was determined relative deviation to illustrate dispersion values between several measurements.

Table 4 Size of measurement relative deviations in sample 3

Sample 3	0.5T		1T		1.5T	
	Maximum relative deviation	Average relative deviation	Maximum relative deviation	Average relative deviation	Maximum relative deviation	Average relative deviation
B_{max}	3.18 %	1.42 %	0.29 %	0.19 %	0.47 %	0.38 %
H_{max}	1.78 %	0.73 %	0.72 %	0.38 %	3.94 %	3.65 %
P_s	4.97 %	1.84 %	1.86 %	1.51 %	4.38 %	2.56 %
μ_a	1.51 %	1.24 %	0.4 %	0.32 %	3.57 %	3.25 %

Prepared samples were measured in different frequencies with the same excitation 0.1 T. Amount of losses is more increasing with frequencies and mechanical damage of samples. Higher deviation would be at higher excitation, but there were a limits of measuring device.

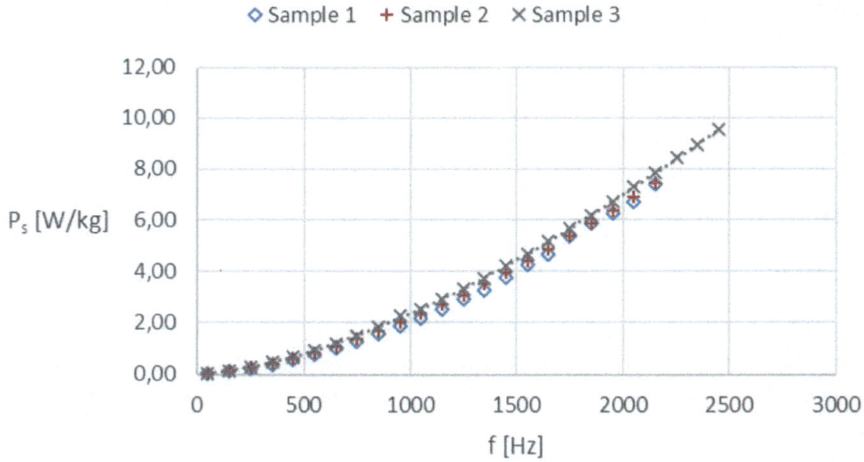

Fig. 1. Dependence of magnitude of the losses at the frequency at excitation B = 0.1T

Measurement at 50 Hz was provided with increasing excitation. As excitation of the sample was growing so the amount of losses was increasing exponentially. The worst results gave Sample 3 as in previous measurement, but with higher excitation there can be seen greater difference between individual samples.

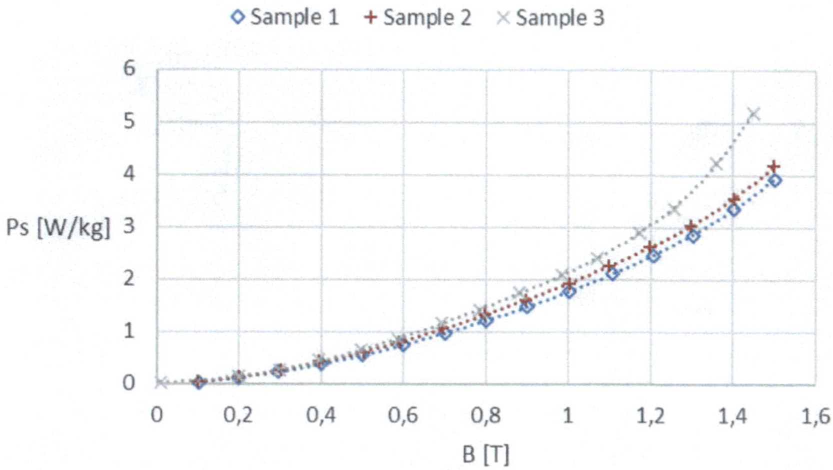

Fig. 2. Dependence of magnitude of the losses on the size of excitation

Same result as in Figure 2 can be found on Figure 3 because permeability is equal to amount of losses. The highest permeability has the Sample 1, because there are the smallest losses. The difference between samples is best seen on Figure 3. Where permeability is displayed.

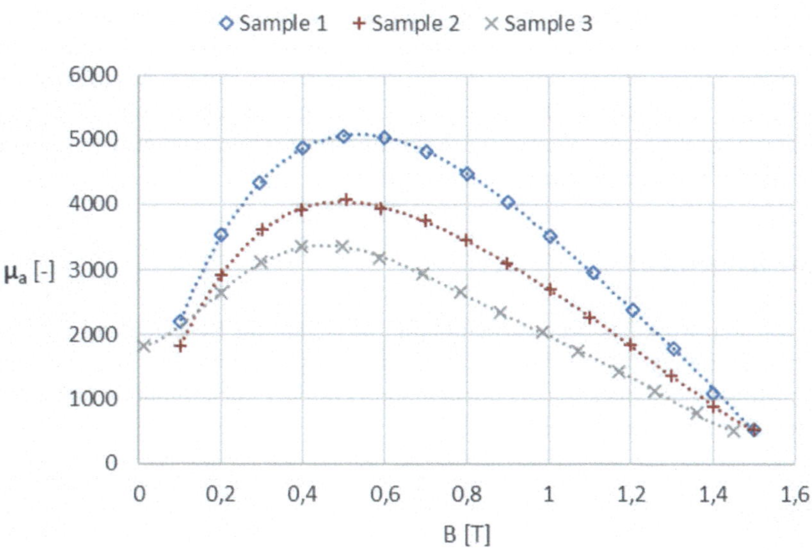

Fig. 3. Dependence of permeability on the size of excitation

On the Figure 4. and Figure 5. there can be seen simulation and real lamination of rotor of synchronous reluctance motor. This shape of laminations is specific with very narrow sheets, where deviation of magnetic parameter could be quite large.

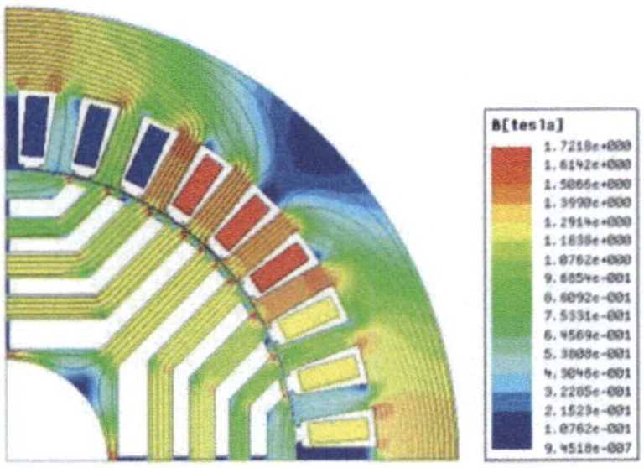

Fig. 4. Simulation of excitation in reluctance motor [3]

Fig. 5. Rotor of reluctance motor [3]

3 Conclusion

Magnetic properties of steel M470-50A, which was changed after machining, were investigated. Value of magnetic losses was increasing with narrower annulus, but it was still smaller than maximum losses indicating by manufacturer. Measured value of minimum magnetic polarization, at magnetic field strength 2500 A/m, was smaller than in catalogue and its value was decreasing with increasing losses. Irregular cross-section in sample 3 caused significant deterioration of magnetic parameters and it proved that it is important to provide measurement on samples with square cross-section. Dependence of magnetic losses on magnetic flux density was investigated, value of losses is exponentially increasing with magnetic flux density. The highest value of permeability was found at 0.5 T for all three samples and curve of magnetic losses value for different frequencies was measured.

Acknowledgement

This research work has been carried out in the Centre for Research and Utilization of Renewable Energy (CVVOZE). Authors gratefully acknowledge financial support from the Ministry of Education, Youth and Sports of the Czech Republic under NPU I programme (project No. LO1210).

References

1. Fiorillo F., "Measurements of magnetic materials," *Metrologia*, vol. 47, no. 2, pp. S114–S142, 2010.
2. Acroni, d.o.o. *Non-oriented electrical steels*. Jesenice, Slovenia, 2013. 15.3.2015. <http://www.acroni.si/assets/Uploads/Neorientirana-elektro-plocevina.pdf>
3. Bárta J., "Design and optimalisation of synchronous reluctance machine," Diploma thesis, Brno, 2014.

Rheology of inks for various techniques of printed electronics

Łucja Dybowska-Sarapuk[1,2], Jerzy Szałapak[1,2], Grzegorz Wróblewski[1], Iwona Wyżkiewicz[1], Marcin Słoma[1,2], Małgorzata Jakubowska[1,2]

[1] Institute of Electronic Materials Technology, Wólczyńska 133, 01-919 Warsaw, Poland
[2] Faculty of Mechatronics Warsaw University of Technology, Andrzeja Boboli 8, 02-525 Warsaw, Poland
l.dybowska@mchtr.edu.pl

Abstract. The inks and pastes with carbon nanoparticles, such as graphene nanoplatelets and carbon nanotubes, for diverse printed electronics techniques were produced. These composite materials, dedicated for screen printing, inkjet printing, spray coating and gravure printing were tested in context of their rheological properties. After comparing the obtained values of viscosity with standard values found in literature, it was confirmed that the tested suspensions may meet rheological requirements of particular techniques and can be transferred correctly on the substrate in the form of good quality pattern.

Keywords: printed electronics inks, graphene inks, rheology of inks

1 Introduction

Printed electronics techniques are a low-cost solution for obtaining layers of different electrical properties and thicknesses from thin, transparent spray coated films to thick, well conductive screen printed patterns. However, all of the printing and coating techniques require inks of different compositions and viscosities which are related to the essence of the deposition.

The ink viscosity is one of the rheological properties, crucial to achieve high quality of printable films [1][2]. It affects the usability of the inks, good ink transfer [3], and the functionality of the resulting films [4]. Moreover, viscosity influences the resolution parameters of the printed patterns and the high aspect ratios [5]. That's why a good layer quality can be achieved only, if the viscosity of the ink is adapted to the particular printing technique.

The ink will be printable and the correct film can be printed only if inks viscosity is appropriate for the selected technique range. For the particular printing techniques, different requirements, in terms of rheological properties, have to be met.

R. Jabłoński and T. Brezina (eds.), *Advanced Mechatronics Solutions*,
Advances in Intelligent Systems and Computing 393,
DOI 10.1007/978-3-319-23923-1_65

Inhalainen Petri et al. [3] reported that for the inkjet printing layer process it is proper (there's no nozzle clogging, no satellite or double droplet) when viscosity is in the range from 1 to 20 mPas. For the gravure printing, ink viscosity is usually higher than 50 and less than 500 mPas. Typical screen printable inks viscosity is between 500 and 50000 mPas [3]. Good quality of spray coated films and good ink transfer can be achieved for low inks viscosity, in the range from 0.7 to 1.5 mPas, since these values allow the proper ink atomization.

Carbon nanomaterials, such as graphene nanoplatelets and carbon nanotubes, became very popular due to its excellent and unique electrical, optical and mechanical properties [7]. These materials are more often used as the functional phase in heterogeneous suspensions precisely tuned for the printed electronics. [6] However, the suspensions dedicated for printed electronic, with mentioned earlier carbon nanomaterials, have not yet been thoroughly tested in the terms of rheology. The aim of this work was to test the viscosities of the carbon nanoparticle inks for screen-, graviure-, inkjet- printing and spray coating and to examine, whether they are adjusted to the selected techniques.

2 Materials and methods

A wide range of inks precisely tuned for diverse printing techniques, such as screen printing, spray coating, gravure printing and ink-jet printing were prepared. During our experiments, various carbon nanomaterials were used as the functional phases, including multiwalled and doublewalled carbon nanotubes and graphene nanoplatelets. These materials are shown below (Fig 1a, b, c and d), on scanning electron microscope (SEM) images. Organic vehicles were based on different types of polymers, solvents and dispersing agents. For the spray coating technique, diethylene glycol n-butyl ether acetate (DBAC) was used as an evaporating solvent, for the screen printing – solution of polymethyl methacrylate (PMMA) dissolved in DBAC and for the ink-jet printing – ethyl alcohol or deionized water based suspensions. Organic vehicle for gravure printing was based on polyurethanes and nitrocellulose in mixture of ethyl acetate and ethyl alcohol.

The viscosity of the prepared low-viscosity suspensions was measured, using the BROOKFIELD DV2T cone plate viscosity meter. The rotational rheometer BROOKFIELD RS CPS+ designed to test high viscosity was used to examine rheology of screen printing paste. The measurements were conducted at 25°C, set by the ultrathermostat Polyscience.

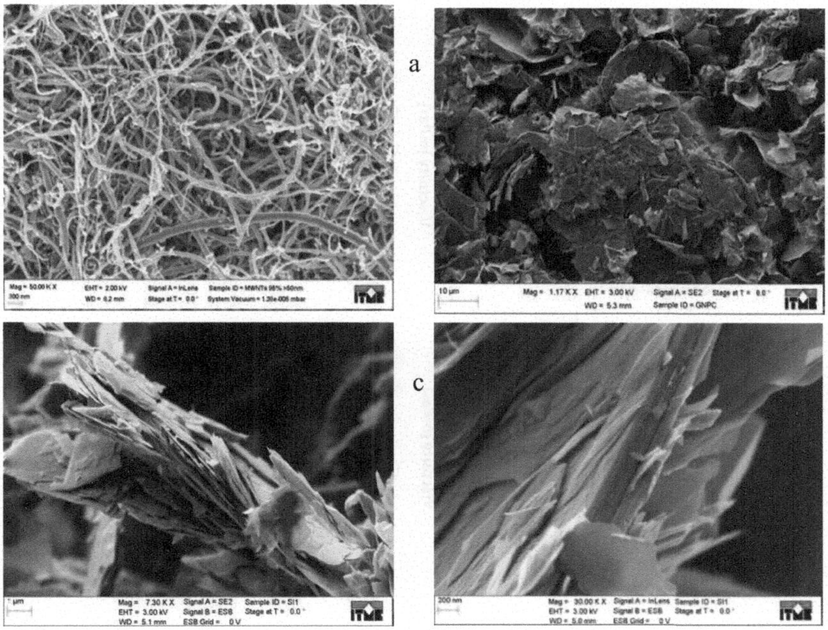

a b c d

Fig. 1. Carbon nanomaterials: a. multiwalled carbon nanotubes, b. graphene nanoplatelets type C, c. and d. graphene nanoplatelets type SI

3 Results and discussion

3.1 Viscosity of the inks

Rheological properties of the inks for the four various techniques are illustrated below in the form of viscosity curves (Fig. 2). All of the suspensions exhibited shear thinning properties, which means that the viscosity value decreased with the increase of the shear rate value. However, the viscosity values over the range of the shear rates are significantly different for each technique.

For the four different inks, the viscosity values are stabilized at the end of the measurement range. Therefore, the viscosity values shown in Table. 1. are readings for selected shear rate from the end of the measurement range of the viscometer.

	Screen printing	Gravure printing	Inkjet printing	Spray coating
Measured viscosity	2070.8	71.3	5.4	0.9
Range of viscosity from literature	500-50000	50-200	1-20	0.7-1.5

Table 1. Viscosity values for the four tested inks in [mPas]

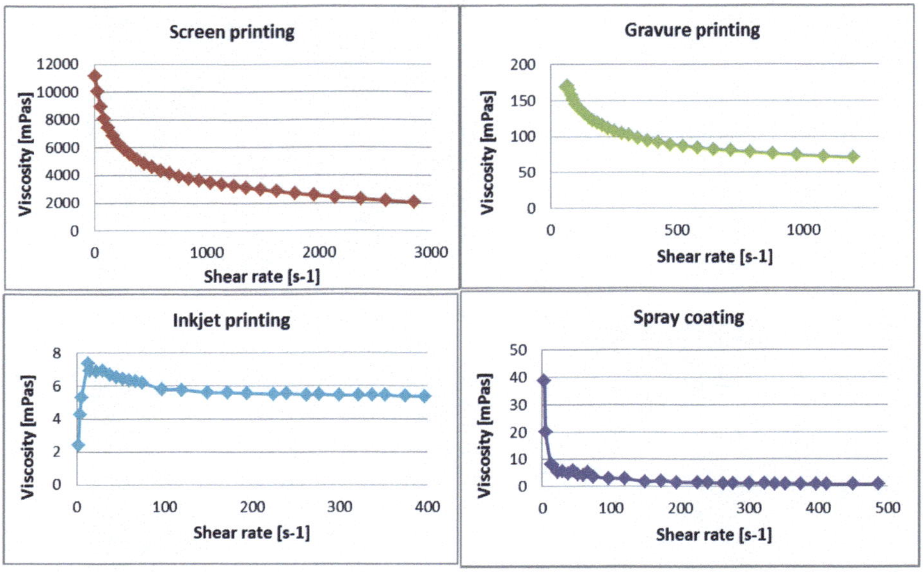

Fig. 2. Viscosity curves of the inks for: screen printing, gravure printing, inkjet printing and spray coating

The screen printing ink has the highest viscosity value. It was nearly 30 times higher than the viscosity of the gravure printing ink and more than 2000 times higher than the lowest, spray coating ink viscosity.

It can be concluded, that the measured values of carbon nanoparticle inks viscosity for screen-, gravure-, inkjet- printing and spray coating and the differences between the viscosities in these techniques are consistent with the guidelines for the inks viscosities, found in the literature, which were presented in the introduction.

The differences in the inks viscosities are caused by the technical differences that occur in the process of applying inks in the particular techniques. For example, pastes for the screen printing must have very high viscosity to stay on the sieve before printing, but the inkjet ink must be characterized by low viscosity. In this case, too high surface tension or ink viscosity value may result in clogging of the printing nozzle.

Of course, the viscosity can be modified by changing the composition of the ink: the type and quantity of the vehicle, content of the functional phase or by changing the temperature of the ink [8,9]. However, after all changes, the final viscosity value must be matched to the used printing technique. It is necessary to achieve a good ink printability and receive a correct layer without defects.

4 Conclusion

During the experiment, the rheological properties of the carbon nanoparticle inks dedicated for the four different printing techniques were tested. Despite all of the inks characterized shear thinning behavior, viscosity values were different for each technique.

The highest viscosity had screen printing paste (2070 mPas), then successively inks for rotogravure (71 mPas), inkjet printing (5 mPas) and spray coating (0.9 mPas). For each technique, the measured viscosity values of the inks with carbon nanoparticles, such as graphene nanoplatelets and carbon nanotubes, are in the operational window in terms of rheological properties, which was found in the literature.

5 Acknowledgements

This work was supported by the Polish National Center for Research and Development (NCBiR) through GRAF-TECH Project no. 513L 1142 0907 000.

6 References

1. Montanino, M., De Girolamo Del Mauro, A., Tesoro, M., Ricciardi, R., Diana, R., Morvillo, P., Nobile, G., Imparato, A., Sico, G., Minarini, C.: Gravure-printed PEDOT:PSS on flexible PEN substrate as ITO-free anode for polymer solar cells. Polym. Compos.36, 1104-1109 (2015)
2. Huang, L., Huang, Y., Liang, J., Wan, X., Chen, Y.:Graphene-based conducting inks for direct inkjet printing of flexible conductive patterns and their applications in electric circuits and chemical sensors. Nano Res. 4, 675–684 (2011)
3. Ihalainen, P., Määttänen, A., Sandler, N.: Printing technologies for biomolecule and cell-based applications. Int. J. Pharm., (2015)
4. Silva-Weiss, A., Bifani, V., Ihl, M., Sobral P. J. A., Gómez-Guillén, M. C.: Structural properties of films and rheology of film-forming solutions based on chitosan and chitosan-starch blend enriched with murta leaf extract. Food Hydrocoll. 31, 458–466 (2013)
5. Pospischil, M., Zengerle, K., Specht, J., Birkle, G., Koltay, P., Zengerle, R., Henning, A., Neidert, M., Mohr, C., Clement, F., Biro, D.: Investigations of thick-film-paste rheology for dispensing applications. Energy Procedia. 8, 449-454 (2011)
6. Wroblewski, G., Kielbasinski, K., Stapinski, T., Jaglarz, J., Marszalek, K., Swatowska, B., Dybowska-Sarapuk, L., Jakubowska, M.: Graphene platelets as morphology tailoring additive in carbon nanotube transparent and flexible electrodes J. Nanomater. (2015)
7. Li, J., Ye, F., Vaziri, S., Muhammed, M., Lemme, M. C., Östling, M.: Efficient inkjet printing of graphene. Adv. Mater. 25, 3985–3992 (2013)
8. Nascimento, Z. C., Reis Pinto, P., da Silva Sirqueira, A.: Rheological studies of conducting acrylic paint. Acta Scientiae et Technicae.2 (2014)
9. Satas, D.Tracton, A.: Coatings technology handbook. CRC Press, 2005.

Specialized MEMS microphone for industrial application

Magdalena A. Ekwińska[1], Tomasz Bieniek[1], Grzegorz Janczyk[2], Jerzy Wąsowski [2], Paweł Janus[1], Piotr Grabiec[1], Grzegorz Głuszko[1], Jerzy Zajac[1], Daniel Tomaszewski, Karina Wojciechowska[1], Rafał Dobrowolski[1], Tadeusz Budzyński[1]

[1] Institute of Electron Technology, Division of Silicon Microsystem and Nanostructure, Piaseczno, Polska
{ekwinska, tbieniek, janus, grabiec, ggluszko, jzajac, dtomasz }@ite.waw.pl
[2] Institute of Electron Technology, Department of Integrated Circuits and Systems, Warszawa, Polska
{janczyk,jwasowski}@ite.waw.pl

Abstract. This article is an attempt to describe design and verification process of the MEMS+IC (micro-electro-mechanical system assisted by an integrated readout circuit) structure designed in Institute of Electron Technology (ITE) dedicated for specific industrial microphone application. Design tools and methods are presented in this paper along with results of numerical simulations compared with real measurements performed for the key steps of device development.

Keywords: MEMS; ASIC; MEMS microphone; Readout electronics, Modelling and simulation; Electric measurements; MEMS+, Cadence, Experimental verification.

1 Introduction

Development of innovative smart systems incorporating micro-electro-mechanical part (MEMS) assisted by some sort of autonomous characteristics supported by dedicated, application specific integrated electronic modules (ASICs) does not fit to any standard design methodology, nor design verification flow. Effectiveness of this process is a measure for company competiveness embracing initial idea specification, design, modeling and simulation stages assisted by virtual fabrication [15], optimization, final verification of the design, device fabrication and measurements. For the device presented in this paper MEMS design and optimization have been performed using CoventorWare and MEMS+ simulation environments. The second one is fully compatible with IC design environment Cadence which supported multidomain simulation of the whole MEMS+IC system yet on the design stage. After the technology guide was set down a set of technological test were performed. Next the MEMS microphone structure was manufactured and its parameter were checked during multi – stage analysis.

R. Jabłoński and T. Brezina (eds.), *Advanced Mechatronics Solutions*,
Advances in Intelligent Systems and Computing 393,
DOI 10.1007/978-3-319-23923-1_66

2 Product Specification and the Design Concept

People should not be exposed to the sound exceeding pain threshold level (120 dB) however it has to be measurable also over that level. One of available solutions is to apply smart microphone based on MEMS module to measure mean and peak values in acoustically harsh environments. Several approaches towards design of microphone are given in literature. There are two main types of MEMS microphones: capacitive [1-6] and piezoresistive [7-9] version. The sensitivity parameters of capacitive microphones are three orders of magnitude better than achieved one for piezoresistive structures. The design concept of the MEMS microphone (Fig.1.) is based on the design of the capacitive microphone and fabrication technologies owned by Institute of Electron Technology (ITE).

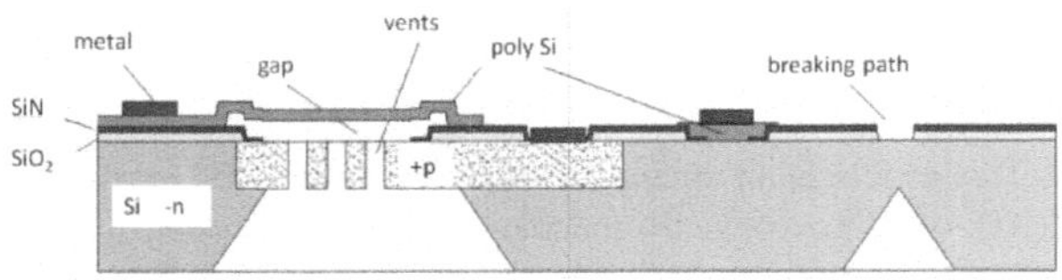

Fig. 1. Schematics of MEMS microphone model

3 Methodology of microphone design

Whole the methodology (fig.2) being developed in frame of SMAC activities is focused on co-simulation profiting from an application of innovative tools developed in frame of the same project or already existing in the EDA market. One of them is MEMS+ environment by Coventor. Microphone incorporates two modules: single MEMS and the dedicated readout electronics (IC's). The model of the MEMS microphone (Fig.3a, b) has been created according to specification and technology requirements. A set of simulations have been performed to optimize [14] the microphone structure. Such a procedure led to automate generation of the electrical model (Fig.3 c) of the MEMS microphone module by MEM+ software by Coventor .
In parallel with MEMS module design IC designers led readout design development
The presented SiP design methodology is a key issue in a still missing efficient development chain for SiPs. The block diagram of the MEMS microphone system being developed in frame of the project (Fig. 4) incorporates:

- The MEMS module supporting detection of dynamic pressure level,
- The analog front-end developed by project partners supporting conversion between varying capacitance to voltage following acoustic wave pressure,
- The Sigma-Delta Modulator supporting conversion from analog voltage to digital domain by means of high resolution Sigma Delta Analog-to-Digital Converter (ADC).
- The digital block supporting conversion from the modulator output to the SPI serial output interface.

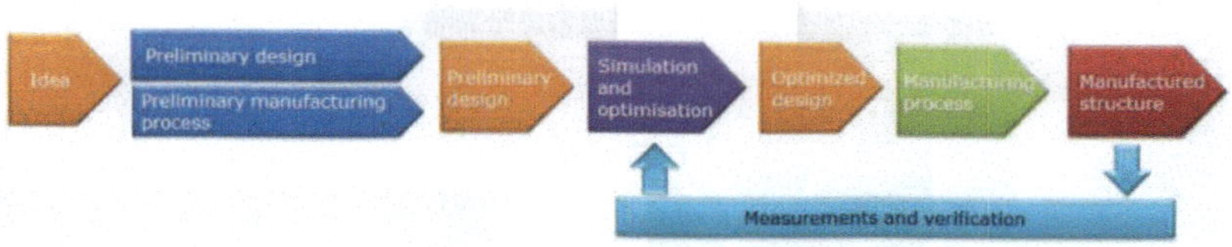

Fig. 2. Design flow

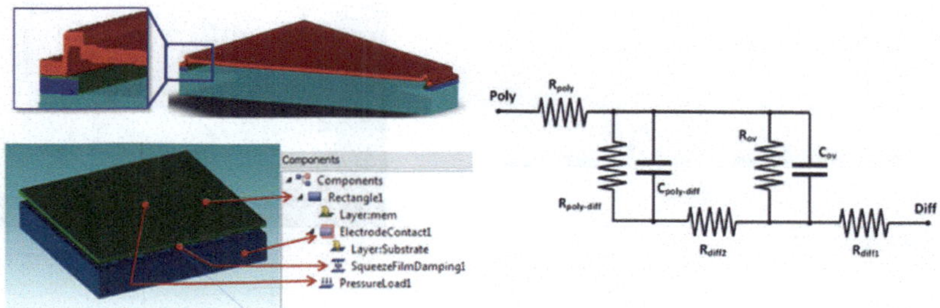

Fig. 3. MEMS microphone model a), FEM model created in CoventorWare™, b) model created in MEMS+, elements of the MEMS+ microphone model, c) electrical model: R_{poly} – resistance of the polysilicon membrane; R_{diff} – resistance of the backplate; $R_{poly\text{-}diff}$ – resistance between membrane and backplate

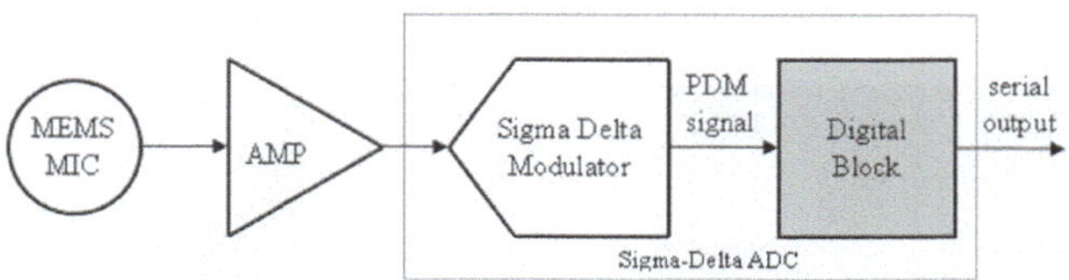

Fig. 4. MEMS microphone system block diagram

4 Experiments

After manufacturing process was finished a set of experiments were made in order to verify the most important parameters of MEMS microphone.

4.1 Electrical experiments on wafer level

For the MEMS microphone structures (Fig. 5) manufactured in ITE typical electrical measurements have been carried out by the Semi-Automatic Wafer Prober - Cascade Microtech Summit 12000 AP and Keithley 4200-SCS Analyzer with CV unit.
The set of CV characteristics have been gathered for 0÷3V membrane-backplate voltage under 100kHz for measurement signal frequency (Fig.6a). The membrane-backplate capacitance in the applied voltage range was kept constant (Fig.6b). Characteristics capacitance change vs. time and phase vs. time have been aggregated estimated and show that the capacitance and phase to not change in time [14].

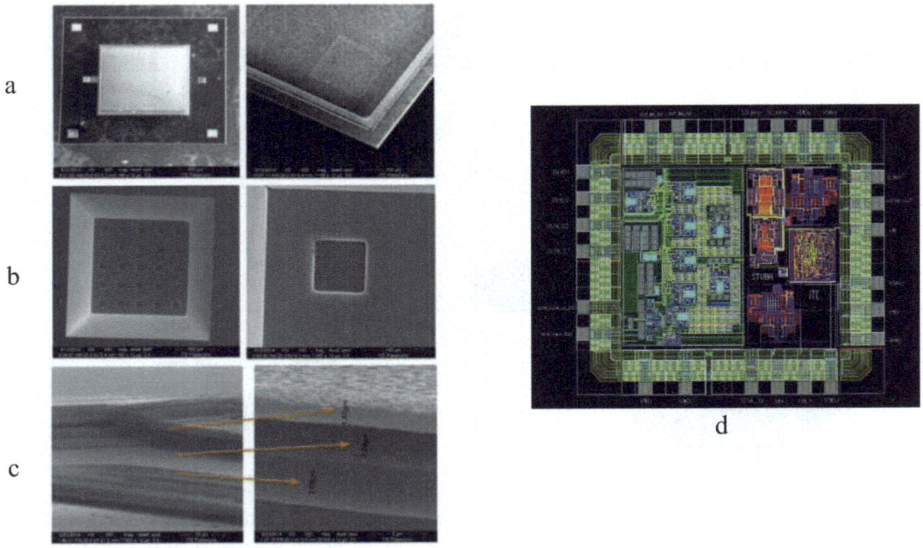

Fig. 5. MEMS microphone: a) membrane topography, b) backplate topography, c) cross section of the structure, thickness of microphone elements presented on the right, d) readout ASIC final layout fabricated in AMSC035um technology.

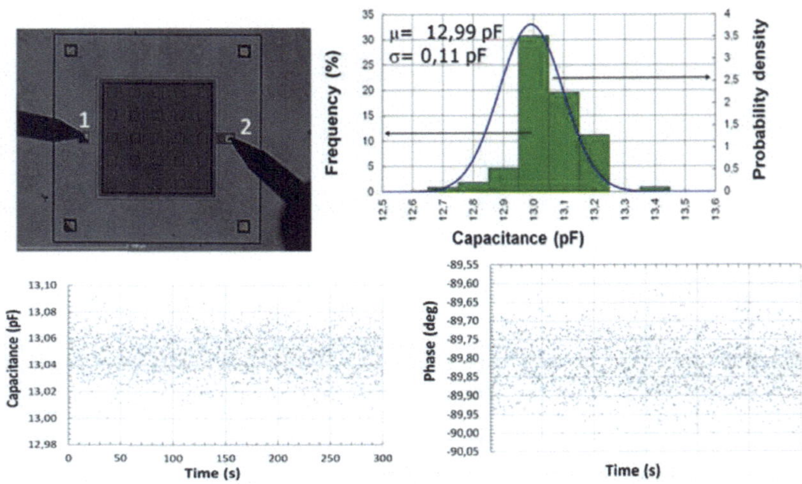

Fig. 6. Electrical measurements a) The schematics of measurement: 1 – contact to backplate, 2 - contact to polysilicon membrane → during the measurements it was connected to ground; b) Distribution of measured capacities, μ– expected value, s - standard deviation, c) The change of the capacitance in time for chosen microphone, d) The change of the phase in time for chosen microphone

4.2 Single structure mechanical experiments

Measurements of resonant frequency have been performed on single structures (Fig.7a). Measurements have been made with with Polytec Scanning Vibrometer MSA 500 within vibration spectrum (Fig. 7b, 7c) for resonance frequency from Table 1.

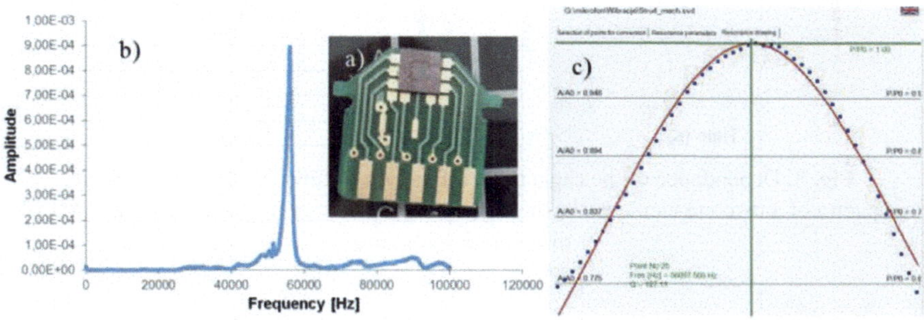

Fig. 7. Resonant frequency of the single MEMS microphone structure a) single MEMS microphone structure; b) vibration spectrum; c) fitted resonance curve, resonance quality Q=127

Table. 1. Resonant frequency of one of investigated structures

Mod	Resonant frequency [kHz]	Displacement [μm]
1	56.1	0.893
2	51.6	0.107
3	74.5	0.034
4	89.7	0.049

4.3 Electrical experiments on single structures

MEMS microphone capacitance has been measured for particular pressure levels imposed to the membrane. The experiment was carried out for very small capacitances to verify the stability of the system as well. The study was conducted during 40000s time slot (Fig. 8a). The average value measured was 8.66pF, with a standard deviation of 0.002pF. It confirmed the proper configuration of the experimental setup. Two types of MEMS microphone structures have been observed (Fig. 9b): the "alpha" - with periodic time-dependent capacitance (green) and the all remaining structures marked as "beta" with electrical behavior similar (black) to the discrete element (red). It was caused by influence of laboratory vibration noise in the on the membrane under test, lowered stiffness of the microphone marked "alpha" resulting in amplified vibration. Concurrent reason could be reduced distance between the microphone covers of "beta" resulting in exaggerated capacitance changes.

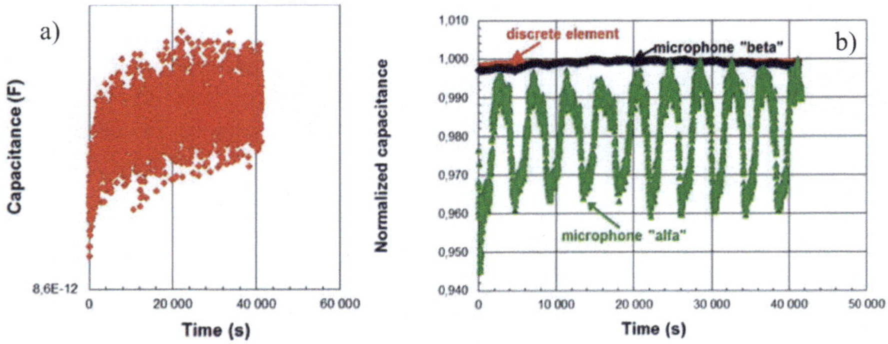

Fig. 8. Dependance of the capacity as a function of time, a) reference capacity (capacity of a discrete element), b) the capacitance measured for the investigated MEMS microphone structures

4.4 Preliminary experiments on the microphone sensitivity

The lowest applied pressure was approx. 1500Pa (equivalent to 158dB) for three series of measurements with pressure gradually increased and reduced down to initial state. Characteristics are coherent with small discrepancies exposed due to deformation of the membrane fixed during preceding series of measurements. For the pressure range considered, the capacity had a logarithmic character (Fig. 10a). It was coherent with the simulation results. The Fig. 10b shows the course of the microphone sensitivity as a function of applied pressure.

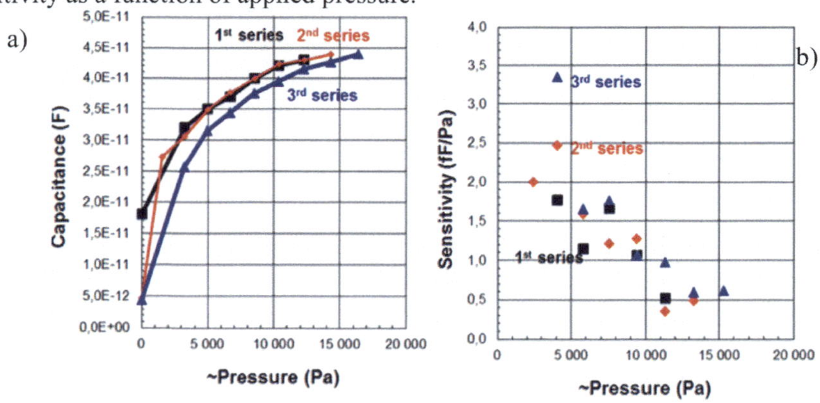

Fig. 9. Dependence of the measured capacitance values vs. applied pressure (a) and dependence of the microphone sensitivity vs. applied pressure (b)

5 Summary

The MEMS microphone simulation results by CoventorWare have been coherent with theoretical analysis based on well-known physical phenomena. The coherence observed lead the designers to conclude that achieved results were satisfactory on the early stage of system development process. Preliminary simulations of the mechanical part of the structure has been performed using MEMS+ environment as well as preliminary simulations of the dedicated readout electronics IC module. Than full co-simulation process of the MEMS module along with dedicated readout circuitry was achievable. This validation process was performed in order to lower design costs and shortening device development time. Based on these results authors team successfully manufactured and characterized microphone structure. Integration of the system with ASIC measurements is scheduled for the close future after full MEMS characterization.

6 Acknowledgements

This work profits from excellent collaboration within the SMAC project. Presented work has been partially supported by the EU FP7 Project SMArt Systems Co-design (SMAC), FP7-ICT-2012-7 288827

References

1. Scheeper P. R., Nordstrand B., Gullov J. O., Bin L.., Clausen T., Midjord L., Storgaard-Larsen T., *A new Measurement Microphone Based on MEMS Technology*" Jurnal of Microelectromechanical systems , Vol 12, No 6, pp.880 – 891 (2003)
2. Torkkeli A., Rusanem O., Saarilahti J., Seppa H., Sipola H., Hietanen J., *Capacitive microphone with low-stress polysilicon membrane and high-stress polysilicon backplate*, Sensors and Actuators A, Physical 01/2000 (2000)
3. Hsu P. C., Mastrangelo H. C., Wise K. D., *A high sensitivity polysilicon diaphragm condenser microphone*, in Digest Tech. Papers IEEE MEMS 1998, pp.580-585
4. Bergqvist J., Rudolf F., Silicon A Condenser Microphone Using Bond and Etch-back Technology, Sensors and Actuators A 45, pp. 115 – 124 (1994)
5. Ning Y. B., Mitchell **A. W.**, Tait R. N., *Fabrication of silicon micromachined microphone using a dry-etch process*, Sensors and Actuators A 53, pp. 237 – 242 (1996)
6. Ganji B. A., Majlis B. Y., *Design and fabrication of a new MEMS capacitive misrophone using a perforated Al diaphragm*, Sensors and Actuators A 149, pp 29- 37 (2009)
7. Schellin R., Hess G., *A silicon subminiature microphone based on piezoresistive polysilicon strain gauges* , Sensors and Actuators A, 32, pp. 555 – 559 (1992)
8. Kaelvesten E., Loefdahl L., Stemme G., *Small piezoresistive silicon microphones specially designed for the characterization of turbulent gas flows*, Sensors and Actuators A; 46, 1/3; 151, Proceedings of Eurosensors VIII (1995)
9. Martins P., Beclin S., Brida S., Mettivet S., Stojanovic O., Malhaire C., Design of bossed silicon membranes for high sensivity microphone application, Microsyst. Technol., 13, pp. 1495 – 15000 (2007)
10. Bieniek T., Janczyk G., Grabiec P., Szynka J., Kaliciński S., Janus P., Domański K., Sierakowski A., Ekwińska M., Szmigiel D., Tomaszewski D., Holzer G., Schropfer G, *Customer-Oriented Product Engineering of Micro and Nano Devices*, Proceedings of the 17th Int. Conf. on Mixed Design of Integrated Circuits and Systems, Wroclaw, Poland, 24-26.06.2010, pp. 81-84 (2010)

11. Janus P., Domański K., Ekwińska M., Bieniek T., Kaliciński S., Sierakowski A., Dobrowolski R., Szmigiel D., Prokaryn P., Grabiec P., *Design and Manufacturing of Heterogeneous Microsystems for Micro- and Nanotechnology Applications*, Elektronika 2010 vol. LI no. 4 pp. 62-71 (2010)
12. Bieniek T., Janczyk G.,Janus P., Szynka J., Grabiec P., Kociubiński A., Ekwińska M., Tomaszewski D., Malinowski A., *Multi-Domain Modeling and Simulations of the Heterogeneous Systems*", Journal of Telecommunications and Information Technology, vol 1, 2010, 34-39 (2010)
13. Bieniek T., Janus P., Kociubiński A., Grabiec P., Janczyk G., Szynka J., *Coupled Thermo-electro-mechanical Modeling and Simulation of 3D Micro- and Nanostructures, Elektronika*, nr. 1, pp. 99-101 (2008)
14. Ekwińska M., Janczyk G., Bieniek T., Grabiec P., *Methodology of MEMS-based Smart System Design, Modelling and Simulation – Development Practice*, Smart System Integration Proceedings, pp. 463 (2014)
15. Ortlof D., Schmidt T., Hahn K., Bieniek T., Janczyk G., Bruck R.: "MEMS Product Engineering. Handling the Diversity of an Emerging Technology. Best Practices for Cooperative Development", Springer (2014)

Template design for craniometric landmarks identification

Iryna Gorbenko, Krzysztof Kałużyński, Krzysztof Mikołajczyk

Warsaw University of Technology, Faculty of Mechatronics, Institute of Metrology and Biomedical Engineering, Św. Andrzeja Boboli 8, 02-525 Warsaw, Poland
{i.gorbenko,k.kaluzynski,k.mikolajczyk}@mchtr.pw.edu.pl

Abstract. An important role in the soft tissue thickness measurement plays the head template. Suggested parameter settings during template creation are: sampling 2:5 mm, no smoothing, use of single image or averaging of maximum 4 images, use of inhomogeneity correction. The algorithm of template creation shows limited sensitivity to these parameters. Use of two templates for low and high Body Mass Index (BMI) is recommended.

Keywords: head template, non-rigid matching, similarity measures, Body Mass Index, facial soft tissue thickness measurement.

1 Introduction

The most important issue for forensic facial reconstruction and soft tissue thickness measurement is the correct identification of craniometric landmarks, which strongly affects results of soft tissue thickness measurement. As a base for soft tissue thickness measurement a method based on coregistration of unknown skull to the reference head model was chosen [1]. It is based on non-rigid matching of individual Magnetic Resonance Imaging (MRI) data to a template and applying a transform to the set of landmarks defined on the template [2, 3]. The use of MRI data is advantageous, because of absence of ionizing radiation and better soft tissue imaging resolution. The important part of the algorithm is the creation of the head template from MRI data.
The aim of this paper is to propose parameters for MRI template creation. Effect of smoothing, sampling rate, intensity inhomogeneity correction, number of averaged images and BMI on template quality was investigated. Image denoising is routinely used in preprocessing of MRI data [4] and for this reason we did not investigate its influence on the template.

2 Materials and methods

The template creation algorithm [2] consists of following steps:
1) Creation of starting templates;
2) Selection of number of MRI images from a database according to the BMI classes;

R. Jabłoński and T. Brezina (eds.), *Advanced Mechatronics Solutions*,
Advances in Intelligent Systems and Computing 393,
DOI 10.1007/978-3-319-23923-1_67

3) Data denoising using non local means algorithm [5];
4) Intensity inhomogeneity correction of data [6];
5) Elastic matching of the first MRI image to the starting template [3];
6) Elastic matching of the following image to the result of previous matching;
7) Averaging of the first two matched images; saving the result to a temporary variable;
8) Repeat points 6 and 7 for subsequent selected images (as template use temporary variable);
9) Smoothing of the template (optional);
10) Mask creation (segmentation and binarization).

For each value of investigated parameters templates were created. During the investigation of one parameter other parameters did not change. Every following stage of experiment used values of parameters selected in previous stages. Each template was assessed using binary masks obtained from non-rigid matching of the set of MRI images to this template. Each mask was compared with template mask and similarity measures were calculated. To assess the matching quality the following measures were used: value difference (VD), signed value difference (SVD), specificity, sensitivity, Dice Similarity Coefficient (DSC) and Jaccard Similarity Coefficient (JSC) [8]. For masks with identical volume, VD and SVD have zero value, with increasing values indicating increasing volume difference between them. Specificity, sensitivity, DSC and JSC for identical masks have a value of 1, decreasing values indicate imperfect overlap.

Three BMI classes were defined in this paper: BMI<25 (low BMI – class L; BMI units [kg/m^2] will be omitted in the text), 25<=BMI<30 (medium BMI – class M) and 30<=BMI<60 (high BMI – class H). T1-weighted images from MRI database IXI [7] recorded in Caucasian adults were used in this study. 15 MRI images obtained in females having BMI in the range 20-22 were used to validate the templates (see subsections 2.1 – 2.4).

For statistic analysis ANOVA, Tukey HSD test and Student test were used and significance level $\alpha = 0.05$ was chosen.

The PMOD software (version 3.6, PMOD Technologies, Zurich, Switzerland) was used as the image processing tool to implement and to test the methodology and for statistical calculations [9].

2.1 Evaluation of the effect of template smoothing

Four templates with different amount of smoothing were created using Gaussian 3D filter with full width at half maximum (FWHM): 0 mm (no smoothing), 1 mm, 2 mm and 3 mm. Higher levels of smoothing led to loss of important features of the MRI images. In each case four MRI images of females with BMI in the range 20-22 were averaged. The sampling rate was 2 mm. Inhomogeneity correction was not used. To validate these templates images from the test group were matched to each template and the similarity measures were computed. For all similarity measures the ANOVA was performed.

2.2 Effect of sampling rate

Six templates using sampling rate 2, 3, 4, 5, 6 and 8 mm were created. All other parameters of template creation process were as in subsection 2.1. Templates were created without smoothing. Images from the test group were matched to each template. For each set of similarity measures the ANOVA, Tukey HSD test and Student test were performed.

2.3 Effect of intensity inhomogeneity Bayesian correction

Two templates with and without intensity inhomogeneity correction were created. All parameters of template creation process except sampling rate were as in subsection 2.1. The sampling rate was 2 mm. Templates were created without smoothing.
Images from the test group without inhomogeneity correction were matched to template without correction; corrected images from the test group were matched to corrected template; images without inhomogeneity correction were matched to corrected template; corrected images were matched to template without inhomogeneity correction. For each set of similarity measures the ANOVA was performed.

2.4 Effect of number of averaged images

Three templates were created with inhomogeneity correction. The first one is a single-image template (without averaging), second template used 4 averaged images, third – 9 averaged images (figure 1). Other parameters of the template creation were as in subsection 2.3. For each set of similarity measures the ANOVA was performed.

2.5 Evaluation of the effect of body mass index

Three templates according to BMI classes L, M and H were created, using a single image with inhomogeneity correction, no smoothing and sampling rate 2 mm.
Three sets of 15 MRI images according to BMI ranges were chosen from the database. Pairs of data from L and M classes, M and H classes and L and H classes were matched to all templates. In each case pairs similarity measure sets were statistically tested. For all sets of similarity measures the ANOVA and Tukey HSD test were performed.

3 Results

All similarity measures show no statistically significant differences between group means obtained for different levels of smoothing. According to results of ANOVA the use of Gaussian smoothing with filter FWHM 1-2 mm is recommended only in the case of templates created using single image.

Only in the case of the SVD obtained for different sampling rates ANOVA shows significant differences between group means (p = 0.0382). The Tukey HSD test and Student test were performed on SVD sets to determine if there are two sets of data significantly different from each other. For chosen significance level there are differences between group means for templates created using sampling rate 3 mm and 8 mm. The p-values decrease when differences of sampling rate increase. Student test shows no significant differences between group means for templates created with sampling rate in the range 2-5 mm. The p-values decrease when differences of sampling rate increase.
The inhomogeneity correction does not have significant influence on quality of the templates, but an improvement in intensity inhomogeneity of templates can be observed.

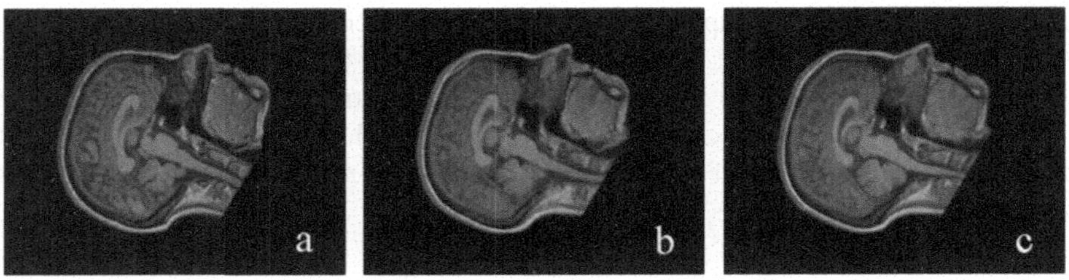

Figure 1. Templates with different number of averaged MRI images: a – single image, b – 4 averaged images, c – 9 averaged images

Only in the case of Sensitivity (p = 0.0173) there are significant differences between group means for templates with different number of averaged images. P-value for SVD measure is close to significance level. Excessive averaging leads to excessively smoothed and poorly matched templates.

Table 1. ANOVA results for test groups with different BMI matched to L class template

	VD	SVD	Specificity	Sensitivity	DSC	JSC
P-value	4.16e-05	8.01e-08	0.041	2.03e-09	3.92e-06	3.4e-06

Table 2. Results of Tukey HSD test for the L class template

BMI	VD	SVD	Specificity	Sensitivity	DSC	JSC
L - H	0.00003	0	0.03214	0	2e-06	2e-06
M - H	0.01563	0.00513	0.30324	0.00249	0.06277	0.06272
M - L	0.0897	0.00174	0.50568	0.00012	0.00347	0.00309

All similarity measures show significant differences between group means for results of matching to the L class template (table 1). To investigate this further Tukey HSD test was performed (table 2). All similarity measures show significant differences between group means for results of matching of L and H class data to the L class template. Specificity, DSC and JSC show no significant differences between group means for results of matching of M and H class data to the L class template. Four measures show significant differences between group means for results of matching of L and M class data to the L class template.

Table 3. ANOVA results for test groups with different BMI matched to M class template

	VD	SVD	Specificity	Sensitivity	DSC	JSC
P-value	0.0618	0.00497	0.034	3.59e-05	0.283	0.268

Table 4. Results of Tukey HSD test for the M class template

BMI	VD	SVD	Specificity	Sensitivity	DSC	JSC
L - H	0.11537	0.00594	0.05024	0.00002	0.46576	0.45635
M - H	0.98921	0.80714	0.98372	0.07598	0.93542	0.92702
M - L	0.08651	0.02999	0.07368	0.01673	0.28379	0.26664

Half of measures show no significant differences between group means for results of matching of test groups obtained for different BMI to the M class template (table 3). Results of Tukey HSD test show that only in the case of SVD and Sensitivity there are significant differences between group means for L-H and L-M class data (table 4). There are no significant differences between group means for matching results of M and H class data to the M class template.

Table 5. ANOVA results for test groups with different BMI matched to H class template

	VD	SVD	Specificity	Sensitivity	DSC	JSC
P-value	0.00012	6.23e-06	7.85e-07	0.386	0.00013	0.00011

Table 6. Results of Tukey HSD test for the H class template

BMI	VD	SVD	Specificity	Sensitivity	DSC	JSC
L - H	0.00084	0.00002	0.000003	0.35843	0.00045	0.00038
M - H	0.94062	0.77491	0.85345	0.8531	0.98957	0.98864
M - L	0.00031	0.00013	0.00002	0.67699	0.00068	0.00059

All similarity measures except Sensitivity show significant differences between group means for matching results of test groups with different BMI to the H class template (table 5). All similarity measures except Sensitivity show significant differences between group means for results of matching L and H class data and L and M class data to the H class template. There are no significant differences between group means for results of matching of M and H class data to the H class template (table 6).

4 Discussion and conclusions

There is no need to use smoothing for templates created from several MRI images. The averaging brings in some smoothing and additional smoothing could result in a loss of important features of the templates and poor matching to these templates. Gaussian smoothing is recommended only in the case of single-image templates. The FWHM of the filter should be 1-2 mm.

Low sampling rate causes deterioration of the result of non-rigid matching, which may result in incorrect position of the craniometric landmarks. High sampling rate slows down the algorithm. Sampling rate of 2-5 mm is recommended.

There is no significant effect of inhomogeneity correction on quality of templates and non-rigid matching process. However, data from IXI database are strongly inhomogeneous what would affect face surface estimation and mask creation. Use of intensity inhomogeneity correction is therefore recommended.

The averaging was expected to provide universal template and better results during matching to this template. However, increased amount of averaging led to excessively smoothed template and poorly matched structures. Use of single-image templates or templates with low number of averaged images is recommended.

Summarizing, a set of parameters was proposed for template creation. However, the algorithm shows limited sensitivity to these parameters in terms of statistical difference of similarity measures.

In a previous paper we stated that BMI is the most important parameter for the outcome of the non-rigid matching [2]. Current results indicate significant differences between group means when matching L-H and L-M class data to the L and H class templates. There are no significant differences when matching M and H class data to the M and H class templates. Therefore, use of the M class template is not necessary.

References

1. Clement, J.G, Marks M.K.: Computer-graphic facial reconstruction" Elsevier, Academic Press, Burlington (2005)
2. Gorbenko, I., Mikołajczyk, K., Iarovyi, I., Kubik, T., Kałużyński, K.: A new method of automatic craniometric landmarks definition and soft tissue thickness measurement based on MRI data. In: 4th International Conference Information Technologies in Biomedicine, Advances in Intelligent Systems and Computing, vol. 283, pp. 115-126. Springer (2014)
3. SPM8 manual: http://www.fil.ion.ucl.ac.uk/spm/doc/spm8_manual.pdf
4. Manjon, J.V., Thacker, N.A., Lull, J.J., Garcia-Marto, G., Marti-Bonmati, L., Robles, M.: Multicomponent MR Image Denoising. Int. J. Biomed. Imaging 10, pp. 1155-1180 (2009)
5. Buades, A., Coll, B., Morel, J.M.: A review of image denoising algorithms, with a new one. SIAM Interdisciplinary Journal 4 (2), pp. 490-530 (2005)
6. Hou, Z.: A Review on MR Image Intensity Inhomogeneity Correction. Int. J. Biomed. Imaging, pp. 1-11 (2006)
7. IXI dataset: http://www.brain-development.org.
8. PFUSEIT manual: http://doc.pmod.com/PDF/PFUSEIT.pdf.
9. PMOD documentation: http://www.pmod.com/technologies/products/products.html.

Stress assessment in steel truss structures on the basis of magnetoelastic effects

Dorota Jackiewicz

Institute of Metrology and Biomedical Engineering, Warsaw University of Technology, sw. A. Boboli 8, 02-525 Warsaw, Poland
d.jackiewicz@mchtr.pw.edu.pl

Abstract. In this paper, magnetoelastic evaluation method of stress assessment in constructional steel samples is presented. For the investigation special test stand was designed and built, along witch the truss structure, which allowed for installation of sample members. Measurements of the hysteresis loops of the sample members were performed under varying mechanical load, resulting in magnetoelastic $B_m(\sigma)_{Hm}$ characteristics. The experimentally obtained characteristics were consistent with the magnetoelastic effect theory, and allow for the on-line monitoring of the structure state. Presented results confirm the feasibility of the magnetoelastic effect based NDT stress assessment in the steel truss constructions.

Keywords: magnetoelastic effects, stress assessment, constructional steel

1 Introduction

Modern steel structures require continuous monitoring of the forces and internal stresses. This is particularly important in the case of bridges, halls, electricity masts and pylons made as truss constructions. In such a case, there are critical components whose failure poses a threat to the entire structure. Current methods for assessment of such stresses, such as the strain gauge method or magnetostrictive method [1], have their major limitations. There is therefore a need to develop new methods for assessing the state of stress in structural elements. Recent trends indicate the rapid development of research in the field of magnetic methods. Currently, magnetic testing methods are widely developed: both for classic materials [2], as well as new methods are developed [3]. A physical phenomenon that creates opportunities for use in such studies is the magnetoelasticity, also known as Villari effect [4]. It involves change of the magnetic properties of the material under stress from external forces. It is currently used in the construction of magnetoelastic force and stress sensors [5, 6, 7]. Due to the nondestructive nature of the measurement, and the possibility of continuous operation, this effect is interesting to study because of its use for on-line monitoring of stresses in the given structure.

The changes of microstructure of constructional steel significantly influence its magnetic properties. These effects were successfully utilized for pipelines monitoring previously [1]. On the other hand, the possibility to utilize the magnetoelastic Villari

R. Jabłoński and T. Brezina (eds.), *Advanced Mechatronics Solutions*,
Advances in Intelligent Systems and Computing 393,
DOI 10.1007/978-3-319-23923-1_68

effect for stress assessment and construction state monitoring in large steel structures (such as trusses) was not previously considered.

In this paper the specially developed magnetoelastic method investigation of steel truss construction is presented. It allows for stress assessment in the preselected members of the truss under mechanical load. Steel trusses are widely used in medium and big scale constructions, such as electric grid power line distribution towers, bridges etc., but without proper maintenance they are prone to catastrophic failures [5,6], which is a significant risk for the local environment and economies. It is especially important for energetic and civil engineering structures.

2 Materials and methods

In the Figure 1a) the truss designed for magnetoelastic testing under varying mechanical load is shown. Three central members of the truss, that is the central bottom chord and the central webs, were planned as the test samples. Their cross-sections are significantly lower than the other elements, in order to carry out tests without any plastic damage nor significant elastic deformation of the rest of the members.

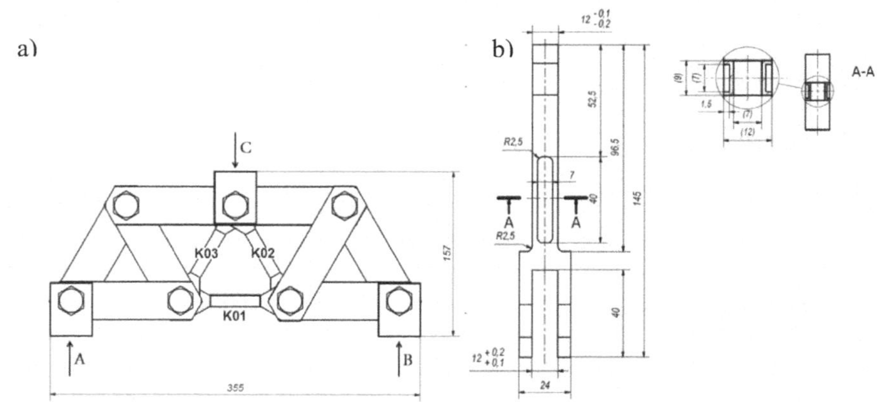

Fig. 1. a) Schematic of the experimental steel truss structure. A,B – support points; C – loading point; K01 – bottom chord, stretched sample; K02,K03 – webs, compressed samples. b) Drawing of the sample member. Middle section consist of two separate columns, on which the measuring and magnetizing windings are wound.

To investigate the basic magnetic properties of the given sample, three requirements have to be fulfilled. Firstly, the closed magnetic circuit in the sample is to be made. Thus the demagnetizing field influence on the measurements is greatly reduced, and the shape influence is nearly eliminated. Secondly, there should be uniform stress distribution along the magnetic circuit in the sample. Acquiring this condition allows for elimination of the stress influences cancelling each other, which happens when there are positive stresses in one part of the sample, and negative in another. The third, important condition is ensuring the distribution of the effective stresses parallel or perpendicular to the magnetic patch direction in the sample.

For the magnetoelastic tests, frame-shaped member sample was developed. The sample is presented on Figure 1 b). It is made of the 13CrMo4-5 constructional steel, widely used in the energetic industry. On the side columns in the middle section of the sample, both sensing and magnetizing windings were made. The magnetizing and sensing windings were wound on both columns. Additionally, to reduce demagnetization effects, sensing winding have to be located under the magnetizing winding. In the presented investigation, each sample was wound by 260 turns of magnetizing winding (130 turns per column), and 100 turns of sensing winding (50 turns per column).

3 Experimental

The hysteresis loops measurements are done on the hysteresisgraph, which is the specially developed test stand for magnetic hysteresis loop measurements (Figure 2). The whole test stand is controlled by the PC equipped with NI USB 6525 data acquisition card. The DAQ is controlled by NI LabView software, for real-time control as well as for the data processing. Voltage signal generated by the data acquisition card drives the Kepco BOP 36 voltage/current converter. The output of Kepco BOP 36 is connected to the magnetizing winding of the investigated member sample. Measuring winding is connected to the Lakeshore 480 fluxmeter, which measures flux density B in the sample. Voltage output of the fluxmeter is connected to the data acquisition card. As a result, the hysteresisgraph presents B(H) hysteresis loops under digitally controlled values of amplitude of magnetizing field, as well as for different frequencies and shapes of magnetizing signals.

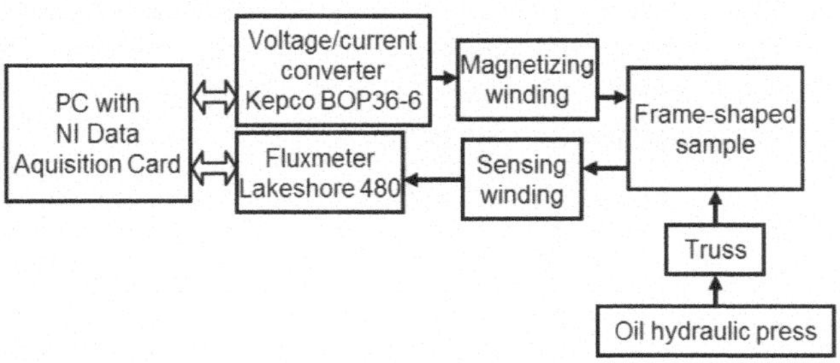

Fig. 2. Schematic block diagram of the hysteresisgraph - magnetic hysteresis loop measurement system

The truss was supported on the bottom edge nodes (Fig. 1.a, A,B). The mechanical force was vertically exerted on the upper central node of the truss (Fig 1.a, C) by the oil hydraulic press.

During the magnetoelastic measurements, three frame-shaped member samples were installed into the truss. The truss was put under the mechanical load. The sample K01 (bottom chord) was stretched with tensile stresses, while the samples K02, K03 (webs) were loaded with compressive stresses (Fig. 1.a). The influence of the stresses on the shapes of the B(H) hysteresis loops was measured for amplitudes of magnetizing field H_m equal to 350 A/m, 435 A/m, 655 A/m, 870 A/m, 1310 A/m and

2180 A/m. The frequency of the magnetizing signal during the measurements was set to 0.1 Hz.

The stresses in the member samples were calculated on the basis of load, truss geometry and sample members dimensions (Table 1). The load was incremented, and hysteresis loop measurements were done for the calculated stresses. Unfortunately, one of the compressed member samples buckled, preventing the further measurements for higher stress values.

Table 1. Mechanical load applied to the truss. F – force exerted by the oil hydraulic press; σ – calculated mechanical stresses in the samples.

F (kN)	σ (MPa)		F(kN)	σ (MPa)	
	tensile	compressive		tensile	compressive
0,6	20	13	5,0	180	120
1,1	40	27	5,5	200	133
1,7	60	40	6,9	250	167
2,2	80	53	8,3	300	200
2,8	100	67	9,7	350	233
3,3	120	80	11,1	400	267
3,9	140	93	12,5	450	300
4,4	160	107			

4 Results and discussion

The experimental results of measurements of the magnetic characteristics stress dependence of the bottom chord sample are presented in Figure 3. Stress dependence of the shape of magnetic hysteresis B(H) loops may be observed for different values of magnetizing field H_m amplitude. Changes of the basic magnetic parameters: flux density, remanence and coercivity are easily seen. From the technical point of view, changes of flux density are the most interesting.

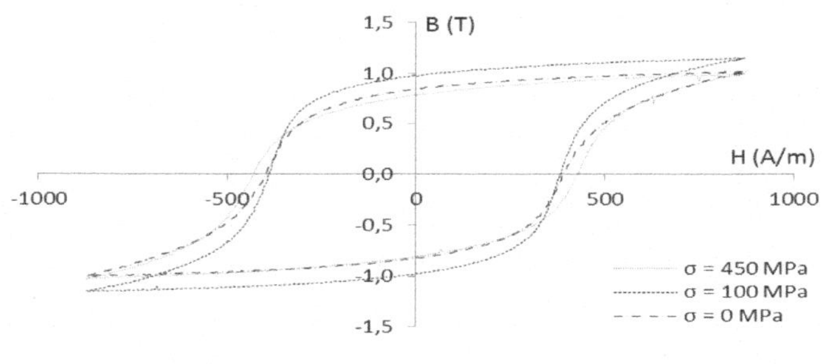

Fig. 3. Magnetic hysteresis loops of the K01 stretched member sample (bottom chord), under various mechanical loads. H - magnetizing field; B - flux density, σ – tensile stress.

Figure 4 presents the magnetoelastic $B_m(\sigma)_{Hm}$ characteristics for the K01stretched sample. Under the tensile stresses, value of the flux density B in the sample first increase, and then, after reaching the Villari point, it starts to decrease. The characteristics are incomplete, because of the compressed member sample buckling. Moreover, these changes are relatively higher for lower values of the amplitude of magnetizing field H_m. This occurs due to the fact, that for lower values of magnetizing field H_m, participation of magnetoelastic energy in the total free energy in the sample's material is significantly higher.

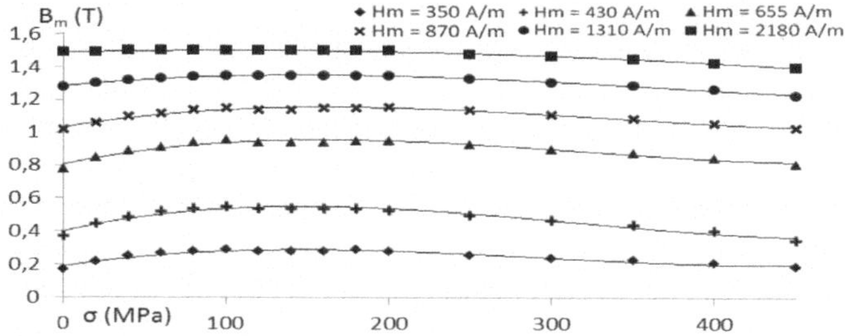

Fig. 4. Influence of the tensile stresses on the flux density of the K01 sample under various magnetizing fields. H - magnetizing field; B - flux density, σ – tensile stress.

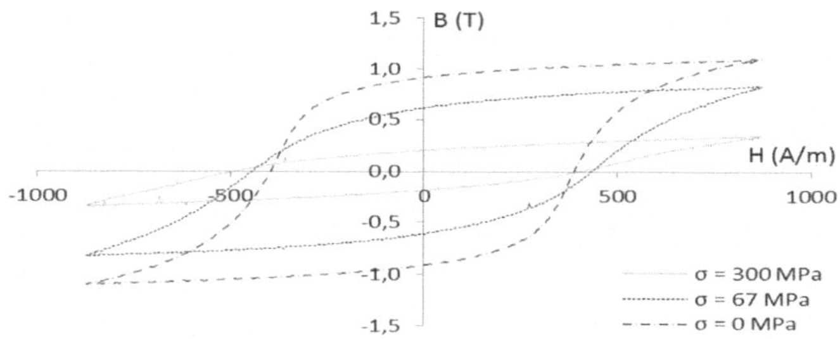

Fig. 5. Influence of the compressive stresses on the flux density of the K02 sample under various magnetizing fields. H - magnetizing field; B - flux density, σ – compressive stress.

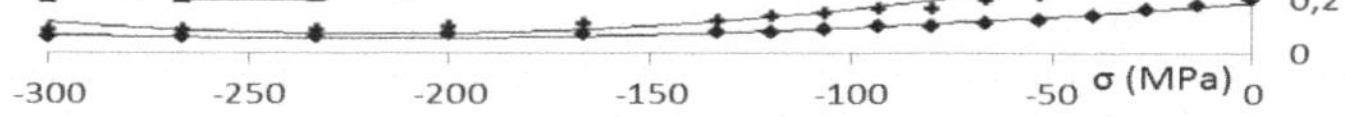

Fig. 6. Influence of the compressive stresses on the flux density of the K02 sample under various magnetizing fields. H - magnetizing field; B - flux density, σ – compressive stress.

Figure 5 presents the results of measurements of stress dependence of magnetic hysteresis loops of the K02 compressed sample. Figure 6 presents the experimental results of the magnetoelastic $B_m(\sigma)_{Hm}$ characteristics for the compressed K02 sample. Under the compressive stresses, value of the flux density B in the sample nonlinearly decreases. These changes are relatively higher for lower values of the amplitude of magnetizing field H_m. The reason is the same as for the tensile stresses. The obtained characteristics are consistent with the theoretical expectations, both for the compressive and tensile stresses.

5 Conclusions

Test stand for the investigation was designed and built, along witch the special truss with interchangeable members, which allowed for installation of specially developed sample members. Magnetoelastic effect based method of stress assessment in constructional steel samples was presented. Measurements of the hysteresis loops of the sample members were performed under varying mechanical loads, which allowed to obtain magnetoelastic $B_m(\sigma)_{Hm}$ characteristics. The experimentally obtained characteristics are consistent with the theory of magnetoelastic effect. The results presented in the paper confirm the possibility of the magnetoelastic effect based measurements for NDT stress assessment in the steel truss structures. The development of new nondestructive evaluation method for industrial applications is feasible. The measurements can be performed continuously for on-line monitoring of the given structure state. Moreover, the method is relatively low-cost, and neutral to environment and operators.

Acknowledgment

This work was partially supported by The National Centre of Research and Development (Poland) within grant PBS1/B4/6/2012.

References

1. Bartels, K., C. Dynes, and H. Kwun: Nondestructive evaluation of prestressing strands with magnetostrictive sensors. SPIE Proceedings Structural Materials Technology III-An NDT Conference 3400, 326-337 (1998)
2. M. Kachniarz, J. Salach, R. Szewczyk, A. Bienkowski: Temperature Influence on the Magnetic Characteristics of Mn-Zn Ferrite Materials. Progress in Automation, Robotics and Measuring Techniques Advances in Intelligent Systems and Computing 352, 121-127 (2015)
3. M. Nowicki, R. Szewczyk. Ferromagnetic Objects Magnetovision Detection System. Materials 6, 5593-5601 (2013)
4. R. Szewczyk, A. Bienkowski, J. Salach, E. Fazakas, L. Varga: The influence of microstructure on compressive stress characteristics of FINEMET-type nanocrystalline sensors. Journal of Optoelectronics and Advanced Materials 5, 705-708 (2003).
5. A. Bieńkowski, R. Szewczyk, J. Salach: Industrial application of magnetoelastic force and torque sensors. Acta Physica Polonica A 118, 1008-1009 (2010).
6. C. Hlenschi, S. Corodeanu, H. Chiriac: Magnetoelastic Sensors for the Detections of Pulse Waves. IEEE Transactions on Magnetics 49, 117-119 (2013).
7. J. Salach, A. Bienkowski, R. Szewczyk: The ring-shaped magnetoelastic torque sensors utilizing soft amorphous magnetic materials. Journal of Magnetism and Magnetic Materials 316, e607-e609 (2007).

Measurement system for investigating magnetic characteristics of soft magnetic materials in Rayleigh region

Maciej Kachniarz

Institute of Metrology and Biomedical Engineering, Warsaw University of Technology, św. A. Boboli 8, 02-525 Warsaw, Poland
m.kachniarz@mchtr.pw.edu.pl

Abstract. The paper presents unique measurement system and methodology for investigating magnetic characteristics of soft magnetic materials in Rayleigh region. Rayleigh region is the first part of the initial magnetization curve, where magnetizing fields with low amplitude are acting on the investigated material. As a part of the paper, the physical processes underlying the phenomenon of magnetization in Rayleigh region are discussed. Outline of the innovative measurement methodology involving elements of mathematical modeling and concept of the measurement system are presented. On the basis of the concept, measurement system was designed and developed with special measurement software included. Functional properties of the system were investigated with experimental sample of soft magnetic material. Finally, the conclusions were formulated, which are included in the last section of the paper.

Keywords: magnetism, soft magnetic materials, Rayleigh region, magnetic characteristics, measurement system.

1 Introduction

Inductive components have wide range of applications in modern electronic including power supplies, resonant circuits, filters and many others. For this reason there is a great need to investigate properties of magnetic materials used for manufacturing magnetic cores of inductive components. Especially important are magnetic characteristics of such materials: initial magnetic curve and hysteresis loop, which are the most comprehensive form of description of material's magnetic properties.

In modernly developed electronic devices one of the main purposes is to reduce the electric energy consumption. In devices designed in this way small electric currents are flowing, which results in small magnetizing fields acting on the magnetic cores of the inductive components. For low magnetizing fields, material exhibits different magnetic properties than for high magnetizing fields. So there is a need to investigate the magnetic properties of materials for low fields, in so-called Rayleigh region [1]. Previous attempts indicates that there is also a possibility of utilizing magnetic cores

R. Jabłoński and T. Brezina (eds.), *Advanced Mechatronics Solutions*,
Advances in Intelligent Systems and Computing 393,
DOI 10.1007/978-3-319-23923-1_69

working in Rayleigh region in many industrial applications, for example in low current transformer [2].

The main obstacle in investigating magnetic properties of the materials in Rayleigh region is the lack of measurement system allowing to investigate magnetic characteristics in low magnetizing fields. Commercially available systems for measuring the magnetic characteristics are designed to work with high magnetizing fields in near saturation region [3, 4]. In this paper, new and innovative measurement system is presented utilizing elements of mathematical modeling, which allows to investigate magnetic materials in Rayleigh region.

2 Rayleigh region

Term Rayleigh region refers to the first part of the initial magnetization curve, where low magnetizing fields are acting on the investigated magnetic material [5]. In the Rayleigh region, initial magnetization curve is a parabolic curve described by second order polynomial equation called Rayleigh law [6]:

$$B(H) = \mu_0 \mu_i H + \alpha_R \mu_0 H^2 \tag{1}$$

where B is magnetics flux density, H is magnetizing field and μ_0 is a vacuum magnetic permeability (physical constant). Symbols μ_i and α_R refer to material parameters respectively initial relative magnetic permeability and so-called Rayleigh coefficient. Dominant mechanism of magnetization in Rayleigh region are reversible elastic deflections of the Bloch domain walls [7], which are described by first, linear part of the equation (1). Second order part of the equation refers to irreversible translations of the domain walls, which are also occurring in Rayleigh region allowing to observe magnetic hysteresis phenomenon. Hysteresis loop in Rayleigh region (so-called Rayleigh hysteresis loop) has lenticular shape, unlike loops in near saturation region. It is also described with second order equations.

The main problem with measurements of magnetic characteristics in Rayleigh region are low values of measured signals, especially voltage induced under the influence of magnetizing field. This creates a need to apply special methodology of measuring.

3 Outline of the measurement methodology

Proposed measurement methodology is based on the oscilloscope method of measuring of magnetic characteristics [1], which is much less complex than other methods (like ballistic or differential). In classic oscilloscope method, magnetizing waveform and integrated measuring waveform are applied to the X and Y input of the oscilloscope working in X-Y mode and hysteresis loop is visualized on the screen. Integration of the measuring waveform is necessary due to the differential dependence between magnetic flux density B and voltage u_B induced in sensing winding of the investigated sample.

To adapt oscilloscope method for measuring magnetic characteristics in Rayleigh region, some modifications are made. The most important change is elimination of the integrator circuit. It was possible due to the specifics of Rayleigh region. When investigated material is magnetized with sinusoidal waveform, induced voltage waveform is also sinusoidal, because reversible elastic deflections of the domain walls does not influence the deformation of the waveform. When material is magnetized with sinusoidal field:

$$H(t) = H_m \sin 2\pi\omega t \tag{2}$$

the magnetic flux density waveform in Rayleigh region is given with the equation:

$$B(t) = B_m \sin(2\pi\omega t + \Delta\varphi) \tag{3}$$

In the equations (2) and (3) H_m and B_m are amplitudes of the waveforms, ω is frequency of the waveforms, t is time and $\Delta\varphi$ is phase shift resulting from delay in irreversible translations of the domain walls, which causes magnetic hysteresis phenomenon to occur.

Measurement of the magnetic flux density B is indirect measurement, where the value measured directly is voltage u_B induced in the sensing winding. The equation connecting these parameters has the form:

$$u_B(t) = -n_s S_e \frac{dB(t)}{dt} \tag{4}$$

where n_s is number of sensing coils and S_e is effective cross-sectional area of the sensing coil. After substitution of the equation (3), u_B is given as:

$$u_B(t) = -n_s S_e B_m \frac{d \sin(2\pi\omega t + \Delta\varphi)}{dt} = -n_s S_e B_m 2\pi\omega \cos(2\pi\omega t + \Delta\varphi) \tag{5}$$

The equation (5) contains cosine function, which can be treated as sine function with phase shift $\pi/2$. So equation (5) can be written as:

$$u_B(t) = n_s S_e B_m 2\pi\omega \sin\left(2\pi\omega t + \Delta\varphi - \frac{\pi}{2}\right) \tag{6}$$

Due to equation (6), there is no need to integrate $u_B(t)$ waveform while measuring magnetic characteristics in Rayleigh region. The same effect will be obtained when $\pi/2$ is added to the phase shift angle of sine function in equation (6). Obtained waveform shifted in phase u_{Bps} is given by the equation:

$$u_{Bps}(t) = n_s S_e B_m 2\pi\omega \sin(2\pi\omega t + \Delta\varphi) = n_s S_e 2\pi\omega B(t) \tag{7}$$

and values of magnetic flux density B can be calculated as:

$$B(t) = \frac{u_{Bps}(t)}{n_s S_e 2\pi\omega} \tag{8}$$

Presented methodology allows to obtain much better results in Rayleigh region than using integrator circuit and eliminates zero drift error, which is important problem in most of the integrator circuits discussed in the literature [8].

4 Conception of the measurement system

The schematic block diagram illustrating the conception of designed measurement system is presented in Fig. 1.

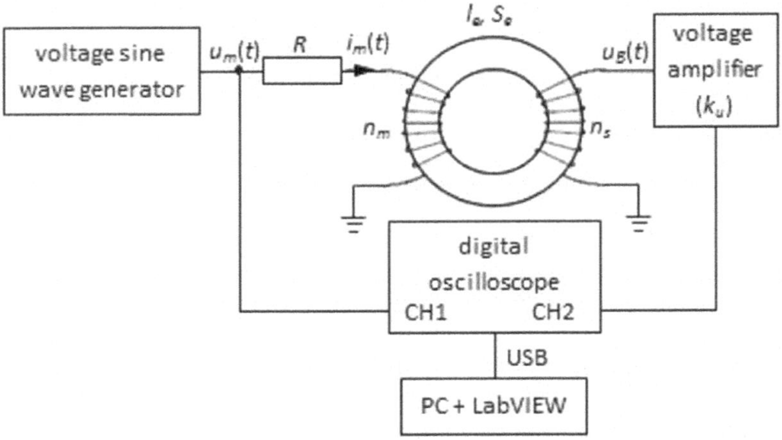

Fig. 1. Schematic block diagram of the conception of the designed measurement system, $u_m(t)$ – magnetizing voltage waveform, $i_m(t)$ – magnetizing current waveform, $u_B(t)$ – induced voltage waveform, n_m – magnetizing coils, n_s – sensing coils, l_e – effective magnetic path in the magnetic core, S_e – effective cross-sectional area of the sensing coil, k_u – gain of the amplifier

As a source of magnetizing waveforms, voltage sinusoidal waveform generator is used. It should be analog RC generator to avoid distortions resulting from quantization of the signal by digital generator. The magnetizing current waveform $i_m(t)$ is determined by resistor R and drives magnetizing winding of the magnetic core. The magnetizing field waveform is described with the formula:

$$H(t) = \frac{n_m i_m(t)}{l_e} = \frac{n_m u_m(t)}{R l_e} \tag{9}$$

Induced voltage waveform $u_B(t)$ is amplified in voltage amplifier with gain k_u. Magnetic flux density B is calculated as in equation (8). Taking into account the gain of the amplifier the equation describing B has the form:

$$B(t) = \frac{u_{Bps}(t)}{k_u n_s S_e 2\pi\omega} \tag{10}$$

Both magnetizing and measuring waveforms are collected by the digital oscilloscope and sent to the PC via USB interface. Computer is processing measurement data: calculates the values of magnetizing field and magnetic flux density utilizing equations (9) and (10) and draw Rayleigh hysteresis loop.

5 Developed measurement system

On the basis of presented outline of measurement methodology and conception, the measurement system for investigating magnetic characteristics in Rayleigh region was designed, including electronics, PCB and measurement software. The view of the entire developed system operating is shown in Fig. 2.

Fig. 2. The view of the developed measurement system operating

For generating magnetizing field, magnetizing current generator circuit was designed utilizing ICL8038 analog RC generator integrated circuit. The electric scheme of the designed circuit is presented in Fig. 3.

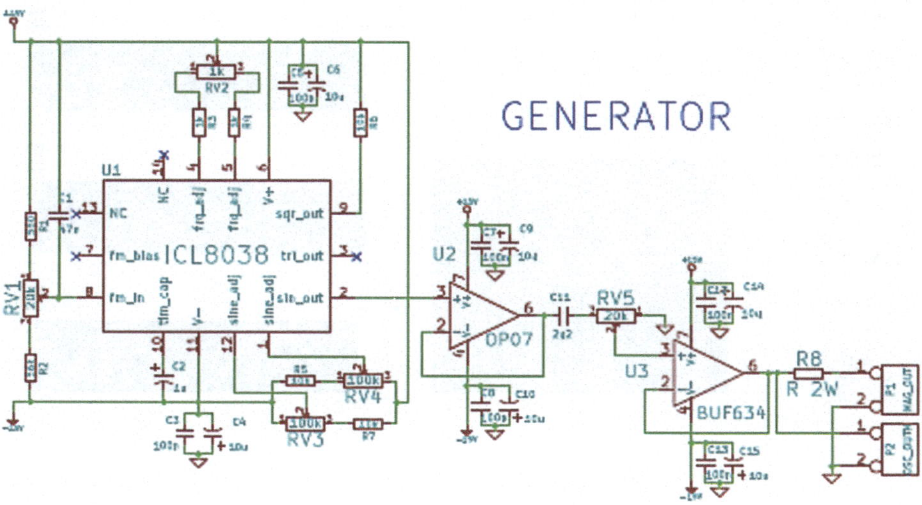

Fig. 3. Electric scheme of magnetizing current generator

Designed circuit allows to modulate amplitude and frequency of the magnetizing current waveform with potentiometers RV1 and RV5. It is also possible to reduce sinusoidal wave distortions with potentiometers RV3 and RV4. Operational amplifier BUF634P with high output current allows to generate current waveforms with amplitudes high enough to obtain magnetizing fields corresponding to the Rayleigh region. Magnetizing current is forced by resistor R8 with nominal power of 1 W. The resistor is replaceable, so it is possible to choose the value of its resistance to the individual properties of each investigated material.

Voltage waveform induced in the sensing winding is amplified with two-stage voltage amplifier. Each stage is non-inverting voltage amplifier with gain of 34, which results in total gain of the measuring signal amplifier of 1156. The output of the amplifier is connected to the Y channel of the digital oscilloscope.

On the basis of developed electric schemes, the PCB (Printed Circuit Board) layout was designed with respect for the rules of electromagnetic compatibility. The PCB with all electronic components mounted was placed inside the aluminum shielding enclosure which protected the electronic circuits from the influence of external electromagnetic noises.

For magnetizing and measuring data acquisition, dual channel digital oscilloscope Tektronix TDS 1002B was chosen. The device allows to collect the voltage waveforms and measure its basic parameters like amplitude, frequency, etc. Oscilloscope is equipped with USB interface and the manufacturer provides drivers for LabVIEW environment. This allows to easily connect the device to the PC and send measurement data.

To acquire and process measurement data, special software was developed in National Instruments LabVIEW environment. The user interface of the application is presented in Fig. 4.

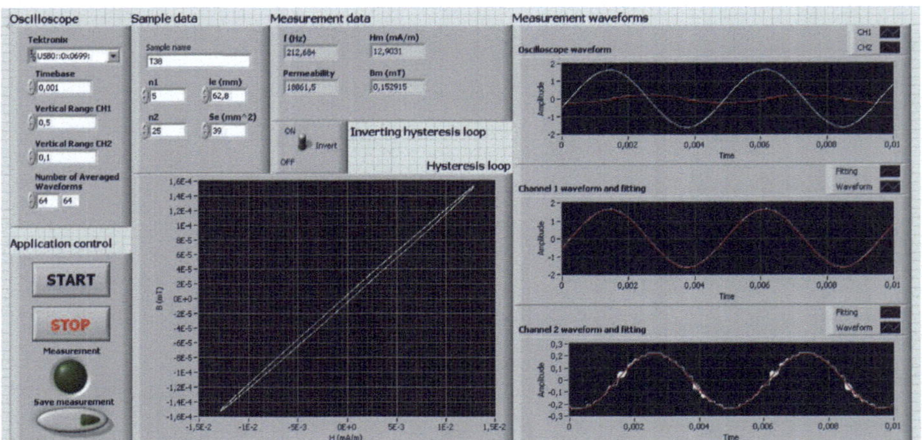

Fig. 4. User interface of the developed measurement application

Both magnetizing and sensing waveforms are collected from the digital oscilloscope and then processed in the program. Each waveform is fitted with sinusoidal function of three parameters: amplitude, frequency and phase angle. Measuring waveform is also shifted in phase by $\pi/2$ according to the equations (6) and (7). In the program, user defines the values of basic parameters of the investigated magnetic core: number of the coils and geometrical parameters. This parameters are used to calculate the waveforms of magnetizing field and magnetic flux density in the material, as presented in equations (9) and (10). On the basis of obtained waveforms program draws the graph of the Rayleigh hysteresis loop and calculates its basic parameters. After processing there is a possibility of saving data in the text file.

6 Verification of functional properties of the developed system

To verify if developed measurement system is working correctly, the measurement of the exemplary magnetic core was performed. Investigated magnetic core was made of manganese-zinc ferrite material of high magnetic permeability. During the investigation the family of Rayleigh hysteresis loops for the material was measured. The results are presented in Fig. 5.

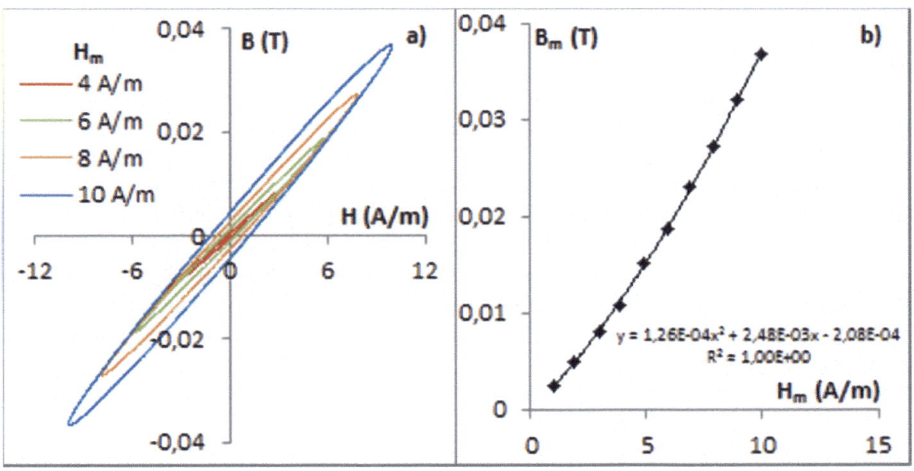

Fig. 5. Measurement results obtained for the exemplary magnetic core investigated with developed system: a) the family of Rayleigh hysteresis loops, b) initial magnetization curve

Presented results indicates that developed measurement system is working correctly. Obtained Rayleigh hysteresis loops has lenticular shape described in the literature [5]. The points of maximum magnetic flux density forming initial magnetization curve are distributed with second-order dependence, as described by Rayleigh in equation (1). After fitting points with second order polynomial the R^2 coefficient reaches the value of 0.9995, which indicates high second order determination of the characteristic.

7 Conclusion

Developed measurement system is working properly and allows to measure magnetic characteristics of soft magnetic materials in Rayleigh region. New and original approach to the problem of magnetic measurements was necessary to elaborate to solve the problem of measurements in Rayleigh region. Unique measurement methodology with elements of mathematical modeling presented in the paper provides resistance for the electromagnetic interferences and allows to measure even low-amplitude signals for low magnetizing fields. During the designing of electronic circuits much effort was put into providing electromagnetic compatibility to obtain signals as little distorted as possible. The results obtained from the system complies with the theoretical description of Rayleigh hysteresis phenomenon available in the literature [5, 6].

Presented system allows not only to measure magnetic characteristics of the materials. It is also possible, utilizing additional equipment, to measure the influence of temperature and mechanical stresses on the magnetic properties of the materials. Especially measurements of magnetomechanical characteristics of magnetic materials in Rayleigh region are very interesting due to the possibility of utilizing this phenomenon in magnetoelastic sensors of strength and mechanical stress working in Rayleigh region.

References

1. Nałęcz, M., Jaworski, J.: Miernictwo magnetyczne. WNT, Warsaw (1986)
2. Salach, J., Hasse, L., Szewczyk, R., et al.: Low Current Transformer Utilizing Co-Based Amorphous Alloys. IEEE Transactions on Magnetics, 48(4), 1493-1496 (2012)
3. Laboratorio Elettrofisico Walker LDJ Scientific measuring systems (2015), on-line source, http://www.laboratorio.elettrofisico.com/measuring-equipment/soft-magnetic-materials/ [cited on 08.05.2015]
4. MAGNET-PHYSIC Dr. Steingroever GmbH (2015), on-line source, http://www.magnet-physik.de/messtechnik.html?&L=1 [cited on 08.05.2015]
5. Bertotti, B.: Hysteresis in Magnetism For Physicists, Materials Scientists and Engineers. Academic Press, San Diego (1998)
6. Jiles, D.C.: Introduction to Magnetism and Magnetic Materials. CHAPMAN&HALL, London (1998)
7. Novák, L., Kováč, J.: Rayleigh Region in Amorphous and Nanocrystaline FINMET Alloy. Acta Physica Polonica A, 126(1), 126-127 (2014)
8. Horowitz, P., Hill, W.: The Art of Electronics. Cambridge University Press, New York (1989)

Fast alignment procedure for MEMS accelerometers

Sergiusz Łuczak

Faculty of Mechatronics, Warsaw University of Technology, Warsaw, Poland
s.luczak@mchtr.pw.edu.pl

Abstract. A procedure of physical alignment of sensitive axes of various types of microelectromechanical accelerometers (mono-, bi- and tri-axial) is proposed. Owing to the procedure it is possible to align the axes in a relatively short time with a satisfactory accuracy of even hundredths of degree arc. Results of an experimental application of the procedure are illustrated. Special precise instrument simplifying realization of the procedure is presented. The related limitations, advantages and disadvantages are discussed.

Keywords: accelerometer · MEMS · tilt · alignment · sensitive axis

1 Introduction

Microelectromechanical (MEMS) accelerometers can be employed in measurements of both variable and constant accelerations. A special case of detecting a constant acceleration is tilt measurement, which is one of the most typical applications of accelerometers [1]. Some common applications of MEMS accelerometers (photo cameras, cellular phones, etc.) do not require a precise alignment of the embedded MEMS accelerometers, unlike in the case of more sophisticated devices, like e.g. mobile robots, including micro-robots: whether simple like in [2] or complicated like in [3]. The required accuracy of the acceleration measurements, depending on alignment precision, may be determined using e.g. real-time control prototyping [4].

Another interesting field of applications are orthotic robots, proposed e.g. in [5]. Such robots may be considerably developed and comprise e.g. a dedicated safety module [6] employing various sensors for tilt measurements, e.g. accelerometers or gyroscopes [7], in order to control the actuators of the robot in emergency situations [8] according to a predefined scenario. Still another example are various kinds of vehicles, especially untypical solutions, like the screw-propelled vehicle proposed in [9], which can operate under a considerable tilt.

Since acceleration is a vectorial quantity, no matter what the application, accelerometer must be aligned with direction of the detected acceleration. Precision of the alignment, whose required value is different in various applications, significantly affects accuracy of the future measurements [10]. Especially in the case of tilt measurements, misalignments may be very detrimental, resulting in a significant decrease of the accuracy of measuring the component tilt angles (i.e. pitch and roll), as proved in [11].

R. Jabłoński and T. Brezina (eds.), *Advanced Mechatronics Solutions*,
Advances in Intelligent Systems and Computing 393,
DOI 10.1007/978-3-319-23923-1_70

2 Alignment methods

There are two basic methods of dealing with the misalignment problem:

- to estimate values of the existing misalignments related to particular sensitive axes of the accelerometer and then numerically compensate for them,
- to physically align the sensitive axes with direction of the acting acceleration as precisely as necessary.

The first method makes the mathematical transformations necessary for determination of accelerations, calculated on the basis of sensor indications, more complicated. First, the applied accelerometer must be subjected to initial calibration [12] or auto-calibration [13-14] in order to estimate the existing misalignments. Then, various approaches are employed to compensate for them. The most typical one is introduction of matrixes containing values of the estimated misalignments [15-18]. Other ideas are based on affine coordinate systems [19] or even empirical formulas [20]. Each approach is characterized by a different accuracy, which sometimes is hard to be estimated due to a lack of data in the related publications.

The second method, which involves a physical alignment, makes it possible to realize precise and simple measurements. However, accelerometer must have been physically aligned beforehand, what is connected with a toilsome work of experimental nature. In order to make the work easier, the proposed alignment procedure was elaborated.

As it is well known, one of the most stable reference sources of acceleration is the gravitational acceleration [21], so the proposed procedure employs this kind of reference. Necessary changes of the reference acceleration measured by the accelerometer to be aligned are obtained by rotating it with respect to the gravitational acceleration, i.e. applying its pitch or roll respectively. The rotations should be applied using some instrument. However, a manual handling is also possible (using e.g. appropriate accelerometer casing), yet the result is a lower precision of the alignment.

3 Experimental application of the alignment procedure

In order to verify the proposed procedure, a test rig, presented in [22], based on two rotary stages powered by stepper motors, and a tri-axial MEMS accelerometer AXDL 330 were used.

To make the procedure more time-efficient and less laborious, an aligning instrument presented in Fig. 1 was employed, which makes it possible to apply small angular displacements (in a range of few degrees arc) of the accelerometer PCB about x-, y- and z-axis with respect to the casing of the instrument, which is rigidly fixed to the test rig.

The procedure basically consists in applying either roll or pitch angles of the accelerometer (with a step of e.g. 5°) and recording series of its output voltages at each angular position. In the case of an ideal alignment, respective sensitive axis should generate almost constant value of the recorded output voltage, with some variations resulting only from the accelerometer noise. A misalignment, having two angular components, results in a more or less sinusoidal course of the alignment characteristics. A key idea of the proposed procedure is to observe such sinusoidal courses and diminish their amplitude as much as possible.

Fig. 1. Top view of the aligning instrument in horizontal position

With regard to spatial arrangement of the aligning instrument at the initial (horizontal) position, the following were assumed: *x*-axis approximately overlaps roll axis, *y*-axis approximately overlaps pitch axis and *z*-axis approximately overlaps direction of the gravitational acceleration.

First, the procedure was applied for aligning *x*-axis of the accelerometer. The rotary stage responsible for application of roll angle was activated, setting successive values of pitch with a step of 5°. At each angular position series of 30 recordings of the output voltage assigned to *x*-axis was performed (variations within each series are represented by small vertical sections on each alignment characteristic presented in Fig. 2 and 3, which can be observed every 5°). These steps had been repeated, until a full rotation (360°) was realized. Course of the recorded voltage (Alignment #1) is illustrated in Fig. 2 and represents initial misalignment of the related sensitive axis.

Then, the accelerometer was set at horizontal position (pitch and roll of 0°) and physically aligned by means of the aligning instrument (as minutely described in sec. 4). Successively, the accelerometer was set at vertical position (pitch of 0° and roll of 90°) and analogically aligned. Exceptionally, in order to demonstrate two components of the observed initial misalignment, additional rotation was performed after the first alignment at roll of 0° (Alignment #2). After the second alignment at roll of 90° the accelerometer was rotated again (Alignment #3). Courses Alignment #1, #2 and #3 clearly prove that the misalignment had two components in perpendicular directions: maximal amplitude of Alignment #1 at roll of ca. -180°, 0° or 180° and in the case of Alignment #2 at roll of -90° or 90°.

Prior to the 4th and 5th rotation, the accelerometer was aligned both at horizontal as well as vertical position. As can be seen in Fig. 2 (Alignment #5), repeating the procedure 3 times (rotation related to course Alignment #2 was performed only for demonstration purposes and thus should not be counted) made it possible to obtain almost a perfect alignment, where variations of the recorded signal result almost only from the accelerometer noise. However, a considerable improvement (see Alignment #3) can be observed already after two alignments, at roll of 0° and 90°, each corresponding to another angular component of the initial misalignment.

The above procedure can be applied to all kinds of accelerometers: mono-, bi- and tri-axial. One sensitive axis, dealt with as the first, can be aligned very precisely. In this case, the precision can be evaluated at few hundredths of degree arc.

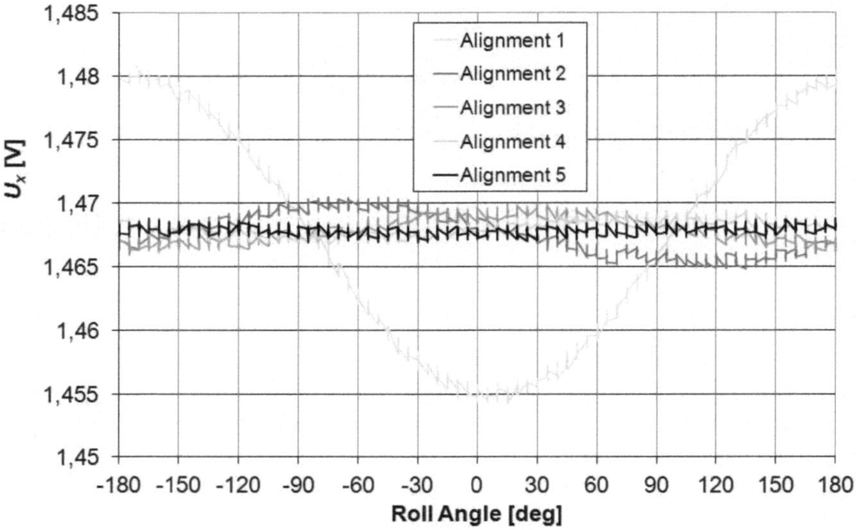

Fig. 2. Alignment of *x*-axis

Analogical steps apply to alignment of *y*-axis. Yet, in this case pitch is applied instead of roll. However, if *x*- and *y*-axes are not orthogonal (what is often true in the case of bi- and tri-axial MEMS accelerometers), the accelerometer can be aligned only at pitch of 0°, omitting the alignment at 90°. The results are illustrated in Fig. 3.

As can be observed, the procedure should be apparently continued, until variations of the output voltage become as small as in the case of Alignment #5 (Fig. 2). However, in such case the *x*-sensitive axis would get misaligned (unless we use two separate accelerometers), i.e. amplitude of variations of the respective output voltage would increase to the level corresponding to Alignment #2 (Fig. 3).

So, at this point (in the case of bi- and tri-axial MEMS accelerometers) it must be decided which sensitive axis should be aligned more accurately, or divide the misalignment equally between both axes. If the *x*-sensitive axis should not be misaligned, alignment precision of the *y*-sensitive axis would be lower. In this case it can be evaluated at few tenths of degree arc, what results from inherent misalignment (non-orthogonality) of the accelerometer sensitive axes.

For the same reasons, the *z*-axis of a tri-axial accelerometer cannot be aligned, as its alignment results from alignment of *x*- and *y*-axis. However, in the case of using a separate accelerometer for measurements in *z*-axis, the axis can be aligned determining appropriate angular phase shifts of a nonlinear regression models created on the basis of data recorded while aligning *x*- and *y*-axis. The idea is minutely discussed in [23].

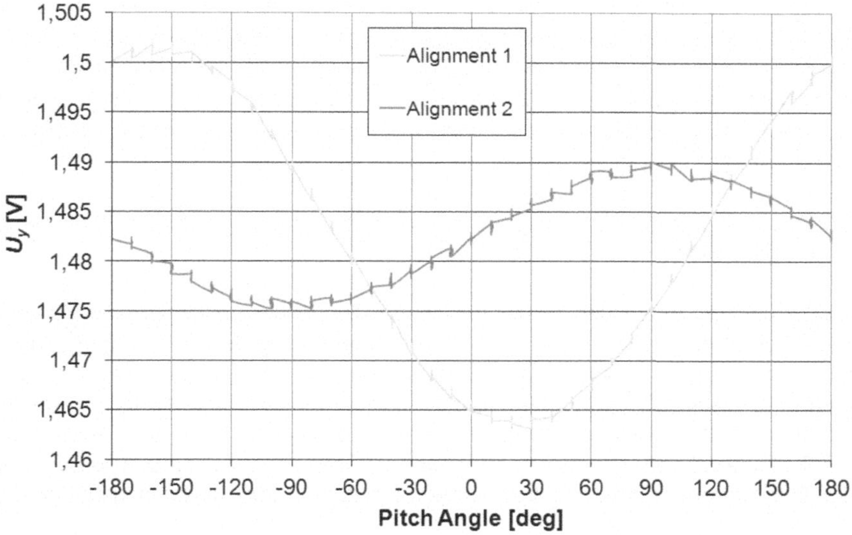

Fig. 3. Alignment of *y*-axis

4 A simplified procedure

At the cost of a lower accuracy of the alignment, the procedure can be simplified. A test rig may be applied, but not necessary. Since only 4 specific values of pitch and roll angles must be applied, a cubical casing of the accelerometer is sufficient. Then, the following steps should be realized while aligning the *x*-axis:

- (fixing the accelerometer in the test rig),
- leveling the accelerometer: pitch and roll angle of ca. 0°,
- successive applying roll angle of 0°, 90°, 180° and 270°,
- recording series of output voltages (e.g. 30) assigned to *x*-axis at each roll angle,
- calculating a total average value of all the recorded output voltages,
- applying roll angle of 0° again,
- aligning the accelerometer PCB (angular displacements around *y*-axis),
- applying roll angle of 90° again,
- aligning the accelerometer PCB (angular displacements around *z*-axis).

Aligning the accelerometer PCB is realized by applying its small angular displacements around respective sensitive axis (in this case *y* or *z*) until an average value of series of the output voltages recorded after each angular displacement equals the calculated total average value. It is important not to compare a single indication of the accelerometer with the calculated total average value - an average value must be used instead.

The same steps should be repeated in the case of *y*-axis, substituting roll angle with pitch angle and *y*-axis with *x*-axis respectively. It must not be overlooked, that realizing the last step will increase misalignment of the *x*-axis in the case of an inherent misalignment between *x*- and *y*-sensitive axes of the accelerometer. This kind of misalignment, experimentally evaluated at few tenths of a degree arc [23], is typical for bi- and tri-axial MEMS accelerometers, especially the tri-axial, as in fact the fabrication techniques of MEMS devices are semi-three-dimensional [24].

5 Conclusions

Physical alignment of accelerometers makes it possible to simplify the related computations. Therefore, two versions of a fast and simple procedure for aligning MEMS accelerometers are proposed: a more accurate one or a faster one. However, some equipment or a special casing of the aligned accelerometer must be used in order to realize the procedure. Besides, it must be taken into account that in the case of multi-axial MEMS accelerometers a full alignment is impossible due to considerable perpendicularity errors (usually of few tenths of a degree arc) between their sensitive axes.

Acknowledgments

The author would like to thank a retired employee of the Institute of Micromechanics and Photonics, Warsaw University of Technology – Mr. Marian Kowalski, for designing and building a prototype of the aligning instrument presented in Fig. 1.

References

1. Wilson, J.S.: Sensor Technology Handbook. Newnes, Burlington (2005)
2. Bodnicki, M., Kamiński, D.: In-pipe Microrobot Driven by SMA Elements. In: Březina, T., Jabloński, R. (eds.) Mechatronics 2013. Recent Technological and Scientific Advances, pp. 527-533. Springer International Publishing (2014)
3. Vlachý, D., Zezula, P., Grepl, R.: Control unit architecture for biped robot. In: Jabłoński, R., Turkowski, M., Szewczyk, R. (eds.) Recent Advances in Mechatronics, pp. 6-10. Springer, Berlin Heidelberg (2007)
4. Grepl, R.: Real-time control prototyping in MATLAB/simulink: Review of tools for research and education in mechatronics. In: 2011 IEEE International Conference on Mechatronics (ICM), pp. 881-886. IEEE (2011)
5. Bagiński, K., Jasińska-Choromańska, D., Wierciak, J.: Modelling and simulation of a system for verticalization and aiding the motion of individuals suffering from paresis of the lower limbs. Bull. Pol. Ac.: Tech. 61, 919-928 (2013)
6. Jasińska-Choromańska, D., Kabziński, B., Matyjewicz-Maciejewicz, M., et al: Safety module for the system of verticalization and aiding motion of the disabled. In: Březina, T.,

Jabłoński, R. (eds.) Mechatronics 2013. Recent Technological and Scientific Advances, pp. 79-86. Springer International Publishing (2014)
7. Łuczak, S.: Guidelines for Tilt Measurements Realized by MEMS Accelerometers. Int. J. Precis. Eng. Manuf. 15, 489-496 (2014)
8. Bagiński, K., Wierciak, J.: Forming of operational characteristics of an orthotic robot by influencing parameters of its drive systems. In: Szewczyk, R., Zieliński, C., Kaliczyńska, M. (eds.) Progress in Automation, Robotics and Measuring Techniques Vol. 2 Robotics, pp. 1-9. Springer International Publishing (2015)
9. Osiński, D., Szykiedans, K.: Small remotely operated screw-propelled vehicle. In: Szewczyk, R., Zieliński, C., Kaliczyńska, M. (eds.) Progress in Automation, Robotics and Measuring Techniques Vol. 2 Robotics, pp. 191-200. Springer International Publishing (2015).
10. Syed, Z., Aggarwal, P., Goodall, C., et al: A new multi-position calibration method for MEMS inertial navigation systems. Meas. Sci. Technol. 18, 1897-1907 (2007)
11. Łuczak, S.: Effects of Misalignments of MEMS Accelerometers in Tilt Measurements. In: Březina, T., Jabłoński, R. (eds.) Mechatronics 2013. Recent Technological and Scientific Advances, pp. 393-400. Springer International Publishing (2014)
12. Won, S.P., Golnaraghi, F.: A Triaxial Accelerometer Calibration Method Using a Mathematical Model. IEEE Trans. Instrum. Meas. 59, 2144-2153 (2010)
13. Luinge, H.J., Veltink, P.H.: Inclination Measurement of Human Movement Using a 3-D Accelerometer with Autocalibration, IEEE Trans. Neural Syst. Rehabil. Eng. 12, 112-121, (2004)
14. Frosio, I., Pedersini, F., Borghese, N.A.: Autocalibration of Triaxial MEMS Accelerometers with Automatic Sensor Model Selection. IEEE Sensors J. 12, 2100-2108 (2012)
15. Stančin, S., Tomažič, S.: Time- and Computation-Efficient Calibration of MEMS 3D Accelerometers and Gyroscopes. Sensors 14, 14885-14915 (2014)
16. Olivares, A., Ruiz-Garcia, G., Olivares, G., et al: Automatic Determination of Validity of Input Data Used in Ellipsoid Fitting MARG Calibration Algorithms. Sensors 13, 11797-11817 (2013)
17. Parsa, K., Lasky, T. A., Ravani, B.: Design and Implementation of a Mechatronic All-Accelerometer Inertial Measurement Unit. IEEE/ASME Trans. Mechatr. 12, 640-650 (2007)
18. Ang, W.T., Khosla, P.K., Riviere, C.N.: Nonlinear regression model of a low-g MEMS accelerometer. IEEE Sensors J. 7, 81-88 (2007)
19. Šipoš, M., Pačes, P., Roháč, J., et al: Analyses of triaxial accelerometer calibration algorithms. IEEE Sensors J. 12, 1157-1165 (2012)
20. Qian, J., Fang, B., Yang, W., et al: Accurate Tilt Sensing with Linear Model. IEEE Sensors J. 11, 2301-2309 (2011)
21. Model 900 Biaxial Clinometer, Data Sheet. Applied Geomechanics, Santa Cruz (1995)
22. Łuczak, S.: Dual-Axis Test Rig for MEMS Tilt Sensors. Metrol. Meas. Syst. 21, 351-362 (2014)
23. Łuczak, S.: Experimental Studies of Hysteresis in MEMS Accelerometers: a Commentary. IEEE Sensors J. 15, 3492-3499 (2015)
24. Kaajakari, V.: Practical MEMS. Small Gear Publishing, Las Vegas (2009)

Contact Area Evaluation Experiment Verified by Computational Model in MBS

Ondřej Maršálek[1], Jozef Dlugoš[1], Pavel Novotný[1], Lubor Zháňal[1]

[1] Brno University of Technology, Brno, Czech Republic
{marsalek,dlugos,novotny,zhanal}@fme.vutbr.cz

Abstract. This article focuses on the experimental contact area evaluation. Test rig is designed in order to investigate the piston/cylinder liner contact. The conducted experiment is based on the principle of electrical contact resistivity between two tested machinery parts. The measured results are compared with different contact model theories. The computational simulation uses precomputed database of contact pressure and contact area depending on separation distance of surfaces. This simulation is solved by Multibody System (MBS) enhanced by the user-written FORTRAN subroutine. The results of experiment and simulations are discussed.

Keywords: Contact pressure·Multibody System·rough surface·contact resistivity

1 Introduction

Monitoring of the contact area of two surfaces during dynamic simulation is a relatively complicated task. Especially if it is case of piston/cylinder liner interaction. With respect to the crawl-walk-run approach the following procedure was executed.

Authors of [5], [6], [7] and [8] describe the methodology for experimental estimation of the oil film thickness in a roller bearing. In all these sources the principle when the electrical current goes through the lubricated contact pair is used. The thickness of oil film layer is then given by electrical contact resistance. In this case the resistance becomes relatively high due to the oil insulating properties. When this principle is applied to a dry contact the electrical current goes through a high number of micro-contacts which can be represented by collaterally connected resistors. Thus, the total value of electrical resistance is very low. The aim of this paper is to apply the above mentioned experimental methodology on a dry contact, so the contact area can be estimated and compared with the different analytical contact models.

2 Analytical contact models

The first analytical model is Greenwood & Tripp [1]. It is very effective and widely used, nowadays and also in the past. In this model, the parameter $(\eta\beta\sigma)$ has the

R. Jabłoński and T. Brezina (eds.), *Advanced Mechatronics Solutions*,
Advances in Intelligent Systems and Computing 393,
DOI 10.1007/978-3-319-23923-1_71

essential impact on the contact pressure values. This parameter represents the surface pattern information: η is the surface density of asperities [m^{-2}], β is the average radius of asperity curvature [m] and σ is the combined Root Mean Square (RMS) of both surfaces [m]. Next, probably the most well-known Hertz [2] contact model was implemented. The contact model published by Volker Lagemann in his dissertation thesis [3] was used to represent modified Hertz pressure formulae. Finally, the last simulated contact model is the one published by Passaribu & Shipper [4]. It takes into consideration all three possible states of material behaviour – fully elastic, elasto-plastic, and fully plastic. Detailed information of each of the presented contact models can be found in the information sources.

3 Resistivity prediction

Following relations were used for design of measuring circuit [7]:

$$R_T = \frac{\zeta h_R}{A_R}, \quad I = U_R / \left(R_T + R_{Piston} + R_{Liner} \right). \tag{1}$$

Where R_T is resistivity of the contact [Ω], R_{Piston} is resistivity of the piston [Ω], R_{Liner} is resistivity of the cylinder liner [Ω], A_R is contact area [m^2], h_R is height of electrical contact [m], ζ is resistance of material [Ω·m] and I is the electrical current [A].

The experimental device was designed according to Fig. 1. Crankshaft motion was constrained so the experiment could be performed as the simplest scenario. Set of weights was then gradually applied on the experimental piston as loads.

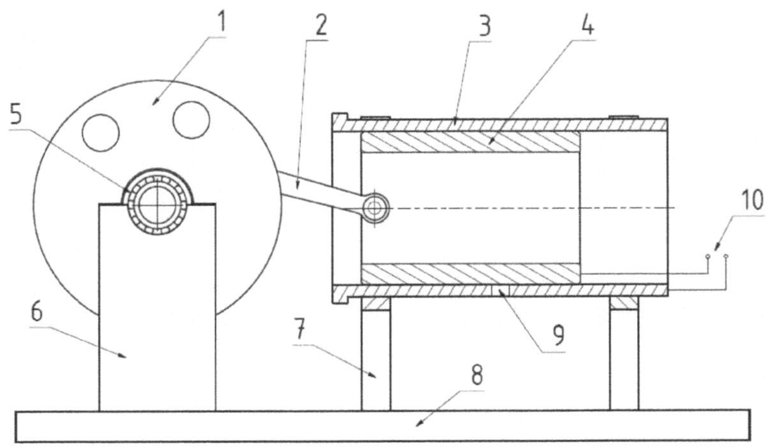

Fig. 1. Test rig scheme: 1 – crankshaft powered by electric engine, 2 – connecting rod, 3 – piston liner, 4 – experimental engine, 5 – crankshaft mount, 6 – crankshaft bearing housing, 7 – piston liner support, 8 – base, 9 – oil intake, 10 – cables for electric circuit connection.

4 MBS model

From the computational algorithms described above a database of contact pressures and contact areas, depending on separation distance of analysed surfaces, was created for the MBS simulation. Trends of all contact curves are shown in Fig. 2. Because the values of contact pressure given by the Greenwood & Tripp model were too different (lower), this curve was plotted on the secondary axis of the graph. The reason is the deviation of the parameter ($\eta\beta\sigma$) which should be in a range between 0.03 and 0.05 [1]. If the tested surfaces are not within this range, we can expect biased results. In this case measured parameter ($\eta\beta\sigma$) was 0.023. Volker Lagemann [3] also confirmed the lower contact pressure values of Greenwood & Tripp model.

According to the low loads applied, the low elastic deformations of each body contact region are considered as negligible and the plastic deformations are not expected at all. Therefore, the MBS model with rigid bodies is sufficient. Commercially available MSC ADAMS enhanced by FORTRAN user-written subroutine was used as MBS software. MBS-subroutine solution loop is shown in Fig. 3.

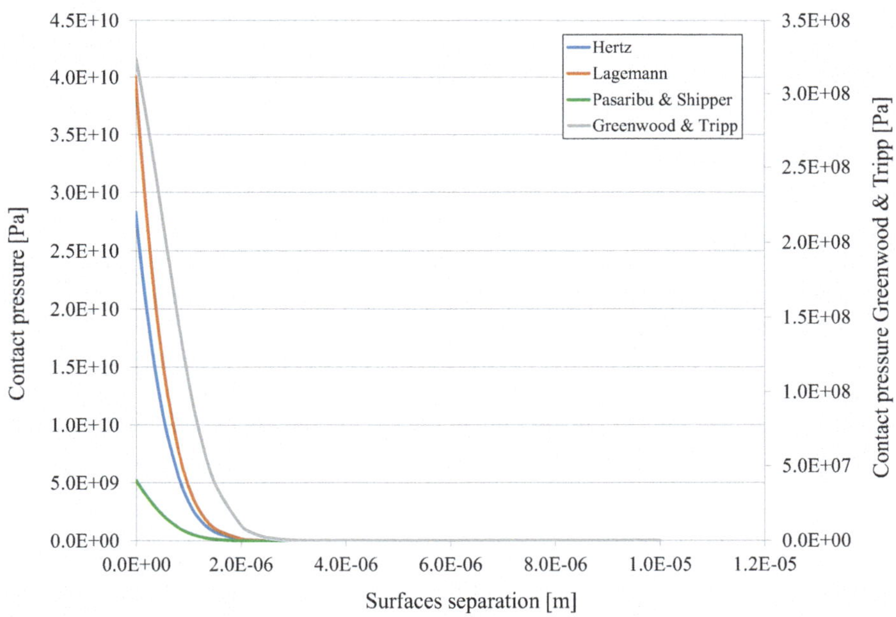

Fig. 2. Contact pressure dependence on surfaces separation

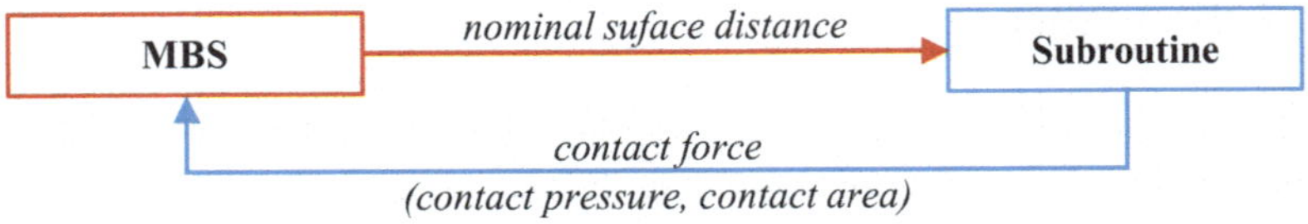

Fig. 3. Flowchart of piston/cylinder liner simulation

In order to increase solution accuracy and to speed-up the simulation the variable density of the mesh grid with respect to the expected location of contact is used. This approach provides acceptable speed of simulation with accurate results. To suppress the unwanted fluttering of the piston, the damping force is introduced. Overall contact force then takes the following form:

$$F_{c\_MBS} = P_c A_c + B_{MBS} v_{MBS} . \tag{2}$$

Where F_{c_MBS} is contact force [N], P_c is contact pressure [Pa], A_c is contact area [m^2], B_{MBS} is damping [N·s·m^{-1}] and v is the relative speed of both surfaces [m·s^{-1}]. The computational model simulates a static equilibrium of the forces between the piston and the cylinder liner actuated by the gravity (no motion in the axis of the cylinder liner is applied).

5 Results discussion

5.1 Experiment

Measurement was done by device Keithley 6221/2182A Delta mode system with AC and DC current source and nanovoltmeter.

Measured values are very different from the ones predicted by equations listed in chapter 3. The main causes are: differences in resistance of both materials; impact of wires resistance; impact of transition resistance of contacts and connectors; different resistance of surface layer (oxidation, contamination), etc. See Fig. 4.

5.2 Comparison

Since the measurement was done for different weight loads applied, measured and calculated contact area values can be compared to the values of nominal load 53 N (5.440 kg), therefore the dimensionless ratios are available as depicted in Fig. 4.

Test rig bore was 76 mm (Fig. 1), while the maximal load force applied was 255N (25.966 kg). These values are insufficient to cause any plastic deformations. In addition, only very low elastic deformations are present. All the implemented models provide under these conditions very similar results – surface separation above 3 μm (Fig. 2). Therefore, the graph in Fig. 5 does not clearly state which of the observed contact models is the most accurate.

To highlight the differences between the contact models, the large elasto-plastic deformations would have to be tested. That means the use of the hydraulic press to cause higher loads applied. MBS model would have to be modified as well – the use of flexible bodies.

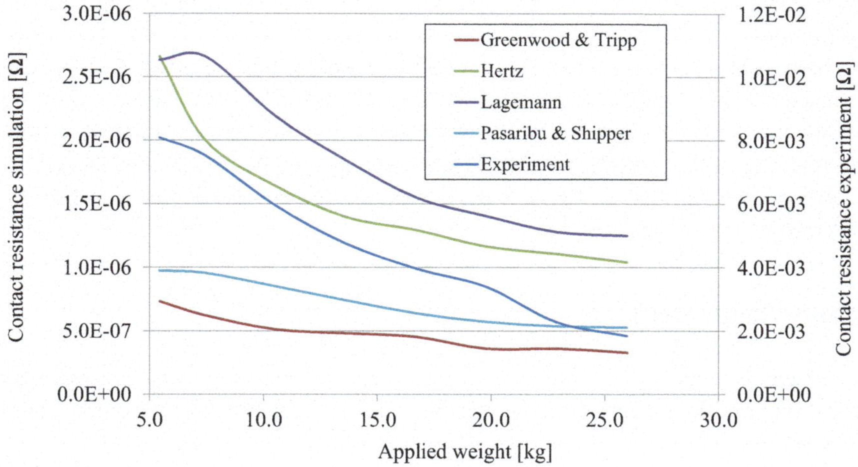

Fig. 4. Comparison of simulated and experimental results - absolute values

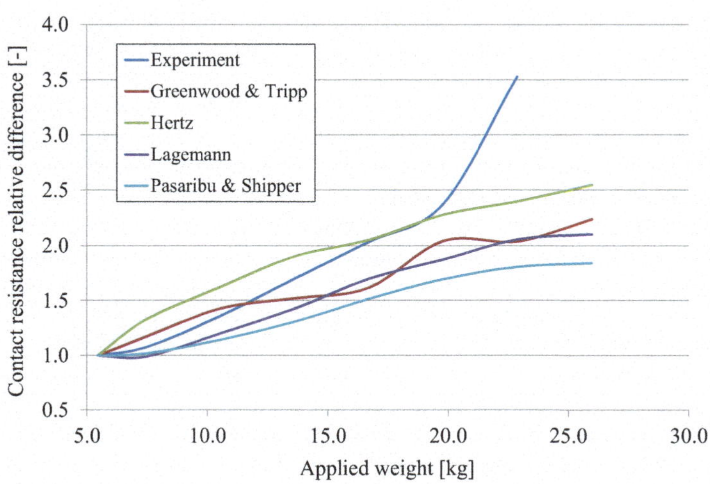

Fig. 5. Comparison of simulated and experimental results – relative values

6 Conclusions

The designed test rig was, in this case, oriented at the dry contact of piston/cylinder liner. The change in electrical contact resistance was able to record the differences in the contact area due to various loads applied, even when only very low elastic deformations were present – the measuring principle using contact resistance was proven.

Experimental and simulated data have similar trends, but the offset is significant. It is caused by the unknown resistance between the individual measuring circuit items. At this moment, the best from the selected analytical contact models cannot be determined. For this purpose, higher deformations would have to be introduced.

Further development of the test rig leading to the successful continuation of the experiment is suggested.

Acknowledgement

The research leading to these results has received funding from the Ministry of Education, Youth and Sports under the National Sustainability Programme I (Project LO1202).

References

1. Greenwood, J. & Tripp, J.. 1970. The contact of two nominally flat rough surfaces. *ARCHIVE: Proceedings of the Institution of Mechanical Engineers 1847-1982 (vols 1-196)*, vol. 185, issue 1970, pp. 625-634.
2. Stachowiak, G. 2014. *Engineering tribology*. 4th ed. Oxford: Butterworth-Heinemann.
3. Lagemann, V. 2000. *Numerische Verfahren zur tribologischen Charakterisierung bearbeitungsbedingter rauher Oberflächen bei Mikrohydrodynamik und Mischreibung*. Dissertation. Kassel: Univ.-Bibliothek.
4. Pasaribu, H. & Schipper, D.. 2005. Application of a Deterministic Contact Model to Analyze the Contact of a Rough Surface Against a Flat Layered Surface. *Journal of Tribology*, vol. 127, issue 2, pp. 451-455. Available from: http://Tribology.asmedigitalcollection.asme.org/article.aspx?articleid=1467327
5. Prashad, H. 2006. *Tribology in electrical environments*. Boston: Elsevier.
6. Jang, Y. & Barber, J.. 2003. Effect of contact statistics on electrical contact resistance. *JOURNAL OF APPLIED PHYSICS*, vol. 94, no. 11.
7. Matharu, S., Sanyal, S. & Bal, D.. 2010. Development of a multipurpose, efficient and inexpensive bearing test rig. *Journal of Engineering and Technology Research*, no. 2, pp. 44-49.
8. Peng-shun, Z., Shu-guang, L. & Wen-jie, Z.. 1991. A new measurement method of oil film thickness in the EHL condition: the R-C oscillation technique. *Wear*, vol. 148, no. 1, pp. 39-46.

Validation of Analytical Calculation of Contact Pressure

Ondřej Maršálek[1], Jan Vopařil[1], Pavel Novotný[1]

[1] Brno University of Technology, Brno, Czech Republic
{marsalek,voparil,novotny}@fme.vutbr.cz

Abstract. This paper discusses the selection of the most appropriate analytical approach for computational modelling of contact pressure using commercial software tool ANSYS. All the analytical relations, characterizing each computational model useful for detailed determination of the contact pressure between two rough surfaces, are presented here. Furthermore, the process of creation of the FEM model of rough surface along with the description and analysis of all the challenges associated with it are described in this paper. In the conclusion are listed results of simulations together with identification of the chosen analytical approach for fast and also accurate calculations.

Keywords: Contact pressure·FEM·rough surface·surface deflection

1 Introduction

For the purpose of making fast and still sufficiently accurate computational software tool for detailed analysis of contact pairs of modern combustion engines the most appropriate analytical model for determining the contact pressure between two rough surfaces had to be chosen. And because the experimental verification of contact pressure in the scale of surface roughness of real machinery parts is, even in these modern times, something unfeasible, one of the most sophisticated commercial FEM software tools ANSYS was used for this purpose. Introduction of each analytical relation, description of development process of FEM model of rough surfaces and all challenges associated with it, are described in the following sections.

2 Analytical models

2.1 Contact model Greenwood & Tripp

This model was published in 1970 by authors Greenwood and Tripp [1]. The basic equation for the contact pressure calculation has the following form:

$$p_c(h) = \frac{8\pi}{5}(\eta\beta\sigma)KF_{5/2}\left(\frac{h}{\sigma}\right). \tag{1}$$

R. Jabłoński and T. Brezina (eds.), *Advanced Mechatronics Solutions*,
Advances in Intelligent Systems and Computing 393,
DOI 10.1007/978-3-319-23923-1_72

Where:

$$K = \frac{2\sqrt{2}}{3}(\eta\beta\sigma)E'\sqrt{\left(\frac{\sigma}{\beta}\right)},\ \sigma = \sqrt{Rq_1^2 + Rq_2^2},\ \frac{1}{E'} = \frac{1-\upsilon_1^2}{E_1} + \frac{1-\upsilon_2^2}{E_2}. \qquad (2)$$

Here η is the average surface density of asperity peaks [m^{-2}], β is average radius of peak curvature [m], σ is combined RMS for both surfaces [m], v is Poisson constant [-], Rq is the RMS of each surface [m], E' is the combined Young's modulus [Pa], $E_{1(2)}$ is the Young's modulus of each material [Pa] and $F_{5/2}$ is the function of the contact pressure increase [-].

Function $F_{5/2}(h/\sigma)$ is in the original information source [1] defined solely for values of ratio (h/σ) in the range between 0 and 4. For values of the ratio (h/σ) higher than 4 the situation is not so complicated. Function $F_{5/2}(h/\sigma)$ has here zero values (surfaces are separated enough from each other and the contact pressure does not occur), therefore also the contact pressure is zero. In the case that EHD approach is not considered, a situation where this ratio (h/σ) drops under zero value may occur (penetration of contact surfaces occurs). Thus, this situation also has to be covered in the numerical solution. Function $F_{5/2}(h/\sigma)$ in the value $(h/\sigma)=0$ tangentially goes into linear function.

2.2 Contact model Hertz

In our case it is convenient to replace the real shape of asperities by the spherical shape and to solve the Hertz pressure as a contact of two spherical bodies.

$$W = \sqrt{\frac{\delta_R^{\ 3} E'^2 R'}{1.0397^3}},\ a = \left(\frac{3WR'}{E'}\right)^{1/3},\ P_{\max} = \frac{3W}{2\pi a^2},\ P_{av} = \frac{W}{\pi a^2} \qquad (3)$$

Here the variable W represents contact force [N], δ_R is the deflection (penetration) of surfaces [m], E' is the combined Young's modulus [Pa], R' is the combined radius of curvature [m] and a [m] is radius of contact area, P_{max} is the relation for the maximal and P_{av} is the relation for the average contact pressure [Pa].

Input values for this computational model are: the deflection (penetration) of contact surfaces δ_R [m], material characteristic Young's module E' [Pa], and combined radius of asperity curvature R' [m].

For all the contact models (except for model Greenwood and Tripp) the same way of calculation of the combined radius of asperity curvature is used. It is the determination of the osculation circle radius of the surface roughness curve in the given point. For each point of the surface roughness profile is this radius calculated for direction x and y.

2.3 Contact model Lagemann

Volker Lagemann has described in his dissertation thesis [3] slightly modified equations for the Hertz pressure calculation. For completeness that also the Hertzian pressure equations can have few shapes and modifications are these equations also implemented as one of the possibilities to determine the contact pressure.

$$W = \sqrt{\frac{16\delta_R^{\ 3} E'^2 R'}{9}},\ a = \left(\frac{3WR'}{4E'}\right)^{1/3},\ P_{\max} = \frac{1}{\pi}\sqrt{\frac{6E'^2 W}{R'^2}}. \tag{4}$$

Combined Young's modulus and combined (reduced) radius of asperities curvature of contact bodies are given by the same equations as in the previous case (classic Hertzian pressure).

It was necessary to implement also contact model which considers elasto-plastic behavior of contact bodies, while preserving the simplicity and calculation speed. As turned out later, for this purpose it is possible to use the computational model described below.

2.4 Contact model Pasaribu & Shipper

This computational model [4] takes into consideration all the three possible states of material behavior – fully elastic, elasto-plastic, and fully plastic behavior. When compared to all previous models, here one additional variable is added – hardness of the softer material. Transitions between the individual material behaviors are controlled by depth of the penetration. Boundary values of penetration (deflection) (ω_c [-]) are given by following equations:

$$\omega_{c1} = 0.89R'\left(\frac{H}{E'}\right),\ \omega_{c2} = 54\omega_{c1}. \tag{5}$$

If the condition that $\delta_R < \omega_{c1}$ is met, materials are in the area of elastic contact. In the case when $\omega_{c1} < \delta_R < \omega_{c2}$, materials are in the elasto-plastic state. If $\omega_{c2} < \delta_R$, materials behave fully plastically.

In the case of fully plastic behaviour of materials is the calculation given by following equations:

$$A_{el} = \pi R'\delta_R,\ F_{el} = \frac{4}{3}E'R'^{0.5}\delta_R^{\ 1.5}. \tag{6}$$

If materials fulfil the second condition (are in the area of elasto-plastic behaviour), following equations are valid:

$$A_{ep} = \pi R'\delta_R\left[1 - 2\left(\frac{\delta_R - \omega_{c1}}{\omega_{c2} - \omega_{c1}}\right)^3 + 3\left(\frac{\delta_R - \omega_{c1}}{\omega_{c2} - \omega_{c1}}\right)^2\right], \tag{7}$$

$$F_{ep} = \left[H - 0.6H\frac{\ln\omega_{c2} - \ln\delta_R}{\ln\omega_{c2} - \ln\omega_{c1}}\right]A_{ep}. \tag{8}$$

Lastly, equations for fully plastic behaviour of materials are:

$$A_p = 2\pi R'\delta_R, \; F_p = HA_p. \tag{9}$$

Here A_{el}, A_{ep} a A_p are contact areas [m^2] for given material states, H is the hardness of softer material [Pa], and F_{el}, F_{ep} a F_p are contact forces [N].

3 FEM modeling

The only input for the finite element method (FEM) modelling were matrixes of values describing a roughness in one axis (height) and value indicating the distance between the individual points (step). In the final calculations the grids with the number of points 400 x 400 were used.

3.1 Material model

Different materials (material properties) were adjusted for each body with surface roughness, but the same material model was used for both: multilinear isotropic hardening. Multilinear model was necessary for better capturing of the stress-strain curve, especially in areas with a larger deformation where the stress-strain curve is approximated by horizontal line. Isotropic model was chosen because of the better tendencies towards the convergence.

3.2 Boundary conditions

Pushing of surfaces to each other was simulated by moving the upper side of one body towards the other, see Fig. 1.

For each computed surface were solved about twenty loading states with different distance of zero planes. It enabled a sufficiently accurate evaluation of the behaviour of the rough surfaces with different degrees of deformation of individual peaks of roughness.

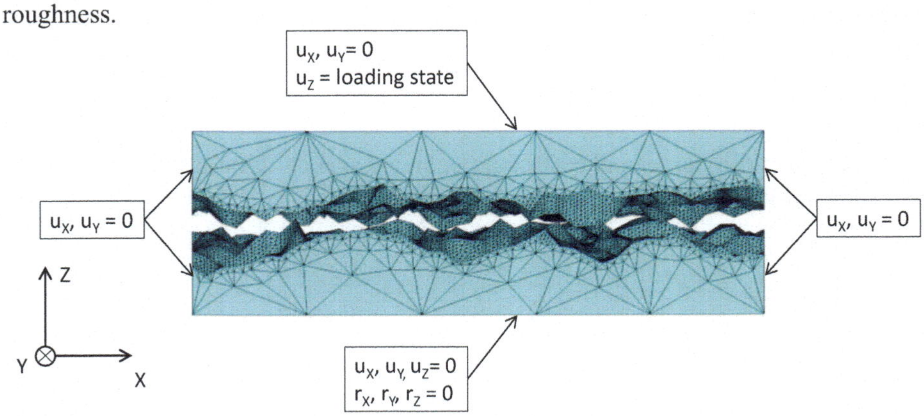

Fig. 1. Boundary conditions

3.3 Correction of model deformations

The loading state for comparison with the analytical computational method was defined as the distance between zero planes of rough surfaces. Zero plane is defined as a place where arithmetic mean of all heights of roughness's inequality is zero.

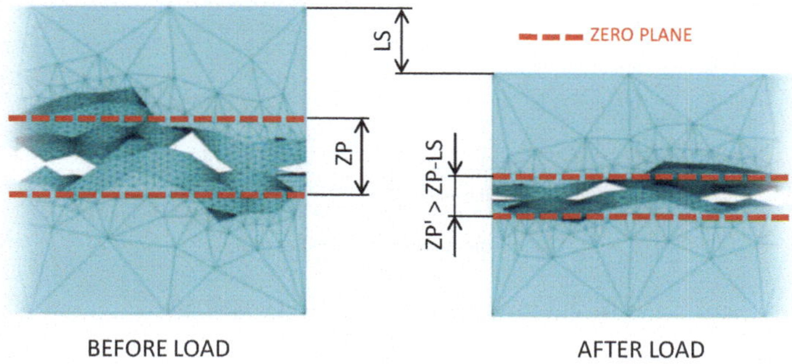

Fig. 2. Scheme of explanatory needs of model deformation corrections

3.4 Convergence challenges

Calculation of contact problem of real rough surfaces proved to be very problematic in terms of convergence. The reason is that the individual gradients of roughness's raising are relatively steep and the contact with opposite surface is realized in a very small area. Load state is therefore composed from many "point load" states. This way of the loading is usually referred to as divergent without any further conditions.

4 Results discussion

For the results of computational models Lagemann and Hertz minor axis of the graph had to be selected. The results of various computational models correspond with the results published by the information source [3]. Computational models Lagemann and Hertz give relatively similar values of contact pressures. This is logical – both approaches are based on the Hertz theory.

Computational model Greenwood & Tripp gives the lowest values of contact pressures. For this reason, plotting the values had to be multiplied by 10. The reason is the deviation of the parameter ($\eta\beta\sigma$) which should be, based on information source [1], in a range between 0.03 and 0.05. If the tested surfaces are not within this range, we can expect biased results.

Closest to the results of FEM analysis is the model according to the authors Pasaribu & Shipper. This computational model considers all possible states of the material behaviour, as well as the FEM model in ANSYS.

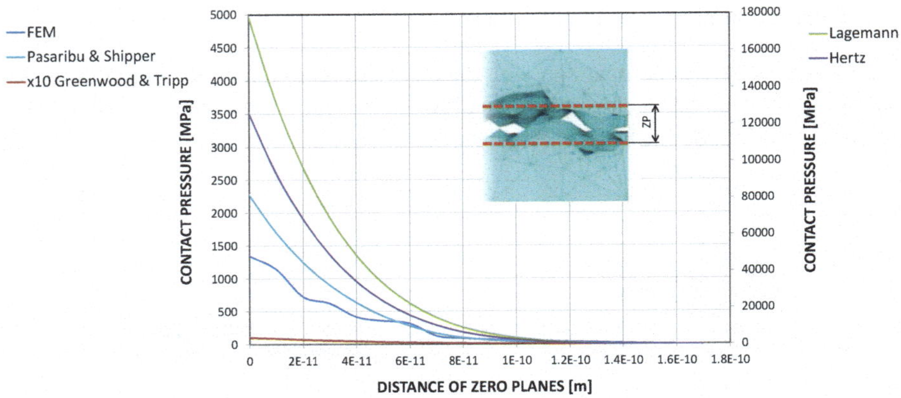

Fig. 3. Computational methods comparison

5 Conclusions

The purpose of this analysis using FEM was, by all means, fulfilled. From the selected computational models the best one that will be used for the fast solution of contact problems of real rough surfaces of machinery parts was determined. Detailed solution using FEM is not applicable to online contact task solutions, using the MBS, due to its high demands, and not only its time-inefficiency.

Acknowledgement

The research leading to these results has received funding from the Ministry of Education, Youth and Sports under the National Sustainability Programme I (Project LO1202).

References

1. Greenwood, J. & Tripp, J.. 1970. The contact of two nominally flat rough surfaces. *ARCHIVE: Proceedings of the Institution of Mechanical Engineers 1847-1982 (vols 1-196)*, vol. 185, issue 1970, pp. 625-634.
2. Stachowiak, G. 2014. *Engineering tribology*. 4th ed. Oxford: Butterworth-Heinemann.
3. Lagemann, V. 2000. *Numerische Verfahren zur tribologischen Charakterisierung bearbeitungsbedingter rauher Oberflächen bei Mikrohydrodynamik und Mischreibung*. Dissertation. Kassel: Univ.-Bibliothek.
4. Pasaribu, H. & Schipper, D.. 2005. Application of a Deterministic Contact Model to Analyze the Contact of a Rough Surface Against a Flat Layered Surface. *Journal of Tribology*, vol. 127, issue 2, pp. 451-455. Available from: http://Tribology.asmedigitalcollection.asme.org/article.aspx?articleid=1467327

Impact of graphene coatings on nanoscale tribological properties of miniaturized mechanical objects

Marcin Michałowski[1], Jan Tomasik[2], and Marta Wiśniewska[2]

[1]Institute of Micromechanics and Photonics, Faculty of Mechatronics, Warsaw University of Technology, Warsaw, Poland

m.michalowski@mchtr.pw.edu.pl

[2]Institute of Metrology and Biomedical Engineering, Faculty of Mechatronics, Warsaw University of Technology, Warsaw, Poland

{j.tomasik,martwisn}@mchtr.pw.edu.pl

Abstract. Impact of graphene coatings on nanoscale tribological properties of miniaturised mechanical objects has been examined. In order to do so, a set of steel,copper-platted specimens has been used, upon which graphen coating has been deposited. Their mechanical and tribological properties such as friction coefficient, hardness and Young's modulus, before and after covering of these samples with graphene layer, have been compared. The results are outlined in the paper, some interesting properties have been found.

Keywords: Graphene·nanotechnology·tribology·friction

1. Introduction

Due to its unique features [1,2], graphene has been staying in scientific limelight since its discovery in 2002. It is claimed that the deposition of graphene layer on surfaces of industrial elements may significantly improve their mechanical properties and lead to decline in friction coefficient or increase in hardness and elasticity [3]. Therefore, graphene coatings may become an attractive alternative to the ones that are already used in the manufacturing. However, no attempt has been made to investigate, and quantitatively describe, graphene performance in practical industrial applications yet. In order to overcome this barrier and, in effect, make graphene coatings conducive to manufacture on a mass scale, the GRAPH-TECH research project was introduced. Its aim is to evaluate impact of graphene coatings on tribological properties of miniaturized industrial objects, i.e. bearings or gearwheels. To do this, values of

R. Jabłoński and T. Brezina (eds.), *Advanced Mechatronics Solutions*,
Advances in Intelligent Systems and Computing 393,
DOI 10.1007/978-3-319-23923-1_73

hardness, Young's modulus, friction coefficient and roughness parameters [4], both before and after graphene deposition, have been estimated and compared.
The core findings of the research, referring to nanoscale mechanical properties of these objects, are presented in the paper.

2. Materials and Methods

2.1 Test Specimens

Three sets of test specimens, differing in the steel grade they were made of (40HM, 35HGS and MS45), have been used to assess an influence of graphene layer on nanoscale tribological properties of mechanical elements. All these sets consist of three samples, each with test surface of (20x20) mm. First, the test surfaces were milled and grinded. Then, there was an electrolytic copper plated on them. Only the surfaces prepared this way were graphene-covered.

2.2 Measurement Equipment

The test specimens have been measured after copper- and graphene- plating respectively.
In order to evaluate their friction coefficient values the NT-206 AFM by Microtestmachines was used. AFM's are often used for a wide range of nano-scale studies in recent years [5, 6]. This microscope is equipped with a beryllium bronze cantilever with a tip of a steel bearing ball that has a diameter of 0.35 mm. Before conducting the measurements the NT-206 system had been calibrated with stiff reference surfaces, silicon wafer and Nanoidea membranes consecutively. Then, the friction loops (Fig. 1) were created for each and every test specimen before and after graphene deposition. The measurement data obtained this way was used to estimate friction coefficient values.
In the same time, the hardness-indentation depth characteristics were obtained using Hystiron UBi-1 with Berkovi nanoindenter. Then, the Oliver and Pharr formula [7] was used to identify hardness and Young's modulus of the test surfaces.

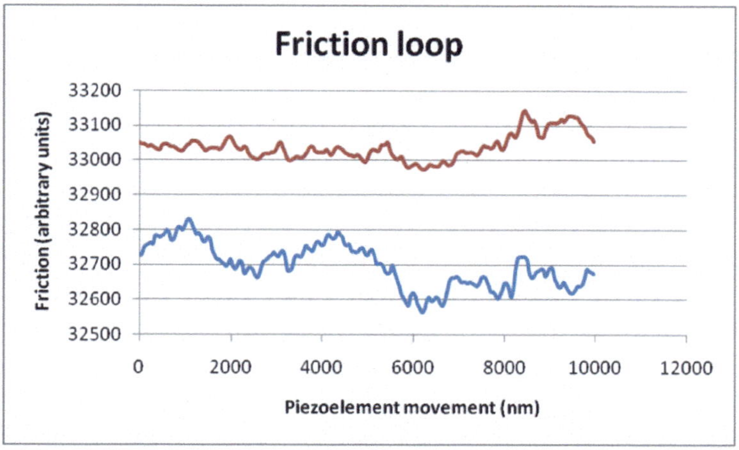

Fig. 1. Sample friction loop obtained using AFM (forward movement – red line; backward movement – blue line)

3. Results

3.1 Hardness

In order to evaluate impact of graphene coating on a nanoscale hardness of industrial elements, 30 indentations of every specimen were done, both before and after graphene deposition, performed. The maximum load of 1500 μN has been applied for two seconds each time. It is also worth mentioning that both loading and unloading the specimens have lasted for five seconds.

Sample hardness measurement results are presented in Fig. 2 and Fig. 3. They clearly show that hardness is strongly dependent on indentation depth. It may be due to the fact that during experiments indentation process is strongly affected by the geometrical features of the investigated surface (peak, valley or slope).

Also, there are two levels of hardness for the same indentation depth value acquired during the measurements of the test specimens without graphene coating. However, when the graphene-coated samples considered only one level of hardness is visible.

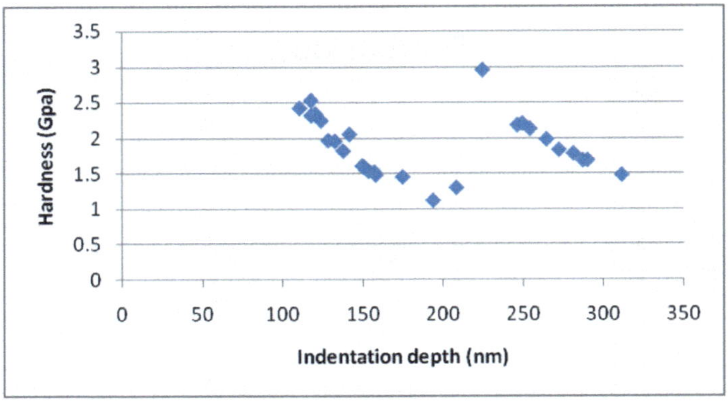

Fig. 2. Sample hardness vs. indentation depth -specimen before graphene deposition (40HM-1)

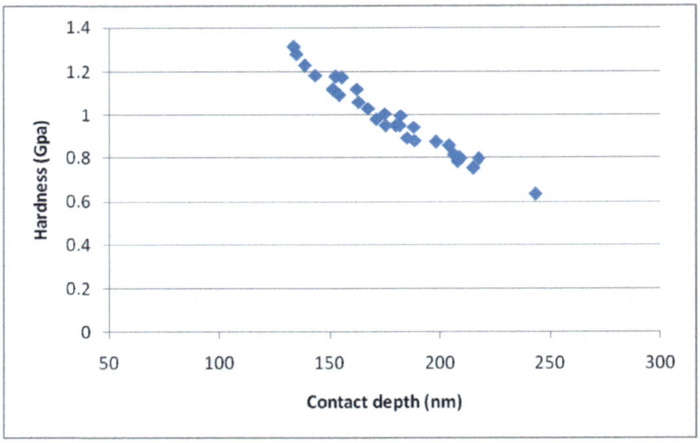

Fig. 3. Sample hardness vs. indentation depth - graphene-coated specimen (MS45-1)

Sample hardness measurement results for constant indentation depth are presented in Fig. 4. The chart proves that hardness of measured surface becomes stable after deposition of a graphene layer, but the values of hardness are smaller than the ones obtained before graphenecoating. However, it must be taken into account that during hardness measurements of layers as thin as graphene ones, an impact of the base surface on the results is crucial. Therefore, the decrease of nanoscale hardness is not a result of a graphene layer itself, but is dependent on a choice of an electrolytic copper as a base material.

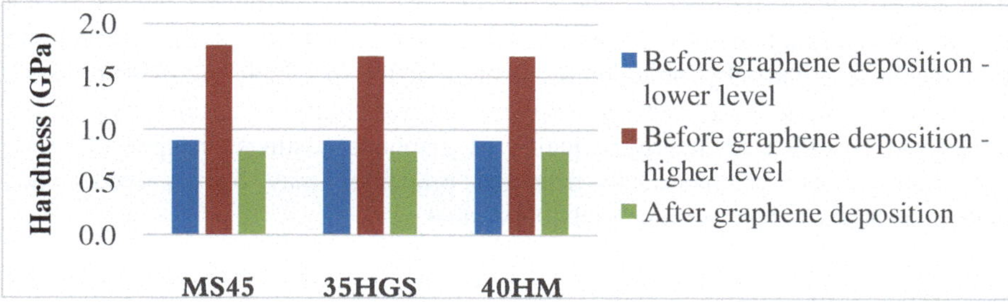

Fig. 4. Hardness measurement results obtained for indentation depth of 200 nm

3.2 Friction Coefficient

Friction coefficient values of the specimens before and after graphene deposition are presented in Fig. 5.

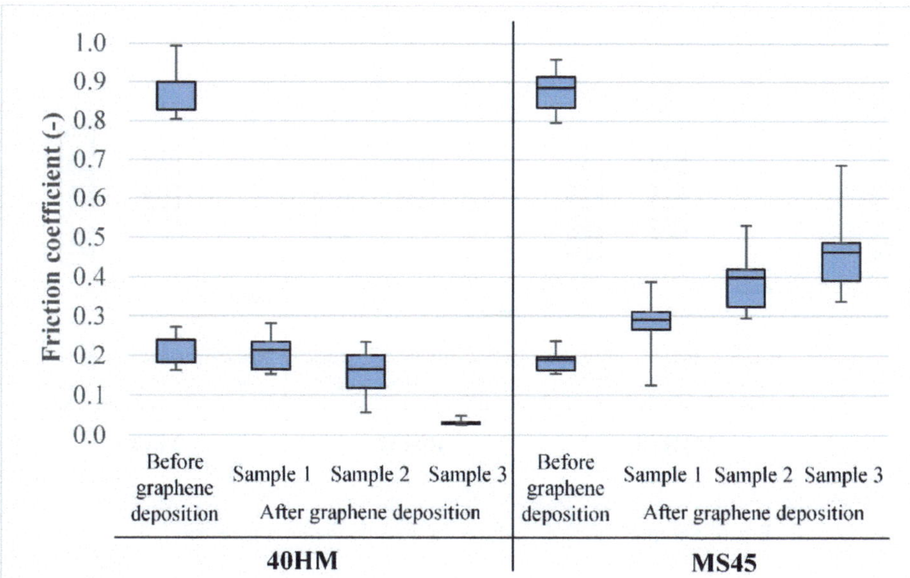

Fig. 5. Sample friction coefficient measurement results

Each data set in box-whiskers chart refers to ten measurement results. Every measurement consisted of 256 scanning lines with 256 measurement points each, along the specimen surface. Similarly to the hardness measurement results, there had been a significant variability of measurement results observed before the specimens were graphene-covered. The results can even be divided in two separate groups with medians of about 0.2 and 0.9 respectively, as it is shown in Fig. 5.

Such an extreme variability of the estimated friction coefficient values is not discerned when graphene-plated surfaces are considered. However, the repeatability of the results is slightly dependent on the steel grade the specimens are made of. Also, even with the same base material applied, the variability of results may differ, probably as an effect of divergent quality of graphene coatings. In spite of these, slight improvement of nanoscale frictional properties may be observed, when graphene is deposited on the coppered 40HM steel.

3.3 Young's Modulus

Young's modulus was measured by thirty indentations of each specimen. Sample results given in the research are presented in Fig. 6.

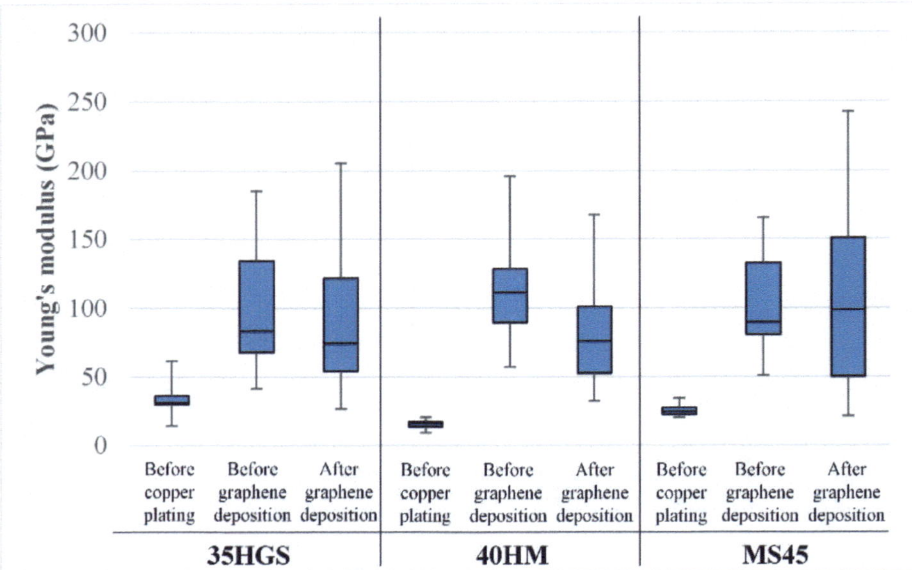

Fig. 6. Sample Young's modulus values obtained in the research

There is a significant improvement of nanoscale Young's modulus of the test samples just after plating copper on the steel base. However, there are no differences between values of this parameter that are important from practical point of view.

4. Conclusion

The results achieved in the research indicate that graphene coatings may improve the nanoscale tribological properties of steel, copper-plated mechanical objects. Firstly, in spite of their initial variability, hardness and friction coefficient values become uniform within range of the whole graphene-plated surface. However, depending on the choice of base material, Young's modulus values are not improved by using graphene coatings. In order to minimise an influence of base material on the assessment of tribological features of graphene-coated industrial elements lower loads can be used in the future researches.

Acknowledgements

This work was supported by grant Graf-Tech/NCBR/05/13/2012 from the National Centre for Research and Development, Poland.

References

1. A. Geim "Graphene:status and prospects" Science 324 (2009).
2. U. K. Sur "Graphene: a rising star on the horizon of materials science" International Journal of Electrochemistry 2012.
3. A. Klemenz, L. Pastewka, S. G. Balakrishna[et al.]"Atomic Scale Mechanisms of Friction Reduction and Wear Protection by Graphene" Nano Lett.14 (2014).
4. J. Tomasik, M. Wiśniewska "Badania wpływu powłok grafenowych na mikrogeometrię powierzchni" Mechanik 88 (2015).
5. R. Voicu, Z. Rymuza, M. Michałowski [et al.] "Investigations of surface properties of SiO2 and Si3N4 thin layers used for MEMS vibrating structures applications" CAS 2013 Proceedings / Conference Committee, 2013.
6. D. M. Jarząbek, A. N. Kaufmann, H. Schift [et al.] "Elastic modulus and fracture strength evaluation on the nanoscale by scanning force microscope experiments", Nanotechnology 25 (2014).
7. W.C. Oliver, G.M. Pharr „Measurement of hardness and elastic modulus by instrumented indentation: Advances in understanding and refinements to methodology", J. Mater. Res. 19 (2004).

Distance determination from the magnetic dipole with magnetometry and Levenberg-Marquardt algorithm

Michał Nowicki

Institute of Metrology and Biomedical Engineering, Warsaw University of Technology, Poland
nowicki@mchtr.pw.edu.pl

Abstract. Paper presents the analysis of distance measurements from an unknown magnetic dipole. Simulations of axial magnetic field distribution were performed in open-source software Elmer FEM based on finite element method. The results analysis was performed in Matlab, and possibility of distance determination is presented. Moreover, validation of the results on the laboratory test stand was performed.

Keywords: magnetovision, finite element method, magnetostatic simulations, hidden objects detection.

1 Introduction

The paper presents the measurements of influence of the high magnetic permeability objects (mostly ferromagnetic) on the natural Earth's magnetic field. The ferromagnetic objects influence surrounding magnetic field mainly because of their own remanent magnetization, which in effect causes them to act like a magnetic dipole, which creates distortions in otherwise homogenous (in small scale) Earth's natural magnetic field.

There is an ongoing research to connect said distortions with hidden objects presence and localization [1]. Existing passive gradiometer systems are designed for deep-level search of relatively big targets, such as unexploded aircraft bombs. They are used for magnetic imaging, using data logging and GPS systems, but they have very low resolution [2, 3]. Therefore a high-resolution scanning system was designed for search of small targets from small distance, such as potentially dangerous objects in baggage.

The 'planar' (XY coordinates) localization of the distortion on the measurement plane is quite obvious [1], giving us information about where to search for an object. There is an open question however, about the method of distance from the object determination, which would reveal the information how deep it is giving us full 3D localization of the object (coordinate Z). In this paper, one of the possible solutions is presented.

Moreover, the paper presents an application of magnetostatic calculations based on the free open-source Elmer FEM software for the 3D modelling of the magnetic field distribution caused by small, ferromagnetic objects. The overall effect would be hard

R. Jabłoński and T. Brezina (eds.), *Advanced Mechatronics Solutions*,
Advances in Intelligent Systems and Computing 393,
DOI 10.1007/978-3-319-23923-1_74

to calculate using analytical methods, therefore finite element method was implemented. The distance determination method was mainly developed basing on the numerous modelling results. Furthermore, validation of the method using laboratory measurements is shown.

2 Modelling of the magnetic field distribution

Analytically, the magnetic field of the magnetic dipole is given by the formula [4,5]:

$$|\vec{B}| = \frac{\mu_0 M}{4\pi r^3}(1 + 3\cos^2\theta_m)^{\frac{1}{2}} \tag{1}$$

Where:
B – magnetic induction value, μ_0 – magnetic vacuum permeability, M - magnetic dipole moment per unit volume, r – distance from the object, θ_m – angle between r and magnetic dipole (object) axis.
For real world magnetized objects, the equation (1) is valid for distances r significantly greater than the objects dimensions. The same can be observed for the modelled results.
The magnetic field distribution of the small iron cylinder (Ø70 x 10 mm) was calculated in 3-dimensional space using the finite element method and Whitney elements. Figure 1 presents the results of simulation utilizing both NETGEN and ELMER FEM open source software [6,7,8,9]. As could be expected, the axial magnetic induction values distribution is inversely proportional to the cube of the distance from the object.

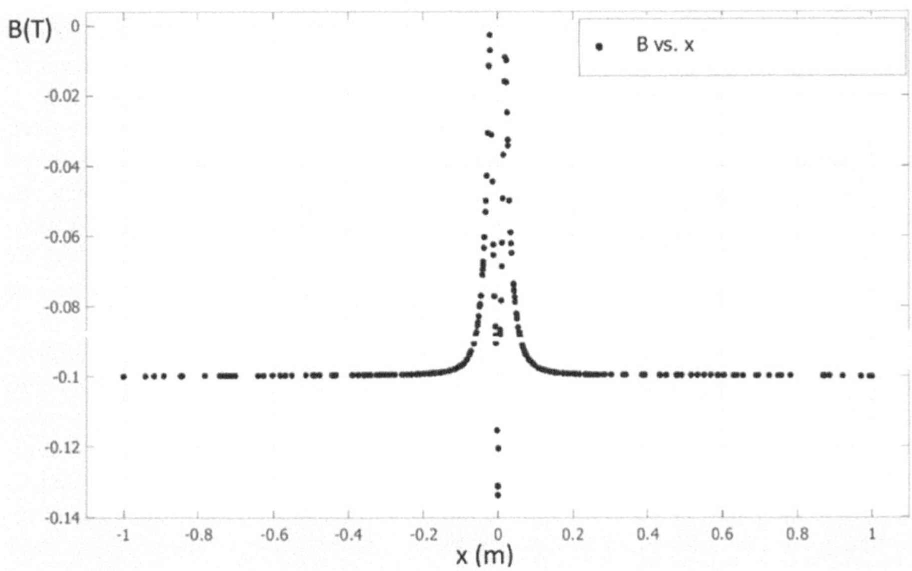

Fig. 1. Modelled axial magnetic field distribution of the homogenously magnetized iron cylinder

In order to utilize this known behaviour of magnetic field distribution in order to find the distance from the unknown magnetic dipole object, the following equation was derived from the (1) :

$$B = \frac{a}{(x+b)^3} + c \tag{2}$$

Where a – replacement magnetic moment parameter, varying in value between M/5 (θ_m=0°) to M/10 (θ_m=90°), depending on the θ_m angle; b –distance of the first measurement point from the object; c – constant, background field value.

If we assume a, b and c parameters as unknowns, and :

- know where (coordinates X and Y) is the hidden object, for example by magnetovision method [1],
- measure the magnetic induction in vertical line pointed at the object,
- measure the distances between the measurement points,

Then, it is possible to determine the a, b and c parameters using the Leveberg-Marquardt algorithm to fit the equation (2) to the measurement results. Of these, b is of utmost interest, as it determines the distance of the first measurement point from the hidden object, thus giving us missing Z coordinate of the object.

In figure 2 the curve fitting of the (2) to the measurement results modelled with FEM method is shown. The distance of the first measurement point in the simulation was set to 100 mm ; the calculations returned the 103 mm value, which is a 3% relative error, probably due to the differences between simplified analytical equation valid for point objects, and 3D object simulation with determined dimensions.

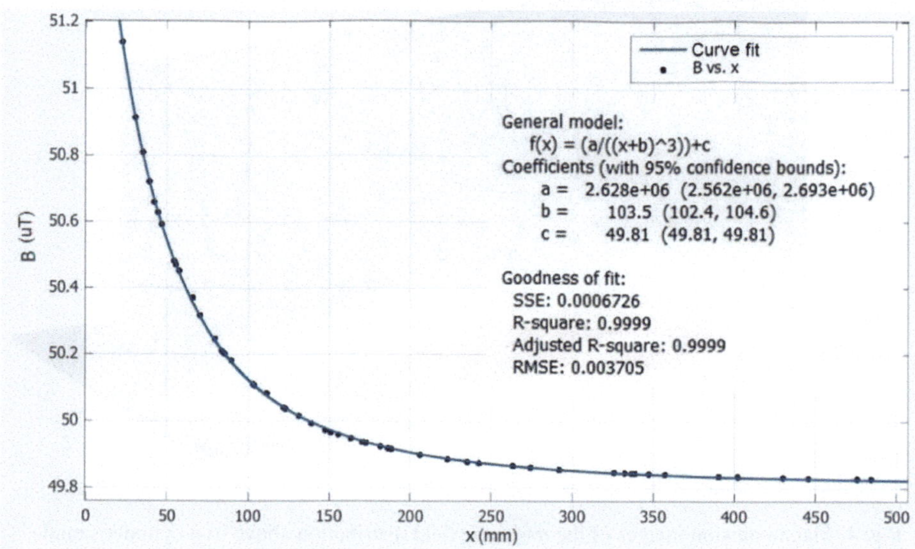

Fig. 2. Curve fit of the modelled axial field distribution, with dipole magnetic moment, distance from the object and background field treated as unknowns.

3 Experimental results

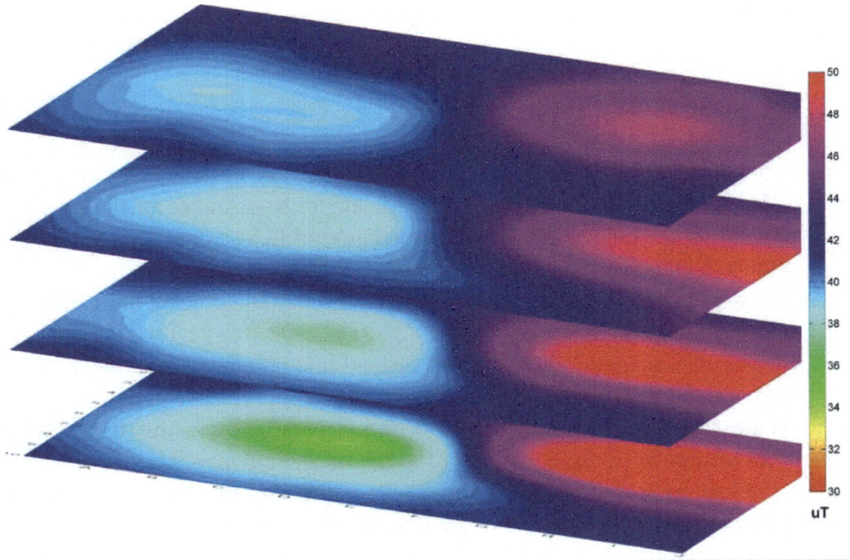

Fig. 3. Magnetovision images of the magnetic field distribution above iron cylinder sample. Distance between the measurement planes is 10 mm.

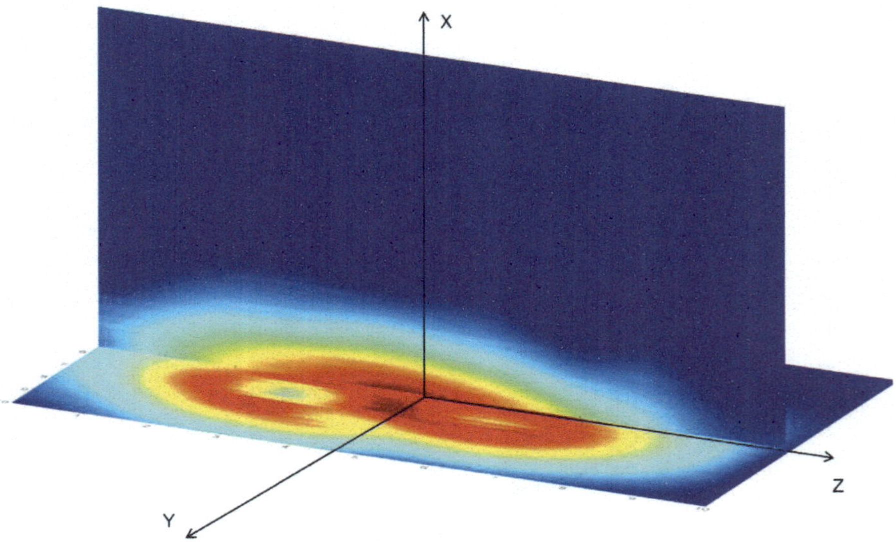

Fig. 4. Magnetovision images of the magnetic field distribution above iron cylinder sample. First plane, used for the sample localization, is horizontal. Second, used for distance measurement, is vertical above the magnetic distortion centre.

For the validation of the method, iron cylinder sample magnetic field distribution was measured on specialized magnetovision scanner, described in detail in [1]. The system utilized high accuracy HMR2300 magnetoresistive magnetometer [10,11]. In figure 3 magnetic induction distribution on 200x200mm measurement planes above the sample can be seen. The planes are 10 mm apart, the first one is 80 mm above the sample. The angle θ_m is 45°, thus the dual minimum and maximum distortions values can be seen.

In Figure 4 the magnetic induction distribution on the horizontal and vertical planes above the sample can be seen. The angle θ_m is 0°. The horizontal plane serves for the XY localization of the object, the vertical one allows for localization refinement, and determination of the object distance with the method described above, for the most distinctive vertical measurement line.

Figure 5 shows the application of the developed method. The real distance of the magnetometer from the object was set to 80 mm, the calculated one was 79,5 mm, with about 1 mm uncertainty. Unfortunately, the uncertainty raises significantly with the distance from the object, because the measured induction B value decreases inversely proportional to the cube of the distance. The theoretical limit for this method is determined by the magnetometer accuracy, there are other effects to take into account however, such as AC noise, and local background field inhomogeneity.

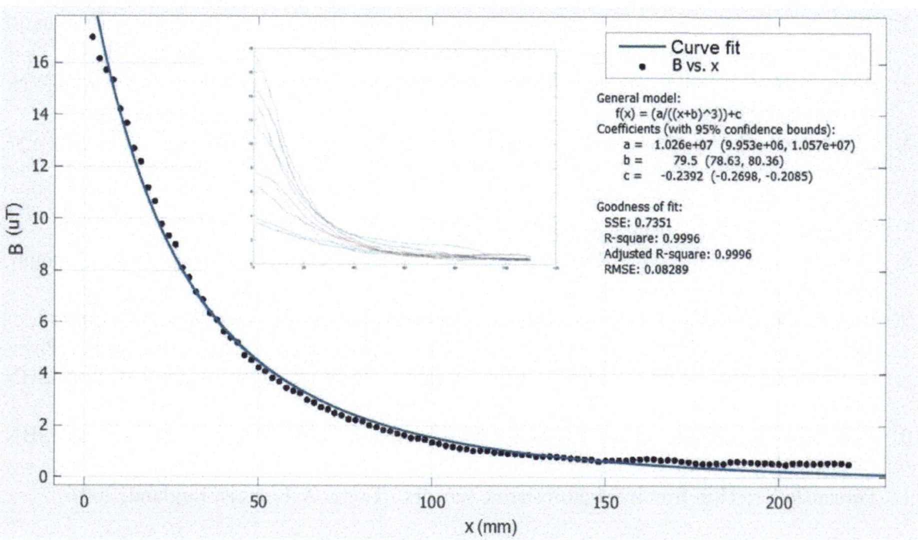

Fig. 5. Curve fit of the measured magnetic field distribution, in the axis of the iron cylinder sample. Miniature shows the vertical measurement lines values from the Fig.4

4 Conclusions

Presented in the paper, method of measurement of distance from the unknown magnetic dipole using magnetometry and signal processing allows for more precise localization of hidden, ferromagnetic objects. The b parameter in the model equation is largely independent from the object θ_m angle relative to the measurement line, as it affects mainly the a parameter and hinders the object's magnetic moment value determination. As long as the measurement line is pointed in the object centre (with about 5° accuracy), the presented method allows for the distance determination, from the unknown object, with unknown magnetic moment value, and with unknown magnetic background value (parameter c). Interestingly, additional measurement of the c parameter and apparent simplification of the (2) by removing one of the unknowns, hampers the method. One of the possible reasons of this behaviour is constraining of the curve fitting algorithm 'flexibility'.

References

1. Nowicki, M., Szewczyk, R. Ferromagnetic Objects Magnetovision Detection System. Materials, 6, 2013, pp 5593
2. Guelle D., Smith A., Lewis A., Bloodworth T., Metal detector handbook for humanitarian demining. Office for Official Publications of the European Communities, Luxembourg, 2003.
3. Billings S. D. Discrimination and classification of buried unexploded ordnance using magnetometry. IEEE Transactions on Geoscience and Remote Sensing, 6,2004, pp 1241
4. Halliday D., Resnick R., Walker J.: Podstawy Fizyki. T. 3. Wydawnictwo Naukowe PWN, Warszawa 2009.
5. Feynman R. P., Leighton R.B., Sands M.: Feynmana wykłady z fizyki. T. 2, cz. 1. Wydawnictwo Naukowe PWN, Warszawa 2004.
6. http://www.hpfem.jku.at/netgen
7. http://www.csc.fi/english/pages/elmer
8. P. Råback, M. Malinen, J. Ruokolainen, A. Pursula, T. Zwinger: Elmer Models Manual, CSC – IT Center for Science, 2014.
9. Nowicki M., Szewczyk R.: Modelling Of The Magnetovision Image With The Finite Element Method, In PROCEEDINGS of the 20 th International conference on Applied Physics Of Condensed Matter (APCOM2014) / Vajda J., Jamnicky I., 2014, pp. 131-134, ISBN 978-80-227-4179-8.
10. Tumanski S.: Handbook of Magnetic Measurements. CRC Press, 2011. 404 p. ISBN 1439829527, 9781439829523
11. Tumański S. , Thin film magnetoresistive sensors ; Taylor & Francis, England, 2001.

Magnetoelastic effect for three type magnetic materials under torque load

Jacek Salach

Institute of Metrology and Biomedical Engineering, Warsaw University of Technology
Warsaw, Poland
j.salach@mchtr.pw.edu.pl

Abstract. The paper presents the comparison of magnetoelastic properties of three new magnetic materials under torque load. The results of investigation on the influence of a torsion on the magnetic properties of different amorphous Fe-based ring cores is also presented. The results of tests on $Fe_{78}Si_{13}B_9$ and $Fe_{40}Ni_{38}Mo_4B_{18}$ amorphous alloys in as quenched and annealed state as well as $Fe_{73.5}Nb_3Cu_1Si_{13.5}B_9$ nanocrystalline alloy indicate, that torque sensitivity is connected with both alloy composition and state of the alloy. Moreover, results confirm that amorphous materials after annealing get better sensitivity than in as-quenched state.

Keywords: : magnetoelastic effect, torque sensors, amorphous alloys.

1 Introduction

New magnetic materials provide possibilities to use them to develop new sensors [1,2,3]. The results of compression and tensile forces impact on magnetoelastic effect for amorphous materials are commonly known [4,5]. Physical models have already been developed as well [6,7]. On this basis the new force sensors have been developed [8]. However, research on the impact of torque on the magnetic properties is not as well developed, despite the fact that both materials and research methodology are known. The main barrier in the methodology was the lack of possibility to apply a torque to the core of wound ring sample. This barrier has already been broken [9].

2 Method of investigation

The main problem connected with the development of magnetoelastic ring-shaped torque sensors is lack of the method, which enable applying the uniform, torsional axial shearing stress to the core.

To overcome this problem, a new method of applying torsional axial shearing stress to the ring-shaped magnetoelastic sensing element was developed [9]. In this method the core under investigation is mounted to the base plates and the torque is applied in direction of axis of the ring core. Special nonmagnetic backings with radial

R. Jabłoński and T. Brezina (eds.), *Advanced Mechatronics Solutions*,
Advances in Intelligent Systems and Computing 393,
DOI 10.1007/978-3-319-23923-1_75

grooves are attached to the base plates of the ring shaped core. Due to these grooves the core can be wound and changes of its magnetic parameters under influence of torsion can be measured. The special backings (3) are necessary to obtain uniform torsional shearing stress in the ring-shaped core. The specialised mechanical system for application of this method is presented in figure 1.

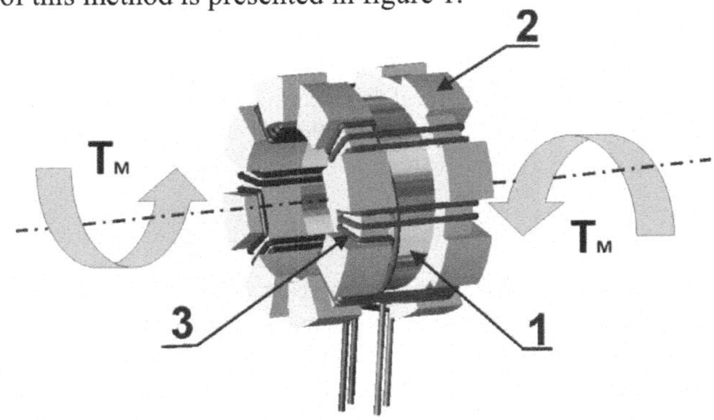

Fig. 1 Mechanical system for application of the uniform torsional axial shearing stress to the ring-shaped amorphous alloy. 1-ring-shaped core under investigation, 2-magnetaizing and measuring windings, 3-nonmagnetic backings with grooves.

This method of applying torsion to the ribbon ring cores can be utilized in the new type of magnetoelastic torque sensors.
During the investigation the ring-shaped cores were subjected to the torsion in the axis of the ring, and next were magnetized from + H_m to - H_m, and the $B(T_M)_{Hm}$ characteristics were determined.

3 Results

The results of investigation of the influences of a torque on the flux density of investigated cores ($B(T_M)_{Hm}$ characteristics) are presented in figure 2 and figure 3. It should be indicated that investigated cores exhibit different values of coercive field H_c. Moreover, the most significant changes of value of flux density B was observed in the case of annealed $Fe_{78}Si_{13}B_9$ amorphous alloy. In the case of this alloy, value of flux density have most significant changes for magnetic field H_m equal to coercive force.

Figure 4, figure 5, figure 6 and figure 7 present the influence of torsion T_M on the flux density B_m for three Fe based materials : $Fe_{78}Si_{13}B_9$ and $Fe_{40}Ni_{38}Mo_4B_{18}$ amorphous alloys in as quenched and annealed state, as well as $Fe_{73.5}Nb_3Cu_1Si_{13.5}B_9$ in as quenched state and nanocrystalline alloy state. Characteristics on figure 4 and figure 6 are for magnetic field H_m equal to coercive field H_c, and characteristics on figure 5 and figure 7 are for magnetic field H_m equal to one and a half of coercive field H_c. The core is subjected to influence of torque T_M and then magnetized at constant value of magnetizing field H_m. It should be highlighted, that so big changes (over 80 percent) of the amplitude of magnetic permeability μ_a can be easily converted to output signal from torque measuring transducer. As it was indicated in previous publications, for conversion of permeability to useful signal both resonant circuit and transformer configuration can be utilized.

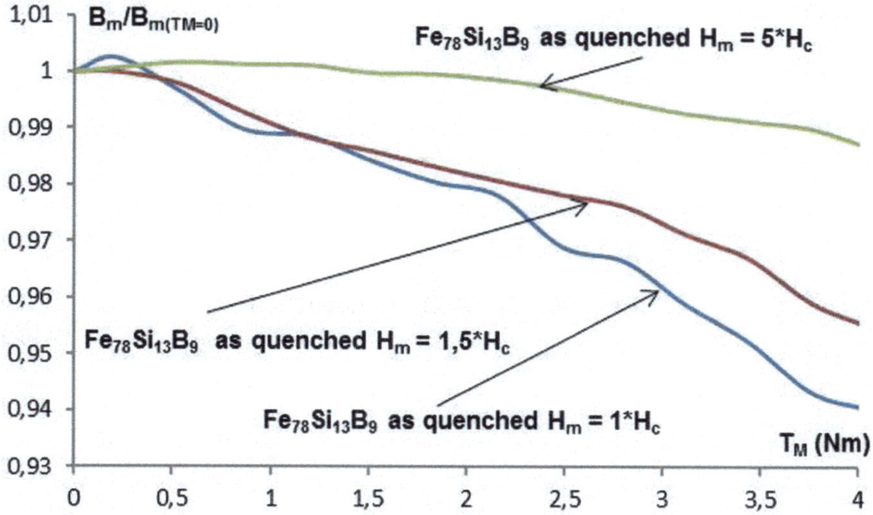

Fig. 2. Magnetoelastic $B_m/B_{m(TM=0)}(T_M)$ characteristics of $Fe_{78}Si_{13}B_9$ amorphous alloy in as quenched state

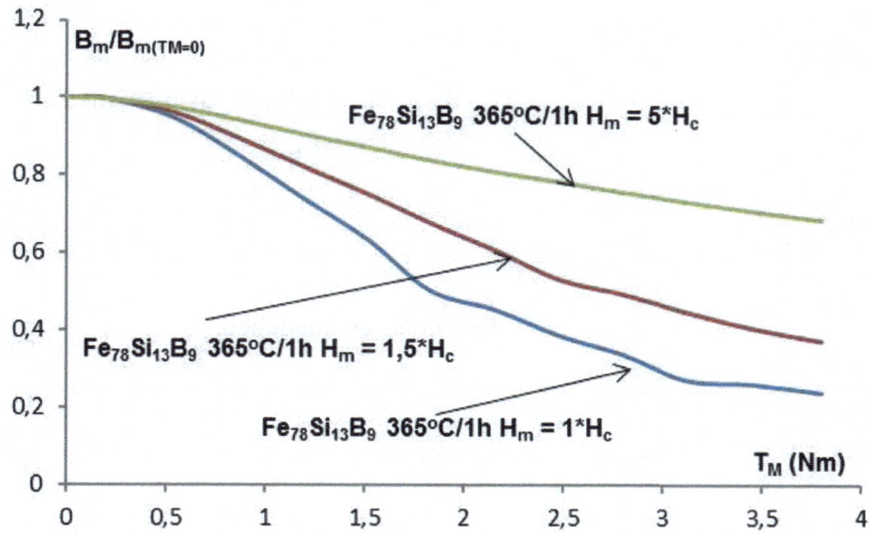

Fig. 3. Magnetoelastic $B_m/B_{m(TM=0)}(T_M)$ characteristic of $Fe_{78}Si_{13}B_9$ amorphous alloy after 1h annealing in 365°C

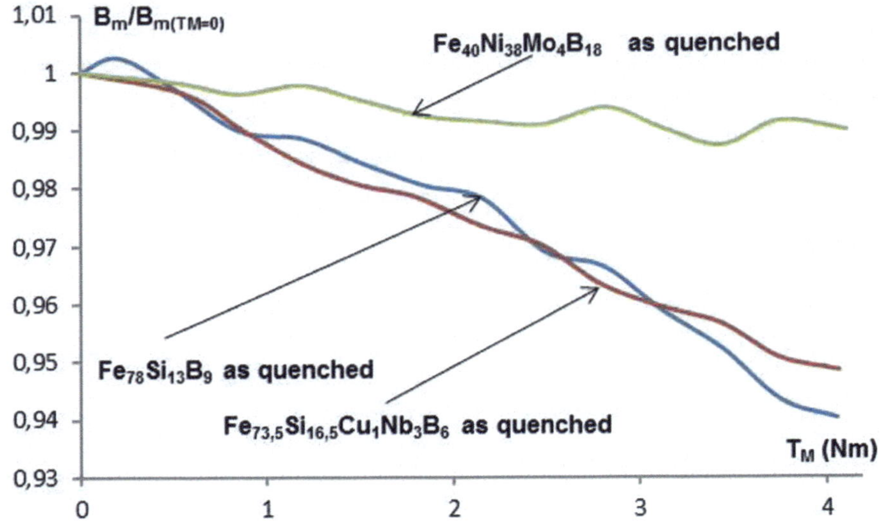

Fig. 4. Magnetoelastic $B_m/B_{m(TM=0)}(T_M)$ characteristics of $Fe_{78}Si_{13}B_9$, $Fe_{40}Ni_{38}Mo_4B_{18}$ and $Fe_{73.5}Nb_3Cu_1Si_{13.5}B_9$ amorphous alloys in as quenched state, $H_m = H_c$.

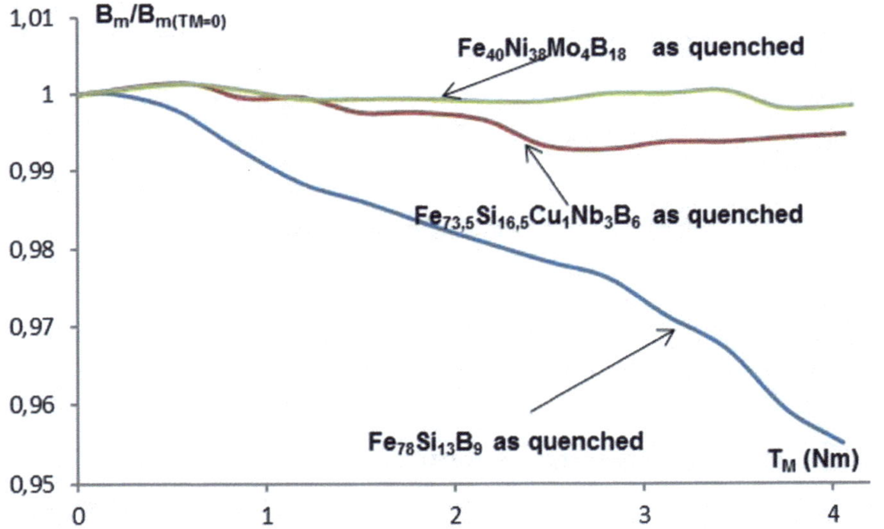

Fig. 5. Magnetoelastic $B_m/B_{m(TM=0)}(T_M)$ characteristics of $Fe_{78}Si_{13}B_9$, $Fe_{40}Ni_{38}Mo_4B_{18}$ and $Fe_{73.5}Nb_3Cu_1Si_{13.5}B_9$ amorphous alloys in as quenched state, $H_m = 1{,}5\ H_c$

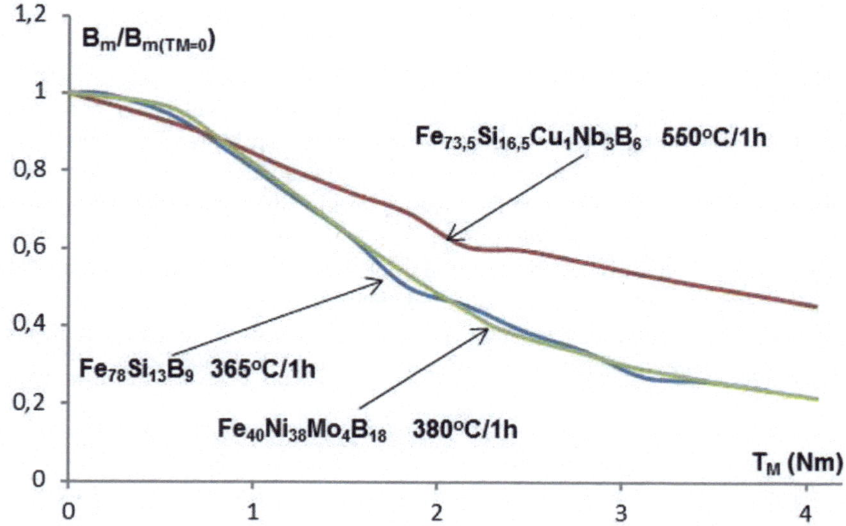

Fig. 6. Magnetoelastic $B_m/B_{m(TM=0)}(T_M)$ characteristic of $Fe_{78}Si_{13}B_9$, $Fe_{40}Ni_{38}Mo_4B_{18}$ and $Fe_{73.5}Nb_3Cu_1Si_{13.5}B_9$ amorphous alloys after annealing, $H_m = H_c$.

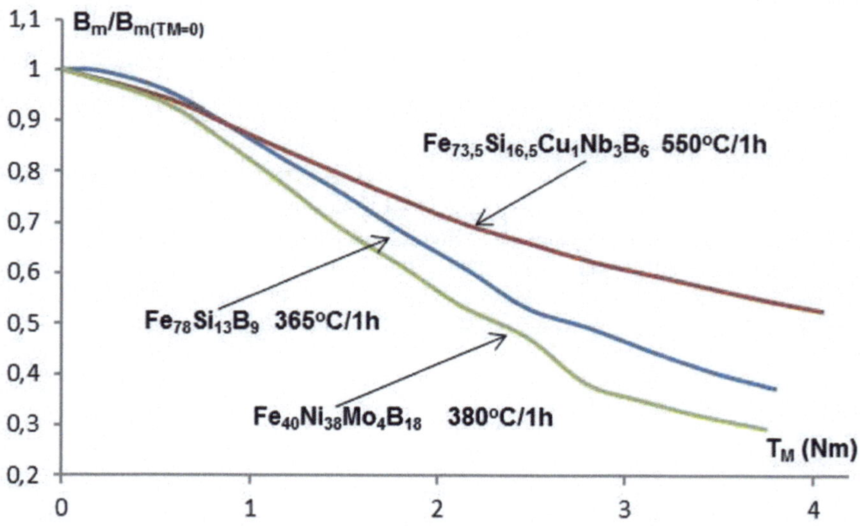

Fig. 7. Magnetoelastic $B_m/B_{m(TM=0)}(T_M)$ characteristics of $Fe_{78}Si_{13}B_9$, $Fe_{40}Ni_{38}Mo_4B_{18}$ and $Fe_{73.5}Nb_3Cu_1Si_{13.5}B_9$ amorphous alloys after annealing, $H_m = 1{,}5\ H_c$.

4 Conclusions

The test results confirm that amorphous materials exhibit the magnetoelastic characteristics with respect to torque T_M. Impact of torque is greatest for cores made of amorphous materials after heat treatment. The most stable trend exhibits the iron-based material $Fe_{78}Si_{13}B_9$. In both the as quenched state and after thermal relaxation a significant effect of the torque on the magnetoelastic characteristics is observed. Also, a large torque impact can be observed for the material based on iron-nickel $Fe_{40}Ni_{38}Mo_4B_{18}$, but only for the material after thermal relaxation. This material in as quenched state is practically not sensitive. The material of the lowest sensitivity is nanocrystalline iron-based $Fe_{73.5}Nb_3Cu_1Si_{13.5}B_9$ alloy, despite the very high permeability. Furthermore, it is brittle and has low mechanical strength.

The experimental results show that the annealed material, especially $Fe_{78}Si_{13}B_9$, are suitable for torque sensors, because of their high magnetoelastic sensitivity. However, the material in as-quenched state is well suited for the DC current transformer core construction, working under mechanical load. .

References

1. O'Handley, R.,Modern magnetic materials – principles and applications,:John Wiley and sons New York, 2000
2. Hristoforu, E. J. Magn. Magn. Mater. **249** (2002)387-392.
3. Herzer, G. Scripta Metallurgica **33** (1995)1741-55.
4. Szewczyk, R., Bienkowski, A., Kolano, R. Cryst. Res. Technol. **38**, (2003) 320-324.
5. Bieńkowski A., Kolano R., Szewczyk R. J Magn Magn Mater. 254-255 (2003) 67-69.
6. Szewczyk R. J. Phys. D: Appl. Phys. 40 (2007)4109-11
7. Szewczyk, R. (2014). Materials (MDPI) 7 (2014)5109-16.
8. Bieńkowski A., Szewczyk R. Sensors and Actuators A: Physical 113 (2004), 270-276
9. Salach J., Bienkowski A. Szewczyk R. Acta. Phys. Pol. A Vol. 113 (2008) 115-118

Tribological behavior of graphene-coated mechanical elements

Jan Tomasik[1], Marta Wiśniewska[1], and Marcin Kamiński[2]

[1] Institute of Metrology and Biomedical Engineering, Faculty of Mechatronics, Warsaw University of Technology, Warsaw, Poland
{j.tomasik,martwisn}@mchtr.pw.edu.pl
[2] Industrial Research Institute for Automation and Measurements, Warsaw, Poland
mkaminski@piap.pl

Abstract. An impact of graphene coatings on tribological and wear properties of mechanical elements was evaluated. In the research, the graphene layer was deposited on two different base materials: electro- or cyanide copper. To quantitatively assess wear of the investigated surfaces, their microgeometry was measured both before and after tribological tests. Differences between surface roughness parameters obtained this way are presented in the paper. Also, limitations of the graphene coatings usability, dependent on the applied base material, are outlined.

Keywords: Graphene·surface metrology·roughness·tribology

1 Introduction

Graphene's properties are astonishing: in spite of being the thinnest material in the known universe, it remains the strongest one ever discovered, with an ultimate tensile strength exceeding one hundred gigapascals [1, 2]. Therefore, the proliferation of patents referring to potential applications of graphene [3] cannot be surprising. However, neither wide array of graphene applications suggested by the number of scientific papers referring to this material, nor its apparent interest to media, is reflected by use of graphene on industrial scale.

In order to bridge the gap between potential and practical uses of graphene, the GRAPH-TECH research project was introduced. Its main aim is to devise graphene coatings that improve the mechanical properties of miniaturized industrial elements. Findings of the research, concerning an impact of graphene-deposition on wear, and, in effect, changes of the microgeometry of the mechanical elements, are presented in the paper.

R. Jabłoński and T. Brezina (eds.), *Advanced Mechatronics Solutions*,
Advances in Intelligent Systems and Computing 393,
DOI 10.1007/978-3-319-23923-1_76

2 Materials and methods

2.1 Test Samples

In order to assess an impact of graphene coatings on tribological and wear properties of mechanical objects, numerous specimens with the test surfaces of (20 x 20) mm, differing in the steel grade they were made of (MS45, 40H or 35HGS), have been used. Each of these three groups consisted of 36 samples that were milled and grinded. Then, half of them was covered with electrolytic copper and another half – with cyanide copper. Nine specimens from the sets obtained this way were used as the reference samples, whereas the graphene layer was deposited on the test surfaces of remaining ones.
The test surfaces of the specimens were measured three times with a use of Form Talysurf PGI 830 stylus profilometer before and after performing tribological tests. The measurement settings have already been listed in [4]. In order to quantitatively describe the changes of the surface microgeometry, *Ra, Rz, Rp* and *Rv* parameters were estimated for each of the measured surface profiles with the Gaussian linear filter applied (λ_c = 0.8 mm). These parameters have been chosen due to their popularity and vulnerability to change value during wear of investigated surface.

2.2 Tribological Tests

The test stand for tribological investigations that is used in the research is presented in Fig. 1 [5]. This stand consists of the replaceable track (1) and three specimen holders (2). The investigated samples (3) are pressed to the track by shield (4), pin (5) and loads (6). Then, with a use of force sensor (7), force, which is an effect of friction between samples and track, is measured. Data given this way, as well as signal from rotation sensor (8) is used to obtain friction characteristics. In Table 1 there is a detailed information concerning the test conditions presented.

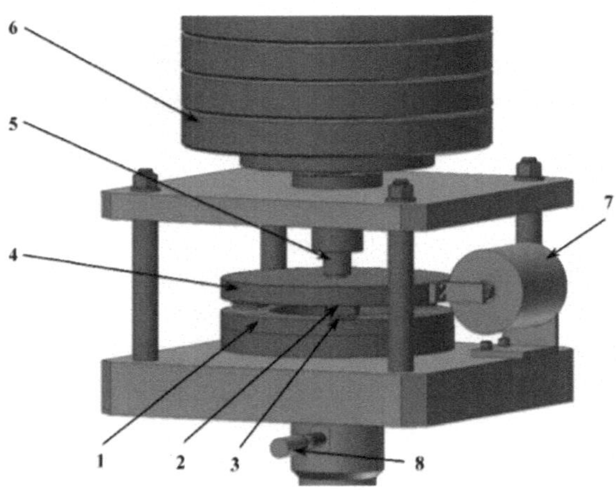

Fig. 1. Test stand used in the research [5]

Table 1. Testing conditions applied in the research

Type of grease	Lithium
Kinematic viscosity of grease	100 mm^2/s (in 40°C)
NLGI consistency number	2
Load	0.54 MPa
Speed	0.53 m/s
Friction distance	40 000 m
Track material	40H steel

3 Results

3.1 Electrocoppered Samples

The *Ra* and *Rz* parameter values obtained in the research both before and after testing specimens are outlined in Fig. 2 and Fig. 3 respectively. The box-whiskers charts made for other samples and roughness parameters reveal the same tendencies, so they have not been presented.

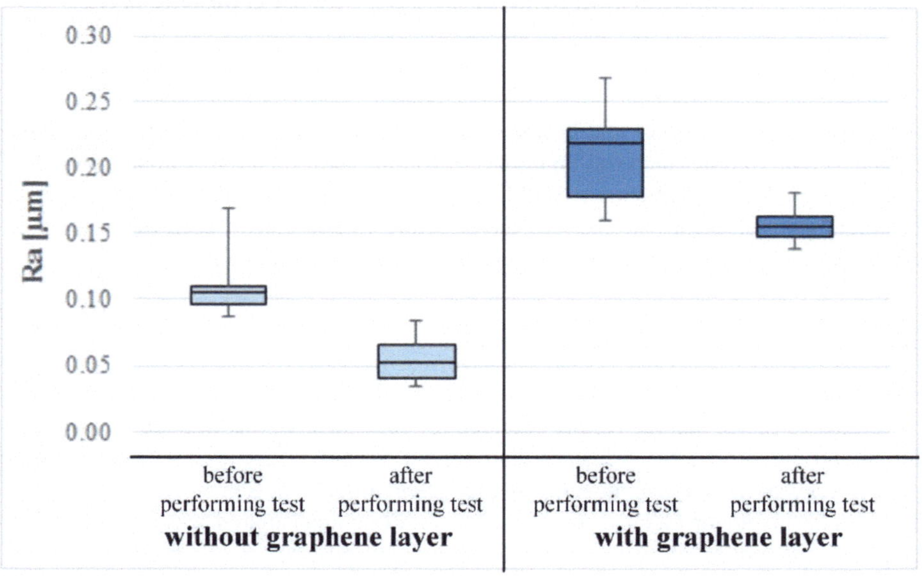

Fig. 2. *Ra* parameter values obtained in the research (electrocoppered MS45 specimens)

Firstly, these pictures clearly indicate that there are some differences between surface microgeometry of the specimens that had been covered with graphene layer and the ones that had not. This phenomena has already been investigated and explained in [4]. Furthermore, the values of surface texture parameters significantly decrease after conducting tribological tests. To prove so, Kruskal-Wallis test ($\alpha = 0.05$) was used, when necessary. However, the relative differences between these values were visibly smaller, if performance of graphene-coated specimens was investigated.

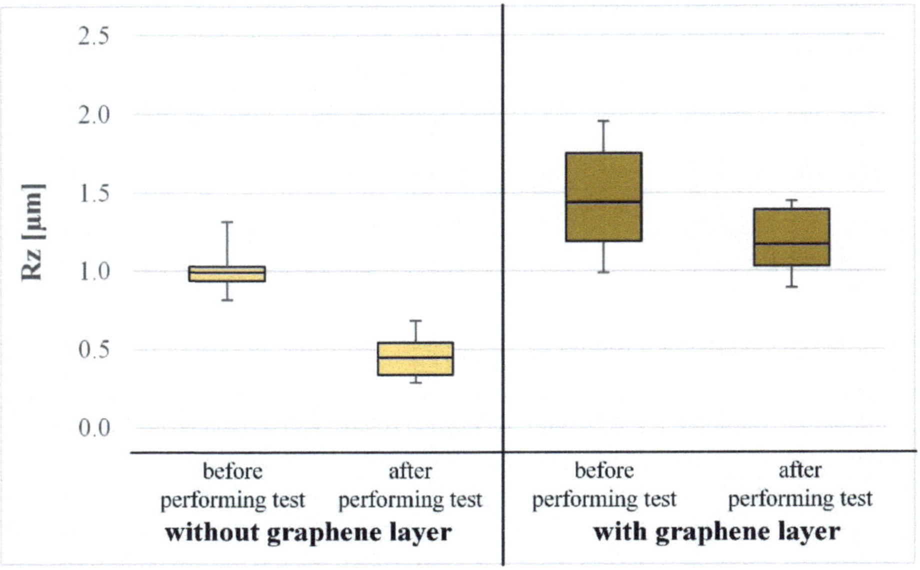

Fig. 3. *Rz* parameter values obtained in the research (Electrocoppered MS45 specimens)

Also, material ratio curves and height distributions obtained for each specimen have been analyzed in order to assess an impact of friction wear on the functional characteristics of mechanical elements. It was found out that neither graphene deposition, nor conducting tribological test changes the structure of surface. In each of these cases the height distributions of surface irregularities resemble normal ones.

3.2 Cyanide Copper Plated Samples

The values of surface roughness parameters have also been compared for cyanide copper plated surfaces. Sample results obtained for specimens of the kind are presented in Fig. 4 and Fig. 5. The box-whiskers charts made for other samples and roughness parameters reveal the same tendencies, so they have not been presented.

The results given for electrocoppered and cyanide copper plated specimens without graphene are similar. However, there are significant differences observed for the graphene-covered samples. When samples with cyanide copper plating considered, the values of roughness parameters increase dramatically after performing tribological test, due to severe damage of the measured surfaces. In the extreme cases the graphene and copper coating may even be torn off, as it is shown in Fig. 6. It may be an effect of the specimen form deviations introduced during graphene deposition.

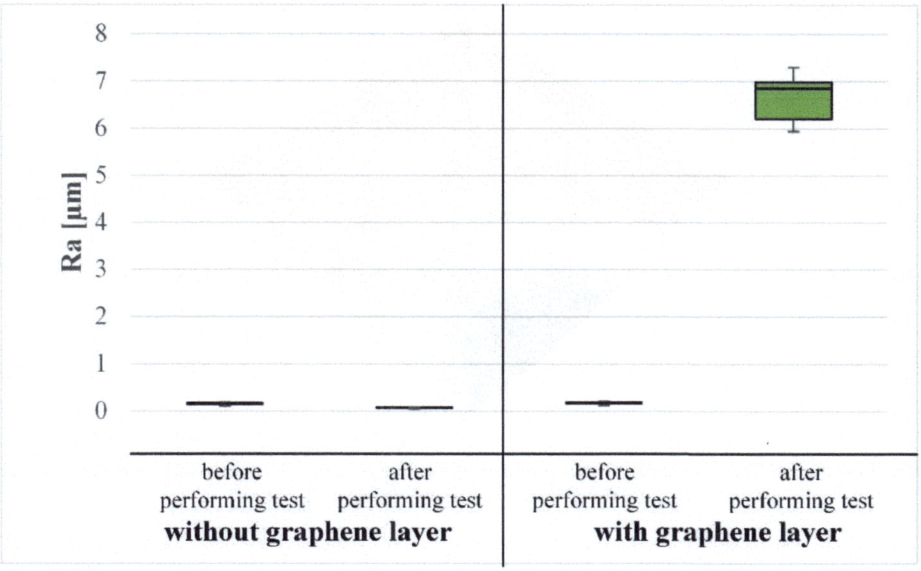

Fig. 4. *Ra* parameter values obtained in the research (cyanide copper plated 40H specimens)

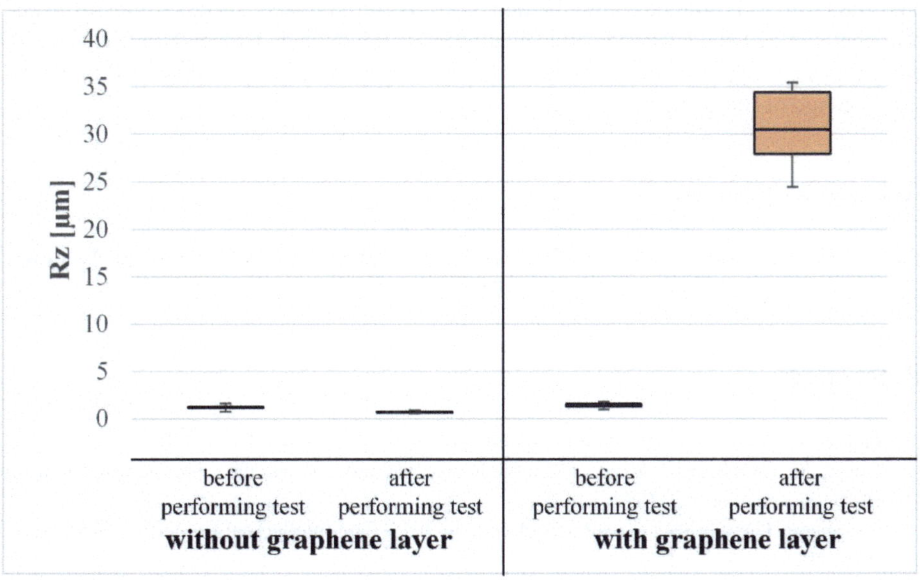

Fig. 5. *Rz* parameter values obtained in the research (cyanide copper plated 40H specimens)

Fig. 6. Image of the damaged surface obtained with the CCI SunStar white-light interferometer (field of view: 6.6 mm x 6.6 mm)

4 Conclusion

The research has demonstrated that graphene coatings can improve wear resistance of mechanical elements. However, in order to achieve these, it is necessary to use proper base material, as graphene cannot be deposited directly on steel surfaces. One of materials that can be used successfully is electrodeposited copper. However, a use of some other types of copper coatings, such as the ones of cyanide copper, may lead to extensive damage of the surface.

Acknowledgements

This work was supported by grant Graf-Tech/NCBR/05/13/2012 from the National Centre for Research and Development, Poland.

References

1. Geim, A.: Graphene: status and prospects. Science 5934, 1530 – 1534 (2009)
2. Sur, U.K.: Graphene: a rising star on the horizon of materials science. International Journal of Electrochemistry 237689, 12 (2012)
3. Graphene. The worldwide patent landscape in 2013. The Intellectual Property Office, Newport (2013)
4. Tomasik, J., Wiśniewska M.: Badania wpływu powłok grafenowych na mikrogeometrię powierzchni. Mechanik 88, 234-237 (2015)
5. Missala, T., Szewczyk, R., Winiarski, W. et al.: Study on Tribological Properties of Lubricating Grease with Additive of Graphene. Progress in Automation, Robotics and Measuring Techniques. Advances in Intelligent Systems and Computing 352, 181-187 (2015)

Part VI
Photonics, Vision Systems and Image Processing

Workshop tomographic system for 3D refractive index investigations in optical fibers

Michał Dudek[1], Nikolai Suchkov[1], Michał Józwik[1], and Małgorzata Kujawińska[1]

[1] Warsaw University of Technology, Institute of Micromechanics and Photonics, Warsaw, Poland
{m.dudek,n.suchkov,m.jozwik,m.kujawinska}@mchtr.pw.edu.pl

Abstract. In this work we present the development of the tomographic phase microscopy system for 3D refractive index measurement with application for optical fibers and their accessories in workshop or factory conditions. The holographic projections are captured at the output of a compact shearing interferometer module with a significant shear of an object beam. The system can work in vertical and horizontal position. It is versatile and less sensitive to the external conditions, maintaining similar accuracy as the conventional Mach-Zehnder interferometer based system.

Keywords: interferometric tomography · tomographic phase microscopy · optical diffraction tomography · refractive index

1 Introduction

With the expansion and development of the fiber optics communication and sensing there appeared a big demand for accurate measurement devices for various parameters of optical fibers. One of the most crucial optical parameter is the 3D refractive index distribution. There are simple ways of determining the 1D refractive index profile of optical fibers [1]. It is valid only for axially-symmetrical structures like singlemode or multimode fibers. It is not enough for optical fibers with more complex internal structure such as polarization maintaining fibers or microstructured fibers and fiber optics accessories [2]. Recently this problem is solved by employing tomographic reconstruction to determine 3D refractive index distribution in arbitrary optical fibers [3-5]. Although the technology for tomographic measurements of phase microobjects is well established, it is not widely used in workshops or for quality control in optical fiber manufacturing process. Most of the interferometric based solutions provide measurements in laboratory conditions only.

The main goal of this work was to provide a compact tomographic system for full tomographic investigations of 3D refractive index distributions of optical fibers in a workshop or factory environments.

R. Jabłoński and T. Brezina (eds.), *Advanced Mechatronics Solutions*,
Advances in Intelligent Systems and Computing 393,
DOI 10.1007/978-3-319-23923-1_77

2 Tomographic systems and measurement methodology

2.1 Mach-Zehnder interferometer based tomographic system

The first tomographic setup we had designed and used extensively [3,4] was based on the modified Mach-Zehnder interferometer with an accurate rotational stage (Fig. 1). The typically achieved accuracy of refractive index determination was 10^{-4}, which was sufficient for most applications, however its biggest disadvantage was high sensitivity to vibrations and other environmental changes.

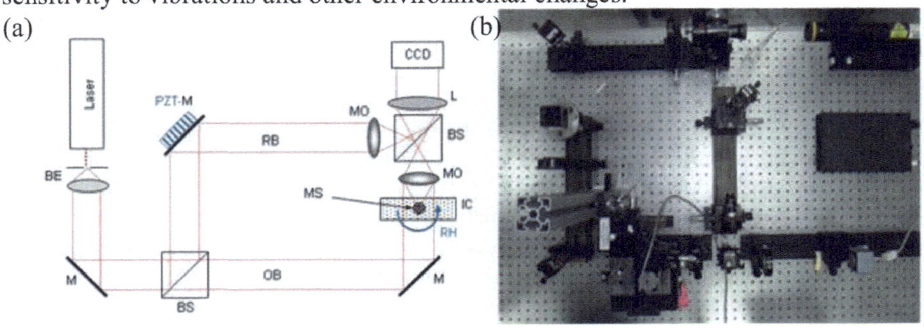

Fig. 1. Tomographic system based on Mach-Zehnder interferometer scheme (a) and its photo (b); BE – beam expander, M – mirrors, BS – beam splitters, MO – microscope objectives, PZT-M – mirror mounted on PZT drive, L – lens, IC – cuvette with immersion oil, RH – rotational holder, MS – measured sample, CCD – camera sensor, OB – object beam, RB – reference beam

2.2 Shearing interferometer based tomographic system

The most important modification of the above tomographic system was to change the configuration by adding the shearing interferometric module after the object. In the modified system (Fig. 2) the object was illuminated by a plane wavefront created by a collimator – the beam was radiated from an optical fiber placed at the focal length of a collimating lens (f = 100 mm). After passing through an object submerged in a cuvette with immersion liquid, the beam entered a telecentric imaging system composed of the microscopic objective (20x, NA 0.6) and lens (f = 200 mm), providing the magnification matched to the expected field of view (350 × 260 μm) and size of the CCD matrix (4.4 × 4.4 μm pixel size). Next, the object beam entered compact Michelson interferometer. The reciprocal tilt of the mirrors M produced shear of the two object beams, which interfered and the interferogram was captured by the CCD camera. The amount of shear was significant, so that the area of the beam modified by a fiber interferes with the unmodified part of the beam. Similar solution was already implemented for biological cells investigations in holographic microscopes [6].

The new system was versatile and enabled to perform measurements in workshop conditions due to greatly improved mechanical stability. Additionally this system might work with two different cuvettes (Fig. 2c,d) in: vertical configuration enabling measurements of optical fibers without the need to cut them, and horizontal configuration for investigations of fiber accessories such as optical fiber microtips [2].

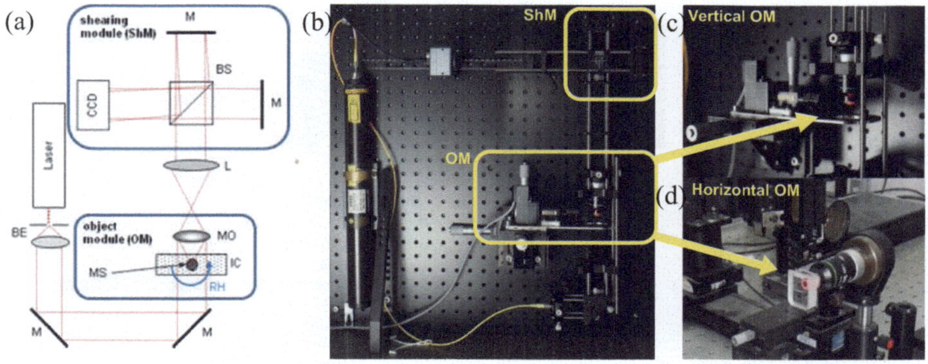

Fig. 2. Tomographic system based on shearing interferometry scheme (a) and its photo (b), also presented with the object module for vertical (c) and horizontal (d) configuration

2.3 Tomographic measurement methodology

Measurement methodology for tomographic reconstruction using Mach-Zehnder interferometer based system was already reported in [3,4]. Here, we present it with the modifications required for implementation of data obtained in the shearing interferometer configuration (Fig. 3).

The set of interferograms was captured for several angular position of an optical fiber. The interferograms were analyzed by means of Fourier Transform algorithm. This solution sped up the data capture procedure and it was very efficient for workshop version of tomographic system. Later the phase mod(2π) was unwrapped using a fast two-dimensional phase-unwrapping algorithm based on sorting by reliability following a noncontinuous path (2D-SRNCP) [7].

These integrated phases might be corrupted by optical system aberrations due to interference of an object wavefront with its replica at the regions in which we theoretically assumed to have an ideal plane wavefront. Therefore it was necessary to consider subtraction of the integrated phase without any object in the optical path from the phases obtained for the sequential projections. Also background normalization algorithm was used for each integrated phase image and from the finally corrected phases sinograms were calculated. Additional improvement of the data was achieved by applying run-out error correction, both for lateral shift and axial shift [8,9].

For the tomographic reconstruction of the refractive index distribution, 360 projections with angular step $\Delta\alpha = 1°$ were captured and used to create sinograms, which were later reconstructed by means of hybrid filtered backprojection, as reported in [8]. Finally, the phase difference values were scaled to the refractive index values $n(x,y,z)$:

$$n(x,y,z_i) = \frac{\Phi(x,y,z_i)\lambda}{2\pi d} + n_0 \tag{1}$$

where $\Phi(x,y,z)$ is the phase difference calculated for layer z_i, d is the thickness of the layer z_i, which depends on the dimension of the CCD's pixel and magnification, and n_0 is the refractive index of the immersion liquid.

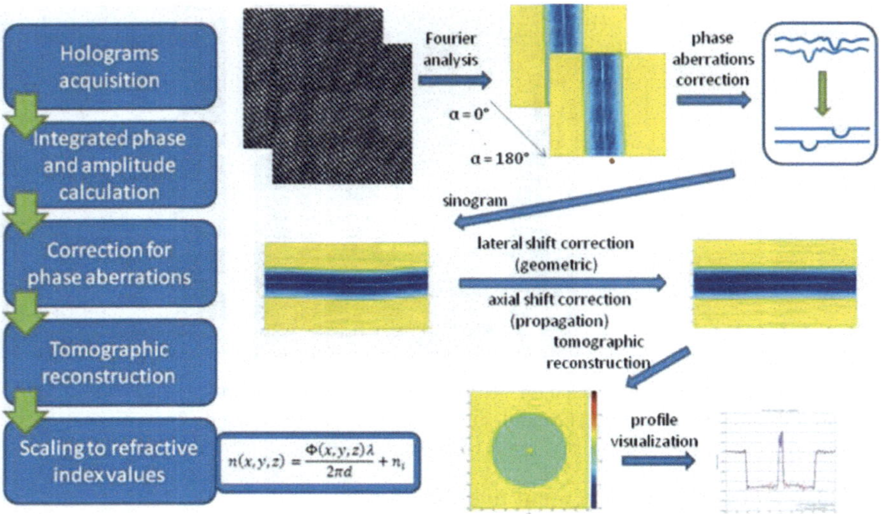

Fig. 3. Tomographic process reconstruction scheme for shearing interferometer based setup

2.4 Wavefront analysis

As indicated in Section 2.3, it was necessary to measure the quality of the wavefront W (beam without an object) entering the shearing module. This can be done using the shearing module present in the system. When a small shear S was introduced in x and y directions [10], the two interferograms obtained sequentially provided the information about the first derivatives of the wavefront in those directions:

$$\frac{\partial W(x,y)}{\partial x(or\ \partial y)} S = n\lambda \tag{2}$$

After retrieving the first derivatives in x and y directions, the phase representing aberrations of the system can be determined through integration of the sum of the derivatives [10]. Usually, iterative least-square minimization or noniterative complex plane integration algorithms are used, but these methods, although fast, usually produce a boundary error [11]. Here, the finite difference method was used for numerical integration. It does not have the boundary error problem, but it is slower than Fourier Transform based methods. However, the data for the phase correction were taken once for the given setup, so the calculation speed should not pose a problem.

3 Results

The measurements of the wavefront W were performed in the shearing system by means of capturing x and y first derivatives of the wavefront and finally integrating them [10]. The procedure was performed for two cases: the system with and without the object cuvette with the immersion oil. The shear angle was around 2°, which was small enough to assume that interferograms represented the derivatives of wavefront, but also to introduce sufficient spatial carrier fringes to allow implementation of the Fourier Transform method of the fringe pattern analysis. The reconstructed wavefronts presented in Fig. 4 are in parts of the wavelength.

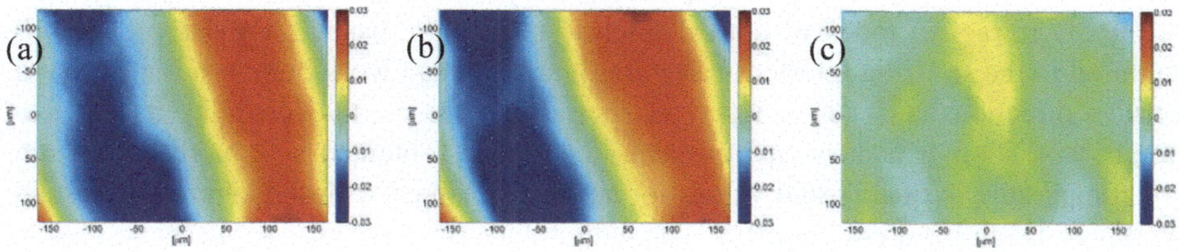

Fig. 4. The integrated wavefronts representing aberrations of the system without the cuvette (a), with a cuvette (b), and the wavefronts difference (c)

The aberrations in the resulting wavefronts in both cases were approximately 1/20 of the wavelength (Fig. 4a,b), which indicated good quality of the system. However in order to eliminate the wavefront systematic error, the phase map may be used for the integrated phase data correction. The aberrations introduced by the cuvette with the immersion oil are negligible, as can be seen in Fig. 4c.

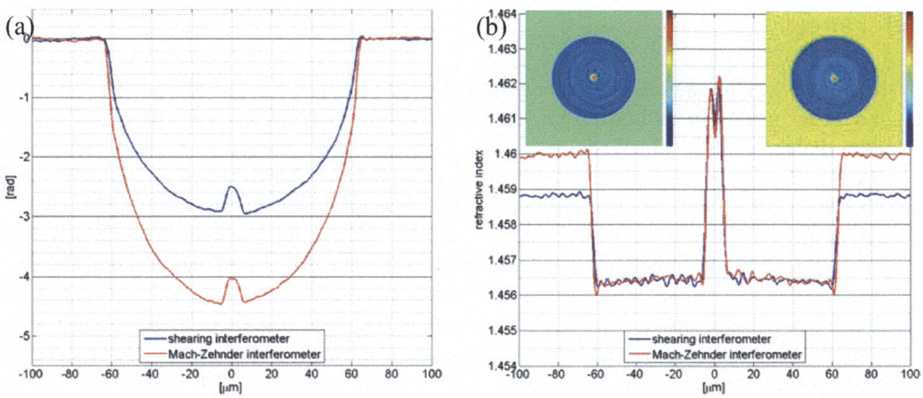

Fig. 5. Comparison of integrated phase (a) and refractive index (b) profiles together with inset of 2D refractive index distribution for different setups (the same scale of the size and the refractive index represented as color) – blue lines for shearing interferometer (left inset) and red lines for Mach-Zehnder interferometer (right inset)

For comparison of the results obtained in the laboratory, Mach-Zehnder interferometer based setup and shearing interferometer based setup, a single-mode fiber G.652 was chosen. The presented results (Fig. 5) show the example integrated phase and refractive index profiles averaged through the 100 μm of the fiber's axis.

As can be clearly seen, the profiles of the refractive index for optical fiber measured in both setups are in a good agreement with each other. The only significant difference, specifically observed for the integrated phase values (Fig.5a), was due to the change of the refractive index value of the immersion oil. The outside temperature during both experiments was different and it caused the change of $\Delta n_0 = 0.0012$ (see Eq.1). This fact indicates strong requirement to monitor the actual temperature during workshop measurements and provide the appropriate corrections for the final results.

4 Conclusions

We have developed and tested the integrated system for tomographic investigations of the 3D refractive index of optical fibers. The system, based on phase projections captured in the shearing interferometer configuration, can work both in horizontal and vertical directions and it is resistant to external conditions, thus enabling its application in workshop or factory environments. The results obtained with modified system are comparable in quality and values to the ones obtained with the laboratory system working in stabilized environment. The output wavefront analysis suggests negligible influence of the cuvette with immersion oil on the results, however special attention have to be paid to monitor the temperature in which measurements are performed.

This work was realized within the program TEAM/2011-7/7 of Foundation for Polish Science, co-financed from the European Regional Development Fund, also by statutory funds of Warsaw University of Technology and within the project financed by the Dean of the Faculty of Mechatronics, Warsaw University of Technology.

References

1. Dasch, C. J.: One-dimensional tomography: a comparison of Abel, onion-peeling, and filtered backprojection methods. Appl. Opt. 31(8), 1146-1152 (1992)
2. Pura, P., Szymański, M., Wiśniowski, B., Dudek, M., Jaroszewicz, L. R., Napierała, M., Spadło, A., Marć, P., Kujawińska, P.: Polymer microtips fabricated at the extremity of photonic crystal fibers. Journal of Materials Science and Eng. B 3(12), 749-758 (2013)
3. Kujawińska, M., Parat, V., Dudek, M., Siwicki, B., Wójcik, S., Baethge, G., Dahmani, B.: Interferometric and tomographic investigations of polymer microtips fabricated at the extremity of optical fibers. Proc. SPIE 8494, 849404 (2012)
4. Dudek, M., Kujawińska, M., Parat, V., Baethge, G., Michalska, A., Dahmani, B., Ottevaere, H.: Tomographic and numerical studies of polymer bridges between two optical fibers for telecommunication applications. Opt. Eng. 53(1), 016113 (2014)
5. Yablon, A. D.: Multifocus tomographic algorithm for measuring optically thick specimens. Opt. Lett. 38(21), 4393-4396 (2013)
6. Kemper, B., Vollmer, A., Rommel, C. E., Schnekenburger, J., von Bally, G.: Simplified approach for quantitative digital holographic phase contrast imaging of living cells. J. Biomed. Opt. 16(2), 026014 (2011)
7. Herráez, M. A., Burton, D. R., Lalor, M. J., Gdeisat, M. A.: Fast two-dimensional phase-unwrapping algorithm based on sorting by reliability following a noncontinuous path. Appl. Opt. 41(35) 7437-7444 (2002)
8. Kostencka, J., Kozacki, T., Dudek, M., Kujawińska, M.: Noise suppressed optical diffraction tomography with autofocus correction. Opt Express 22(5), 5731-5745 (2014)
9. Kostencka, J., Kozacki, T., Dudek, M., Kujawińska, M.: Time efficient method for defocus error compensation in tomographic phase microscopy. Photonics Lett. Pol. 6(3), 102-104 (2014)
10. Malacara, D., Servin, M., Malacara, Z.: Interferogram Analysis for Optical Testing (Second Edition). CRC Press, Boca Raton (2005)
11. Bon, P., Monneret, S., Wattellier, B.: Noniterative boundary-artifact-free wavefront reconstruction from its derivatives. Appl. Opt. 51(23) 5698-5704 (2012)

Holographic cameras for active 3D data capture

Michał Józwik, Jakub Żak, Małgorzata Kujawińska

Institute of Micromechanics and Photonics, Warsaw University of Technology, Warsaw, Poland

{m.jozwik, m.kujawinska}@mchtr.pw.edu.pl, jjacobzak@gmail.com

Abstract. In this paper the new approach towards 3D holographic data capture systems is presented. An active holocamera configurations designed for imaging and characterization of engineering objects are discussed. Two designs of holocameras with extended imaging and measurement capabilities are presented. The first one includes a spatial light modulator allowing for active forming of a reference beam. The second configuration is based on simultaneous capturing of an object beam by 2 cameras in order to extend the captured field of view. The performance of those configurations is illustrated with exemplary results.

Keywords: digital holography· holocamera · spatial light modulator· extended field of view

1 Introduction

The imaging and metrology based on the digital holography (DH) has been developing together with the development of CCD and CMOS detectors. DH allows to register in digital form the full-field three-dimensional (3D) amplitude and phase information of the object. Registered digital hologram can be numerically reconstructed on the computer or optoelectronically in the holographic display. The measured phase contains the information about the shape of the object under test or its deformation after application of load. DH as robust measurement technique offers new possibilities for optical metrology especially in the material science, bioengineering or MEMS technology [1]. During last decade the different approaches to develop DH systems have been proposed [2,3]. Part of those solutions, simple opto-mechanical setups named holocameras, are of highest interest for mechanical elements testing. The main requirements for holocamera are: simple and low cost design, compact construction with low sensitivity to vibrations and temperature changes, robustness and portability, quick preparation for measurement, usage in various applications with the high quality of output data. However, due to limited size and resolution of matrix detector (when compared with classical holography) the angular field of view of an object is significantly limited. To enhance this feature in recording system the synthetic aperture technique was implemented [4]. In other hand, for holographic display the multi-SLM display designs have been proposed [5]. Such ideas can also be implemented in holocamera construction. This paper describes two holocamera configurations de-

R. Jabłoński and T. Brezina (eds.), *Advanced Mechatronics Solutions*, Advances in Intelligent Systems and Computing 393, DOI 10.1007/978-3-319-23923-1_78

signed for characterization of engineering objects. The possible working modes and exemplary results are discussed.

2 Registration and reconstruction process in DH

The hologram is created due to the interference of the object and reference waves. The reference wavefront can be plane (Fig. 1a) or spherical. When the point source of the spherical reference wave is located in the plane of the object, then the setup is known as lensless Fourier holography (Fig. 1b). Another existing configurations with asymmetrical position of the object towards detector and the tilted reference beam illumination were not considered.

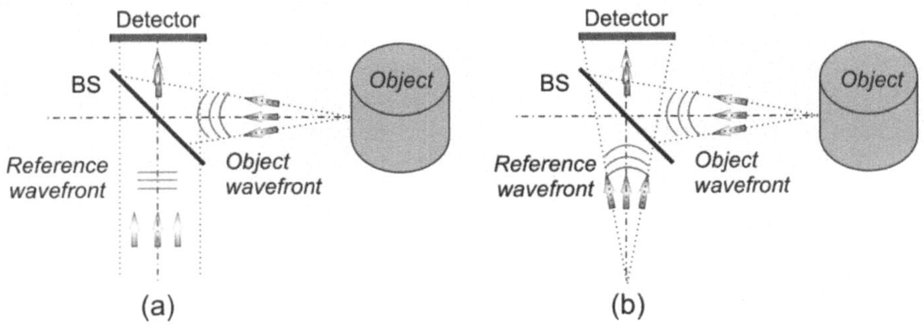

Fig. 1. Configurations of the registration setups: a) Fresnel and b) Fourier holograms

For recording of a hologram by a digital camera, the sampling theorem states, that every interference fringe must be recorded at least by two pixels of the camera. The camera registers an intensity distribution of the interference pattern I(x,y) converted into a two-dimensional array of discrete signals. Resulting image I_{DH} is modified by the parameters of camera and both - sampling and quantization process:

$$I_{DH}(x,y)=\gamma_C\left(\frac{1}{ab}comb\left(\frac{x}{a},\frac{y}{b}\right)*\Pi\left(\frac{x}{\Delta x},\frac{y}{\Delta y}\right)\right)I(x,y)t\,, \tag{1}$$

where γ_C is a characteristic parameter of camera, a,b – camera resolution in x,y direction, $\Delta x, \Delta y$ – dimension of a pixel, t – acquisition time.
The sampling criteria and illumination wavelength λ determine the acceptance angle θ for a camera sensor used in registration process:

$$\theta_x \le \frac{\lambda}{2\Delta x} \text{ or } \theta_y \le \frac{\lambda}{2\Delta y}\,. \tag{2}$$

This allow to calculate the maximal dimension of the object D_{max} in the function of distance d from detector:

$$D_{x\max} = \frac{\lambda d}{\Delta x} + a\Delta x \text{ or } D_{y\max} = \frac{\lambda d}{\Delta y} + b\Delta y. \quad (3)$$

In DH method the reconstruction of the wavefield is performed by numerical methods. The numerical reconstruction depends on the capture configuration and can be achieved by different methods (Fourier or Fresnel transformation, convolution approach) [4]. The amplitude and phase of an object wave can be reconstructed from a single hologram or from a set of holograms with phase-shifting algorithm [5]. The advantage of phase-shifting digital holography is that the reconstructed image is free from a zero-order and a conjugate image terms.

3 Configurations of holocamera

The laboratory version of holocamera is set on the vibration isolated table to assure mechanical stability. The beam radiated from the pigtailed Nd-YAG laser (λ = 532 nm) is collimated by lens (f = 100 mm). A plane wave is polarized by a polarizer P, adjusted by a half-wavelength plate HP and divided by a beam splitter BS into a reference and an object beam.

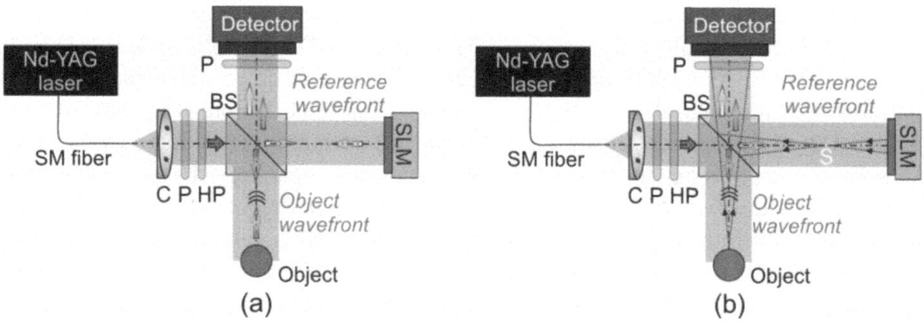

Fig. 2. The scheme of holocamera with active modification of reference wavefront: (a) flat reference wavefront, (b) spherical reference wavefront.

In the system presented in Fig. 2 the reference wavefront is created by reflection from the spatial light modulator SLM (HOLOEYE Photonics AG, model Pluto, reflective phase only Liquid Crystal on Silicon microdisplay resolution 1920 x 1080, 8 μm pixel size). The use of an active SLM allows for electronic control of the phase shifting and active manipulation of the radius of the reference wavefront. In order to capture Fresnel holograms (Fig. 2a) SLM acts as a flat mirror, reflecting the incoming plane wavefront and introducing required phase shifts. The system can be easily transformed to capture Fourier holograms (Fig. 2b) by creating a point source S. Here SLM is programmed to change the incoming plane wavefront into a spherical one. This creates very compact and versatile tool for capturing holograms of different types of objects. After reflections/scattering from reference and object surface both beams are directed to the detector (CMOS camera, Pixelink B781, resolution: 2208 x 3000, pixel size: 3.5 μm x 3.5 μm). The acceptance angle θ for this detector is 4.35°.

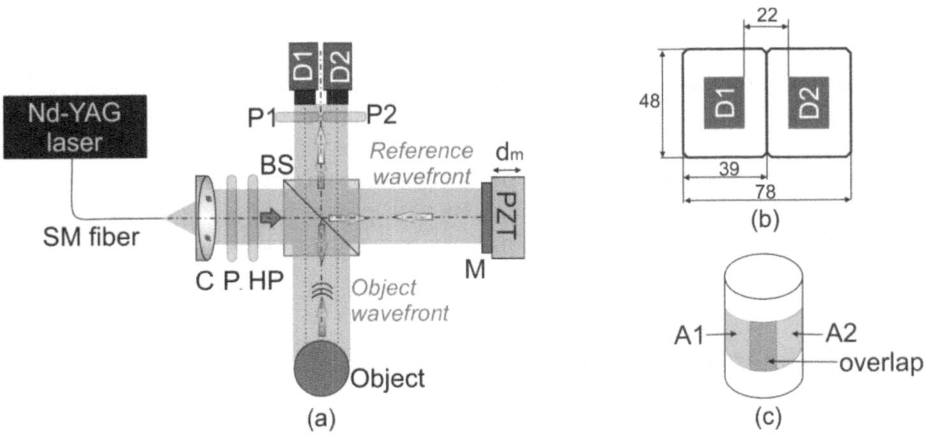

Fig. 3. The holocamera with two detectors: (a) schematic of system for Fresnel holography, (b) dimensions of sensors, (c) object with an indication of areas measured by detectors.

In the system shown in Fig. 3 the plane reference wavefront is created by reflection from the mirror M. The M is mounted on PZT translator allowing to introduce the discrete displacements d_m for the phase shifting technique. The interfering beams are registered by the module including two CMOS cameras D1, D2 (Pixelink B781) and polarizers P1,P2. As CMOS are board level type they can be mounted on the same platform with minimal distance between imager sensors of 22 mm (Fig. 3b). In fact D1 and D2 register the data from areas A1 and A2 with an overlap region (Fig. 3c). Actually the processing of captured data is performed separately. However the two captured individual wavefields can be used to calculate an object wavefield coherently assembled from the captured data. Reconstruction of such synthetic aperture hologram will enable to measure objects with extended size. In contrast to classically captured synthetic holograms [4], the hologram does not need numerical stitching and can be captured in a simple setup of holocamera.

4 Experimental results

Figures 4 and 5 show the initial results of numerical reconstruction of the intensity from holograms recorded with systems described in the previous section. The reconstructions were obtained by means of commercial software *Fringe Processor*. The first object was a coin with a diameter of 19.5 mm, which results in minimal distance between object and camera of 77.45 mm (see Eq. 3). In holographic camera with SLM (see Fig. 2a) the holograms of the object were recorded at a distance 275 mm. The object was fully illuminated and in the reconstruction process the image of the whole object was obtained (Fig. 4a). Due to the precise phase shift introduced by SLM the zero order has been nearly completely eliminated and a good quality image was obtained. In case of Fourier holography setup (see Fig. 2b) the spherical reference wavefront was created by introducing mod(2π) phase distribution simulating Fresnel lens with a focal distance 254.56 mm. This distance was matched actively to virtually place the reference source in the object plane. During the reconstruction the image of an object, the conjugate image and zero order beams are placed in the same plane (Fig. 4b). In order to properly reconstruct object intensity and phase, the object position have to be slightly offset from the optical axis of the camera system.

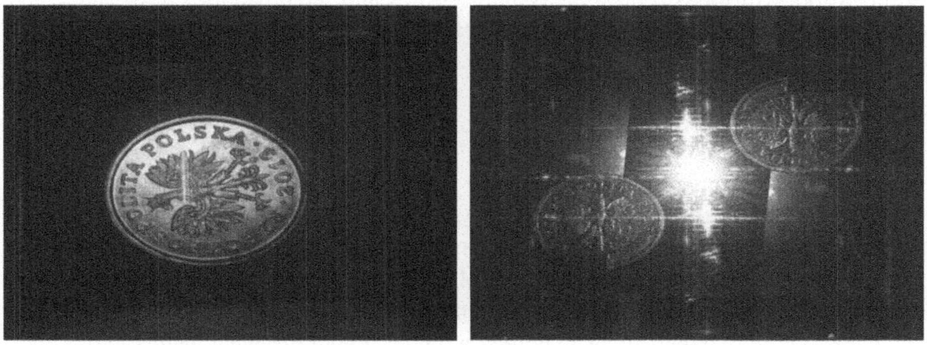

(a) (b)

Fig. 4. The results obtained from the holocamera with SLM: intensity images of the coin reconstructed from: (a) Fresnel hologram, and (b) Fourier hologram.

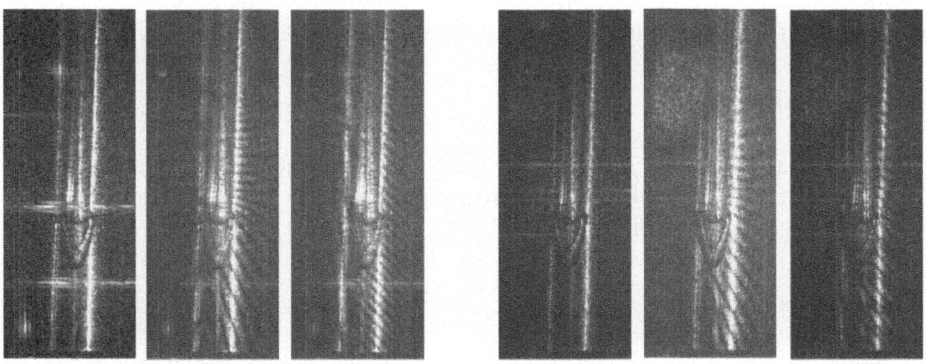

(a) (b)

Fig. 5. Results from holocamera with two detectors registered by: (a) D1, (b) D2.

The interpretation of results is more complicated in the case of holographic camera with two detectors. The object under test was a metal pen cap. The intensity reconstructions at the first photo of Fig. 5a and b show that the object is observed from detectors under different angles. For the reconstruction distance of 445 mm the difference of angles was equal of 7°. These angles can be modified by adjustment of distance between the cameras and by change of an object position. However, with modification of that distances the overlap region of data will be changed as well. To prove the system functionality in the digital holographic interferometry mode the object was heated by hot air from right side. The fringes visible on reconstructions at the second and third photo of Fig. 5a and b indicate the changes in the object shape due to heating. Observed differences in images quality of Fig. 5a and b are connected with different inclination of cameras D1, D2 and polarizers P1, P2. The quality of registered holograms is acceptable, but for the metrological tasks need to be improved. Different artifacts appearing in reconstructions will be limited by better adjustment of the ratio between object and reference beams and elimination of unwanted reflections. The future works will focus on testing system settings and the development of shape and out-of-plane displacement measurement methodology for technical objects.

5 Conclusions

A new configurations of holocamera have been introduced that offer improved performance and flexibility in the registration scheme. The use of an active SLM device that controls and modifies the reference wavefront will dramatically increase the system utility in different imaging and metrology tasks. Future experiments with holocamera will prove possibility of measurements for different types of objects. The holocamera with two detectors can deliver a bigger hologram, what results in possibility of testing of bigger objects and increased viewing angle. After solution of the problem with the quality of reconstructions the algorithms that allow assembly of separated object wavefields will be tested. The future plan is to combine functionality of an active reference beam modification by SLM and multi-detector registration in one system of an active, multifunctional holocamera.

Acknowledgements

This work was financially supported by the National Science Centre through project 2011/02/A/ST7/00365 realized within program MAESTRO and by statutory funds of Warsaw University of Technology.

References

1. Osten W.: Optical inspection of Microsystems. Taylor and Francis, New York (2006)
2. Singh V.R., Liansheng S., Asundi A.: Compact handheld digital holographic microscopy system development. Proc. of SPIE 7522, 75224L (2010)
3. Michałkiewicz A., Kujawińska M., Stasiewicz K.: Digital holographic camera and data processing for remote monitoring and measurements of mechanical parts. Opto-Electron. Rev. 16, 68–75 (2008)
4. Makowski P.L., Kozacki T., Zdankowski P., Zaperty W.: Synthetic aperture Fourier holography for wide-angle holographic display of real scenes. Appl. Opt. 54(12), 3658-3665 (2015)
5. Kozacki T., Kujawinska M., Finke G., Hennelly B., Pandey N.: Extended viewing angle holographic display system with tilted SLMs in a circular configuration. App. Opt. 51, 1771-1780 (2012)
6. Schnars U., Juptner W.: Digital recording and numerical reconstruction of holograms. Meas. Sci. Technol. 13, R85-R101 (2002)
7. Yamaguchi I., Zhang T.: Phase-shifting digital interferometry. Opt. Lett. 22, 1268-1270 (1997)

Accurate DHM method for topography characterization of reflective microoptics

Marta Mikuła[1], Tomasz Kozacki[1]

[1] Warsaw University of Technology, Institute of Micromechanics and Photonics

8 Św. Andrzeja Boboli St., 02-525 Warsaw, Poland

Abstract: In this paper the new measurement method for topography characterization of reflective microoptics is presented. The proposed measurement method is based on digital holographic microscope (DHM) system utilizing spherical wave illumination. The method permits to obtain full characterization of shape parameters of high numerical aperture (NA) microstructures such as micromolds [1]. For numerical reconstruction of topography of high NA micromolds modified Spherical Local Ray Approximation (SLRA) algorithm has been employed. We present results of topography reconstruction of triangular reflective micromolds.

Keywords: digital holographic microscopy, topography reconstruction, microoptics

1 Introduction

Nowadays, microlenses have a wide range of applications, for instance: in optical sensors, medicine and communication systems. Due to the application of focusing microoptics in many fields of technology, accurate control of their topography is extremely important. For noncontact quality control the interferometric and Digital Holographic Microscopy (DHM) techniques are used, because of the ability to capture full field phase distribution of measuring microelements. On the other hand, in many cases the numerical aperture (NA) of measured microobjects is larger than the NA of DHM system, hence the full characterization of high NA microoptics cannot be performed.

In DHM configuration for quantitative measurements two different types of illumination are usually employed: plane wave [2-3] or spherical wave [4-6]. When the high NA microobject is illuminated by the normal plane wave and the measurement setup has insufficient NA, the total topography characterization cannot be obtained. Recently, new approaches with inclined plane wave have been developed [7,8], which theoretically permit to obtain the accurate topography of microoptics. However, to obtain full topography several measurements are necessary. Furthermore, the numerical propagation and shape merging procedures is needed for merging and conversion of wavefronts into the final shape [8]. Moreover, the micromolds are very demanding sample to measure, due to the fact that they are the objects with high frequency spatial phase, and there is no clear position of sharp imaging reference plane (IRP).

To overcome the problem of insufficient NA of the optical system, we apply the most optimal configuration with a spherical wave illumination in Twyman-Green

R. Jabłoński and T. Brezina (eds.), *Advanced Mechatronics Solutions*,
Advances in Intelligent Systems and Computing 393,
DOI 10.1007/978-3-319-23923-1_79

interferometer. When the point source is placed in the center of curvature of the object surface, it is possible to capture the wavefront corresponding to the highest NA of microelement. This method permits to increase the total NA of the DHM system, due to spherical illumination and it allows to use an accurate non-paraxial Spherical Local Ray Approximation (SLRA) algorithm for shape calculation [9]. The capabilities of SLRA algorithm provide accurate reconstruction and characterization of focusing reflective microoptics with a condition of precise knowledge about location of IRP and point source. Calibration process presented in [9] allows obtaining accurate determination of IRP only in the case when investigated microelements possess highly distinctive substrate. However, for many microobjects (such as microelements with high fill factor) substrate surface is indistinguishable and inaccessible. Hence, such calibration method and SLRA algorithm [9] cannot be applied.

In this paper, for the first time to the best of our knowledge, we present the measurement method for accurate topography characterization of reflective microstructures with point source object illumination. In our work we apply modified SLRA algorithm, which for accurate results of shape reconstruction requires accurate calibration of the position of the IRP and illuminating point source. In this article we develop method, which allows for accurate determination of the IRP position and point source without need for a substrate surface. Developed calibration method is based on the mutual control of the position of the focal point of the sample, the bottom of the sample and position of CCD camera.

This work is organized as follows: Section 2 presents the experimental DHM system in reflective configuration with a point source illumination; Section 3 presents developed measurement method for obtaining parameters for topography reconstruction; Section 4 presents experimental results of measured triangular micromold.

2 Measurement setup

Fig. 1. presents the experimental system of DHM utilizing the Twyman-Green configuration. He-Ne laser is generating a monochromatic ($\lambda = 632.8$ nm) coherent and linearly polarized beam. The beam is filtered by a spatial filter (SP) and collimated by lens (C). Then a plane wave passes through a polarizing beam splitter (PBS1) and is divided into an object and a reference beam. The reference beam is reflected from a mirror (M1) mounted on piezoelectric transducer (PZT). Then, it passes PBS1 again, PBS2, polarizer (P) and is captured by CCD camera. The object beam is reflected from mirrors M2 and M3 then passes through the microscope objective (MO1), mounted on a 100 mm range motorized linear stage. The application of the linear stage allows to accurate determination of the position of the point source illumination. The spherical wave is passing through the second polarizing beam splitter (PBS2) and an afocal imaging system consisting of a lens (S200) and a microscope objective (MO2, Mitutoyo, NA = 0.42, 20x). The object beam is reflected from the tested micromold and comes back to the PBS2. Then, the reference beam and the object beam are interfering. The interferogram is acquired by the CCD camera.

The magnification of our afocal system, which conjugates the CCD camera with IRP, is 22x. The CCD detector (resolution 2456 x 2058, pixel size 3.45) is mounted on a precise motorized linear stage, which provides accurate determination of the object plane. This setup can work in two configurations: with plane wave illumination and spherical wave illumination.

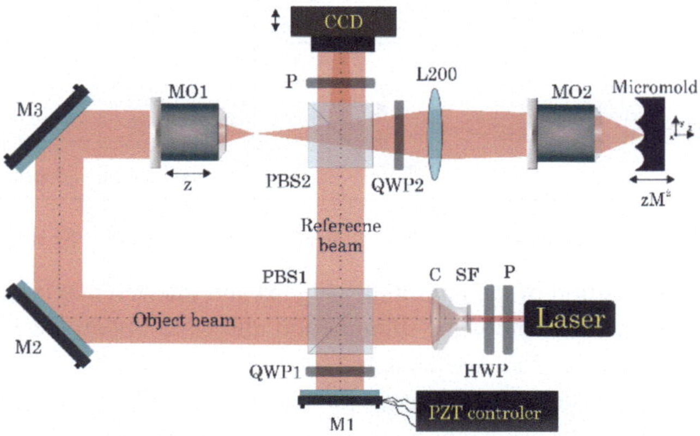

Fig. 1. Optical scheme of DHM system.

3 Measurement method

SLRA algorithm allows to calculate topography reconstruction of microelements using equation [9]:

$$z(x + x_s) = -CB^{-1}, \quad (1)$$

where coefficients: $B = -2z_0 + 2(x - x_0)\tan u' + 2\sec u'(Rn + [NM])$, $C = z_0^2 + (z - x_0)^2 - (R - [NM])^2$, the x_0 is the location of the point source, the radius of the illumination wave $R = (z_0^2 + (x_P - x_0)^2)^{1/2}$ and $[NM] = -\int_P^N \sin u' \, dx$. The transverse shift of the ray $x_s = Z\tan u'$, where u' is calculated from the phase of the hologram, which is captured on the IRP position, $u' = \sin^{-1}(\varphi'/k_0)$, where φ' is a transverse derivative of the measured phase.

Accurate topography measurement is possible only when the position of the IRP and point source are known. In this Section method for calibration of these parameters is showed. The calibration method is based on the determination of the position of the sample in relation to the generated point source. Moreover the microelements with the highest NA can be measured when the point source coincides with the center of curvature of the measured surfaces. Thus the calibration method provides such conditions. The next steps of the calibration method are presented in the table on the Fig. 2.

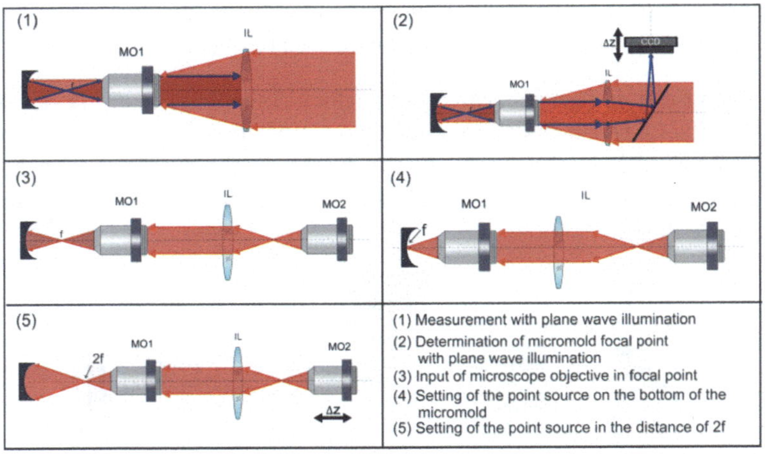

Fig. 2. Steps of the calibration method of DHM system required for SLRA algorithm.

- Step 1 – acquisition of the amplitude (Fig. 3) and interference images with plane wave illumination. Plane wave illumination allows to obtain the sharp and clear edges of the measured microelements.
- Step 2 – movement of the camera until the focal point of the testing objects is found. In order to determine the focal point of the microobject we use the algorithm based on the full width at half maximum (FWHM) parameter. The image and cross-section is presented on Fig. 5 and Fig. 6.
- Step 3 – the point source is placed in the focal point of microobject; the microscope objective (MO1), which is used to generate a point source illumination, is inserted into the object beam. When the interference image is nulled, it means that we are in the focal point of the tested object.
- Step 4 – the point source is imaged at the bottom of the sample. For this purpose, an axial movement of the object is performed; the distance, by which the test object has to be moved to the position of the point source is a focal length (f). In this arrangement the CCD camera and IRP are conjugated in focal point of the sample.
- Step 5 – the point source is placed at the double focus (2f) distance from measured micromold. In this arrangement the CCD camera and IRP are conjugated in on the bottom of the microobject. The final amplitude (Fig. 4) and interference images with spherical wave illumination are acquired.

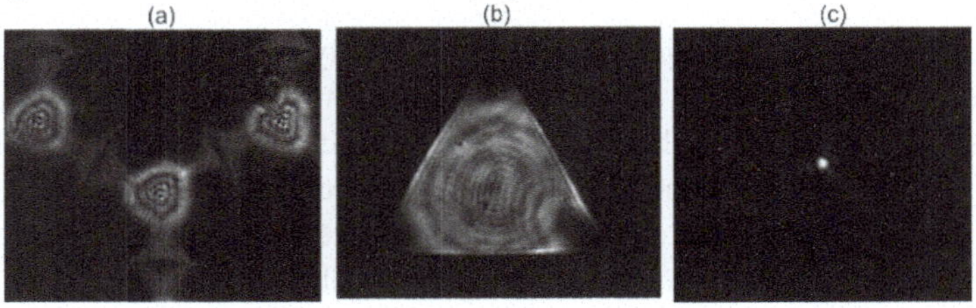

Fig. 3. (a) Amplitude image of the sample illuminated by plane wave, (b) amplitude image of the sample illuminated by spherical wave, (c) focus point of the sample.

4 Experimental results

To obtain topography reconstruction the SLRA algorithm has been used. The reflective triangular micromold, with edge length = 69.39 μm was measured. In Fig. 7 the measurement results of high NA microobject are illustrated. Fig. 7a presents interference pattern, Fig. 7b the results of topography reconstruction of the sample. Cross-sections are presented on Fig. 7c and Fig. 7d. Maximum height of the reconstructed topography of the micromold is 12.52 μm, edge length = 63.08 μm, NA = 0.39. According to the calibration method focal length is 24 μm. The obtained edge length varies from the value supplied by the micromold matrix manufacturer. This may be due to the chemical etching process of monocrystalline silicon [9].

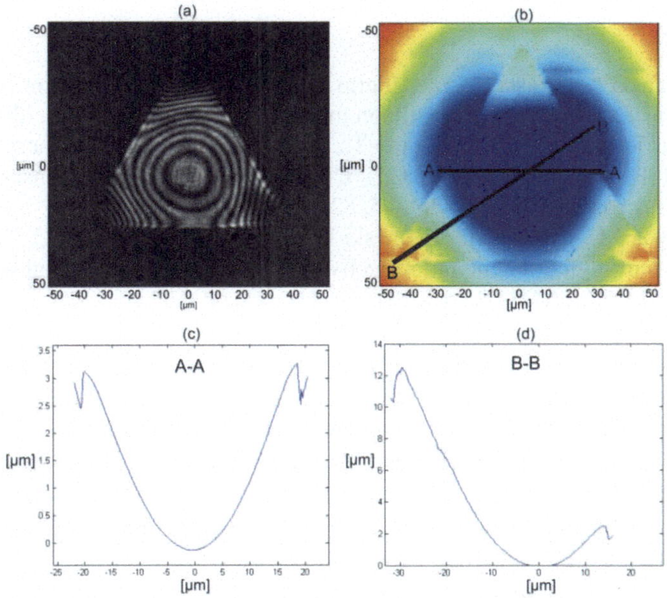

Fig. 4. (a) Interferogram, (b) topography reconstruction of the measured micromolds (edge length = 69.39μm), (c) cross-section A-A, (d) cross-section B-B.

5 Conclusion

In this article, the new measurement method, based upon a DHM system in reflective configuration with spherical wave illumination arrangement, has been used to obtain topography reconstruction of reflective microstructures. Through configuration with spherical wave illumination the phase of the entire area of measured high NA microobject can be captured. We have measured and presented results of topography reconstruction of reflective microobject with NA up to 0.4. The measurements were possible due to application of developed here calibration method, which allows to obtain accurate position of IRP, which is crucial in obtaining accurate results of topography reconstruction. Moreover the calibration method sets the illuminating point source in the center of curvature of the measured surface optimizing measurement system NA; only with such a setting it is possible to obtain measurement of entire topography. Presented results confirm that developed measurement method in DHM system and application of SLRA algorithm provide topography reconstruction of reflective micromolds with a very high precision.

Acknowledgments

The research leading to the described results is realized TEAM/2011-7/7 of the Foundation for Polish Science, co-financed by the European Funds of Regional Development and Faculty of Mechatronics Dean's Grant of 2015. We acknowledge the support of the statutory funds of Warsaw University of Technology.

References

1. Baranski M., Passilly N., Alberto J., Gorecki C., "Fabrication of 100% fill factors arrays of microlenses from silicon molds", Proc. of SPIE, 8428, (2014).
2. Charrière F., Kühn J., Colomb T., Montfort F., Cuche E., Emery Y., Weible K., Marquet P., and Depeursinge C., "Characterization of microlenses by digital holographic microscopy," Appl. Opt. **45**, 829-835 (2006).
3. Józwik M., Kozacki T., Liżewski K., Barański M., Gorecki C. "Topography measurements of high gradient and reflective micro-structures by digital holography," Proceedings of Fringe, 623-628, (2013).
4. Reichelt S. and Zappe H., "Combined Twyman-Green and Mach-Zehnder interferometer for microlens testing," Appl. Opt. **44**, 5786-5792 (2005).
5. Gomez V., Ghim Y.-S., Ottevaere H., Gardner N., Bergner B., Medicus K., Davies A., and Thienpont H., "Micro-optic reflection and transmission interferometer for complete microlens characterization," Meas. Sci. Technol. 20, 025901 (2009).
6. Kim M.-S., Scharf T., and Herzig H. P., "Small-size microlens characterization by multiwavelength high-resolution interference microscopy," Opt. Express 18, 14319-14329 (2010).
7. Kozacki T., Liżewski K., and Kostencka J., "Holographic Method for Topography Measurement of Highly Tilted and High Numerical Aperture Micro Structures," Opt. Laser Technol. 49, 38-46 (2013).
8. Józwik M., Kozacki T., Liżewski K., Kostencka J., "Digital holography with multi-directional illumination by LCoS SLM for topography measurement of high gradient reflective microstructures," Applied Optics, 54 (9), 2283-2288 (2015).
9. Kozacki T., Liżewski K., Kostencka J., "Absolute shape measurement of high NA focusing microobjects in digital holographic microscope with arbitrary spherical wave illumination," Opt. Express, 22, 16991 – 17005, (2014).

Control of the End Effector Position Based on Motion Capture

Vít Ondroušek[1]

[1] Mendel University in Brno, Department of Informatics, Brno, Czech Republic
vit.ondrousek@mendelu.cz

Abstract. This paper is focused on the design, development and testing complex human-robotic arm interaction system. This system should offer an intuitive control of the end effector of the robot arm to the operator. The designed solution allows to perform several main tasks, especially grasp, move and place objects in the working space of the manipulator. Used approach consists of tracking operator's palm, calculating configuration of the robotic arm with five DOF using an algorithm of inverse kinematics and performing high level commands through graphical user interface.

Keywords: manipulator control • end effector • motion capture • robotic arm

1 Introduction

The main aim of this issue is to design and implement complex human-robotic arm interaction system. This system should offer an intuitive control of the end effector of the robot arm to the operator. It is assumed that the manipulator and the manipulated objects are in the field of view of the human operator. The designed solution allows to perform several main tasks, especially grasp, move and place objects in the working space of the manipulator. A lot of attention was also paid to the design of the graphical user interface (GUI) of the control application. The GUI was redesign several times based on the user experience of the different human operators, with or without previous experiences with the control of the robotic arm.

Using the human motion and gestures for the control of robotic arm is still in the forefront of the scientific research in the fields of Human-robot interaction and Robots teleoperation. Various problems have to be solved to achieve stated goal of controlling the position of the end effector via tracking human motions: tracking movements of the whole body or upper limbs, mapping the movements of the selected parts or joints of the body to the movements of the drives, detecting selected gestures to perform special tasks (e.g. pause of operating or emergency stop) and implement high level optimization of the resulting trajectory. These issues where solved in different ways in last few years.

Authors of the paper [1] are focused on the design of the gesture based control system intended for the ad-hoc robotic arm. The movement of the human arm is captured

R. Jabłoński and T. Brezina (eds.), *Advanced Mechatronics Solutions*,
Advances in Intelligent Systems and Computing 393,
DOI 10.1007/978-3-319-23923-1_80

in 3D space, processed and replicated by the robotic arm. In fact, only two different angles are calculated using the tracked positions of three joints (hand, elbow, shoulder) of the operator's right hand. It means, that the robotic arm with three degrees of freedom (DOF) is controlled using two calculated values. The human arm has much more DOF, but these degrees are not fully independent. Therefore it is assumed, that the direct mapping of the joints to servo drives positions cannot be used for robotic arm with five or more DOF. The implementation of suitable method of inverse kinematics is inevitable.

Such approach is shown in [2]. In this case, the human-hand movement was used to control the electro-hydraulic manipulator with servo drives. The human posture recognition is also implemented using the Kinect sensor. The necessary positions of joints (shoulder, elbow and hand) are used to scale and compute inverse kinematic of the manipulator. The analytical calculation of the inverse problem is performed, because the robotic arm has only two DOF. Authors of this paper stated that the biggest issue occurred, when the operator's hand freeze while the recognized skeleton points were still changing in time.

Entirely different approach is shown in [3]. The full-body inertial motion capture suit Xsens MVN is used for the design of a complex human-robot interaction system. The Xsens MVN suit consists of seventeen inertial motion trackers attached to the body capable of providing real-time full body kinematic data (position, velocity, acceleration). The body center of mass and the right hand kinematic data of the operator are used to generate position and orientation references for the robot base and manipulator with five DOF, respectively. The left arm is employed to provide high-level user commands such as "Manipulator On/Off", "Base On/Off" and "Manipulator Pause/Resume".

2 Used Approach

2.1 Overall concept of the Solution

In our approach we have focused on the design of the intuitive control system of the robotic arm Katana HD300s with five DOF equipped with the standard gripper. The robot is placed into the protective cage [4] with controllable lighting to avoid any injuries. The human operator should be able to use the designed solution to perform pick and place tasks with objects placed inside the protective cage. The Kinect sensor is used for tracking the position of the operator. However, in opposite to the solutions mentioned in the previous chapter, we do not track the whole body of the operator but only the palm of the hand. Therefore, it is possible to replace the low cost Kinect sensor with any professional vibration resistant tracker with MEMS based gyroscopes, e.g. Xsens MTi-100, and thus achieve higher accuracy, if required. The tracked position of the palm is used as an input of the calculation of the inverse kinematics problem. The whole concept is shown on Fig. 1 and Fig. 3. The operator stands in front of the protective cage equipped with the robotic arm and Kinect sensor. The movement of the palm should be mirrored by the end effector of the robotic arm in the real time. It is also necessary to find appropriate way, how to process high level commands from the operator, e.g. open the gripper, grasp the object, pause or stop tracking operator's palm.

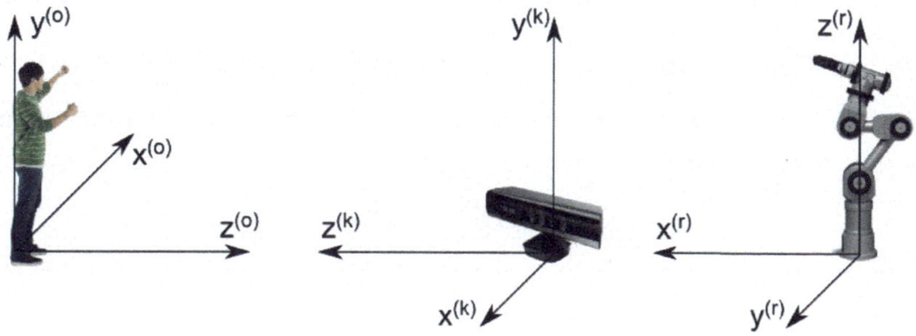

Fig. 1. Overall view of the designed concept

The manipulator and Kinect sensor are connected to the control PC, which runs the main designed software with special graphical user interface. This software reads data from the sensor and controls movements of the robotic arm. Whole program is implemented in the C# language. Only the low level routines, which provide control of five drives of the robotic arm, are implemented in C++ language. These routines are wrapped by the Visual C++ methods and deployed using the dynamic linked library. This approach allowed to develop complex solution, which uses sensory data and controls movements of the manipulator, in the C# language.

2.2 The Designed Method

The designed method reproduces movements of the operator by the robotic arm in the real time. This issue consists of several partial problems:

- sensing position of the operator's palm
- calculating desired position of the gripper
- calculating target positions of five drives
- processing high level commands

The position of the palm $H^{(k)}$ in the coordinate system k is obtained using the Kinect sensor and standard software developing kit. The information on position is updated thirty times per second. The obtained position of the operator's hand is used as an input for calculation of the desired position of the gripper. Let's denote $H_0^{(k)}$ as the known starting position of the operators hand in the coordination system k , and $G_0^{(r)}$ as the known starting position of the gripper in the coordination system r . The position of the operator's hand and the position of the gripper are linked together only once, when the operator selects the start command. All following positions of the gripper $G_i^{(r)}$, are calculated using equation (1).

$$G_i^{(r)} = M^{RotX}\left(\frac{\pi}{2}\right) \cdot M^{RotY}\left(\frac{\pi}{2}\right) \cdot M^{Trans}(\Delta) \cdot H_i^{(k)} \tag{1}$$

Where $H_i^{(k)}$ denotes new position of the operator's hand, M^{RotX} is well known rotation matrix along x axis, M^{Trans} is a transition matrix and Δ is distance between $H_0^{(k)}$ and $G_0^{(r)}$. Thus the equation (1) can be simplified, because both rotations, about x axis and y axis, are performed by ninety degrees:

$$\Delta = H_0^{(k)} - G_0^{(r)} \tag{2}$$

$$G_i^{(r)} = \begin{bmatrix} 0 & 0 & 1 & \Delta z \\ 1 & 0 & 0 & \Delta x \\ 0 & 1 & 0 & \Delta y \\ 0 & 0 & 0 & 1 \end{bmatrix} \cdot H_i^{(k)} \tag{3}$$

Once the new desired position of the gripper is obtained, it is possible to calculate new configuration of the robotic arm. It is necessary to implement an algorithm of the inverse kinematics problem for positions. The iterative calculation of the numerical inversion of the Jacobian was selected, [5]. Joint velocities $\dot{q}$ can be obtained using the inversion of the Jacobian matrix J, see equation (4). Joint positions can be computed by integrating velocities over time. Analytical integration can be avoid using numerical integration method. Once the joint positions and velocities at time t_k are known, the joint positions at time t_{k+1} can be computed according equation (5).

$$\dot{q} = J^{-1}(q) v_e \tag{4}$$

$$q(t_{k+1}) = q(t_k) + \dot{q}(t_k)\Delta t \tag{5}$$

Consider that the initial joint position and velocity are known in the described approach. Moreover, in the numerical implementation of (5) computation of joint velocities are obtained using the inverse of the Jacobian evaluated with the joint variables at the previous instant of time:

$$q(t_{k+1}) = q(t_k) + J^{-1}(q(t_k)) v_e(t_k)\Delta t \tag{6}$$

Using this calculation leads to the difference between the computed joint variables and the desired one. However, practical experience showed, that this error is negligibly small against the disturbances caused by the Kinect sensor, even after a small number of iterations of the algorithm.

The last resolved problem is processing of the high level commands by the operator. Whereas operator's right hand is used for controlling position of the manipulator, the left hand can be used for performing special high level commands. Four different commands were implemented so far: open the gripper, grasp the object, pause and stop tracking of the operator's palm. These commands are integrated into special circular menu, see Fig. 2. The graphical user interface is shown to the operator throughout performing the manipulation tasks. The center of this menu is placed into the elbow joint of the operator's left hand. Thus the menu is moving with an operator all the time. Moreover, the size of the circle is set up dynamically according to the size of forearm. This solution enables using the menu by different operators standing in the different distances from the manipulator. It also enables moving of the operator during the pick and place tasks. The selected high level command is performed whenever the operator places his hand over the colored circle, see Fig. 2c.

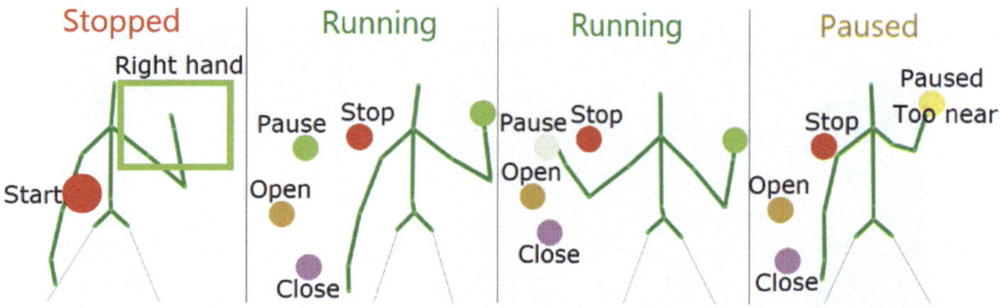

Fig. 2. Graphical user interface of the designed software

3 Achievements and Discussion

The described method of the end effector control was successfully implemented and dozens of experiments with twelve different human operators were performed. These experiments proofed, that the used approach is correct and the designed method can be used for performing pick and place tasks by tracking only the operator's palm. However, the real experiments showed several usability issues, which had to be solved.

At first place, insufficient accuracy of the palm's position sensing was partially solved applying the low pass filter to the data obtained from the Kinect sensor. Using this filter causes freezing of the end effector position, whenever the operator's hand freeze. However, it is necessary to use another type of tracking system, to obtain higher accuracy required in industrial solutions. Another problem was caused by several operators, when they lost their concentration and put their hands down during the tracking. This led to the collision between the end effector and the ground. Therefore the working area of the manipulator was restricted by incorporating the boundary conditions into the numeric calculations to prevent damage of the end effector.

Position of the operator's right hand and the position of the gripper are linked together, when the start command is performed by left hand. So that, it is necessary to prepare right hand into an appropriate position before selecting the start command. Especially untrained operators had problems with this requirement. This problem was solved incorporating restriction on right hand position at the start of tracking, see Fig. 2a. The tracking is paused, when the operator selects the pause command. It is also paused automatically, when the palm's position reaches working envelop or another operator steps in front of the working one. Operator has to place his palm in the last known position to enable further tracking, see yellow point on Fig. 2d.

4 Conclusion

The design and implementation of the complex human-robotic arm interaction system is discussed in this paper. Described solution consists of tracking operator's palm, calculating configuration of the robotic arm with five DOF and performing high level commands through graphical user interface. Dozens of experiments with different operators proofed suitability and correctness of the used approach.

Fig. 3. Real-time experiment

5 Acknowledgement

Special thanks goes to my colleague Josef Kolaja.

6 References

1. Hussein, M. A., Ali, A. S., Elmisery, F. A., Mostafa, R.: Motion Control of Robot by using Kinect Sensor. In: Applied Sciences, Engineering & Technology. Vol. 8, Issue: 11, pp. 1384-1388. (2014)
2. Gośliński, J., Owczarek, P., Rybarczyk, D.: The use of Kinect sensor to control manipulator with electrohydraulic servodrives. In: Pomiary, Automatyka, Robotyka. Vol. 17, Issue: 2, pp. 481-486. (2013)
3. Khassanov, Y., Imaberdiyev, N., Varol, H. A.: Inertial motion capture based reference trajectory generation for a mobile manipulator. In: HRI '14 Proceedings of the 2014 ACM/IEEE international conference on Human-robot interaction. New York: ACM, pp. 202-203. ISBN 978-1-4503-2658-2. DOI: 10.1145/2559636.2559812. (2014)
4. Vytečka, M., Ondroušek, V., Kolomazník J., Hammerschmiedt, M.: Design and Implementation of the Remote Control of the Manipulator. In: Acta Universitatis Agriculturae et Silviculturae Mendelianae Brunensis. Vol. 62, Issue: 6, pp. 1521-1525. (2014)
5. Siciliano, B., Sciavicco, L., Villani, L., Oriolo, G.: Robotics: Modeling, Planning and Control, Springer, ISBN 978-1-84628-641-4. (2009)

3D white light interference microscope with specialized illumination for better sample imaging and observation

Joanna Schmit

Nano Surfaces Division, Bruker Inc., Tucson, Arizona, USA

joanna.schmit@bruker.com

Abstract. 3D microscopes based on white light interferometry, in addition to great precision topography measurement of small objects, can now provide great sample color or grayscale imaging capabilities. In addition to topography, image representation of the sample allows for additional ways of feature or defect characterization of the sample. A simple but very powerful additional illumination source on interference microscope allows for excellent imaging especially of low reflective and diffusive surfaces. The additional illumination enables ease of use features like finding focus and location of area of interest on sample, which are so important in commercial products nowadays. The paper industry may be one area that would significantly benefit from this modification of the white light interferometer.

Keywords: white light interferometry· 3D microscopy · color imaging · grayscale imaging

1 Introduction

3D microscopy based on white light interferometry (WLI) [1] is commonly used as a metrology tool for surface roughness and small-structure measurements because it delivers high resolution and precision topography measurements. It also comes with the convenience of full-field, non-contact and fast measurements. Some system modifications or special algorithms for white light fringes allow also for film thickness measurement, through a protective cover glass or in liquid measurement, large area measurement via a stitching procedure, or measurement of samples in different types of motion. The measured topography is conventionally presented in a color-coded scale; however, in today's digital world a crisp true color image of the sample is expected. These images may provide additional information for the sample measurement like automatic defect or special area identification. This article will explain how a color or grayscale image can be obtained as a result of the interferometric measurement and how some sample's visualization in live observation can be improved.

R. Jabłoński and T. Brezina (eds.), *Advanced Mechatronics Solutions*,
Advances in Intelligent Systems and Computing 393,
DOI 10.1007/978-3-319-23923-1_81

2 3D microscope based on white light interferometry

3D microscopes based on white light interference [1] use specialized interferometric objectives to create fringes that represent an object's shape with a nanometer vertical resolution under any system magnification. Fringes are collected during the vertical scan and then analyzed for their vertical position. Interferometric objectives deliver excellent 3D topography, but good quality color determination is not straight forward. A live image via an interferometric objective as compared to image via a bright field objective is obscured by the light beam from the reference mirror and created fringes.

The sample viewed with interferometric objective is observed in the presence of Aadditional background light reflected from the reference mirror and the ratio of sample and mirror reflectivity determine the clarity of a sample's image. For samples of very low reflectivity in comparison with mirror's reflectivity, the image clarity may be very poor. In Fig. 1 examples are shown of low and high reflective samples as observed with color camera via bright field and via interferometric objective. Fig. 2 shows a schematic of interferometric objective.

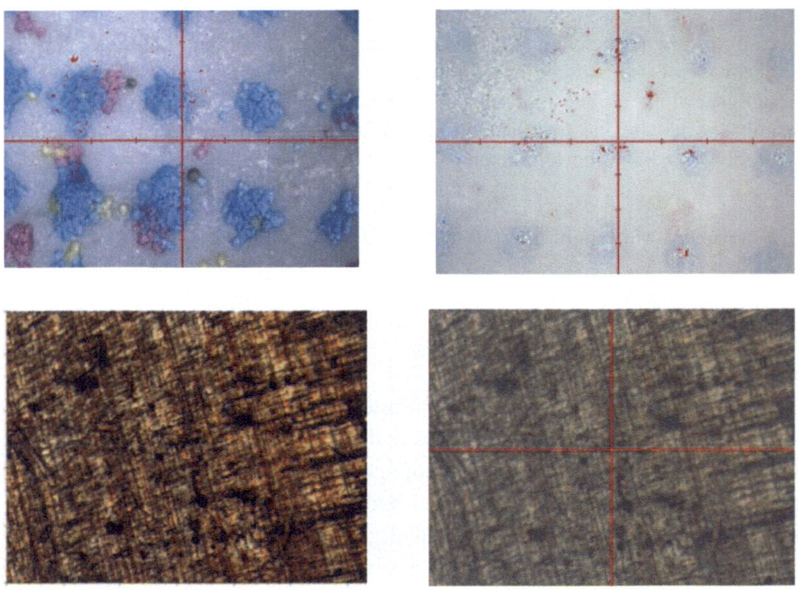

a) b)

Fig. 1. Live image of low reflectivity print and a high reflectivity coin as observed with a) 20X NA=0.4 bright field objective and b) 2X NA=0.3 interferometric Mirau objective.

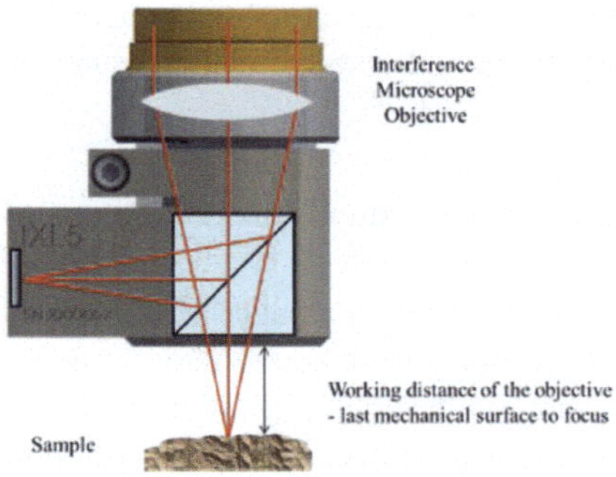

Fig. 2. Interferometric objective – Michelson type with a beamsplitter cube and a reference mirror.

3 Color imaging with white light interferometry

In order to digitally observe a sample in its natural color, white light illumination and a color camera are needed, or a combination of (three) cameras with color filters (red green blue RGB or other combination) and a white light or RGB source; RGB illumination and a B&W camera also can be used. Some considerations need to be done when choosing the illumination-camera combination as different options may somewhat reduce the lateral or vertical system resolution, making the system more expensive or larger in size, or significantly slow down the measurement if, for example, RGB illumination is sequential.

As mentioned, interferometric objectives do not provide the same image of a sample as regular bright-field objectives. Thus, one method to obtain a 3D color image is to do a two-scan measurement: one scan with an interferometric objective delivering 3D data based on finding a best focus position for each pixel, and a second scan with a bright-field objective during which color information is collected only at the previously predetermined best-focus positions and then overlaying the information on the 3D interferometric data.

Because this solution requires two scans, which extends the measurement time, it was worth looking into obtaining a true color image of an object using only interferometric objectives [3,4,5]. There are a few issues that need to be overcome in an image provided using an interferometric objective: presence of fringes and sometimes poor colors and poor object-features visibility. First issue is that fringes present at the best focus obscure the sample and need to be removed by, for example, frames averaging around the best focus. The second issue is that interferometric objectives introduce additional background intensity from the reference mirror causing the image of low reflective samples to have low saturation and low contrast colors. These two issues can be dealt with by image processing, but samples such as paper, foam or shiny surfaces, may benefit from higher numerical aperture illumination or special illumination.

3.1 Side illumination for imaging of low reflective and diffusive samples

In Figure 3a is an example of a paper image as observed by a color camera with a 5X Michelson interferometric objective under regular illumination through the objective conditions using white light LED illumination. Due to the low reflectivity of the sample and the small numerical aperture of the objective, the colors and sample itself are difficult to see making it difficult to find focus and the area needed to be measured in addition to extracting color. Figure 3b shows the live image of the same sample but with additional side illumination. Scattered light from side illumination clearly brings out colors and features of sample.

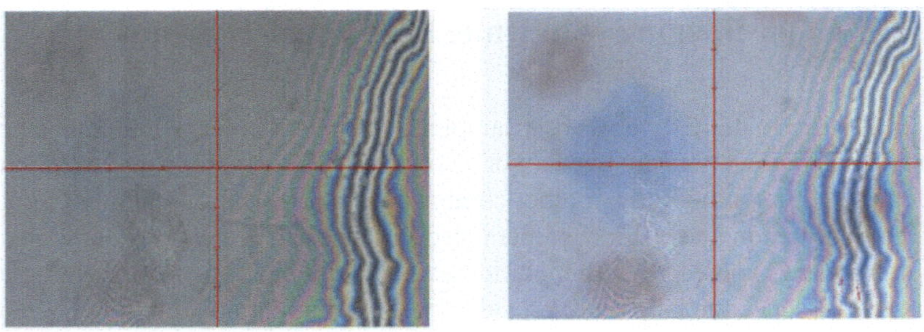

a) b)

Fig. 3 Live image of print as observed with 5X interferometric objective a) without and b) with side illumination.

The intensity (correctly irradiance) of each color (red green and blue: RGB) on the camera can be described, as a function of scanner position z, by the following equation:

$$I^{RGB}(z) = I^{RGB}_{obj} + I^{RGB}_{ref} + I^{RGB}_{side} + g^{RGB}(z)\sqrt{I^{RGB}_{obj} I^{RGB}_{ref}}\cos\left(\frac{2\pi}{\lambda}z + \varphi\right) \quad (1),$$

where I^{RGB}_{obj} represents the red, green or blue interference wavefront intensity returned by the sample, I^{RGB}_{ref} represents the red, green or blue interference wavefront intensity returned by the reference mirror, $g^{RGB}(z)$ is the signal modulation function associated with the objective and light source used by the white-light interferometer, λ is corresponding wavelength and φ is phase.

The value of z varies linearly with the physical separation between the sample and the reference mirror (i.e., scanner motion), while the phase φ is introduced by the relative height of each sample pixel with respect to the reference plane and represents the object's topography. The additional DC term $I_{side}^{RGB}(z)$ is the contribution received from the object as a result of side illumination. Side illumination has been found to contribute advantageously to color detection based on the DC component interferometric irradiance when low-reflectivity samples are involved. Side illumination is used in machine vision to help obtain a desired look of the sample.

Figure 4 shows an example of a different print observation without and with side illumination and the final color image of the print overlaid on the measured topography representing texture of the print.

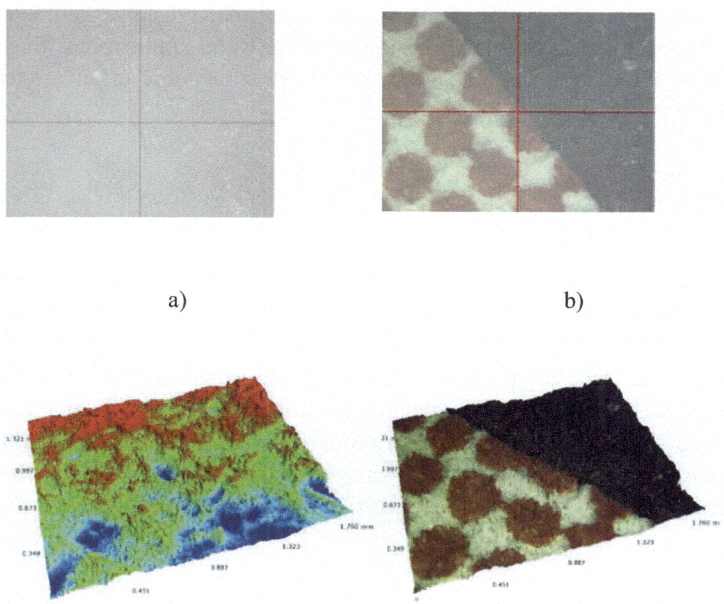

Fig. 4. Paper sample a) as observed with 5X interferometric objective, b) with additional side illumination c) resulting height map in color coded scale, d) height map with overlaid color image.

The most stunning colors are interference colors created and easily obtained with interference or bright-field objectives for thin (0.1-1.4 micron) film layers on samples. Simple sample color visualization is sufficient to determine if there is a large film-thickness variation across the film or just a thin film (less than 1.5 micrometers) present.

Fig. 5. Color fringe image of varying thickness thin film

In addition to precise measurements and great image display there is also an important additional aspect of metrology tools for industrial applications: they provide a high level of automation that excludes manual labor for more efficient and more repeatable results [6].

4 Summary

3D microscopes based on white light interferometry can have color imaging capabilities similar to an ordinary microscope. A simple addition of side illumination significantly aids in observation via interferometric objective of low reflective and diffusive samples. Side illumination helps in finding focus and area of interest in addition to retrieval of sample's true color image. Imaging of low reflective and diffusive samples also significantly benefits interferometric imaging with a monochrome camera. Color imaging and precise surface topography measurement capabilities make 3D microscopes that are based on WLI a very attractive metrology tool for the paper industry.

References

1. Caber, P.J.: Interferometric Profiler for Rough Surfaces, Appl. Opt. 32(19) (1993) 3438.
2. Koschan, A., Abidi, M. A.: "Digital Color Image Processing" WILEY-Interscience, 2008
3. Kim, M. K., Yu, L.: US Patent 7317540, 2008
4. Machleidt, T., Kollhoff, D., Dathe, O., Kapusi, O., Nestler, R., Franke, K.-H.: Application of Color Cameras for 3D Surface Measurements with White Light Interferometers 14th SpectroNet Collaboration Forum AM 05.09.2012 in Jena
5. Beverage J. L., Colonna De Lega X., Fay M. F., Interferometric Microscope with True Color Imaging, SPIE Proc.Vol. 9203, (2014) 92030S
6. Novak E., Guenther B., "Automation in Interferomerty" in Handbook of Optical Dimensional Metrology", Ed. K. Harding, CRC Press, 2013

The WUT database for modeling the visual quality

Rafał Kłoda[1], Anna Ostaszewska-Liżewska[1], Sabina Żebrowska-Łucyk[1]

[1] Institute of Metrology and Biomedical Engineering,
Warsaw University of Technology, Poland
{kloda,ostaszew,szl}@mchtr.pw.edu.pl

Abstract. The article presents ready-to-use database containing a set of objective measures and Mean Opinion Scores (MOS) for the set of 40 video sequences. The format of published data enables to use a wide range of software. Additionally scripts for R (a programming language and software environment for statistical computing and graphics) are published, to use the database in that environment.

Keywords: DVD-Video, modeling, R (programming language), subjective testing, video quality assessment.

1 Introduction

Modern coding standards, widely used in the film industry for full-length movies distribution on optical discs, include: MPEG-2 – used in the production of DVD-Video and Blu-ray, H.264/MPEG-4 (AVC) and SMPTE 421M (VC-1) – used in case of Blu-ray and MVC – for Blu-ray 3D. One of the stages of the production of such discs [1, 2, 3] is video coding, when the source signal is subjected to a lossy compression. Compression parameters are set up earlier, during bit budgeting [1-4] and they establish both the volume of the data file and the probability of compression artefacts. It is an important element in the process of preparing optical discs, as visual information takes the largest part of the disc space and the visual quality is an important element in assessing the final product. Discs, unlike the signals transmitted live, can be played repeatedly, so in case of a poor quality they can be disqualified from the market. Therefore, preventing such failures is important both for producers and customers. Advanced encoders, required in this process, use complex techniques of automatic selection of compression parameters [1, 4] and to improve quality, a coding system enables for segment-based re-encoding [1, 4, 5]. The omnipresence of these standards together with a big and still growing market of DVD-Video and Blu-ray disc implies the improvement of quality control tools, which automatically flag the fragments of video which probably need to be re-encoded.

To develop an algorithm that is able to find low-quality segments in accordance with human visual perception, an appropriate hardware and software facilities as well as deep knowledge in many fields of science and technology are essential. Researcher taking on this task should have a set of test scenes that would be processed in a coding system first. Then, for a set of sequences thus obtained, mean of scores (MOS) from a

R. Jabłoński and T. Brezina (eds.), *Advanced Mechatronics Solutions*,
Advances in Intelligent Systems and Computing 393,
DOI 10.1007/978-3-319-23923-1_82

chosen subjective test should be calculated as well as a set of objective measures (metrics) characterizing the video before the compression process and after its reconstruction. It should be noted that preparing such a set of data for modeling is complicated and time-consuming. For that reason, numerous organizations and research centers built and published video databases, to make researches on objective quality assessment easier. Winkler S. [6, 7] analyzed the available video databases, that are publicly available, and relevant to quality assessment. Unfortunately, available databases contain only test scenes in a source format (e.g. [8]), compressed video sequences and reference test scenes (e.g. [9]), or video sequences with MOS (e.g. [10]). There are no reports on the ready-to-use databases containing aggregated data for test sequences with objective metrics and MOS from subjective experiment. Thus authors decided to prepare and publish such a complete database in a general format, which would enable using any software tools and techniques for modeling purpose.

2 Test scenes

As a source material, a set of video sequences extracted from DVD-Video during production process and scenes published by research centers were used. Collected materials were processed by using measurement station in the Institute of Metrology and Biomedical Engineering at the Faculty of Mechatronics, Warsaw University of Technology (WUT). All of them were in PAL format with the standard resolution of 720 × 576 pixels. When choosing material for the database its content and character was taken into consideration. To describe the character of the video sequences, two parameters *SI* and *TI* (defined in ITU Recommendations [11, 12]) were used (fig. 1).

ITU-T P.910 [11] recommends that the calculations should be performed on a sub-image of the video to avoid unwanted edge effects, and because the extreme edges of a video frame are usually invisible to CRT users. Thus the analyzed area of the frame was 672 × 528 pixels (24 pixels from each side of the frame were discarded). This area was used to calculate the rest of the objective measures.

Scenes with the prefix *ID* come from DVD-Video production process and with *src* – from VQEG[1] resources. The rest of the sequences derive from Tektronix[2].
The analysis revealed a huge variety of film material (fig. 1). In the lower left of the chart there are scenes with very low dynamic and low complexity of the image. In the right upper part of the plot there are scenes with high dynamics and a large number of details in the image.

SI and *TI* help to choose sequences for subjective quality tests optimally, which enables to reduce the length of a test material. It is crucially important because of high costs of such experiments. To build the mentioned database, four scenes were chosen: ID02, ID04, ID09 and ID16, representing a wide range of *TI* and *SI* variability (fig. 1).

---

[1] ftp://vqeg.its.bldrdoc.gov/SDTV/VQEG_PhaseI/TestSequences/Reference/
[2] ftp://ftp.tek.com/tv/test/streams/Element/MPEG-Video/625/

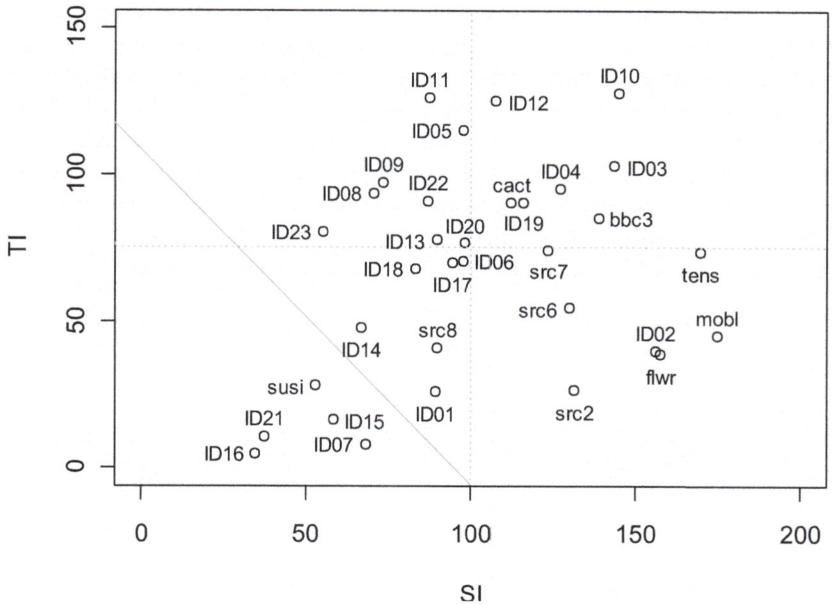

Fig. 1. Characteristics of scenes collected in the library

Scene ID02 is a part of the show and presents a musician playing a violin. Only the violinist's head and shoulders are visible. The background is dark. In the final part of the sequence there is a frame change with a dissolve effect. This scene contains a lot of details and is quite static. Changes of *TI(Fn)* from the beginning to the frame no. 190 are associated with the movement of the music inside the frame (fig. 2 a). From the frame no. 191 there are *TI(Fn)* fluctuations visible. This is where the dissolve effect starts. The effect ends with the image of a musician with a harp. *TI(Fn)* fluctuations can be assigned to the harp strings vibrations. It can be seen that the *SI(Fn)* is also increasing in the final interval (fig. 2 b). It is associated with strings of the instrument gradually appearing in the foreground. The scene was named *viol*.

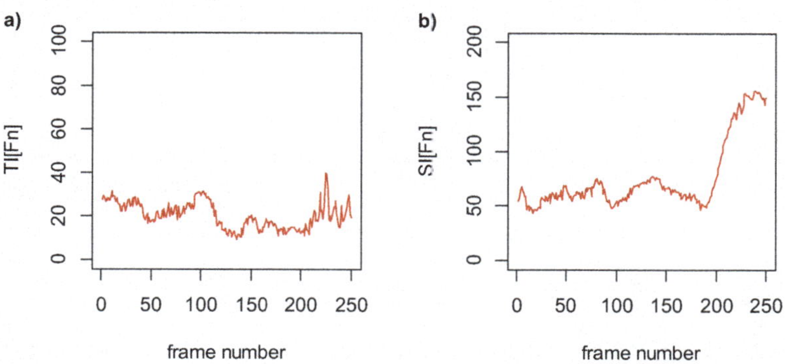

Fig. 2. $TI(F_n)$ and $SI(F_n)$ values for consecutive frames of *viol* scene: a) $TI(F_n)$, b) $SI(F_n)$

Scene ID04 is a GSM network operator advertisement. It consists of several shots. Players run on the pitch – it's a rainy night, stadium lights blind the camera. Hence, there are a large number of details (high *SI* value) and a high dynamic (indicated by the value of *TI*). The scene consists of eight shots - there are peak values of *TI(Fn)* clearly visible (fig. 3 a). Each shot is more dynamic in comparison with the rest of the test sequences chosen (fig. 2 a and fig. 4 a). The change of shots is sometimes accompanied by a significant change in the amount of details in the picture. The smallest values of *SI(Fn)* can be observed in the range of frames 76 - 120 and 190 - 203 (fig. 3 b), when a player is filmed in a medium shot. The scene was named *comm*.

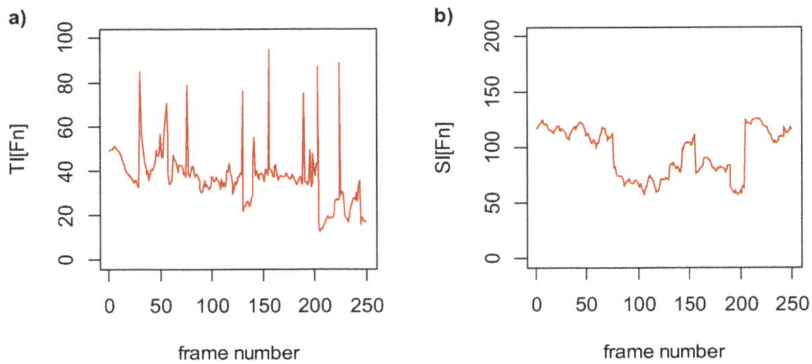

Fig. 3. $TI(F_n)$ and $SI(F_n)$ values for consecutive frames of *comm* scene: a) $TI(F_n)$, b) $SI(F_n)$

ID09 scene is a computer animation. It is full of colors wit high dynamic and the low number of details. It contains a large number of objects moving in various directions on a colorful background. Characteristic peaks on a *TI(Fn)* plot (fig. 4 a) are the result of straight cuts in animation – the sequence consists of five different shots. Shots have a similar number of details (fig. 4 b). The scene was named *pold*.

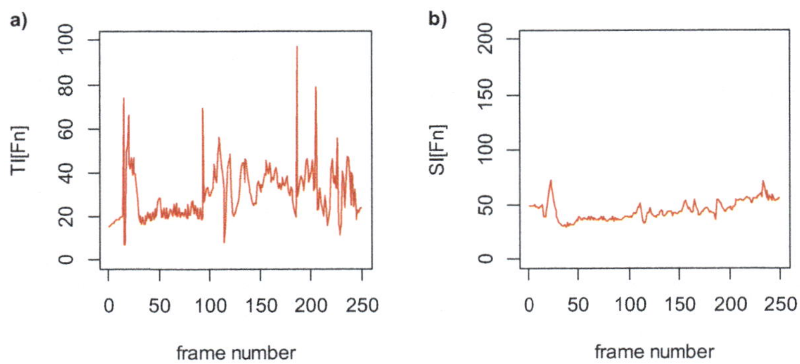

Fig. 4. $TI(F_n)$ and $SI(F_n)$ values for consecutive frames of *pold* scene: a) $TI(F_n)$, b) $SI(F_n)$

The last scene ID16 is a fragment of a feature film in an anamorphic format, in which the actor speaks his lines to a persons located outside the frame. The frame includes only the character actor sitting on a bed in the close-up shot. In the background there are paintings hanging on the wall and a lampshade of a bedside lamp. This scene is an example of a movie with a very small number of details and a very low dynamic range. *SI(Fn)* and *TI(Fn)* values are almost constant (fig. 5 a, b). Sometimes this type of audiovisual material is referred to as a "talking head". The scene was named *shot*.

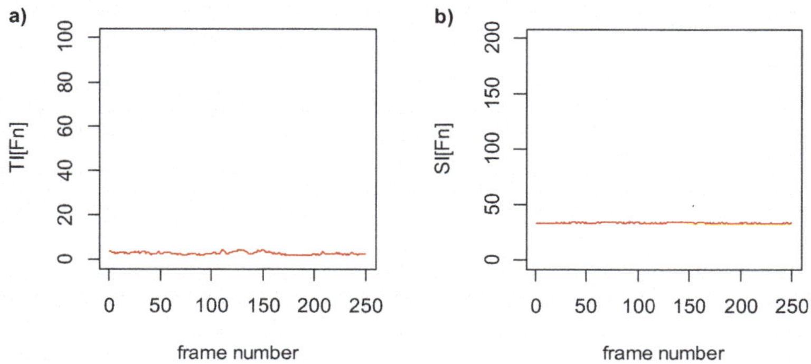

Fig. 5. $TI(F_n)$ and $SI(F_n)$ values for consecutive frames of *shot* scene: a) $TI(F_n)$, b) $SI(F_n)$

The analysis conducted above shows, that the chosen sequences are highly diversified. It can be expected that those scenes will be prone to compression artifacts in different ways. The described set of four scenes was named *build*.

3 Database

3.1 Encoding parameters and calculated metrics

Selected videos have been encoded with hardware encoder MPEG-2 (ISO-13818-2) MP@ML. The encoding process was performed with specialized tool *Encode*, which is a part of a DVD premastering system – Sonic DVD Creator™ [4].

The studies used single-pass encoding with constant bitrate (CBR) and the following set of parameters for group of pictures (GOP): 13 frames long, with two B frames. This configuration is considered to be optimal for encoding most of the footage [4].

Scenes from the *build* set were coded with seven levels of bitrate $S = \{1{,}14; 2{,}0; 3{,}0; 4{,}0; 5{,}0; 7{,}0; 9{,}0\}$ Mb/s, which covers all the useful (for the chosen format) range of this parameter. This way authors got 28 test sequences, which were processed with VImaQ software [13] to get the values of metrics. Among them there are both metrics which use the model of human visual system (HVS) like *SSIM* and the simplest ones, which can even be calculated in real time such as *PSNR*. Among selected metrics, there are also others based on uncomplicated algorithms like *MAE* and metrics for detecting some specific types of impairments, like blurring (*BLUR*) or blocking artifacts (*BLOCK*). Other metrics describe the video content: its temporal (*TI(Fn)*) and spatial (*SI(Fn)*) features. Tab. 1 displays calculated values for extreme values of bitrate *S* for each scene. It may be noted that the ratios for individual scenes are significantly different from each other, and also scenes have different sensitivity to compression artifacts.

Table 1. Metrics for sequences selected from a *build* set with the lowest and the highest values of bitrate *S*.

	scene	comm		pold		shot		viol	
	bitrate *S* (Mb/s)	1,14	9,00	1,14	9,00	1,14	9,00	1,14	9,00
metric	AD	-2,321	-2,950	-2,061	-2,175	-3,180	-2,864	-3,376	-3,764
	BLOCK	0,602	0,206	0,625	0,236	0,195	0,152	0,323	0,168
	BLUR	0,281	0,293	0,316	0,461	0,332	0,260	0,329	0,284
	MAE	10,918	6,056	7,990	4,254	3,669	3,258	7,906	5,686
	MSE	339,328	81,653	163,115	42,690	20,384	15,857	191,048	75,730
	MSSSIM	0,645	0,887	0,762	0,938	0,914	0,938	0,713	0,863
	PSNR-HVS-M	19,011	28,625	22,622	32,110	32,003	33,613	24,046	28,430
	PSNR	23,256	29,308	26,652	34,005	35,047	36,136	26,995	30,235
	SI(Fn)	73,497	95,178	49,926	44,711	33,463	34,176	51,413	74,832
	SSIM	0,556	0,795	0,746	0,916	0,872	0,881	0,742	0,859
	TI(Fn)	32,039	37,255	28,604	30,692	2,318	3,148	16,801	19,907
	MIN.TI(Fn)*	9,743	12,791	5,804	6,342	1,190	2,197	7,829	9,065
	TI**	89,925	94,111	95,112	95,986	4,624	5,485	29,789	39,694
	Q3.TI***	34,943	40,941	34,991	37,666	2,763	3,508	21,103	24,254
	MIN.SI(Fn)*	38,270	56,876	31,138	29,921	32,192	32,912	36,149	43,867
	SI**	112,962	126,709	79,128	71,186	34,354	34,980	71,973	157,898
	Q3.SI***	80,380	115,234	55,421	50,007	33,733	34,390	55,727	73,094

* – value of this metric was calculated as minimum of TI(Fn) or SI(Fn) from all frames
** – value of this metric was calculated as maximum of TI(Fn) or SI(Fn) from all frames [11]
*** – value of this metric was calculated as upper quartile of TI(Fn) or SI(Fn) from all frames [14]

3.2 Subjective experiment

In order to obtain the reliable information on the quality of chosen sequences processed with the set of encoding parameters, the DCR (Degradation Category Rating) method was used in accordance with ITU-T P.910 [11]. The DCR method presents test sequences in pairs, where the first of them is a source one and the second (presented after 2-seconds break) – the same video subjected to lossy compression. After watching each set of two, the impairment of the second stimulus in relation to the reference is rated. The five-level scale for rating the impairment is used: 5 – imperceptible, 4 – perceptible but not annoying, 3 – slightly annoying, 2 – annoying, 1 – very annoying.

The whole experiment with each observer lasted nearly an hour and was divided into initial step, training session and two test sessions. At the initial step the observer got instructions on the plan of the experiment, the way of quality evaluation of compressed video and the way lossy compression may distort a picture. In accordance with ITU-T P.910 Recomendation [11], the visual acuity of observers was tested with Snellen charts and their normal color perception was checked with Ishihara plates. A success in both tests qualified the observer to the rest of the experiment.

The training session was performed to check whether observers cope with video quality assessment process and are able to grade the existing distortions. ID01 and ID23 sequences (fig. 1) were used in this part. These additional scenes were encoded with the same encoder settings and processed in the same way as in case of a *build* set. The only difference was that for the training sequences only six values of bitrate were used $S = \{1,5;\ 2,0;\ 2,5;\ 5,0;\ 7,0;\ 9,0\}$ Mbps. This way 12 test sequences were obtained. Scenes were named *musi* and *fish,* and the whole set of 12 got the name *verify*.

After the training session the final experiment was conducted. The test material was prepared with the use of *build* set. The order of sequences in each test session was different and there was a short break after each session.

The group of observers which take part in the experiment, according to ITU Recomendations [11, 12] should consist of 4 – 40 people. The final number of observers was a compromise between costs of the time-consuming experiment and the acceptable confidence intervals for computed mean of scores.

In this experiment 52 scores were obtained for each of test and training sequence. The observers watched the video from the distance of four the display heights (4H). In the experiment the 21” CRT SONY PVM-20M4E display was used.

The following plots display mean opinion score (MOS) in a function of bitrate S for each sequence from the *build* set (fig. 6).

All curves increase monotonically, but differ from each other in shape. The highest scores got the *shot* scene (fig. 6 c). It was characterized by the lowest TI and SI values. The lowest scores got *pold* scene (fig. 6 b). In general, each scene has two sections in the plot. The first ($1 \le S \le 4$ Mb/s) is a fast monotonic increase in quality. The second ($S > 4$ Mb/s) is the interval where the scores stabilize, which is a result in accordance with previously obtained by the authors [15,16].

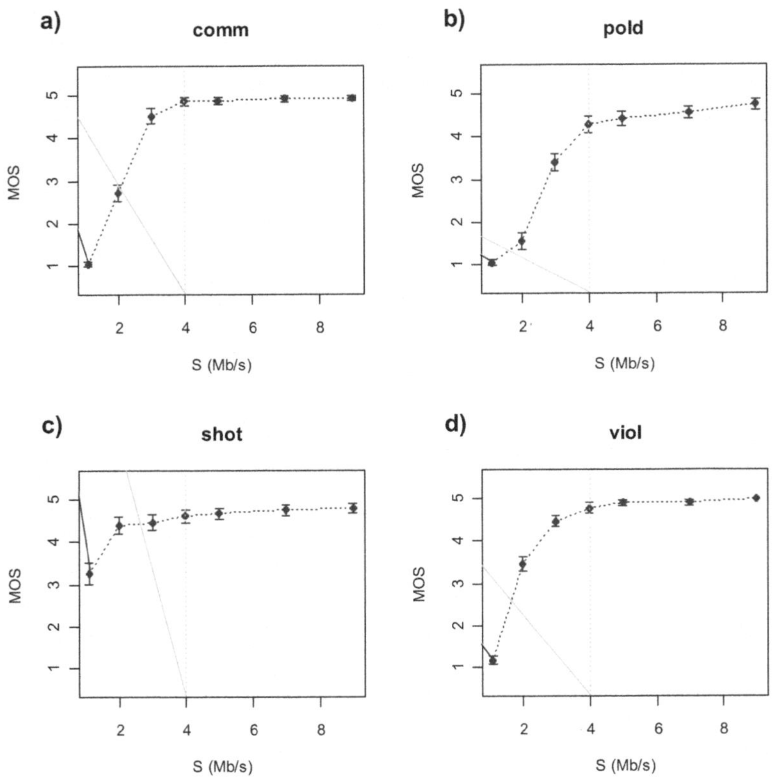

Fig. 6. Average scores and their confidence intervals for scenes from the *build* set

3.3 Download and usage

The WUT database with aggregated datasets *build* and *verify* and scripts are available for download at `http://adam.mchtr.pw.edu.pl/~kloda/wut/`. Datasets can be directly accessed from R statistical environment from this location. Authors also provided R scripts for described database management, which could be helpful for researchers in further analysis. Moreover authors provided all raw data from subjective experiment and computed objective metrics (displayed in Tab. 1). The WUT database contains the following folders:

- `datasets` – aggregated data for *build* and *verify* sets with calculated averages for MOS and 11 objective metrics (values of 6 metrics were calculated in different way – see comment in tab. 1),
- `scripts` – set of R functions useful to manage WUT database,
- `exp_DCR` – raw data from subjective experiments (*test* and *training* sessions). For test session the first section of data (rows from 1 to 26) comes from the first repetition. Second part (rows from 27 to 52) - from second repetition,

- `metrics` – collection of 11 objective metrics computed for each frames for all sequences used in database,
- `VQA` – detailed information about all sequences used in database obtained by using VISUALmpeg analyser. These kind of data could be helpful for testing and comparison of algorithms.

4 Conclusion

The published WUT database includes 17 objective measures and MOS obtained with DCR method for 40 video sequences. It is particularly useful in case of DVD-Video premastering process optimization involving detection of segments with visible compression artifacts. Conclusions and results can be generalized to other standards.

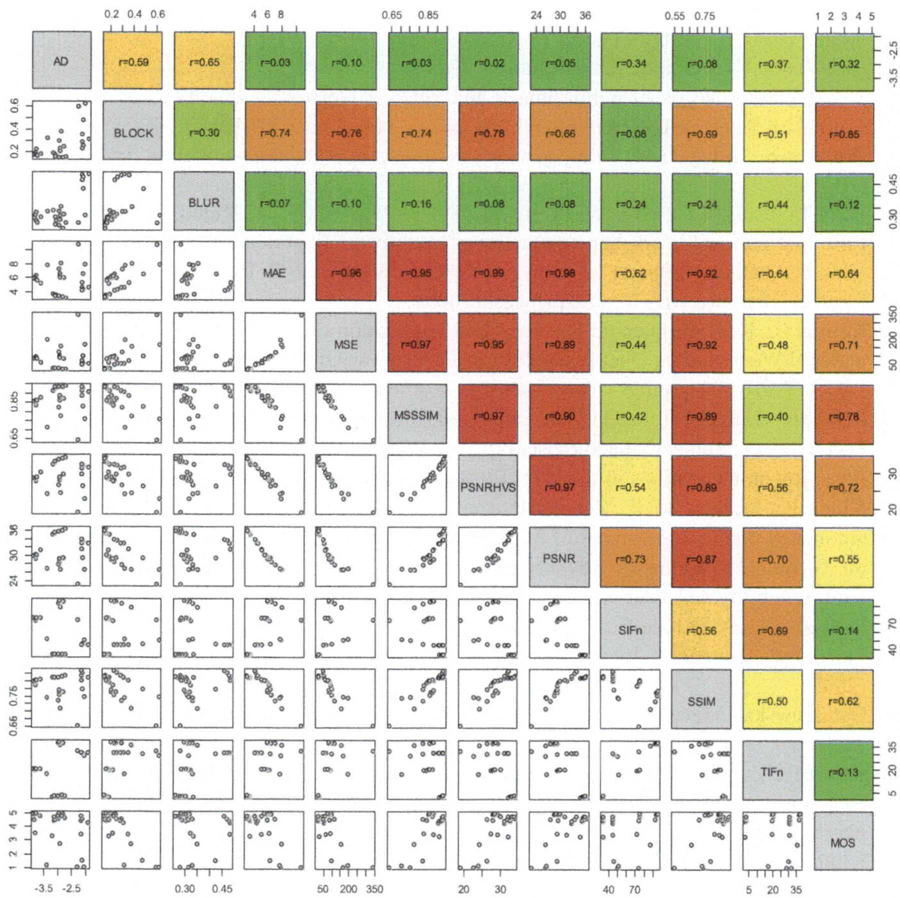

Fig. 7. Correlation analyses for main metrics in the dataset *build*

Correlation analysis of the *build* dataset (fig. 7) presents a practical example of the use of the WUT database to answer the question, how to select the best explanatory variables that predict the assessment of observers. A strong correlation between explanatory variables entering the regression equation and the dependent variable is expected. In that case predictor variables show more power to predict the value of the response variable. The same time the explanatory variables shouldn't be highly correlated with each other to not deprive each other explanatory power [17].

References

1. Ely M., Block D.: Publishing in the Age of DVD, Sonic Solutions (1998)
2. Taylor J., Johnson M.R., Crawford C.G.: DVD Demystified, Third Edition, McGraw-Hill (2006)
3. Taylor J., Zink M., Crawford C.G., Armbrust C.M.: Blu-ray Disc Demystified, McGraw-Hill (2008)
4. Sonic DVD Creator User Guide, Sonic Solutions (1999)
5. CineVision User Guide, Sonic Solutions, (2011)
6. Winkler S.: Analysis of public image and video databases for quality assessment, IEEE Journal on Selected Topics in Signal Processing, vol. 6, no. 6 (2012)
7. Winkler S.: Image and video quality resources, http://stefan.winkler.net/resources.html (2012)
8. Song L., Tang X., Zhang W., Yang X., Xia P.: The SJTU 4K Video Sequence Dataset, Fifth International Workshop on Quality of Multimedia Experience (2013)
9. Video Quality Experts Group (VQEG), VQEG FR-TV Phase I Database, ftp://ftp.crc.ca/crc/vqeg/TestSequences/ (2000)
10. Keimel C., Redl A., Diepold K.: The TUM High Definition Video Data Sets, Fourth International Workshop on Quality of Multimedia Experience (2012)
11. ITU-T P.910, Subjective video quality assessment methods for multimedia applications, Recommendations of the ITU, Telecommunication Standardization Sector (2008)
12. ITU-T P.911, Subjective audiovisual quality assessment methods for multimedia applications, Recommendations of the ITU, Telecommunication Standardization Sector (1998)
13. Kłoda R., Oczkowski P., Żebrowska-Łucyk S., Ostaszewska-Liżewska A.: Software for automated quality assessment of digitally processed video, 13th IMEKO TC10 Workshop on Technical Diagnostics Advanced measurement tools in technical diagnostics for systems' reliability and safety (2014)
14. Ostaszewska A., Kłoda R.: Quantifying the amount of spatial and temporal information in video test sequences, Recent Advances in Mechatronics, Springer Verlag, pp. 11-15, (2007)
15. Ostaszewska-Liżewska A., Kłoda R., Żebrowska-Łucyk S.: Continuous quality evaluation: subjective tests vs. quality analyzers, Recent Advances in Mechatronics, 55-60 (2010)
16. Kłoda R., Żebrowska-Łucyk S., Ostaszewska A.: A dedicated station for MPEG-2 quality evaluation with ACR method, Fifth International PhD Conference on Mechanical Engineering (2007)
17. Kłoda R.: The method of detection and reduction compression artifacts in DVD premastering process of DVD-Video (in polish), PhD Thesis, Warsaw University of Technology, Faculty of Mechatronics (2014)

Multimodal perception in subjective quality evaluation of compressed video

Anna Ostaszewska-Liżewska[1], Rafał Kłoda[1], Sabina Żebrowska-Łucyk[1]

[1] Institute of Metrology and Biomedical Engineering,
Warsaw University of Technology, Poland
{ostaszew,kloda,szl}@mchtr.pw.edu.pl

Abstract. A movie, by its nature, is a combination of two stimuli simultaneously given: video and sound. Although multimodal perception phenomenon is widely known [1, 2, 3], in the area of subjective quality evaluation of compressed video there is still a lack of a coherent opinion on the mutual influence of picture and sound on the final impression of the observer. The aim of this study was to perform an experiment to check the influence of sound on perceptual evaluation of compressed video. Results of the experiment show that both the presence and the type of sound simultaneously given changes the video quality perception.

Keywords: lossy compression, Single Stimulus Continuous Quality Evaluation, multimodal perception

1 Introduction

As the number of digital videos which are created, saved and transmitted each day constantly increases and despite of a growing internet connection speed and the capacity of computer drives, lossy compression still plays an important role. Multimedia companies are constantly working on more efficient algorithms, which would result in getting the smaller files of a better quality video. To accomplish that goal, in some point it is necessary to assess the quality of the final product with a group of observers. There are numerous methods recommended by International Telecommunication

R. Jabłoński and T. Brezina (eds.), *Advanced Mechatronics Solutions*,
Advances in Intelligent Systems and Computing 393,
DOI 10.1007/978-3-319-23923-1_83

Union (ITU), which explain in details the way of planning and conducting the experiment. Despite those methods are broadly used, there is still a question if the video should be evaluated with or without a sound. There have been many studies on the effects of sound on the assessment of image quality, but they approach the issue in different ways and produce different conclusions [4, 5, 6, 7]. Hence the authors decided to perform an experiment to check the actual influence of sound on the perceived quality of compressed video in subjective methods.

2 The experiment

The experiment was conducted on a test station which was designed and set up in the Institute of Metrology and Biomedical Engineering at the Faculty of Mechatronics, Warsaw University of Technology (Fig. 1).

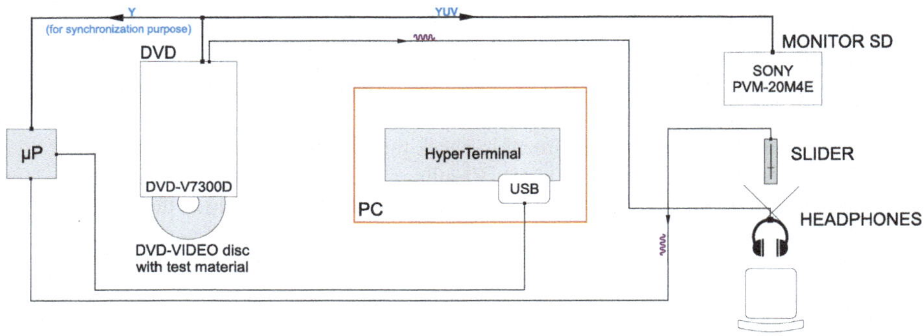

Fig. 1. The scheme of the experimental station

The playback process is started by selecting the appropriate material from DVD-Video. This automatically begins the process of data acquisition from the slider. A microprocessor responsible for synchronizing communicates with the PC via a USB interface. Its configuration and measurement data acquisition from the slider is performed with the use of HyperTerminal application. The test material is displayed on the broadcasting monitor SONY.

Four 15-seconds MPEG-2 video scenes, coded with 13-frame group of pictures (GOP) were prepared both in good (bitrate of 4 Mbps) and poor (bitrate of 2 Mbps) quality. Then each sequence was presented to the human audience in three versions – without sound, with pleasant sound and annoying sound. The accompanying sounds were chosen by sound engineering specialists relying on the effects described in [8, 9], (Tab. 1).

The experiment was conducted in accordance with Single Stimulus Continuous Quality Evaluation method recommended by International Telecommunication Union [10]. A group of 15 people participated in tests. Every configuration was displayed twice, in random sequence. Data acquisition was carried out with a frequency of 5 Hz. The data obtained was processed with dedicated software created in C++.

Table 1. Sequences chosen to build the test material with two types of the acompanying sound (pleasant and annoying): a) bbc3 b) mobl c) cact d) susie

	Scene	Pleasant sound	Annoying sound
a)		G. Rossini wind quartet in F major - Theme & variations	white noise
b)		Tomohito Nishiura „Puzzle”	crying baby
c)		W. A. Mozart Concerto for Flute and Harp in C major, II. Andantino	buzz at the festival
d)		Norah Jones „Painter song”	annoying alien buzzer

3 Results

The first observation was that in general the global mean of scores for 4 Mbps was higher than for 2 Mbps (54, 3 vs. 40,7), which is obvious in case of chosen values and the format of the video. Besides the lowest mean of all grades given characterized the scene 'mobl' (42, 0) and 'bbc3' (44,7), which were the most prone to compression artifacts due to the great amount of details and the high dynamic of the scene. The 'cact' scene (global mean of 49,6) used to get scores close to the easiest to decode – 'susi'(global mean of 53,8), which was in accordance with the results obtained previously [11]. Fig. 1 and Fig. 2 display the mean of scores given to each scene with all three combinations of sound.

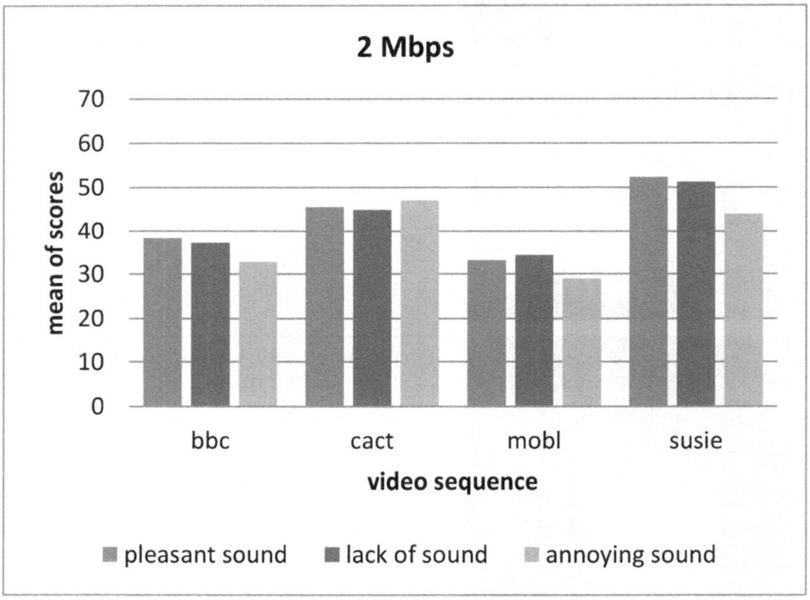

Fig. 2. Mean of scores for a set of four video sequences coded with the bitrate of 2 Mbps with three combinations of sound

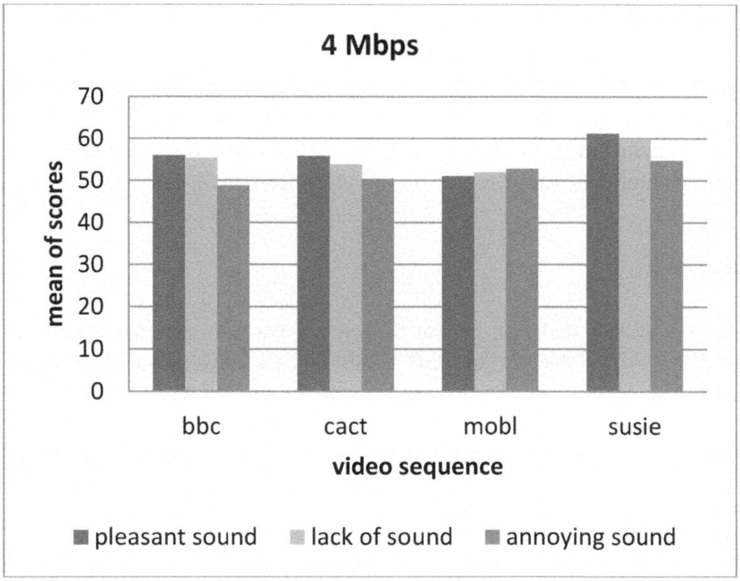

Fig. 3. Mean of scores for a set of four video sequences coded with the bitrate of 4 Mbps with three combinations of sound

In order to check whether there is a statistically significant difference between the average score of videos according to the scene and the type of audio, two-way analysis of variance was performed. Separate analyzes were performed for two values of a binary stream 2 Mbps and 4 Mbps. The factors were: the video sequence and type of the accompanying audio. For both analyzes (significance level $\alpha = 0{,}05$) p-value is: 0,075 for 2 Mbps and 0,084 for 4 Mbps, which means that there was no significant influence of the sound on the mean score.

Then, a two factorial analysis of variance of all data (for both bitrates taken together) was performed. In this case, the p – value is 0,016 ($\alpha = 0{,}05$), which proves the influence of the sound on the mean score. Significant differences between the results of the analysis result from the different number of data (which is twice as high as in the case of the analysis for bitrates analyzed separately).

In the next step the two-way analysis of variance, where the factors are the bitrate and the sound was carried out. In this case, the influence of sound has also proved to be significant (p – value is 0,011; $\alpha = 0{,}05$).

Additionally the three-way analysis of variance for such factors as: the video sequence, the bitrate and the sound was performed. It showed the significance of all three factors. For the sound the p-value is 0,007 ($\alpha = 0{,}05$), which shows it is strongly significant. However, no significant interaction of sound with the video sequence or with the bitrate was shown.

4 Conclusions

Results of the experiment show that both the presence and the type of sound simultaneously given changes the video quality perception. Therefore, in the area of subjective quality evaluation of compressed video, sound should be considered as an additional factor, which has an influence on the measurement.

References

1. Kunka, B.: System śledzenia punktu fiksacji wzroku jako narzędzie wspierające badania korelacji wzrokowo-słuchowych, PhD Thesis, Gdańsk University of Technology, 2011.
2. McGurk, H. , MacDonald, J.: Hearing lips and seeing voices, Nature, vol. 264, 746–748, 1976.
3. Shams, L., Kim, R.: Crossmodal influences on visual perception, Physics of Life Reviews, 2010.
4. Storms, R. L. & Zyda, M.: Interactions in perceived quality of auditory-visual displays. Presence 9, 2000.
5. Hulusic, V., Aranha, M., Chalmers, A.: The influence of cross-modal interaction on perceived rendering quality thresholds. 16th International Conference in Central Europe on Computer Graphics, Visualization and Computer Vision, 2008.
6. Lee, J.-S., De Simone, F., Ebrahimi, T.: Influence of audio-visual attention on perceived quality of standard definition multimedia content. First International Workshop on Quality of Multimedia Experience (QoMEX 2009), San Diego, CA, U.S.A., July 29–31, 2009.
7. Aldridge, R.P., Hans, D.S., Pearson, D.E., Lodge N.K.: Continuous quality assessment of digitally-coded television pictures, IEE Proceedings online no. 19981843.
8. Campbell, The Mozart Effect, Tapping the Power of Music to Heal the Body, Strengthen the Mind and Unlock the Creative Spirit, Harper Paperbacks 1997.
9. Chang, R. S., & Thompson, N. S.: Whines, cries, and motherese: Their relative power to distract. Journal of Social, Evolutionary and Cultural Psychology, 5(2), 131-141, 2011.
10. ITU-T Recommendation P.911 (1996), Subjective audiovisual quality assessment methods for multimedia applications.
11. Ostaszewska, A., Kłoda, R.: Quantifying the amount of spatial and temporal information in video test sequences, Recent Advances in Mechatronics, Springer Verlag, pp. 11-15, 2007.

Part VII
Robotics

A Wiimote 3D Localization Scheme without Channel Constraints

Ting-Hao Li[1], Kuo-Shen Chen[1]

[1] Department of Mechanical Engineering, National Cheng-Kung University, Tainan, Taiwan, 701, R.O.C.

blackstone0711@hotmail.com, kschen@mail.ncku.edu.tw

Abstract. A novel 3D Wiimote localization scheme with no channel constraints has been proposed for smart living or computer animation applications. This scheme utilizes two Wiimotes for performing stereo range finding and an Arduino-based IR LED switching for breaking the constraint on available channels. By controlling the switching of IR LED units sequentially, the equivalent available channels can be extended with a trade off in system bandwidth. This Wiimote-based 3D localization system can detect more IR LEDs than the number of original channels and each IR LED can be identified correctly even if it passes through a shelter. A testing system is also setup for evaluating the performance on the 3D localization and IR LED switching control. The preliminary results indicate that the position of IR LEDs can be accurately detected and the system can identify IR LEDs without channel constraints. This novel localization system is expected to have a great potential on applications such as mobile robots monitoring or human motion capture for animations.

Keywords: Wiimote, 3D localization, Arduino-based IR LED control

1 Introduction

Position sensing is critical for indoor mobile robot navigation [1]. With accurate positioning information, navigation, control, coordinated motion, and task planning can be realized. Typical localization techniques, such as inertial navigations [2], ultrasonics [3], and vision [4], have been used. However, due to integrated consideration in terms of cost, resolution, and signal reliability, a flawless positioning technique has not yet been achieved. IR localization is another possible choice [5]. Particularly, IR-based Wiimotes have been proposed for many applications beyond its original TV game design [6]. Due to simple relationship between the displayed pixel and the sensing distance, it has been used as an alternative choice for indoor localization sensors. In our previous studies, Gu and Chen utilized Wiimotes to develop an accurate 2D localization system for indoor mobile robot control [6]. Fu and Chen proposed a Wiimote 2D localization system with a controllable IR LED array [7]. By the controlled input of the IR LED array, it is possible to extend the sensing area to cover the entire living space and the target can be identified correctly.

R. Jabłoński and T. Brezina (eds.), *Advanced Mechatronics Solutions*,
Advances in Intelligent Systems and Computing 393,
DOI 10.1007/978-3-319-23923-1_84

Recently, 3D indoor localization becomes important for robot control and for computer animation applications. In this work, based on our previous experiences in 2D localization, the 3D Wiimote localization technique is proposed and validated. The system utilizes two parallel Wiimotes for performing stereo range finding. However, Wiimote has a constraint on the number of detectable IR LEDs and this considerably restricts the usage of the scheme. Therefore, similar to the method proposed earlier [7], this work develops a novel IR LED switching technique to extend the detectable IR LEDs from four to virtually no limit. By such an effort, the proposed scheme should be useful in applications such as mobile robots monitoring or motion captures.

2 Wiimote 3D Localization System

As shown in Fig. 1(a), Wiimote is originally designed for TV games. It contains Bluetooth communication and a built-in CMOS image sensor (i.e., Wiimote camera), which is able to detect up to four IR LEDs with visual angles of approximately 45° (X) and 35° (Y) to form a sensing zone with a resolution of 1024×768 pixels. It also has tracking ability with a sampling rate of 100 Hz. Fig. 1(b) shows the concept of the Wiimote 3D localization scheme. When the IR LED is in the overlapped zone, both Wiimotes detect the pixel coordinate of the IR LED and then transmit it to computer for calculating the position of the IR LED. Specifically, the goal is to find the position of the IR LED (i.e., D_X, D_Y, and D_Z) from the readout of these Wiimotes. Here, we define the central position of the front end of two Wiimotes for the origin of coordinate. The distance between Wiimotes (D_W) also needs to be determined in advance. A D_W of 10 cm is chosen for the subsequent development. From Fig. 2, it can be seen that these two Wiimotes and the IR LED form a triangle. By calculating the three side lengths and the angles, the horizontal position (D_X), vertical position (D_Y), as well as the distance between Wiimotes and the IR LED (D_Z), can be determined.

$$D_X = \frac{D_W}{2} - (L_1 \times \cos\theta_{X1}) \tag{1}$$

$$D_Y = D_Z \times \tan(\theta_Y - 17.5°) \tag{2}$$

$$D_Z = L_1 \times \sin\theta_{X1} \tag{3}$$

where

$$L_1 = \frac{D_W \times \sin\theta_{X2}}{\sin(180° - \theta_{X1} - \theta_{X2})} \tag{4}$$

As shown in Fig. 3, based on the detected pixel coordinate (X, Y) of the IR LED and the geometric setting, it is possible to calculate the visual angles θ_{WX1}, θ_{WX2}, and θ_{WY} from geometry as

$$\frac{X_1}{\sin\theta_{WX1}} = \frac{1024 - X_1}{\sin(45° - \theta_{WX1})} \tag{5}$$

$$\frac{X_2}{\sin(45^\circ - \theta_{WX2})} = \frac{1024 - X_2}{\sin\theta_{WX2}} \tag{6}$$

$$\frac{Y_1}{\sin\theta_{WY}} = \frac{768 - Y_1}{\sin(35^\circ - \theta_{WY})} \tag{7}$$

where (X_1, Y_1) and (X_2, Y_2) are the pixel coordinates detected by the Wiimotes. Referring to Fig. 3, it can obtain that $\theta_{X1} = \theta_{WX1} + 67.5^\circ$, $\theta_{X2} = \theta_{WX2} + 67.5^\circ$, and $\theta_Y = \theta_{WY}$. Once θ_{X1}, θ_{X2}, and θ_Y are calculated and substituted into Eq.1 to Eq.4, it is able to determine the position of the IR LED (i.e., D_X, DY, and D_Z).

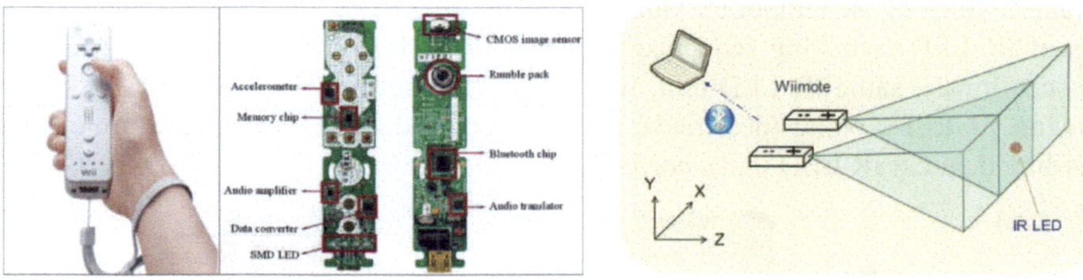

Fig. 1 (a) The Nintendo Wiimote and (b) the schematic plot of the 3D localization scheme

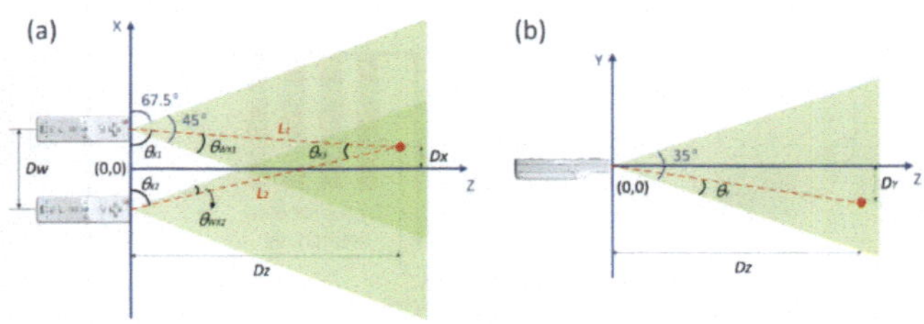

Fig. 2 The coordinated relation between Wiimotes and IR LED (a) top view (b) side view

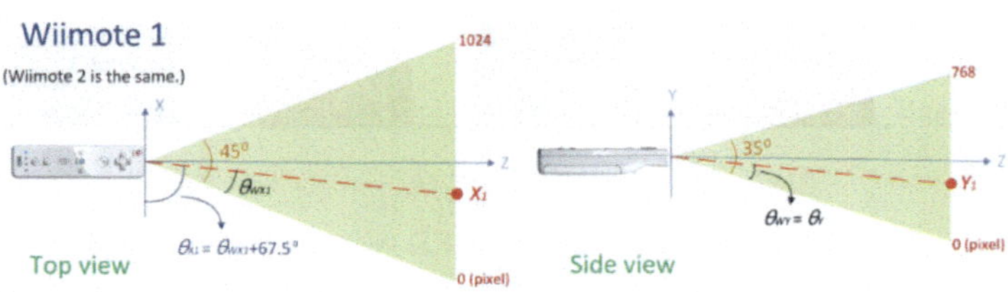

Fig. 3 Geometric relationship between pixel coordinate and visual angles

3 IR LED Switching Technique

Here we develop IR LED switching technique to extend the number of the detectable IR LEDs from 4. Fig. 4(a) shows the schematic concept of the technique. An Arduino is used for rapidly switching the IR LED units. With sufficient switching speed, all LEDs can be treated as being simultaneously detected by the Wiimote. Fig. 4(b) shows the original condition of the Bluetooth channel occupation during localization.

Each channel can only detect one IR LED at one time and four IR LEDs in total for a single Wiimote. On the other hand, by carefully controlling the switching sequence, each channel can detect two IR LEDs and all 8 IR LEDs can be detected in this case. The number can be further extended with a trade off on the equivalent bandwidth.

Meanwhile, for LEDs in the same IR LED unit, how to keep the same channel alignments in each time is also a problem. To keep the desired sequence, we also control the switches for these LEDs in the same unit. Consequently, these IR LEDs will sequentially enter the sensing zone and the channels are aligned correctly in each sampling. This action also avoids misalignment when the IR LEDs pass through a shelter, which cause temporary vanishing of the IR LEDs. Fig. 5 shows the schematics of IR LED switching sequence with two IR LED units. To keep the desired sequence for the same IR LED unit, we need to find both of the time interval, T_{ON} (allowing IR LEDs being identified), and T_{OFF} (providing sufficient reset time for recording the next IR LED unit), correctly.

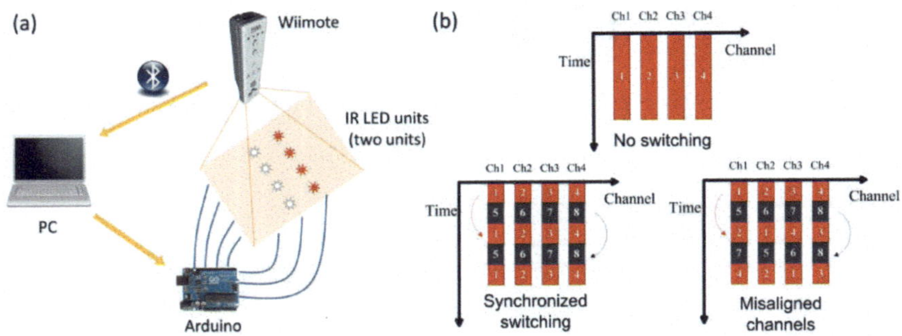

Fig. 4 Schematic of IR LED switching scheme (a) setup and (b) channel sequencing

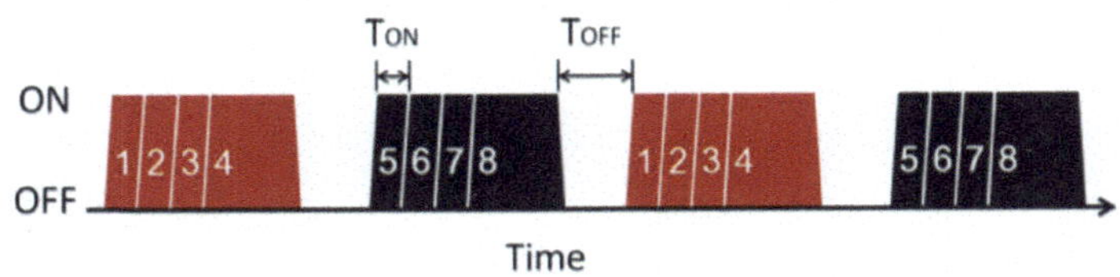

Fig. 5 Schematic plot of IR LED switching time interval

4 System Integration and Experiments

Here we integrate the 3D localization scheme with the switching technique to show the feasibility of detecting more IR LEDs in the localization task. First, we utilize two Wiimotes to monitor an IR LED for examining the feasibility of the 3D localization scheme. Experimental setup for evaluating the performance is shown in Fig. 6(a). Two Wiimotes are placed to monitor the motion of an IR LED mounted on a two-axis servo motor, which is programmed to perform 2D motions with various paths and speeds. It is possible to compare the paths localized by Wiimote versus their actual moving paths. Fig. 6(b) shows the results with speeds of 10cm/s and 20cm/s in a saw tooth path. The maximum position errors in X and Z directions are ±0.37cm and ±4cm, respectively. They correspond to only 3.7% and 3.3% of the nominal distance. This demonstrates the ability of the 3D localization scheme in object tracking. However, we also found that the positioning errors in Z direction become larger as the sensing distance farther increases and this is due to larger pixel length in larger Z.

Meanwhile, experimental setup for evaluating the IR LED switching performance is shown in Fig. 7(a). We utilize two Wiimotes to detect the two IR LED units (4 LEDs/unit) and to find the relationship between switching rate and data integrity. Suitable T_{ON} and T_{OFF} must be determined for keeping correct channel sequence. By extensive testing, as shown in Fig. 7(b), T_{ON} and T_{OFF} are determined as 5 ms and 13 ms and the equivalent bandwidth is restricted by these two parameters. In this case, the switching rate is approximately 16Hz. The time interval between each sampling is approximately 62ms. Half of the interval is due to Bluetooth communication. For two units, the sampling frequency of each unit is cut by a factor of 2 (i.e., 8Hz). We are currently improving the switching technique to enhance the bandwidth by both hardware and software approaches.

Experimental setup of Wiimote 3D localization with two IR LED units is shown in Fig. 8(a) with a 16Hz switching rate. Two Wiimotes are mounted on the motor to monitor the motion relative to the two IRLED units (4 LEDs/unit). The motor is programmed to perform rectangular looping. Fig. 8(b) shows the result with speed of 20cm/s and the relative trajectory of these two units can be monitored excellently. However, if we use a T_{ON} and T_{OFF} pair of 5ms and 8ms, the channel sequence could be wrong and mistaken locations will be reported (shown in Fig. 8(c)).

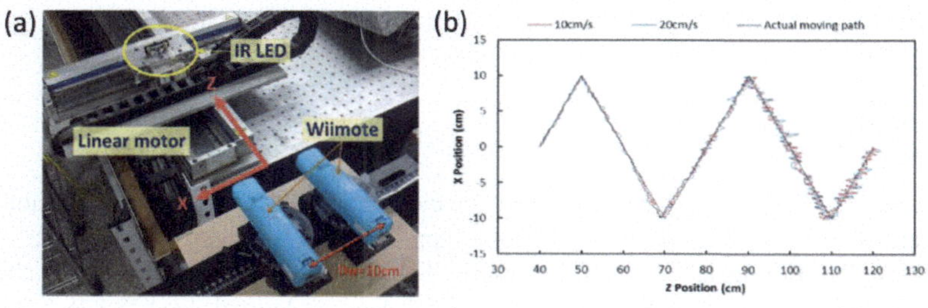

Fig. 6 (a) The experimental setup for 3D localization and (b) results for a saw tooth path

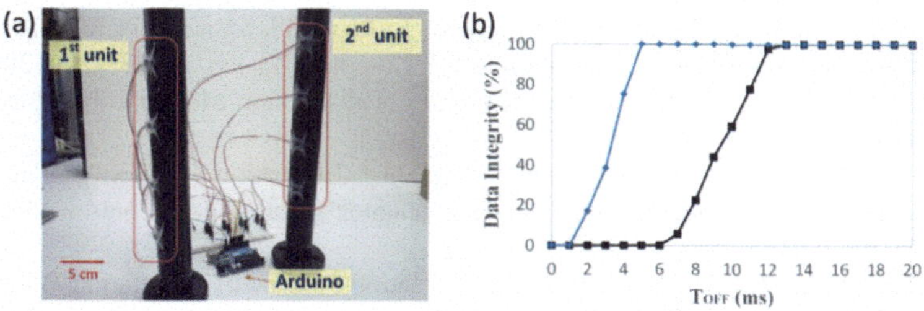

Fig. 7 (a) The setup for IR LED switching and (b) the experimental result for T_{ON} and T_{OFF}

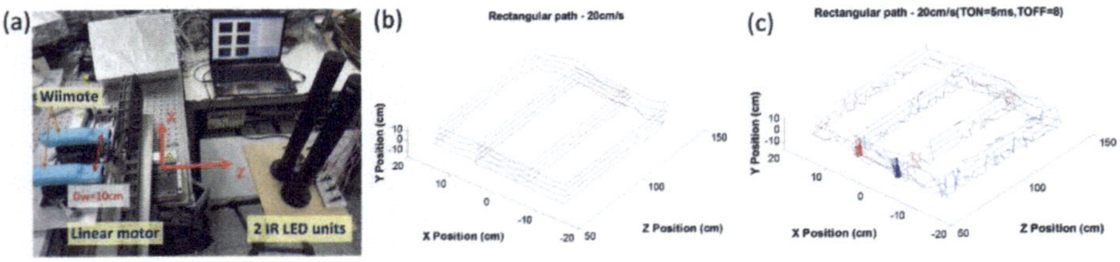

Fig. 8 (a) The experimental setup for Wiimote 3D localization (b) results with perfect channel integrity and (c) poor channel integrity due to unsuitable T_{OFF}.

5 Conclusion

A novel 3D Wiimote localization scheme with no channel constraints is proposed for smart living or computer animation applications. We develop the Wiimote 3D localization system and IR LED switching technique for 3D positioning and breaking the channel constraints. A testing system is setup for evaluating the performance on the 3D localization and IR LED switching. The preliminary results indicate that the position of IR LEDs can be accurately detected and the system can identify IR LEDs without channel constraints. This simple but effective 3D localization scheme is expected to have potential on applications such as robots monitoring or motion captures.

6 Acknowledgements

This work is supported by Ministry of Science and Technology (MOST) of Taiwan (NSC101-2221-E-006-032-MY3.)

7 References

1. Vechet, S., Krejsa, J., Houska, P.: The enhancement of PCSM method by motion history analysis, Recent Advances in Mechatronics, pp. 107-110, Springer, (2007)
2. Zhang, R., Höflinger, F., Reindl L.: Inertial Sensor Based Indoor Localization and Monitoring System for Emergency Responders, IEEE Sensors Journal, 13, 838-848 (2013)
3. Kim, S.J., Kim, B.K.: Dynamic Ultrasonic Hybrid Localization System for Indoor Mobile Robots, IEEE Trans. Industrial Electronics, 60, 4562-4573(2013)
4. Royer, E., Lhuillier, M., Dhome, M., Lavest, J.: Monocular Vision for Mobile Robot Localization and Autonomous Navigation, Int. J. Computer Vision, 74, 237-260 (2007)
5. Krejsa, J., Vechet, S.: Infrared Beacons based Localization of Mobile Robot, Electronika ir Electrotechnika, 1, 17-22 (2012)
6. Gu, D., Chen, K.S.: Design and Performance Evaluation of Wiimote-Based Two-Dimensional Indoor Localization Systems for Indoor Mobile Robot Control, Measurement, 66, 95-108 (2015)
7. Fu, Y.T., Chen, K.S.: A Novel Indoor Localization Scheme by Integrating Wiimote Sensing and A Controllable IR-LED Array, Mechatronics 2013: Recent Technological and Scientific Advances, pp. 567-574, Springer (2013)

Comparative Analysis of Posture Controllers for Tracking Control of a Four-Wheeled Skid-Steered Mobile Robot – Part 1. Theoretical Considerations.

Maciej Trojnacki[1], Przemysław Dąbek[1], Janusz Kacprzyk[1], Zenon Hendzel[2]

[1] Industrial Research Institute for Automation and Measurements PIAP, Warsaw, Poland
{mtrojnacki, pdabek, jkacprzyk}@piap.pl
[2] Rzeszów University of Technology, Rzeszów, Poland
zenhen@prz.edu.pl

Abstract. The paper is concerned with the problem of trajectory tracking control of a four-wheeled PIAP GRANITE mobile robot. All wheels of the robot are non-steered and the servomotors are used for driving the robot. Kinematic structure of the robot and its kinematics are described. Structure of robot motion control system containing posture controller and drive controller is presented. Various solutions of the posture controller which allow realization of tracking control are discussed. Because of limitations of the analyzed solutions their modifications are proposed. Methodology of posture controller tuning is proposed. According to it the controller parameters for particular solutions are determined from conditions for maximum velocities of robot motion and maximum posture errors.

Keywords: wheeled mobile robot, tracking control, posture controller, drive controller, controller tuning.

1 Introduction

Tracking control, which is the kind of control where chosen point of a robot has to move on certain desired motion trajectory, belongs to important problems of control theory and robotics. From the point of view of synthesis of control systems, wheeled mobile robots can be divided in two groups: robots for which in typical operating conditions there is almost no wheel sliding on the ground during motion, and robots with all wheels non-steered for which wheel sliding is an inherent feature of motion (skid-steered robots).

Robots with steered or caster wheels belong to the first group. They are usually intended for use inside buildings (indoor robots). An example of such a design is the popular Pioneer 2-DX robot shown in Fig. 1a. Robots of this kind usually move with

R. Jabłoński and T. Brezina (eds.), *Advanced Mechatronics Solutions*,
Advances in Intelligent Systems and Computing 393,
DOI 10.1007/978-3-319-23923-1_85

minor wheel slips, so their tracking control often takes place based on wheel control only. This approach was applied to the Pioneer 2DX robot in the work [3].

a) b)

Fig. 1. Examples of wheeled mobile robots: a – Pioneer 2-DX, b – IBIS [17]

The second group is comprised of robots with all wheels non-steered. Usually robots like that are intended for use in open terrain (outdoor robots). Example of this kind of design is the IBIS robot shown in Fig. 1b [17], developed by the Industrial Research Institute for Automation and Measurements PIAP. In case of this kind of robots, wheel slip is inherent property of their motion and occurs especially during robot turning thus having significant influence on its motion. For this reason, control of this kind of robots cannot be done based only on wheel control, which was also pointed out in the work [14] dedicated to dynamics of this class of robots. Knowledge of actual robot motion parameters and control based on those known parameters are necessary.

In the present article the problem of trajectory tracking control of a four-wheeled mobile robot with non-steered wheels using posture controller is considered. By robot posture its position and course are meant. Therefore, the task of this kind of controller is determination of robot motion velocities so as to minimize position and course errors, that is the robot posture errors.

As far as mobile robot posture controllers are concerned, the following solutions are usually used: back-stepping method [6], [7], sliding mode method [9] or switching control approach [13], robust control [10], adaptive control [12], using artificial neural networks [1] or fuzzy logic [8] and various hybrid strategies.

Stability of the control systems is usually studied using Lyapunov method whose advantage is that as one of few methods it can be applied in case of non-linear systems.

Within the present work, selected existing position and course (posture) controller solutions are analyzed and their modifications aimed at removing certain limitations are proposed. Additionally, methodology of controller tuning for each solution is proposed for assumed robot boundary velocities and maximum posture errors.

2 Four-Wheeled Skid-Steered Mobile Robot

The object of the research is a small four-wheeled mobile robot called PIAP GRANITE (Ground Robot for ANalyzes of wheels Interaction with various TErrain). The robot has all wheels driven independently by DC servomotors with gear units and encoders. A visualization of the robot is shown in Fig. 2a, and its kinematic structure is presented in Fig. 2b.

a) b)

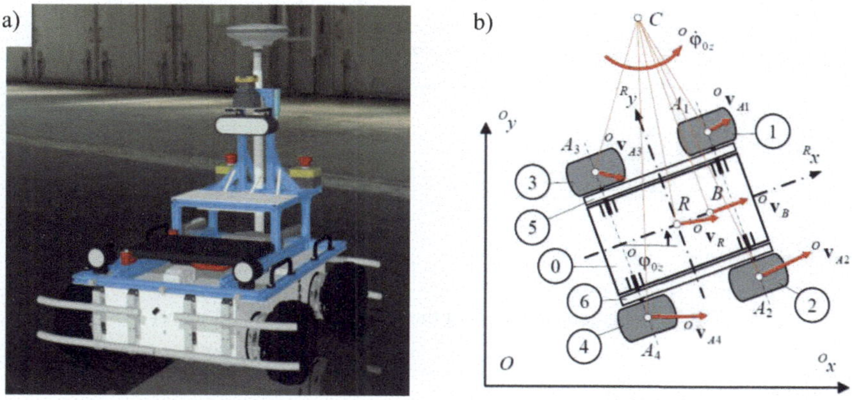

Fig. 2. Four-wheeled mobile robot: a – visualization of the robot, b – kinematic structure of the robot

It is possible to distinguish the following main components of the robot: 0 – body with frame for installation of the research equipment, 1-4 – wheels with toothed belt pulleys, 5-6 – toothed belts.

The drive transmission in each drive unit can be decoupled which permits obtaining the following three configurations of the robot chassis:

- driving of front or rear wheels only,
- driving of front or rear wheels and transmission of the drive to the remaining wheels using the toothed belts,
- independent driving of all wheels.

The following designations for the ith wheel have been introduced in the robot model: A_i – geometrical center, r_i – radius, θ_i – rotation angle ($i = \{1, ..., 4\}$).

The robot is equipped with:

- a laptop computer for control and data acquisition purposes,
- iNEMO sensors module with 3-axis MEMS accelerometer, gyro and magnetometer for determination of motion parameters of the robot [15],
- a GNSS receiver and antenna for robot navigation [11],
- a 2D laser scanner for localization in known environment [4],
- bumper switches for obstacles detection by means of direct contact,
- router and USB modem for Internet connection,
- video cameras and lights for robot teleoperation.

The robot drives are DC servomechanisms. The mechanical power from the motor is transmitted to the axle of the wheel by means of the gear unit, which is illustrated in Fig. 3.

It is assumed that configuration of the robot with independent driving of all 4 wheels is analyzed in this paper.

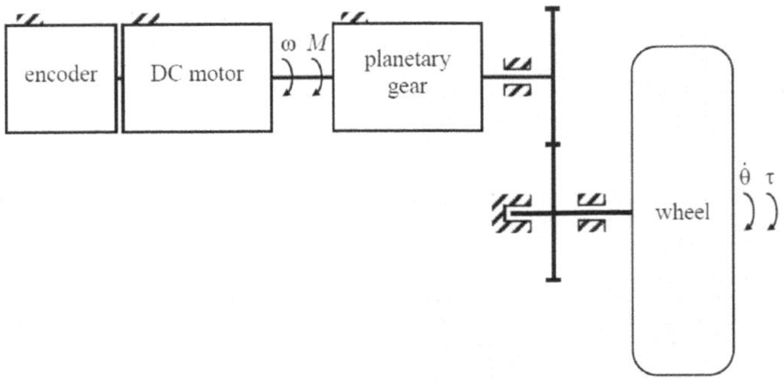

Fig. 3. Schematic diagram of the robot drive unit and wheel

3 Robot Kinematics

It is assumed that robot motion is realized in *Oxy* plane of the fixed coordinate system $\{O\}$. The moving coordinate system, considered as rigidly connected to the robot, is denoted with symbol $\{R\}$. Position and orientation of the mobile platform are described by the vector of generalized coordinates:

$$ {}^{O}\mathbf{q} = [{}^{O}x_R,\ {}^{O}y_R,\ {}^{O}\varphi_{0z}]^T , \tag{1}$$

where: ${}^{O}x_R$, ${}^{O}y_R$ are coordinates of point R belonging to the mobile platform, and ${}^{O}\varphi_{0z} = \psi$ denotes angle of spin of mobile platform about z axis with respect to fixed coordinate system $\{O\}$, which is also named the course angle.

In turn, vectors of generalized velocities respectively in $\{O\}$ and $\{R\}$ coordinate systems can be written as:

$$ {}^{O}\dot{\mathbf{q}} = \left[{}^{O}\dot{x}_R,\ {}^{O}\dot{y}_R,\ {}^{O}\dot{\varphi}_{0z}\right]^T,\ {}^{R}\dot{\mathbf{q}} = \left[{}^{R}v_{Rx},\ {}^{R}v_{Ry},\ {}^{R}\dot{\varphi}_{0z}\right]^T, \tag{2}$$

where: ${}^{O}v_{Rx} = {}^{O}\dot{x}_R$, ${}^{O}v_{Ry} = {}^{O}\dot{y}_R$, ${}^{O}\omega_{0z} = {}^{O}\dot{\varphi}_{0z}$.

Those two vectors satisfy the relationship:

$$ {}^{O}\dot{\mathbf{q}} = {}^{O}\mathbf{J}^{R}\ {}^{R}\dot{\mathbf{q}} , \tag{3}$$

where matrix ${}^{O}\mathbf{J}^{R}$ has the following form:

$$ {}^{O}\mathbf{J}^{R} = \begin{bmatrix} \cos\left({}^{O}\varphi_{0z}\right) & -\sin\left({}^{O}\varphi_{0z}\right) & 0 \\ \sin\left({}^{O}\varphi_{0z}\right) & \cos\left({}^{O}\varphi_{0z}\right) & 0 \\ 0 & 0 & 1 \end{bmatrix}. \tag{4}$$

If one makes assumption that ${}^{R}v_{Ry} \approx 0$ and motion is realized in *Oxy* plane then vector of generalized velocities ${}^{O}\dot{\mathbf{q}}$ can be defined on the basis of kinematic equations of motion in the form:

$$ {}^{O}\dot{\mathbf{q}} = \begin{bmatrix} {}^{O}\dot{x}_R \\ {}^{O}\dot{y}_R \\ {}^{O}\dot{\varphi}_{0z} \end{bmatrix} = \begin{bmatrix} \cos({}^{O}\varphi_{0z}) & 0 \\ \sin({}^{O}\varphi_{0z}) & 0 \\ 0 & 1 \end{bmatrix} \begin{bmatrix} {}^{R}v_{Rx} \\ {}^{R}\dot{\varphi}_{0z} \end{bmatrix}, \tag{5} $$

where vector ${}^{R}\mathbf{v} = [{}^{R}v_{Rx}, {}^{R}\dot{\varphi}_{0z}]^T$ contains respectively component of velocity of the point R of the robot on the x direction of $\{R\}$ coordinate system and yaw rate of the mobile platform.

In case of plane motion of the mobile platform, velocity of the point R depends on angular velocity ${}^{R}\dot{\varphi}_{0z}$ and radius of curvature R_z of the path according to the formula:

$$ {}^{R}v_R = {}^{R}\dot{\varphi}_{0z} R_z, \tag{6} $$

where ${}^{R}v_R$ is the velocity of the point R motion with respect to the stationary coordinate system $\{O\}$, expressed in the coordinates of the reference frame $\{R\}$ associated with the robot.

In addition, in a general case the acceleration vector ${}^{R}\mathbf{a}_R$ of the point R, expressed in the robot coordinate system $\{R\}$, has both tangential and normal components, that is, the following dependencies are valid:

$$ {}^{R}\mathbf{a}_R = {}^{R}\mathbf{a}_{R\tau} + {}^{R}\mathbf{a}_{Rn} \quad => \quad {}^{R}\mathbf{a}_{R\tau} = {}^{R}\mathbf{a}_R - {}^{R}\mathbf{a}_{Rn}, \tag{7} $$

where ${}^{R}\mathbf{a}_{R\tau}$ and ${}^{R}\mathbf{a}_{Rn}$ denote respectively vectors of tangential and normal accelerations of the point R.

Thus, for determination of velocity components of the point R, based on the known acceleration vector ${}^{R}\mathbf{a}_R$, the following equations resulting from projecting acceleration vectors on axes of $\{R\}$ coordinate system should be used:

$$ {}^{R}\dot{v}_{Rx} = {}^{R}a_{Rx} + {}^{R}v_{Ry}\,{}^{R}\dot{\varphi}_{0z}, \quad {}^{R}\dot{v}_{Ry} = {}^{R}a_{Ry} - {}^{R}v_{Rx}\,{}^{R}\dot{\varphi}_{0z}. \tag{8} $$

Moreover, projections of velocity of the point A_i, which is the center of the i-th wheel, on Rx and Ry axes of $\{R\}$ coordinate system depend on angular velocity of spin of the wheel $\dot{\theta}_i$ and the velocity of slip, that is, the following equations are satisfied:

$$ {}^{R}v_{Aix} = \dot{\theta}_i r_i + {}^{R}v_{Six}, \quad {}^{R}v_{Aiy} = {}^{R}v_{Siy}, \tag{9} $$

where ${}^{R}v_{Six}$, ${}^{R}v_{Siy}$ are components of slip velocity, that is, velocity of motion of points of wheel which are in contact with the ground with respect to the ground.

Coordinates of the instantaneous center of rotation C (Fig. 2b) can be calculated from dependencies:

$$ {}^{R}x_C = {}^{R}x_B = -{}^{R}v_{Ry} / {}^{R}\dot{\varphi}_{0z}, \quad {}^{R}y_C = {}^{R}v_{Rx} / {}^{R}\dot{\varphi}_{0z}. \tag{10} $$

4 Posture Controller

4.1 Desired motion and errors

In the control of posture of the robot, one assumes that motion of the robot is realized based on the desired vector of its posture (i.e. position and course), which has the form:

$$^{O}\mathbf{q}_{d}=\left[^{O}x_{Rd},\ ^{O}y_{Rd},\ ^{O}\varphi_{0zd}\right]^{T}, \tag{11}$$

where: $^{O}x_{Rd}$, $^{O}y_{Rd}$ – desired coordinates of characteristic point R of the robot in the $\{O\}$ coordinate system in (m), $^{O}\varphi_{0zd}=\psi_{d}$ – desired course of the mobile platform with respect to z-axis of $\{O\}$ coordinate system in (rad).

Desired trajectory of robot motion can be also represented in the form of vector of desired generalized velocities $^{O}\mathbf{v}_{d}=[^{O}v_{Rd},\ ^{O}\omega_{0zd}]^{T}$, which corresponds to the vector of desired generalized coordinates $^{O}\mathbf{q}_{d}$, where: $^{O}v_{Rd}$, $^{O}\omega_{0d}=\ ^{O}\dot{\varphi}_{0zd}$ – respectively desired linear velocity of the characteristic point R of the robot in (m/s) and desired angular velocity of its mobile platform in (rad/s), in the stationary coordinate system $\{O\}$.

Moreover, it is assumed that $^{O}v_{Rd}=\ ^{R}v_{Rd}=\ ^{R}v_{Rxd}$ and $^{O}\dot{\varphi}_{0zd}=\ ^{R}\dot{\varphi}_{0zd}$, that is as a result of the fact that the robot moves on a horizontal plane, values of the linear velocity and angular yaw velocity are independent of the chosen reference frame and that the lateral component of the desired linear velocity vector $^{R}v_{Ryd}\approx 0$.

In order to define the problem of tracking control, let us define desired parameters of motion of the robot in the form of equation:

$$^{O}\dot{\mathbf{q}}_{d}=\begin{bmatrix}^{O}\dot{x}_{Rd}\\ ^{O}\dot{y}_{Rd}\\ ^{O}\dot{\varphi}_{0zd}\end{bmatrix}=\begin{bmatrix}\cos(^{O}\varphi_{0zd}) & 0\\ \sin(^{O}\varphi_{0zd}) & 0\\ 0 & 1\end{bmatrix}\begin{bmatrix}^{O}v_{Rd}\\ ^{O}\omega_{0d}\end{bmatrix}. \tag{12}$$

In the problem of tracking control one should determine vector of control of position and course of the robot $\mathbf{u}_{s}=[^{R}v_{s},\ ^{R}\omega_{s}]^{T}$, such that $^{O}\mathbf{q}\rightarrow\ ^{O}\mathbf{q}_{d}$ for $t\rightarrow\infty$.

The errors of robot position and course in the coordinate system associated with the robot $\{R\}$ and in the stationary system $\{O\}$ can be determined from the relationship:

$$^{R}\mathbf{q}_{e}=\begin{bmatrix}^{R}e_{F}\\ ^{R}e_{L}\\ ^{R}e_{O}\end{bmatrix}=\begin{bmatrix}\cos(^{O}\varphi_{0z}) & \sin(^{O}\varphi_{0z}) & 0\\ -\sin(^{O}\varphi_{0z}) & \cos(^{O}\varphi_{0z}) & 0\\ 0 & 0 & 1\end{bmatrix}{}^{O}\mathbf{q}_{e},\quad ^{O}\mathbf{q}_{e}=\begin{bmatrix}^{O}e_{Rx}\\ ^{O}e_{Ry}\\ ^{O}e_{\psi}\end{bmatrix}=\begin{bmatrix}^{O}x_{Rd}-{}^{O}x_{R}\\ ^{O}y_{Rd}-{}^{O}y_{R}\\ ^{O}\varphi_{0zd}-{}^{O}\varphi_{0z}\end{bmatrix}, \tag{13}$$

where $^{R}e_{F}$, $^{R}e_{L}$, $^{R}e_{O}$ are respectively longitudinal position error in (m), lateral position error in (m), and course error in (rad).

In Fig. 4, desired and actual postures and robot posture errors are illustrated.

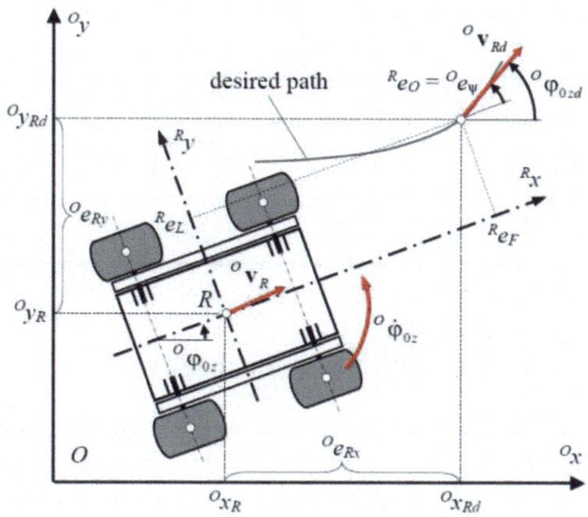

Fig. 4. Desired and actual postures and errors of posture of a robot

Vector of robot velocity errors is defined after [2], [7], [9] and [16] in the form:

$$ {}^{R}\dot{\mathbf{q}}_e = \begin{bmatrix} {}^{R}\dot{e}_F \\ {}^{R}\dot{e}_L \\ {}^{R}\dot{e}_O \end{bmatrix} = \begin{bmatrix} {}^{O}\dot{\varphi}_{0z}\,{}^{R}e_L - {}^{O}v_R + {}^{O}v_{Rd}\cos({}^{R}e_O) \\ -{}^{O}\dot{\varphi}_{0z}\,{}^{R}e_F + {}^{O}v_{Rd}\sin({}^{R}e_O) \\ {}^{O}\dot{\varphi}_{0zd} - {}^{O}\dot{\varphi}_{0z} \end{bmatrix}. \tag{14} $$

Moreover, it is assumed that ${}^{O}v_R = {}^{R}v_R \approx {}^{R}v_{Rx}$, that is motion in the ${}^{R}y$ axis direction with velocity ${}^{R}v_{Ry}$ can occur, but $|{}^{R}v_{Ry}| \ll |{}^{R}v_{Rx}|$.

In turn, the distance of the characteristic point R of the robot from the desired position of this point is described by the relationship:

$$ d = e_R = \|{}^{O}\mathbf{e}_R\| = \|{}^{R}\mathbf{e}_R\| = \sqrt{{}^{O}e_{Rx}^{\,2} + {}^{O}e_{Ry}^{\,2}} = \sqrt{{}^{R}e_F^{\,2} + {}^{R}e_L^{\,2}}, \tag{15} $$

where ${}^{O}\mathbf{e}_R = [{}^{O}e_{Rx}, {}^{O}e_{Ry}]^T$, ${}^{R}\mathbf{e}_R = [{}^{R}e_F, {}^{R}e_L]^T$.

It should be noted that because of kinematic constraints: ${}^{O}\dot{\varphi}_{0z} = \dot{\varphi}_{0zmin} = 0$ for $|{}^{O}v_R| = v_{Rmax}$ and ${}^{O}v_R = v_{Rmin} = 0$ for $|{}^{O}\dot{\varphi}_{0z}| = \dot{\varphi}_{0zmax}$.

4.2 Structure of analyzed control system

A schematic diagram of the robot tracking control system containing posture controller is shown in Fig. 5.

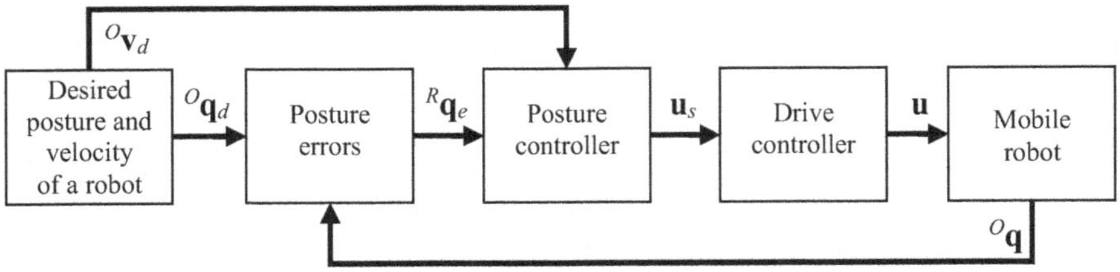

Fig. 5. Schematic diagram of control system of robot motion

The following blocks are distinguished in the diagram:

- Desired posture of a robot – responsible for generation of a vector of desired generalized coordinates $^O\mathbf{q}_d = [^O x_{Rd}, {}^O y_{Rd}, {}^O \varphi_{0zd}]^T$ and corresponding vector of desired generalized velocities $^O\mathbf{v}_d = [^O v_{Rd}, {}^O \omega_{0zd}]^T$.
- Posture controller – on the basis of desired and actual vectors of generalized coordinates, i.e., vectors $^O\mathbf{q}_d = [^O x_{Rd}, {}^O y_{Rd}, {}^O \varphi_{0zd}]^T$ and $^O\mathbf{q} = [^O x_R, {}^O y_R, {}^O \varphi_{0z}]^T$ respectively, it determines control signals $\mathbf{u}_s = [^R v_s, {}^R \omega_s]^T$, using one of the solutions discussed below.
- Drive controller – determines a vector of control signals for motors of driven wheels $\mathbf{u} = [u_l, u_r]^T$ on the basis of vector of control signals from posture controller.
- Mobile robot – dynamics model, including the drive model, of the investigated skid-steered four-wheeled PIAP GRANITE mobile robot with measurement and control equipment described in section 2.

The control signal vector $\mathbf{u}_s = [^R v_s, {}^R \omega_s]^T$ contains generalized velocities of motion of the mobile platform expressed in the robot coordinate system $\{R\}$, that is, linear velocity of characteristic point R in (m/s) and angular velocity of the mobile platform in (rad/s), which minimize posture errors. In a sense, it is a modification of the desired generalized velocities vector as a result of the occurrence of posture errors.

4.3 Velocity limits for posture controllers

It is assumed that each posture controller should generate maximum velocities respectively equal to $v_{smax} = \dot{\theta}_{max} r / 2$, and $\omega_{smax} = v_{smax} / W$. The assumed velocity v_{smax} corresponds to half of the maximum velocity of longitudinal motion of the robot, which in the analyzed case is equal to $v_{Rmax} = 1.525$ m/s. As far as $\dot{\varphi}_{0zmax}$ and ω_{smax} are concerned, in the case when the wheel slips were absent during robot turning, it would be possible to assume $\dot{\varphi}_{0zmax} = 2\, v_{Rmax} / W$ and $\omega_{smax} = 2\, v_{smax} / W$.

Maximum angular velocity of turning of the robot is chosen on assumption that longitudinal slips of wheels can be as high as 50%, which follows from the fact that in the analyzed case robot wheels are not steered. The chosen maximum velocities v_{smax} and ω_{smax} generated by the controller guarantee possibility of realization of those velocities for their arbitrary combination, which is illustrated in Fig. 6, on assumption that $\dot{\varphi}_{0zmax} = \dot{\theta}_{max} r / W$ (rad/s). In other words, in the "velocity space" is included "control space", in which in turn should be included the "space of desired trajectory of robot motion".

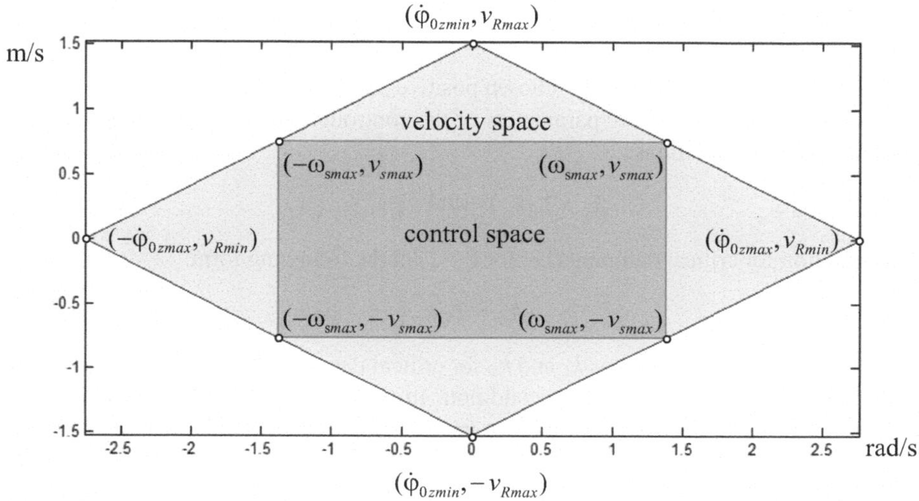

Fig. 6. Boundary velocities resulting from robot kinematic constraints and boundary velocities limiting the control signals of the posture controller

In the process of designing a posture controller, it is also necessary to make decisions as to the following questions:

- Is the robot body allowed to go backwards, that is, if the ${}^{O}v_{Rd}$ and ${}^{R}v_{s}$ velocities can have negative values?
- Are the robot wheels allowed to rotate both forwards and backwards, that is if the angular velocities $\dot{\theta}_i$ can have negative values?

Answers to those questions have important influence on the chosen law of control.

For instance, if the robot cannot move backwards, the sat(${}^{R}v_{s}$, 0, v_{smax}) function should be introduced for the ${}^{R}v_{s}$ velocity which additionally bounds the "control space", and if it can, then sat(${}^{R}v_{s}$, $-v_{smax}$, v_{smax}).

By analogy, in case of control of the angular velocities of spin of wheels, the sat() function with appropriate limits should be introduced.

4.4 Selected solutions of posture controllers

In the present work the selected solutions of posture controller are considered and their modifications aimed at solving problems occurring in case of nonzero initial and final errors and in case of errors associated with large course error are proposed.

Finite values of errors of longitudinal position, lateral position and course are assumed. Also it is assumed that the course error is in the range of $\langle -\pi, \pi \rangle$.

Solution no. 1

The control signal of posture controller can be determined based on the following relationship [7]:

$$\mathbf{u}_s = \begin{bmatrix} {}^R v_s \\ {}^R \omega_s \end{bmatrix} = \begin{bmatrix} k_F \ {}^R e_F + {}^O v_{Rd} \cos({}^R e_O) \\ {}^O \dot{\varphi}_{0zd} + k_L \ {}^O v_{Rd} \ {}^R e_L + k_O \ {}^O v_{Rd} \sin({}^R e_O) \end{bmatrix}, \tag{16}$$

where: k_F (s^{-1}), k_L (m^{-2}), k_O (m^{-1}) – chosen positive parameters.

For the values of the above parameters of the controller, authors of the work [7] propose the following relations:

$$\zeta = k_O / (2\sqrt{k_L}) \quad \text{and} \quad \xi = {}^O v_{Rd} \sqrt{k_L} \tag{17}$$

and condition for critical damping (i.e. for $\zeta = 1$) in the following form:

$$k_O = 2\sqrt{k_L} \,. \tag{18}$$

They suggest choice of gains k_L and k_O for critical damping, because the convergence is the fastest under non-oscillatory condition. In turn, in case of the ξ parameter, although its larger value makes convergence faster, too large ξ demands an excessive rotational velocity from the robot.

The proposed control law of posture controller can be modified in a certain way, so as to enforce robot stopping when the desired ${}^O v_{Rd}$ velocity is equal to zero despite the ${}^R e_F$ error being different than zero.

To this end the following formula can be used:

$$\mathbf{u}_s = \begin{bmatrix} {}^R v_s \\ {}^R \omega_s \end{bmatrix} = \begin{bmatrix} k_F \ {}^R e_F \,|\, \mathrm{sgn}({}^O v_{Rd}) \,| + {}^O v_{Rd} \cos({}^R e_O) \\ {}^O \dot{\varphi}_{0zd} + k_L \ {}^O v_{Rd} \ {}^R e_L + k_O \ {}^O v_{Rd} \sin({}^R e_O) \end{bmatrix}, \tag{19}$$

in which the expression $|\mathrm{sgn}({}^O v_{Rd})|$ was added to enforce robot stopping.

It can be noticed that in case of occurrence of the course error with the absolute value larger than $\pi/2$ rad, the proposed controller will misbehave, because in this case the $\sin({}^R e_O)$ function returns value of opposite sign with respect to ${}^R e_O$ which results in robot turning in wrong direction and further increase of the course error. Therefore, instead of the $\sin({}^R e_O)$ function, for instance the $\sin(k_a \ {}^R e_O)$ with positive nondimensional parameter $0 < k_a \leq 0.5$ or $\tanh(k_b \ {}^R e_O)$ with $k_b > 0$ parameter can be introduced. In the latter case, in particular $k_b = 1$ can be chosen, which will remove the mentioned problem.

Similarly, for the course error of absolute value greater than $\pi/2$ and smaller than π, the $\cos({}^R e_O)$ function will return a negative value, thus changing the sign of the desired velocity ${}^R v_s$, so it can be replaced for instance by $\cos(k_a \ {}^R e_O)$ or $\mathrm{sech}(k_b \ {}^R e_O) = 1/\cosh(k_b \ {}^R e_O)$ functions.

To sum up, the control vector of the posture controller can be calculated from the relationship:

$$\boxed{\mathbf{u}_{s1} = \begin{bmatrix} {}^R v_{s1} \\ {}^R \omega_{s1} \end{bmatrix} = \begin{bmatrix} k_F \ {}^R e_F \,|\, \mathrm{sgn}({}^O v_{Rd}) \,| + {}^O v_{Rd} f_v({}^R e_O) \\ {}^O \dot{\varphi}_{0zd} + k_L \ {}^O v_{Rd} \ {}^R e_L + k_O \ {}^O v_{Rd} f_\omega({}^R e_O) \end{bmatrix}}, \tag{20}$$

where: $f_v = \cos(k_a\,{}^R e_O)$ or $f_v = \mathrm{sech}(k_b\,{}^R e_O)$, and $f_\omega = \sin(k_a\,{}^R e_O)$ or $f_\omega = \tanh(k_b\,{}^R e_O)$.

In Fig. 7 plots of particular $f_v({}^R e_O)$ and $f_\omega({}^R e_O)$ functions are compared for $k_a = 0.5$ and $k_b = 1$. In the further considerations $f_v = \mathrm{sech}(k_b\,{}^R e_O)$ and $f_\omega = \tanh(k_b\,{}^R e_O)$ are assumed.

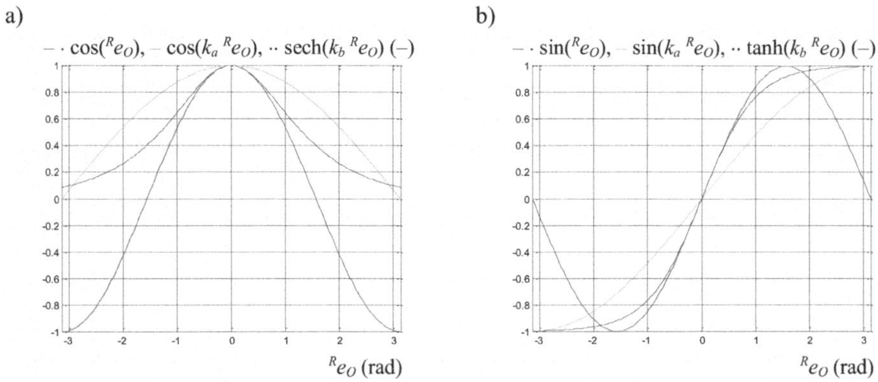

Fig. 7. Comparison of plots of various functions: a – $f_v({}^R e_O)$, b – $f_\omega({}^R e_O)$

Solution no. 2

In the work [2] the control law for posture controller is proposed in the form:

$$\mathbf{u}_s = \begin{bmatrix} {}^R v_s \\ {}^R \omega_s \end{bmatrix} = \begin{bmatrix} k_F\ {}^R e_F + {}^O v_{Rd} \cos({}^R e_O) \\ {}^O \dot{\varphi}_{0zd} + k_\psi |{}^R e_O|^a \operatorname{sgn}({}^R e_O) \end{bmatrix}, \tag{21}$$

where k_F (s^{-1}), $k_\psi > 0$ (s^{-1}), $0 < a < 1$ (–) – chosen positive parameters.

As far as the k_ψ parameter is concerned, authors of the work [2] notice that the larger k_ψ is, the sooner ${}^R e_O$ converges, but larger k_O can lead to controller saturation. Therefore, k_ψ should be chosen properly so as to satisfy the relationship:

$$\left| {}^O \dot{\varphi}_{0zd} + k_\psi |{}^R e_O|^a \operatorname{sgn}({}^R e_O) \right| \le {}^O \omega_{0max}, \tag{22}$$

where ${}^O \omega_{0max}$ is maximum angular velocity of the robot.

Moreover, they bring attention to the fact that the angular velocity control law will be singular when ${}^R e_O = 0$, therefore, when ${}^R e_O$ tends to zero the following control law can be used:

$${}^R \omega_s = k_\psi\,{}^R e_O + {}^O v_{Rd}\,{}^R e_F\,. \tag{23}$$

The above controller in comparison to the solution proposed in the work [7] contains identical control law for the linear velocity v_s, but it is different in terms of the control law for the angular velocity ${}^R\omega_s$. As a result, it has analogous limitation in the case when the desired velocity ${}^O v_{Rd}$ is equal to zero and the ${}^R e_F$ error is different than zero. In this case the problem of loss of controller stability occurs for the course error of absolute value greater than π. One may try to solve this problem by choosing appropriate parameter k_ψ. Moreover, this controller does not directly remove the lateral position error ${}^Re_L$.

Taking the above considerations into account, for further studies the following modified version of this controller is proposed:

$$\mathbf{u}_{s2} = \begin{bmatrix} {}^Rv_{s2} \\ {}^R\omega_{s2} \end{bmatrix} = \begin{bmatrix} k_F\ {}^Re_F\,|\operatorname{sgn}({}^Ov_{Rd})| + {}^Ov_{Rd} f_v({}^Re_O) \\ {}^O\dot{\varphi}_{0zd} + k_\psi |{}^Re_O|^a \tanh(k_\omega\ {}^Re_O)|\operatorname{sgn}({}^Ov_{Rd})| \end{bmatrix}, \tag{24}$$

where, similarly as in the previous solution, the expression $|\operatorname{sgn}({}^Ov_{Rd})|$ aimed at enforcing robot stopping and $f_v({}^Re_O)$ function instead of $\cos({}^Re_O)$ function which eliminates problems for the course error of absolute value greater than $\pi/2$ are added. Additionally, instead of $\operatorname{sgn}({}^Re_O)$ expression the $\tanh(k_\omega\ {}^Re_O)$ function is introduced with $k_\omega > 0$, which leads to less oscillations generated by the controller in case of the ${}^Re_O$ error close to zero.

Solution no. 3

In turn in the work [9] authors propose the sliding-mode controller. For the switching function the expression ${}^Re_O = -\operatorname{atan}({}^Ov_{Rd}\ {}^Re_L)$ is chosen, and then the following switching surface is proposed:

$$\mathbf{s} = \begin{bmatrix} s_v \\ s_\omega \end{bmatrix} = \begin{bmatrix} {}^Re_F \\ {}^Re_O + \operatorname{atan}({}^Ov_{Rd}\ {}^Re_L) \end{bmatrix}. \tag{25}$$

Eventually, using a reaching law approach, they assumed the control law in the following form:

$$\mathbf{u}_s = \begin{bmatrix} {}^Rv_s \\ {}^R\omega_s \end{bmatrix} = \begin{bmatrix} {}^Re_L\ {}^O\dot{\varphi}_{0z} + {}^Ov_{Rd}\cos({}^Re_O) + k_S\operatorname{sat}(s_v) \\ {}^O\dot{\varphi}_{0zd} + \dfrac{\partial\alpha}{\partial^O v_{Rd}}{}^O\dot{v}_{Rd} + \dfrac{\partial\alpha}{\partial^R e_L}\left({}^Ov_{Rd}\sin({}^Re_O)\right) + k_\psi\operatorname{sat}(s_\omega) \end{bmatrix}, \tag{26}$$

where $\dfrac{\partial\alpha}{\partial^O v_{Rd}} = \dfrac{{}^Re_L}{1+({}^Ov_{Rd}\ {}^Re_L)^2}$ after rescaling is in (s/m), $\dfrac{\partial\alpha}{\partial^R e_L} = \dfrac{{}^Ov_{Rd}}{1+({}^Ov_{Rd}\ {}^Re_L)^2}$ after rescaling is in (m^{-1}), k_S (s^{-1}), k_ψ (s^{-1}) are chosen positive parameters and sat() is a saturation function.

Unfortunately, the authors do not give any guidance concerning selection of controller parameters.

The present controller contains more complex expressions in comparison to the previously described solutions and additionally it is relatively demanding when it comes to required knowledge of robot motion parameters.

In particular, it requires knowledge of acceleration ${}^O\dot{v}_{Rd}$ and angular velocity ${}^O\dot{\varphi}_{0z}$, but in case of the latter, one may notice that:

$${}^O\dot{\varphi}_{0z} = {}^O\dot{\varphi}_{0zd} - {}^Re_O\,. \tag{27}$$

With the above control law are associated similar problems as described previously. Moreover, instead of the sat(s_x) function it is beneficial to introduce tanh(k_x s_x) function with appropriately chosen positive parameter k_x, where $x = \{v, \omega\}$.

One may also notice that use of this controller is troublesome in case of large ${}^Re_L$ error, large value of acceleration ${}^O\dot{v}_{Rd}$, and large value of velocity ${}^Ov_{Rd}$ as compared to the angular velocity ${}^O\dot{\varphi}_{0z}$. Then the terms ${}^Re_L\,{}^O\dot{\varphi}_{0z}$ and ${}^O\dot{v}_{Rd}$ and ${}^Ov_{Rd}\sin({}^Re_O)$ for respective control signals ${}^Rv_s$ and ${}^R\omega_s$ can dominate the control law leading to algorithm instability, therefore limitation of their influence on the total control signal should be considered by means of introduction of additional controller parameters.

After assuming that:

$$k_F|{}^Re_L\,{}^O\dot{\varphi}_{0z}|\le{}^Ov_{Rd},\; k_L\frac{\partial\alpha}{\partial^Ov_{Rd}}|{}^O\dot{v}_{Rd}|\le{}^O\dot{\varphi}_{0zd} \text{ and } k_O\frac{\partial\alpha}{\partial^Re_L}|{}^Ov_{Rd}|\le{}^O\dot{\varphi}_{0zd}, \tag{28}$$

where k_F (–), k_L (m^{-2}) and k_O (m^{-2}) are chosen positive parameters, ultimately the following modified version of the control law is proposed:

$$\boxed{\mathbf{u}_{s3}=\begin{bmatrix}{}^Rv_{s3}\\ {}^R\omega_{s3}\end{bmatrix}=\begin{bmatrix}\begin{array}{c}k_F\,{}^Re_L\,{}^O\dot{\varphi}_{0z}\,|\mathrm{sgn}({}^Ov_{Rd})|+{}^Ov_{Rd}f_v({}^Re_O)+\\ +k_S\,|\mathrm{sgn}({}^Ov_{Rd})|\tanh(k_vs_v)\end{array}\\ \begin{array}{c}{}^O\dot{\varphi}_{0zd}+k_L\frac{\partial\alpha}{\partial^Ov_{Rd}}{}^O\dot{v}_{Rd}+k_O\frac{\partial\alpha}{\partial^Re_L}\left({}^Ov_{Rd}f_\omega({}^Re_O)\right)+\\ +k_\psi\,|\mathrm{sgn}({}^Ov_{Rd})|\tanh(k_\omega s_\omega)\end{array}\end{bmatrix}}. \tag{29}$$

Solution no. 4

In the work [16] a controller of the following form proposed in [6] is used:

$$\mathbf{u}_s=\begin{bmatrix}{}^Rv_s\\ {}^R\omega_s\end{bmatrix}=\begin{bmatrix}k_F\,{}^Re_F+{}^Ov_{Rd}\cos({}^Re_O)\\ {}^O\dot{\varphi}_{0zd}+k_\psi\,{}^Re_O+{}^Ov_{Rd}\,{}^Re_L\sin({}^Re_O)/{}^Re_O\end{bmatrix}, \tag{30}$$

where $k_F > 0$ (s^{-1}), $k_\psi > 0$ (s^{-1}), and authors focused on the problems associated with estimation of position errors.

Unfortunately authors do not give any guidance concerning choice of the controller parameters.

In the analyzed case, during simulation numerical problems can occur for ${}^Re_O = 0$. For this reason, one may assume a small allowable course error ${}^Re_{Omin}$, and then instead of $\sin({}^Re_O)/{}^Re_O$ introduce $|\sin({}^Re_O)|/(|{}^Re_O|+{}^Re_{Omin})$.

One may also notice that in the case of the analyzed controller, for large errors ${}^Re_L$ and ${}^Re_O$ and for large value of velocity ${}^Ov_{Rd}$ as compared to the angular velocity ${}^O\dot{\varphi}_{0z}$, the expression ${}^Ov_{Rd}\,{}^Re_L\sin({}^Re_O)/{}^Re_O$ can dominate the control signal ${}^R\omega_s$ leading to algorithm instability.

As a result, introduction of additional k_O parameter, modification of this expression to the form $k_O\,{}^O v_{Rd}\,{}^R e_L\,|\,f_\omega({}^R e_O)\,|\,/(|{}^R e_O|+{}^R e_{Omin})$ and acceptance of the following assumption are proposed:

$$k_O\,{}^O v_{Rd}\,{}^R e_L\,|\,f_\omega({}^R e_O)\,|\,/(|{}^R e_O|+{}^R e_{Omin}) \le {}^O\dot{\varphi}_{0zd}\,. \tag{31}$$

After introducing similar as previously modifications resulting from analogous as previously limitations, one may eventually assume the following control law:

$$\mathbf{u}_{s4} = \begin{bmatrix} {}^R v_{s4} \\ {}^R \omega_{s4} \end{bmatrix} = \begin{bmatrix} k_F\,{}^R e_F\,|\,\mathrm{sgn}({}^O v_{Rd})\,| + {}^O v_{Rd} f_v({}^R e_O) \\ {}^O\dot{\varphi}_{0zd} + k_\psi\,{}^R e_O\,|\,\mathrm{sgn}({}^O v_{Rd})\,| + \\ + k_O\,{}^O v_{Rd}\,{}^R e_L\,|\,f_\omega({}^R e_O)\,|\,/(|{}^R e_O|+{}^R e_{Omin}) \end{bmatrix}, \tag{32}$$

where $k_O > 0$ (m^{-2}).

It can be noticed that the discussed solution does not eliminate the ${}^R e_L$ error when ${}^R e_O = 0$, because $f_\omega(0) = 0$.

Solution no. 5

In the work [5] the following control law of the posture controller is proposed:

$$\mathbf{u}_s = \begin{bmatrix} {}^R v_s \\ {}^R \omega_s \end{bmatrix} = \begin{bmatrix} k_S \tanh({}^R e_F) + {}^O v_{Rd} \cos({}^R e_O) \\ {}^O\dot{\varphi}_{0zd} + k_\psi \tanh({}^R e_O) + {}^O v_{Rd}\,{}^R e_L \sin({}^R e_O)/\left({}^R e_O(1+{}^R e_F^2+{}^R e_L^2)\right) \end{bmatrix}, \tag{33}$$

which for the reasons discussed earlier (problems for large values of ${}^R e_O$ errors and for ${}^R e_O = 0$, necessity of limitation of the ${}^O v_{Rd}\,{}^R e_L$ value, etc.) may be modified to the form:

$$\mathbf{u}_{s5} = \begin{bmatrix} {}^R v_{s5} \\ {}^R \omega_{s5} \end{bmatrix} = \begin{bmatrix} k_S \tanh(k_v\,{}^R e_F)\,|\,\mathrm{sgn}({}^O v_{Rd})\,| + {}^O v_{Rd} f_v({}^R e_O) \\ {}^O\dot{\varphi}_{0zd} + k_\psi \tanh(k_\omega\,{}^R e_O)\,|\,\mathrm{sgn}({}^O v_{Rd})\,| + \\ + k_O\,{}^O v_{Rd}\,{}^R e_L\,|\,f_\omega({}^R e_O)\,|\,/\left((|{}^R e_O|+{}^R e_{Omin})(1+{}^R e_F^2+{}^R e_L^2)\right) \end{bmatrix}. \tag{34}$$

where k_S (m/s), k_ψ (s^{-1}), k_O (–) – chosen positive parameters.

In the proposed solution, by analogy to the solution no. 3, also k_v (m^{-1}) and k_ω (–) parameters are introduced.

Unfortunately authors do not give guidance for choice of controller parameters. It can be also noticed that the presented solution does not eliminate ${}^R e_L$ error, when ${}^R e_O = 0$.

Solution no. 6

In the work [13] a bit different approach is applied in comparison to previously described controllers, because of the proposed definition of the position and course errors. The position error is defined not as two separate longitudinal and lateral position errors but as a single error in the form of a distance from the actual to desired position. In this case, the posture error vector has the following form:

$$^{R}\mathbf{q}_e = \begin{bmatrix} ^{R}d \\ ^{R}e_O \end{bmatrix} = \begin{bmatrix} \sqrt{(^{R}e_F)^2 + (^{R}e_L)^2} \\ ^{R}\varphi^{O}_{0zd} - ^{R}\varphi^{O}_{0z} \end{bmatrix} = \begin{bmatrix} \sqrt{(^{O}e_{Rx})^2 + (^{O}e_{Ry})^2} \\ ^{O}\varphi_{0zd} - ^{O}\varphi_{0z} \end{bmatrix}. \tag{35}$$

Then, the control law for the posture controller has the form:

$$\mathbf{u}_s = \begin{bmatrix} ^{R}v_s \\ ^{R}\omega_s \end{bmatrix} = \begin{bmatrix} \dfrac{d}{1+d}\, ^{O}v_{Rdmax} \cos(^{R}e_O) \\ \dfrac{^{O}v_{Rd}}{d} \sin(^{R}e_O) + k_\psi \tanh(k_\omega\, ^{R}e_O) \end{bmatrix}, \tag{36}$$

Unfortunately, in their work authors use identical symbols v in completely different contexts, so it is difficult to tell if in the analyzed case for the course controller the correct meaning is $^{O}v_{Rd}$ or $^{O}v_R$ or $^{R}v_s$. Moreover, in the work there is no guidance concerning choice of the parameters.

One disadvantage of the proposed approach is the occurrence of division by d, and as a result, as the authors also point out, the robot has to stop in the desired position with assumed inaccuracy d_{min}. Moreover, because of the previously discussed problems instead of the $\cos(^{R}e_O)$ and $\sin(^{R}e_O)$ functions the $f_v(^{R}e_O)$ and $f_\omega(^{R}e_O)$ functions are introduced.

Additionally, in the law of control of the longitudinal velocity in the denominator the value 1 is present, which can result in excessive limitation of the linear velocity and consequently larger error of robot position, therefore it is beneficial to introduce certain value d_{min} in place of 1.

Finally, the control law may have the following form:

$$\boxed{\mathbf{u}_{s6} = \begin{bmatrix} ^{R}v_{s6} \\ ^{R}\omega_{s6} \end{bmatrix} = \begin{bmatrix} k_F \dfrac{d}{d+d_{min}}\, ^{O}v_{Rdmax} f_v(^{R}e_O) |\mathrm{sgn}(^{O}v_{Rd})| \\ \dfrac{^{O}v_{Rd}}{d+d_{min}} f_\omega(^{R}e_O) + k_\psi \tanh(k_\omega\, ^{R}e_O) |\mathrm{sgn}(^{O}v_{Rd})| \end{bmatrix}}, \tag{37}$$

where k_F (–), k_ψ (s^{-1}) – chosen positive parameters.

It can be noticed that described solution does not eliminate the $^{R}e_L$ error when $^{R}e_O = 0$.

4.5 Drive controller

Having control signal $\mathbf{u}_s = [^{R}v_s, ^{R}\omega_s]^T$ containing generalized velocities in (m/s) and (rad/s) respectively, the control signal for motors of driven wheels in (V) can be determined using the relationship which results from robot kinematics:

$$\mathbf{u} = \begin{bmatrix} u_l \\ u_r \end{bmatrix} = \mathrm{sat}\left(\frac{k_u}{r} \begin{bmatrix} 1 & -W/2 \\ 1 & W/2 \end{bmatrix} \mathrm{sat}(\mathbf{u}_s, \mathbf{u}_{smin}, \mathbf{u}_{smax}), \mathbf{u}_{min}, \mathbf{u}_{max} \right), \tag{38}$$

where u_l and u_r are control signals for respectively left-hand side and right-hand side wheels of the robot, the parameter $k_u = 1$ Vs, and $\mathbf{u}_s = \{\mathbf{u}_{s1}, \mathbf{u}_{s2}, \mathbf{u}_{s3}, \mathbf{u}_{s4}, \mathbf{u}_{s5}, \mathbf{u}_{s6}\}$ is a control vector from one of the earlier described solutions of the posture controller, $\mathbf{u}_{smax} = [v_{smax}, \omega_{smax}]^T$, $\mathbf{u}_{smin} = [v_{smin}, -\omega_{smax}]^T$, $\mathbf{u}_{max} = [u_{max}, u_{max}]^T$, $\mathbf{u}_{min} = [u_{min}, u_{min}]^T$, the sat() function is saturation function aimed at bounding the signal, and $u_{min} = -u_{max}$ or $u_{min} = 0$ and $v_{smin} = -v_{smax}$ or $v_{smin} = 0$, depending on chosen assumptions.

In real implementation, to the hardware drive controller are passed integer control values within the range of ±970, and it directly controls the drives using pulse width modulation technique (PWM), where the amplitude of this control is equal to u_{max}.

One should also notice that the expression

$$\frac{1}{r}\begin{bmatrix} 1 & -W/2 \\ 1 & W/2 \end{bmatrix}\mathbf{u}_s \tag{39}$$

has dimension of (rad/s), that is of angular velocity of spin of wheels, which due to the properties of drive units cannot rotate with velocities higher than $\dot{\theta}_{max}$.

4.6 Posture controller tuning methodology

During posture controller tuning it is assumed that for any time instant t, the following conditions are satisfied:

$$|{}^O v_{Rd}| \le v_{smax} \le v_{Rmax} \text{ and } |{}^O\dot{\varphi}_{0zd}| = |{}^O\omega_{0d}| \le \omega_{smax} \le \dot{\varphi}_{0zmax}, \tag{40}$$

$$|{}^O v_R| \le v_{smax} \text{ and } |{}^O\dot{\varphi}_{0z}| = |{}^O\omega_0| \le \omega_{smax}. \tag{41}$$

In case of using $f_v = \cos(k_a\,{}^R e_O)$ and $f_\omega = \sin(k_a\,{}^R e_O)$ functions the parameter $k_a = 1/2$ is assumed, which eliminates the problems described earlier associated with change of $\sin({}^R e_O)$ and $\cos({}^R e_O)$ functions for arguments of absolute values greater than $\pi/2$, which leads to losing the controller stability. In turn, for the tanh() function occurring in solutions no. 3, 5 and 6 it is assumed that it should saturate for the argument of $\pm\pi$ value. After assuming that maximum absolute values of longitudinal position error and course error are respectively equal to ${}^R e_{Fmax} = 1$ m and ${}^R e_{Omax} = \pi/2$ rad, the gains k_v and k_ω for the analyzed solutions of controllers should be respectively equal to π and 2.

Values of the critical errors should be chosen for a given object in such way that the controller does not generate errors larger than critical. Therefore, they should not be too small, to avoid exceeding the limits easily.

During choice of the remaining parameters for posture controller it is assumed that the controller should generate maximum velocities of v_{smax} or ω_{smax} for minimum or critical posture errors respectively equal to ${}^R e_{Fmin} = 0$, ${}^R e_{Lmin} = 0$, ${}^R e_{Omin} = 0$ and ${}^R e_{Fmax} = 1$ m, ${}^R e_{Lmax} = 1$ m, ${}^R e_{Omax} = \pi/2$ rad. Moreover, it is assumed that in case of their surpassing, the controller should still generate maximum velocities v_{smax} or ω_{smax} thanks to use of the saturation function. The case when maximum errors are assumed too large leads to situation where controller gains are relatively small, and as a result the controller is less effective, that is larger tracking errors are generated.

After taking into account the above considerations, values of parameters for the analyzed solutions of the posture controller can be assumed as initial approximation according to the rules described below.

Solution no. 1
From the condition for maximum linear velocity:

$$k_F \; {}^R e_{Fmax} | \operatorname{sgn}({}^O v_{Rdmax}) | + {}^O v_{Rdmax} f_v({}^R e_{Omin}) \leq v_{smax} \tag{42}$$

the following constraint for the k_F parameter is obtained:

$$\boxed{k_F \leq (v_{smax} - {}^O v_{Rdmax}) / {}^R e_{Fmax}}, \tag{43}$$

whereas from the condition for maximum angular velocity:

$${}^O \dot{\varphi}_{0zdmax} + k_L \; {}^O v_{Rdmax} \; {}^R e_{Lmax} + k_O \; {}^O v_{Rdmax} f_\omega({}^R e_{Omax}) \leq \omega_{smax} \tag{44}$$

it is possible to define constraints for k_L and k_O parameters.

After assuming, following in this respect the work [7], that values of those parameters should be chosen for critical damping, that is, that the relationship below is satisfied:

$$\boxed{k_L = (k_O / 2)^2}. \tag{45}$$

Then, based on the condition:

$$(k_O)^2 \; {}^R e_{Lmax} / 4 + k_O - (\omega_{smax} - {}^O \dot{\varphi}_{0zdmax}) / {}^O v_{Rdmax} \leq 0 \tag{46}$$

one of two possible solutions, that is for $k_O > 0$, has the form:

$$\boxed{k_O \leq 2\left(\sqrt{1 + {}^R e_{Lmax} (\omega_{smax} - {}^O \dot{\varphi}_{0zdmax}) / {}^O v_{Rdmax}} - 1\right) / {}^R e_{Lmax}}. \tag{47}$$

Solution no. 2
As previously, from the condition for maximum linear velocity one obtains:

$$\boxed{k_F \leq (v_{smax} - {}^O v_{Rdmax}) / {}^R e_{Fmax}}, \tag{48}$$

whereas from the condition for maximum angular velocity:

$${}^O \dot{\varphi}_{0zdmax} + k_\psi \, | {}^R e_{Omax} |^a \leq \omega_{smax}, \tag{49}$$

one gets:

$$\boxed{k_\psi \leq (\omega_{smax} - {}^O \dot{\varphi}_{0zdmax}) / | {}^R e_{Omax} |^a}, \tag{50}$$

where $0 < a < 1$ (–), e.g. $a = 0.5$ (–).

Solution no. 3

In case of the present solution for the controller, determination of boundary parameter values is more complex.

From the condition:

$$k_F\, {}^R e_{Lmax}\omega_{smax} + {}^O v_{Rdmax} f_v({}^R e_{Omin}) + k_S \tanh(k_v s_{vmax}) \le v_{smax}, \tag{51}$$

assuming that ${}^O\dot{\varphi}_{0zmax} = {}^R\omega_{smax}$, it is possible to determine parameters k_F and k_S by introducing additional assumptions.

After assuming that expression $k_F\, {}^R e_{Lmax}\omega_{smax}$ should have smaller influence on the control of robot longitudinal velocity relatively to other expressions, and assuming maximum value of this expression equal to $\eta\, {}^O v_{Rdmax}$, where η is a parameter such that $0 < \eta \ll 1$, it is possible to determine k_F parameter:

$$k_F\, {}^R e_{Lmax}\omega_{smax} = \eta\, {}^O v_{Rdmax} \quad => \quad \boxed{k_F = \eta\, {}^O v_{Rdmax} / ({}^R e_{Lmax}\omega_{smax})}, \tag{52}$$

and next the parameter k_S depending on the parameter k_F:

$$\boxed{k_S \le v_{smax} - k_F\, {}^R e_{Lmax}\omega_{smax} - {}^O v_{Rdmax}}, \tag{53}$$

where in the further studies $\eta = 0.1$ (–) will be assumed.

Similarly, it is possible to choose boundary values of parameters for control of angular velocity of robot turning.

Then, based on the condition:

$${}^O\dot{\varphi}_{0zdmax} + k_L \frac{\partial\alpha}{\partial {}^O v_{Rd}} {}^O\dot{v}_{Rdmax} + k_O \frac{\partial\alpha}{\partial {}^R e_L}\left({}^O v_{Rdmax} f_\omega({}^R e_{Omax})\right) + k_\psi \tanh(k_\omega s_{\omega max}) \le \omega_{smax}, \tag{54}$$

after assuming, similarly as previously, that maximum values of expressions containing parameters k_L and k_O should be equal to $\eta\, {}^O\dot{\varphi}_{0zdmax}$, one obtains:

$$\frac{k_L\, {}^R e_{Lmax}\, {}^O\dot{v}_{Rdmax}}{1 + ({}^O v_{Rdmax}\, {}^R e_{Lmax})^2} = \eta\, {}^O\dot{\varphi}_{0zdmax} \quad => \quad \boxed{k_L = \eta\, {}^O\dot{\varphi}_{0zdmax} \frac{1 + ({}^O v_{Rdmax}\, {}^R e_{Lmax})^2}{{}^R e_{Lmax}\, {}^O\dot{v}_{Rdmax}}}, \tag{55}$$

$$\frac{k_O ({}^O v_{Rdmax})^2}{1 + ({}^O v_{Rdmax}\, {}^R e_{Lmax})^2} = \eta\, {}^O\dot{\varphi}_{0zdmax} \quad => \quad \boxed{k_O = \eta\, {}^O\dot{\varphi}_{0zdmax} \frac{1 + ({}^O v_{Rdmax}\, {}^R e_{Lmax})^2}{({}^O v_{Rdmax})^2}}, \tag{56}$$

based on which, the last missing parameter k_ψ can be determined, as a function of k_L and k_O parameters, in the form:

$$\boxed{k_\psi \le \omega_{smax} - {}^O\dot{\varphi}_{0zdmax} - k_L \frac{\partial\alpha}{\partial {}^O v_{Rd}} {}^O\dot{v}_{Rdmax} - k_O \frac{\partial\alpha}{\partial {}^R e_L} {}^O v_{Rdmax}}, \tag{57}$$

where in this case $\dfrac{\partial\alpha}{\partial {}^O v_{Rd}} = \dfrac{{}^R e_{Lmax}}{1 + ({}^O v_{Rdmax}\, {}^R e_{Lmax})^2}$, $\dfrac{\partial\alpha}{\partial {}^R e_L} = \dfrac{{}^O v_{Rdmax}}{1 + ({}^O v_{Rdmax}\, {}^R e_{Lmax})^2}$.

Solution no. 4

Because of analogous form of the control law for the linear velocity as in solutions no. 1 and 2, the same constraint for the k_F parameter is obtained:

$$\boxed{k_F \le ({v_{smax}} - {}^O v_{Rdmax}) / {}^R e_{Fmax}}. \tag{58}$$

In turn, boundary values of parameters for the control of angular velocity of robot turning can be determined for the assumed maximum angular velocity of robot turning, from the following condition:

$$\begin{gathered} {}^O\dot{\varphi}_{0zdmax} + k_\psi\, {}^R e_{Omax}\, |\operatorname{sgn}({}^O v_{Rdmax})| + \\ + k_O\, {}^O v_{Rdmax}\, {}^R e_{Lmax}\, |f_\omega({}^R e_{Omax})| / (|{}^R e_{Omax}| + {}^R e_{Omin}) \le \omega_{smax}. \end{gathered} \tag{59}$$

Assuming that the expression $k_O\, {}^O v_{Rdmax}\, {}^R e_{Lmax}$ should have smaller influence on robot angular velocity control as compared to the other expressions, and then after assuming the maximum value of this expression equal to $\eta\, {}^O\dot{\varphi}_{0zdmax}$ (by analogy to the previous solution), it is possible to determine the k_O parameter:

$$k_O\, {}^O v_{Rdmax}\, {}^R e_{Lmax} = \eta\, {}^O\dot{\varphi}_{0zdmax} \quad => \quad \boxed{k_O = \eta \frac{{}^O\dot{\varphi}_{0zdmax}}{{}^O v_{Rdmax}\, {}^R e_{Lmax}}}, \tag{60}$$

and then the k_ψ parameter as a function of the k_O parameter:

$$\boxed{k_\psi \le \left(\omega_{smax} - {}^O\dot{\varphi}_{0zdmax} - k_O\, {}^O v_{Rdmax}\, {}^R e_{Lmax} / (|{}^R e_{Omax}| + {}^R e_{Omin})\right) / {}^R e_{Omax}}. \tag{61}$$

Solution no. 5

From the condition of maximum linear velocity of the robot for linear velocity control:

$$k_S \tanh(k_v\, {}^R e_{Fmax})\, |\operatorname{sgn}({}^O v_{Rdmax})| + {}^O v_{Rdmax} f_v({}^R e_{Omin}) \le v_{smax} \tag{62}$$

the constraint for the k_S parameter is obtained:

$$\boxed{k_S \le v_{smax} - {}^O v_{Rdmax}}. \tag{63}$$

In the case of the angular velocity of robot turning, boundary values of controller parameters are determined based on the relationship:

$$\begin{gathered} {}^O\dot{\varphi}_{0zdmax} + k_\psi \tanh(k_\omega\, {}^R e_{Omax})\, |\operatorname{sgn}({}^O v_{Rdmax})| + \\ + k_O\, {}^O v_{Rdmax}\, {}^R e_{Lmax}\, |f_\omega({}^R e_{Omax})| / \left((|{}^R e_{Omax}| + {}^R e_{Omin})(1 + {}^R e_{Fmax}^2 + {}^R e_{Lmax}^2)\right) \le \omega_{smax}. \end{gathered} \tag{64}$$

As previously, it is possible to assume the maximum value of the expression containing the k_O parameter, for example:

$$k_O\, {}^O v_{Rdmax}\, {}^R e_{Lmax} / \left((|{}^R e_{Omax}| + {}^R e_{Omin})(1 + {}^R e_{Fmax}^2 + {}^R e_{Lmax}^2)\right) = \eta\, {}^O\dot{\varphi}_{0zdmax}, \tag{65}$$

hence:

$$\boxed{k_O = \eta \frac{{}^O\dot{\varphi}_{0zdmax}}{{}^O v_{Rdmax}\ {}^R e_{Lmax}} \left((|{}^R e_{Omax}| + {}^R e_{Omin})(1 + {}^R e_{Fmax}^2 + {}^R e_{Lmax}^2) \right)} \tag{66}$$

and next for this value, determine the critical value of the second parameter, that is:

$$\boxed{k_\psi \le \omega_{smax} - {}^O\dot{\varphi}_{0zdmax} - k_O\, {}^O v_{Rdmax}\ {}^R e_{Lmax} / \left((|{}^R e_{Omax}| + {}^R e_{Omin})(1 + {}^R e_{Fmax}^2 + {}^R e_{Lmax}^2) \right)}. \tag{67}$$

Solution no. 6
In case of the last analyzed solution of the posture controller, from the condition for maximum linear velocity:

$$k_F \frac{d_{max}}{d_{max} + d_{min}} {}^O v_{Rdmax} f_v({}^R e_{Omin}) \le v_{smax}, \tag{68}$$

the boundary value of the k_F parameter is obtained in the following form:

$$\boxed{k_F \le v_{smax}(d_{max} + d_{min}) / (d_{max}\, {}^O v_{Rdmax})}. \tag{69}$$

In turn, for the control of angular velocity of robot turning, from the condition:

$$\frac{{}^O v_{Rdmax}}{d_{max} + d_{min}} f_\omega({}^R e_{Omax}) + k_\psi \tanh(k_\omega\, {}^R e_{Omax}) \,|\, \mathrm{sgn}({}^O v_{Rdmax}) \,| \le \omega_{smax} \tag{70}$$

boundary value of the second parameter is obtained:

$$\boxed{k_\psi \le \omega_{smax} - \frac{{}^O v_{Rdmax}}{d_{max} + d_{min}}}, \tag{71}$$

where d_{min} should not be too small, for the term ${}^O v_{Rdmax} / (d_{max} + d_{min})$ not to dominate the control of robot turning.

Therefore, d_{min} should be chosen so that the following condition is satisfied:

$$\frac{{}^O v_{Rdmax}}{d_{max} + d_{min}} \le {}^O \omega_{0dmax} \quad => \quad \boxed{d_{min} \ge \frac{{}^O v_{Rdmax}}{{}^O \omega_{0dmax}} - d_{max}}, \tag{72}$$

on the assumption that ${}^O \omega_{0dmax} \ne 0$.

5 Summary and Future Works

In the present work, selected solutions of posture controller from the literature were presented and their modifications were proposed aimed at eliminating problems associated with nonzero initial and final errors as well as with large values of the course error.

The structure of the robot tracking control system was presented and methodology of choosing parameters of particular posture controller solutions was proposed, assuming maximum robot posture errors and limits of the control signals following from robot kinematic constraints.

The second part of the work will be concerned with comparative analysis of the described solutions of posture controller by means of simulation studies using the robot dynamics model. Effectiveness of particular solutions will be evaluated using the introduced quality indexes.

Moreover, experimental investigations of robot tracking control using posture controller of different forms are planned to support the theoretical analysis.

Directions of future works will also include development of solutions of the posture controller where:

- artificial intelligence techniques like artificial neural networks or systems with fuzzy logic will be used,
- change of controller parameters will be possible depending on the actual robot motion parameters and posture errors,
- "control space" will be extended to cover the whole "velocity space" and intelligent saturation will be used.

Acknowledgements. The work has been realized as a part of the project entitled “Dynamics modeling of four-wheeled mobile robot and tracking control of its motion with limitation of wheels slip”. The project is financed from the means of National Science Centre of Poland granted on the basis of decision number DEC-2011/03/B/ST7/02532.

References

1. Bayar G., Konukseven E.I., Koku A.B., Mobile Robot Heading Adjustment Using Radial Basis Function Neural Networks Controller and Reinforcement Learning, Proceedings of the 4th WSEAS/IASME International Conference on Dynamical Systems and Controls, 169–174 (2008).
2. Cao M., Algorithms research of autonomous navigation and control of planetary exploration rover, 2010 Chinese Control and Decision Conference, IEEE, 4359–4364 (2010).
3. Hendzel Z., Szuster M., Discrete model-based adaptive critic designs in wheeled mobile robot control, Artifical Intelligence and Soft Computing, Springer Berlin Heidelberg, 264-271 (2010).
4. Jaroszek P., Trojnacki M., Localization of the wheeled mobile robot based on multi-sensor data fusion, Journal of Automation, Mobile Robotics & Intelligent Systems, (submitted).
5. Jiang Z.-P., Lefeber E., Nijmeijer H., Saturated stabilization and tracking of a nonholonomic mobile robot, Systems & Control Letters 42.5, 327-332 (2001).
6. Jiang Z.P., Nijmeijer H., Tracking control of mobile robots: A case study in backstepping, Automatica, Vol. 33, 1393–1399 (1997).
7. Kanayama Y., Kimura Y., Miyazaki F., Noguchi T., A stable tracking control method for an autonomous mobile robot, IEEE International Conference on Robotics and Automation, Cincinnati, OH, USA, 384-389 (1990).

8. Lakehal B., Amirat Y., Pontnau J., Fuzzy steering control of a mobile robot, International IEEE/IAS Conference on Industrial Automation and Control: Emerging Technologies, Proceedings, 383–386 (1995).
9. Lee J. H., Lin C., Lim H., Lee J. M., Sliding mode control for trajectory tracking of mobile robot in the RFID sensor space, International Journal of Control, Automation and Systems, Vol. 7, No. 3, 429–435 (2009).
10. Pei X.-Z., Liu Z.-Y., Pei R., Robust trajectory tracking controller design for mobile robots with bounded input, Zidonghua Xuebao/Acta Automatica Sinica, Vol. 29, No. 6, 876–882 (2003).
11. Perski A., et al., GNSS receivers in engineering practice. Introduction to Global Navigation Satellite Systems, Pomiary, Automatyka, Robotyka, Vol. 17, No. 3, 103–111 (in Polish: "Odbiorniki GNSS w praktyce inżynierskiej. Wprowadzenie do systemów GNSS") (2013).
12. Sun D., Dong H.N., Tso S.K., Tracking stabilization of differential mobile robots using adaptive synchronized control, Proceedings of IEEE International Conference on Robotics and Automation, Vol. 3, 2638-2643 (2002).
13. Toibero J.M., Roberti F., Carelli R., Fiorini P., Switching control approach for stable navigation of mobile robots in unknown environments, Robotics and Computer-Integrated Manufacturing, Vol. 27, No. 3, 558–568 (2011).
14. Trojnacki M.: Dynamics Model of a Four-Wheeled Mobile Robot for Control Applications – a Three-Case Study, IEEE Intelligent Systems IS'14, Series: Advances in Intelligent Systems and Computing, Springer International Publishing, 99-116 (2014).
15. Trojnacki M., Dąbek P.: Determination of Motion Parameters with Inertial Measurement Units. Part 2: Algorithm Verification with a Four-Wheeled Mobile Robot and Low-Cost MEMS Sensors, Mechatronics: Ideas for Industrial Applications, Series: Advances in Intelligent Systems and Computing, Springer International Publishing, 253-267 (2015).
16. Wang K., Liu Y., Li L., Visual servoing trajectory tracking of nonholonomic mobile robots without direct position measurement, IEEE Transactions on Robotics, Vol. 30, No. 4, 1026–1035 (2014).
17. Mobile robots for counter-terrorism (PIAP), `http://www.antiterrorism.eu`.

Comparative Analysis of Posture Controllers for Tracking Control of a Four-Wheeled Skid-Steered Mobile Robot – Part 2. Dynamics Model of the Robot and Simulation Research of Posture Controllers.

Maciej Trojnacki[1], Przemysław Dąbek[1], Janusz Kacprzyk[1], Zenon Hendzel[2]

[1] Industrial Research Institute for Automation and Measurements PIAP, Warsaw, Poland
{mtrojnacki, pdabek, jkacprzyk}@piap.pl
[2] Rzeszów University of Technology, Rzeszów, Poland
zenhen@prz.edu.pl

Abstract. The paper is the second part of the work concerned with the problem of trajectory tracking control of a four-wheeled PIAP GRANITE mobile robot. The first part of the work was devoted to theoretical considerations. Research object and its kinematics were described. Robot motion control system structure comprising posture controller and drive unit controller was presented. Various solutions for posture controller were discussed and their modifications proposed. A methodology of posture controller tuning was introduced in which controller parameters for particular solutions are determined from conditions for maximum velocities of robot motion and maximum posture errors. In the present work dynamics model of the robot is described. It takes into account tire-ground contact conditions and wheel slips. The tire-ground contact conditions are characterized by coefficients of friction and rolling resistance. A simple form of the tire model, which includes only the most important effects of tire-ground interaction, is used. The robot dynamics model also contains the electromechanical model of a servomotor drive unit. The developed model of robot dynamics is used in the simulation studies in which the effectiveness of particular solutions of posture controller is benchmarked. Evaluation of the analyzed solutions is carried out using the introduced quality indexes.

Keywords: wheeled mobile robot, tracking control, posture controller, dynamics model, drive model, wheel slips, comparative analysis, simulation research.

1 Dynamics Model of the Robot

Within the first part of the work, that is [4], the PIAP GRANITE robot which is the object of the study is presented and its kinematics is described. In this work model of that robot, including wheel slips and developed within work [3] will be used. This model will be employed in simulation-based research of posture controllers.

R. Jabłoński and T. Brezina (eds.), *Advanced Mechatronics Solutions*,
Advances in Intelligent Systems and Computing 393,
DOI 10.1007/978-3-319-23923-1_86

It is assumed that the robot is under action of the following external forces:

- ground reaction forces ${}^R\mathbf{F}_{Ai} = [{}^RF_{Aix}, {}^RF_{Aiy}, {}^RF_{Aiz}]^T$ acting on each wheel,
- gravity force ${}^R\mathbf{G} = m_R\, {}^R\mathbf{g}$, where m_R denotes total mass of the robot.

Components of ground reaction forces ${}^R\mathbf{F}_{Aix}$ and ${}^R\mathbf{F}_{Aiy}$, $i = \{1, 2, 3, 4\}$ acting in plane of wheel-ground contact are shown in Fig. 1. For simplicity, in further considerations the subscript A in designations of reaction forces and reaction moments of force will be omitted.

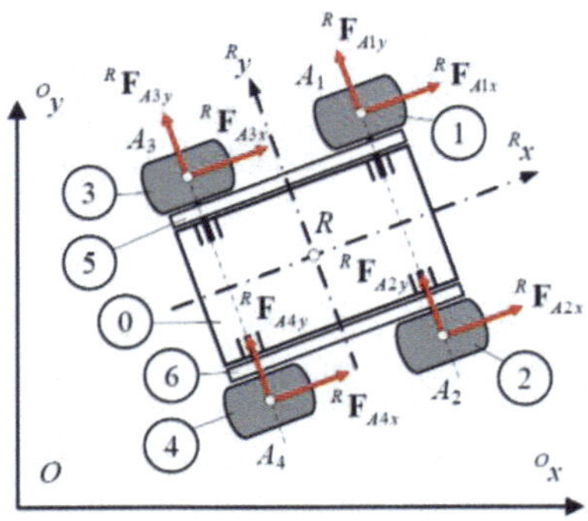

Fig. 1. Reaction forces acting on the PIAP GRANITE robot in the plane of wheel-ground contact

As mentioned in the first part of the paper, the robot configuration with independent drive of all 4 wheels is analyzed. In this configuration, the toothed belts 5 and 6 connecting front and rear wheels on either side of the robot are not present.

The gravity force vector ${}^R\mathbf{G}$ is a function of gravitational acceleration vector ${}^R\mathbf{g} = [{}^Rg_x, {}^Rg_y, {}^Rg_z]^T$ and it is applied at the robot mass center, whose position is described by the vector ${}^R\mathbf{r}_{CM} = [{}^Rx_{CM}, {}^Ry_{CM}, {}^Rz_{CM}]^T$.

Under action of the mentioned forces, according to the Newton's 2nd law, the robot moves with acceleration ${}^R\mathbf{a}_{CM} = [{}^Ra_{CMx}, {}^Ra_{CMy}, {}^Ra_{CMz}]^T$ and the following dynamic equations of motion of the robot mass centre are valid:

$$m_R\, {}^Ra_{CMw} = \sum_{i=1}^{4} {}^RF_{iw} + m_R\, {}^Rg_w, \tag{1}$$

where $w = \{x, y, z\}$.

Taking into account the assumptions that:

- the robot mobile platform is a rigid body,
- the ground surface is rigid, horizontal and even,
- all wheels have contact with this surface,
- points of wheel-ground contact lie in one plane,
- wheel tires deflect directly proportional to reaction forces acting along direction normal to the ground,
- the robot does not rotate about Rx and Ry axes,

- the gravitational acceleration vector expressed in the robot coordinate system $\{R\}$ is equal to ${}^R\mathbf{g} = [0, 0, g]^T$, where $g = 9.81$ (m/s^2),

it is possible to determine normal components of ground reactions (on the assumption that accelerations of the robot mass center are known and ${}^R a_{CMz} = 0$) in the form:

$$ {}^R F_{1z} = m_R\big((-{}^R a_{CMx}/(2L) - {}^R a_{CMy}/(2W))(r + {}^R z_{CM}) + g(1/4 + {}^R x_{CM}/(2L))\big), \tag{2} $$

$$ {}^R F_{2z} = m_R\big((-{}^R a_{CMx}/(2L) + {}^R a_{CMy}/(2W))(r + {}^R z_{CM}) + g(1/4 + {}^R x_{CM}/(2L))\big), \tag{3} $$

$$ {}^R F_{3z} = m_R\big(({}^R a_{CMx}/(2L) - {}^R a_{CMy}/(2W))(r + {}^R z_{CM}) + g(1/4 - {}^R x_{CM}/(2L))\big), \tag{4} $$

$$ {}^R F_{4z} = m_R\big(({}^R a_{CMx}/(2L) + {}^R a_{CMy}/(2W))(r + {}^R z_{CM}) + g(1/4 - {}^R x_{CM}/(2L))\big). \tag{5} $$

In the calculations of those components, values of motion parameters of the robot mass center CM and of geometric centers of wheels Ai from the previous calculation step are used, that is, from time instant $t - \Delta t$, where Δt is the adopted time step. In the first calculation step, some known initial conditions are taken into account.

On the assumption that the robot wheels neither move nor rotate backwards, the current value of longitudinal slip ratio for the i-th wheel is determined from the formula:

$$ \lambda_i = \begin{cases} 0 & \text{for } {}^{Ai}v_x = 0 \text{ and } v_{oi} = 0, \\ ({}^{Ai}v_x - v_{oi})/\max({}^{Ai}v_x, v_{oi}) & \text{for other } {}^{Ai}v_x \text{ and } v_{oi}, \end{cases} \tag{6} $$

where $v_{oi} = \dot{\theta}_i r_i$ and ${}^{Ai}v_x = {}^R v_{Aix}$ are respectively velocity at the wheel circumference and longitudinal component of velocity of wheel geometric center.

Current value of the tire adhesion coefficient on longitudinal direction for this wheel is calculated using Kiencke tire model [2] modified within work [3] to the form:

$$ \mu_{ix} = \begin{cases} \dfrac{2\mu_p \lambda_p \lambda_i}{\lambda_p^2 + \lambda_i^2} & \text{for } |\lambda_i| \le \lambda_p, \\ a_{\lambda x}\lambda_i + b_{\lambda x}\,\mathrm{sgn}(\lambda_i) & \text{for } |\lambda_i| > \lambda_p, \end{cases} \tag{7} $$

where λ_p denotes the value of longitudinal slip corresponding to the value of maximum tire adhesion coefficient μ_p and sgn() is a signum function.

In the classical approach to dynamics modeling of wheeled mobile robots, components of ground reaction forces are unknown quantities and their determination is problematic, because there are more unknown quantities than available equations. In the proposed approach, thanks to knowledge of normal components of ground reactions, the longitudinal component of ground reaction force for the i-th wheel can be calculated based on determined current value of the adhesion coefficient on longitudinal direction, according to relationship:

$$ {}^R F_{ix} = \mu_{ix}\, {}^R F_{iz}. \tag{8} $$

In turn, the current value of the lateral slip angle for i-th wheel is determined from the formula:

$$\alpha_i = \begin{cases} 0 & \text{for} \quad {}^{Ai}v_y^O = 0, \\ \operatorname{atan2}({}^{Ai}v_y^O, |{}^{Ai}v_x^O|) & \text{for} \quad |{}^{Ai}v_y^O| > 0, \end{cases} \tag{9}$$

where ${}^{Ai}v_x = {}^{R}v_{Aix}$ and ${}^{Ai}v_y = {}^{R}v_{Aiy}$ are respectively longitudinal and transversal velocity of the geometric center of the i-th wheel.

Knowing the lateral slip angle it is possible to calculate current value of the adhesion coefficient on lateral direction for i-th wheel. To this end the following approximate relationship (as compared to H.B. Pacejka model [1, 5]) is introduced:

$$\mu_{iy} = -\mu_{ymax} \sin(\alpha_i), \tag{10}$$

where μ_{ymax} denotes maximum value of adhesion coefficient on lateral direction (it is assumed that $\mu_p \geq \mu_{ymax} \geq \mu_k$).

Hence, lateral component of the ground reaction force for i-th wheel is calculated from the formula:

$${}^{R}F_{iy} = \mu_{iy} {}^{R}F_{iz}. \tag{11}$$

For the known components of longitudinal and lateral components of ground reaction forces for particular wheels, it is possible to calculate values of accelerations on longitudinal and lateral direction using equations:

$${}^{R}a_{CMx} = \sum_{i=1}^{4} {}^{R}F_{ix} / m_R, \quad {}^{R}a_{CMy} = \sum_{i=1}^{4} {}^{R}F_{iy} / m_R. \tag{12}$$

For the robot mobile platform, it is also possible to write the dynamic equation of motion resulting from its rotation about Rz axis with angular acceleration ${}^{R}\ddot{\varphi}_{0z}$:

$$\begin{aligned} {}^{R}\ddot{\varphi}_{0z} = \Big(&-({}^{R}F_{1x}+{}^{R}F_{3x})(W/2-{}^{R}y_{CM})+({}^{R}F_{2x}+{}^{R}F_{4x})(W/2+{}^{R}y_{CM})+ \\ &+({}^{R}F_{1y}+{}^{R}F_{2y})(L/2-{}^{R}x_{CM})-({}^{R}F_{3y}+{}^{R}F_{4y})(L/2+{}^{R}x_{CM})\Big)/I_{Rz}, \end{aligned} \tag{13}$$

where I_{Rz} is mass moment of inertia of the robot about the Rz axis.

Then one can calculate, by integration, the value of angular velocity of the robot mobile platform about this axis, that is ${}^{R}\dot{\varphi}_{0z}$.

Next, it is possible to calculate values of velocities ${}^{R}v_{Rx}$ and ${}^{R}v_{Ry}$ and then values of velocities of characteristic points A_i. For simplicity, one may assume that ${}^{R}x_C = {}^{R}x_B = 0$, that is the x coordinate describing position of the instantaneous centre of rotation for the mobile platform in the robot coordinate system $\{R\}$ is equal to 0.

For each of the robot wheels it is then possible to write dynamic equation of motion associated with wheel spin:

$$I_{Wy}\ddot{\theta}_i = \tau_i - {}^{R}F_{ix} r - {}^{R}F_{iz} r f_r \operatorname{sgn}(\dot{\theta}_i), \tag{14}$$

where: I_{Wy} – mass moment of inertia of the wheel about its spin axis, τ_i – a driving torque,

f_r – coefficient of rolling resistance, $\dot{\theta}_i$ and $\ddot{\theta}_i$ – angular velocity and acceleration of spin of that wheel, respectively.

In place of the signum function $\mathrm{sgn}(\dot{\theta}_i)$, one may introduce a $\tanh(k_\theta \dot{\theta}_i)$ function, which is better from the point of view of simulation, where the coefficient $k_\theta > 0$ should be the larger, the more the function should be similar to $\mathrm{sgn}(\dot{\theta}_i)$.

After taking into account the above relationships, it is possible to determine angular parameters of wheel spin, i.e., $\dot{\theta}_i$ and $\ddot{\theta}_i$ for driving torques τ_i acting on the driven wheels (forward dynamics problem) based on the equation:

$$\ddot{\theta}_i = \left(\tau_i - {}^R F_{ix} r - {}^R F_{iz} r\, f_r\, \mathrm{sgn}(\dot{\theta}_i)\right) / I_{Wy}. \tag{15}$$

The described simplified model of robot dynamics can be also enhanced with the model of its drive units. It is particularly important, when one wants to connect the model of robot dynamics (treated as a multi-body system) with a control system. In such case the drive unit model is an intermediate link between the control system and the robot dynamics model.

It is assumed that:

- each of the robot drive units consists of identical DC motor, encoder, and transmission system (according to the scheme in Fig. 3 in [4]),
- robot drive units are not self-locking,
- mass moments of inertia of the rotating elements of the servomechanisms (DC motor, encoder and gear unit) are small in comparison to mass moments of inertia of the driven parts of the robot (wheels), that is why they are neglected.

The DC motor model of the i-th drive unit is described by the following dependences:

$$\frac{\mathrm{d}i_i}{\mathrm{d}t} = \left(u_i - k_e n_d \dot{\theta}_i - R_d i_i\right) / L_d, \tag{16}$$

$$\tau_i = \eta_d n_d k_m i_i, \tag{17}$$

where: u_i – motor voltage input, i_i – rotor current, L_d, R_d – respectively inductance and resistance of the rotor, k_e – electromotive force constant, k_m – motor torque coefficient, n_d – gear ratio of the transmission system, η_d – efficiency factor of the transmission system.

2 Simulation Research

The main part of the present work are simulation studies aimed at evaluation of several solutions of the posture controller described in the first part of the work, for the controller parameters chosen according to the methodology proposed there.

2.1 Desired trajectory and robot parameters

It is assumed that robot motion consists of three phases: accelerating with maximum acceleration a_{Rmax} on the distance of l_r, steady motion with constant velocity ${}^{O}v_{Rd} = v_{Ru}$ and braking with maximum acceleration (deceleration) a_{Rmax} on the distance of l_h. In turn, the desired path of motion contains straight line segment of length L_p, circular arc of radius R_z and the second straight line segment of length 2 L_p. As a result of turning with maximum angular velocity equal to $\omega_{0zu} = v_{Ru}/R_z$, the robot should turn through the angle of φ_{zmax}.

For the simulation, the maximum values of desired linear and angular velocities are assumed respectively equal to $v_{Ru} = 0.3$ m/s and $\omega_{0zu} = v_{Ru}/R_z$, maximum acceleration during accelerating and braking phases $a_{Rmax} = 0.7$ m/s^2, maximum angular acceleration $\varepsilon_{0zmax} = \pi/4$ rad/s^2 and length $L_p = 1$ m.

In the simulation the following two cases of robot motion are considered:

1. soft turning to the right with radius $R_z = 1$ m and angle of rotation $\varphi_{0zmax} = 2/3\ \pi$ rad, with zero initial conditions, that is, position, course and velocities of the robot,
2. rapid turning to the right with radius $R_z = 0.5$ m and angle of rotation $\varphi_{0zmax} = \pi$ rad, with zero initial conditions.

Lengths of acceleration and braking distances, l_r and l_h respectively, are determined for the given velocity profile based on v_{Ru} velocity and maximum acceleration a_{Rmax}.

Next, based on the knowledge of v_{Ru} and ω_{0zu} velocities, a_{Rmax} and ε_{0zmax} accelerations and the L_p length, characteristic time instants are determined.

For the simulation studies the following values of the basic design parameters of the PIAP GRANITE robot are assumed:

- dimensions: $L = 0.425$ m, $W = 0.553$ m (where: $L = A_1A_3 = A_2A_4$, $W = A_1A_2 = A_3A_4$, see Fig. 1), $r_i = r = 0.0965$ m,
- masses of the components: $m_0 = 40$ kg, $m_i = 1$ kg, $m_5 = m_6 = 0.18$ kg,
- robot mass center coordinates: ${}^{R}x_{CM} = -0.04$ m, ${}^{R}y_{CM} \approx 0$ m, ${}^{R}z_{CM} = 0.14$ m,
- mass moments of inertia: $I_{Wy} = 0.01$ kg m^2, $I_{Rz} = 2.8$ kg m^2,
- parameters of drive units: $L_d = 0.0823$ mH, $R_d = 0.317\ \Omega$, $k_e = 0.0301$ Vs/rad, $k_m = 0.0302$ Nm/A, $n_d = 53$, $\eta_d = 0.8$, $\dot{\theta}_{max} = 15.807$ rad/s, $\dot{\theta}_{min} = 0$, $u_{max} = 32$ V, $u_{min} = 0$.

Moreover, the following environment and tire-ground contact parameters are assumed: $g = 9.81$ m/s^2, $\mu_p = 0.85$, $\mu_s = 0.75$, $f_r = 0.03$, $\lambda_p = 16.5\%$.

The desired motion path and time histories of desired velocities and accelerations of robot motion are illustrated in Fig. 2.

It should be noted that the transition from the rectilinear motion to motion along an arc and vice-versa is associated with change of the angular velocity of robot turning (yaw rate) from zero to ω_{0zu}, and then from ω_{0zu} to zero, which takes place in a finite time and results in appearance of transition curves.

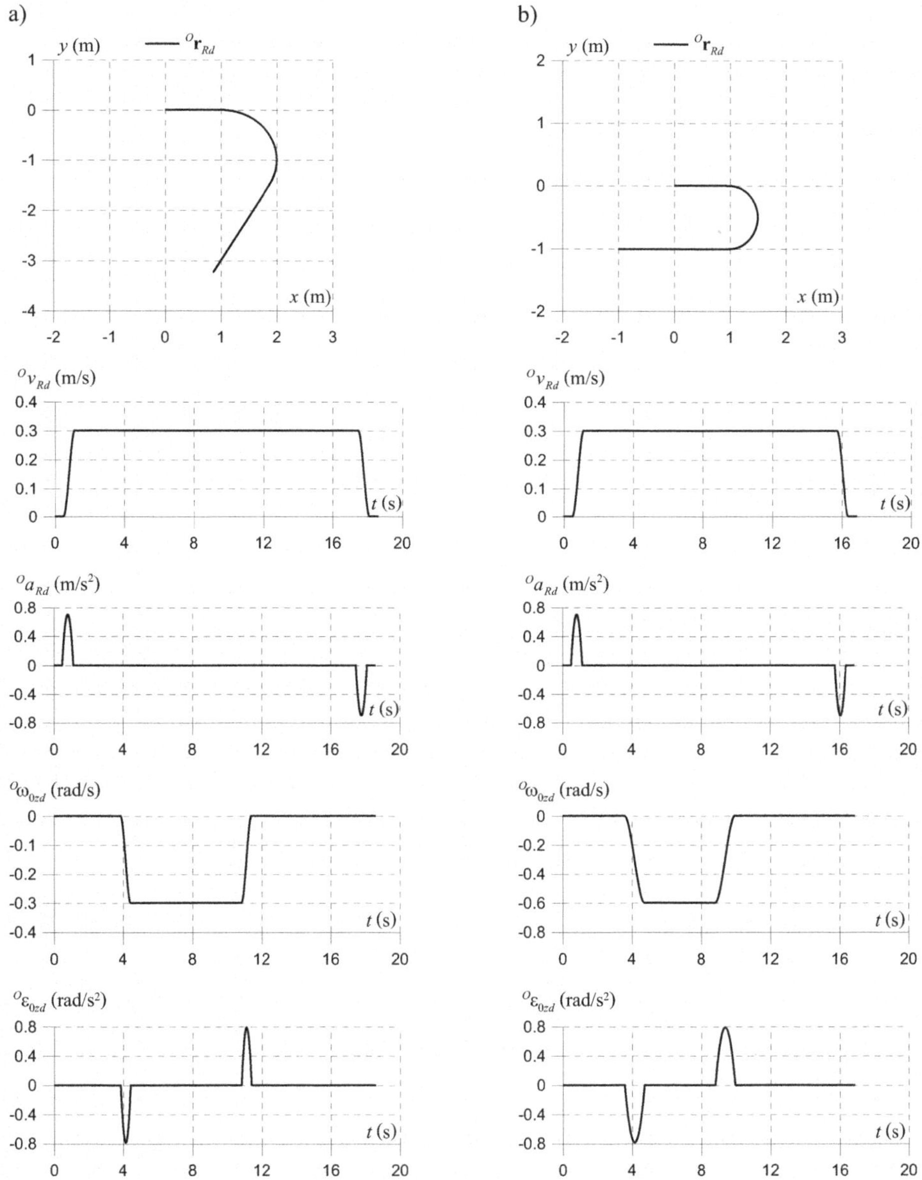

Fig. 2. Desired path of robot motion and time histories of desired velocities and accelerations: a – soft turning to the right with radius $R_z = 1$ m and angle of turn $\varphi_{0zmax} = 2/3\ \pi$ rad, b – rapid turning to the right with radius $R_z = 0.5$ m and angle of turn $\varphi_{0zmax} = \pi$ rad

2.2 Quality indexes

In order to compare in a comprehensive way accuracy of realization of motion by the robot for the analyzed posture control systems, the following quality indexes were introduced during the simulation studies:

- maximum posture errors of the robot:

$$ {}^{R}e_{Fmax} = \max_{t\in\langle 0,T\rangle}\left(|{}^{R}e_{F}|\right),\quad {}^{R}e_{Lmax} = \max_{t\in\langle 0,T\rangle}\left(|{}^{R}e_{L}|\right),\quad {}^{R}e_{Omax} = \max_{t\in\langle 0,T\rangle}\left(|{}^{R}e_{O}|\right), \tag{18} $$

- square root of integral of the squared error for each robot posture error:

$$ E_F = \sqrt{\frac{1}{T}\int_0^T \left({}^{R}e_F\right)^2 \mathrm{d}t},\quad E_L = \sqrt{\frac{1}{T}\int_0^T \left({}^{R}e_L\right)^2 \mathrm{d}t},\quad E_O = \sqrt{\frac{1}{T}\int_0^T \left({}^{R}e_O\right)^2 \mathrm{d}t}, \tag{19} $$

- integral of absolute error multiplied by time for each robot posture error:

$$ Q_F = \frac{2}{T^2}\int_0^T (|{}^{R}e_F|\,t)\,\mathrm{d}t,\quad Q_L = \frac{2}{T^2}\int_0^T (|{}^{R}e_L|\,t)\,\mathrm{d}t,\quad Q_O = \frac{2}{T^2}\int_0^T (|{}^{R}e_O|\,t)\,\mathrm{d}t, \tag{20} $$

where T is the analyzed time period of robot motion.

All above quality indexes have units analogous to the posture errors. The first quality index enables finding the maximum posture errors. In case of the second index, average errors in the assumed time interval are obtained. The third quality index is designed so as to determine the average errors as well, but this time in the averaging process the time plays a key role, that is the errors occurring closer to the final phase of motion are more important than initial errors, which allows for assessment of how the analyzed controller solution deals with reduction of errors with time.

2.3 Values of controller parameters

For simulation studies, maximum (boundary) values of controller parameters calculated according to the methodology discussed earlier are adopted. Values of those parameters are summarized in Tab. 1 for the first desired motion trajectory and in Tab. 2 for the second trajectory.

Table 1. Assumed values of posture controller parameters for the first motion trajectory

Posture controller	k_F	k_L	k_O	k_S	k_ψ
No. 1	0.4627	1.3090	2.2883	–	–
No. 2	0.4627	–	–	–	0.8611
No. 3	0.0218	0.0467	0.3633	0.2163	1.0192
No. 4	0.4627	–	0.1000	–	0.6750
No. 5	–	–	0.4729	0.4627	1.0492
No. 6	2.7221	–	–	–	1.1811

Table 2. Assumed values of posture controller parameters for the second motion trajectory

Posture controller	k_F	k_L	k_O	k_S	k_ψ
No. 1	0.4627	0.8040	1.7933	–	–
No. 2	0.4627	–	–	–	0.6217
No. 3	0.0218	0.0934	0.7267	0.2163	0.6592
No. 4	0.4627	–	0.2000	–	0.4720
No. 5	–	–	0.9529	0.4627	0.7192
No. 6	2.7221	–	–	–	1.1811

It can be noticed that in case of the posture controller solution no. 6, for the constant ${}^{O}v_{Rdmax}$ value and various ${}^{O}\omega_{Odmax}$ values, the controller parameters do not change for the analyzed desired motion trajectories.

Unless it is explicitly stated otherwise, $f_v = \text{sech}(k_b\, {}^{R}e_O)$ and $f_\omega = \tanh(k_b\, {}^{R}e_O)$ functions with parameter $k_b = 1$ are adopted. Moreover, for the solutions no. 3, 4 and 5 parameter $\eta = 0.1$ is assumed, for the solution no. 2, $a = 0.5$, and finally in the case of solution no. 6, $d_{min} = 0.1$ m.

2.4 Illustration of problem with non-zero final errors

In the further part of the work, the earlier mentioned problem associated with non-zero final errors will be discussed on example. This problem is connected with occurrence of non-zero control signals in the case when desired velocities of robot motion are equal to zero, but non-zero final errors are present. In the described solutions this problem exists mainly in the case of robot linear velocity control law and is solved by introduction of additional $|\text{sgn}({}^{O}v_{Rd})|$ expression aimed at generation of zero control signals, and as a result at enforcing the robot stopping. The problem will be illustrated on example of the solution no. 1 in which the term $|\text{sgn}({}^{O}v_{Rd})|$ is not included.

In Figs. 3-4, the considered problem is illustrated for the first of the earlier discussed motion trajectories, for $f_v = \cos(k_a\, {}^{R}e_O)$, $f_\omega = \sin(k_a\, {}^{R}e_O)$ functions with $k_a = 0.5$ parameter. From the obtained time histories (Fig. 4) it is clearly visible that unwanted motion of the robot takes place after desired motion is finished, because the ${}^{R}v_s$ control signal has non-zero value.

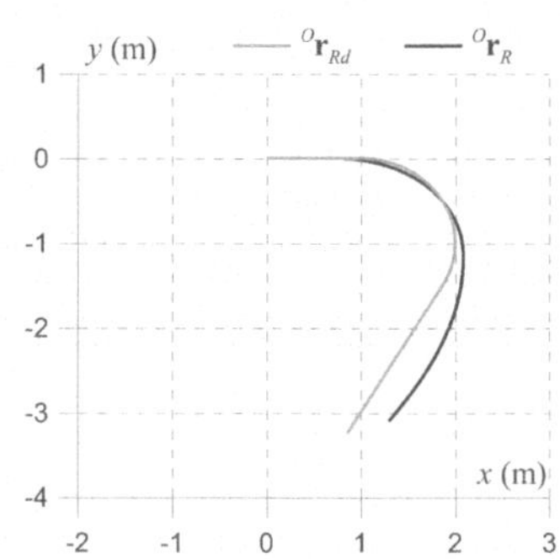

Fig. 3. Illustration of the problem connected with non-zero final errors (original solution no. 1, $f_v = \cos(k_a\, {}^{R}e_O)$, $f_\omega = \sin(k_a\, {}^{R}e_O)$, $k_a = 0.5$) – motion paths of the characteristic point R of the robot

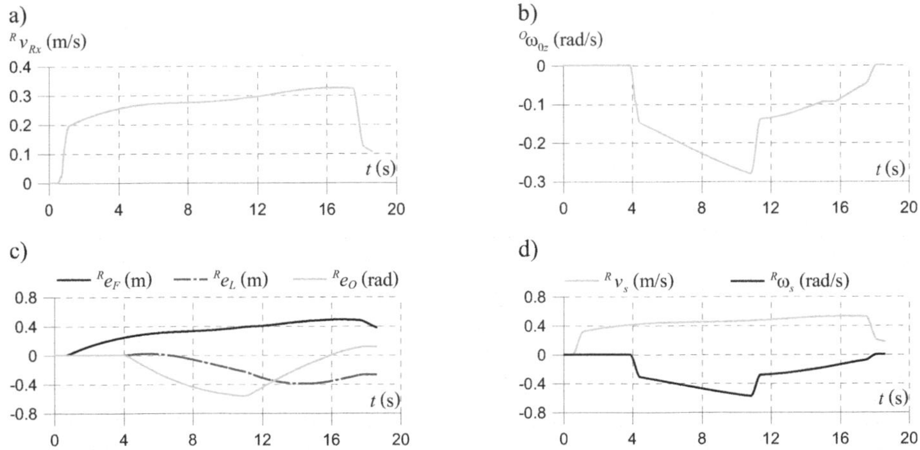

Fig. 4. Illustration of the problem connected with non-zero final errors (original solution no. 1, $f_v = \cos(k_a\, {}^Re_O)$, $f_\omega = \sin(k_a\, {}^Re_O)$, $k_a = 0.5$): a – linear velocity, b – angular velocity, c – posture errors, d – control signals

2.5 Comparative analysis of posture controllers

For the earlier discussed controller solutions and the adopted robot motion trajectories, simulation studied were carried out. Quality indexes obtained for particular simulation cases are shown in Tables 3-5.

In the case of the first desired motion trajectory and zero initial conditions (Tab. 3), the best results were obtained for the solution no. 5 of the posture controller. Quite good results were also obtained for the solution no. 6. The worst results are for solutions no. 4 and 1.

Table 3. Quality indexes obtained in simulation studies for particular controllers, for the first desired motion trajectory and zero initial conditions

Posture controller	${}^Re_{Fmax}$ (m)	${}^Re_{Lmax}$ (m)	${}^Re_{Omax}$ (rad)	E_F (m)	E_L (m)	E_O (rad)	Q_F (m)	Q_L (m)	Q_O (rad)
No. 1	0.4180	0.1490	0.4307	0.3327	0.0813	0.2248	0.3708	0.0872	0.1679
No. 2	0.3963	0.1131	0.3551	0.3176	0.0529	0.1949	0.3520	0.0574	0.1562
No. 3	0.3315	0.1550	0.1991	0.2476	0.0970	0.1072	0.2718	0.1040	0.0808
No. 4	0.4146	0.1971	0.4281	0.3300	0.1077	0.2254	0.3676	0.1152	0.1794
No. 5	0.1391	0.0318	0.1536	0.1248	0.0200	0.0867	0.1311	0.0220	0.0514
No. 6	0.1585	0.0489	0.1798	0.1371	0.0348	0.1040	0.1386	0.0392	0.0584

For the second desired motion trajectory and zero initial conditions unquestionably best results were obtained for the solution no. 6 of the posture controller, which is clearly visible in Tab. 4. As far as the ${}^{R}e_F$ and ${}^{R}e_O$ errors are concerned, good results are obtained for solution no. 5, and considering the ${}^{R}e_L$ error, for solution no. 3. The worst results were obtained again for solutions no. 4 and 1.

Table 4. Quality indexes obtained in simulation studies for particular controllers, for the second desired motion trajectory and zero initial conditions

Posture controller	${}^{R}e_{Fmax}$ (m)	${}^{R}e_{Lmax}$ (m)	${}^{R}e_{Omax}$ (rad)	E_F (m)	E_L (m)	E_O (rad)	Q_F (m)	Q_L (m)	Q_O (rad)
No. 1	0.5688	0.5651	0.9960	0.3765	0.3348	0.4915	0.4308	0.3629	0.3871
No. 2	0.4649	0.5001	0.8384	0.3327	0.2693	0.4195	0.3772	0.2897	0.3519
No. 3	0.3649	0.1172	0.5933	0.2610	0.0696	0.2772	0.2935	0.0760	0.1798
No. 4	0.5568	0.7318	0.9872	0.3667	0.4068	0.5074	0.4195	0.4396	0.4485
No. 5	0.2032	0.2423	0.5080	0.1457	0.1616	0.2404	0.1539	0.1796	0.1438
No. 6	0.1981	0.0868	0.4283	0.1431	0.0627	0.2113	0.1404	0.0699	0.1111

In order to choose the optimal solution of the controller for desired motion trajectories and zero initial conditions, a total score for each solution was calculated as total sum of the individual quality indexes for particular controller solution divided by the number of performed simulations (Tab. 5). According to this measure, the solution no. 6 is optimal. The second best one is the solution no. 5. The third best solution is the solution no. 3. The worst alternatives are the solutions no. 4, 1 and 2 (in this order).

Table 5. Total scores for each controller solution calculated based on all simulations

Posture controller	No. 1	No. 2	No. 3	No. 4	No. 5	No. 6
TOTAL SCORE	3.39	2.92	1.91	3.61	1.37	1.18

Motion paths of the characteristic point R of the robot for all analyzed solutions and cases of robot motion are shown in Fig. 5. Conclusions from their analysis are similar to those from the analysis of the obtained quality indexes.

At the end, as the results of the simulation studies of the analyzed solutions of the posture controller for the considered robot motion cases, in Figs. 6-11 are presented time histories of velocities, posture errors and control signals for selected solutions. The results for presentations were selected according to the rule that for each motion case, at first two best solutions are presented and then the worst one.

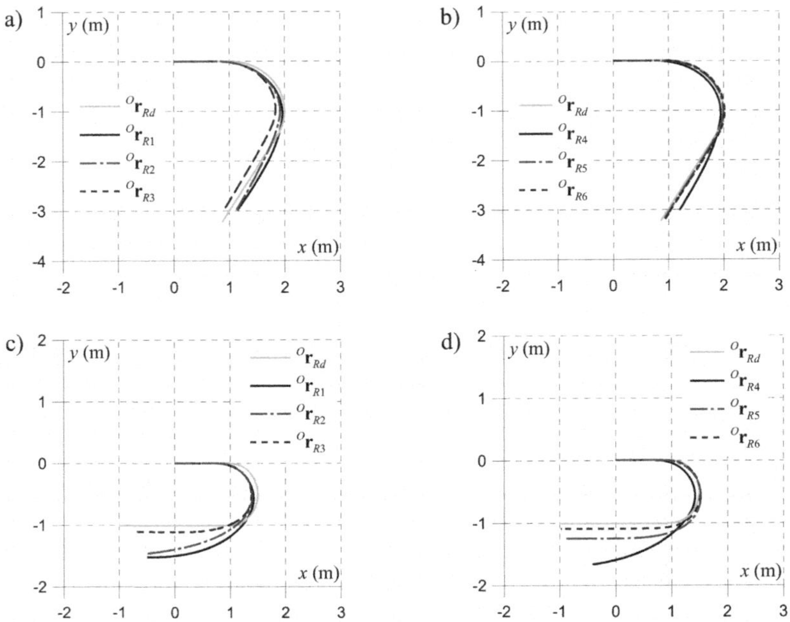

Fig. 5. Motion paths of the characteristic point R of the robot for particular solutions: a, b – for the first desired motion trajectory and zero initial conditions, c, d – for the second desired motion trajectory and zero initial conditions.

In Fig. 6 are shown results for the first desired motion trajectory and zero initial conditions and the posture controller solution no. 5. The robot moves with nearly constant linear velocity (Fig 6a), non-zero angular velocity occurs only during turning (Fig. 6b).

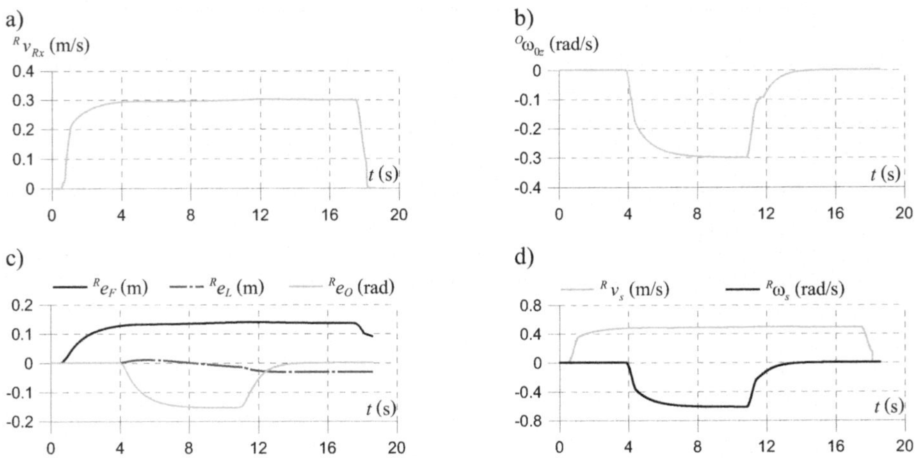

Fig. 6. Time histories of velocities, errors and control signals for the first desired motion trajectory and zero initial conditions and the posture controller solution no. 5

The course error ${}^{R}e_O$ is reduced to zero, and the remaining posture errors eventually attain non-zero values. Control signals ${}^{R}v_s$ and ${}^{R}\omega_s$ change softly and do not exceed the assumed boundary values.

In turn, in Fig. 7 are shown analogous results for solution no. 6 of the posture controller. The obtained time histories of quantities are similar to results for the solution no. 5.

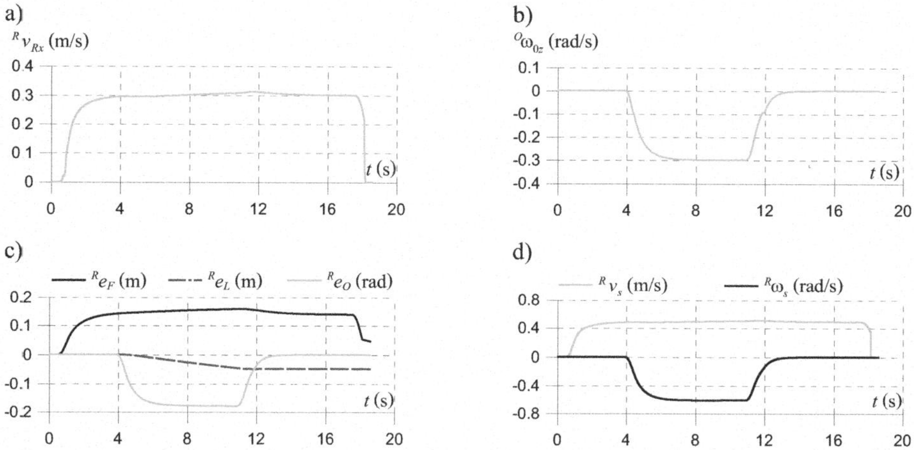

Fig. 7. Time histories of velocities, errors and control signals for the first desired motion trajectory and zero initial conditions and the posture controller solution no. 6

In the case of the posture controller solution no. 4 and analogous case of motion, the robot moves with slightly too small linear velocity (Fig. 8a), which results in increasing ${}^{R}e_F$ error (Fig. 8c). Moreover, after turning begins a large value of ${}^{R}e_L$ error appears (Fig. 8c).

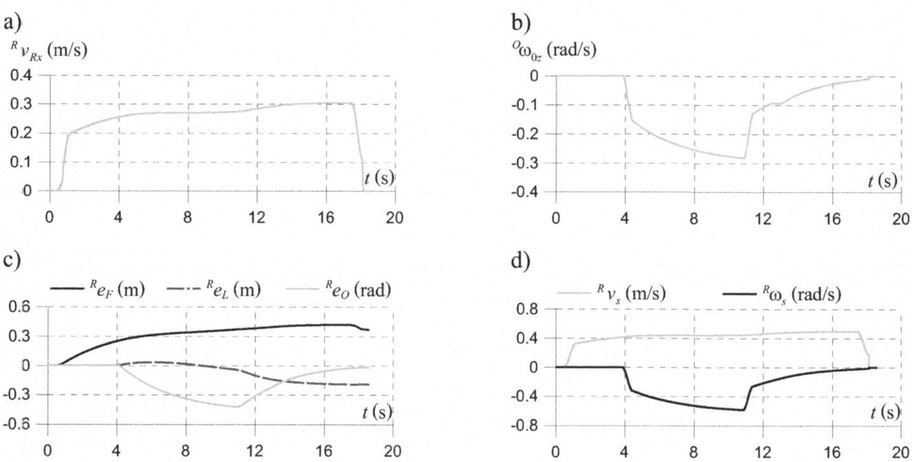

Fig. 8. Time histories of velocities, errors and control signals for the first desired motion trajectory and zero initial conditions and the posture controller solution no. 4

In the next three figures are shown results of simulation studies for the second desired trajectory of motion and zero initial conditions. In Fig. 9 results for solution no. 6 of the posture controller are presented. This time during robot motion small changes in linear velocity can be noticed (Fig. 9a). As previously, the course error ${}^{R}e_{O}$ is reduced to zero, whereas the ${}^{R}e_{L}$ error is not eliminated. The control signals ${}^{R}v_{s}$ and ${}^{R}\omega_{s}$ do not exceed the assumed boundary values.

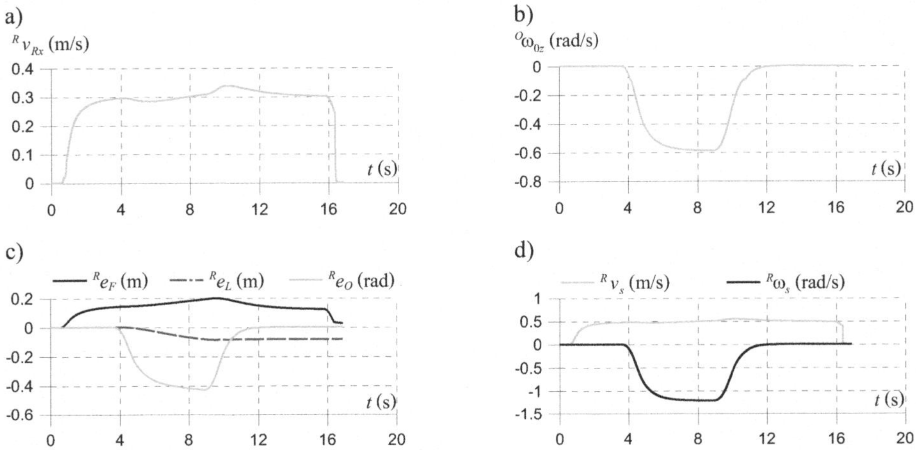

Fig. 9. Time histories of velocities, errors and control signals for the second desired motion trajectory and zero initial conditions and the posture controller solution no. 6

In the case of posture controller solution no. 5, the problem with too low value of the robot linear velocity occurs (Fig. 10a), which results in increasing ${}^{R}e_{F}$ error (Fig. 10c). Additionally, quite large ${}^{R}e_{L}$ error is present (Fig. 10c).

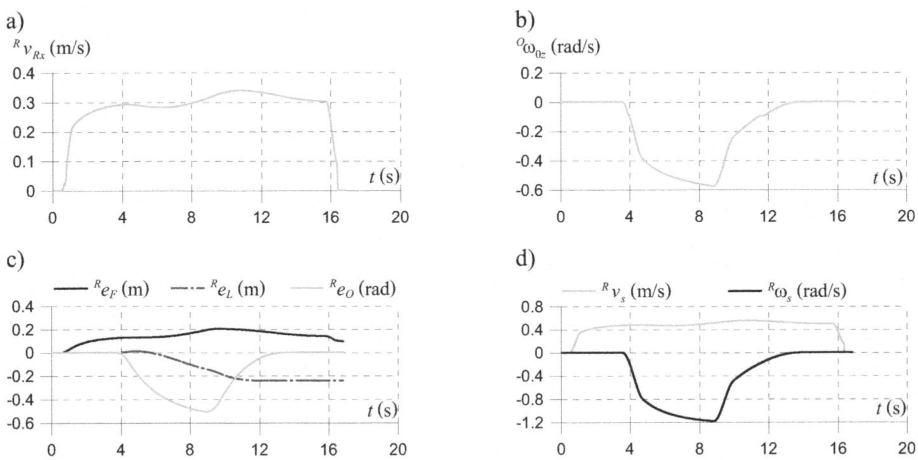

Fig. 10. Time histories of velocities, errors and control signals for the second desired motion trajectory and zero initial conditions and the posture controller solution no. 5

Also this time, the worst results for the solution no. 4 were obtained (Fig. 11). This controller eliminates mainly the course error ${}^{R}e_{O}$ (Fig. 11c), but poorly deals with the other types of errors.

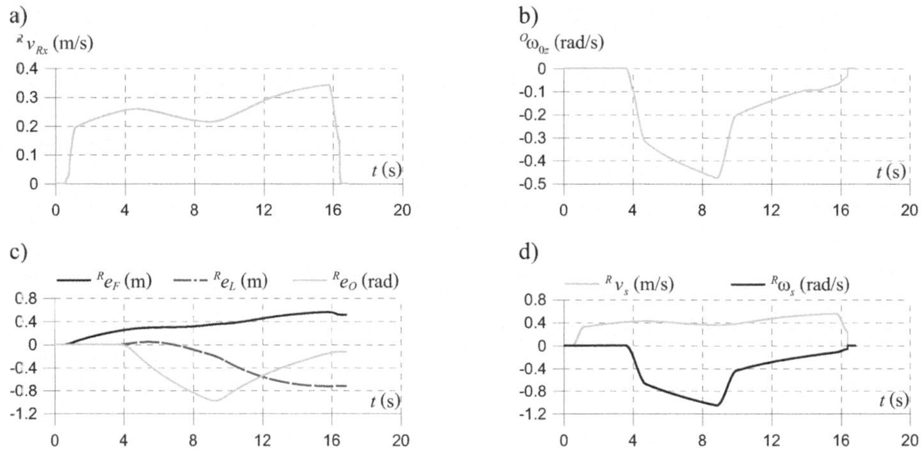

Fig. 11. Time histories of velocities, errors and control signals for the second desired motion trajectory and zero initial conditions and the posture controller solution no. 4

3 Conclusions and Future Works

Within the present, second part of the work, the model of robot dynamics including occurrence of wheel slips and model of drive units were described. This model was adopted in the conducted simulation studies of robot motion for several solutions of posture controller described in the first part of the article and for various cases of desired robot motion. Motion paths obtained from simulations for all analyzed cases as well as time histories of velocities, posture errors and control signals for selected cases were presented. Control quality indexes were introduced and calculated based on results of the simulations.

The conducted simulation studies demonstrated that the proposed methodology of posture controller tuning allows choosing the parameters which guarantee stable and correct operation of the analyzed solutions of controller.

The simulation studies also allowed evaluation and comparison of particular solutions of posture controller and yielded the following conclusions:

- The best performance was observed for solutions no. 6 and 5 of the posture controller in terms of the values of quality indexes.
- The solution no. 3 of the posture controller is the most complex and requires the largest number of both input quantities and parameters. Its performance is satisfactory and can be classified third best according to the proposed evaluation scheme.

- In the analyzed cases of motion and considering the adopted methodology of choosing the controller parameters, the worst solutions are the following: no. 4, 1 and 2 (in this order).

It should be emphasized that the obtained results of robot tracking control with the analyzed solutions of posture controller depend on the introduced modifications and the adopted values of parameters, therefore the presented results should be deemed preliminary results. Further investigations involving more advanced tuning of controllers, possibly using some alternative approach, are necessary and may lead to different quantitative results of comparative analysis of the controller solutions than presented in the present work.

As continuation of the present work are planned:

- simulation studies concerning robot tracking control for analyzed solutions of posture controller in case of non-zero initial conditions,
- experimental investigations of robot tracking control with posture controller aimed at conducting analogous comparative analysis, including the same solutions of the posture controllers as in the present work.

Acknowledgements. The work has been realized as a part of the project entitled "Dynamics modeling of four-wheeled mobile robot and tracking control of its motion with limitation of wheels slip". The project is financed from the means of National Science Centre of Poland granted on the basis of decision number DEC-2011/03/B/ST7/02532.

References

1. Pacejka H.B.: Tire and Vehicle Dynamics, 2nd Edition, SAE International and Elsevier (2005).
2. Ping L.Y., Slip Modelling, Estimation and Control of Omnidirectional Wheeled Mobile Robots with Powered Caster Wheels, Doctorial Thesis, National University of Singapore, Singapore (2009).
3. Trojnacki M.: Dynamics Model of a Four-Wheeled Mobile Robot for Control Applications – a Three-Case Study, IEEE Intelligent Systems IS'14, Series: Advances in Intelligent Systems and Computing, Springer International Publishing, 99-116 (2014).
4. Trojnacki M., Dąbek P., Kacprzyk J., Hendzel Z., Comparative Analysis of Posture Controllers for Tracking Control of a Four-Wheeled Skid-Steered Mobile Robot – Part 1. Theoretical Considerations (2015).
5. Wong J.Y.: Theory of Ground Vehicles, 3rd Edition, Wiley-Interscience (2001).

Adaptive Control of Electro-Mechanical Actuator using Receptive Field Weighted Regression

Robert Grepl[1], Vaclav Sova[1], Jan Chalupa[1]

[1] Mechlab, Faculty of Mechanical Engineering, Brno University of Technology
{grepl, sova, chalupa}@fme.vutbr.cz

Abstract.
For control of strongly nonlinear electromechanical actuator, it is useful to use composite regulator consisting of a PID and feedforward and/or feedback nonlinearities compensator. In this article, we show the application of Receptive Field Weighted Regression for approximation of reduced inverse dynamic model of the actuator. This model neglects the effect of inertia in the model of the actuator, which solves the problem of difficulty related to practical measurement/estimation of the acceleration from the measured positions. The resulting controller is able to self-tune and further adapt to the changes of parameters due to wear, temperature and other environmental conditions.

Keywords: nonlinear control · adaptive control · electromechanical actuator · receptive field weighted regression

1 Introduction

One of the typical mechatronic applications is the position control of electromechanical actuator (EMA), which are used in the aerospace, automotive and military. These actuators are often strongly nonlinear due to dry friction, return spring and other factors and for their management, a simple PID controller would not suffice. It is necessary to use feedforward and/or feedback compensation based on a knowledge model. Another important factor is the change of properties (parameters) of EMA during operation due to temperature, wear, lubrication quality, etc. - usable controller must therefore be at least somewhat adaptive.
In this article, we describe the application of local linear models (Receptive Field Weighted Regression - RFWR) on the problem of composite feedforward controller design. We show, that the results of the controller give very good results even when using a quasi-static approximation of the system dynamics and that the RFWR structure is able to adapt to the change of the parameters.

1.1 Modelling and nonlinear control of an EMA

A typical EMA consists of a DC or BLDC motor, gearbox and nonlinear spring. Model after reduction can usually be formulated as follows:

R. Jabłoński and T. Brezina (eds.), *Advanced Mechatronics Solutions*,
Advances in Intelligent Systems and Computing 393,
DOI 10.1007/978-3-319-23923-1_87

$$I\ddot{\varphi} = u - b\dot{\varphi} - u_S(\varphi) - u_F(\dot{\varphi}), \tag{1}$$

where I is system inertia, b is reduced linear damping (includes also motor EMF), u_F is dry friction and u_S is nonlinear spring model [7][6].

For such a model, it is possible to estimate the parameters using experimental data and propose a feedback and / or compensation of nonlinearities. Example of regulator is in Fig. 1. Dry friction is compensated by the feedback based on position deviation measurement, the spring nonlinearity is compensated in feedforward. PID regulator compensates for the linear part of the model (1) and also the deviation of nonlinear models parts and the real system.

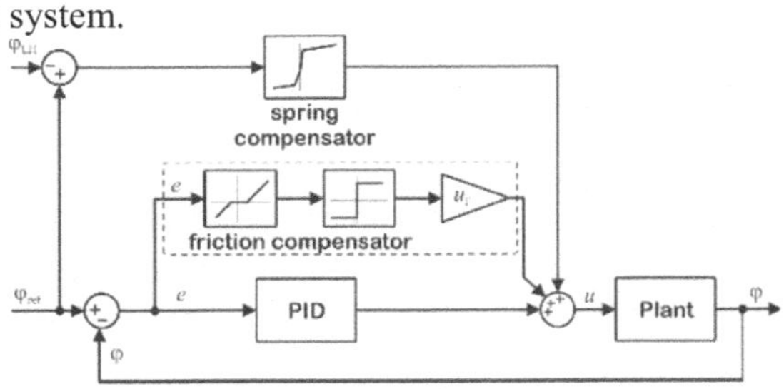

Fig.1.Schema of composite controller for an EMA

1.2 Local linear models and RFWR

The purpose of this article is to show that feedforward and feedback compensators shown in Fig. 1 can be created automatically and also further adapted using local linear approximators and online measured data.

Local linear models are a very interesting alternative to conventional global approximators eg. multilayer ANN. The advantage is that during the learning only the local portion of the approximated function adapts, which guarantees a considerably higher reliability and stability of approximation.

RFWR [1] is memory free method with possibility of incremental learning. It adaptively creates / modifies a set of receptive fields with a linear model. First results in this area were published in [8, 9]. This article provides further simulation experiments and demonstrated adaptivity of RFWR network during the change of the parameters.

2 Implementation of feedforward compensator using RFWR

2.1 Approximation of the model discontinuity due to dry friction

In the model (1) occurs dry friction which is inherently discontinuous function which is difficult to approximate accurately. Elegant solution to this problem is to use two independent structures RFWR. The principle can be demonstrated on a simple task of 1D approximation function:

$$y = \text{sign}(x) + 0,3x + 0,8x^3 \tag{2}$$

Result for 1 structure and a 2 independent structures can be seen in Fig. 2.

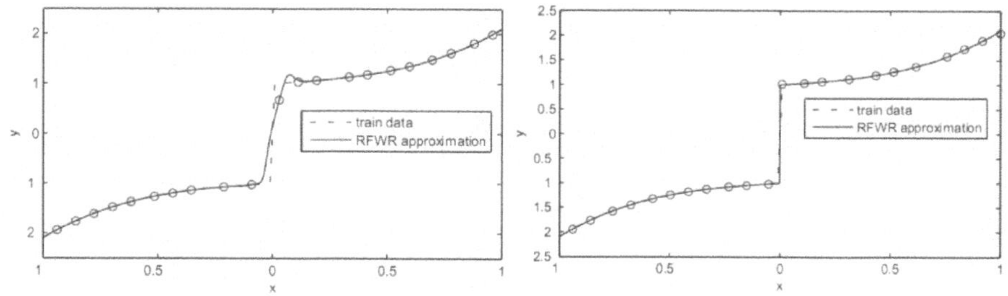

Fig.2.Approximation of non-smooth function: one RFWR with 21 receptive fields (left); two RFWR structures with 10+11 receptive fields (right)

2.2 Feedforward compensator using full inverse dynamic model

Inverse dynamic model of the system (1) can be formulated as a nonlinear problem of static approximation function [5,12]:

$$u = b\dot{\varphi} + u_S(\varphi) + u_F(\dot{\varphi}) + I\ddot{\varphi} = f(\varphi, \dot{\varphi}, \ddot{\varphi}) \tag{3}$$

Application of this full inverse model is depicted in Fig. 3. The objective of first simulation experiments was to verify the design concept and the learning rate of RFWR. However, in practical terms, it is difficult to obtain acceleration from the measured position for learning/adaptivity phase.

In Fig. 4 are the simulation results of the experiment, when RFWR learning was under way for 10 seconds and system was controlled using only the PID. Subsequently, the feedforward compensator was turned on. Looking at the input into the system u it can be seen, that after the learning practically all action interference of compensator component is carried out, the PID is active only during the step changes of the reference value.

2.3 Feedforward compensator using reduced dynamic model

Practical implementation of the previous concept gets into the problem of finding out acceleration from noisy position data measurements. Direct use of numerical derivation is not possible, the filtration always brings undesirable delays into the system and the use of additional sensors is not generally possible due to economical reasons and often also due to the limited building space. However, micro-machined devices, e.g. accelerometers, can be an exception, as they are miniature [10] and inexpensive at the same time [11]. The solution may be the neglecting of the influence of the drive inertia, problem of approximation is then formulated as follows:

$$u = f(\varphi, \dot{\varphi}) \tag{4}$$

Fig. 5 are shows the results of approximation of the reduced dynamic model. Two RFWR network are used for the reasons described in 2.1 Fig. 6 shows the result of position control simulation with the use of approximator.

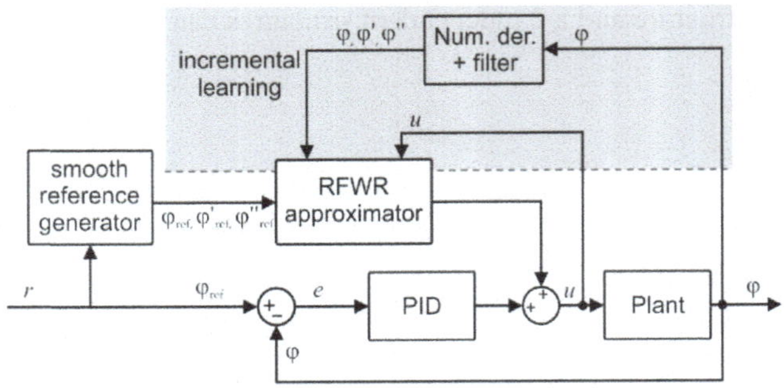

Fig.3.Schema of composite controller based on RFWR

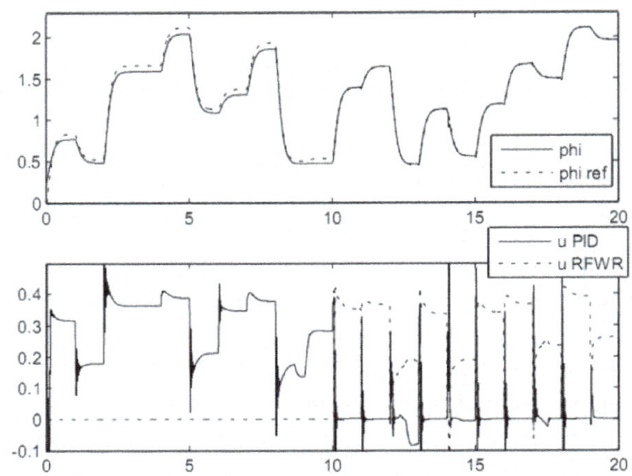

Fig.4.Control of EMA using full feedforward inverse model (simulation results)

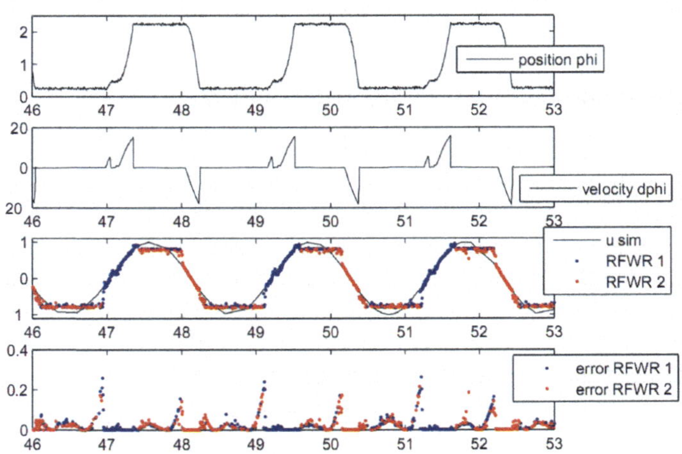

Fig.5. Approximation of reduced inverse model using two RFWR structures (simulation results)

2.4 Adaptivity of RFWR approximator

To demonstrate the possibilities of RFWR network adaptability, experiment was conducted:

Two RFWR structures were trained for 240 sec. by data generated in a closed loop. The result is a network with 84 resp. 82 receptive fields.

• Subsequently, friction value was changed abruptly. During the 60 sec. approximated function was adjusted shown in Fig. 7

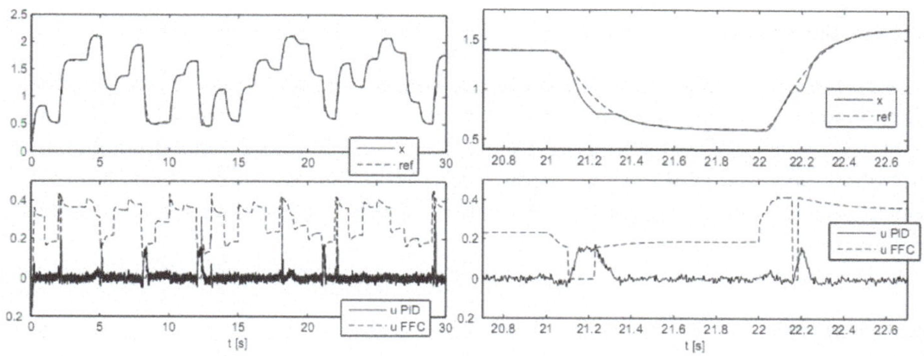

Fig.6. Control of EMA using reduced feedforward inverse model (simulation results)

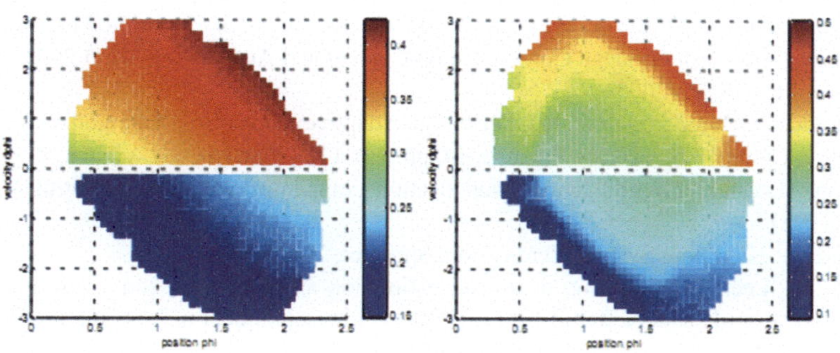

Fig.7. Adaptivity of reduced model approximation (left: original dry friction; right: friction reduced by 50%)

3 Conclusion

This article shows the automatic design of feedforward compensator for a specific class of nonlinear system (EMA) using the local linear models. Two separate RFWR structures are used for separated approximation of two parts of reduced inverse dynamic model EMA.

PID controller as part of the composite controller is set for a relatively small increase, and its task is to compensate for the linear part of the system model, as well as deviations of approximated model from reality.

During the system control, input values are measured and the position. By the numerical derivation with filtration speed is calculated. These values are used online for the adaptation of receptive fields. On simulation examples we demonstrated that the proportion of PID regulator action with respect to the overall steering intervention is relatively small and the ability to adapt is very fast.

Acknowledgement

This work was supported by the European Commission within the FP7 project Efficient Systems and Propulsion for Small Aircraft "ESPOSA", contract No. ACP1-GA-2011-284859-ESPOSA, by NETME Centre, Reg. No. CZ.1.05/2.1.00/01.0002 and, in the follow-up sustainability stage, through NETME CENTRE PLUS (LO1202) by financial means from the Ministry of Education, Youth and Sports under the „National Sustainability Programme I.

References

1. Schaal, S. & Atkeson, C. G. (1998). "Constructive incremental learning from only local information." Neural Computation, 10, 8, pp.2047-2084.
2. Deur, J., Pavkovic, D., Jansz, M., Peric, N.: Automatic Tuning of Electronic Throttle Control Strategy, 11th Mediterranean Conference on Control and Automation MED 2003, Rhodes, Greece, June 18-20 (2003)
3. Pavkovic, D.,Deur, J., Jansz, M., Peric, N.: Adaptive Control of Automotive Electronic Throttle, Control Engineering Practice, 14, pp. 121 136 (2006)
4. Olsson, H. et al.: Friction models and friction compensation. Eur. J. Control 4(3):176-195.(1998)
5. Nelles, O.: Nonlinear System Identication. Springer, 2001.
6. Grepl, R., Lee, B.: Modeling, parameter estimation and nonlinear control of automotive electronic throttle using a Rapid-Control Prototyping technique, Int. J. of Automotive Technology, 2010, Volume 11, Number 4, pp. 601-610
7. Grepl, R.: Composite Controller for Electronic Automotive Throttle with Self-tuning Friction Compensator, 9th International Conference on Mechatronics Location: Warsaw,
8. Grepl, R.: Modelling and control of electronic throttle of a passenger car – present and trends in the design of mechatronics systems, Brno University of Technology, 2010.
9. Grepl, R.: Adaptive Composite Control of Electronic Throttle using Local Learning Method, IEEE International Symposium on Industrial Electronics (ISIE 2010), Bari, 2010
10. S. Łuczak "Specific Measurements of Tilt with MEMS Accelerometers". In "Mechatronics. Recent Technological and Scientific Advances" Springer-Verlag, Berlin Heidelberg, 2011, 705.
11. Łuczak, S.: Guidelines for Tilt Measurements Realized by MEMS Accelerometers. Int. J. Precis. Eng. Manuf. 15, 489-496 (2014)
12. Honek, M, Wojnar, S, Seman, P, Juhas, M, Rohal'-Ilkiv, B; Electronic Throttle Identification and Modeling, 2013 INTERNATIONAL CONFERENCE ON PROCESS CONTROL, 2013

A .NET application searching for deleted, modified and added statements into control programs of the KUKA industrial welding robot during its testing

Igor Košťál[1]

[1] University of Economics in Bratislava, Faculty of Economic Informatics, Bratislava, Slovakia
igor.kostal@euba.sk

Abstract. During testing of a certain activity of a KUKA industrial robot equipped with some tool, the operator many times deletes or modifies an existing motion statement or adds a new statement into the relevant robot program for various reasons. All modifications in robot programs carried out through the KCP (KUKA Control Panel) teach pendant are irreversible. If the operator from some reason wants to return to the deleted motion statement or to the original version of modified motion statement, for example, the precision of the TCP (Tool Center Point) of a robot tool motion along the prescribed path gets worse after carrying out a program modification, it is not possible. We have created a .NET search application that can find deleted, modified and added statements in control programs of a robot during its testing in the *Logbuch.txt* file. It allows for the operator, among other things, to restore/reverse carried out modifications in robot programs.

Keywords: Tool Center Point; robot program, log file; testing and retesting robot; motion statement

1 Introduction

KUKA industrial robots are the part of a robotic workstation which mostly, beside them, consists of manipulators and of a control of whole robotic workstation. The design and creation of the robotic workstation is a relatively a large-scaled and complicated process. A very important and necessary part of it is testing of particular robots in the designed and assembled robotic workstation in the developers company and their testing in the assembled workstation at a customer. This fundamentally affects the work quality of particular robots in a serial production in robotic workstation. We have created the .NET search application that makes retesting of certain activities of the KUKA industrial welding robot more efficient and makes this process faster. This robot is the part of robotic workstation welding a cooling system radiator bracket of a passenger car. The paper deals with the functioning of our .NET search application and outputs that this application provides to the robot operator during retesting of certain activities of this robot.

R. Jabłoński and T. Brezina (eds.), *Advanced Mechatronics Solutions*,
Advances in Intelligent Systems and Computing 393,
DOI 10.1007/978-3-319-23923-1_88

2 Testing a robot, motion programming a robot

As we mentioned above, during testing of a certain activity of a robot the operator many times deletes or modifies an existing statement or adds a new statement into the corresponding robot program for various reasons. In the case of motion statements this is mostly carried out to improve the precision of the TCP motion along the prescribed path. However by carrying out motion statement modifications he does not always achieve this goal.

For this reason, searching for deleted statements, modified statements, their original versions and added statements has a great meaning to the robot operator testing the robot in the test operation mode. For example, if the operator modifies a motion statement in some robot program and the precision of the TCP motion gets worse after this modification, than the operator wants to return to the original statement. However, this statement is stored nowhere, because any modifications made in SRC files of the robot program by the operator through the KCP (KUKA Control Panel) teach pendant are irreversible. The same situation occurs when the operator deletes a motion statement in some robot program and the precision of the TCP motion along the prescribed path gets worse after deletion of this statement. The operator wants to recover the deleted motion statement, but it is not possible, because it is stored nowhere for the same reason as the original version of the modified statement is stored nowhere.

Such modifications of statements, deleting statements and also adding statements are recorded only by the KUKA System Software (KSS) and can be stored in the robot log file *Logbuch.txt*. The only way to find the original version of the modified statement, a deleted statement or an added statement, is to search the robot log file *Logbuch.txt* that was created by the operator. However, if the file has e.g. 5281 logs, which are written in 27531 lines, it is very time consuming work which is prone to errors. In this case can help our .NET search application that is able to find logs of modified, but also deleted and added non-motion and motion statements in the robot log file *Logbuch.txt* according to the user requirements. For modified statements it is also able to find and display their original versions, as well as find and display a full deleted statement and a full added statement. Besides that, our .NET search application stores particular results of searches into disk files with a time stamp.

Because our .NET search application searching especially for a detailed information about the motion statements PTP and LIN, we shortly describe these statements.

Control programs of KUKA industrial robots are created in the KUKA Robot Language (KRL). Its syntax allows developers motion programming (e.g. PTP, LIN statements), a program execution control, programming inputs/outputs (e.g. OUT, ANOUT statements), programming subprograms and functions, etc. A KRL motion program generally consists of an SRC file and a DAT file of the same name [1]. An SRC file contains the program code and a DAT file contains permanent data and point coordinates [1].

PTP statement executes a point-to-point motion to the end point. The coordinates of the end point are absolute [1].

Syntax [1]: PTP *End_Point* <C_PTP <*Approximate_Positioning*>>

LIN statement executes a linear motion to the end point. The coordinates of the end point are absolute [1].
Syntax [1]: LIN *End_Point <Approximate_Positioning>*

3 A .NET application searching for deleted, modified and added statements into control programs of the KUKA industrial welding robot during its testing

Our .NET search application was developed in the C# language in the development environment Microsoft Visual Studio 2013 for the Microsoft .NET Framework version 4. This .NET application searches for deleted, modified and added non-motion and motion statements into three KRL programs with names *servisna_pozicia* (*service_position*), *frezovanie_capic* (*milling_caps*) and *vymena_capic* (*replacement_caps*) in the log file *Logbuch.txt*. These KRL programs are part of the KUKA industrial welding robot control program. Using these programs the robot carries out activities related to the programs names. It means, if the cover caps of robot welding pliers spikes are worn, the TCP of the robot moves to the specified position, where these cover caps are milled (the program *frezovanie_capic* (*milling_caps*)). If these cover caps are worn more than the permissible value, they are exchanged for new ones (the program *vymena_capic* (*replacement_caps*)) in the TCP service position (the program *servisna_pozicia* (*service_position*)).
Immediately after the start-up our .NET search application attempts to connect through the Intranet to the robot control PC and searches for the directory with the *Logbuch.txt* file and the directory with DAT and SRC files of above mentioned programs on its hard drive. If the .NET search application does not find these directories for some reasons, then it calls the user to search for them through a dialog box. When the .NET search application is connected to the correct directories on the robot control PC hard drive, it will create items of both its combo boxes dynamically. The left combo box will contain items with SRC and DAT files names from the found directory. The .NET application is ready to full use now.
For each group of founded statements independently, for ***deleted*** **non-motion** and **motion statements**, for ***modified*** **non-motion** and **motion statements** and for ***added*** **non-motion** and **motion statements** our .NET search application provides the following two kind of independent outputs to the user:

- *Basic statistics* about ***founded statements*** of a given group in the SRC files of **all 3 programs**, which the .NET search application found in the *Logbuch.txt* file: the total number of all logs in the *Logbuch.txt* file, the number of founded statements of a given group in the SRC files of all 3 programs, the number of founded statements of a given group in the particular SRC files of all 3 programs. The .NET search application displays these statistics in a text box.
  It displays as items in a control checked list box ***all found complete logs*** about founded statements of a given group in the SRC files of relevant programs and in the case of searching for deleted or added statements it also displays above men-

tioned information together ***with full notations of*** these ***deleted*** or ***added statements***.

Besides that the .NET application displays its search results in the form of basic statistics and complete found requested logs in its two controls, it also creates 6 ***disk files*** with the names, for example,
ALL_DELETED_nonmotion_stat_in_ALL_SRC_20150503_114522.DTX,
ALL_DELETED_motion_stat_in_ALL_SRC_20150503_113551.DTX (Fig. 1),
ALL_MODIFIED_nonmotion_stat_in_ALL_SRC_20150503_103925.DTX (Fig. 2),
ALL_MODIFIED_motion_stat_in_ALL_SRC_20150503_102209.DTX,
ALL_nonmotion_stat_ADDED_into_ALL_SRC_20150503_114234.DTX or
ALL_motion_stat_ADDED_into_ALL_SRC_20150502_093450.DTX containing the date and time when the files were created (2015-05-03 11:45:22, 2015-05-03 11:35:51, 2015-05-03 10:39:25, 2015-05-03 10:22:09, 2015-05-03 11:42:34, or 2015-05-02 09:34:50). These files will contain the ***complete search results***.

```
The results of searching for ALL DELETED motion statements in ALL programs
(SERVISNA_POZICIA.SRC, FREZOVANIE_CAPIC.SRC, VYMENA_CAPIC.SRC)
(from 2015-05-03 11:35:51Z)

The TOTAL number of ALL logs in the 'Logbuch.txt' file                  : 5281
The number of DELETED motion statements in ALL SRC files                : 6
The numbers of DELETED motion statements in particular SRC files        :
SERVISNA_POZICIA.SRC          : 0      FREZOVANIE_CAPIC.SRC             : 2
VYMENA_CAPIC.SRC              : 4
--------------------------------------------------------------------------------
        The list of ALL full logs with DELETED motion statements in ALL SRC files:
Log 5224 (User action, Error)
2014-01-17 11:47:12'129
Delete statement: SN 174: PTP SG0000032 Vel= 100 % PDAT10 TipDress  ProgNr= 31
ServoGun= 1  Part= 0 mm  Force= 2 kN  Comp= 1  Trigger= 0 mm Tool[1]:SpotGun
Base[0] (KRC:\R1\PROGRAM\FREZOVANIE_CAPIC.SRC)
Module: TPBASIS   MsgNo: 0
        [full statement PTP SG0000032]
;FOLD PTP SG0000032 CONT Vel=100 % PDAT0000032 Tool[1]:SpotGun Base[0];
%{PE}%R 8.2.24,%MKUKATPBASIS,%CMOVE,%VPTP,%P 1:PTP,
2:SG0000032, 3:C_DIS, 5:100, 7:PDAT0000032
$BWDSTART=FALSE
PDAT_ACT=PPDAT0000032
FDAT_ACT=FSG0000032
BAS(#PTP_PARAMS,100)
PTP XSG0000032 C_DIS
;ENDFOLD
```

Fig. 1. The part of the file *ALL_DELETED_motion_stat_in_ALL_SRC_20150503_113551.DTX* with search results (the .NET search application displays the same outputs in its two controls)

```
The results of searching for ALL MODIFIED non-motion statements in ALL programs
(SERVISNA_POZICIA.SRC, FREZOVANIE_CAPIC.SRC, VYMENA_CAPIC.SRC)
(from 2015-05-03 10:39:25Z)

The TOTAL number of ALL logs in the 'Logbuch.txt' file                  : 5281
The number of MODIFIED non-motion statements in ALL SRC files           : 8

The numbers of MODIFIED non-motion statements in particular SRC files :
SERVISNA_POZICIA.SRC        : 4
FREZOVANIE_CAPIC.SRC        : 0
VYMENA_CAPIC.SRC            : 4
------------------------------------------------------------------------------------------------------------
The list of ALL full logs with MODIFIED non-motion statements in ALL SRC files:

Log 5267 (User action, Error)
2014-01-17 12:09:18'120
Modify: SN 132: From:  OUT 27 'OUT027_EL_CHANGE_REQUEST' State= TRUE  ,
To: OUT 26 'OUT026_DRESSING_REQUEST' State= TRUE  (KRC:\R1\PROGRAM\
VYMENA_CAPIC.SRC)
Module: TPBASIS   MsgNo: 0

Log 3934 (User action, Error)
2014-01-16 11:58:11'360
Modify: SN 49: From:  ; servisna_pozicia ,  To: ; PTP_servisna_pozicia  (KRC:\R1\
PROGRAM\SERVISNA_POZICIA.SRC)
Module: TPBASIS   MsgNo: 0
```

Fig. 2. The part of the file *ALL_MODIFIED_nonmotion_stat_in_ALL_SRC_20150503_103925.DTX* with search results (the .NET search application displays the same outputs in its two controls)

- *Basic statistics* about ***founded statements*** of a given group in the SRC file of **each particular program**, which the .NET search application found in the *Logbuch.txt* file: the total number of all logs in the *Logbuch.txt* file, the number of founded statements of a given group in the SRC file of every particular program. The .NET search application displays these statistics in a text box.
  It displays as items in a checked list box control ***all found complete logs*** about founded statements of a given group in the SRC file of every particular program and in the case of searching for deleted or added statements it also displays above mentioned information together ***with full notations of*** these ***deleted*** or ***added statements***.
  Besides that the .NET application displays its search results in the form of basic statistics and complete found requested logs in its two controls, it also creates 12 ***disk files*** with the names, for example,

DELETEnonmotion_stat_inFREZOVANIE_CAPIC_SRC_20150503_082238.DTX, *DELETEDmotion_stat_inFREZOVANIE_CAPIC_SRC_20150503_082549.DTX*, *MODIFIEDnonmotion_stat_inSERVISNAPOZICIA_SRC_20150503_104022.DTX* (Fig. 3), *MODIFmotion_statinFREZOVANIECAPIC_SRC20150503_083128.DTX*, *nonmotion_statADDEDintoSERVISNA_POZICIA_SRC_20150503_083206.DTX* or *motion_statADDEDintoVYMENACAPIC_SRC_20150503_113511.DTX* (Fig. 4) containing the date and time when files were created (2015-05-03 08:22:38, 2015-05-03 08:25:49, 2015-05-03 10:40:22, 2015-05-03 08:31:28, 2015-05-03 08:32:06, or 2015-05-03 11:35:11). These files will contain the ***complete search results***. The .NET search application will not create a DTX file for SRC files, in which it did not find statements of a given group.

The results of searching for MODIFIED non-motion statements in
the SERVISNA_POZICIA.SRC file (from 2015-05-03 10:40:22Z)

The TOTAL number of ALL logs in the 'Logbuch.txt' file : 5281
The numb. of MODIFIED non-motion statemen. in the SERVISNA_POZICIA.SRC file : 4

The list of ALL full logs with MODIFIED non-motion statements in
the SERVISNA_POZICIA.SRC file:

Log 3934 (User action, Error)
2014-01-16 11:58:11'360
Modify: SN 49: From: ; servisna_pozicia , To: ; PTP_servisna_pozicia (KRC:\R1\PROGRAM\SERVISNA_POZICIA.SRC)
Module: TPBASIS MsgNo: 0

Log 3290 (User action, Error)
2014-01-16 09:16:18'102
Modify: SN 65: From: OUT 17 'OUT017_SERVICE_POS' State= FALSE CONT , To: OUT 17 'OUT017_SERVICE_POS' State= FALSE (KRC:\R1\PROGRAM\SERVISNA_POZICIA.SRC)
Module: TPBASIS MsgNo: 0

Log 3288 (User action, Error)
2014-01-16 09:16:11'132
Modify: SN 59: From: OUT 17 'OUT017_SERVICE_POS' State= TRUE CONT , To: OUT 17 'OUT017_SERVICE_POS' State= TRUE (KRC:\R1\PROGRAM\SERVISNA_POZICIA.SRC
Module: TPBASIS MsgNo: 0

Fig. 3. The part of the file *MODIFIEDnonmotion_stat_inSERVISNAPOZICIA_SRC_20150503_104022.DTX* with search results (the .NET search application displays the same outputs in its two controls)

```
                 The results of searching for motion statements ADDED into
                 the VYMENA_CAPIC.SRC file (from 2015-05-03 11:35:11Z)

The TOTAL number of ALL logs in the 'Logbuch.txt' file                          : 5281
The number of motion statements ADDED into the VYMENA_CAPIC.SRC file            : 2
(added statements: PTP P13, PTP P12)

The motion statements were ADDED into the VYMENA_CAPIC.SRC file for
2 NEW-FOUNDED points in VYMENA_CAPIC.DAT : XP12, XP13
The list of ALL points in VYMENA_CAPIC.DAT   : XP1, XP2, XP3, XP4, XP5, XP6, XP7,
XP8, XP9, XP10, XP11, XP12, XP13, XP14, XP15, XP16, XP17, XP18, XP19, XP20
------------------------------------------------------------------------------------------
        The list of ALL full logs and related full motion statements ADDED into
                          the VYMENA_CAPIC.SRC files:

Log 329 (User action, Error)
2014-01-15 05:45:32'128
Lines 16 to 23 inserted   into file /R1/VYMENA_CAPIC.SRC [statement PTP P13]
Module: XEdit   MsgNo: 3
        [full statement PTP P13]
;FOLD PTP P13 CONT Vel=100 % PDAT13 Tool[1]:SpotGun Base[0];%{PE}%R 8.2.24,
%MKUKATPBASIS,%CMOVE,%VPTP,%P 1:PTP, 2:P13, 3:C_DIS, 5:100, 7:PDAT13
$BWDSTART=FALSE
PDAT_ACT=PPDAT13
FDAT_ACT=FP13
BAS(#PTP_PARAMS,100)
PTP XP13 C_DIS
;ENDFOLD

Log 303 (User action, Error)
2014-01-15 05:43:23'235
Lines 10 to 17 inserted   into file /R1/VYMENA_CAPIC.SRC [statement PTP P12]
Module: XEdit   MsgNo: 3
        [full statement PTP P12]
;FOLD PTP P12 CONT Vel=100 % PDAT12 Tool[1]:SpotGun Base[0];%{PE}%R 8.2.24,
%MKUKATPBASIS,%CMOVE,%VPTP,%P 1:PTP, 2:P12, 3:C_DIS, 5:100, 7:PDAT12
$BWDSTART=FALSE
PDAT_ACT=PPDAT12
FDAT_ACT=FP12
BAS(#PTP_PARAMS,100)
PTP XP12 C_DIS
;ENDFOLD
```

Fig. 4. The file *motion_statADDEDintoVYMENACAPIC_SRC_20150503_113511.DTX* with search results (the .NET search application displays the same outputs in its two controls)

The .NET search application provides *18 different outputs* in total to the user. It displays these outputs into its two controls and stores each one from these outputs into a DTX disc file with a time stamp. By means of it is possible to chronologically restore all modifications of non-motion and motion statements made in SRC files of robot programs by the operator during testing of a robot in a given time period.
The output displayed *in Fig. 1* provides an useful information about deleted motion statements in all 3 robot programs during testing of a robot to the operator. It displays clearly in which program and when the operator deleted a given motion statement together with a full notation of deleted statement. The output displayed in Fig. 1 containing full notations of all deleted motion statements is the part of a DTX file, which was created by the .NET search application. From this DTX file the operator can get anytime a chronologically ordered list of all deleted motion statements and if he would like to restore whichever deleted motion statement then he has its full notation in this DTX file at disposal.
The output displayed *in Fig. 4* provides an useful information about the points XP12 and XP13, which were added to the motion path of TCP of a robot tool through motion statements PTP P13 and PTP P12 the robot program *vymena_capic* (*replacement_caps*).
These supplemented points provide important information about the TCP path sections for the operator, on which were problems during testing of the precision of the motion of the TCP along the prescribed path. During retesting of this activity of the robot it is therefore necessary to focus specially on these sections of the path, i.e. on the points XP12 and XP13 and their close vicinity. This is very important information, which the operator will use during retesting of this activity of the robot. He will test the precision of the motion of the TCP of a robot tool along the prescribed path only on its parts formed by the points XP12 and XP13 and their close vicinity. Such procedure of retesting makes it more efficient.

4 Conclusion

The most valuable outputs that our .NET search application provides to the robot operator are:

- the lists of deleted and added non-motion and motion statements into all 3 programs and into each one individually, together with full notations of deleted and added statements,
- the lists of modified non-motion and motion statements into all 3 programs and into each one individually,
- additional information to the lists of added motion statements into all 3 programs and into each one individually about new points created to added motion statements.

The first mentioned list provides very important information about all deleted statements in given robot programs to the operator with the possibility to return deleted statements, because he has full notations of deleted statements at disposal.

Additional information to the lists of added motion statements about new points created to these statements enable the operator to quickly identify problematic spots of a motion path of the robot TCP. He then focuses on these spots during retesting the precision of the robot TCP motion along the prescribed path.
The results of all searches stored in the corresponding disk files make it easier for the operator to archive and a time recapitulate all changes done in the SRC files of relevant programs during testing of a robot particular activities. Thus, our .NET search application can be a very effective software support for the operator during testing and retesting of robot particular activities.

References

1. KUKA Roboter GmbH "KUKA System Software, KUKA System Software 5.5, Operating and Programming Instructions for System Integrators" KUKA Roboter GmbH, Augsburg, Germany, 2010.
2. KUKA Roboter GmbH "KUKA System Technology, KUKA.RobotSensorInterface 2.3 for KUKA System Software 5.4, 5.5, 5.6, 7.0" KUKA Roboter GmbH, Augsburg, Germany, 2009.
3. KUKA Roboter GmbH "Controller, KR C4; KR C4 CK, Specification" KUKA Roboter GmbH, Augsburg, Germany, 2013.
4. KUKA Roboter GmbH "KUKA System Technology KUKA.Load 5.0 KUKA.Load Pro 5.0 Valid for KSS 5.5, 5.6, 8 and VSS 8" KUKA Roboter GmbH, Augsburg, Germany, 2013.
5. Microsoft Corporation 2015 "MSDN Library on-line" http://msdn.microsoft.com.

A Task Planner for Autonomous Mobile Robot Based on Semantic Network

Petr Mašek[1], Michal Růžička[2]

Brno University of Technology, Faculty of Mechanical Engineering: Institute of Automation and Computer Science, Technická 2896/2, 616 69, Brno, Czech Republic
{y70232, y110384}@stud.fme.vutbr.cz

Abstract. Autonomous mobile robots movement in shared indoor environment, where are people, needs to have natural communication interface. Presented paper describes a formal representation of robots environment by semantic network. Such representation allows us to communicate with autonomous robot with terms defined in semantic network. This kind of human-machine interface is highly necessary in areas of health-care mobile robots or rapidly growing field of social robotics. Presented semantic network can handle relations of positions between places in the map, objects placed in defined places, interaction with objects, etc.

Keywords: Autonomous mobile robot, Semantic network, Map tagging problem, Task planning, Indoor robots, Human-machine interface.

1 Introduction

The rapidly growing field of social robotics brings a new kind of issues into mapping representation problem together with advanced human-machine interface for mobile robots. The field of social robotics addresses problems, which arises from the close cooperation of mobile robots with people. Especially the indoor autonomous mobile robots which shares the environment with people, moreover when cooperate with people. Such kind of robot needs to have advanced human-machine interface for natural communication. Typical application is health-care robots, which interact with patients or seniors. This kind of robots needs to have speech based human-machine interface.

Human-machine interface is not novelty. Many speech recognition engines exist. These engines are able to recognize voice commands or convert text to speech [1, 2]. The two most popular voice recognition engines are Google API and Microsoft SAPI.

Presented human-machine interface is focused to improvement of recognition accuracy. The robot has to precisely recognize to the user's natural commands [3, 4]. The basic level of understanding can be realized as a simple "map tagging" [5, 6] which adds a semantic information into the environment map. In this case the robot can react to the commands like "go to the kitchen" or "clean the living room". This

R. Jabłoński and T. Brezina (eds.), *Advanced Mechatronics Solutions*,
Advances in Intelligent Systems and Computing 393,
DOI 10.1007/978-3-319-23923-1_89

can be realized with adding tags "kitchen" and "living room" into the map, so the map is not just a set of x, y coordinates which can be reached. [7]

The higher level of "understanding" to the human commands can be realizes with semantic map or semantic network which brings semantic information into the map as described above and furthermore it can address the relation between known objects ("fridge is in kitchen", "carpet is in bed room", "apple is in larder", etc.). The higher knowledge improvement of the environment can be defined by the possible actions, which can be executed with the presented objects in environment e.g. "fridge cannot be moved", "apple can be moved", "living room can be cleaned", "stairs are not feasible". [8, 9]

The Fig. 1 below demonstrates communication of presented human machine interface, which uses semantic network. The human utters some command and robot returns voice response back to human. The robot is equipped by specialized software, which contains semantic network engine, syntax and context module, which are able to recognize human command context and return voice feedback to human operator.

This paper is organized as follows. The section 2 deals with semantic network representation. The section 3 contains description of semantic network implementation. The last section contains conclusions and future work.

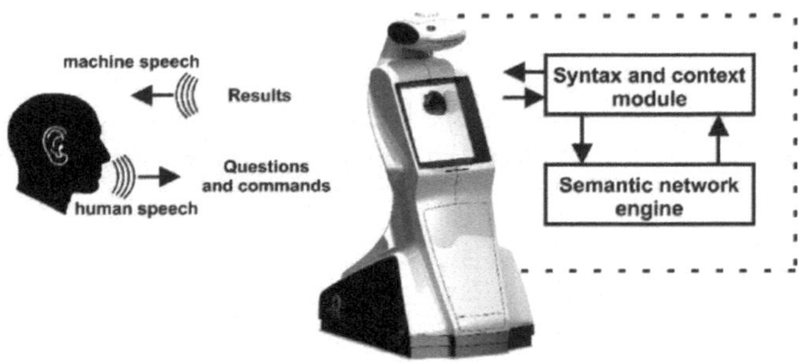

Fig. 1. Basic function

2 Semantic network representation

We use different kinds of mobile robots for real-world experiments. The most advance autonomous mobile robot Advee [12, 13] is already equipped by human-machine interface for intuitive interaction with bystanders. This mobile robot is designed for movement in dynamic environment, where are many people. The Advee is ideal robot for experiments with advanced human-machine interface. [3]

The part of semantic network, which was designed for autonomous mobile robot Advee is shown on figure 1 [10, 11]. It represents all three levels of abstraction, which are described above:

- map tagging,
- object relations,
- possible actions.

The semantic network is created by user before the mobile robot is placed into the environment. There are a few rules which are used to create a semantic network and also which keeps the semantic network consistent:

- The semantic network is represented as oriented graph structure.
- This semantic network makes no difference between classes and objects.
- We assume that everything is the Tag (Object) [14].
- The relation between Tags is given by oriented edges.
- The possible actions are properties of given Tag and can be recognized as a leafs of the network.

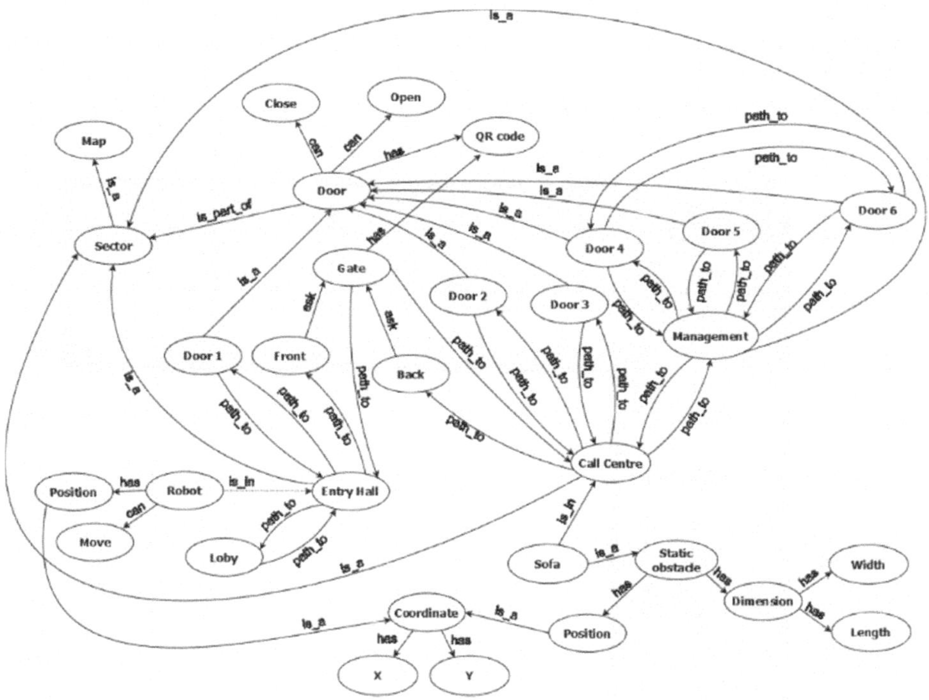

Fig. 2. Example of semantic network

The semantic network is formed via different kinds of links between objects which can be used for knowledge inference (see figure 2). This links are follows:

- The ***is_a*** link means that the connected object share the same properties/actions. This link can be used for sharing proper-ties/action over the network.
- The ***can*** link means that the object (or with the object) can be performed some action (eg. move the object, take or drop it).
- The ***has*** links means simple properties of given objects.
- The ***is_in*** links means position of given objects.

- The ***path_to*** link indicates that objects connected via these links are connected and path can be planned from one location to each other [6].
- The ***ask*** link means calling authorized person.
- The ***is_part_of*** link means that the object is part of another object.

3 Semantic network usage

The semantic network from figure 2 is bounded to the map from figure 3, both are simplified for the purposes of this paper. The presented map and semantic network are based on real application for service robot inside of an office building. The robots main task is to travel from starting point to the goal [5]. In our experiments the goal is not defined by exact coordinates *x*, *y*, but by simple natural language sentence. Such sentence can be formed by tags and action from semantic network as "Move to the laboratory" or "Find path to the corridor".

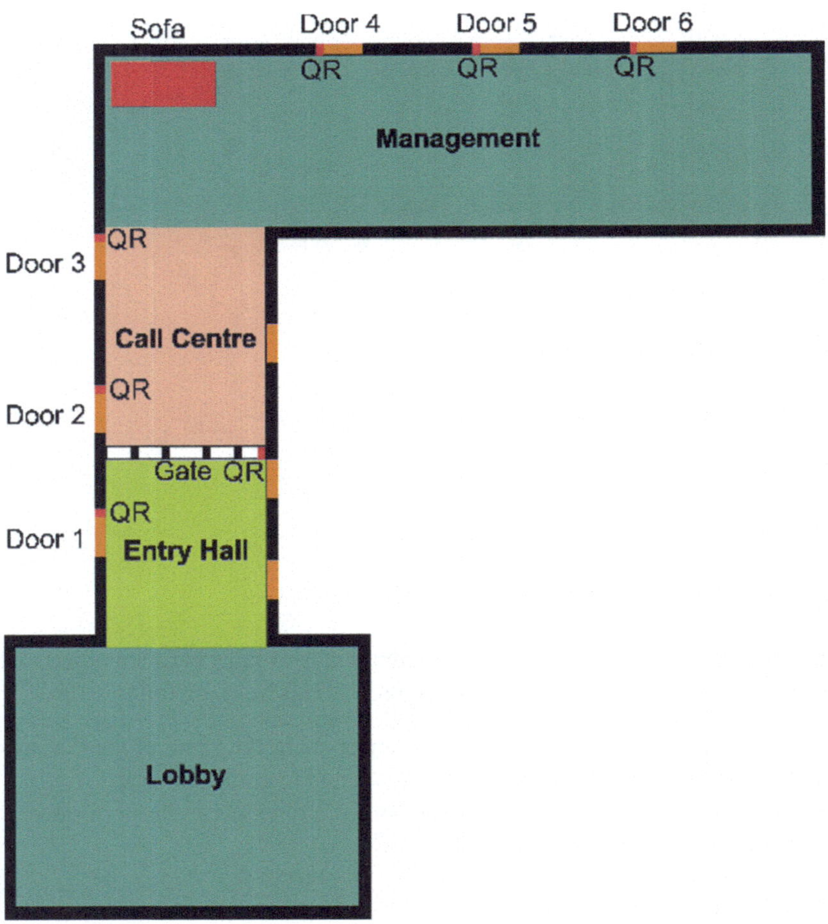

Fig. 3. Map of office environment

The map is divided into several sections. This section was tagged by four tags [6, 7] ("Lobby", "Entry Hall", "Call Centre" and "Management"). These environments are represented by open space except transition between the "Entry Hall" and "Call Centre" where is "Gate". This transition is solved as follows: The robot has to go to object "Front". When object is reached, then the robot has to perform activity "ask". This activity tries to call gateway operator. If a time limit for open "Gate" is expired, the robot tries to call gateway operator again.

The semantic network can be used to check whether the command is correct and the solution is feasible. This could be done by searching in the graph structure of semantic network via commonly known searching methods e.g. depth first search. In case that user calls wrong command (e.g. "Move room to the Lobby") the searching result is false due to the room has no property move. In the case, that user calls inaccurate command such as "move cap to the room", the recursive search finds four possible rooms and asks human operator for more specification.

4 Conclusion

The semantic network can bring more intuitive interaction with autonomous mobile robots. Those robots can have semantic information related to the commonly used data structures for map representation and path planning. This can be used in many cases from laboratory experiments to health-care robots, etc.

Current experiments are focused to effective implementation of semantic network, which is directly connected to planning engine of the robot. The robot can perform simple tasks as path planning to given places. The commands are entered via voice commands e.g. "Move to the Entry Hall".

Current state of relation between places and map raw representation is defined by ranges of dimensions. If the semantic network includes a tag "Lobby" the connection into the map representation is defined by geometric primitives as rectangles (x=<0, 10>[m], y=<0, 10>[m]). This exacts definition could be improved in the future works e.g. by fuzzy sets, where the location will be defined smoother.

Even such simple semantic network was described in this paper. Semantic network contains wide range of knowledge about environment, thus the actual research is focused to the effective representation and fitting of the semantics into the robots internal model of the world.

Acknowledgement: This work was supported by grant of BUT IGA No. FSI-S-14-2533: "Applied Computer Science and Control".

References

1. Mattys, S. L., Davis, M. H., Bradlow, A. R. Scott, S. K.: Speech recognition in adverse condi-tions: A review, Language and Cognitive Processes, 953-978, (2012)

2. Katz, S.: Estimation of probabilities from sparse data for the language model component of a speech recognizer, Acoustics, Speech and Signal Processing, IEEE Transactions, vol. 35, no. 3, pp. 400-401, DOI 10.1109/TASSP.1987.1165125, (1987)
3. Al-Moadhen, A. Packianather, M., Setchi, R.: Integrating Robot Task Plan-ner with Com-mon-sense Knowledge Base to Improve the Efficiency of Planning. Procedia Computer Sci-ence, vol. 22, pp. 211-220. ISSN 1877-0509, (2013)
4. Al-Moadhen, A., Packianather, M., Setchi, R.: Automation in Handling Uncertainty in Se-mantic-knowledge based Robotic Task-planning by Using Markov Logic Networks. Proce-dia Computer Science, vol. 35, pp. 1023-1032, ISSN 1877-0509, (2014)
5. Galindo, C., Fernández-Madrigal, J., González, J., Saffiotti, A.: Robot task planning using semantic maps. Robotics and Autonomous Systems, vol. 56, pp 955-966, ISSN 0921-8890, (2008)
6. Lakemeyer, G., Nebel, B.: Exploring artificial intelligence in the new millennium, Morgan Kaufmann (2003)
7. Kuipers, B.: Modeling spatial knowledge, vol. 2, Cognitive Science, pp. 129–153, Ablex Publishing Corporation, New Jersey (1978)
8. Galindo, C., Saffiotti, A., Coradeschi, S., Buschka, P., Fernandez-Madrigal, J.A., Gonza-lez, J.: Multi-hierarchical semantic maps for mobile robotics, Intelligent Robots and Sys-tems, 2005 IEEE/RSJ International Conference, pp.2278-2283, DOI 10.1109/IROS.2005.1545511, (2005)
9. Wolf, D.F., Sukhatme, G.: Semantic Mapping Using Mobile Robots, Robotics, IEEE Trans-actions on , vol. 24, no. 2, pp. 245-258, DOI 10.1109/TRO.2008.917001, (2008)
10. Vechet, S., Ondrousek, V.: Motion Planning of Autonomous Mobile Robot in Highly Pop-ulated Dynamic Environment. 9th International Conference on Mechatronics, no. 9, pp. 453-461, Warsaw (2011)
11. Vechet, S., The Rule Based Path Planner for Autonomous Mobile Robot. Proceedings of 17th International Conference on Soft Computing – MENDEL 2011, no. 17 in MENDEL, pp. 546–551. Brno University of Technology, VUT Press, Brno (2011)
12. Krejsa, J., Vechet, S.: Infrared Beacons based Localization of Mobile Robot., ELEKTRONIKA IR ELEKTROTECHNIKA, pp. 17-22, DOI 10.5755/j01.eee.117.1.1046, Kaunas (2012)
13. Vechet, S., Krejsa, J., Houska, P.: The enhancement of PCSM method by motion history analysis, International Conference on Mechatronics, pp. 107-110, DOI 10.1007/978-3-540-73956-2_22, Warsaw (2007)
14. Krejsa, J.: Reduction of Face Detection False Positive in Mobile Robot Interaction Using Proximity Sensors. Proceedings of 17th International Conference on Soft Computing – MENDEL 2011, no. 17 in MENDEL, pp. 540–545. Brno University of Technology, VUT Press, Brno (2011)

Visual environment mapping based on computer vision

Michal Růžička[1] and Petr Mašek[2]

[1] Brno University of Technology, Institute of Automation and Computer Science, Brno, Czech Republic
Y110384@stud.fme.vutbr.cz
[2] Brno University of Technology, Institute of Automation and Computer Science, Brno, Czech Republic
Y70232@stud.fme.vutbr.cz

Abstract. This paper presents method, which can be used for localization of autonomous mobile robots inside of buildings. The method is able to map environment, where the mobile robot operates. For this purpose was used phase correlation and image stitching method. Presented method uses continuously created map of environment for localization. Localization uses particle filters for mobile robot position estimation.

Keywords: Phase correlation, particle filters, localization, mapping, stitching, computer vision, image processing, mobile robot, discrete Fourier transform, inverse discrete Fourier transform and Hanning window.

1 Introduction

This paper presents method for environment mapping to create map of environment, where mobile robot operates. This map of environment is used for mobile robot localization purpose. Localization is realized by probability approach, specifically by particle filters. Continuous mapping runs parallel with location estimation. Image data are grabbed by one camera, which is located perpendicular to the ceiling.
The state of the art solution for environment mapping is Project Tango from Google. It is software and hardware solution, which is able to map the environment in 3D space. Disadvantage of this solution is low invariance to sun lightening of the scene. This solution is based on sterovision and infrared projector. Presented solution is focused to 2D problem mapping by single camera only. 2D map representation is completely sufficient for indoor mobile robot mapping and localization. Specialized hardware cannot be used for mapping tasks.
This paper is organized as follows. The section 2 contains description of environment mapping and localization. Practical experiment is described in the section 3. Section 4 is reserved for conclusions and future works.

R. Jabłoński and T. Brezina (eds.), *Advanced Mechatronics Solutions*,
Advances in Intelligent Systems and Computing 393,
DOI 10.1007/978-3-319-23923-1_90

2 Mapping and localization

This section deals with design of mapping and localization system for mobile robot. This system was designed for environment inside of buildings, where the mobile robot operates. The ceiling represents map of environment. Presented system uses single camera with low frame resolution 320x240. Image data are sequentially grabbed and processed by phase correlation method. After that the image data are stitched to one frame, which represents map of environment. Localization estimation uses environment map and runs parallel with mapping of environment.
The next subsection explains the concept of phase correlation. The image stitching method is described in subsection 2.2. The subsection 2.3 describes probabilistic localization method.

2.1 Phase correlation

Phase correlation is image processing method, which is able to register affine transformation between two similar images. Affine transformations are translation, rotation and scale of the image. The base of phase correlation is mainly composed from three methods. The first one is two dimensional discrete Fourier transformations (1) [6, 7, 11]. The second one is two dimensional inverse discrete Fourier transformations (2) [6, 7, 11]. The last one is correlation formula (3) [1, 2, 3, 4, 11].

$$F(\xi,\eta)=\sum_{x=0}^{N-1}\sum_{y=0}^{N-1} f(\mathrm{x},\mathrm{y})\,\mathrm{e}^{-\frac{2\pi i}{N}(\mathrm{x}\,\xi+\mathrm{y}\,\eta)} \tag{1}$$

$$D^{-1}\{\mathrm{F}\}(\mathrm{x},\mathrm{y})=\frac{1}{N^2}\sum_{\xi=0}^{N-1}\sum_{\eta=0}^{N-1} F(\xi,\eta)e^{\frac{2\pi i}{N}(\mathrm{x}\,\xi+\mathrm{y}\,\eta)} \tag{2}$$

$$P_{f_1,f_1}^{p,q}(\mathrm{x},\mathrm{y})=\mathrm{D}^{-1}\left\{\frac{F_1(\xi,\eta)\cdot F_2^*(\xi,\eta)}{\left(\left|F_1(\xi,\eta)\right|+p\right)\cdot\left(\left|F_2(\xi,\eta)\right|+q\right)}\right\} \tag{3}$$

Hanning window function has to be applied to image dada from camera to remove edges around the perimeter of image.

Two dimensional discrete Fourier transformation compute magnitude-frequency spectrum $F(\xi,\eta)$ from original image, which is defined by image function $f(\mathrm{x},\mathrm{y})$ Correlation formula operates with logarithmic magnitude-frequency image spectrums *F1* and *F2*. If the rotation and scale will be registered, then logarithmic magnitude-frequency image spectrums have to be converted to logarithmic polar coordinate system. This transformation converts image rotation character to translation character on the vertical axis. The scale is defined on horizontal axis. Logarithmic polar conversion is not applied for translation estimation. Correlation formula contains constants p

and q, which define type of correlation and thus the character of correlation formula result (3). If variables p and q converge to zero, than we are talking about phase correlation. On the other side when variables p and q converge to one, we are talking about cross correlation.

The main advantage of phase correlation, against to cross correlation, is sharpen global peak. This means the global peak is easier to locate against to peak of cross correlation. Ideal case for phase correlation would be, than variable p and q equal zero, but convergence to zero arranges invariance against to division by zero.

Two dimensional inverse discrete Fourier transformation converts spectrum $\mathrm{P}(\xi,\eta)$, which was computed by correlation formula back to image $\mathrm{p}(\mathrm{x},\mathrm{y})$.

This algorithm is able to compute translation with sub-pixel accuracy in hundredths of pixel. The estimation of rotation accuracy is approximately in tenths of degree. The estimation accuracy of scale is in hundredths pixels. This is valid for high resolution frames. The accuracy for low resolution frames, which presented method uses, is worse but it still acceptable. Two images, that will be registered, have to be overlapped to each other. The higher the image resolution can thus be smaller overlap.

2.2 Map stitching

Phase correlation outputs are affine transformation, between previously and currently grabbed images, which have to be applied to currently grabbed image frame. Firstly is applied rotation, after that is applied change scale by scale factor. Finally is applied translation to stitch rotated and scaled currently grabbed image into map of environment, where was stitched previously grabbed image as well.

2.3 Localization

Localization process uses continuously stitched map of environment and probabilistic method, specifically particle filters, which contain mobile robot motion model (4) (5) [8, 9, 10].

$$x = d \cdot \cos(\varphi) \tag{4}$$

$$y = d \cdot \sin(\varphi) \tag{5}$$

The motion model variable φ represents mobile robot orientation; d is travelled distance and variables x, y represents mobile robot location. Particle generator is described by expressions (6) and (7) [10].

$$x_i = d_N \cdot \cos(\varphi_N) \tag{6}$$

$$y_i = d_N \cdot \sin(\varphi_N) \tag{7}$$

Where dN was randomly generated with parameters: $\mu = d$, $\sigma d = 10$ and φN for $\mu = \varphi$, $\sigma\varphi = 5$. Index i represents particle index. N particles on the algorithm start and after re-sampling, which uses weight for each particle. Weights evaluation is based on the phase correlation for translation registration only. Idea is as follows. What would be seen on camera frame if the robot was located at the place where this particle is located? We create image by this idea from the map. Basically cut of part the map image, where a new image centre represents particles location. This image is compared with real camera image to get shift between both of them. The smaller the Euclidean distance between these images, the higher the weight of particles will be evaluated. After re-sampling are left just 25 particles. The robot location is evaluated as weighted mean of re-sampled particles locations.

3 Practical experiment

This section contains practical experiment, which proves correct functionality presented system.

Source video was captured by non calibrated low resolution camera with high noise level. The resolution is 320x240. Presented system is very compute intensive and therefore all experiments run in offline mode yet. System was written in C++ language with support OpenCV libraries. Software runs on computer configuration with Intel i7 processor, supported by 4 GB ram.

This practical experiment demonstrates result the stitched map from many frames. There are evident a very significant boundary between stitched images in the map. This was caused by non-homogenous light condition by fluorescent bulb during mobile robots movement. Despite this fact mapping works well. Phase correlation is robust image processing method, which is suitable for application, where lighting conditions are variable. For example, optical flow method would fail in this case. Blue circle represents mobile robots start position. Green circles equal estimated mobile robots locations. Red points mean particles, which were rejected from calculation during re-sampling process. Finally green points serve particles, which were selected for next computation during re-sampling process. Result is shown in Fig. 1. During experiment mobile robot did not go completely flat. At one point slightly change direction to the left and after that robot returns to origin direction. This fact is reflected also in determining the mobile robot position by presented localization method.

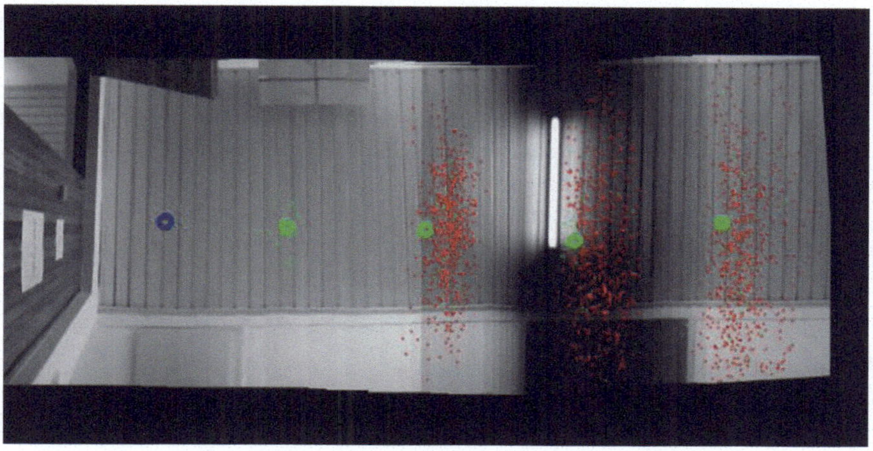

Fig. 1. Mapping and localization practical experiment

4 Conclusions

This paper deals with design suitable method for mapping of environment and localization. Mapping is based on phase correlation. Localization uses continuously created map and probability approach, specifically particle filters, which contain mobile robot motion model. Re-sampling rules are based on phase correlation for translation registration only.

The mapping and localization methods were described in the section 2. Practical experiment, which confirms function of proposal system, was described in section 3.

Phase correlation robustness can be increased by using low pass filter before applying two dimensional inverse discrete Fourier transformation. Low pass filter parameters setting determining is very problematic in this case due to dynamic changes in the context of scene mapping. Adaptive low pass filter is solution of this problem. Low pass filter was not used in presented mapping method. Despite this solution works very well.

This solution improves commonly used mapping methods, which are based on using only parts of the image for the registration of an affine transformation. Presented solution works with all frequencies in the image data. Used method is highly invariant to high speed changes in lightening of scene compared to conventionally used methods for environment mapping.

The next development steps will be mainly focused to reduce the demand for computer power. The reduction can be solved by converting calculations to multiple processor cores.

The next development step will solve loop enclosure problem. During mapping process the mobile robot holds the movements, such as a circular trajectory, must return to the starting position. The same applies for creating maps of the environment. First and last frames shall be continuous. If not, then we are talking about open loop. Then map of environment have to be transformed to get closed loop.

The next problem that has to be solved is rotation estimation range by phase correlation. Rotation can be estimated in interval from 0 to π. Estimation will be possible for interval from 0 to 2π in the future.

One of essential advantage of presented system is possibility to apply this system to any type of mobile robot chassis. Just change the mobile robot motion model, which is part of particle filters algorithm.

In the present system, there are still many problematic issues that will be resolved in the further development. This is a new method for mobile robot localization and environment mapping and development still continues.

Acknowledgement

This research was supported by grant of BUT IGA No. FSI-S-14-2533: "Applied Computer Science and Control".

References

1. Druckmüller, M.: Phase Correlation Method for the Alignment of Total Solar Eclipse Images, ASTROPHYSICAL JOURNAL, 2009(706), vol. 2, pp. 1605-1608, ISSN 0004-637X, (2009)
2. Berjak, J., Druckmüller, M.: Automatic analysis and recognition of moving objects in the picture by method of phase correlation, pp. 35, (2004)
3. Konecny, Z., Druckmüllerová, H.: Improvement of time-periodical production schedule of the group of products in the group of workplaces through the lot sizes alteration. In: R. Matousek (ed.) Proceedings of 19th International Conference on Soft Computing – MENDEL 2013, no. 19 in MENDEL, pp. 331–336. Brno University of Technology, VUT Press, Brno (2013)
4. Druckmüllerová, H.: Registration of real images by means of phase correlation, In: R. Matousek (ed.) Proceedings of Mendel 16th International Conference on Soft Computing – MENDEL 2010, no. 16 in MENDEL, pp. 578-583. Brno University of Technology, VUT Press, Brno (2010)
5. Argyriou, V., Vlachos, T.: A study of sub-pixel motion estimation using phase correlation, In: BMVC, pp. 387-396, (2006)
6. Bracewell, R. N.: Affine Theorem for Two-Dimensional Fourier transform, Electronics Letters, vol. 29, no. 3, pp. 304, (1993)
7. De Castro, E., Morandi, C.: Registration of Translated and Rotated Images Using Finite Fourier Transforms, IEEE Transactions on Pattern Analysis and Machine Intelligence,, vol. PAMI-9, no. 5, pp. 700–703. (1987)
8. Vechet, S., Ondrousek, V.: Motion Planning of Autonomous Mobile Robot in Highly Populated Dynamic Environment. 9th International Conference on Mechatronics, no. 9, pp. 453-461, Warsaw (2011)
9. Vechet, S.: The Rule Based Path Planner for Autonomous Mobile Robot. In: R. Matousek (ed.) Proceedings of 17th International Conference on Soft Computing – MENDEL 2011, no. 17 in MENDEL, pp. 546–551. Brno University of Technology, VUT Press, Brno (2011)

10. Krejsa, J., Vechet, S.: Infrared Beacons based Localization of Mobile Robot., ELEKTRONIKA IR ELEKTROTECHNIKA, pp. 17-22, DOI 10.5755/j01.eee.117.1.1046, Kaunas (2012)
11. Druckmüllerová, H.: Phase-correlation based image registration, Brno: Brno University of Technology, Faculty of Mechanical Engineering. (100 pages). Supervisor Mgr. Jana Procházková, PhD., (2010)

Design & FEA and Multi Body System Analysis of Human Rescue Robot Arm

H. Saygin Sucuoglu, Ismail Bogrekci, Pinar Demircioglu, and Ogulcan Turhanlar

Mechanical Engineering Department, Faculty of Engineering, Adnan Menderes University
{hilmisucuoglu,ibogrekci,pinar.demircioglu,121802011}@adu.edu.tr

Abstract. In this study, human rescue robot arm was designed using an integrated application on both Computer Aided Design (CAD) and Computer Aided Engineering (CAE) methods. The components and assembly model of the system were created using solid state modelling technique with CAD that enables the structural and Multi Body System (MBS) analysis with CAE methods. The strength of the design was performed by Finite Element Analysis (FEA). The maximum stress value in initial static case was calculated about 8 MPa. Then, MBS analysis consisted of kinematic and transient structural was conducted to find the torque requirements, velocity and acceleration and to check the mechanical safety of the system under dynamic conditions. Transient structural analysis showed that the yield strength of the selected material (Aluminum 6061) higher than maximum stress. The results indicated that designed robot arm is mechanically safe (safety factor=15) to execute to human rescue operation.

Keywords: CAD and CAE methods · FEA · MBS analysis · Robot arm

1 Introduction

Robots are really mechatronic devices. Their performance can be managed with interaction of mechanics, electronics, control and software [1]. Designing and developing robotic system is required taking a holistic approach that covers the disciplines mentioned above. A type of today's robotic research robot manipulator also known as robot arm is serial chain of rigid parts [2]. Robot arms are designed and manufactured to perform special tasks such as welding, palletizing, painting etc. with end effector due to their power, speed and accuracy.

In the design process of the robot arm several parameters and stages are taken into consideration: (1) Specifications (payload, dimensions, material for body, etc.), (2) CAD, (3) Determination of driving system, (4) FEA of the components and assembly, (5) Simulation and analysis of the robot as multibody system (6) Power management, (7) Creation of the prototype using Computer Aided Manufacturing Methods (CAM) [3, 4].

Today, many computer-based tools have been available to help design and analysis cycle of the robot arm [5]. Solid modelling using CAD methods assists the robot de-

R. Jabłoński and T. Brezina (eds.), *Advanced Mechatronics Solutions*,
Advances in Intelligent Systems and Computing 393,
DOI 10.1007/978-3-319-23923-1_91

signers to define the parts and assembly of the system and to utilize the geometry in applications such as simulation, analysis and prototyping. Virtual prototype simulations, static, kinematic and dynamic analysis can be conducted with CAE methods [6].

In this article we focused on the design and FEA analysis of human rescue robot arm. The design parameters that ensure the mechanical safety of robot under given conditions were determined. MBS analysis was also conducted to estimate the behavior of the robot in dynamic conditions and to determine the required torque at each joint, total velocity and acceleration of the system.

2 Geometry and 3D Model

In the mechanical design process of the human rescue robot arm, several issues were considered such as compact size, sufficient stiffness and toughness, management of time and cost. Components of the robot arm were drawn using parametric solid modelling technique. Subsequently, created parts were assembled to build up the 3D model. Design details are summarized in Table 1. Although MBS analysis was carried out using Ansys Workbench, the 3D model was created in Autodesk Inventor because of the Ansys's poor modelling performance [7]. The robot arm designed for the purpose of human rescue operation is shown in Fig. 1.

Table 1. Details of the human rescue robot arm design

Number	Part	Length (mm)	Weight (kg)
1	Base	650	275
2	Upper Arm	1000	100
3	Lower Arm	850	75
4	Servo Motor x 5	-	75
5	Lifting Arm	2000	200
	Total	4175	725

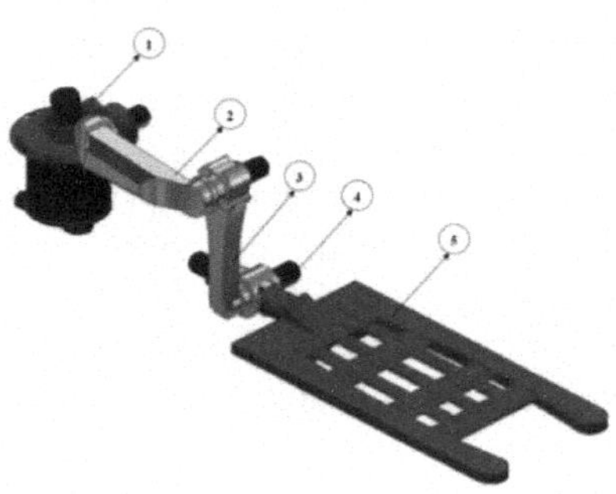

Fig. 1. Human rescue robot arm

3 Static Analysis

In order to determine the structural behavior of the robot arm on the model characteristic static analysis was carried out for the initial position using FEA. This approach increased the design process progress and efficiency [8, 9]. The area of the lifting arm was 1.5 m^2 and the maximum lifting capacity of the robot was assumed as 150 kg. Therefore, the calculated pressure force 1000 Pa was applied to lifting arm surface.
FEA was done with Autodesk Inventor. The parameters obtained from design environment such as dimensions, component weights etc. were used as input for analysis. The material of the components was defined as Aluminum 6061. The mechanical properties of Aluminum 6061 are given in Table 2. Before the calculations the model was simplified to increase the analysis speed and computation precision. The finite element model of the robot arm was composed of 50,357 nodes and 29,112 elements. FEA are shown in Fig. 2.

Table 2. Mechanical properties of Aluminum 6061

Property	Content
Yield Strength (MPa)	276 MPa
Elastic Modulus (GPa)	68.9 GPa
Poisson's Ratio	0.33
Density (g/cm^3)	2.71

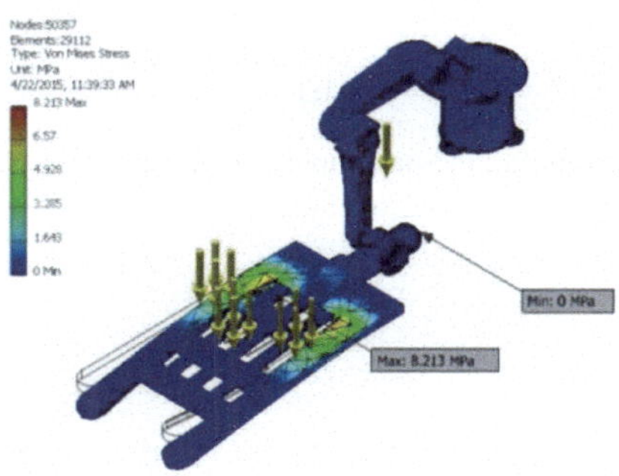

Fig. 2. FEA with Autodesk Inventor

4 Multi Body System Analysis

MBS analysis was carried out in two steps: (1) Kinematic (2) Transient Structural. In the kinematic analysis, all of the components were assumed as rigid to decrease the number and time of computations [10]. The required torque, total velocity and acceleration values were calculated. Equivalent (von Mises) and Maximum Principal Stresses were computed by means of transient structural analysis to predict the behavior of the system under dynamic conditions.

4.1 Kinematic

Kinematic analysis was conducted using Ansys Workbench Rigid Dynamic Module. The connections in the model were defined. According to the rescue scenario, rotary joints that had relative motion assigned as the revolute connection. In this research kinematic analysis was divided into two stages: (1) Initial to rescue position, (2) Rescue operation. The supposed scenario includes rescue arm motion from initial to rescue position in 20 seconds firstly and the arm lifts the patient in 10 seconds subsequently. The defined rotation angles to fulfill the operation are shown in Fig. 3.

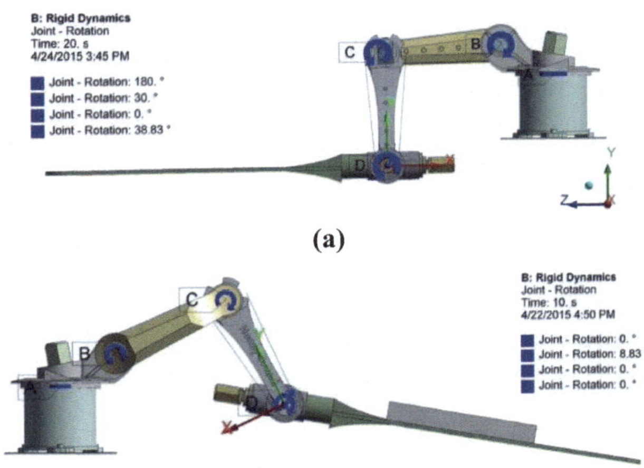

(b)

Fig. 3. Joint motion definition in kinematic analysis (a) initial to rescue position (b) rescue operation

4.2 Transient Structural

Components of the robot were defined as flexible bodies in transient structural analysis. The values for rotational angles given in kinematic study were applied to joints. In the first part of the scenario system was driven from initial to rescue position and then weight (patient) was lifted and held. The strength of the robot under dynamic loads was checked.

5 Results and Discussion

5.1 Static Analysis

The total calculated displacements were about 3 mm and maximum von Mises stress was 8.213MPa in initial static case. According to Mott's recommendation for machine member's deflection ranges (from 0.0005 to 0.003 m/m) [11] displacements are within allowable limits. Maximum stress doesn't have high value that could affect the operation of the robot. The factor of safety also indicates that system is 15 times stronger for the intended loads.

5.2 Kinematic Analysis

Fig. 4 plots the variation of torque values for the A, B, C and D joints. Joint A which is the most distant from the center of gravity and has the maximum joint load requires the largest torque as much as 8726 Nm. Maximum velocity and acceleration values of the robot were found as 400 mm/s and 67 mm/s^2, respectively.

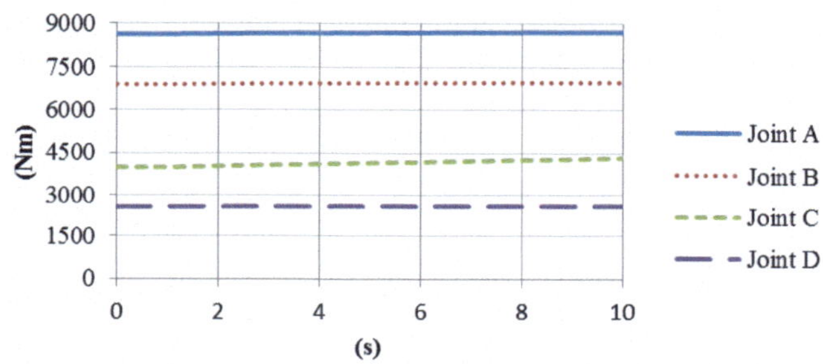

Fig. 4. Calculated torque values at joints

5.3 Transient Structural Analysis

Fig. 5 shows the variation of the von Mises stress. The maximum von Mises and Principal stresses were about 153 MPa and 156 MPa, respectively. Although values are much higher than those in the static analysis, stresses are almost twice lower than yield strength. Transient structural analysis showed that the design of the robot could ensure the safety under dynamic loads.

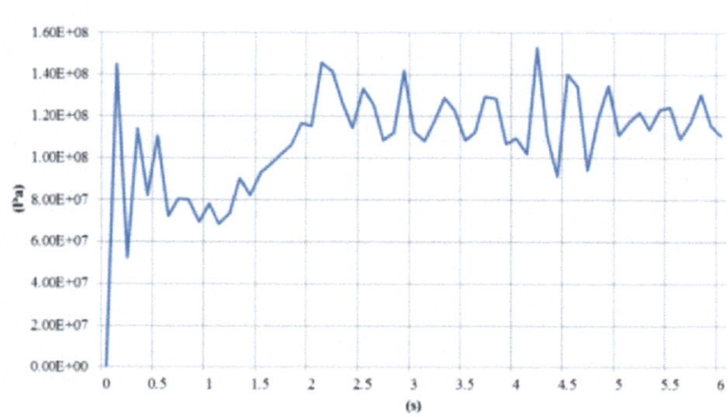

Fig. 5. Variation of von Mises stress

6 Conclusions and Future Works

In this study, robot arm for human rescue operation was proposed. An integrated application of CAD and CAE was systematically performed at the design and analysis stages. Design of the human rescue robot arm was accomplished. FEA and MBS analysis were conducted to check the mechanic safety of the system and to find torque, velocity and acceleration values. The selected results obtained from analysis were presented. The results from the static analysis showed that maximum stress was about 8 MPa. The safety factor was found 15 for the whole robot's body. The strength of the robot under intended dynamic conditions was checked by means of transient structural analysis. It can be concluded that the robot is compact and strong enough for human rescue operation. In the future, a conveyor belt mechanism will be developed and integrated into robot arm.

References

1. Kazi, A., Merk, G., Otter, M., Fan, H.: Design Optimization of Industrial Robots Using The Modelica Multi-Physics Modeling Language. In: Proceedings of the 33rd International Symposium on Robotics (2002)
2. Garcia, E., Jimenez, M.A., De Santos, P.G., Armada, M.: The Evolution of Robotics Research. IEEE Robot. Autom. Mag. 14(1), 90-103 (2007)
3. Bogrekci, I., Demircioglu, P., Kayhan, E.B.: Mechanical Design and Fea Analysis for Vineyard Robot. In: International Conference on Production Research - Regional Conference Africa, Europe and the Middle East (ICPR-AEM) / 3rd International Conference on Quality and Innovation in Engineering and Management (QIEM), pp. 33-36. Romania (2014)
4. Pupăză, C., Constantın, G., Negrılă, Ş.: Computer Aided Engineering of Industrial Robots. Proc in MS. 9(2), 87-92 (2014)
5. Gelgele, H.L.: Study of CAD Integrated Analysis for Complex Structures. In: Wang K., Kovacs G., Wozny M., Fang M. (eds.): International Federation for Information Processing (IFIP), vol. 207, pp. 673-678. Springer, Boston (2006)
6. Park, K., Kim, Y. S., Kim, C. S., Park, H. J.: Integrated Application of CAD/CAM/CAE And RP for Rapid Development of a Humanoid Biped Robot. J. Mater. Process. Technol. 187, 609-613 (2007)
7. Zhou, J., Yang, Z., Chen, S.: Analysis of the Harvesting Robot Arm Modal Based on CAE. J. Chem. Pharm. Res. 6(11) (2014)
8. Adams, V., Askenazi, A.: Building Better Products with Finite Element Analysis. Cengage Learning (1999)
9. Choong, W.H., Yeo, K. B.: Structural Design for a 3DOF Robot Lower-Arm via Computer Aided Engineering. Centre of Materials & Minerals, Universiti Malaysia Sabah. 88999, 8-18 (2007)
10. Zglińska, M., Orzechowski, G., Frączek, J.: Dynamical Analysis of the Deformable Power Transmission Mechanism Using Mixed FEM and MBS Method. Journal of KONES. 15, 235-245 (2008)
11. Mott, R.L.: Machine Elements in Mechanical Design. Digital Designs (1992)

Enhancement of Autonomous Robot Navigation via Sensor Failure Detection

Stanislav Vechet[1], Jiri Krejsa[1], and Kuo-Shen Chen[2]

[1] Brno University of Technology, Faculty of Mechanical Engineering, Technicka 2, 616 69, Brno, Czech Republic
{vechet.s, krejsa}@fme.vutbr.cz

[2] National Cheng Kung University, Department of Mechanical Engineering, No.1, Ta-Hsueh Road, Tainan 701, Taiwan
kschen@mail.ncku.edu.tw

Abstract. Sensor Failure Detection and Identification (FDI) is a standard method in many technical applications. Presented paper describes the usage of such approach in autonomous mobile robot navigation tasks. The FDI method uses pattern matching principle based on Markov chain theory. Such approach is capable to avoid any miss-readings or sensor failures during navigation task. This method significantly improves the robustness of mobile robot navigation in crowded environment shared with people. Practical verification experiments were performed on autonomous mobile robot Advee, which is designed to safely move among people.

Keywords: Autonomous mobile robots navigation. Sensor failure detection and identification. Pattern matching. First order Markov chain.

1 Introduction

Autonomous mobile robots are more and more commonly used in environment shared with people. This brings new kind of problems to mobile robotics. Moving machines which are not directly controlled by human operator has to be extremely safe in case they are operated in crowded places such as exhibition halls or shopping malls.

During last five years we have developed an autonomous mobile robot Advee [1] which is suitable for safe operation among moving people in exhibition halls [2]. An example of a typical environment can be seen on figure 1. It is clear that such environment is full of moving obstacles/people and they cannot be harmed by moving robot.

Once the navigation methods were solved (localization, path planning, reaching the goal position, etc.), the problem of safety arises. The problem of robust navigation task cannot be solved completely without having additional information about the relevance of sensor readings [3]. The robot can move through nearby bystanders only if the sensor readings are valid, which means that the navigation method works with correct data. If the sensor readings are incorrect or untrustworthy the robot should

R. Jabłoński and T. Brezina (eds.), *Advanced Mechatronics Solutions*,
Advances in Intelligent Systems and Computing 393,
DOI 10.1007/978-3-319-23923-1_92

immediately stop or run in emergency mode [4]. Such methodology can be seen in real-time failure detection of autonomous helicopters [5] or large scale distributed systems [6]. The most common approach uses the probability calculation or histogram methods described in [7].

Presented paper describes a method which uses a calculation of join probability via Markov chain method for time-series data. The principle of this method is described in chapter 2. The join probability is used as input to log odds ratio calculation, which is used as final result of presented method to decide whether the sensor measurement is correct or not. The verification of the method by practical experiments is described in chapter 3.

Fig. 1. Autonomous mobile robot Advee in its typical environment surrounded by bystanders.

2 Time series data monitoring method

The method used for "health" monitoring of given sensor is based on first/second order Markov chain. Markov chain allows us to calculate the join probability of time series data which is consequently used for decision if such data are trustworthy or not.

The first order Markov chain is given by equation 1:

$$P(X) = P(X_1)\prod_{t=2}^{n} P(X_t \mid X_{t-1}) \tag{1}$$

which takes into account two consequent measurements, the actual X_t, and previous X_{t-1}. The $P(X_t \mid X_{t-1})$ is transition probability (the probability of transition from state X_{t-1} to subsequent state X_t) and can be used repeatedly for calculation of join probability of given measurements in process history. The number of measurement samples in history which is used for calculation is given by number n.

The second order Markov chain calculates the join probability based on the current and two previous measurements and it's given by equation 2:

$$P(X)=P(X_1)P(X_2\mid X_1)\prod_{t=3}^{n}P(X_t\mid X_{t-1},X_{t-2}) \tag{2}$$

where the join probability is calculated based on two previous measurement X_{t-1} and X_{t-2}. The higher order Markov chain can be used, however the increased computational complexity doesn't bring adequate benefit in more reliable results.

In further text we compare described first and second order Markov chain on real data measured by autonomous mobile robot Advee.

3 Practical experiments

The real world experiments were performed on autonomous mobile robot Advee which is equipped with ultrasonic sensors for navigation purposes. Presented method uses measured data for training and separates all measurements to two groups: Correct and Incorrect. These pre-trained data are lately used for evaluation of measurement in real time and for making the decision whether the navigation can continue or emergency navigation procedures must be called. Typical measurements from ultrasonic sensors during navigation are shown on figure 2 and 3.

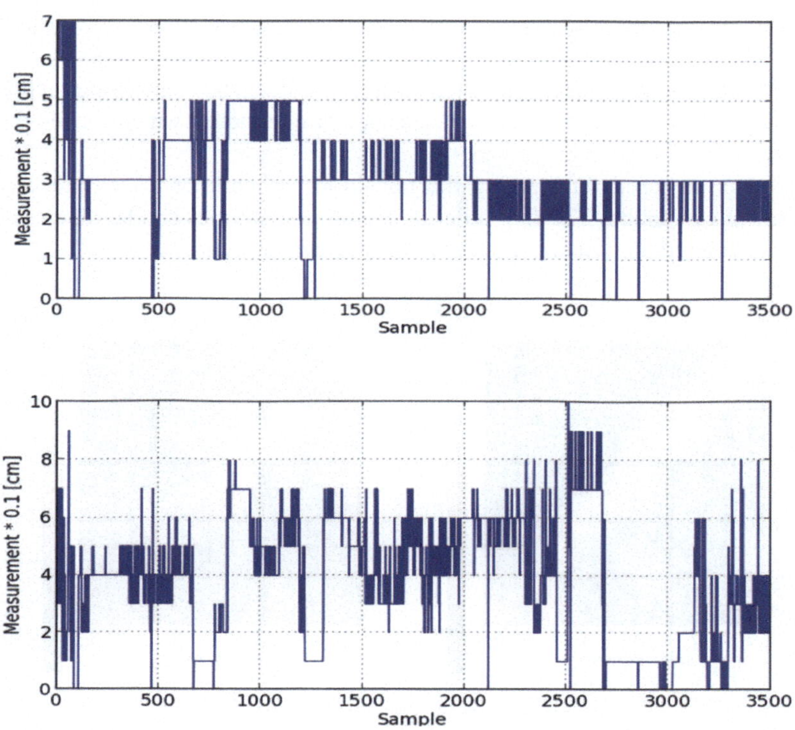

Fig. 2. Typical correct ultrasonic measurements from sensors used for navigation

Figure 2 shows two correctly working sensors and figure 3 incorrect measurements. Presented measurements shown on figures 2 and 3 were measured on different sensors placed in mobile robots chassis. This data shows actual sensor readings from ultrasonic sensors during navigation task. The erroneous measurement was caused by malfunction of one from used sensor. Such malfunction does not allow reliable navigation. The data differs from correct measurement (compare to figure 2) in two ways: there is many zero distances measured, the zero distance is measured frequently. However the difference is obvious for human operator the detection method for usage during navigation is needed.

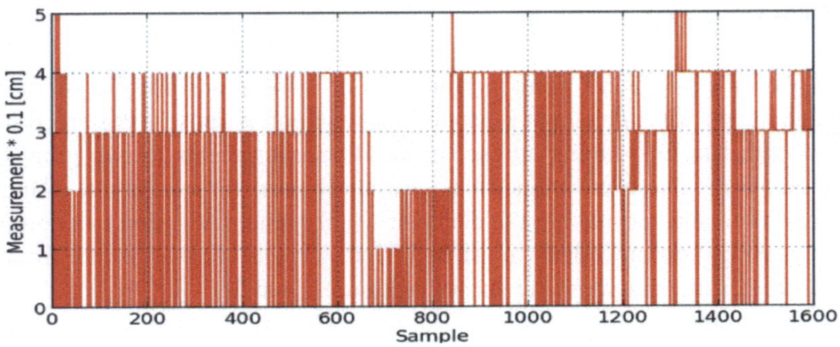

Fig. 3. Typical erroneous ultrasonic measurements from sensors used for navigation:

The measured data are discretized to ten discrete values, where each value represents 10 cm range. The maximum measured distance is one meter, which means that the values on all figures are in range 0 to 9.

These data are used for learning the transition probabilities which are used for measurement evaluation. The first order transition probabilities can be represented as table or 2D colored plot. Such representation is shown on figure 4.

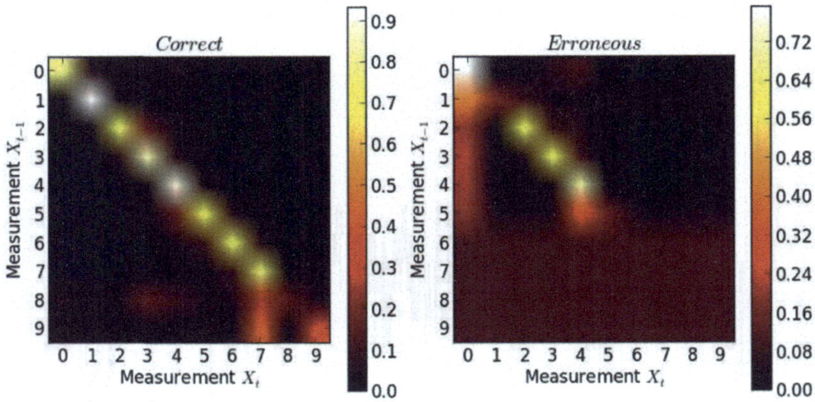

Fig. 4. Transition probabilities represented as 2D color map for correct measurement (left) and erroneous (right)

The left graph represents the correct measurement patterns and the right the erroneous. The colors represent the transition probabilities as described on side color bar. The erroneous probabilities shown on figure 4 (right) exhibit the lack of measurement above 0.6m which was caused by the fact, that the malfunction sensor wasn't capable to detect higher distances, as is can be also seen on figure 3.

Such pre-learned transition probabilities are used for data evaluation. In the verification phase, the measurement was initially taken from the valid sensor and at sample 9000 the signal was switched to invalid sensor. When a measurement is taken, the log odds ratio (see equation 3) is calculated from probabilities of incorrect to correct measurement. The resulted log odds ratios are shown on figure 5.

$$L(X) = \log \frac{P(X \mid Correct)}{P(X \mid Erroneous)} \tag{3}$$

where, X is the calculated join probability via equation 2 from *Correct* and *Erroneous* transition probabilities (figure 4), the logarithm of resulted quotient $L(X)$ represents the log odds ratio shown on figure 5. The positive value of log odds means the measurement is correct, the negative value indicates erroneous one. The higher the number is, the more reliable the detection is.

Figure 5 shows the comparison of first and second order Markov chain. The comparison shows that the first order Markov chain is more certain about the measurement in absolute values, however in disputable parts (e.g. around steps 10300 and 10600) the performance is similar.

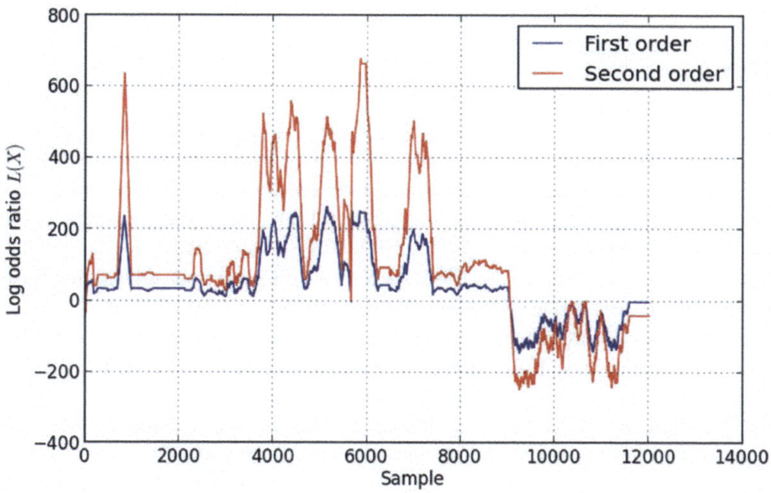

Fig. 5. Log odds ratios for first and second order Markov chain as a result of online measuring time series data from ultrasonic sensors

4 Conclusions

The practical experiments showed that the Second order Markov model has better performance in typical operational conditions; however in disputable cases the performance of both models is similar. The figure 5 clearly shows that the method quickly identifies the change from the correct measurements to erroneous one, as can be seen around measurement step 9000 where the log odds ration drops to negative numbers.

The drawback of this method can be seen in local misdetections (around steps 10300, 10600) where the method identified the erroneous measurement as correct one. On the other hand, the overall trend in data shows that the method is able to decide whether the measurement is correct or not.

The difference between first and second order Markov chain is minor and due to the fact that the first order method is faster and easier to implement it will be used for further experiments in the future.

5 Acknowledgement

Presented outputs of this project NETME CENTRE PLUS (LO1202) were created with financial support from the Ministry of Education, Youth and Sports under the "National Sustainability Programme I".

References

1. Krejsa J., Věchet S., et.al. “Presentation robot Advee”, Engineering Mechanics, 2012, pp. 307-322, ISSN 1802-1484
2. Vechet S., Chen K. S., Krejsa J. “Hybrid navigation method for dynamic indoor environment based on mixed potential fields”, In Recent Technology and Scientific Advances, Mechatronics 2013, Springer, pp. 575-582, ISBN 978-3-319-02293-2, 2013
3. Willsky A. S. “A Surway of Design Methods for Failure Detection in Dynamic Systems,” Automatica, vol. 12, pp. 601-611. Pergamon Press, 1976
4. Soika M. “A sensor failure detection framework for autonomous mobile robots,” In Intelligent Robots and Systems, 1997. IROS'97., Proceedings of the 1997 IEEE/RSJ International Conference on, vol. 3, pp. 1735-1740. IEEE, 1997.
5. Heredia G., Ollero A., Bejar M., and Mahtani R. "Sensor and actuator fault detection in small autonomous helicopters." Mechatronics 18, no. 2 (2008): 90-99.
6. Lavinia A., Dobre C., Pop F., Costan A., Andreica M. I., Cristea V. “Robust Failure Detection Architecture for Large Scale Distributed Systems,” In Complex, Intelligent and Software Intensive Systems (CISIS), 2010 International Conference on (pp. 482-489). IEEE.
7. Sharma A. B., Golubchik L., Govindan R. “Sensor Faults: Detection Methods and Prevalence in Real-World Datasets,” ACM Transactions on Sensor Networks (TOSN) 6, no. 3, 2010, vol. 23.

The Development of the Autonomous Indoor Robot

Vít Ondroušek[1], Jiří Lýsek[1], Marcel Vytečka[1] and Ondřej Švehla[1]

[1] Mendel University in Brno, Department of Informatics, Brno, Czech Republic
{vit.ondrousek, jiri.lysek, marcel.vytecka, ondrej.svehla}@mendelu.cz

Abstract. This paper is focused on the development of the autonomous wheeled indoor robot of a small size. The autonomous robot K3 is designed as a universal platform for developing new algorithms of path planning, localization and mapping. Detailed description of the hardware, sensory system and high level control software is given. The design of the image processing method intended for solving the Bear Rescue discipline of the autonomous robot competition is discussed in detailed to prove correctness and suitability of the introduced solution.

Keywords: autonomous robot · indoor robot · machine vision · bear rescue

1 Introduction

Mobile indoor robots are used in a wide range of applications in today's world. Companies can utilize mobile robots for public relations activity. Also, much effort has been devoted to build security indoor robots or healthcare robots, e.g. assistant robot for the elderly [1]. The most challenging problems of the indoor mobile robotics are still navigation and mapping of the unmanaged dynamic environment [2, 3 and 4] as well as the object recognition based on image processing [5].

Authors of this paper are focused on developing new algorithms for solving selected problems of the SLAM and computer vision in the indoor environment. For that reason, we have decided to build our own universal mobile robot, which can be used as a testing platform of the mentioned algorithms.

The main aim of this paper is introducing our own design of the autonomous mobile indoor robot K3. The second chapter describes the designed construction and integrated hardware, including the sensory system. Next chapter describes the design of the control software unit of the robot. We participated with the described robot in the competition of autonomous indoor robots held in the city of Prague in 2015. Our solution of the Bear Rescue discipline is described in the last chapter as a proof of concept.

2 The Construction Design and Sensory System

The K3 robot is designed for testing various algorithms of path planning, localization and mapping of the inner space of office buildings. The development team also intend

R. Jabłoński and T. Brezina (eds.), *Advanced Mechatronics Solutions*,
Advances in Intelligent Systems and Computing 393,
DOI 10.1007/978-3-319-23923-1_93

to use this robot in various disciplines of autonomous robots competitions, e.g. the Bear Rescue discipline. These requirements define the resulting dimensions of the robot, namely, 260 x 260 x 320 mm. The configuration with two actively driven wheels and one passive wheel was chosen. Such holonomic type of construction is convenient especially in the limited areas of various robotic playgrounds. The robot has a square shape and the base frame is manufactured from aluminum. Several auxiliary parts and both main wheels are printed on the 3D printer, see Fig. 1. The wheels are driven using the Microcon SX stepper motors. The outer part of the wheel is connected to the shaft of the motor through the wheel hub. The motor is mounted to the side plate of the chassis using a flange with a bearing inside. As you can see, the third motor of the same type is used for opening and for closing of the gripper mechanism as well.

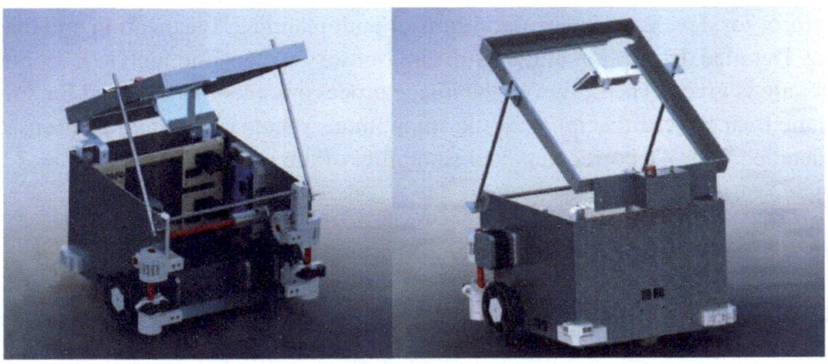

Fig. 1. The CAD model of the developed robot K3

The tablet PC HP ElitePad 900 with Windows 8 operating system is used as a main control unit of the robot. We assume that the battery-powered device with sufficient computing power and well documented platform would be beneficial for robot software development. This tablet is placed on the top of the robot using a frame bed, which is made of aluminum bars. The operator can easily execute commands using the control tablet, because it has a touch screen and the front side is not covered by any other components. The bed is adjustable by long screws placed on both sides. Unfortunately, the integrated camera of the control tablet faces almost directly into the ground. This issue is solved using the special bracket with a mirror, which is placed under the camera to change the field of view.

The K3 robot is also equipped with the special gripper that enables robot to pick and carry various objects. The figures Fig. 1 and Fig. 7 show the second improved generation of the gripper construction, which consists of two movable claws, system of gears and shafts. The propelled shaft is placed inside the robot body and the stepper drive is placed on the outer side. This shaft is mounted with gears that transmit the movement to the pivots using the worm gears. The gear transmission enables synchronous opening and closing using one stepper drive only. All gears were fabricated using a 3D printer.

The low level control unit consists of two Arduino boards, see Fig. **2**. These boards provide sufficient performance for the stepper drives control and for reading data from the IR proximity sensors. The information from IR sensors and stepper drives is sent to the control computer on demand only. Main control unit sends the message over the interface and waits for the response of the microcontroller. The custom communication protocol between tablet and both Arduino microcontrollers was designed. For example, only one byte of message is used for storing information about distance during the communication with proximity sensors. In this case, the data byte contains the address of the sensor, which value is requested by the main control unit. The stepper motors are connected to the Arduino microcontroller through the custom boards with four inputs that serve for bipolar control of the motors.

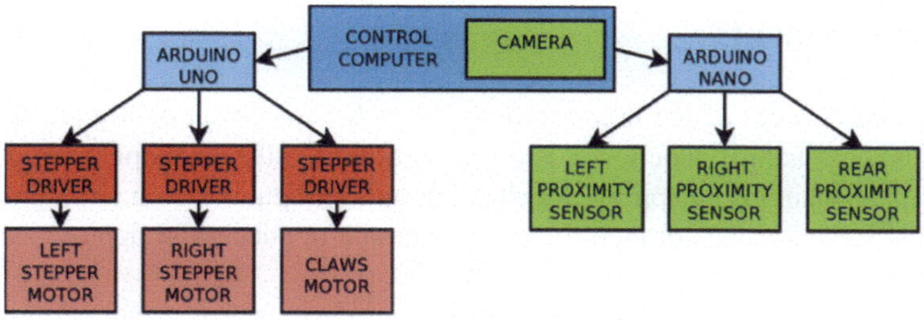

Fig. 2. Diagram of the low level communication

3 Main Control Software

Main control software ensures the autonomous movements of the robot, i.e. it solves problems of path planning, localization and mapping. Moreover the program has to perform sensory data acquisition, image processing and stepper motors control at the same time. Therefore it is necessary to use the multithreaded architecture. The control software is implemented using the C# language and .Net framework, since this high level programming language enables easy and rapid prototyping of new algorithms. Moreover the standard .Net Task Parallel Library (TPL) is used for parallel processing of required tasks. The simplified schema of the control software of the robot is shown on Fig. 3.

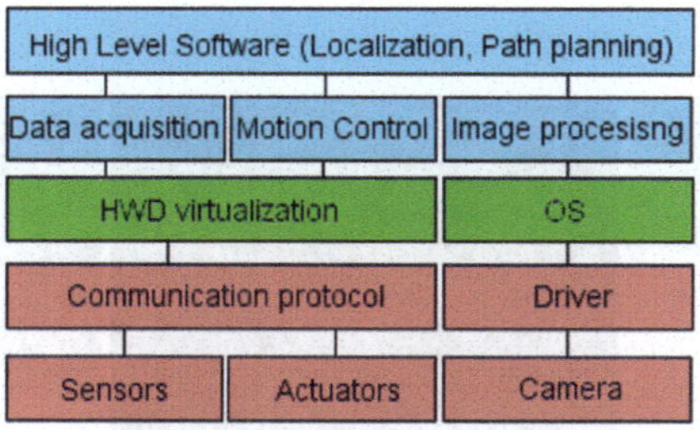

Fig. 3. Simplified schema of the control software

The designed architecture consists of three main layers. The highest layer is represented by the graphical user interface, which informs operator about actual state of the robot and enables to perform some basic operations, e.g. starting operations, setting up parameters of the tested algorithms, emergency shutdown, etc. The XAML language and windows presentation foundation (WPF) was used for developing the user interface. The designed and developed algorithms of image processing, path planning or localization are implemented in the middle layer of the architecture, e.g. the method of object detection described in the next chapter. Let's note, that the AForge .Net library is used in our approach for the design of methods of computer vision. The lowest layer of the architecture consists of communication routines, which transform and send middle layer commands over the communication protocol to the subordinate hardware.

4 Achievements

The described concept of the indoor mobile robot K3 was successfully produced and tested. This robot is fully functional, and is currently used for developing various algorithms of path planning, mapping, localization and computer vision. It is possible to mention the Bear Rescue competition as an example of successful deployment of the designed robot.

The autonomous robot is placed into the known structured area that is divided into two sections. The teddy bear is placed into the second section. The main aim for the control unit of the robot is to navigate through the first section and to find and pick the teddy bear in the second section. Consequently, it is necessary to return back to the starting position. The problem of path planning through the known first section is solved using the predefined checkpoints in the inner representation of the map. The localization of the robot is managed using the odometry information only. This simple approach is permitted, because the length of the path through the first section is very short and thus the cumulative error does not take effect.

The biggest challenge of the whole task is to find the teddy bear in the second section. Consider that the teddy bear is of an unknown color, size and shape. Experimental tests showed, that the use of common proximity sensors, e.g. sonic sensors or IR sensors, do not produce satisfactory results. The best results were achieved using the image processing methods. The whole method of detecting the teddy bear in the image can be divided into three main steps: border detection, inner area detection and the goal object detection.

Fig. 4. Border detection using edge detection and Hough transformation

The image obtained from the robot's camera is transformed to the binary image and the Canny edge detection algorithm is applied in the first step. After that, the Hough transformation is performed to find straight lines in the image, see Fig. 4. If the detected straight line is corrupted by the goal object, then the approximation of the edge of the border is performed. Resulting line leads from the left side of the image to the right side without any disruption, see Fig. 5.

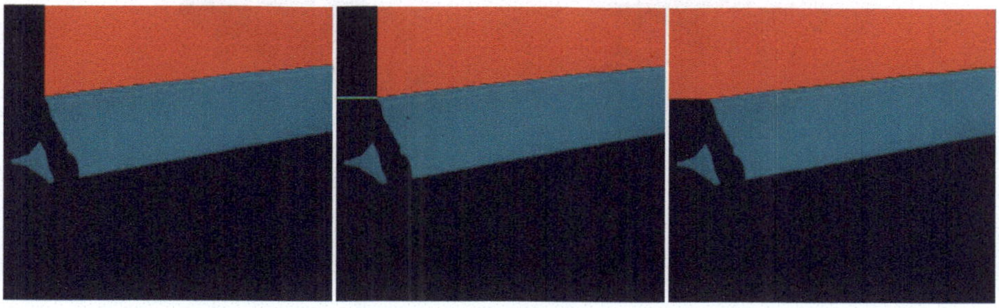

Fig. 5. Approximation of the border (green) and outer space detection (red)

The outer area and the border are successfully detected. Consequently, the inner area is detected in the second step. The inner area is found using the subtraction of pixels in the original image (Fig. 4a) and the image with known outer area (Fig. 5c). Resulting values, that are smaller then allowed minimum, are truncated to that minimum value, see Fig. 6a. Once are the border, the outer and inner areas detected, teddy bear object is detected as well. The center of gravity of the object is found in the next step, see the yellow point in the Fig. 6c. This information is used by the control unit to move the robot to the suitable position, see the pink point in the Fig. 6c, and the object is picked up. For more information on the designed algorithm see [6].

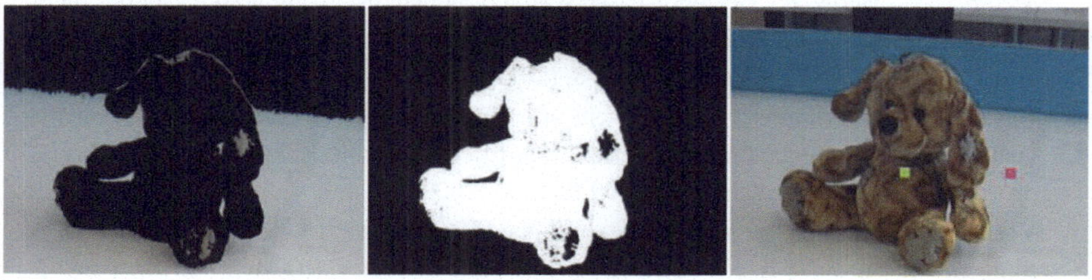

Fig. 6. Inner space and object detection, center of gravity (yellow), desired position (pink)

5 Conclusion

The complex development of an autonomous indoor robot K3, see Fig. 7 is described in this paper. The design of the construction, low level control hardware and sensory system was under discussion. The implementation of the software control unit was described as well. Successful participation in the competition, see chapter four, of the autonomous indoor robots proved the correctness of the whole concept.

Fig. 7. The developed autonomous indoor robot K3

Acknowledgement

Published results were acquired using the subsidization of the Ministry of Education, Youth and Sports of the Czech Republic, research plan IGA MENDELU MP PEF_TP_2015_006 "Design and Implementation of an Intelligent Control for Robotic Systems". Special thanks goes to our colleagues Martin Pánek, Robert Čížek and Robin Antonič.

References

1. Mehdi, S. A., Berns, K.: Behavior-based search of human by an autonomous indoor mobile robot in simulation. In: Universal Access in the Information Society. Vol. 13, Issue: 1, pp. 45-58. DOI: 10.1007/s10209-013-0301-8. (2013)
2. Hayashi, E., Kinoshita, T.: Development of an indoor navigation system for a monocular-vision-based autonomous mobile robot. In: Artificial Life and Robotics. Vol. 14, Issue 3, pp. 324-328. DOI: 10.1007/s10015-009-0671-4. (2009)
3. Vechet, S., Chen, K. S., Krejsa, J.: Hybrid Navigation Method for Dynamic Indoor Environment Based on Mixed Potential Fields. In: MECHATRONICS 2013: RECENT TECHNOLOGICAL AND SCIENTIFIC ADVANCES. Springer, pp. 575-582, DOI: 10.1007/978-3-319-02294-9_73, WOS:000345341700073 (2014)
4. Stachniss, C., Mozos, O. M., Burgard, W.: Efficient exploration of unknown indoor environments using a team of mobile robots. In: Annals of Mathematics and Artificial Intelligence. Vol. 52, Issue: 2-4, pp. 205-227. DOI: 10.1007/s10472-009-9123-z. (2009)
5. Minarik, M., Stastny, J.: Recognition of Randomly Deformed Objects. In: MENDEL, 14th International Conference on Soft Computing. Brno University of Technology, pp. 275-280, ISBN 78-80-214-3675-6. (2008)
6. Antonič, R.: Using the Mobile Robot for Objects Manipulation Based on Image Processing. Bachelor thesis. Mendel University in Brno. Brno. (2015)

The manufacturer's authorised representative in the EU is Springer Nature Customer Service Centre GmbH, Europaplatz 3, 69115 Heidelberg, Germany. If you have any concerns regarding our products, please contact ProductSafety@springernature.com

Printed and bound by CPI Group (UK) Ltd, Croydon, CR0 4YY
02/10/2026
02226631-0001